Lecture Notes in Computer Science   16445

Founding Editors

Gerhard Goos
Juris Hartmanis

The series Lecture Notes in Computer Science (LNCS), including its subseries Lecture Notes in Artificial Intelligence (LNAI) and Lecture Notes in Bioinformatics (LNBI), has established itself as a medium for the publication of new developments in computer science and information technology research, teaching, and education.

LNCS enjoys close cooperation with the computer science R & D community, the series counts many renowned academics among its volume editors and paper authors, and collaborates with prestigious societies. Its mission is to serve this international community by providing an invaluable service, mainly focused on the publication of conference and workshop proceedings and postproceedings. LNCS commenced publication in 1973.

Neeldhara Misra · Arti Pandey

Editors

Algorithms and Discrete Applied Mathematics

12th International Conference, CALDAM 2026
Dharwad, India, February 12–14, 2026
Proceedings

 Springer

Editors
Neeldhara Misra ⓘ
Indian Institute of Technology Gandhinagar
Gandhinagar, Gujarat, India

Arti Pandey ⓘ
Indian Institute of Technology Ropar
Rupnagar, Punjab, India

ISSN 0302-9743   ISSN 1611-3349 (electronic)
Lecture Notes in Computer Science
ISBN 978-3-032-17155-9   ISBN 978-3-032-17156-6 (eBook)
https://doi.org/10.1007/978-3-032-17156-6

This Springer imprint is published by the registered company Springer Nature Switzerland AG
The registered company address is: Gewerbestrasse 11, 6330 Cham, Switzerland

If disposing of this product, please recycle the paper.

Preface

This volume contains the papers presented at CALDAM 2026: the 12th International Conference on Algorithms and Discrete Applied Mathematics, held during February 12–14, 2026, in Dharwad. CALDAM 2026 was organized by the Department of Mathematics, Indian Institute of Technology Dharwad.

The 92 submissions had authors from various countries. Each submission received detailed single-blind reviews, with submissions reviewed on average by 2.4 reviewers. The committee decided to accept 35 submissions as full papers and 8 submissions as posters. The posters appear in these proceedings as two-page abstracts, and are considered non-archival.

The program also included three invited talks: Sandip Das gave a talk on Facility Location Problems, Andrzej Dudek gave a talk on Homogeneous substructures in (random) ordered matchings, and Géza Tóth gave a talk on Generalizations of the Crossing Lemma. We are grateful to the invited speakers for accepting our invitation and giving these wonderful talks.

We would like to thank all the authors for contributing high-quality research papers to the conference. We express our sincere thanks to the Program Committee members and the external reviewers for reviewing the papers within a very short period of time. We thank Springer for publishing the proceedings in the Lecture Notes in Computer Science series. We thank the Organizing Committee, chaired by Sagnik Sen, from the Indian Institute of Technology Dharwad, for the smooth functioning of the conference.

We are also grateful to the Steering Committee for their active help, support, and guidance throughout, with special thanks to the chair, Subir Kumar Ghosh, for helping us navigate the process of putting the program together. We thank Google for their financial support and Springer for its support for the Best Student Presentation Awards. We would also like to acknowledge the important role of the EasyChair conference management system, which we used to run the reviewing process end to end, from submissions to proceedings.

February 2026

Neeldhara Misra
Arti Pandey

Organization

Steering Committee

Subir Kumar Ghosh (Chair)	Ramakrishna Mission Vivekananda Educational and Research Institute, India
Gyula O. H. Katona	Alfréd Rényi Institute of Mathematics, Hungarian Academy of Sciences, Hungary
János Pach	Alfréd Rényi Institute of Mathematics, Hungary
Swami Sarvattomananda	Ramakrishna Mission Vivekananda Educational and Research Institute, India
Chee Yap	Courant Institute of Mathematical Sciences, New York University, USA
Anil Maheshwari	Carleton University, Canada

Program Committee

Aida Abiad	Eindhoven University of Technology, Netherlands
Niranjan Balachandran	Indian Institute of Technology Bombay, India
Aritra Banik	National Institute of Science Education and Research, India
Sasmita Barik	Indian Institute of Technology Bhubaneswar, India
Ahmad Biniaz	University of Windsor, Canada
Manoj Changat	University of Kerala, India
Sandip Das	Indian Statistical Institute, Kolkata, India
Palash Dey	Indian Institute of Technology Kharagpur, India
Christopher Duffy	University of Melbourne, Australia
Florent Foucaud	LIMOS - Université Clermont Auvergne, France
Daya Gaur	University of Lethbridge, Canada
Sathish Govindarajan	Indian Institute of Science, India
Joachim Gudmundsson	University of Sydney, Australia
Wilfried Imrich	Montanuniversität Leoben, Austria
Subrahmanyam Kalyanasundaram	Indian Institute of Technology Hyderabad, India
Rogers Mathew	Indian Institute of Technology Hyderabad, India
Neeldhara Misra	Indian Institute of Technology Gandhinagar, India
Wolfgang Mulzer	Freie Universität Berlin, Germany
Rahul Muthu	Dhirubhai Ambani University, India

Bhawani Panda	Indian Institute of Technology Delhi, India
Arti Pandey	Indian Institute of Technology Ropar, India
Iztok Peterin	University of Maribor, Slovenia
Dömötör Pálvölgyi	Eötvös Loránd University, Hungary
Indhumathi Raman	PSG College of Technology, India
Bodhayan Roy	Indian Institute of Technology Bombay, India
Maria Saumell	Czech Technical University in Prague, Czech Republic
Rishi Ranjan Singh	Indian Institute of Technology Bhilai, India
André van Renssen	University of Sydney, Australia

Organizing Committee

Zin Mar Myint	Indian Institute of Technology Dharwad, India
Supraja D. K.	Indian Institute of Technology Dharwad, India
Kumud Singh Porte	Indian Institute of Technology Dharwad, India
Sandeep R. B.	Indian Institute of Technology Dharwad, India
Sagnik Sen (Chair)	Indian Institute of Technology Dharwad, India

Additional Reviewers

Adiga, Shreyas	Haslegrave, John
Alvarez Montaner, Josep	Huang, Zijin
Ashok, Pradeesha	Huemer, Clemens
Bandopadhyay, Susobhan	Inamdar, Tanmay
Basavaraju, Manu	Jana, Satyabrata
Basu Roy, Aniket	Jany, Ben
Berthe, Gaétan	Jung, Mook Kwon
Bhyravarapu, Sriram	Kalb, Antonia
Bhyravarapu, Sriram	Kare, Anjeneya Swami
Chakraborty, Dipayan	Kasthurirangan, Prahlad Narasimhan
Deryckere, Lindsey	Khare, Niraj
Dhar, Amit Kumar	Khramova, Antonina
Dutta, Madhura	Kielak, Bartek
Fiala, Jiří	Kuppusamy, Lakshmanan
Francis, Mathew	Lorieau, Lucas
Gahlawat, Harmender	M. A. Shalu
Gorain, Barun	M. Mitra, Ritam
Gupta, Shiwali	Madireddy, Raghunath Reddy
Guzman Pro, Santiago	Majumder, Atrayee

Manna, Bubai
Matamala, Martín
Mishra, Tapas Kumar
Mondal, Kaushik
Nasre, Meghana
P. Ragukumar
Padinhatteeri, Sajith
Paul, Kaustav
Pradhan, Dinabandhu
Rachuri, Anirudh
Ramdas, Omkar
Reddy, Vinod
Reijnders, Luuk
S. Kavitha
Sahu, Abhishek
Sankarnarayanan, Brahadeesh
Sanpui, Md Azharuddin

Sarma Sarkar, Rohit
Saurabh, Nitin
Shankar, Umesh
Simoens, Robin
Singh, Anurag
Singh, Karamjeet
Singh, Ranveer
Sinha, Deepa
Szabó, Balázs
Tarsissi, Lama
Tiwary, Hans Raj
Tompkins, Casey
Tripathi, Vikash
van de Berg, Nils
Venkitesh, S.
Wong, Sampson
Zólomy, Kristóf

Abstracts of Invited Talks

Facility Location Problems

Sandip Das

Indian Statistical Institute, Kolkata
sandipdas@isical.ac.in

Abstract. Facility location problems are mathematical optimization problems used to determine the most strategic locations for facilities to achieve specific objectives, such as minimizing costs, maximizing service coverage, or balancing equitable access, while adhering to constraints like capacity, budget, or distance. These problems play a critical role in logistics, supply chain management, and public service planning. Center, Median, Coverage are some important variations of these problems. These problems are considered in geometry as well as graph domain. We will discuss those in this talk.

Homogeneous Substructures in (Random) Ordered Matchings

Andrzej Dudek

Western Michigan University

A hypergraph on a linearly ordered vertex set is called *ordered*, and it is *r-uniform*, $r \geq 1$, if all its edges have the same size r. Two ordered hypergraphs are *order-isomorphic* if there is an isomorphism between them preserving their linear orders. An *ordered r-matching* of *size n* is an ordered r-uniform hypergraph consisting of n pairwise disjoint edges (and no isolated vertices). For fixed r and n, there are precisely

$$\frac{(rn)!}{(r!)^n n!}$$

distinct ordered r-matchings of size n. The family of all of them will be denoted by $M_n^{(r)}$.

A natural and convenient way to represent ordered r-matchings is in terms of *words*. One simply fixes an alphabet of size n and assigns different letters to different edges (so the resulting word contains each letter exactly r times). For instance, if $r = 3$, $n = 4$, and

$$M = \{1, 2, 4\}, \{3, 10, 12\}, \{5, 7, 11\}, \{6, 8, 9\},$$

then two examples of words representing M are

$$AABACDCDDBCB \text{ and } CCECDFDFFEDE.$$

Obviously, there are many such words, differing only in the choice of letters in the alphabet and their assignment to the edges.

Ordered r-matchings consisting of two edges only (i.e., $n = 2$) are called *r-patterns*. For each $r \geq 2$ there are exactly $\frac{1}{2}\binom{2r}{r}$ of them. For example, when $r = 3$ the pattern $AAABBB$ forms a *line*. For a fixed pattern P, a *P-clique* is an ordered matching in which every pair of edges forms pattern P. Hence, if $P = AAABBB$, then the P-clique of size five is of the form

$$AAABBBCCCDDDEEE.$$

In this talk, we present an Erdős--Szekeres type result guaranteeing, in every $M_n^{(r)}$, the presence of a P-clique of a prescribed size for some pattern P. In addition, for each pattern P (as well as for several classes of sets of patterns), we determine with high probability the largest possible size of a P-clique in a *random* ordered r-uniform hypergraph.

This is joint work based on several projects carried out together with Jarosław Grytczuk, Jakub Przybyło, and Andrzej Ruciński.

Keywords: Ordered hypergraphs · Ordered matchings · Erdős--Szekeres · Random hypergraphs

Generalizations of the Crossing Lemma

Géza Tóth[1,2] (iD)

[1] HUN-REN Alfréd Rényi Institute of Mathematics, Budapest, Hungary
[2] Department of Computer Science and Information Theory, Budapest University
of Technology and Economics, Budapest, Hungary
geza@renyi.hu

Abstract. The *crossing number* $CR(G)$ of G is the minimum number of edge crossings over all drawings of G on the plane. It has a fascinating history and lots of applications in theory and practice. It started with Turán's Brick Factory problem in 1944, also known as Zarankiewicz's problem [8]. Determining or even estimating the crossing number is a notoriously difficult problem [7, 11], and it has a huge number of practical and theoretical applications [5, 13] in VLSI design, geographic information systems, graph drawing, computer graphics, and many other fields. Even the crossing numbers of complete graphs are not known, despite the huge number of results, attempts, improvements [1–3, 9]. The Crossing Lemma of Ajtai, Chvátal, Newborn, Szemerédi and independently Leighton, 40 years ago was a real breakthrough. It is a general lower bound on the crossing number, it states that any simple graph with n vertices and $e \geq 4n$ edges has crossing number at least ce^3/n^2 for some $c > 0$.

Székely [13] and then others observed that crossing numbers, in particular the Crossing Lemma, also has important applications in the theory of incidences [12], additive number theory [6], in the analysis of geometric algorithms [4] and many other fields.

In this talk we survey some results related to the Crossing Lemma and its generalizations. First we review the history and improvements of the Crossing Lemma. It is easy to see that it does not hold for multigraphs in general, not even in any weaker form. We investigate its generalizations for multigraphs under different conditions on the drawing. Then we try to figure out, how far can we go with such a generalization, that is, what could be the weakest condition which implies some Crossing Lemma type statement. Finally, we introduce some other versions of the crossing number, the pair-crossing number and the odd-crossing number, their relationships, and generalizations of the Crossing Lemma for these crossing numbers. This talk and abstract is based on the survey [10].

Keywords: Crossing number · Crossing Lemma

References

1. Abrego, B., Cetina, M., Fernández-Merchant, S., Leaños, J., Salazar, G.: On $\leq k$-edges, crossings, and halving lines of geometric drawings of k_n. Discrete Comput. Geom. **48**(1), 192–215 (2012)

2. Aichholzer, O., Duque, F., Fabila-Monroy, R., García-Quintero, O., Hidalgo-Toscano, C.: An ongoing project to improve the rectilinear and the pseudolinear crossing constants. J. Graph Algorithms Appl. **24**(3), 421–432 (2020)

3. Balogh, J., Lidicky, B., Salazar, G.: Closing in on hill's conjecture. SIAM J. Discrete Math. **33**(3), 1261–1276 (2019)

4. Dey, T.K.: Improved bounds for planar k-sets and related problems. Discrete Comput. Geom. **19**, 373–382 (1998)

5. Didimo, W., Liotta, G., Montecchiani, F.: A survey on graph drawing beyond planarity. ACM Comput. Surv. **52**(1), 1–37 (2019)

6. Elekes, G.: On the number of sums and products. Acta Arithmetica **81**(4), 365–367 (1997)

7. Garey, M.R., Johnson, D.S.: Crossing number is np-complete. SIAM J. Algebraic Discrete Methods **4**, 312–316 (1983)

8. Guy, R.K.: The decline and fall of zarankiewicz's theorem. In: Proof techniques in graph theory, pp. 63–69. Academic Press, New York (1969)

9. Harary, F., Hill, A.: On the number of crossings in a complete graph. Proc. Edinburgh Math. Soc. **13**(2), 333–338 (1962)

10. Katona, G.O.H., Patkós, B., Tompkins, C.: Surveys in Extremal Combinatorics and Combinatorial Geometry. Bolyai Society Mathematical Studies, Springer, Berlin. to appear

11. Schaefer, M.: Crossing Numbers of Graphs. CRC Press, Boca Raton (2018)

12. Solymosi, J., Tóth, C.D.: Distinct distances in the plane. Discrete Comput. Geom. **25**, 629–634 (2001)

13. Székely, L.: Crossing numbers and hard erdős problems in discrete geometry. Comb. Probab. Comput. **6**(3), 353–358 (1997)

Contents

Abstracts for Posters

Localization: A Framework to Generalize Extremal Graph Problems

Rajat Adak[(✉)] and L. Sunil Chandran[(✉)]

Department of Computer Science and Automation, Indian Institute of Science, Bangalore, India
{rajatadak,sunil}@iisc.ac.in

Abstract. Extremal graph theory studies the maximum or minimum number of subgraphs isomorphic to a prescribed graph under given constraints. *Localization* has recently emerged as a framework that refines such problems by assigning extremal quantities locally (to vertices or edges) and then aggregating them. This perspective not only recovers classical results but also leads to sharper bounds. A classical result states that a connected planar graph with a finite girth g satisfies

$$m \le \frac{g}{g-2}(n-2)$$

Wood [19] derived upper bounds on the number of K_t-cliques in graphs of bounded maximum degree, expressed in terms of both the number of vertices and the number of edges:

$$ex(n, K_t, K_{1,d+1}) \le \frac{n}{d+1}\binom{d+1}{t}$$

$$mex(m, K_t, K_{1,d+1}) \le \frac{m}{\binom{d+1}{2}}\binom{d+1}{t}$$

More recently, Chakraborty and Chen [9] established a similar upper bound for graphs with bounded path length:

$$mex(m, K_t, P_{r+1}) \le \frac{m}{\binom{r}{2}}\binom{r}{t}$$

In this paper, we employ the localization framework to improve these bounds and provide structural characterizations of the extremal graphs attaining them.

Keywords: Extremal graph theory · Localization · Planar graphs · Cliques · Turán number

1 Introduction

Extremal graph theory, initiated by Turán, studies the maximum or minimum number of copies of a certain subgraph in a graph that does not contain any

N. Misra and A. Pandey (Eds.): CALDAM 2026, LNCS 16445, pp. 1–15, 2026.
https://doi.org/10.1007/978-3-032-17156-6_1

member of a prescribed family as a subgraph. Some of the earliest foundational results in extremal graph theory are the following classical theorems.

Theorem 1. (Turán [17]) *Let G be a simple graph on n vertices with clique number at most r. Then*

$$|E(G)| \leq \frac{n^2(r-1)}{2r},$$

with equality if and only if G is a regular Turán graph on n vertices with r classes.

Definition 1. *A planar graph G is called a k-angulation if it admits a planar embedding in which every face (including the outer face) is a cycle of length k.*

Theorem 2. *Let G be a simple planar connected graph on n vertices with finite girth g, where girth is the length of the smallest cycle in the graph. Then*

$$|E(G)| \leq \frac{g}{g-2}(n-2)$$

with equality if and only if G is a g-angulation.

Let $\mathcal{F}$ be a family of graphs. A graph is considered $\mathcal{F}$-free if it contains no subgraph that is isomorphic to any graph in the family $\mathcal{F}$. The Turán number of an $\mathcal{F}$-free graph on n vertices is denoted by $ex(n, \mathcal{F})$, which counts the maximum number of edges in such a graph. The generalized Turán numbers $ex(n, H, \mathcal{F})$ and $mex(m, H, \mathcal{F})$ extend this framework by counting the maximum number of copies of a fixed graph H that can appear in $\mathcal{F}$-free graphs on n vertices or with m edges, respectively. For the special case $\mathcal{F} = \{F\}$, we simplify the notation to $ex(n, H, F)$ (or $mex(m, H, F)$). These parameters have been widely studied when $\mathcal{F}$ is a natural family of graphs, such as paths, cycles, and stars and H is a clique.

Notation. Let $N(G, K_t)$ denote the number of subgraphs of G isomorphic to K_t.

Theorem 3. (Wood [19]) *Let $t \geq 1$ and G be a graph on n vertices with maximum degree, $\Delta(G) = d$. Then*

$$N(G, K_t) \leq ex(n, K_t, K_{1,d+1}) \leq \frac{n}{d+1}\binom{d+1}{t} = \frac{n}{t}\binom{d}{t-1}$$

If $t = 1$, we always get equality; otherwise, equality holds if and only if G is a disjoint union of copies of K_{d+1}.

Corollary 1. (Wood [19]) *Let $t \geq 1$ and G be a graph on n vertices with maximum degree $\Delta(G) = d$. Then*

$$N(G, K_{\geq 1}) \leq \frac{n}{d+1}(2^{d+1} - 1).$$

Theorem 4. (Wood [19]) *Let $t \geq 2$ and G be a graph with m edges and maximum degree, $\Delta(G) = d$. Then*

$$N(G, K_t) \leq mex(m, K_t, K_{1,d+1}) \leq \frac{m}{\binom{d+1}{2}}\binom{d+1}{t} = \frac{m}{\binom{t}{2}}\binom{d-1}{t-2}$$

If $t = 2$, we always get equality; otherwise, equality holds if and only if G is a disjoint union of copies of K_{d+1} with any number of isolated vertices.

Corollary 2. (Wood [19]) *Let $t \geq 2$ and G be a graph with m edges and maximum degree $\Delta(G) = d$. Then*

$$N(G, K_{t \geq 2}) \leq \frac{m}{\binom{d+1}{2}}(2^{d+1} - d - 2).$$

Theorem 5. (Chakraborti-Chen [9]) *Let $t \geq 2$ and G be a P_{r+1}-free graph on m edges, then*

$$N(G, K_t) \leq mex(m, K_t, P_{r+1}) \leq \frac{m}{\binom{r}{2}}\binom{r}{t}$$

If $t = 2$, equality always holds; otherwise, equality holds if and only if G is a disjoint union of copies of K_r with any number of isolated vertices.

1.1 Localization

A classical lower bound on the independence number $\alpha(G)$ of a graph G is, $\alpha(G) \geq \frac{n}{\Delta+1}$, where Δ denotes the maximum degree of G. While this inequality depends only on a global parameter, namely the maximum degree, stronger bounds can be obtained by incorporating local structural information. In this direction, Caro [8] and Wei [18] independently established a celebrated refinement by replacing Δ with the degree of each individual vertex. Their result asserts that

$$\alpha(G) \geq \sum_{v \in V(G)} \frac{1}{d(v) + 1},$$

where $d(v)$ denotes the degree of vertex $v \in V(G)$. This formulation highlights the power of localized parameters in strengthening classical extremal results, and naturally, this led to the question of whether other graph parameters, when localized, could provide comparable extensions of classical extremal theorems.

Bradač [7] and Malec–Tompkins [16] recently introduced the notion of *localization* as a tool to strengthen and generalize classical extremal bounds. Observe that the constraint in Theorem 1 relies on a global graph parameter—specifically, the maximum clique size. In contrast, the localization framework assigns *local weights* to graph elements (such as vertices or edges), thereby transforming global restrictions into locally defined conditions.

Bradač [7] and Malec–Tompkins [16] defined a weight function $w : E(G) \to \mathbb{Z}$, where the value of $w(e)$ is determined by the particular extremal bound under consideration. In the case of Theorem 1, $w(e)$ is defined as the order of the largest clique in G containing the edge e. They proved the following result:

Theorem 6. (Bradač [7], Malec–Tompkins [16]) *For a simple graph G with n vertices,*

$$\sum_{e \in E(G)} \frac{w(e)}{w(e) - 1} \leq \frac{n^2}{2},$$

with equality if and only if G is a regular Turán graph.

It is easy to verify that Theorem 6 implies the bound of Theorem 1, when we assume G to be K_{r+1}-free. These localized weights allow for the improvement of the classical bound in Theorem 1 by replacing the global parameter, maximum clique size, with these edge weights.

In [20], the localization framework was further applied to generalize the Erdős–Gallai theorem for cycles [10] and the corresponding path theorem was localized in [16]. Kirsch and Nir [13] established a localized version of Zykov's generalization [22] of Turán's theorem, which was subsequently further improved in [6], the authors provided a localization for the Graph Maclaurin Inequalities [12]. Moreover, the localization framework has also been extended to hypergraphs [21] and to spectral extremal bounds [11,14]. This further illustrates the versatility and potential of the localization framework to extend across different domains within extremal graph theory.

While edge-based localization often produces distribution-type inequalities reminiscent of the LYM or Kraft inequalities, a more direct alternative was proposed in [3], where localization is applied to vertices rather than edges.

Adak and Chandran [3] introduced the concept of *vertex-based localization.* They provided a vertex-based localization of the Erdős–Gallai theorems for paths and cycles. In subsequent work [5], the same authors established a vertex-based localization of a popular form of Turán's theorem. For a simple graph G they proved that,

$$|E(G)| \leq \frac{n}{2} \left\lfloor \sum_{v \in V(G)} \frac{c(v) - 1}{c(v)} \right\rfloor$$

where the weight function $c : V(G) \to \mathbb{Z}$ is defined such that $c(v)$ is the order of the largest clique containing the vertex v.

More recently, in [4], they extended this localization framework to improve the bounds of generalized Erdős–Gallai theorems due to Luo [15] and in [1] they provided a vertex-based localization of Zykov's Theorem.

In this paper, we use the localization framework to generalize Theorem 2, 3, 4 and 5, and characterize the extremal graphs for these strengthened bounds.

2   Main Results

We provide a vertex-based localization of Theorem 3, and generalize Theorems 2, 4 and 5, by assigning suitable weights to the edges of the graph and applying the localization framework.

Definition 2. *Define $g : E(G) \to \mathbb{Z}$, such that $g(e)$ is the length of the smallest cycle edge e is part of, and if e does not participate in any cycle, set $g(e) = 2$.*

Theorem 7. *Let G be a planar connected graph on n vertices. Then*

$$\sum_{e \in E(G)} \frac{g(e) - 2}{g(e)} \le n - 2$$

equality holds if and only if G is either K_2 or is a k-angulation for some k.

Theorem 8. *Let $t \ge 1$ and G be a simple graph then,*

$$N(G, K_t) \le \sum_{v \in V(G)} \frac{\binom{d(v)+1}{t}}{d(v) + 1} = \frac{1}{t} \sum_{v \in V(G)} \binom{d(v)}{t - 1}$$

Let $X = \{v \in V(G) \mid d(v) \ge t - 1\}$. If $t = 1$, equality always holds; otherwise, equality holds if and only if all the components of $G[X]$ are cliques.

Corollary 3. *Let $t \ge 1$ and G be a simple graph. Then*

$$N(G, K_{t \ge 1}) \le \sum_{v \in V(G)} \sum_{t=1}^{d(v)+1} \frac{\binom{d(v)+1}{t}}{d(v) + 1} = \sum_{v \in V(G)} \frac{2^{d(v)+1} - 1}{d(v) + 1}.$$

Definition 3. *A graph G is called a diamond-graph if it is isomorphic to K_4 minus one edge. A graph is diamond-free if it has no diamond as an induced subgraph.*

Definition 4. *Define $w : E(G) \to \mathbb{Z}$, such that $w(e) = $ Number of common neighbors of the endpoints of e. In other words, $w(e)$ is the number of triangles that edge e is part of in G. This can be seen as the degree of edge e.*

Theorem 9. *Let $t \ge 2$ and G be a simple graph then,*

$$N(G, K_t) \le \sum_{e \in E(G)} \frac{\binom{w(e)+2}{t}}{\binom{w(e)+2}{2}} = \frac{1}{\binom{t}{2}} \sum_{e \in E(G)} \binom{w(e)}{t - 2}$$

Let $Y = \{e \in E(G) \mid w(e) < t - 2\}$. If $t = 2$, equality always holds; otherwise, equality holds if and only if $G \setminus Y$ is a diamond-free graph.

Corollary 4. *Let $t \ge 2$ and G be a simple graph. Then*

$$N(G, K_{t \ge 2}) \le \sum_{e \in E(G)} \sum_{t=2}^{w(e)+2} \frac{\binom{w(e)+2}{t}}{\binom{w(e)+2}{2}} = \sum_{e \in E(G)} \frac{2^{w(e)+2} - w(e) - 3}{\binom{w(e)+2}{2}}.$$

Corollaries 3 and 4 can be seen as the localized counterparts of Corollaries 1 and 2 respectively.

Definition 5. *Define* $p : E(G) \rightarrow \mathbb{Z}$*, such that* $p(e)$ *is the length of the longest path in* G *containing the edge* e*.*

Theorem 10. *Let* $t \geq 2$ *and* G *be a simple graph, then*

$$N(G, K_t) \leq \sum_{e \in E(G)} \frac{\binom{p(e)+1}{t}}{\binom{p(e)+1}{2}} = \frac{1}{\binom{t}{2}} \sum_{e \in E(G)} \binom{p(e) - 1}{t - 2}$$

Let $Z = \{e \in E(G) \mid p(e) < t - 1\}$*. If* $t = 2$*, equality always holds; otherwise, equality holds if and only if all the components of* $G \setminus Z$ *are cliques.*

The recovery of the classical bounds from our results is deferred to the full version of the paper [2].

3   Proof of Results

3.1   Proof of Theorem 7

Proof. Since G is planar and connected, from Euler's formula, we get that

$$|F(G)| = 2 - |V(G)| + |E(G)| \tag{1}$$

Let $F(G) = \{f_1, f_2, \ldots, f_{|F(G)|}\}$. If an edge $e \in E(G)$ is incident to a face f_i for some $i \in [\,|F(G)|\,]$, then $g(e) \leq d(f_i)$, where $d(f_i)$ denotes the degree of the face f_i, that is the number of edges in the boundary walk of f_i (thus cut-edges are counted twice).

Observe that every edge of G which is not a cut edge is incident to exactly two faces, contributing once to the degree of each of them. In contrast, a cut-edge is incident to a single face but contributes twice to its degree.

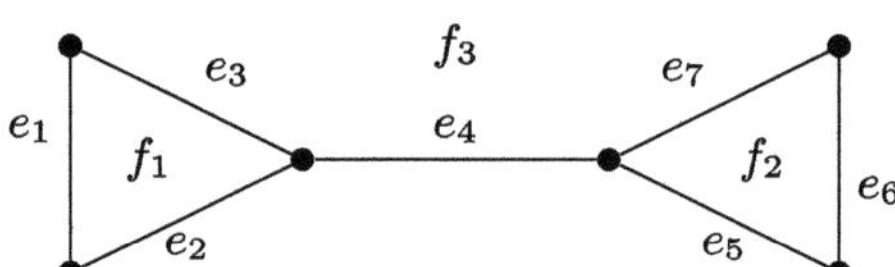

Fig. 1. $E(f_1) = \{e_1, e_2, e_3\}$, thus $d(f_1) = 3$. But $E(f_3) = \{e_1, e_2, e_4, e_5, e_6, e_7, e_4, e_3\}$, note that e_4 comes twice, and therefore $d(f_3) = 8$.

Let $E(f_i)$ be the multiset of edges in the boundary walk of f_i, where the cut-edges in the boundary of f_i, appear twice. Refer to Fig. 1 for an example. Note that $|E(f_i)| = d(f_i)$. Thus, we get

$$2 \sum_{e \in E(G)} \frac{1}{g(e)} \geq \sum_{i=1}^{|F(G)|} \sum_{e \in E(f_i)} \frac{1}{d(f_i)} = \sum_{i=1}^{|F(G)|} d(f_i) \frac{1}{d(f_i)} = |F(G)| \tag{2}$$

Thus from Eqs. (1) and (2) we get that

$$2 \sum_{e \in E(G)} \frac{1}{g(e)} \geq 2 - |V(G)| + |E(G)|$$

$$\implies n - 2 \geq \sum_{e \in E(G)} \left(1 - \frac{2}{g(e)}\right) = \sum_{e \in E(G)} \left(\frac{g(e) - 2}{g(e)}\right)$$

For the extremal analysis, we first consider the case when G is either K_2 or a k-angulation. If $G \cong K_2$, then $E(G) = \{e\}$ and $g(e) = 2$. In this case, both sides of the bound evaluate to 0. If G is a k-angulation, it follows that $g(e) = k$ for all $e \in E(G)$ and $d(f_i) = k$ for all $f_i \in F(G)$. Hence, equality holds in Eq. (2), and consequently, in the bound of Theorem 7.

For the converse, assume equality holds in the bound of Theorem 7 with $G \not\cong K_2$. This implies equality in Eq. (2), and therefore $d(f_i) = g(e)$ for every $f_i \in F(G)$ and every $e \in E(f_i)$. Since $g(e) = 2$ if and only if e is a cut-edge, while $d(f_i) > 2$ for all $f_i \in F(G)$ (as $G \not\cong K_2$), it follows that G contains no cut-edges and therefore, each face in G is bounded by a cycle.

Two faces are *adjacent* if they share at least one edge. In the dual graph, this corresponds to the adjacency of the associated vertices. Similarly, two faces are *connected* if the corresponding vertices in the dual graph are connected.

If f_i and f_j are adjacent and share an edge e, then we have $d(f_i) = d(f_j) = g(e)$. Since the dual graph of any planar graph is connected, every face is connected to every other face. By propagating this equality of $d(\cdot)$ along paths in the dual, it follows that all faces share the same degree. Hence $d(f)$ is constant for all $f \in F(G)$.

Since every face is bounded by a cycle, the existence of a face f whose boundary cycle has length strictly less than $d(f)$ would imply the presence of an edge e on the boundary of f with $g(e) < d(f)$. This is a contradiction. Hence, each face must be bounded by a simple cycle of length exactly $d(f)$. Consequently, G is a k-angulation with $k = d(f)$ for any $f \in F(G)$. $\qquad\square$

3.2 Proof of Theorem 8

Notation. If $v \in V(G)$, then $N(G, K_t, v)$ denotes the number of copies of K_t in G containing the vertex v.

Recall that the set X consists of all vertices in G whose degrees are at least $t - 1$.

Lemma 1. *Theorem 8 holds for G if and only if it holds for $G[X]$.*

For the proof of the above lemma, please refer to the full version of the paper [2]. Therefore, without loss of generality, we may assume that G contains only vertices of degree at least $t - 1$. Now we are ready to prove Theorem 8.

8   R. Adak and L. S. Chandran

Proof. For $t = 1$, the bound and the equality case can both be verified trivially; thus, we assume $t \geq 2$. Since we assumed $d(v) \geq t - 1$ for all $v \in V(G)$, we get that G does not contain any isolated vertex; therefore, $\delta(G) > 0$.

Proof by induction on the number of vertices.

Base Case: Since $\delta(G) > 0$, $|V(G)| \geq 2$. For $n = 2$, let $V(G) = \{u, v\}$ and $E(G) = \{\{u, v\}\}$ (since $\delta(G) > 0$). For $t > 2$, the bound is trivially true since both sides are zero. For $t = 2$, $N(G, K_t) = 1$ and the right-hand side also sums up to 1.

Induction Hypothesis: Suppose the claim in Theorem 8 is true for all graphs with fewer than $|V(G)|$ vertices.

Induction Step: Let $x \in V(G)$ such that $0 < d(x) = \delta(G)$. Now define $G' = G \setminus \{x\}$. Recall that $N(G, K_t, x)$ denotes the number of copies of K_t in G containing the vertex $x \in V(G)$. Clearly

$$N(G, K_t) = N(G', K_t) + N(G, K_t, x) \tag{3}$$

Note that for every K_t containing x, there exists a unique copy of K_{t-1} in $G[N(x)]$. Therefore

$$N(G, K_t, x) = N(G[N(x)], K_{t-1}) \leq \binom{d(x)}{t-1} \tag{4}$$

Let $d_{G'}(v)$ be the degree of vertex v in G'. Now note that $d_{G'}(v) = d(v)$ for all $v \in V(G') \setminus N(x)$, where $N(x)$ denotes the set of vertices adjacent to x in G. Clearly, $d_{G'}(v) = d(v) - 1$ for all $v \in N(x)$. From the induction hypothesis, we know that

$$N(G', K_t) \leq \frac{1}{t} \sum_{v \in V(G')} \binom{d_{G'}(v)}{t-1} = \frac{1}{t} \sum_{v \in V(G') \setminus N(x)} \binom{d(v)}{t-1} + \frac{1}{t} \sum_{v \in N(x)} \binom{d(v) - 1}{t-1} \tag{5}$$

Thus from Eqs. (3), (4) and (5) we get that

$$N(G, K_t) \leq \frac{1}{t} \sum_{v \in V(G') \setminus N(x)} \binom{d(v)}{t-1} + \frac{1}{t} \sum_{v \in N(x)} \binom{d(v) - 1}{t-1} + \binom{d(x)}{t-1} \tag{6}$$

Note that

$$\frac{1}{t} \binom{d(v) - 1}{t-1} = \frac{(d(v) - t + 1)}{t d(v)} \binom{d(v)}{t-1} = \frac{1}{t} \binom{d(v)}{t-1} - \frac{1}{d(v)} \binom{d(v)}{t-1} + \frac{1}{t d(v)} \binom{d(v)}{t-1} \tag{7}$$

Thus from Eqs. (6) and (7) we get

$$N(G, K_t) \leq \binom{d(x)}{t-1} + \frac{1}{t} \sum_{v \in V(G') \setminus N(x)} \binom{d(v)}{t-1}$$

$$+ \sum_{v \in N(x)} \left(\frac{1}{t} \binom{d(v)}{t-1} - \frac{1}{d(v)} \binom{d(v)}{t-1} + \frac{1}{t d(v)} \binom{d(v)}{t-1} \right) \tag{8}$$

$$= \binom{d(x)}{t-1} + \frac{1}{t} \sum_{v \in V(G')} \binom{d(v)}{t-1} + \sum_{v \in N(x)} \left(\frac{1}{t d(v)} \binom{d(v)}{t-1} - \frac{1}{d(v)} \binom{d(v)}{t-1} \right) \tag{9}$$

Claim 1. *For $d(v) \geq d(x) > 0$ and $t \geq 2$ we have*

$$\sum_{v \in N(x)} \left(\frac{1}{td(v)} \binom{d(v)}{t-1} - \frac{1}{d(v)} \binom{d(v)}{t-1} \right) + \binom{d(x)}{t-1} \leq \frac{1}{t} \binom{d(x)}{t-1}$$

Proof. Note that $\delta(G) = d(x) \leq d(v)$ for all $v \in V(G)$. Therefore, we get $\frac{1}{t-1}\binom{d(v)-1}{t-2} \geq \frac{1}{t-1}\binom{d(x)-1}{t-2} \implies \frac{1}{d(v)}\binom{d(v)}{t-1} \geq \frac{1}{d(x)}\binom{d(x)}{t-1}$. Thus we get

$$\binom{d(x)}{t-1} = \sum_{v \in N(x)} \frac{1}{d(x)} \binom{d(x)}{t-1} \leq \sum_{v \in N(x)} \frac{1}{d(v)} \binom{d(v)}{t-1} \tag{10}$$

$$\implies \frac{t-1}{t} \binom{d(x)}{t-1} \leq \sum_{v \in N(x)} \frac{t-1}{td(v)} \binom{d(v)}{t-1}$$

$$\implies \binom{d(x)}{t-1} - \frac{1}{t} \binom{d(x)}{t-1} \leq \sum_{v \in N(x)} \left(-\frac{1}{td(v)} \binom{d(v)}{t-1} + \frac{1}{d(v)} \binom{d(v)}{t-1} \right)$$

$$\implies \sum_{v \in N(x)} \left(\frac{1}{td(v)} \binom{d(v)}{t-1} - \frac{1}{d(v)} \binom{d(v)}{t-1} \right) + \binom{d(x)}{t-1} \leq \frac{1}{t} \binom{d(x)}{t-1}$$

$\square$

From Eq. (8) and Claim 1 we get that

$$N(G, K_t) \leq \frac{1}{t} \binom{d(x)}{t-1} + \frac{1}{t} \sum_{v \in V(G')} \binom{d(v)}{t-1} = \frac{1}{t} \sum_{v \in V(G)} \binom{d(v)}{t-1}$$

Thus, we get the required bound. Now we will characterize the extremal graphs for Theorem 8.

First, suppose that all the components of G are cliques. Let C be one such component with order k. Clearly $d(v) = k - 1$ for all $v \in V(C)$. And C contains $\binom{k}{t}$ copies of K_t. The contribution by the vertices of C to the right side of the bound in Theorem 8 is,

$$\frac{1}{t} \sum_{v \in V(C)} \binom{k-1}{t-1} = \frac{k}{t} \binom{k-1}{t-1} = \binom{k}{t}$$

Thus, we get equality in the bound.

Now, assume that we have equality in the bounds of Theorem 8. Thus, we must have equality in Eqs. (4), (5) and (10). From equality in Eq. (4), we get that $G[N[x]]$ is a clique, where $N[x]$ denotes the closed neighborhood of x. Using the induction hypothesis and the equality in Eq. (5), we get that all the components of G' are cliques.

Assume that C is a connected component of G, then it is enough to show that C is a clique. If C does not contain x, clearly C does not contain any vertex from $N(x)$. Thus, C is a component in G' as well, and therefore, by induction hypothesis, C is a clique. Now suppose C contains the vertex x, therefore C contains all the vertices of $N(x)$. Since $N[x]$ induces a clique in G, $C \setminus \{x\}$ must be connected and therefore, $C \setminus \{x\}$ must be a component of G', and thus is a clique. Suppose there exists $y \in V(C) \setminus N[x]$. Clearly y is adjacent to $v \in N(x)$ but not adjacent to x. Therefore, $d(v) > d(x)$. But from equality in the bound of Eq. (10) we get that, $d(x) = d(v)$ for all $v \in N(x)$. We get a contradiction. Therefore, $V(C) = N[x]$, which induces a clique. $\qquad \square$

3.3   Proof of Theorem 9

Recall that $Y = \{e \in E(G) \mid w(e) < t - 2\}$

Lemma 2. *Theorem 9 holds for G if and only if it holds for $G \setminus Y$*

For the proof of the above lemma, please refer to the full version of the paper [2]. Therefore, without loss of generality, we may assume that G contains only edges with weights at least $t - 2$.

Proof. If $t = 2$, clearly the bound holds with equality; thus, assume $t \geq 3$. Let $e \in E(G)$, if $e = \{u, v\}$ where $u, v \in V(G)$. If A is a K_t containing the edge e, then $u, v \in V(A)$ and $V(A) \setminus \{u, v\} \subseteq N(u) \cap N(v)$. Note that $|N(u) \cap N(v)| = w(e)$. Thus, the number of copies of K_t containing the edge e is at most $\binom{w(e)}{t-2}$. If we add up the contribution of each edge, that is the number of copies of K_t, containing an edge, then each copy is counted $\binom{t}{2}$ times, since it is counted once for each edge participating in that K_t copy. Thus, we get;

$$N(G, K_t) \leq \frac{1}{\binom{t}{2}} \sum_{e \in E(G)} \binom{w(e)}{t - 2}$$

Thus, we get the required bound. Now we will characterize the extremal graphs for Theorem 9.

First, assume we have equality in the bound of Theorem 9. Therefore, we must have the number of copies of K_t containing the edge e to be exactly $\binom{w(e)}{t-2}$ for all $e = \{u, v\} \in E(G)$. Thus, the vertices in $N(u) \cap N(v)$ must induce a clique. Now suppose G contains an induced diamond as shown in Fig. 2. Note that $\{c, d\} \in E(G)$ and $a, b \in N(c) \cap N(d)$, but a and b are not adjacent. Thus, we get a contradiction; therefore, G is diamond-free.

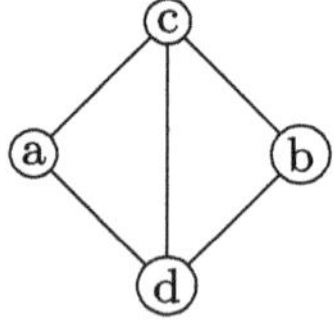

Fig. 2. Diamond in the extremal graph G.

Now suppose G is diamond-free. Suppose there exists an edge $e = \{x, y\} \in E(G)$ such that $N(x) \cap N(y)$ does not induce a clique in G. Since $w(e) \geq t-2 \geq 1$, $N(x) \cap N(y)$ is non-empty, thus, there exists $u, v \in N(x) \cap N(y)$ such that u and v are not adjacent. Therefore, we get a diamond induced by u, v, x, and y. Thus, the common neighbors of the endpoints of any edge in G must induce a clique. Therefore, the number of copies of K_t containing some $e \in E(G) = \binom{w(e)}{t-2}$. Therefore, we get G must be extremal for Theorem 9. $\qquad\square$

An alternative proof for Theorem 8, which utilizes Theorem 9, is provided in the full version of the paper [2].

3.4 Proof of Theorem 10

Recall that $Z = \{e \in E(G) \mid p(e) < t - 1\}$.

Lemma 3. *Theorem 10 holds for G if and only if it holds for $G \setminus Z$.*

We defer the proof of the lemma to the full version of the paper [2]. Therefore, without loss of generality, we may assume $p(e) \geq t - 1$ for all $e \in E(G)$.

Proof. If $t = 2$, clearly the bound holds with equality; thus, assume $t \geq 3$.
Proof by induction on the number of vertices of the graph.

Base Case: Since $p(e) \geq t - 1 \geq 2$, $|V(G)| \geq 3$. For $n = 3$, let $V(G) = \{u, v, w\}$ and $\{\{u, v\}, \{v, w\}\} \subseteq E(G)$ (since $p(e) \geq 2$). If $t > 3$, the bound is trivially true as both sides are zero. If $t = 3$, $N(G, K_t) \leq 1$, with equality if and only if $G \cong K_3$. Clearly, the right-hand side sums up to 1 when $t = 3$.

Induction Hypothesis: Suppose the claim in Theorem 10 is true for all graphs with fewer than $|V(G)|$ vertices.

Induction Step: If G is disconnected, by the induction hypothesis, the statement of the theorem is true for each connected component and thus for G. Therefore, assume G to be connected. Let $k = max\{p(e) \mid e \in E(G)\}$ and P be a k-length path in G. Let $P = v_0, v_1, \ldots, v_k$. Now consider the following two cases;

Case 1. There exists a $(k + 1)$-length cycle in G.
Without loss of generality assume the endpoints of P, v_0 and v_k, are adjacent thus forming a $(k + 1)$-length cycle.

Claim 1. $V(P) = V(G)$.

Proof. We know that v_0 is adjacent to v_k. Suppose $V(G) \setminus V(P) \neq \emptyset$. Since G is connected, there exists a vertex $v \in V(G) \setminus V(P)$, adjacent to some vertex in P. Since the vertices of P form a cycle, without loss of generality, we can assume v is adjacent to v_0. Now consider the path;

$$Q = v, v_0, v_1, \ldots, v_k$$

Thus Q is a path in G with length $k + 1$. Thus, we arrive at a contradiction to the assumption that the length of the longest path in G is k. $\qquad\square$

From Claim 1 we get that $p(e) = |V(G)| - 1 = n - 1$, for all $e \in E(G)$. Therefore;

$$\frac{1}{\binom{t}{2}} \sum_{e \in E(G)} \binom{p(e) - 1}{t - 2} = \frac{1}{\binom{t}{2}} \sum_{e \in E(G)} \binom{n - 2}{t - 2} \tag{11}$$

Recall from the bound of Theorem 9

$$N(G, K_t) \leq \frac{1}{\binom{t}{2}} \sum_{e \in E(G)} \binom{w(e)}{t - 2}$$

where $w(e)$ is the number of common neighbors of the endpoints of $e \in E(G)$. Clearly $w(e) \leq n - 2$ for all $e \in E(G)$. Thus we get, $\frac{1}{\binom{t}{2}} \binom{w(e)}{t-2} \leq \frac{1}{\binom{t}{2}} \binom{n-2}{t-2}$. Therefore from Eq. (11) we get that;

$$N(G, K_t) \leq \frac{1}{\binom{t}{2}} \sum_{e \in E(G)} \binom{w(e)}{t - 2} \leq \sum_{e \in E(G)} \frac{1}{\binom{t}{2}} \binom{n - 2}{t - 2} = \frac{1}{\binom{t}{2}} \sum_{e \in E(G)} \binom{p(e) - 1}{t - 2} \tag{12}$$

Clearly, for equality $w(e) = n - 2$ for all $e \in E(G)$. Thus G must be a clique.

Case 2. There does not exist any $(k + 1)$-length cycle in G.

From the assumption of the case, v_0 and v_k are non-adjacent. Without loss of generality assume $d(v_0) \geq d(v_k)$.

Claim 2. $d(v_k) \leq \frac{k}{2}$

Proof. Suppose not, that is $d(v_k) \geq \frac{k+1}{2}$. Thus we get $d(v_0) \geq \frac{k+1}{2}$. Since P is a longest path in G and v_0 is not adjacent to v_k, we get, $N(v_0), N(v_k) \subseteq \{v_1, v_2, \ldots, v_{k-1}\}$. Let $N = N(v_k) \setminus \{v_{k-1}\}$ and $N^+ = \{v_{j+1} \mid v_j \in N\}$. Clearly, $|N| = |N^+| \geq \frac{k+1}{2} - 1 = \frac{k-1}{2}$. Note that $v_0, v_k \notin N^+$. Therefore, $|\{v_1, v_2, \ldots, v_{k-1}\} \setminus N^+| \leq \frac{k-1}{2}$. Since $|N(v_0)| \geq \frac{k+1}{2}$, by pigeonhole principle, we get that, there exists $v_i \in N^+$ such that $v_i \in N(v_0)$ and $v_{i-1} \in N(v_k)$. Consider the cycle, $K = v_0, v_1, \ldots, v_{i-1}, v_k, v_{k-1}, \ldots, v_i, v_0$. Clearly K is a $(k + 1)$-length cycle in G. Thus, we get a contradiction. $\square$

Define $G' = G \setminus \{v_k\}$. Now, using the induction hypothesis, we get that

$$N(G', K_t) \leq \frac{1}{\binom{t}{2}} \sum_{e \in E(G')} \binom{p_{G'}(e) - 1}{t - 2}$$

where for $e \in E(G')$, $p_{G'}(e)$ is the weight of the edge e restricted to G'. Clearly $p_{G'}(e) \leq p(e)$ for all $e \in E(G')$. Thus we get that

$$N(G', K_t) \leq \frac{1}{\binom{t}{2}} \sum_{e \in E(G')} \binom{p_{G'}(e) - 1}{t - 2} \leq \frac{1}{\binom{t}{2}} \sum_{e \in E(G')} \binom{p(e) - 1}{t - 2}$$

Note that the number of copies of K_t containing the vertex v_k is at most $\binom{d(v_k)}{t-1}$. Therefore,

$$N(G, K_t) \leq \binom{d(v_k)}{t - 1} + \frac{1}{\binom{t}{2}} \sum_{e \in E(G')} \binom{p(e) - 1}{t - 2} \tag{13}$$

Claim 3. $p(e) = k$, *for all* $e \in E(G) \setminus E(G')$.

Proof. Recall that, since P is the longest path in G, $N(v_k) \subseteq \{v_1, v_2, \ldots, v_{k-1}\}$. Define $e_i = \{v_i, v_k\} \in E(G) \setminus E(G')$ where $v_i \in N(v_k)$. Thus $E(G) \setminus E(G') = \{e_i \mid v_i \in N(v_k)\}$. Define P_i to be the path formed by removing the edge $\{v_i, v_{i+1}\}$ from P and adding edge e_i. Clearly, P_i is also a k-length path. Therefore, $p(e) = k$ for all $e \in E(G) \setminus E(G')$. $\qquad\square$

Claim 4. *For $t \geq 3$ we have*

$$\binom{d(v_k)}{t-1} < \frac{1}{\binom{t}{2}} \sum_{e \in E(G) \setminus E(G')} \binom{p(e) - 1}{t - 2}$$

Proof. If $\binom{d(v_k)}{t-1} = 0$, the bound is trivially true, since $p(e) \geq t - 1$ for all $e \in E(G)$. Thus we assume $\binom{d(v_k)}{t-1} > 0$. From Claim 3 we have that

$$\frac{1}{\binom{t}{2}} \sum_{e \in E(G) \setminus E(G')} \binom{p(e) - 1}{t - 2} = \frac{d(v_k)}{\binom{t}{2}} \binom{k - 1}{t - 2}$$

Recall that from Claim 2, $d(v_k) \leq \frac{k}{2}$. Therefore, we get

$$\frac{d(v_k)}{\binom{t}{2}} \binom{k - 1}{t - 2} \geq \frac{d(v_k)}{\binom{t}{2}} \binom{2d(v_k) - 1}{t - 2} = \frac{2d(v_k)}{t(t-1)} \binom{2d(v_k) - 1}{t - 2} = \frac{1}{t} \binom{2d(v_k)}{t - 1}$$

Since we assumed $t \geq 3$, we get that

$$\frac{\frac{d(v_k)}{\binom{t}{2}} \binom{k-1}{t-2}}{\binom{d(v_k)}{t-1}} \geq \frac{\frac{1}{t} \binom{2d(v_k)}{t-1}}{\binom{d(v_k)}{t-1}} = \frac{2d(v_k)(2d(v_k) - 1) \ldots (2d(v_k) - t + 2)}{td(v_k)(d(v_k) - 1) \ldots (d(v_k) - t + 2)} \geq \frac{2^{t-1}}{t} > 1$$

$$\qquad\square$$

From Eq. (13) and Claim 4 we get that

$$N(G, K_t) < \frac{1}{\binom{t}{2}} \sum_{e \in E(G')} \binom{p(e) - 1}{t - 2} + \frac{1}{\binom{t}{2}} \sum_{e \in E(G) \setminus E(G')} \binom{p(e) - 1}{t - 2}$$

$$= \frac{1}{\binom{t}{2}} \sum_{e \in E(G)} \binom{p(e) - 1}{t - 2}$$

Clearly, we can not get equality in the bounds of Theorem 10 for Case 2.

Equality in Theorem 10 is possible only for Case 1. Thus, we must have equality in Eq. (12) and therefore $w(e) = n - 2$ for all $e \in E(G)$; thus, G must be a clique. $\qquad\square$

4 Concluding Remarks

In this paper, we developed localized versions of classical extremal bounds, focusing on the number of edges in planar graphs and clique bounds for graphs with bounded maximum degree and bounded path length. By introducing specifically designed local parameters, our results refine and extend the classical bound for planar graphs and improve the bounds of Wood [19] and Chakraborty–Chen [9]. We also characterized the extremal graphs that attain these bounds. These results demonstrate the strength of the localization framework: by shifting the analysis from global to local parameters, it not only recovers known extremal results but also provides sharper bounds and new structural insights, underscoring its potential as a powerful tool for generalizing a wide range of extremal problems.

References

1. Adak, R., Sunil Chandran, L.: Generalized Zykov's theorem. arXiv preprint arXiv:2512.02958 (2025)
2. Adak, R., Sunil Chandran, L.: Localization: a framework to generalize extremal problems. arXiv preprint arXiv:2508.20946 (2025)
3. Adak, R., Sunil Chandran, L.: Vertex-based localization of Erdős-Gallai theorems for paths and cycles. arXiv preprint arXiv:2504.01501 (2025)
4. Adak, R., Sunil Chandran, L.: Vertex-based localization of generalized Turán Problems. arXiv preprint arXiv:2508.20936 (2025)
5. Adak, R., Sunil Chandran, L.: Vertex-based localization of Turán's theorem. arXiv preprint arXiv:2504.02806 (2025)
6. Aragão, L., Souza, V.: Localised graph Maclaurin inequalities. Ann. Comb. **28**(3), 1021–1033 (2024)
7. Bradač, D.: A generalization of Turán's theorem. arXiv preprint arXiv:2205.08923 (2022)
8. Caro, Y.: New results on the independence number. Technical report, Technical Report, Tel-Aviv University (1979)
9. Chakraborti, D., Chen, D.Q.: Exact results on generalized Erdős-Gallai problems. Eur. J. Comb. **120**, 103955 (2024)
10. Erdős, P., Gallai, T.: On maximal paths and circuits of graphs. Acta Math. Acad. Sci. Hungar. **10**, 337–356 (1959)
11. Rajesh Kannan, M., Kumar, H., Pragada, S.: Localization of spectral Turán-type theorems. arXiv preprint arXiv:2512.01409 (2025)
12. Khadzhiivanov, N.: Inequalities for graphs. CR Acad. Sci. Bul **30**(6), 793 (1977)
13. Kirsch, R., Nir, J.D.: A localized approach to generalized Turán problems. Electron. J. Comb. **31**(3), (2024)
14. Liu, L., Ning, B.: Local properties of the spectral radius and perron vector in graphs. J. Comb. Theor. Ser. B **176**, 241–253 (2026)
15. Luo, R.: The maximum number of cliques in graphs without long cycles. J. Comb. Theor. Ser. B **128**, 219–226 (2018)
16. Malec, D., Tompkins, C.: Localized versions of extremal problems. Eur. J. Comb. **112**, 103715 (2023)

17. Turán, P.: On an extremal problem in graph theory. Matematikai és Fizikai Lapok **48**, 436–452 (1941)
18. Wei, V.K.: A lower bound on the stability number of a simple graph (1981)
19. Wood, D.R.: On the maximum number of cliques in a graph. Graphs Comb. **23**(3), 337–352 (2007)
20. Zhao, K., Zhang, X.-D.: A localized approach for turán number of long cycles. J. Graph Theor. **108**(3), 582–607 (2025)
21. Zhao, K., Zhang, X.-D.: Localized version of hypergraph Erdős-Gallai theorem. Discret. Math. **348**(1), 114293 (2025)
22. Zykov, A.A.: On some properties of linear complexes. Matematicheskii Sbornik, Novaya Seriy

Complexity Results on Paired Domination in Cubic Graphs

Deepak M. Bakal[1,2]([✉]) and Y. M. Borse[2]

[1] Department of Applied Sciences, COEP Technological University, Pune, India
iamdeepakbakal@gmail.com
[2] Department of Mathematics, Savitribai Phule Pune University, Pune, India

Abstract. A (total) dominating set D of a graph G with no isolated vertices is called a paired dominating set, if the subgraph induced by D in G contains a perfect matching. The decision problem PAIRED DOMINATION takes as input a graph G and a positive integer k, and asks whether G contains a paired dominating set of size at most k. PAIRED DOMINATION is known to be NP-complete for planar bipartite subcubic graphs. In this paper, we prove that PAIRED DOMINATION remains NP-complete on planar cubic graphs and, assuming the Exponential Time Hypothesis (ETH), the problem cannot be solved in time $2^{o(\sqrt{|V(G)|})}$ in planar cubic graphs. We also show that PAIRED DOMINATION is NP-complete on triangle-free d-regular graphs for each fixed $d \geq 3$, and it does not admit a $2^{o(|V(G)|)}$ time algorithm unless ETH fails. Additionally, we establish that determining if the paired domination number coincides with the total domination number is NP-hard across planar bipartite subcubic graphs, planar cubic graphs, and triangle-free d-regular graphs.

Keywords: Paired domination · Total Domination · Cubic graphs · Planar graphs · Triangle-free graphs

1 Introduction

We consider finite, simple, and undirected graphs without isolated vertices. Let G be a graph with vertex set $V(G)$ and edge set $E(G)$. For terminology and notation related to graphs and computational complexity, we follow [13] and [11] respectively.

A *dominating set* in G is a subset $D \subseteq V(G)$ such that every vertex in $V(G) \setminus D$ is adjacent to at least one vertex in D. The *domination number* of G, denoted by $\gamma(G)$, is the minimum cardinality of a dominating set in G. Given a graph G and an integer k as input, DOMINATING SET is the decision problem of determining whether G contains a dominating set of size at most k. A dominating set D of G is a *total dominating set* if the subgraph induced by D in G, denoted by $G[D]$, is isolate-free. A dominating set D is called a *paired dominating set* if the subgraph $G[D]$ contains a perfect matching. For a paired dominating set D of G, we say that two vertices $u, v \in D$ are *paired* or *partners* if they belong to a

N. Misra and A. Pandey (Eds.): CALDAM 2026, LNCS 16445, pp. 16–28, 2026.
https://doi.org/10.1007/978-3-032-17156-6_2

perfect matching in $G[D]$. It is easy to see that a paired dominating set of G is, by definition, a total dominating set of G. The *paired (resp. total) domination number* of G, denoted by $\gamma_{pr}(G)$ (resp. $\gamma_t(G)$), is the minimum cardinality of a paired (resp. total) dominating set of G.

The decision problem PAIRED DOMINATION takes as input a graph G and a positive integer k, and asks whether G contains a paired dominating set of size at most k. This problem is formally stated as follows. The TOTAL DOMINATION problem is defined similarly.

PAIRED DOMINATION

Instance: A graph G and a positive integer k, such that $k \leq |V|$.

Question: Does there exist is a paired dominating set D of G, such that $|D| \leq k$?

The concept of paired domination was introduced by Haynes and Slater [14], motivated by the area monitoring problem, where each guard must have a unique adjacent guard assigned as a backup. They showed that PAIRED DOMINATION is NP-complete for general graphs. As a result, the complexity of the problem has been studied for various restricted classes of graphs. Chen et al. [6] proved that the PAIRED DOMINATION problem is NP-complete for bipartite graphs. This was strengthened by Panda et al. [26] for perfect elimination bipartite graphs and by Chen et al. [4] for star convex and comb convex bipartite graphs. Chen et al. [6] established that the problem is NP-complete in split graphs. Mu et al. [24] showed that PAIRED DOMINATION is NP-complete in circle graphs. Chang et al. [3] established the NP-completeness in undirected path graphs. Hanaka et al. [12] proved the NP-completeness of PAIRED DOMINATION in planar bipartite graphs with maximum degree at most three.

Chen et al. [5] showed that the MINIMUM PAIRED DOMINATION problem is APX-complete for subcubic graphs. Recently, Wang et al. [30] established the inapproximability result for generalized convex graphs, a subclass of bipartite graphs that extends convex bipartite graphs. On the positive side, Panda et al. [25] and Hung [16] gave linear time algorithm for convex bipartite graphs. Moreover, various other graph classes such as trees, cographs, AT-free graphs, chordal bipartite graphs, distance hereditary graphs, strongly chordal graphs admit polynomial time algorithms for computing the exact value of paired domination number.

Due to interest in applications, the impact of graph modifications such as deleting or subdividing an edge on the paired domination number has also been studied [2, 27]. For a comprehensive treatment of the results on paired domination, we refer the reader to [9]. For a detailed overview of algorithmic results, we refer readers to [19, 29], and the references therein.

A paired dominating set is a more restrictive version of a total dominating set. Total domination is a well-studied topic in structural and algorithmic graph theory [15]. Zhu [31] established that the TOTAL DOMINATION problem remains NP-complete even when restricted to planar bipartite subcubic graphs via a linear time reduction from the VERTEX COVER problem. Under the Exponential Time Hypothesis (abbreviated ETH), this implies that TOTAL DOMINATION cannot be solved in $2^{o(|V(G)|)}$ time. Mohar [23] proved that the VERTEX COVER problem remains NP-complete even when restricted to 2-connected planar cubic graphs. Consequently, TOTAL DOMINATION is NP-complete for planar bipartite subcubic graphs. It is also known to be NP-complete for cubic graphs [11]. Antony et al. [1] further strengthened this result by showing that TOTAL DOMINATION remains NP-complete for triangle-free cubic graphs. They extended this further by establishing the NP-completeness of TOTAL DOMINATION in triangle-free d-regular graphs for any fixed integer $d \geq 3$ Moreover, they proved that, assuming the ETH, TOTAL DOMINATION does not admit subexponential time algorithms in this graph class. Shalu et al. [28] proved that TOTAL DOMINATION remains NP-complete for planar cubic graphs.

Due to the close relationship between the two notions, PAIRED DOMINATION is known to inherit many of the known hardness results of TOTAL DOMINATION. However, the complexity of the PAIRED DOMINATION problem remains unresolved for planar cubic graphs and triangle-free cubic graphs.

Since, the total domination number $\gamma_t(G)$ and the paired domination number $\gamma_{pr}(G)$ are related by the inequality $\gamma_t(G) \leq \gamma_{pr}(G)$, it is of interest to determine the complexity of deciding when these two parameters are equal. Dettlaff et al. [10] showed that determining whether a graph G satisfies $\gamma_t(G) = \gamma_{pr}(G)$ is NP-hard for bipartite graphs. In this work, we the NP-hardness result due to Dettlaff et al. [10] for planar bipartite subcubic graphs, planar cubic graphs, and triangle-free d-regular graphs, for each fixed integer $d \geq 3$.

Structure of the paper

In this paper, we first determine the algorithmic complexity of PAIRED DOMINATION on planar cubic graphs.

Theorem 1. *PAIRED DOMINATION is NP-complete for planar cubic graphs. Moreover, assuming ETH, there is no algorithm that solves it in $2^{o(\sqrt{|V(G)|})}$ time.*

Then we show that PAIRED DOMINATION remains NP-complete on triangle-free cubic graphs and triangle-free d-regular graphs, for each fixed $d \geq 4$. We also obtain the the complexity lower bounds under the Exponential Time Hypothesis.

Theorem 2. *PAIRED DOMINATION is NP-complete on triangle-free cubic graphs and assuming ETH, the problem cannot be solved in $2^{o(|V(G)|)}$ time.*

Theorem 3. *PAIRED DOMINATION is NP-complete on triangle-free d-regular graphs, for each fixed integer $d \geq 4$. Moreover, assuming ETH, the problem cannot be solved in $2^{o(|V(G)|)}$ time.*

Finally, we present the following strengthening of Dettlaff et al. [10].

Theorem 4. *Deciding if a graph G satisfies $\gamma_t(G) = \gamma_{pr}(G)$ is NP-hard, even for (i) planar bipartite subcubic graphs, (ii) planar cubic graphs, and (iii) triangle-free d-regular graphs, for each fixed integer $d \geq 3$.*

The paper is organised as follows. In Sect. 2 we provide Basic notations and definition used in the paper. We also explain the Exponential Time Hypothesis. The proof of Theorem 1 is given in Sect. 3. We establish Theorem 2 and Theorem 3 in Sect. 4. We give the proof of Theorem 4 in Sect. 5. Finally, we close the paper with some open problems.

2 Preliminaries

2.1 Basic Notations and Definitions

The *degree* of a vertex is the number of edges incident to it. A vertex of degree zero is an *isolated vertex*. A graph is *isolate-free* if it contains no isolated vertex. The *maximum degree* of a graph G, denoted by $\Delta(G)$, is the largest degree of any vertex in G. A vertex of degree one is called a *pendant* vertex and its unique adjacent vertex is called *support* vertex. A graph G is called *d-regular* if every vertex in G has degree d. We say that G is *cubic* if the degree of each vertex in $V(G)$ is three and if the maximum degree of G is three then G is *subcubic*. We refer to vertices of degree one and two in a subcubic graph as *non-cubic* vertices.

We use the standard notation $[k]$ to denote the set $\{1, 2, \ldots, k\}$. Let P be a path in a graph G, defined as a sequence of vertices $v_1, v_2, \ldots, v_k$ such that each consecutive pair $(v_i, v_{i+1}) \in E(G)$ for $i \in [k-1]$. The *length* of the path P, denoted $\ell(P)$, is the number of edges in the path. Given two vertices x and y in G, the *distance* between x and y is the length of a shortest path between x and y in G.

2.2 Exponential Time Hypothesis:

A theorem establishing the NP-hardness of a decision problem does not offer insight into how efficiently the problem can be solved, aside from ruling out polynomial time algorithms unless P = NP. To address this gap and provide fine-grained lower bounds on the time complexity of solving NP-complete problems, Impagliazzo, Paturi, and Zane [18] formulated the *Exponential Time Hypothesis* (ETH). Let us recall the classical NP-complete problem 3-SATISFIABILITY, commonly referred to as 3-SAT.

> 3-SAT
>
> *Instance:* A collection $\mathscr{C} = \{C_1, C_2, \ldots, C_m\}$ of clauses over a finite set $U = \{u_1, u_2, \ldots, u_n\}$ of variables such that $|C_j| = 3$ for each $j \in [m]$.
> *Question:* Is there a satisfying truth assignment for U that satisfies all the clauses in $\mathscr{C}$?

ETH conjectures that there is no deterministic algorithm that solves every instance of 3-SAT with n variables in time $2^{o(n+m)}$. In addition to formulating ETH, Impagliazzo, Paturi, and Zane proved the celebrated *Sparsification Lemma*, which has the following implication [17] that 3-SAT can be solved in time $2^{o(n)}$ if and only if it can be solved in time $2^{o(m)}$.

In order to transfer this lower bound to rule out an algorithm for a problem X running in time $2^{o(f(|x|))}$, it is enough to give a polynomial-time reduction that maps every 3-SAT instance with n variables and m clauses to an instance x of X of size $|x| = O\big(g(n+m)\big)$, where f denotes an inverse of g. For a comprehensive discussion of these notions, we direct the reader to Chap. 14 of the book [7].

We next state the PLANAR EXACTLY 3-BOUNDED 3-SAT problem, abbreviated PE-3B-3SAT introduced by Dahlhaus et al. [8] with following restrictions on 3-SAT instance.

(i) Each clause $C_j \in \mathscr{C}$ satisfies $|C_j| = 2$ or $|C_j| = 3$.

(ii) For each variable u_i, one of the literals among u_i, $\overline{u_i}$ appears in exactly two clauses and the other appears in exactly one clause.

(iii) The bipartite incidence graph $H = \{U \cup \mathscr{C}, E\}$ with vertex set $U \cup \mathscr{C}$ and edge set $E = \{(u, c) \mid u \in U, c \in \mathscr{C},$ and either $u \in C$ or $\overline{u} \in C\}$ is planar. Note that, the maximum degree of the incidence graph H is at most three.

PLANAR EXACTLY 3-BOUNDED 3-SAT (PE-3B-3SAT)

Instance: A collection $\mathscr{C} = \{C_1, C_2, \ldots, C_m\}$ of clauses over a finite set $U = \{u_1, u_2, \ldots, u_n\}$ of variables such that each clause $C_j \in \mathscr{C}$ satisfies conditions (i), (ii), and (iii) stated above.

Question: Does there exist a satisfying truth assignment for $\mathscr{C}$?

The PE-3B-3SAT problem was introduced and shown to be NP-complete by Dahlhaus et al. [8] via a reduction from 3-SAT that has quadratic blow-up due to $O(m^2)$ crossovers.

Theorem A ([8]). *The* PE-3B-3SAT *problem is* NP-*complete. Moreover, assuming the Exponential Time Hypothesis, it cannot be solved in time* $2^{o(\sqrt{n})}$ *for instances with n variables.*

3   Hardness Results in Planar Cubic Graphs

This section focuses on the complexity of PAIRED DOMINATION in planar cubic graphs. This is in contrast to the classical DOMINATING SET problem, which is known to be NP-complete for planar 4-regular graphs and planar subcubic graphs [11]. To the best of our knowledge, complexity of DOMINATING SET problem is not known for planar cubic graphs.

Lin et al. [20,21] posed the question of determining the complexity of the problem in planar graphs. Tripathi et al. [29] answered this by proving the NP-completeness of PAIRED DOMINATION in planar graphs with maximum degree at most five. Hanaka et al. [12] improved their result by establishing the NP-completeness in planar graphs with maximum degree at most three. Their result follows from a reduction via PE-3B-3SAT, which implies that there is no algorithm that solves PAIRED DOMINATION in time $2^{o(\sqrt{|V(G)|})}$ for planar graphs, unless ETH fails. We complement Hanaka et al.'s result by proving NP-hardness of PAIRED DOMINATION in planar cubic graphs and assuming ETH, there is no algorithm that solves it in time $2^{o(\sqrt{|V(G)|})}$ for planar cubic graphs. Since

PAIRED DOMINATION is easy on graphs with maximum degree at most two, we get the following complexity dichotomy: the problem is polynomial time solvable for maximum degree two graphs, but becomes NP-complete once the degree increases to three; even when restricted to planar bipartite or planar regular graphs.

Theorem 1. *PAIRED DOMINATION is NP-complete for planar cubic graphs. Moreover, assuming ETH, there is no algorithm that solves it in $2^{o(\sqrt{|V(G)|})}$ time.*

Proof. Consider a planar bipartite subcubic graph G. We construct a planar cubic graph G', by applying a local transformation to each non-cubic vertex in G as follows:

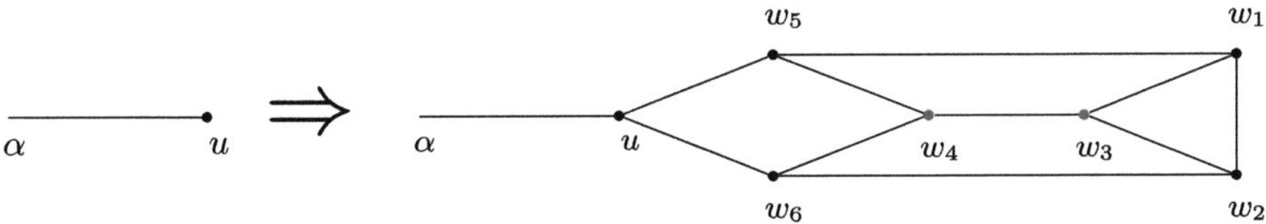

(a) Transformation of the vertices of degree one in the proof of Theorem 1

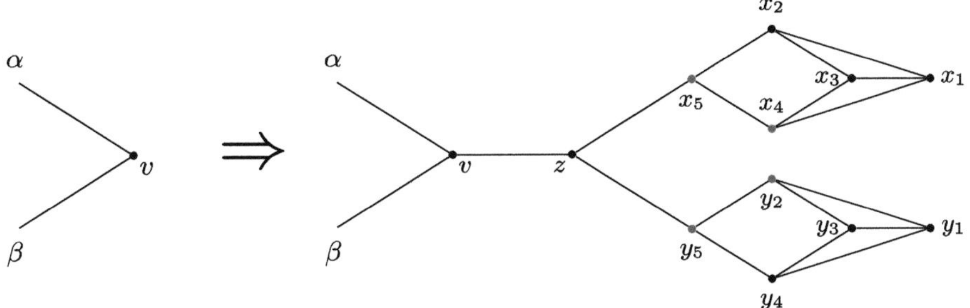

(b) Transformation of the vertices of degree two in the proof of Theorem 1

Fig. 1. Transformations for vertices of degree one (top) and degree two (bottom) used in the proof of Theorem 1.

The above attachments preserve planarity and ensure that all vertices in G' have degree exactly three. If G has r vertices of degree one and s vertices of degree two then the resultant graph G' has $6r+11s$ new vertices from $G'-G$. Figure 2(a) shows an example of a planar bipartite subcubic graph G, and Fig. 2(b) shows the corresponding planar cubic graph G' obtained by the above construction (Fig. 1).

Observe that any paired dominating set of the resultant graph G' must include at least two vertices from each gadget W_1 corresponding to a degree

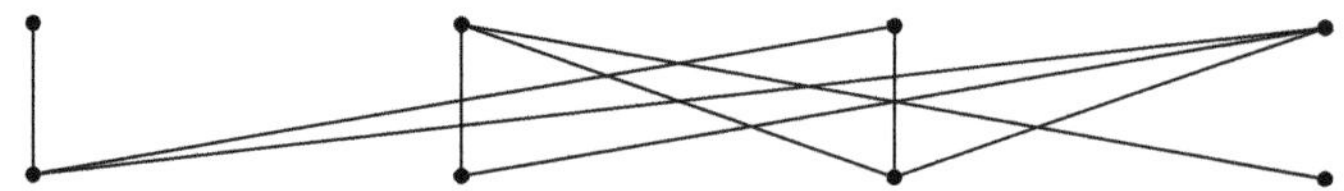

(a) An example of a planar bipartite subcubic graph G.

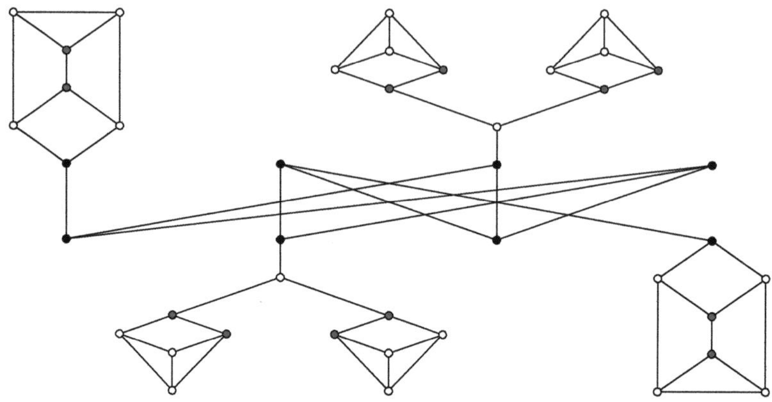

(b) The corresponding planar cubic graph G' obtained by the construction in the proof of Theorem 1.

Fig. 2. Planar bipartite subcubic graph G and resultant planar cubic graph G' obtained from G.

one vertex, and at least four vertices from each gadget W_2 corresponding to a degree two vertex. Furthermore, the two vertices w_3 and w_4 paired dominate the gadget W_1, while the four vertices x_4, x_5, y_2, and y_5 paired dominate the gadget W_2. These vertices are shown in the red color in the above figures. Without loss of generality, we may assume that the vertices w_5, w_6 and z are not included in a minimal paired dominating set of G'. If a gadget vertex among w_5, w_6 and z is included in a minimal paired dominating set of G' as a partner of a vertex in G, then we can replace it with another neighbour of that vertex within G.

Claim. The graph G has a minimal paired dominating set of size at most k if and only if the graph G' has a minimal paired dominating set of size at most $k + 2r + 4s$.

Suppose G has a minimal paired dominating set D with $|D| \leq k$. We construct a minimal paired dominating set D' of G' as follows. Let $D' = D \cup R_1 \cup R_2$, where R_1 consists of set of all vertices w_3, w_4 from each gadget W_1 corresponding to each vertex of degree one in G, and R_2 consists of four vertices x_4, x_5, y_2, and y_5 from each gadget W_2 corresponding to each vertex of degree two in G. As there are with r vertices of degree one and s vertices of degree two, D' is a minimal paired dominating set of G' of size at most $k + 2r + 4s$. On the other hand, let the graph G' has a minimal paired dominating set D' of size at most $k + 2r + 4s$. By previous discussion we assume that vertices w_5, w_6 and z are not included in D'. By the construction of G', each gadget W_1, corresponding to a degree one vertex of G, requires at least two of its internal vertices to be included in any paired dominating set of G'. Similarly, each gadget W_2, corresponding to a degree two vertex of G, requires at least four of its internal vertices for paired domination. It follows that D' must contain at least $2r + 4s$ vertices from $G' - G$. Since the gadget vertices w_5, w_6 and z are not included in the paired dominating set D' of G', the set of remaining vertices in $V(G) \cap D'$ has size at most k and they must form a minimal paired dominating set of G. This proves the Claim 3.

Notice that G' has $|V(G)| + 6r + 11s$ vertices where r and s denote the vertices of degree 1 and 2 in G, respectively. This is a linear reduction from PAIRED DOMINATION in planar bipartite subcubic graphs to PAIRED DOMINATION in planar cubic graphs. since the earlier problem is NP-complete, the NP-completeness follows and the linearity of the reduction establishes that assuming ETH, there is no algorithm that solves it in $2^{o(\sqrt{|V(G)|})}$ time.

4 Hardness Results in Triangle-Free Regular Graphs

In this section, we establish the NP-completeness of PAIRED DOMINATION on triangle-free d-regular graphs, for every fixed integer $d \geq 3$.

Merouane et al. [22] showed that for any triangle-free graph G, total domination number $\gamma_t(G)$ and CD-chromatic number $\chi_{cd}(G)$ are equal. Antony et al. [1] showed that the CD-COLORING problem is NP-complete on triangle-free d-regular graphs for every fixed integer $d \geq 3$, and does not admit subexponential time algorithm unless ETH fails, by describing a reduction from TOTAL DOMINATION on bipartite subcubic graphs. As a consequence, they established that the TOTAL DOMINATION problem is NP-complete on triangle-free d-regular graphs for every fixed integer $d \geq 3$, and does not admit subexponential time algorithm unless ETH fails.

We establish that the analogous results for PAIRED DOMINATION follow by using the constructions of Antony et al. [1] together with the NP-completeness of the PAIRED DOMINATION problem on bipartite subcubic graphs [12]. These constructions (Constructions 1âĂŞ3 in [1]) are incorporated directly into the proof and since the proof closely parallels as Theorem 1, we omit the details.

Theorem 2. PAIRED DOMINATION *is* NP-*complete on triangle-free cubic graphs and assuming* ETH, *the problem cannot be solved in* $2^{o(|V(G)|)}$ *time.*

Proof. Let G be a bipartite subcubic graph with r number of degree one and s number of degree two vertices. We obtain the triangle-free cubic graph G' from G by the Construction 1 in [1]. We form a graph G' by attaching to G a gadget X, defined as follows. The gadget X consists of a star $K_{1,2}$ with center p and leaves x and y, together with two copies of $K_{2,2}$, say $D_1 = (U_1, V_1)$ and $D_2 = (U_2, V_2)$. Vertex x (respectively y) is adjacent to all vertices of U_1 (respectively U_2), and each vertex of V_1 is adjacent to exactly one vertex of V_2. To construct G', for each vertex of degree 2 in G, attach one copy of X by joining its root p to that vertex; for each vertex of degree 1, attach two copies of X. It is easy to prove that, $\gamma_{pr}(G) = k$ if and only if $\gamma_{pr}(G') = k + 4r + 8s$. Since the arguments mirror those used in the proof of Theorem 1, the details are omitted. As PAIRED DOMINATION is NP-complete on bipartite subcubic graphs and underlying construction is linear, the theorem follows.

Now using Theorem 2 we prove the next result.

Theorem 3. *PAIRED DOMINATION is NP-complete on triangle-free d-regular graphs, for each fixed integer $d \geq 4$. Moreover, assuming ETH, the problem cannot be solved in $2^{o(|V(G)|)}$ time.*

Proof. Now, we prove the following result on triangle-free $(d-2)$-regular graphs, for any odd integer $d \geq 5$.

Claim. PAIRED DOMINATION is NP-complete on triangle-free d-regular graphs, for any odd integer $d \geq 5$.

Proof. Let G be a triangle-free $(d-2)$-regular graph with n vertices, for any odd integer $d \geq 5$. We construct d-regular graph G' from G by the Construction 2 in [1]. We can prove that $\gamma_{pr}(G) = k$ if and only if $\gamma_{pr}(G') = k + 4n(d-1)$. Thus the Claim follows from Theorem 2.

Now, we prove the following result on triangle-free $(d-1)$-regular graphs, for any even integer $d \geq 4$.

Claim. PAIRED DOMINATION is NP-complete on triangle-free d-regular graphs, for any even integer $d \geq 4$.

Proof. Let G be a triangle-free $(d-1)$-regular graph with n vertices, for any even integer $d \geq 4$. We construct d-regular graph G' from G by the Construction 3 in [1]. We can prove that $\gamma_{pr}(G) = k$ if and only if $\gamma_{pr}(G') = k + 2n(d-2)$. Thus the Claim follows from Theorem 2.

Therefore, by the claims established above, we proved that PAIRED DOMINATION is NP-complete on triangle-free d-regular graphs for every fixed integer $d \geq 4$ and the linearity of the reductions, implies that it does not admit subexponential time algorithm unless ETH fails. Hence, theorem is proved.

5   Hardness of Equality

Graph theorists have long been drawn to investigate the interplay between related graph parameters and their structural properties, particularly focusing on the algorithmic consequences for graph classes in which these parameters coincide. Recall that for any graph G without an isolated vertex we have, $\gamma_t(G) \leq \gamma_{pr}(G)$. In this section we discuss the hardness of determining the equality $\gamma_t(G) = \gamma_{pr}(G)$. Now we prove Theorem 4 which strengthens the earlier hardness result on bipartite graphs due to Dettlaff et al. [10]. We note that statements *(ii)* and *(iii)* of Theorem 4 uses the constructions in the previous sections, making them natural consequences of the earlier reductions.

Theorem 4. *Deciding if a graph G satisfies $\gamma_t(G) = \gamma_{pr}(G)$ is NP-hard, even for (i) planar bipartite subcubic graphs, (ii) planar cubic graphs, and (iii) triangle-free d-regular graphs, for each fixed integer $d \geq 3$.*

Proof. (i) To prove the result, we present a reduction from the PLANAR EXACTLY 3-BOUNDED 3-SAT problem. Let $(U, \mathscr{C})$ be an instance of PE-3B-3SAT. We construct a bipartite graph G from the instance $(U, \mathscr{C})$, as follows. For each variable u_i, we introduce the literal vertices $u_i, \overline{u_i}$ and two additional vertices p_i, q_i and add the edges $u_i q_i, \overline{u_i} q_i, p_i q_i$. Let us call this literal gadget as H_i. For each clause C_j, we introduce the clause vertex c_j and add the edge $c_j u_i$ if $u_i \in C_j$, and the edge $c_j \overline{u_i}$ if $\overline{u_i} \in C_j$ to the edge set of G. These completes the construction of G. It is immediate that the graph G is a planar bipartite subcubic graph.

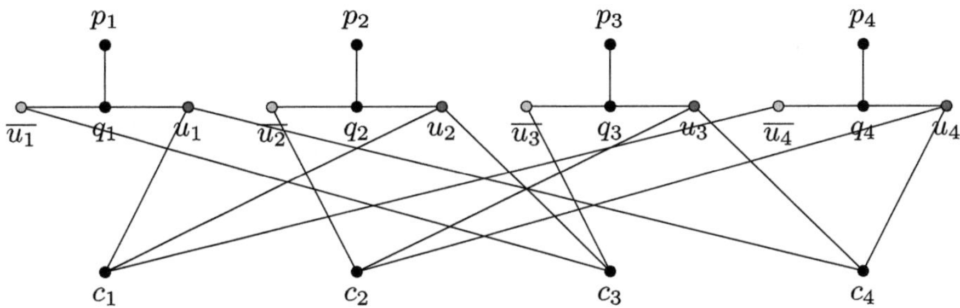

Fig. 3. An illustration of the construction of G in the proof of Theorem 4.

It is easy to observe that a support vertex must be included in every paired (resp. total) dominating set and thus vertices q_i are included in every (resp. total) dominating set of G. However, any two of these support vertices are at distance at least 4 from each other. Therefore any paired (resp. total) dominating set must include one additional vertex from the literal gadget H_i, as a partner of q_i leading to, $\gamma_{pr}(G) \geq 2n$ (resp. $\gamma_t(G) \geq 2n$) (Fig. 3).

We claim that $\gamma_{pr}(G) = 2n$ if and only if the collection $\mathscr{C}$ is satisfiable. Suppose D is a paired dominating set of G with $|D| = \gamma_{pr}(G) = 2n$. Without loss of generality, we may assume that for each $i \in [n]$, the vertex q_i and exactly one of the literal vertices from $\{u_i, \overline{u_i}\}$ is included in D. Define a truth assignment $t : U \to \{T, F\}$ by setting $t(u_i) = T$ if $u_i \in D$, and $t(u_i) = F$ otherwise. Since no clause vertex c_j is in D, each clause vertex must be dominated by some literal vertex that appears in it. By our construction of t, the dominating literal vertex corresponds to a literal that is assigned true. Therefore, every clause is satisfied, and t is a satisfying assignment of $\mathscr{C}$.

Conversely, suppose $\mathscr{C}$ is satisfiable with satisfying truth assignment t. For each variable u_i, include the vertex q_i and the literal vertex corresponding to the true literal in the satisfying assignment in D. Now $\{u_i, q_i\}$ as well as $\{\overline{u_i}, q_i\}$ are dominating sets for the gadget H_i. As t is a satisfying truth assignment for $\mathscr{C}$, the clause vertex c_j in G is adjacent to at least one vertex in D. Therefore, each clause vertex c_j with $j \in [m]$ is dominated by a vertex in D. Also, the vertex q_i is paired with the true literal vertex. This gives us a paired dominating set D of size $2n$, and hence, $\gamma_{pr}(G) = 2n$. Similarly we can argue that $\gamma_t(G) = 2n$ if and only if the collection $\mathscr{C}$ is satisfiable.

Now, we prove that $\gamma_{pr}(G) = \gamma_t(G)$ if and only if $\gamma_{pr}(G) = \gamma_t(G) = 2n$. One direction is straightforward. For the other direction, suppose that $\gamma_{pr}(G) = \gamma_t(G)$, and let D be a minimum paired dominating set of G. Since each support vertex q_i must be included in any paired dominating set and the support vertices are at distance at least 4 from one another, it follows that $|D \cap H_i| \geq 2$ for every $i \in [n]$. Without loss of generality, we may assume that D contains q_i and one of its corresponding literal vertices, either u_i or $\overline{u_i}$. We claim that no clause vertex c_j belongs to D. Suppose, for contradiction, that $c_j \in D$ for some $j \in [m]$. Then c_j could be paired with one of its adjacent literal vertices, say $w_i \in \{u_i, \overline{u_i}\}$, for some $i \in [n]$. In that case, the set $D' = D \setminus \{c_j\}$ is a total dominating set of G with strictly smaller cardinality than D, implying that $\gamma_t(G) < \gamma_{pr}(G)$, which contradicts our assumption. Therefore, no clause vertex is included in D.

Next, we claim that exactly one literal vertex from each H_i is present in D. Suppose, for contradiction, that both u_i and $\overline{u_i}$ belong to D for some $i \in [n]$. Since no clause vertex is present in D and both u_i and $\overline{u_i}$ are adjacent only to the vertex q_i, only one of them can be paired with q_i. However, then the other vertex would remain unpaired, contradicting the definition of a paired dominating set. Therefore, exactly one literal vertex from each H_i must be in D. This implies that $\gamma_{pr}(G) = \gamma_t(G) = 2n$. Hence, it follows that, deciding whether a planar bipartite subcubic graph G satisfies $\gamma_t(G) = \gamma_{pr}(G)$ is NP-hard.

(ii) Let G be an instance of planar bipartite subcubic graph and let G' be obtained by attaching the gadgets W_1 and W_2 to vertices of degree one and degree two respectively, as in the proof of Theorem 1. We have established that, for each positive integer k, $\gamma_{pr}(G) = k$ if and only if $\gamma_{pr}(G') = k + 2r + 4s$. Similarly, $\gamma_t(G) = k$ if and only if $\gamma_t(G') = k + 2r + 4s$.

Therefore, the equality $\gamma_{pr}(G') = \gamma_t(G')$ holds if and only if $\gamma_{pr}(G) = \gamma_t(G)$ holds. Since deciding $\gamma_{pr}(G) = \gamma_t(G)$ is NP-hard by Theorem 4(i), the same holds for $\gamma_{pr}(G') = \gamma_t(G')$.

(iii) Now, to prove the NP-hardness of deciding if a graph G satisfies $\gamma_t(G) = \gamma_{pr}(G)$ in triangle-free d-regular graphs for every fixed integer $d \geq 3$ we use the constructions and supporting claims from Theorem 2 and Theorem 3. The proof proceeds as in Theorem 4(ii). The details are omitted as they are analogous.

We have established that it is NP-hard to decide the equality of $\gamma_t(G)$ and $\gamma_{pr}(G)$. The equality problem is not known to be in NP. It does not seem to admit a polynomial-time verifiable certificate for equality since verifying equality of these parameters requires checking both lower and upper bounds. Hence the problem is unlikely to be in NP unless NP = coNP.

6   Concluding Remarks

In this work, we proved that PAIRED DOMINATION remains NP-complete even on planar cubic graphs. We further proved that it is NP-complete for triangle-free d-regular graphs, for any fixed integer $d \geq 3$. We have established that, the problem of comparing the paired domination number and the total domination number is NP-hard for planar bipartite subcubic graphs, planar cubic graphs and triangle-free d-regular graphs.

It is an interesting open problem is to determine the complexity of PAIRED DOMINATION and TOTAL DOMINATION in planar bipartite cubic graphs. Similarly, It is worth investigating the complexity of these problems in planar d-regular and bipartite d-regular graphs for each fixed integer $d \geq 4$.

Acknowledgments. The authors gratefully acknowledge the anonymous reviewers for their careful reading and constructive feedback. This work was carried out while the first author was visiting the Institute of Mathematical Sciences (IMSc), Chennai, during JanuaryâĂŞJuly 2025. The first author expresses sincere gratitude to Professor Sushmita Gupta for the opportunity and for her kind mentorship.

Disclosure of Interests. The authors confirm that they have no conflicts of interest, competing financial interests, or personal relationships that could have influenced the work presented in this paper.

References

1. Antony, D., Chandran, L.S., Gayen, A., Gosavi, S., Jacob, D.: Total domination, separated-cluster, CD-coloring: algorithms and hardness. In: Latin Amer. Symp. Theoret. Informatics, pp. 97–113. Springer (2024). arXiv preprint: arXiv:2307.12073
2. Bakal, D.M., Borse, Y.M., Waphare, B.N.: Algorithmic complexity of three domination subdivision number problems in graphs. Commun. Comb. Optim. (2025). https://doi.org/10.22049/cco.2025.29896.2213. to appear
3. Chang, G.J., Panda, B.S., Pradhan, D.: Complexity of distance paired-domination problem in graphs. Theoret. Comput. Sci. **459**, 89–99 (2012)
4. Chen, H., Lei, Z., Liu, T., Tang, Z., Wang, C., Xu, K.: Complexity of domination, Hamiltonicity and treewidth for tree convex bipartite graphs. J. Combin. Optim. **32**(1), 95–110 (2016)
5. Chen, L., Lu, C., Zeng, Z.: Hardness results and approximation algorithms for (weighted) paired-domination in graphs. Theoret. Comput. Sci. **410**(47–49), 5063–5071 (2009)
6. Chen, L., Lu, C., Zeng, Z.: Labelling algorithms for paired-domination problems in block and interval graphs. J. Combin. Optim. **19**(4), 457–470 (2010)
7. Cygan, M., et al.: Parameterized Algorithms, vol. 5. Springer, Berlin (2015)
8. Dahlhaus, E., Johnson, D.S., Papadimitriou, C.H., Seymour, P.D., Yannakakis, M.: The complexity of multiterminal cuts. SIAM J. Comput. **23**(4), 864–894 (1994)
9. Desormeaux, W.J., Haynes, T.W., Henning, M.A.: Paired domination in graphs. In: Topics in Domination in Graphs, pp. 31–77. Springer (2020)

10. Dettlaff, M., Gözüpek, D., Raczek, J.: Paired domination versus domination and packing number in graphs. J. Combin. Optim. **44**(2), 921–933 (2022)
11. Garey, M.R., Johnson, D.S.: Computers and Intractability: A Guide to the Theory of NP-Completeness. W. H. Freeman, San Francisco (1979)
12. Hanaka, T., Ono, H., Otachi, Y., Uda, S.: Grouped domination parameterized by vertex cover, twin cover, and beyond. Theoret. Comput. Sci. **996**, 114507 (2024)
13. Haynes, T.W., Hedetniemi, S.T., Henning, M.A.: Domination in Graphs: Core Concepts. Springer, Berlin (2023)
14. Haynes, T.W., Slater, P.J.: Paired-domination in graphs. Networks **32**(3), 199–206 (1998)
15. Henning, M.A., Yeo, A.: Total Domination in Graphs. Springer, Berlin (2013)
16. Hung, R.W.: Linear-time algorithm for the paired-domination problem in convex bipartite graphs. Theory Comput. Syst. **50**, 721–738 (2012)
17. Impagliazzo, R., Paturi, R.: On the complexity of k-SAT. J. Comput. System Sci. **62**(2), 367–375 (2001)
18. Impagliazzo, R., Paturi, R., Zane, F.: Which problems have strongly exponential complexity? J. Comput. System Sci. **63**(4), 512–530 (2001)
19. Lin, C.C., Hsieh, C.Y., Mu, T.Y.: A linear-time algorithm for weighted paired-domination on block graphs. J. Combin. Optim., 1–18 (2022)
20. Lin, C.C., Ku, K.C., Hsu, C.H.: Paired-domination problem on distance-hereditary graphs. Algorithmica **82**(10), 2809–2840 (2020)
21. Lin, C.C., Tu, H.L.: A linear-time algorithm for paired-domination on circular-arc graphs. Theoret. Comput. Sci. **591**, 99–105 (2015)
22. Merouane, H.B., Haddad, M., Chellali, M., Kheddouci, H.: Dominated colorings of graphs. Graphs Comb. **31**(3), 713–727 (2015)
23. Mohar, B.: Face covers and the genus problem for apex graphs. J. Combin. Theory Ser. B **82**(1), 102–117 (2001)
24. Mu, T.Y., Lin, C.C.: Paired-domination problem on circle and k-polygon graphs. arXiv Preprint (2024)
25. Panda, B.S., Pradhan, D.: A linear time algorithm for computing a minimum paired-dominating set of a convex bipartite graph. Discrete Appl. Math. **161**(12), 1776–1783 (2013)
26. Panda, B.S., Pradhan, D.: Minimum paired-dominating set in chordal bipartite graphs and perfect elimination bipartite graphs. J. Combin. Optim. **26**(4), 770–785 (2013)
27. Rad, N.J., Kamarulhaili, H.: On the complexity of some bondage problems in graphs. Australas. J. Combin. **68**, 265–275 (2017)
28. Shalu, M.A., Kirubakaran, V.K.: Total domination and open packing in two subclasses of triangle-free graphs. In: Conf. Algorithms Discrete Appl. Math, pp. 319–330. Springer (2025)
29. Tripathi, V., Kloks, T., Pandey, A., Paul, K., Wang, H.L.: Complexity of paired domination in at-free and planar graphs. Theoret. Comput. Sci. **930**, 53–62 (2022)
30. Wang, P.Y., Kitamura, N., Izumi, T., Masuzawa, T.: Approximation hardness of domination problems on generalized convex graphs. Theoret. Comput. Sci. **1028**, 115035 (2025)
31. Zhu, J.: Approximation for minimum total dominating set. In: Proc. 2nd Int. Conf. Interaction Sci.: Information Technology, Culture and Human, pp. 119–124 (2009)

Monotone Decontamination of Arbitrary Dynamic Graphs with Mobile Agents

Rajashree Bar[1], Daibik Barik[2], Adri Bhattacharya[1], and Partha Sarathi Mandal[1(✉)]

[1] Indian Institute of Technology Guwahati, Guwahati 781039, India
`psm@iitg.ac.in`
[2] Indian Statistical Institute, Bangalore 560059, India

Abstract. Network decontamination is a well-known problem, in which the aim of the mobile agents should be to decontaminate the network (i.e., both nodes and edges). This problem comes with an added constraint, i.e., of *monotonicity*, in which whenever a node or an edge is decontaminated, it must not get recontaminated. Hence, the name comes *monotone decontamination*. This problem has been relatively explored in static graphs, but nothing is known yet in dynamic graphs. We, in this paper, study the *monotone decontamination* problem in arbitrary dynamic graphs. We designed two models of dynamicity, based on the time within which a disappeared edge must reappear. In each of these two models, we proposed lower bounds as well as upper bounds on the number of agents, required to fully decontaminate the underlying dynamic graph, monotonically. Our results also highlight the difficulties faced due to the sudden disappearance or reappearance of edges. Our aim in this paper has been to primarily optimize the number of agents required to solve monotone decontamination in these dynamic networks.

Keywords: Mobile agent · Network Decontamination · Dynamic Graphs

1 Introduction

We consider a connected network where nodes and edges are infected by various attacks, including but not limited to, viruses, spam, and malware. In real life scenarios, the concern is to keep the network safe from such harmful contamination. In particular, we work on dynamic graphs where the edges can disappear and reappear at any time, provided the underlying graph remains connected. This phenomenon of connectivity is termed as *1-interval connectivity*. In literature, the decontamination has been achieved by using a team of mobile agents,

D. Barik–A portion of the work was completed during a summer internship from May to July 2025 at the Indian Institute of Technology Guwahati.
A. Bhattacharya–Supported by CSIR, Govt. of India, Grant Number: 09/731(0178)/2020-EMR-I.

N. Misra and A. Pandey (Eds.): CALDAM 2026, LNCS 16445, pp. 29–42, 2026.
https://doi.org/10.1007/978-3-032-17156-6_3

called searchers. Each node of a graph can be in one of two states: contaminated or decontaminated. There are certain variations of contamination which has been studied in literature. A brief survey of which can be found in [15]. Some of these variations are: *Edge-search* [17], *Node-search* [10] and *Mixed-search* [3]. The *edge-search* model is a decontamination strategy where an agent traverses through an edge and then decontaminates. The *node-search* model is slightly different from the edge-search model. Here, an edge is clean when both of its endpoints are occupied by some agents. And lastly, the *mixed-search* is an edge-search strategy with the difference that an edge becomes clean when either an agent traverses through that edge or its endpoints are guarded by at least one agent. Network decontamination is well studied in static graph settings such as hypercubes [7], tori, and chordal rings [6]. Networks nowadays are prevalently dynamic in nature, so to capture the complexity of this evolving topology is what interested us in studying this problem of decontamination in dynamic graph networks. As far as we know, we are the first ones to explore this decontamination problem on dynamic graphs. We consider two models of dynamicity, in the first one, the adversary can disappear or reappear certain edges (adhering to 1-interval connectivity), but those edges that are disappeared can only remain disappeared for a finite time. We term this model as *Finite time edge appearance* model or FTEA. In the second model, there is no bound on disappearing time, i.e., in other words, an edge can remain disappeared for an infinitely long time. We term this model as *Indefinite edge disappearance* model or IDED. FTEA is a special case of IDED. In this paper, we design algorithms and lower bounds on the number of agents for both these models of dynamicity.

Related Works: Network decontamination, often referred to as graph searching, is a problem that has been studied for decades, beginning with Parsons et al. in the year 1976 [16]. The basic idea is simple: a group of mobile agents (searchers) tries to clean up a contaminated network, using as few agents as possible. Without any knowledge of the underlying graph, identifying the minimum number of searchers needed is actually NP-complete for general graphs [14]. Luccio et al. [13] introduced a different model of decontamination, termed as m-immunity. It is defined to be as follows: any decontaminated node with no agent on it is further recontaminated, only if at least m of its neighbors are contaminated. They solved these problem for trees, tori, mesh networks and for graphs with maximum degree 3. Note that in this paper, the value of m is limited to $m = 1$ or to the strict majority of the neighbor of the nodes. Further, Flocchini et al. [8] extended this problem for any integer value $m \geq 1$. They solved this problem for the underlying graphs to be mesh networks, trees and hypercubes. All these works so far studied in literature, assume the network to be static (i.e., does not change over time). Accordingly, monotone strategies (i.e., once something is cleaned, it stays clean) is designed to decontaminate the underlying graph.

In contrast to static networks, dynamic graphs allow edges to disappear and reappear over time, subject to a connectivity constraint (referred to as 1-interval connectivity, meaning the graph remains connected at every round). Trivially, in comparison to static networks, solving these problems in dynamic networks adds more complexity. Recently, there has been a lot of interest in studying the fundamental problems, such as exploration [4], gathering [5], dispersion [11], black hole search [2,9], etc., in dynamic graphs. Accordingly, in this paper, we explored this decontamination problem for dynamic networks. As per our knowledge, this is the first paper to establish bounds on monotone decontamination in dynamic networks.

Our Contribution: In this paper, our contributions are as follows. Let $\mathcal{G}$ be any dynamic graph with n nodes, and let $\mu(G) = k$ (where $\mu(G) = |E| - |V| - 1$) be the cyclomatic number (Definition 2) of the initial static graph $G = (V, E)$ of $\mathcal{G}$. Based on the two models of dynamicity, we obtain the following results.

- **Finite time edge reappearance Model (FTEA):** In this model, each edge that disappears must reappear within T time. Based on the relation between k and n, we obtain the following results.
  1. For $k < n$, we show that if d is the diameter of G then at least d agents are required to solve monotone network decontamination. Further, we proposed an algorithm that requires $d + k$ agents to solve monotone decontamination on $\mathcal{G}$.
  2. For $k \geq n$, we show that at least $n - 1$ agents are required to solve monotone decontamination. Accordingly, we proposed an algorithm that solves monotone decontamination in this model of dynamicity with n agents.
- **Indefinite Edge Disappearance Model (IDED):** In this model, edges may disappear for an unbounded period of time. This adds more complexity in comparison to the earlier FTEA model, since FTEA is a special case of the IDED model. Based on the relation between k and n, we have the following results.
  1. For $k < n$, we have shown that at least $k + 1$ agents are required to monotonically decontaminate a graph in this model. But, since FTEA is a special case of IDED, the bound for FTEA also holds for IDED. This means, if $d > k + 1$, then as per the result obtained in the FTEA model, we can say that at least d agents are required to solve monotone decontamination in this model. Hence, this gives our lower bound to be $\max\{d, k + 1\}$.
  2. For $k \geq n$, again as per the result obtained in the FTEA model, we propose that at least $n - 1$ agents are required to solve network decontamination under IDED model as well.

Table 1. Summary of the results. The underlying graph is $G = (V, E)$ with $|V| = n$, diameter d, and cyclomatic number k.

Model	Case	Upper Bound	Lower Bound
FTEA	$k < n$	$d + k$ (Theorem 1)	d (Lemma 1)
	$k \geq n$	n (Theorem 3)	$n - 1$ (Theorem 1)
IDED	$k < n$	$d + 2k$ (Theorem 4)	$\max\{d, k + 1\}$ (Lemmas 1,2)
	$k \geq n$		$n - 1$ (Theorem 1)

Lastly, irrespective of $k < n$ or $k \geq n$, we propose a monotone network decontamination algorithm in the IDED model that requires $d + 2k$ agents. Summary of the results show in Table 1.

Several technical details, pseudo codes and proofs are omitted due to lack of space, and they can be found in the full version of the paper [1].

2   Model and Preliminaries

Dynamic Graph Model: The dynamic graph is modeled as a *time-varying graph* termed as TVG, where it is denoted by $\mathcal{G} = (V, E, \mathbb{T}, \rho)$, where V indicates the set of vertices, E indicates the set of edges, $\mathbb{T}$ is said to be the *temporal domain*, which is symbolized to be $\mathbb{Z}^+$, as we consider discrete time steps. Further we define $\rho : E \times \mathbb{T} \to \{0, 1\}$ as the *presence* function, which indicates whether an edge is present or absent at any given time. The graph $G = (V, E)$ is the initial underlying static graph of the *time varying graph* $\mathcal{G}$. This underlying graph $\mathcal{G}$, with $|V|$ many vertices and $|E|$ many edges, is stated to be the footprint of $\mathcal{G}$. The degree of a node $v \in V$ is denoted by δ_v, where Δ denotes the maximum degree in G. The footprint graph is an undirected anonymous graph (i.e., the nodes in G have no IDs). Each edge with respect to a node $v \in V$ is uniquely labeled by a port numbers which ranges from $\{0, 1, \ldots, \delta_v - 1\}$. The adversary has the ability to disappear or reappear any edge from G at any round. This disappearance or reappearance of edge(s) is done by the adversary, keeping in mind that the underlying graph remains connected. This connectivity property is termed as *1-interval connectivity*.

In this paper, we consider two dynamic models based on the time the adversary can make an edge disappear. The first model is defined as *finite time edge reappearance model* (FTEA), where there exists a constant T (> 0) for which at most, the adversary can make any edge disappear, at time $T + 1$, it has to make that edge reappear. The second model is defined as *indefinite edge disappearance* time (IDED), where there is no existence of such a T, i.e., the adversary can make an edge disappear for any amount of time (i.e., even not finite time). Note that, in both case, the adversary must maintain 1-interval connectivity property in the underlying graph.

Agent Model: The agents are initially co-located at a node $r \in V$, chosen by the adversary. This node r is also termed as *Home*. They work synchronously. So, time is calculated in terms of rounds. The agents have distinct IDs, and they do not have any knowledge about any graph parameters, such as $|V|$, $|E|$, Δ, etc. When an agent visits any node $v \in V$, it can know the degree of that node in the footprint graph G. The agents can see the presence or absence of the edges at that round. For each adjacent edge present, the agents can see whether it is contaminated or not. The agents communicate via the *face-to-face* model of communication, where they can share information with another agent when they are at the same node in the same round.

Decontamination Model: Initially, we define all the edges and vertices to be *contaminated*. Whenever an agent visits a node, or traverses through an edge, that specific node or edge becomes *decontaminated*. *Recontamination* can occur if there exists a path, unguarded by agents, from a contaminated component to the decontaminated component. For example, in a cycle graph C_3, an agent starts from a node v_1, making it initially decontaminated. It goes to v_2, decontaminating the edge (v_1, v_2). But since the vertex v_3 is still contaminated and it's directly joined with v_1 and hence with the edge (v_1, v_2), it recontaminates both of them, but not the node v_2 because the agent is still present on it.

Here, we define the required definitions and problem definition.

Definition 1 (Contamination Degree). *The set $\{C_t(v)\}$ is the collection of incident edges of the node v, which are contaminated at time t. We define it as $\{C_t(v)\} = \{(u, v) \in E_t \mid (u, v)$ is contaminated at time t$\}$. Moreover, $C_t(v)$ denotes the contamination degree of v at time t.*

Definition 2 (Cyclomatic Number). *The cyclomatic number of a connected undirected graph $G = (V, E)$ is given by*

$$\mu(G) = |E| - |V| + 1.$$

Equivalently, $\mu(G)$ is the number of independent cycles in G, and equals the number of edges that must be removed to obtain a spanning tree.

Definition 3 (Feedback Edges [12]). *Let $G = (V, E)$ be a static connected graph and $T = (V, E') \subseteq G$ be a fixed spanning tree of G. A feedback edge set is $\tilde{E} = E \setminus E'$, and $|\tilde{E}| = k$ equals the cyclomatic number $\mu(G)$.*

Definition 4 (Separator Vertex). *A vertex $v \in V$ is called a separator vertex, if say at round t, it is incident to a contaminated edge. The set of all separator vertices is denoted by Σ. Moreover, the agents present at these separator vertices can be called separators or separator agent.*

Definition 5 (Problem Definition). *Given an anonymous, port-labeled dynamic graph $\mathcal{G}$, an initial node Home, determine the minimum number of agents required to solve the following protocols:*

- *Decontaminates all vertices and all existing edges of the underlying dynamic network.*
- *Ensures monotonicity by preventing any recontamination.*

Remark 1. It may be noted that, in all our algorithms we assume the initial deployment of agents to be sufficient in number. This sufficient number depends on the underlying graphs diameter (i.e., d), total number of nodes (i.e., n) and the cyclomatic number (i.e., k). But, this knowledge is only confined to the initial deployment on the number of agents, but nowhere in our algorithms do the agents require the help of any of these global graph parameters.

3 Lower Bound

In this section we propose the lower bound results obtained. We divide this section in to two parts, based on the dynamicity model.

3.1 Lower Bound for Finite Time Edge Reappearance Model

This argument gives the lower bound for the case of the finite-time edge reappearance model.

Theorem 1 (Lower Bound). *There exists a graph G with n nodes, and maximum degree $\frac{n}{2}$, on which no deterministic algorithm can solve network decontamination monotonically with $n - 2$ initially co-located agents, in the finite edge reappearance model of dynamicity, where $n > 4$.*

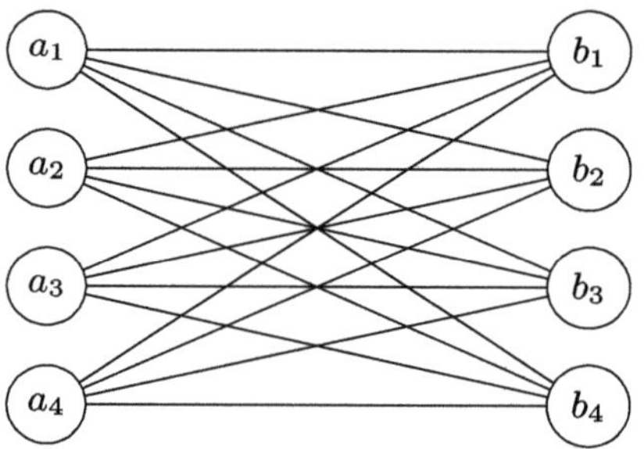

Fig. 1. The complete bipartite graph $K_{4,4}$ with partitions $\{a_1, \ldots, a_4\}$ and $\{b_1, \ldots, b_4\}$.

Proof. Let us consider our underlying graph to be $K_{n/2,n/2}$, where A and B are two disjoint partitions, each of size $\frac{n}{2}$, where $n > 4$. Let the group of $n - 2$ mobile agents, start from any node, we term that node as *Home*. Without loss of generality (WLOG), let us assume that our *Home* node is $a_1 \in A$ (refer to Fig. 1). A node is said to be *clean*, when all its adjacent edges have been decontaminated by the agents. Now, as per the characteristics of $K_{n/2,n/2}$, the vertex a_1 has degree $\frac{n}{2}$. Let $\mathcal{A}$ be any algorithm that claims to solve the network decontamination problem monotonically.

Let r (> 0) be the round at which, during the execution of $\mathcal{A}$, an agent first visits the $\frac{n}{2}$-th adjacent node of *Home*. Let us call this node b_t (where $t \in \{1, 2, \ldots, \frac{n}{2}\}$). Now, for $\mathcal{A}$, there can be two strategies. First, within round r there exists at least one node (say, b_{t1}) in B that is *fully decontaminated*, i.e., in other words b_{t1} along with all its adjacent edges are decontaminated, implying that its neighboring nodes are also decontaminated. Second, within round r, no such node of the form b_{t1} exists. For each of these two strategies, we infer a contradiction in the following manner.

Strategy-1: Let, without loss of generality, b_{t1} be the first node to be fully decontaminated, and let this happen at round r_0 $(< r)$. This means all neighbour nodes of b_{t1}, i.e., each node in A contains at least one agent. If any one of these agents holding a node in A (say, that node is a_{t1}), decides to leave a_{t1}, then a_{t1} gets recontaminated and monotonicity fails. The reason being, there exist at least two adjacent edges of a_{t1} which are contaminated. As otherwise there exists at least one node in A, say b_{t2} which is fully decontaminated before round r. It is because at round r_0, $\frac{n}{2}$ agents are holding a node each in A. Remaining, $\frac{n}{2} - 2$ agents cannot fully decontaminate any node in B, as those nodes must have exactly $\frac{n}{2} - 1$ contaminated edge. This concludes that none of the agents in A can move after b_{t1} is fully decontaminated. This also contradicts that $\mathcal{A}$ solves the problem, as no other node in A can be fully recontaminated with the remaining vertices after round r_0.

Strategy-2: After round r, all nodes in B are occupied by at least one agent. So, before r, no node in A can be fully decontaminated. So, each visited node in A till round r must contain at least one agent. Let there be x many such nodes. Observe that at round r, one of these x many nodes must be fully decontaminated (since each node in A contains at least one vertex). So, at this moment, at least $x-1$ agents remain occupying the remaining $x-1$ yet to be fully decontaminated nodes in A. Now, observe that any node in B must have at least $\frac{n}{2} - x$ many contaminated edges at round r. But since $\mathcal{A}$ is said to achieve monotonicity, so the remaining agents, that are free to move (or not occupying any contaminated node) is $\frac{n}{2} - x - 1$ $(= n - 2 - (\frac{n}{2} + x - 1))$, and these remaining agents cannot fully decontaminate any more node in B, which leads to a contradiction.

This shows that $\mathcal{A}$ fails to decontaminate $K_{n/2,n/2}$ with $n - 2$ co-located agents, monotonically.   □

3.2   Lower Bound for Indefinite Edge Disappearance Model

In this section, we show that there exists a graph, with cyclomatic number $\mu(G) = k$, on which any monotone decontamination protocol in the indefinite edge disappearance model with k agents fails.

Theorem 2. *There exists a graph G with n (with $n > 4$) nodes and maximum degree $n - 1$, with $\mu(G) = n - 1$, on which no deterministic algorithm can solve network decontamination monotonically with $\mu(G)$ many initially co-located agents, in the infinite edge disappearance model.*

Proof. In contradiction, let $\mathcal{A}$ be any deterministic monotone decontamination strategy using $n-1$ agents. Let us consider G to be a wheel graph with the vertex set $V = \{v_0, v_1, \ldots, v_{n-1}\}$, where v_0 is the centre vertex and the remaining are outer vertices (for reference see Fig. 2). Suppose that the adversary selects v_0 to be the initial node for these $n - 1$ agents. In addition, suppose the adversary removes each edge of the form (v_i, v_{i+1}) $(i \in \{1, \cdots, n - 2\})$ and (v_{n-1}, v_1). In this situation, to clean any vertex v_i $(i \in \{1, \ldots, n - 1\})$, an agent must visit it through v_0. Let r_0 be the round, when any agent from v_0 first visits any node of the form v_i, say that the first node visited is v_t. Now, we claim that from round r_0 onward, there must exist at least one agent guarding the node v_t. As otherwise, the adversary can reappear the contaminated edges (v_t, v_{t-1}) or (v_{t+1}, v_t) and hence monotonicity fails. Since, this argument is true in general for each v_i. Hence, this proves that eventually $n - 1$ agents get stuck at each of

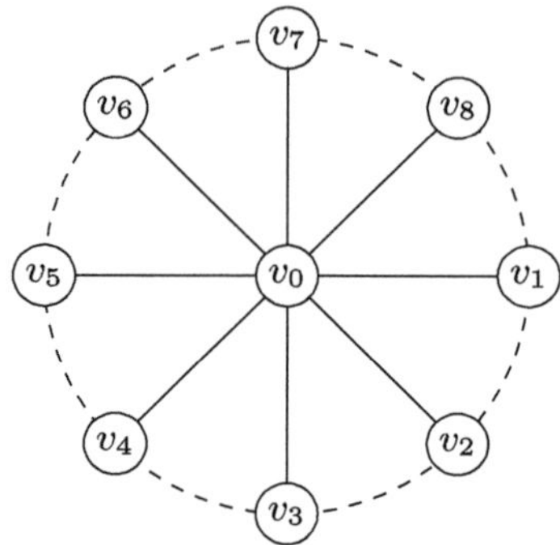

Fig. 2. The wheel graph G on 9 vertices with cyclomatic number $\mu(G) = 8$, where the dotted edges imply the disappeared edges.

these vertices. Say that the round is r'_0 ($> r_0$), when each of the $n - 1$ nodes is guarded by one agent. From $r'_0 + 1$ round onwards, if the adversary reappears on each edge of the form (v_i, v_{i+1}) ($i \in \{1, \cdots, n - 2\}$) and (v_{n-1}, v_1), then none of these agents can move from their respective vertices. It is because the degree of contaminated edges from each v_i is exactly 2, whereas there is only a single agent guarding v_i. This means, G cannot be fully decontaminated using the strategy $\mathcal{A}$. Hence, this leads to a contradiction. $\qquad\square$

Corollary 1. *In the worst case, over all dynamic graphs with cyclomatic number $\mu(G) = k$, at least $k + 1$ agents are necessary for monotone decontamination under adversarial edge dynamics.*

Next, we discuss a lower bound result, that shows that on graphs with diameter d, at least d agents are required to decontaminate those graphs under the *finite time edge appearance* model.

Lemma 1. *There exists a graph G consisting of n nodes with diameter d, on which there does not exist any deterministic monotone decontamination strategy that works with $d - 1$ co-located agents, irrespective of the finite edge reappearance or indefinite edge disappearance model.*

Proof. We prove this claim by considering the finite edge reappearance model, but it can be easily modified to the indefinite edge disappearance model. Let $v_1 \in V$ be an initial vertex where $d - 1$ agents are placed. We assume that, initially, G is contaminated, i.e., all vertices and edges of G are contaminated. Let $\mathcal{A}$ be such a strategy that claims to solve network decontamination with $d - 1$ agents, starting from v_1 (for reference see Fig. 3). Note that, in the first round, all the agents can't leave v_1 to go, say v_2, because the contaminated edge (v_3, v_1) recontaminates the unguarded node v_1. This prevents the *monotonicity* of our process.

So, the only possibility that remains is for the agents to split up, such that one of the 2 agent groups (say $\{A_1\}$) visits v_2 and for the other agent group (say

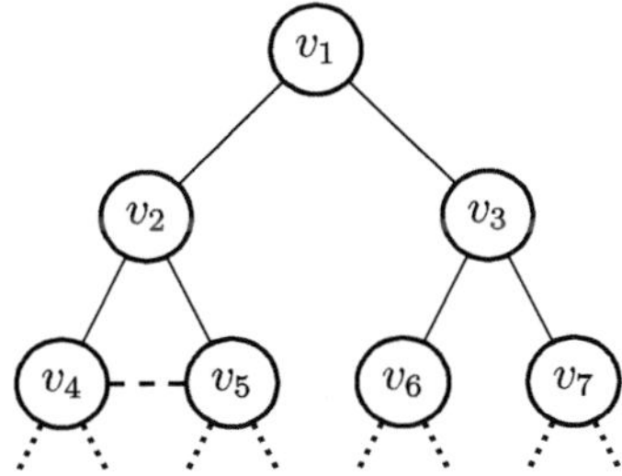

Fig. 3. This shows a graph G where d is the diameter and a possible $k = 1$. The dotted edge between v_4 and v_5 denotes the possibility for $k = 1$.

$\{A_2\}$) to wait or move forward to v_3. Following the recursion again, the agent group $\{A_1\}$ has to split into smaller groups to move forward among v_4 and v_5. Similarly, whenever $\{A_2\}$ moves forward along (v_1, v_2), they have to split off to mover forward among v_6 and v_7 and this continues recursively.

As the size of the groups continue decreasing, and since $d < n/2$, there would be a round r at which every agent, group is of size 1, visits a new parent node. From this point onwards, none of these agents, can move further. It is because, the degree of contaminated edges at parent nodes is 2 each, but a single agent is guarding them. Hence, $\mathcal{A}$ fails, which leads to a contradiction. $\square$

Remark 2. Here in Lemma 1 we took the case of $k = 0$. In fact, we can increase k by connecting children up to $d - 1$ bounded by the result in Lemma 3.2. We can still prove Lemma 1 with some modifications.

4   Algorithm for Network Decontamination

In this section, we propose network decontamination algorithms for both models of dynamicity. We start with the *finite edge reappearance model*.

4.1   Finite Edge Reappearance Model

Under this section, we propose two monotone decontamination algorithms, first for the case $k \geq n$ and next for the case $k < n$.

Case ($k \geq n$): The following algorithm (termed as UNI-DECONTAMINATION) discusses a universal monotone decontamination strategy for the case $k \geq n$, where $\mu(G) = k$, $|V| = n$ and $G = (V, E)$ is the underlying static graph of the dynamic graph $\mathcal{G}$.

High-Level Idea of UNI-DECONTAMINATION: We present a deterministic strategy satisfying monotonic decontamination of a dynamic graph $\mathcal{G}$ using n agents. The agents are initially co-located at a node, termed as *Home*. The algorithm proceeds in two main phases: the first phase is called *dispersal* and the second phase is called *cleaning*. In the *dispersal* phase, the agents starting from

Home perform *breadth first search* (or BFS). The moment a group of agent arrive at an empty node (i.e., not already occupied by any other agent), the lowest ID agent settles, and the rest continue to perform BFS. These stationary agents serve as guards, ensuring that once a vertex is cleaned, it cannot be recontaminated due to changes in the dynamic edge set. During this strategy, whenever an agent finds that the edge towards its destination node has disappeared, it waits for that edge to reappear. Now, once the *Home* node is *fully decontaminated*, i.e., *Home* and its adjacent edges are decontaminated, then the agent at *Home*, say a_1, acts as a *cleaner* agent and starts the second phase of the algorithm.

In the cleaning phase, a_1 begins to explore the network to identify and clean every remaining contaminated edge. The moment it encounters a disappeared contaminated edge, it waits for that edge to reappear, i.e., for at most T times (refer to dynamic graph model in Sect. 2). So, eventually every edge and vertex of $\mathcal{G}$ gets decontaminated, and the algorithm also maintains monotonicity.

Theorem 3. *Let $G = (V, E)$ be the initial underlying static graph of the dynamic graph $\mathcal{G}$ on $|V| = n$ vertices, then n initially co-located agents are sufficient to achieve monotone decontamination of $\mathcal{G}$.*

Time Complexity: The algorithm UNI-DECONTAMINATION takes time proportional to the size of the graph, plus the waiting delays caused by the finite time disappearing edges. The BFS strategy after incorporating the delays due to the disappearing edge requires $O(T \cdot (V + E))$ time, where $G = (V, E)$ is the underlying static graph, T is the maximum time for which any edge can be disappeared.

In the cleaning phase, the cleaner may need to wait at each vertex for up to $(\Delta - 1)T$, where Δ is the maximum degree of G. For all vertices, this adds up to $O(V \cdot \Delta \cdot T)$. Thus, the total running time is $O(T \cdot (V + E) + V \cdot \Delta \cdot T) = O(n^2 T)$, where $|V| = n$.

Remark 3. This trivial bound of n agents is only meaningful as a universal upper bound. In the remainder of this section, we focus on more refined results for the case $k \ll n$, where k measures the number of dynamic threats (e.g., chords or feedback edges) in the underlying static topology.

Remark 4. Even if we are given the condition that the agents are unable to figure out whether the neighbouring nodes are contaminated or not, it doesn't matter. As the edge-reappearing time is finite, the agent on a node v can simply wait until all the edges appear once and check them.

Case ($k < n$): In this section, we discuss a monotone decontamination strategy for the case $k < n$, termed as MODIFIED-DECONTAMINATION, where $\mu(G) = k$, $|V| = n$ and $G = (V, E)$ is the underlying static graph of the dynamic graph $\mathcal{G}$.

High-Level Idea of MODIFIED-DECONTAMINATION: We now present an algorithm MODIFIED-DECONTAMINATION that proposes to solve a monotonic decontamination of any arbitrary unknown graph G, with only $d + k$ ($k \geq 1$)

agents where $k = \mu(G)$ is the cyclomatic number and d is the diameter of the graph. So, if lets say x (where suppose x is sufficiently large) many agents are present, then we claim that among them only $d + k$ agents are sufficient to solve monotone decontamination, using MODIFIED-DECONTAMINATION strategy. Unlike the previous strategy, this approach does not require that each vertex be permanently guarded. Instead, this method relies on a comparatively small team of agents working together to clean the entire graph while ensuring that no cleaned area ever gets recontaminated. The idea is: all agents begin at a certain root vertex (termed as *Home*), which is immediately decontaminated and guarded. The decontaminated area increases as agents traverse the network. At every step, a few agents stay behind to guard the boundary between the cleaned and contaminated parts of the graph. The set of these boundary nodes, are defined to be as the *separator* nodes, and the agents guarding these nodes by remaining stationary are termed as *separator* agents. At each separator vertex, the agent stationed there identifies the contaminated incident edges, including those that have temporarily disappeared but may reappear in the future. It may be noted that, a separator agent only remains stationary at a separator node, say v, if $C_t(v) > 1$, i.e., when it finds the degree of adjacent contaminated edges (including the disappeared contaminated edges) is more than 1. Otherwise, for $C_t(v) = 1$, it continues to sweep through, i.e., in turn decontaminating the separator node. These separator agents prevent contamination from spreading back into the clean zone, even if edges reappear due to the dynamic nature of the graph. For the case, when $C_t(v) > 1$, the separator agent waits till $t' > t$, when $C_{t'}(v) = 1$.

Meanwhile, the free agents (i.e., the ones that are not guarding any separator node as a separator agent) sweep through the graph carefully exploring and cleaning new vertices and edges, until they encounter a separator vertex, u say, with $C_{t''}(u) > 1$. The main task of all the agents, starting from *Home* is to reduce the separator nodes in the graph, until it becomes null. In order to move along the graph, they perform *depth first search* (DFS) traversal. During this traversal, whenever an agent encounters a node, which is fully decontaminated (i.e., the node and all its neighbors are decontaminated), then it backtracks.

In the following part, we show that MODIFIED-DECONTAMINATION monotonically decontaminates the dynamic graph $\mathcal{G}$.

Theorem 1. *Let $\mathcal{G} \subseteq G$ be a dynamic connected graph of the static graph G, with cyclomatic number $\mu(G) = k$ and diameter d. Then, $d + k$ agents are sufficient to perform monotone decontamination starting from any initial node.*

Lemma 1. *Whenever a node, say $v \in V$ is visited for the first time by an agent at round t, our algorithm ensures that within the time interval $[t, t']$ (where $t < t'$), the node v is fully decontaminated.*

Lemma 2. *Every node in G is visited by at least one agent, during the execution of our algorithm.*

Corollary 2 (Time Complexity). MODIFIED-DECONTAMINATION *takes time proportional to the size of the graph, plus the waiting delays caused by the finite time disappearing edges. Incorporating those delays into DFS yields $O(T \cdot (V + E))$ time, where $G = (V, E)$ is the underlying static graph, T is the maximum time for which any edge can disappear.*

4.2 Indefinite Edge Disappearance Model

In this section we propose a monotone decontamination algorithm, that requires $d + 2k$ agents, initially co-located at a node of the underlying dynamic graph $\mathcal{G}$. We define $G = (V, E)$ to be the initial static graph of $\mathcal{G}$, with diameter d and $\mu(G) = k$. We term the algorithm as INFINITE-DECONTAMINATION.

High Level Idea of INFINITE-DECONTAMINATION: We present the algorithm that proposes to decontaminate an underlying unknown dynamic graph $\mathcal{G}$ with $d + 2k$ agents, where k is the cyclomatic number of the initial static graph G of $\mathcal{G}$, and d is the diameter of G. In other words, if x (where x is sufficiently large) many agents are present at the initial node, then we claim that among them $d + 2k$ are sufficient to solve monotone decontamination using the strategy INFINITE-DECONTAMINATION.

Here, similar to the previous algorithm MODIFIED-DECONTAMINATION we rely on a DFS strategy to decontaminate the graph. The exact strategy follows, but unlike the earlier strategy, there are certain changes. Let us suppose v be the node where a set of x agents arrive at time t. Based on this we can have the following cases: $C_t(v) \geq 1$ or $C_t(v) = 0$.

If $C_t(v) = 0$ and there are $M_t(v)$ many missing edges adjacent to v, where $\{M_t(v)\}$ indicates the collection of ports corresponding to these missing edges at v at time t and $M_t(v)$ indicates its cardinality. Then at most $M_t(v)$ agents remain at v, guarding these particular missing edges until they reappear. Remaining agents at v find an alternate route to move along, since at any round $\mathcal{G}$ must be connected.

If $C_t(v) \geq \alpha$ (where $\alpha \geq 1$) and $M_t(v)$ many missing edges are adjacent to v. Then atmost $M_t(v)$ agents remain at v. Among these $M_t(v)$ edges, some may be contaminated and some may be decontaminated. The agents waiting for decontaminated missing edges wait until their respective missing edges appear. On the other hand, the agents waiting for contaminated missing edges not only wait until their respective edges appear but also until the time, say t', when there exists one agent at v, that remains stationary for v to be fully decontaminated. Only if both these conditions hold, then they are allowed to move and decontaminate, after their respective edges has reappeared.

Now, among the remaining $\delta_v - M_t(v)$ many adjacent edges, there are further three sub-cases: **SubCase-1**: none edge is contaminated, **SubCase-2**: a single edge is contaminated, **SubCase-3**: more than one edges are contaminated. **SubCase-1**: If no edge is contaminated, then all agents, except $M_t(v)$ many agents, find a route from v, since $\mathcal{G}$ is connected.

SubCase-2: If a single edge is contaminated, then all the agents, except $M_t(v)$, move along this contaminated edge.

SubCase-3: If more than one edges are contaminated, then the lowest Id agent (say a_1), among all the agents at v, except the $M_t(v)$ many waiting agents, remain stationary, and the remaining agents move along the lowest port labeled contaminated non-missing edge. The agent a_1 remains stationary at v until either **SubCase-1** or **SubCase-2** occurs.

Theorem 4. *$\mathcal{G}$ be a dynamic graph, with underlying initial static graph $G = (V, E)$ with cyclomatic number $\mu(G) = k$ and diameter d. Our algorithm* INFINITE-DECONTAMINATION *monotonically decontaminates $\mathcal{G}$ with $d + 2k$ agents.*

Lemma 2. *Whenever a node, say $v \in V$ is visited for the first time by an agent at time t, our algorithm ensures that from time t onwards, at least one agent remains stationary at v until v is fully decontaminated.*

Lemma 3. *Every node in G is visited by at least one agent, during the execution of* INFINITE-DECONTAMINATION.

Remark 5. Our algorithm guarantees that all the vertices are decontaminated, but it did not guarantee all the edges to be decontaminated, because the adversary may not reappear some edges, after the agents guard those adjacent nodes. So, it is impossible to decontaminate those edges.

Lemma 4. *Algorithm* INFINITE-DECONTAMINATION *ensures that each node in the underlying dynamic graph $\mathcal{G}$ is decontaminated within $O(n^2)$ rounds.*

5   Conclusion

In this paper, we study the monotone decontamination on two models of dynamicity *finite time edge appearance* (i.e., FTEA) and *indefinite edge disappearance* (i.e., IDED) models. The FTEA model is a special case of IDED, accordingly the bounds obtained by us reflects the added complexities for the IDED model. In all our algorithms as well as lower bounds, we assume the underlying graphs to be arbitrary and the agents have no initial knowledge about it. Accordingly, the bounds obtained in this paper are among the first results to be obtained for network decontamination in dynamic networks. There are many directions of future work that can be explored. Among them, the most important fact is that, our results are not tight, so an important future work must be to obtain tight bounds. Another possible direction is to explore different other models of dynamicity, in each of these models this problem can be investigated.

References

1. Bar, R., Barik, D., Bhattacharya, A., Mandal, P.S.: Monotone decontamination of arbitrary dynamic graphs with mobile agents (2025). https://arxiv.org/abs/2511.18315
2. Bhattacharya, A., Italiano, G.F., Mandal, P.S.: Black hole search in dynamic cactus graph. In: International Conference and Workshops on Algorithms and Computation (WALCOM 2024), pp. 288–303. Springer (2024)
3. Bienstock, D., Seymour, P.: Monotonicity in graph searching. J. Algorithms **12**(2), 239–245 (1991)
4. Das, S.: Graph explorations with mobile agents. In: Distributed Computing by Mobile Entities: Current Research in Moving and Computing, pp. 403–422. Springer (2019)
5. Di Luna, G.A., Flocchini, P., Pagli, L., Prencipe, G., Santoro, N., Viglietta, G.: Gathering in dynamic rings. Theoret. Comput. Sci. **811**, 79–98 (2020)
6. Flocchini, P., Huang, M.J., Luccio, F.L.: Decontamination of chordal rings and tori using mobile agents. Int. J. Found. Comput. Sci. **18**(03), 547–563 (2007)
7. Flocchini, P., Huang, M.J., Luccio, F.L.: Decontamination of hypercubes by mobile agents. Netw. Int. J. **52**(3), 167–178 (2008)
8. Flocchini, P., Luccio, F., Pagli, L., Santoro, N.: Network decontamination under m-immunity. Discret. Appl. Math. **201**, 114–129 (2016)
9. Kaur, T., Saxena, A., Mandal, P.S., Mondal, K.: Black hole search in dynamic graphs. In: Proceedings of the 26th International Conference on Distributed Computing and Networking, ICDCN 2025, pp. 221–230. ACM, New York, NY, USA (2025)
10. Kirousis, L.M., Papadimitriou, C.H.: Interval graphs and seatching. Discret. Math. **55**(2), 181–184 (1985)
11. Kshemkalyani, A.D., Molla, A.R., Sharma, G.: Efficient dispersion of mobile robots on dynamic graphs. In: 2020 IEEE 40th International Conference on Distributed Computing Systems (ICDCS), pp. 732–742. IEEE (2020)
12. Kudelić, R.: Feedback arc set: a history of the problem and algorithms. Springer Briefs in Computer Science, Springer (2022)
13. Luccio, F., Pagli, L., Santoro, N.: Network decontamination in presence of local immunity. Int. J. Found. Comput. Sci. **18**(03), 457–474 (2007)
14. Megiddo, N., Hakimi, S.L., Garey, M.R., Johnson, D.S., Papadimitriou, C.H.: The complexity of searching a graph. J. ACM (JACM) **35**(1), 18–44 (1988)
15. Nisse, N.: Network Decontamination, pp. 516–548. Springer Int. Pub., Cham (2019)
16. Parsons, T.D.: Pursuit-evasion in a graph. In: Alavi, Y., Lick, D.R. (eds.) Theory and Applications of Graphs, pp. 426–441. Springer, Berlin, Heidelberg (1978)
17. Parsons, T.D.: The search number of a connected graph. In: Proceedings of the 9th South-Eastern Conference on Combinatorics, Graph Theory, and Computing, pp. 549–554 (1978)

New Lower Bounds for Permutation Codes in the Ulam Metric

Sergey Bereg$^{(\boxtimes)}$, Md Sanaullah Miah, and Md Saiful Islam

Department of Computer Science, University of Texas at Dallas, Box 830688,
Richardson, TX 75083, USA
besp@utdallas.edu

Abstract. Permutation codes in the Ulam metric address translocation errors that arise in flash memories and powerline communications. We investigate the structure and construction of such codes, focusing on the maximum size $P(n, d)$ of an (n, d)-permutation array. Some exact values of $P(n, d)$ are determined. To improve scalability, we develop greedy constructions enhanced by prefix pruning and refined longest common subsequence analysis. These techniques improve the Gilbert—Varshamov lower bound and provide efficient algorithms for constructing larger codes.

Keywords: Error-correcting codes · permutation codes · Ulam metric · Gilbert—Varshamov bound · greedy constructions · clique algorithms

1 Introduction

Permutation codes, introduced by Slepian in 1965 [19], are error-correcting codes that have been widely studied in both combinatorics and information theory [1,4–6,11]. They have applications in flash memory storage and powerline communication. Recent study of permutation codes under the Ulam metric is motivated by their role as rank-modulation codes resilient to translocation errors [10,12,22].

Let S_n be the set of permutations (the symmetric group) on the alphabet $\Sigma = \{1, 2, \ldots, n\}$. For a permutation $\pi \in S_n$ and $i \in \{1, \ldots, n\}$, we write π_i (or $\pi(i)$) for the element in position i of π.

The *Ulam distance* between two permutations π and σ in S_n, is defined as $d_U(\pi, \sigma) = n - \mathrm{LCS}(\pi, \sigma)$ where $\mathrm{LCS}(\pi, \sigma)$ denotes the length of the longest common subsequence between π and σ. It is the number of symbols that must be deleted from each permutation so that the remaining subsequences agree. Equivalently, it is the minimum number of single-symbol moves (translocations) needed to transform π to σ. For example, $d_U(\pi, \sigma) = 2$ for permutations $\pi = (4, 1, 5, 2, 3)$ and $\sigma = (1, 3, 5, 2, 4)$. First, move the symbol 4 so that it appears between 2 and 3. Next, move the symbol 3 so that it appears between 1 and 5.

$$\pi = (4, 1, 5, 2, 3) \rightarrow (1, 5, 2, 4, 3) \rightarrow (1, 3, 5, 2, 4) = \sigma.$$

N. Misra and A. Pandey (Eds.): CALDAM 2026, LNCS 16445, pp. 43–56, 2026.
https://doi.org/10.1007/978-3-032-17156-6_4

For an array A of permutations, the pairwise Ulam distance of A, denoted by $d_U(A)$, is $\min\{\, d_U(\sigma, \pi) \mid \sigma, \pi \in A, \sigma \neq \pi \,\}$. An (n,d)-PA is an array A of permutations on Σ with $d_U(A) \geq d$. Let $P(n,d)$ be the maximum size of a (n,d)-PA under the Ulam metric.

Error-correcting codes using the Ulam metric. A permutation array A is called a *t-correcting code* if, after deleting any t symbols from any permutation $\pi \in A$, π can be uniquely recovered from the subsequence of $n - t$ remaining symbols. It is known [10,17] that a permutation array A is a t-correcting code if and only if $d_U(A) \geq t + 1$. A central objective in coding theory is to construct large permutation codes, since their size is directly related to the amount of information that can be encoded.

Levenshtein [17] designed a perfect code capable of correcting a single deletion. For all integers $n \geq 2$,

$$P(n, 2) = (n - 1)! \tag{1}$$

Mathon and van Trung [18], investigated directed t-packings and in particular of directed t-Steiner systems. They proved $P(6, 3) = 14$ and established $P(7, 4) = 12$ by an exhaustive search.

Farnoud *et al.* [9] proved that for all integers $n \geq d \geq 1$,

$$\frac{(n - d + 1)!}{\binom{n}{d-1}} \leq P(n, d) \leq (n - d + 1)! \tag{2}$$

Göloglu *et al.* [13] derived upper bound on $P(n, d)$ for some n, d using an integer programming approach; for example, $P(7, 3), P(8, 4), P(9, 5) < 120$ and $P(8, 5), P(9, 6) \leq 12$. The other exact values of $P(n, d)$ are shown in Table 2 (see also Table 1). Using Eq. 2, they found an asymptotic lower bound for distance $d = \Delta_n + 1$ where $\Delta_n = n - c\sqrt{n}$ for a constant c

$$\lim_{n \to \infty} \frac{1}{\sqrt{n}} \ln \left(\frac{(n - \Delta_n)!}{\binom{n}{\Delta_n}} \right) = 2c \left(\ln c - 1 \right). \tag{3}$$

They presented an asymptotic improvement on this lower bound using the length of a longest increasing subsequence of a random permutation under the uniform distribution.

Recently, Wang *et al.* [23] proved a lower bound on $P(n, d)$. For a fixed integer $t \geq 1$,[1]

$$P(n, t + 1) = \Omega_t \left(\frac{n! \log n}{n^{2t}} \right). \tag{4}$$

The Ulam distance between two permutations is closely related to the Levenshtein distance. The *Levenshtein distance* between two permutations $\pi, \sigma \in S_n$, denoted by $d_V(\pi, \sigma)$, is defined as the minimum number of insertions or deletions

---

[1] The subscript t means that the hidden constant depends on t, i.e. $.\forall t \exists C_t > 0$ such that $P(n, t + 1) \geq C_t \frac{n! \log n}{n^{2t}}$ for sufficiently large n.

that are needed to change π to σ [22]. Levenshtein distance is widely used in several domains, most notably in text retrieval [14], signal processing [20], and computational biology [3].[2]

Table 1. Color legend for Tables 2 and 3.

color	legend		color	legend
Red	by Theorem 1		Olive	by Mathon and van Trung [18]
Magenta	by Göloglu *et al.*[13]		Yellow	by Theorem 2
Orange	computed by the algorithm G		Green	computed by the algorithm G+
Cyan	computed by the algorithm G++		Gray	computed by the random search

Table 2. Bounds for $P(n,d)$. Bold font denotes tight bounds. *Italic* font denotes lower bounds from [15].

$n\backslash d$	2	3	4	5	6	7	8	9	10	11	12	13	14
3	**2**	1											
4	**6**	**2**	1										
5	**24**	4	**2**	1									
6	**120**	**24**	4	**2**	1								
7	**720**	60	**12**	4	**2**	1							
8	**5040**	283	40	*10*	4	**2**	1						
9	**40320**	1658	173	28	8	**2**	**2**	1					
10	**362880**	11274	910	110	22	7	**2**	**2**	1				
11		87536	5600	525	76	17	6	**2**	**2**	1			
12			39795	3012	328	55	14	5	**2**	**2**	1		
13			65575	11059	1734	222	41	11	5	**2**	**2**	1	
14				55971	5447	1062	158	33	10	4	**2**	**2**	1
15					28706	2673	687	115	26	9	4	**2**	**2**

Gabrys et al. [10] developed error-correcting codes capable of handling deletions and erasures, laying foundational work for edit-distance-based correction. Building upon such ideas, Faircloth and Glenn [8] applied Levenshtein distance to barcode design for correcting sequencing errors in DNA.

Our results

- In Sect. 2 we proved (Theorem 2) that every diagonal of table $P(n,d)$ contains infinitely many twos. We also verified that $P(7,4) = 12$ using an algorithm for computing the maximum clique.

---

[2] Levenshtein and Ulam distances are related by $d_V(\pi,\sigma) = 2d_U(\pi,\sigma)$.

46   S. Bereg et al.

- In Sect. 3, we examined the Gilbert–Varshamov bounds for $P(n, d)$ with $n \leq 12$ (see Table 8). We also implemented a greedy algorithm for constructing permutation arrays and observed that it yields substantially better lower bounds. However, the applicability of the greedy algorithm is limited to $n \leq 12$, since beyond this the factorial growth of $n!$ makes exhaustive greedy search infeasible.
- In Sect. 4 we improve the greedy algorithm using a technique called *prefix pruning*. We were able to get lower bounds for $P(n, d)$ for $n \leq 16$. This algorithm is called G+. However, it is computationally expensive for $n = 16$ and $d < 12$.
- In Sect. 5 we further improve the greedy algorithm making algorithm G++ using an enhanced prefix pruning. We were able to get lower bounds for $P(n, d)$ for $n \leq 20$. Tables 2 and 3 summarize best known bounds for $P(n, d)$.

Table 3. Bounds for $P(n, d)$.

$n\backslash d$	7	8	9	10	11	12	13	14	15	16	17	18	19
16	13051	1389	194	86	22	7	4	4	**2**	1			
17	48691	5254	897	132	68	18	7	4	2	**2**	1		
18	237545	25505	4327	595	96	29	16	7	4	2	**2**	1	
19	590575	125605	14089	2425	270	67	19	14	6	4	2	**2**	1
20		552213	68713	8186	1468	192	52	16	12	6	2	2	**2**

2   Largest Possible Permutation Arrays

The redundancy of an encoding is minimized when the set of codewords is as large as possible [2]. Thus, it is important to determine $P(n, d)$. In this section we present some exact values of $P(n, d)$ by proving or computing. The table $P(n, d)$ consists of diagonals $D_k, k \geq 1$ where the kth diagonal D_k has entries $P(n, n - k)$ with $n \geq k + 1$. We found that, for every k, the diagonal D_k has all values equal to two when n is large enough. In other words, *all but a constant number* of exact values on the diagonal (depending on k) are equal to two.

Recall the classical result of Erdős and Szekeres.

Theorem 1 (Erdős–Szekeres [7]). *Let r, s be positive integers. Every sequence of $(r - 1)(s - 1) + 1$ distinct real numbers contains a monotonically increasing subsequence of length r or a monotonically decreasing subsequence of length s.*

Theorem 2. *Let $k \geq 1$ be an integer. If $n \geq k^3 + 1$, then $P(n, n - k) = 2$.*

Proof. The lower bound is immediate: the identity $\iota = (1, 2, \ldots, n)$ and the reverse permutation $\rho = (n, n-1, \ldots, 1)$ have $\mathrm{LCS}(\iota, \rho) = 1$, hence

$$d_U(\iota, \rho) = n - \mathrm{LCS}(\iota, \rho) = n - 1 \geq n - k,$$

so a permutation array of size two with minimum Ulam distance at least $n - k$ exists.

Next, we prove the upper bound $P(n, n-k) \leq 2$. Suppose to the contrary that $P(n, n-k) \geq 3$ for some $k \geq 1$ and $n \geq k^3 + 1$. Therefore there is a permutation array $A \subseteq S_n$ with $|A| \geq 3$ and minimum Ulam distance at least $n - k$, i.e. $d_U(A) \geq n - k$. Using right-invariance of the Ulam metric, we may assume that the identity permutation $\iota = (1, 2, \ldots, n)$ belongs to A.

For any other permutation $\tau \in A \setminus \{\iota\}$ we have $d_U(\tau, \iota) = n - \mathrm{LIS}(\tau) \geq n - k$, where $\mathrm{LIS}(\tau)$ is the length of a longest increasing subsequence of τ. Thus $\mathrm{LIS}(\tau) \leq k$ for all permutations $\tau \in A \setminus \{\iota\}$.

Now choose two distinct non-identity permutations $\pi, \sigma \in A \setminus \{\iota\}$. We will show that the assumption $d_U(\pi, \sigma) \geq n - k$ is impossible when $n \geq k^3 + 1$.

Apply the Erdős–Szekeres theorem to the permutation π, viewed as a sequence of n distinct real numbers, with parameters $r = k + 1$ and $s = k^2 + 1$. Since $n \geq k^3 + 1 = (r-1)(s-1) + 1$, the theorem guarantees that permutation π contains an increasing subsequence of length $k + 1$, or a decreasing subsequence of length $k^2 + 1$.

Case 1: Permutation π contains an increasing subsequence of length $k + 1$.
   This contradicts the fact that $\mathrm{LIS}(\tau) \leq k$ for all $\tau \in A \setminus \{\iota\}$.

Case 2: Permutation π contains a decreasing subsequence of length $k^2 + 1$.
   In this case, there exist indices $1 \leq i_1 < i_2 < \cdots < i_{k^2+1} \leq n$ such that sequence $X = (\pi_{i_1}, \pi_{i_2}, \ldots, \pi_{i_{k^2+1}})$ is strictly decreasing. All entries of X are distinct elements of $\{1, 2, \ldots, n\}$.
   Consider the restriction of σ to the symbol set

$$S_X = \left\{ \pi_{i_1}, \pi_{i_2}, \ldots, \pi_{i_{k^2+1}} \right\}.$$

Reading σ from left to right and keeping only the entries that belong to S_X, we obtain a sequence $Y = (\sigma_{j_1}, \sigma_{j_2}, \ldots, \sigma_{j_{k^2+1}})$ of length $k^2 + 1$ consisting of distinct real numbers. Note that $S_X = S_Y$ where

$$S_Y = \left\{ \sigma_{j_1}, \sigma_{j_2}, \ldots, \sigma_{j_{k^2+1}} \right\}.$$

Apply the Erdős–Szekeres theorem to Y with $r = s = k + 1$. Since $|Y| = k^2 + 1 = (k + 1 - 1)^2 + 1$, the theorem implies that Y contains a subsequence of length $k + 1$ which is either increasing or decreasing.

Subcase 2a: Sequence Y contains an increasing subsequence of length $k + 1$.
   Then permutation σ has an increasing subsequence of length $k + 1$. This contradicts the fact that $\mathrm{LIS}(\tau) \leq k$ for all $\tau \in A \setminus \{\iota\}$.

Subcase 2b: Sequence Y contains a decreasing subsequence of length $k + 1$.

Let $Z = (z_1, z_2, \ldots, z_{k+1})$ be such a decreasing subsequence of Y. By construction, all elements of Z belong to the set S_Y. Since $S_Y = S_X$ and X is a decreasing subsequence of π, any subsequence of X (in particular, the elements of Z) also appears in decreasing order in π. On the other hand, sequence Z is decreasing in σ by definition. Thus Z is a common subsequence of π and σ of length $k + 1$, and hence $\mathrm{LCS}(\pi, \sigma) \geq k + 1$. Therefore

$$d_U(\pi, \sigma) = n - \mathrm{LCS}(\pi, \sigma) \leq n - (k + 1) = n - k - 1,$$

which contradicts $d_U(\pi, \sigma) \geq n - k$.

In every possible case we reach a contradiction. Thus our initial assumption $|A| \geq 3$ is false, and no permutation array of size three can have minimum Ulam distance $n - k$ when $n \geq k^3 + 1$. Combining this with the lower bound, we conclude that $P(n, n - k) = 2$ for all integers $k \geq 1$ and $n \geq k^3 + 1$. $\qquad\square$

2.1 A Graph Approach

In this section we construct graphs that can be used to compute $P(n, d)$.

Table 4. The number of edges in the graph $G(n, d)$.

$n\backslash d$	1	2	3	4	5
3	15	3	0	0	0
4	276	168	12	0	0
5	7140	6180	2520	60	0
6	258840	249840	184680	47520	360
7	12698280	12607560	11546640	6957720	1081080

Ulam Distance Graph on Permutations. We define the graph $G(n, d)$ as follows. Each vertex corresponds to a permutation $\pi \in S_n$, represented as a string. Two vertices π and σ are connected by an undirected edge if and only if their Ulam distance satisfies $d_U(\pi, \sigma) \geq d$. Thus, computing $P(n, d)$ is equivalent to finding the maximum clique in $G_1(n, d)$. This is a challenging problem, since the size of the graph grows rapidly with n, see Table 4. (Table 5)

We create a new (reduced) graph $G'_{n,d}$ by taking the identity permutation $\pi = (1, 2, .., n)$ and all permutations at distance at least d from π. We use program for computing maximum clique MaxCliqueDyn [15]. It has found the maximum clique of size 12, see Table 6. Thus, $P(7, 4) = 12$.

Table 5. Number of edges in $G_{n,d}$.

$n\backslash d$	1	2	3	4	5
3	10	0	0	0	0
4	253	50	0	0	0
5	7021	4516	205	0	0
6	258121	232032	90791	612	0
7	12693241	12422969	9524012	1918186	1116
8				455736279	40735668

Table 6. Results for $P(7,4)$ using program `MaxCliqueDyn` [15].

1234567	7214365	6174523
7654321	3254761	5274163
4375126	4267315	5634127
4625137	4157362	3164725

3 Greedy Algorithm and Gilbert—Varshamov Bound

Greedy algorithms are well known in the construction of error-correcting codes
and they have been used to produce optimal or near-optimal codes in various
settings, including permutation arrays. See Algorithm 1, where a permutation
array A is created for given n and d.

Algorithm 1: Greedy Construction of a Permutation Array

---

1   $A \leftarrow \emptyset$ and $S \leftarrow S_n$;   // Set of all permutations of $\{1,\ldots,n\}$
2   **while** $S \neq \emptyset$ **do**
3   Pick any $\sigma \in S$ and add it to A;
4   Remove from S all π such that $d_U(\sigma, \pi) < d$;
5   **return** A;

The memory required to store all $n!$ permutations is prohibitive. We implement a version of the greedy algorithm where we generate the permutations
recursively one by one and check if a permutation σ can be added to the current
PA A. The greedy algorithm adds a permutation σ to A if $d_U(\sigma, A) \geq d$ where

$$d_U(\sigma, A) := \min\{\ d_U(\sigma, \pi) \mid \pi \in A\ \}.$$

Let $B(n,r)$ be the number of permutations at Ulam distance at most $r \in \{0, 1, \ldots, n-1\}$ from the identity permutation. By the Gilbert—Varshamov
bound, we have

$$P(n,d) \geq \frac{n!}{B(n,d-1)}. \tag{5}$$

Kong [16] found a new method to calculate $B(n,r)$ using Young tableaux. Let
$\Lambda := \{\lambda \vdash n \ : \ \lambda_1 \geq n-r\}$ be the set of partitions of n whose first part is at least

$n - r$. For example, partitions of 5 are (5), (4,1), (3,2), (3,1,1), (2,2,1), (2,1,1,1), (1,1,1,1,1). Therefore $\Lambda = \{\lambda \vdash 5 : \lambda_1 \geq 3\} = \{(5), (4,1), (3,2), (3,1,1)\}$.

Theorem 3 ([16, **Theorem III.1**]). $B(n,r) = \sum_{\lambda \in \Lambda} \left(f^\lambda \right)^2$, where f^λ denotes the number of standard Young tableaux of shape λ.

It implies $B(n,1) = O(n^2), B(n,2) = \Theta(n^4), B(n,3) = \Theta(n^6)$, see [16] for the exact formulas.

We compute values of $B(n,r)$ for all $3 \leq n \leq 12, 1 \leq r \leq 6$ by generating all permutations $\pi \in S_n$ and computing $d(\pi, id)$. The volumes $B(n,r)$ and the Gilbert—Varshamov bounds are shown in Table 7.

Table 7. The volumes $B(n,r)$.

$n \backslash r$	1	2	3	4	5	6
3	5	-	-	-	-	-
4	10	23	-	-	-	-
5	17	78	119	-	-	-
6	26	207	588	719	-	-
7	37	458	2,279	4,611	5,039	-
8	50	891	6,996	24,553	38,890	40,319
9	65	1,578	18,043	101,072	268,521	358,018
10	82	2,603	40,884	337,210	1,438,112	3,042,210
11	101	4,062	83,923	958,809	6,081,686	20,598,112
12	122	6,063	159,404	2,410,291	21,353,634	108,469,917

Table 8. The Gilbert—Varshamov bounds for $P(n,d)$ (Eq. 5).

$n \backslash d$	2	3	4	5	6	7
3	2	-	-	-	-	-
4	3	2	-	-	-	-
5	8	2	2	-	-	-
6	28	4	2	2	-	-
7	137	12	3	2	2	-
8	807	46	6	2	2	2
9	5583	230	21	4	2	2
10	44254	1395	89	11	3	2
11	395216	9827	476	42	7	2
12	3926243	79005	3005	199	23	5

While the Gilbert–Varshamov (GV) bound provides a general theoretical lower bound for $P(n,d)$, the values it produces are considerably weaker than those obtained through constructive and computational approaches (see Table 2). For example, Table 8 shows that for $(n, d) = (7, 3)$, the GV bound guarantees only 12 permutations, whereas Table 2 reports an array of size 60. Likewise, for $(n, d) = (8, 3)$, the GV bound suggests 46, yet the achieved value is 281. The disparity widens as n grows larger: for $(n, d) = (10, 3)$, the GV bound is only 1395, while Table 2 gives 11212, nearly an order of magnitude greater. At higher dimensions, the gap becomes even more pronounced; for $(n, d) = (12, 4)$ the GV bound is 3005, while Table 2 provides 2981 for $d = 4$ and substantially larger numbers, such as 39795, for nearby parameters.

These comparisons illustrate that although the GV bound is useful as a theoretical baseline, it significantly underestimates the achievable size of permutation arrays in the Ulam metric. Constructive methods such as greedy search, prefix-pruning, and clique-based algorithms consistently yield permutation arrays far larger than those suggested by the GV bound, demonstrating the weakness of this classical estimate in the present setting.

4   Improving Greedy Algorithm by Prefix Pruning

The greedy algorithm is quite expensive since it checks $n!$ permutations. For example, for $n = 13$, it is more than $6 \cdot 10^9$ permutations. We show that it can be reduced significantly using partial permutations. A new permutation σ is generated by extending a prefix of the previous permutation. Let σ' be the current prefix in this process. We denote by $\mathrm{LCS}(\sigma', \pi)$ the length of the longest common subsequence of σ' and π.

Lemma 1. *Let σ' be a prefix of a permutation $\sigma \in S_n$. If $LCS(\sigma', \pi) > n - d$ then $d_U(\sigma, \pi) < d$.*

Proof. We have $d_U(\sigma, \pi) = n - \mathrm{LCS}(\sigma, \pi)$. Note that $\mathrm{LCS}(\sigma, \pi) \geq \mathrm{LCS}(\sigma', \pi)$ since σ is an extension of σ'. Then $d_U(\sigma, \pi) = n - \mathrm{LCS}(\sigma, \pi) \leq n - \mathrm{LCS}(\sigma', \pi) < n - (n - d) = d$. $\qquad\square$

Note that permutation σ is unknown. If σ' is a prefix of length m and π is a permutation in A with $\mathrm{LCS}(\sigma', \pi) > n-d$, then we can skip all $(n-m)!$ extensions of σ' in the greedy algorithm by Lemma 1. We refer to this pruning algorithm as G+ and have implemented it. Compared with the greedy algorithm, G+ is more efficient. For instance, when $n = 13$ and $d = 14$, the greedy algorithm requires 22,487 s, while G+ requires only 1,864 s, making it approximately 12 times faster.

5   Enhanced Pruning

The computation of Ulam (and Levenstein) distance $d_U(\pi, \sigma)$ includes $\mathrm{LCS}(\pi, \sigma)$. The classical dynamic programming algorithm (DP) for the longest common subsequence (LCS) problem was introduced by Wagner and Fischer [21].

52   S. Bereg et al.

Table 9. LCS table for σ' (rows) vs. π (columns). Yellow = diagonal match chosen on Algorithm Y backtrack; cyan = UP/LEFT step on the backtrack.

π	5	10	3	7	11	8	6	9	4	1	2	12	13
σ' 0	0	0	0	0	0	0	0	0	0	0	0	0	0
5 0	1↖	1	1	1	1	1	1	1	1	1	1	1	1
10 0	1	2↖	2←	2	2	2	2	2	2	2	2	2	2
7 0	1	2	2	3↖	3←	3	3	3	3	3	3	3	3
8 0	1	2	2	3	3	4↖	4←	4	4	4	4	4	4
9 0	1	2	2	3	3	4	4	5↖	5←	5←	5←	5	5
12 0	1	2	2	3	3	4	4	5	5	5	5	6↖	6
13 0	1	2	2	3	3	4	4	5	5	5	5	6	7↖
4 0	1	2	2	3	3	4	4	5	6	6	6	6	7↑
1 0	1	2	2	3	3	4	4	5	6	7	7	7	7↑

Consider an example of a prefix $\sigma' = (5, 10, 7, 8, 9, 12, 13, 4, 1)$ and a permutation $\pi = (5, 10, 3, 7, 11, 8, 6, 9, 4, 1, 2, 12, 13)$ in the current PA for $n = 13$ and $d = 6$. The DP algorithm computes the LCS matrix shown in Table 9. The LCS length is 7 and $d_U(\sigma', \pi) = n - 7 = d$. Thus, algorithm G+ will not prune at the prefix σ' when checking π in the current PA. The backtracking algorithm [21] (see the pseudode below) computes an LCS $\alpha = (5, 10, 7, 8, 9, 12, 13)$, see Table 9. In fact, every extension of σ' to a permutation in S_{13} has an LCS with π of length at least 8. This can be seen by taking another LCS $\beta = (5, 10, 7, 8, 9, 4, 1)$. Note that 2 is the remaining symbol to complete σ'. It also appears after 1 in π. Therefore, $\mathrm{LCS}(\sigma, \pi) \geq 8$ for every σ. Thus, the prefix σ' requires modification in the greedy algorithm.

We define a problem MINPOSLCS. Let σ' be a prefix of a permutation in S_n and let π be a permutation in S_n. For any element $x \in \Sigma$, $\mathrm{pos}_\pi(x)$ denotes the index j with $\pi[j] = x$. For a sequence β, let $\mathrm{last}(\beta)$ be the last element of β. Let $\mathcal{S}(\sigma', \pi)$ be the set of all longest common subsequences of σ' and π. Define

$$\mathrm{mpl}(\sigma', \pi) = \min_{\beta \in \mathcal{S}(\sigma', \pi)} \mathrm{pos}_\pi(\mathrm{last}(\beta)). \tag{6}$$

MINPOSLCS
Input: $\pi \in S_n$ and σ', a prefix of a permutation in S_n
Output: Compute $\mathrm{mpl}(\sigma', \pi)$

We show that
(i) problem MINPOSLCS can be solved efficiently, and
(ii) an algorithm for MINPOSLCS can be used for a new pruning method.

5.1   Algorithm for MinPosLCS

We show that the MinPosLCS can be solved using dynamic programming. Let
$\text{len}[1..n][1..m]$ be the standard 2D array in computing LCS of σ' and π. We
create a new array $\text{pos}[1..n][1..m]$ such that

$$\text{pos}[i][j] = \text{mpl}(\sigma'[1..j], \pi[1..i]).$$

The algorithm is described below, see Algorithm 2 MinPosLCS (σ', π). Its running time is $O(nm)$.

5.2   New Pruning

First, we solve the problem MinPosLCS for a given prefix σ' and a permutation $\pi \in S_n$, i.e. $(L, p) = \text{MinPosLCS}(\sigma', \pi)$. The pruning rule in G+ applies
whenever $L > n - d$; that is, the extensions of σ' can be skipped in the greedy
algorithm. As in the above example, we would like to make a new pruning rule
when $L = n - d$.

Again, note that permutation σ is unknown. If, for some $t \in \Sigma \setminus \sigma'$,
$\text{pos}_\pi(t) > \text{mpl}(\sigma', \pi)$, then we can skip all extensions of σ' in the greedy algorithm by Lemma 2. We refer to this pruning algorithm as G++, see Algorithm
3
PRUNE_CHECK(σ', π) below for pruning prefix σ'. Its output is either
PRUNE or KEEP. PRUNE indicates that the current prefix σ' is discarded
because it cannot be extended to a valid PA at distance d, while KEEP indicates
that σ' remains feasible and may still extend to a valid PA.

Lemma 2. *Let σ' be a prefix of a permutation $\sigma \in S_n$. If $LCS(\sigma', \pi) = n - d$
and $pos_\pi(t) > mpl(\sigma', \pi)$ for some $t \in \Sigma \setminus \sigma'$ then $d_U(\sigma, \pi) < d$.*

Proof. Let $L = \text{LCS}(\sigma', \pi)$ and $\pi_{i_1}, \pi_{i_2}, \ldots, \pi_{i_L}$ be the longest common subsequence of σ' and π such that $i_L = \text{mpl}(\sigma', \pi)$. Let t be a number in $\Sigma \setminus \sigma'$ such
that $\text{pos}_\pi(t) > i_L$. Then $\pi_{i_1}, \pi_{i_2}, \ldots, \pi_{i_L}, t$ is a common subsequence of σ and π.
Then $\text{LCS}(\sigma, \pi) \geq L + 1$ and

$$d_U(\sigma, \pi) = n - \text{LCS}(\sigma, \pi) \leq n - (L + 1) < n - L = d.$$

The lemma follows.   □

Algorithm 2: $\textsc{MinPosLCS}(\sigma', \pi)$

Input: $\sigma' = (\sigma'_1, \ldots, \sigma'_m)$ (prefix), $\pi = (\pi_1, \ldots, \pi_n)$ (permutation)
Output: $(L, \mathrm{mpl}(\sigma', \pi))$ where $L = \mathrm{LCS}(\sigma', \pi)$

1 **for** $i \leftarrow 0$ **to** n **do**
2     $\mathrm{len}[i][0] \leftarrow 0; \quad \mathrm{pos}[i][0] \leftarrow +\infty$
3 **for** $j \leftarrow 0$ **to** m **do**
4     $\mathrm{len}[0][j] \leftarrow 0; \quad \mathrm{pos}[0][j] \leftarrow +\infty$
5 **for** $i \leftarrow 1$ **to** n **do**
6   **for** $j \leftarrow 1$ **to** m **do**
7   **if** $\pi_i = \sigma'_j$ **then**
8             $\mathrm{len}[i][j] = \mathrm{len}[i-1][j-1] + 1$
9             $\mathrm{pos}[i][j] = i$
10   **else**
11   **if** $len[i-1][j] > len[i][j-1]$ **then**
12                 $\mathrm{len}[i][j] = \mathrm{len}[i-1][j]$
13                 $\mathrm{pos}[i][j] = \mathrm{pos}[i-1][j]$
14   **else if** $len[i-1][j] < len[i][j-1]$ **then**
15                 $\mathrm{len}[i][j] = \mathrm{len}[i][j-1]$
16                 $\mathrm{pos}[i][j] = \mathrm{pos}[i][j-1]$
17   **else**
18                 $\mathrm{len}[i][j] = \mathrm{len}[i-1][j]$
19                 $\mathrm{pos}[i][j] = \min\{\mathrm{pos}[i-1][j], \mathrm{pos}[i][j-1]\}$
20 **return** $(\mathrm{len}[n][m], \mathrm{pos}[n][m])$

Algorithm 3: $\mathrm{PRUNE\_CHECK}\ (\sigma', \pi)$

Input: $\sigma' = (\sigma'_1, \ldots, \sigma'_m)$ (prefix), $\pi = (\pi_1, \ldots, \pi_n)$ (permutation)
Output: PRUNE or KEEP

1 $n \leftarrow |\pi|;$ `// length of reference permutation`
2 $(L, p) = \textsc{MinPosLCS}(\sigma', \pi);$
3 **if** $L > n - d$ **then**
4   **return** $PRUNE;$ `// G+ pruning rule`
5 **if** $L = n - d$ **then**
6   Compute $T = \Sigma \setminus \sigma'$, the set of unused symbols in σ';
7   **if** $\exists\, t \in T$ *with* $\mathrm{pos}_\pi(t) > p$ **then**
8   **return** $PRUNE;$ `// G++ pruning rule`
9 **return** $KEEP;$

We implemented the algorithm G++, and our experiments show that it is significantly faster than G+. For instance, when $(n, d) = (13, 7)$, the greedy algorithm (G) required 22,487.37 s, while the improved algorithm G+ reduced the runtime to 1,864.16 s. The enhanced pruning method G++ was even faster, finishing in only 181.20 s. Then, in this case, G++ is 124.1 times faster than G, and G++ is 10.28 times faster than G+.

We obtained new lower bounds for $P(n, d)$ for $n \leq 20$. Tables 2 and 3 summarize the best known bounds for $P(n, d)$.

Disclosure of Interests. The authors have no competing interests to declare that are relevant to the content of this article.

References

1. Barg, A., Mazumdar, A.: Codes in permutations and error correction for rank modulation. IEEE Trans. Inf. Theor. **56**(7), 3158–3165 (2010)
2. Barta, J., Montemanni, R., Smith, D.H.: Permutation codes: a branch and bound approach. In: Proceedings of the 2014 International Conference on Pure Mathematics, Applied Mathematics, Computational Methods (PMAMCM 2014), pp. 86–90. WSEAS Press, Santorini, Greece (2014)
3. Berger, B., Waterman, M.S., Yu, Y.W.: Levenshtein distance, sequence comparison and biological database search. IEEE Trans. Inf. Theor. **67**(6), 3287–3305 (2021)
4. Blake, I.F., Cohen, G.D., Deza, M.: Coding with permutations. Inf. Control **43**(1), 1–19 (1979)
5. Cameron, P.J.: Permutation codes. Eur. J. Comb. **31**(2), 482–490 (2010)
6. Chu, W., Colbourn, C.J., Dukes, P.: Constructions for permutation codes in powerline communications. Des. Codes Cryptogr. **32**(1–3), 51–64 (2004)
7. Erdős, P., Szekeres, G.: A combinatorial problem in geometry. Compos. Math. **2**, 463–470 (1935)
8. Faircloth, B.C., Glenn, T.C.: Levenshtein error-correcting barcodes for multiplexed DNA sequencing. BMC Bioinform. **14**(1), 1–15 (2013)
9. Farnoud, F., Skachek, V., Milenkovic, O.: Error-correction in flash memories via codes in the Ulam metric. IEEE Trans. Inf. Theor. **59**(5), 3003–3020 (2013)
10. Gabrys, R., Yaakobi, E., Farnoud, F., Sala, F., Bruck, J., Dolecek, L.: Codes correcting erasures and deletions for rank modulation. IEEE Trans. Inf. Theor. **62**(1), 136–150 (2016)
11. Ganesan, G.: Recursive Methods for Some Problems in Coding and Random Permutations. In: Mudgal, A., Subramanian, C.R. (eds.) CALDAM 2021. LNCS, vol. 12601, pp. 373–384. Springer, Cham (2021). https://doi.org/10.1007/978-3-030-67899-9_30
12. Goldenberg, E., Habib, M., Karthik, C.S.: Explicit good codes approaching distance 1 in Ulam metric. IEEE Trans. Inf. Theor. **71**(7), 5082–5088 (2025)
13. Göloglu, F., Lember, J., Riet, A., Skachek, V.: New bounds for permutation codes in Ulam metric. In: IEEE International Symposium on Information Theory, ISIT 2015, Hong Kong, China, 14-19 June 2015, pp. 1726–1730. IEEE (2015)
14. Hudi, R., Mitra, A.R.: Implementation of Levenshtein distance algorithm for library book search system. In: IEEE International Conference on Consumer Electronics - Taiwan (ICCE-Taiwan), pp. 653–654. IEEE (2025)
15. Konc, J., Janezic, D.: An improved branch and bound algorithm for the maximum clique problem. MATCH Commun. Math. Comput. Chem. **58**, 569–590 (2007)

16. Kong, J.: Ulam ball size analysis for permutation and multipermutation codes correcting translocation errors. IEEE Trans. Inf. Theor. **65**(12), 7806–7828 (2019)
17. Levenshtein, V.I.: On perfect codes in deletion and insertion metric. Discret. Math. Appl. **2**(3), 241–258 (1992)
18. Mathon, R., van Trung, T.: Directed t-packings and directed t-Steiner systems. Des. Codes Cryptogr. **18**, 187–198 (1999)
19. Slepian, D.: Permutation modulation. Proc. IEEE **53**(3), 228–236 (1965)
20. Todd, A., Nourian, M., Becchi, M.: A memory-efficient GPU method for hamming and Levenshtein distance similarity. In: Proceedings of the 2017 IEEE 24th International Conference on High Performance Computing (HiPC), pp. 408–418. IEEE (2017)
21. Wagner, R.A., Fischer, M.J.: The string-to-string correction problem. J. ACM **21**(1), 168–173 (1974)
22. Wang, S., Chee, Y.M., Vu, V.K.: Permutation codes in Levenshtein, Ulam and generalized Kendall-tau metrics. In: IEEE International Symposium on Information Theory, ISIT 2024, pp. 43–48. IEEE (2024)
23. Wang, S., Nguyen, T., Chee, Y.M., Vu, V.K.: Permutation and multi-permutation codes correcting multiple deletions (2024). arXiv e-prints, abs/2406.16656

Constructing Permutation Codes Under the Generalized Kendall-τ Metric

Sergey Bereg$^{(\boxtimes)}$ and Jack N. Truong

Department of Computer Science, University of Texas at Dallas, Box 830688, Richardson, TX 75083, USA
besp@utdallas.edu

Abstract. Permutation codes under the generalized Kendall-τ metric are error-correcting codes with applications in rank modulation and related storage systems. Constructing permutation arrays is a challenging problem since even computing the distance between two permutations is NP-hard. This paper investigates (n, d)-permutation arrays (PAs) with prescribed length of permutations n and minimum generalized Kendall-τ distance d. We determine exact values of $T(n, d)$, the maximum size of a PA, for $n = 4$ and $d = 2, 3$. We develop algorithms for computing Kendall–τ distances and constructing permutation arrays for $n \leq 11$. An efficient testing procedure for $(n, 2)$-PAs is also presented.

Keywords: permutation code · generalized Kendall-τ · sorting by transpositions · cycle graph

1 Introduction

Introduced by Slepian in 1965 [26], permutation codes form an important class of error-correcting codes and have been extensively studied in both combinatorial and information-theoretic contexts [4,6,10,12,17]. Permutation codes have recently become a focus of research, driven by their applications in flash memory technologies [4,15,19,20]. In particular, permutation codes under the Kendall-τ metric have attracted increasing attention [1,9,20,23,27,29,30].

The *Kendall tau distance*, or *Kendall tau rank distance*, measures how different two rankings (or sequences) are by counting the number of pairwise disagreements between them. A larger distance indicates greater dissimilarity. This distance is also known as the *bubble-sort distance*, since it equals the minimum number of adjacent swaps required by the bubble sort algorithm to transform one ranking into another. The distance was originally introduced by Maurice Kendall [21].

Example. For permutations $\pi = (1, 2, 3, 4)$ and $\sigma = (4, 3, 2, 1)$. The Kendall tau distance is 6 using swaps (1,2),(1,3),(1,4),(2,3),(2,4),(3,4).

Chee and Vu [11] introduced the generalized transposition as follows.

© The Author(s), under exclusive license to Springer Nature Switzerland AG 2026
N. Misra and A. Pandey (Eds.): CALDAM 2026, LNCS 16445, pp. 57–70, 2026.
https://doi.org/10.1007/978-3-032-17156-6_5

Definition. Let S_n denote the set of permutations of $\Sigma = \{1, 2, \ldots, n\}$. Let $\pi = (\pi_1, \pi_2, \ldots, \pi_n)$ be a permutation in S_n. An (i, j, k)-*transposition*,[1] where $1 \le i < j \le k \le n$, is an operation that swaps the subsequences $\pi_i, \pi_{i+1}, \ldots, \pi_{j-1}$ and $\pi_j, \pi_{j+1}, \ldots, \pi_k$. yielding a permutation π',

$$\pi' = (\pi_1, \ldots, \pi_{i-1}, \pi_j, \ldots, \pi_k, \pi_i, \ldots, \pi_{j-1}, \pi_{k+1}, \ldots, \pi_n).$$

The *generalized Kendall-τ distance* between two permutations $\pi, \sigma \in S_n$, denoted by $d_T(\pi, \sigma)$, is defined as the minimum number of transpositions required to transform π into σ. We call it the *T-distance* for short, and the corresponding metric the *gK-metric*. For the example above, we have $d_T(\pi, \sigma) = 3$, since $1234 \to 3412 \to 4312 \to 4321$.

Chee and Vu [11] introduced gK-metric for permutation codes. They proved general bounds $(n-3d)! \le T(n, d+1) \le (n-d)!$. They found a relation between T-distances and Hamming distances. The *Hamming distance* $hd(\pi, \sigma) = d$ for $\pi, \sigma \in S_n$ if $\pi_i \ne \sigma_i$ or exactly d positions. Let $M(n, d)$ be the maximum size of a PA in S_n with pairwise Hamming distance at least d. Chee and Vu [11] proved that

$$T(2p - 1, d) \ge T(p, d) \cdot M(p - 1, \lceil 3d/2 \rceil). \tag{1}$$

Wang *et al.* [28] studied the relationship between the generalized Kendall-τ distance and the Hamming distance. They also developed decoding algorithms.

In this paper, we observe that the generalized Kendall-τ distance is related to genome rearrangement, in particular to the problem of *sorting by transpositions*, see details in Sect. 2. This problem has attracted considerable attention in bioinformatics. Sorting by transpositions is NP-hard [8], and several approximation algorithms have been developed [2, 3, 16, 18, 24, 25]. In particular, it implies that computing the T-distance between two permutations in S_n is NP-hard.

Table 1. Maximum distance between permutations of length $n \le 15$ [14].

n	4	5	6	7	8	9	10	11	12	13	14	15
$d_{\max}$	3	3	4	4	5	5	6	6	7	8	8	9

Let $d_{\max}(n)$ be the maximum T-distance between two permutations in S_n. It is called the *transposition diameter*. In contrast with the Kendall-τ metric, which attains a maximum distance of $n(n-1)/2$, the transposition diameter is currently established only for $n \le 15$ [13] (see Table 1). Several lower and upper bounds for the transposition diameter have been derived. The first bounds were proved by Bafna and Pevzner [3] (see Sect. 2 for further discussion) $\left\lfloor \frac{n}{2} \right\rfloor + 1 \le d_{\max}(n) \le \left\lfloor \frac{3n}{4} \right\rfloor$.

Recently, Silva *et al.* [25] proved an upper bound (see also [14]) $d_{\max}(n) \le 11 \left\lfloor \frac{n}{16} \right\rfloor + \left\lfloor \frac{(3 \cdot m)}{14} \right\rfloor$ where $m = (n \bmod 16)$.

---

[1] It is common term in [3, 8, 11, 28] and It is also called *block moves* in [7, 14].

Let $A \subseteq S_n$ be an array (set) A of permutations. The pairwise T-distance of A, denoted by $d_T(A)$, is $\min\{\, d_T(\sigma, \pi) \mid \sigma, \pi \in A, \sigma \neq \pi \,\}$. A permutation array (permutation code) A is called an (n,d)-PA if $A \subseteq S_n$ and $d_T(A) \geq d$. Let $T(n,d)$ be the maximum size of a (n,d)-PA under the generalized Kendall τ metric.

Determining the exact values of $T(n,d)$, or establishing lower and upper bounds for it, is an important problem in error-correcting codes. It is a challenging problem since even computing the T-distance is NP-hard. In this paper, we study permutation arrays under the generalized Kendall-τ metric. The results presented in this paper are summarized as follows.

- In Sect. 3, we prove that $T(4,2) = 4$ and $T(4,3) = 2$ (Theorem 4).
- In Sect. 4, we derived general lower bounds for $T(n,d)$ using the method of contraction.
- In Sect. 5, we design an algorithm for computing T-distances and applied a greedy algorithm to compute permutation arrays for $n \leq 11$.
- In Sect. 6, we present an algorithm for testing a $(n,2)$-PA.

2   Preliminaries

Let $id = (1, 2, ..., n)$ be the identity permutation. The T-distance between permutations π and σ in S_n can computed using

$$d_T(\pi, \sigma) = d_T(id, \pi^{-1} \circ \sigma). \tag{2}$$

It follows from $d_T(\pi, \sigma) = d_T(\pi^{-1}(\pi), \pi^{-1}(\sigma)) = d_T(id, \pi^{-1} \circ \sigma)$. By Eq. 2, the computation of $d_T(\pi, \sigma)$ reduces to computing the distance from a permutation to the identity permutation. This can be viewed as a problem of sorting a permutation by transpositions. Let $d(\pi) = d_T(id, \pi)$ for a permutation $\pi \in S$.

A directed edge-colored *cycle graph* of π [3], denoted by $G(\pi)$, is the graph with vertex set $\{0, 1, \ldots, n+1\}$ and edge set defined as follows. For all $1 \leq i \leq n+1$, gray edges are directed from $i-1$ to i and black edges from π_i to π_{i-1} (In Fig. 1(a), black edges are shown by thick lines and gray edges are shown by thin lines).

An *alternating cycle* of $G(\pi)$ is a directed cycle in which the edges alternate colors. Observe that for each vertex in $G(\pi)$ every incoming edge is uniquely paired with an outgoing edge of different color. This implies that there is a unique decomposition of the edge set of $G(\pi)$ into alternating cycles. A cycle in $G(\pi)$ is *odd* if it has an odd number of black edges and *even* otherwise. To establish a better lower bound we analyze odd and even cycles separately. Define $c_{odd}(\pi)$ $(c_{even}(\pi))$ as the number of odd (even) cycles in π. See Fig. 1, for an example of a cycle graph. It has two odd cycles (0,1) (red) and (1,2,3,4,5,6) (blue).

Theorem 1 ([3]). $d(\pi) \geq \frac{n+1-c_{odd}(\pi)}{2}$.

3   Graphs G_n and $G_{n,d}$

We define a graph $G_n = (V, E)$ where $V = S_n$ and two permutations π and σ are connected by an edge if σ can be obtained from π by a transposition. It is a Cayley graph [14]. We also define a graph $G_{n,d} = (V, E)$ where $V = S_n$ and two permutations π and σ are connected by an edge if $d_T(\pi, \sigma) \geq d$. In graph (V, E), *the neighborhood* of $v \in V$ is denoted by $N(v) = \{u \mid (u, v) \in E\}$. For any n, d, $T(n, d)$ is the *clique number* of G, i.e. $T(n, d) = \omega(G_{n,d})$ (the size of the largest clique in graph $G_{n,d}$).

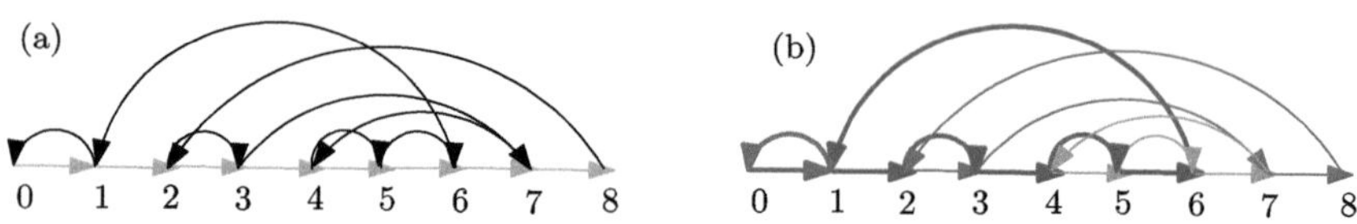

Fig. 1. (a) The cycle graph for $\pi = (1, 6, 5, 4, 7, 3, 2)$. (b) The cycle decomposition of $G(\pi)$. There are two odd cycles $\{0, 1\}$, $\{1, 2, 3, 4, 5, 6\}$. By Theorem 1 the distance $d(\pi)$ is at least $(7 + 1 - 2)/2 = 3$.

Proposition 2 *(i) For all $n \geq 3$, G_n is a regular graph of degree $\binom{n+1}{3}$.*
(ii) For all $n \geq 3$ and $2 \leq d \leq 2n-2$, graph $G_{n,d}$ is the complement of $(G_{n,1})^{d-1}$,
i.e. $G_{n,d} = \overline{(G_n)^{d-1}}$.

Proof. (i) We map every (i, j, k)-transposition to a triple (i, j, m) where $m = k + 1$. Since $1 \leq i < i < m \leq n + 1$, the number of (i, j, k)-transposition is exactly the number of 3-element subsets of $\{1, 2, \ldots, n+1\}$. It is equal to $\binom{n+1}{3}$ and the first claim follows.

(ii) By the definition of $G_{n,1}$, the edges of $(G_{n,1})^{d-1}$ correspond to the pairs (π, σ) of permutations in S_n with $d_T(\pi, \sigma) \leq d - 1$. Therefore $G_{n,d} = \overline{(G_{n,1})^{d-1}}$. Since $G_{n,1} = G_n$, we have $G_{n,d} = \overline{(G_n)^{d-1}}$. $\square$

We will use graphs G_n and $G_{n,d}$ to obtain exact values of $T(n, d)$ or bounds on $T(n, d)$. Observe that $T(n, 1) = n!$ for all $n \geq 1$ since at least one transposition is required to transform a permutation $\pi \in S_n$ to another permutation $\sigma \in S_n$.

Theorem 3. *The clique number of $G_{4,3}$ is 2. Thus, $T(4, 3) = \omega(G_{4,3}) = 2$.*

Proof. Consider graph G_4. By Proposition 2, G_4 is a regular graph of degree $\binom{5}{3} = 10$ with $4! = 24$ vertices. We write a permutation (a, b, c, d) simply as a string $abcd$. Let

$$A = \{3124, 3412, 1342, 1423, 2314\} \quad C = \{2431, 2143, 4213, 4132, 3241\}$$
$$B = \{1243, 1324, 2134, 2341, 4123\} \quad D = \{4312, 4231, 3421, 3214, 1432\}$$

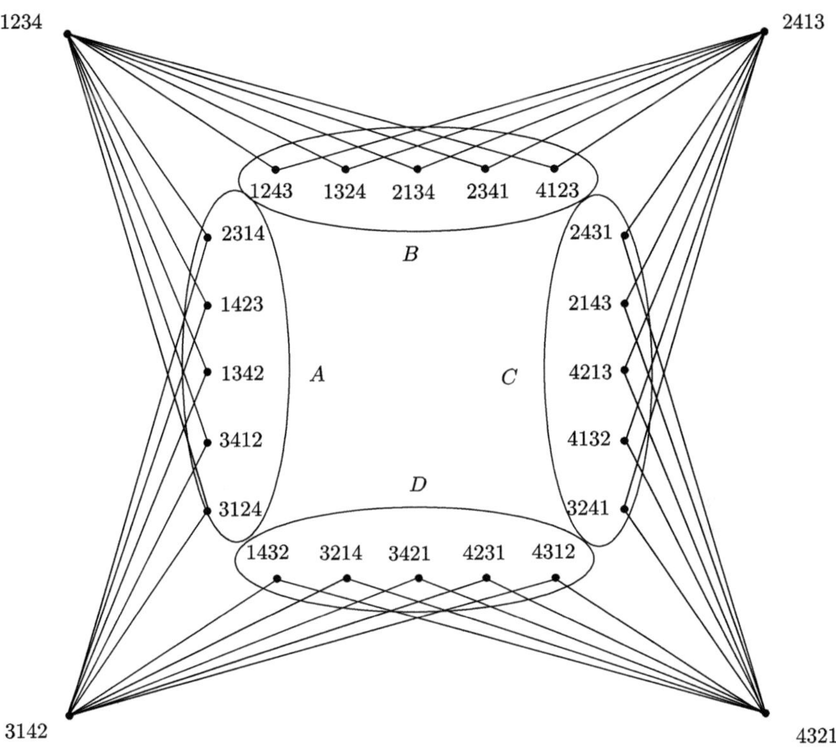

Fig. 2. The graph G_4. The graph is invariant under the rotation $\rho = (2, 4, 1, 3)$. Only the edges incident to permutations in $\{1234, 4321, 2413, 3142\}$ are shown.

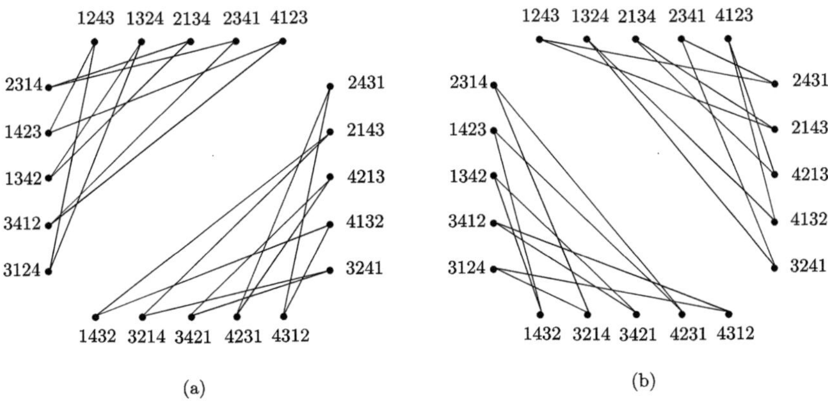

Fig. 3. (a) Edges between groups A, B and between C, D. (b) Edges between groups A, D and between B, C.

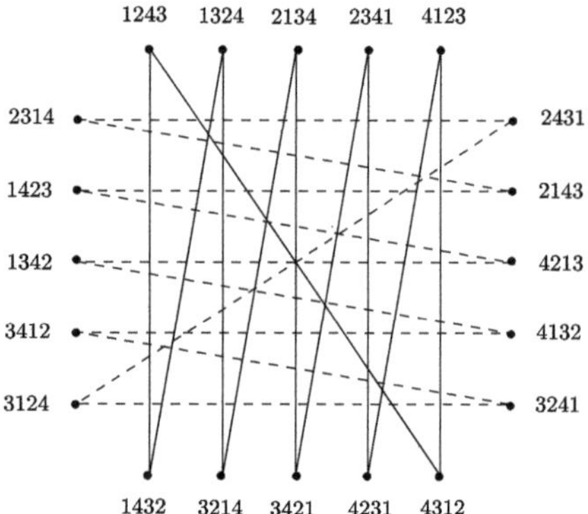

Fig. 4. Edges between groups A, C and between B, D.

The neighborhood of the permutation 1234 is $N(1234) = A \cup B$, and the neighborhood of 4321 is $N(4321) = C \cup D$. The neighborhood of the permutation 2413 is $N(2413) = B \cup C$, and the neighborhood of 3142 is $N(3142) = A \cup D$. See Fig. 2.

Since the neighborhoods $N(1234)$ and $N(4321)$ are disjoint, $d_T(1234, 4321) \geq 3$. On the other hand $d_T(1234, 4321) \leq 3$ using transpositions $1234 \to 2341 \to 3421 \to 4321$. Thus, $d_T(1234, 4321) = 3$. Therefore $T(4, 3) \geq 2$ by taking $P = \{1234, 4321\}$ (Fig. 3 and Fig. 4.).

We show that $T(4, 3) \leq 2$. Consider a $(4, 3)$-array P. Without loss of generality, the identity permutation is in P, i.e. $1234 \in P$. The only vertex in G_4 at distance 3 from the identity permutation is 4321. It follows from $N(1234) = A \cup B$ in G_4 and $N(1234) = C \cup D \cup \{2413, 3142\}$ in G_4^2. Therefore $T(4, 3) = 2$ for $P = \{1234, 4321\}$. $\qquad\square$

Theorem 4. *The clique number of $G_{4,2}$ is 4. Thus, $T(4, 2) = \omega(G_{4,2}) = 4$.*

Proof. By Proposition 2, $\omega(G_{4,2}) = \alpha(G_4)$, the *independence number* of G_4. It can be verified that $\{1234, 4321, 2413, 3142\}$ is a $(4, 2)$-array. Then $T(4, 2) \geq 4$.

To show $T(4, 2) \leq 4$, consider a $(4, 2)$-array P. WLOG, we can assume that $1234 \in P$. If $4321 \in P$ then $S_4 \setminus (\{1234, 4321\} \cup N(1234) \cup N(4321)) = \{2413, 3142\}$ and $|P| \leq 4$. Now, we assume that $4321 \notin P$. Suppose $2413 \in P$. By symmetry we can assume that $3142 \notin P$.

Since $N(1234) \cup N(2413)) = A \cup B \cup C$ and $3142, 4321 \notin P$, we have $P \subseteq \{1234, 2413\} \cup D$, see Fig. 2. Note that the subgraph of G_4 induced by D is a 5-cycle, see Fig. 5. Then at most two vertices of D can be in P. Therefore $|P| \leq 4$.

It remains to consider the case where $1234 \in P$ and $\{4321, 2413, 3142\} \notin P$. Then $P \setminus \{1234\} \subseteq C \cup D$, see Fig. 5 for the subgraph of G_4 induced by $C \cup D$. The subgraph of G_4 induced by C (resp. by D) is 5-cycle. Then $|P \cap C| \leq 2$ and $|P \cap D| \leq 2$. Thus $|P| \leq 5$.

Suppose to the contrary that $|P| = 5$. Then $|P \cap C| = 2$ and $|P \cap D| = 2$. The graph $G_{C \cup D}$ induced by $C \cup D$ is shown in Fig. 5 where the 5-cycles C and D are shown in blue and red, respectively. There are 2 cases.

Case 1. Suppose $3421 \in P$. Then $1342, 3412 \notin P$. Since $|P \cap D| = 2$, we have $1423, 3124 \in P$. Then both 1432 and 4312 are not in P. Then $P \cap C$ contains only one permutation 3421. Contradiction.

Case 2. Suppose $3421 \notin P$. Then $P \cap C$ is either $\{1432, 4231\}$ or $\{4312, 3214\}$. By symmetry, we can assume $P \cap C = \{1432, 4231\}$. Then $1342, 1423, 2314 \notin P$. Then $P \cap D$ contains only one permutation from $\{3412, 3124\}$. □

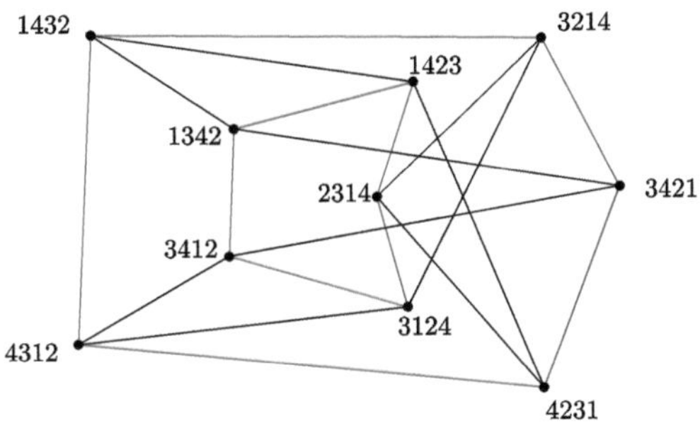

Fig. 5. The subgraph of G_4 induced by $C \cup D$.

4 Contraction

The method of *contraction* is well-known for constructing PAs under the Hamming distance [5]. The contraction of a permutation $\sigma \in S_n$, denoted by σ^{CT}, is the permutation in S_{n-1} defined as follows for $1 \leq i \leq n-1$

$$\sigma^{CT}(i) = \begin{cases} \sigma(i), & \text{if } \sigma(i) \neq n, \\ \sigma(n), & \text{if } \sigma(i) = n. \end{cases} \tag{3}$$

For a PA $A \subseteq S_n$, let $A^{CT} = \{\sigma^{CT} \mid \sigma \in A\}$. The contraction of a PA A decreases the Hamming distance by at most 3 [5]. Surprisingly, we show that the T-distance may decrease by at most 2 under contraction.

Theorem 5. *For any PA $A \subseteq S_n, n \geq 4$, we have $T(n-1, d-2) \geq T(n, d)$.*

Define a map ct : $S_n \to S_{n-1}$ by deleting the symbol n from the one-line notation of a permutation (and closing the gap).[2]

Proof (of Theorem5). Let A be an (n, d)-PA of size $T(n, d)$. Since $A^{CT} \subseteq S_{n-1}$ and $|A^C| = |A|$, it suffices to show that A^{CT} is a $(n-1, d-2)$-PA. In other words, we show that $d_T(A^{CT}) \geq d - 2$.

Take any two permutations $\pi, \sigma \in A$. If $\pi_n = \sigma_n = n$, then $d_T(\pi^{CT}, \sigma^{CT}) = d_T(\pi, \sigma)$ by Lemma 1. If only one of π_n, σ_n is equal n, then $d_T(\pi^{CT}, \sigma^{CT}) \geq d_T(\pi, \sigma) - 1$ by Lemma 2. If $\pi_n \neq n \neq \sigma_n$, then $d_T(\pi^{CT}, \sigma^{CT}) \geq d_T(\pi, \sigma) - 2$ by Lemma 2. Therefore $d_T(A^{CT}) \geq d_T(A) - 2$. □

---

[2] Note that $\text{ct}(\sigma) \neq \sigma^{CT}$ for any $\sigma \in S_n$ such that $\sigma(n) \neq n$.

Lemma 1. *Let $\pi, \sigma \in S_n$ with $\pi_n = \sigma_n = n$. Then $d_T(\pi^{CT}, \sigma^{CT}) = d_T(\pi, \sigma)$.*

Proof. Since $\pi_n = \sigma_n = n$, permutations π^{CT} and σ^{CT} are formed by simply deleting symbol n. Then $d_T(\pi, \sigma) \leq d_T(\pi^{CT}, \sigma^{CT})$ as any sequence of transpositions taking π^{CT} to σ^{CT} can be applied to take π to σ (and keep symbol n fixed).

Next we show $d_T(\pi^{CT}, \sigma^{CT}) \leq d_T(\pi, \sigma)$. Let $\pi = \beta^{(0)}, \beta^{(1)}, \ldots, \beta^{(m)} = \sigma$ be a shortest sequence of permutations in S_n taking π to σ by block transpositions, so $m = d_T(\pi, \sigma)$. For each step $\beta^{(t)} \to \beta^{(t+1)}$, write $\beta^{(t)} = PABQ$, where A, B, P, Q are the contiguous blocks such that $\beta^{(t+1)} = PBAQ$ (A and B are swapping, P and Q are the prefix/suffix). Apply map ct() to both sides:

- If $n \notin A \cup B$, then A and B stay contiguous blocks in $\mathrm{ct}(\beta^{(t)})$, and $\mathrm{ct}(\beta^{(t)}) \to \mathrm{ct}(\beta^{(t+1)})$ is a transposition.

- If $n \in A$ or $n \in B$, then A or B loses one element when we delete n. Then $\mathrm{ct}(\beta^{(t)}) = PA'B'Q$ and $\mathrm{ct}(\beta^{(t+1)}) = PB'A'Q$ where $A' = \mathrm{ct}(A)$ ans $B' = \mathrm{ct}(B)$. If one of A', B' is empty, the two permutations coincide; otherwise they differ by a transposition.

Starting from $\pi^{CT} = \mathrm{ct}(\beta^{(0)})$ and ending at $\sigma^{CT} = \mathrm{ct}(\beta^{(m)})$, we therefore obtain a sequence in S_{n-1} with at most m nontrivial transpositions. Hence $d_T(\pi^{CT}, \sigma^{CT}) \leq m = d_T(\pi, \sigma)$ and lemma follows. $\qquad\square$

Lemma 2. *Let $\pi, \sigma \in S_n$. (i) If $\pi_n = n$ and $\sigma_n \neq n$, then $d_T(\pi^{CT}, \sigma^{CT}) \geq d_T(\pi, \sigma) - 1$. (ii) If $\pi_n \neq n$ and $\sigma_n \neq n$, then $d_T(\pi^{CT}, \sigma^{CT}) \geq d_T(\pi, \sigma) - 2$.*

Proof. First, we prove (ii). Let p be the position of n in π and q the position of n in σ, i.e. $\pi_p = n, \sigma_q = n$, with $1 \leq p, q \leq n-1$ (since $\pi_n \neq n$ and $\sigma_n \neq n$). Consider the $(p, p+1, n)$-transposition. It swaps the blocks (n) and $(\pi_{p+1}, \ldots, \pi_n)$ yielding $\pi^1 = (\pi_1, \ldots, \pi_{p-1}, \pi_{p+1}, \ldots, \pi_{n-1}, n)$. Then $d_T(\pi, \pi^1) = 1$.

Similarly, applying the transposition $(q, q+1, n)$ to σ we obtain a permutation σ^1 with $d_T(\sigma, \sigma^1) = 1$. Since $\pi_n^1 = \sigma_n^1 = n$, $d_T(\mathrm{ct}(\pi^1), \mathrm{ct}(\sigma^1)) = d_T(\pi^1, \sigma^1)$ by Lemma 1. Since $\pi^{CT} = \mathrm{ct}(\pi^1)$ and $\sigma^{CT} = \mathrm{ct}(\sigma^1)$, we have $d_T(\pi^{CT}, \sigma^{CT}) = d_T(\pi^1, \sigma^1)$. By triangle inequality (since d_T is a metric), we have

$$d_T(\pi, \sigma) \leq d_T(\pi, \pi^1) + d_T(\pi^1, \sigma^1) + d_T(\sigma^1, \sigma)$$
$$= 1 + d_T(\pi^{CT}, \sigma^{CT}) + 1 = d_T(\pi^{CT}, \sigma^{CT}) + 2.$$

Therefore $d_T(\pi^{CT}, \sigma^{CT}) \geq d_T(\pi, \sigma) - 2$.

The proof of (i) is slightly different. We can change it by assigning $\pi^1 = \pi$ and using $d_T(\pi, \pi^1) = 0$. The lemma follows. $\qquad\square$

5   Algorithms

In this section we design algorithms for constructing permutations arrays under the generalized Kendall-τ metric. Greedy algorithms play a central role in the design of error-correcting codes and have been successfully applied to obtain optimal or near-optimal codes in many contexts, such as permutation arrays. A commonly used greedy algorithm is presented below, which constructs a permutation code C for specified values of n and d.

Algorithm 1: Find a PA C for given n and d

---

1 $C \leftarrow \emptyset$ and $S \leftarrow S_n$;   `// S is the set of all permutations`
2 **while** $S \neq \emptyset$ **do**
3   Pick any $\pi \in S$ and add it to C;
4   Remove from S all σ such that $d_T(\pi, \sigma) < d$;
5 **return** C;

There are two main drawbacks to this algorithm. First, it requires $O(n!)$ space, even though the resulting permutation code C is typically much smaller. Second, computing the T-distance between two permutations π and σ is costly, as the problem is NP-hard [8]. The first drawback can be resolved by employing an algorithm that systematically generates all permutations and tests each permutation whether it can be included in C. The next section deals with the second drawback.

5.1   Breadth-First Search in G_n

If we apply breadth first search in G_n, the space will be $O(n! \cdot n)$ because there is at most $n!$ permutations with each of the size of n. We reduce it by a factor of $O(n)$ using ranking permutations developed by Myrvold and Ruskey [22]. We store the distance $d_T(id, \pi)$ in $D[rank(\pi)]$. For completeness we provide the pseudocode for computing an array $D[0..(n!-1)]$ and ranking/unranking.

5.2   Computing Permutation Arrays

As the preprocessing step, we compute array D first. Then we apply a version of the greedy algorithm where the permutations are generated recursively one by one and check if a permutation σ can be added to the current PA A. The greedy algorithm adds a permutation σ to A if $d_T(\sigma, A) \geq d$ where

$$d_T(\sigma, A) := \min\{ \, d_T(\sigma, \pi) \mid \pi \in A \, \}.$$

The results are shown in Table 2.

Remark. Note that many of the bounds computed in Table 2 are new. Also, many of them are computationally expensive. For example, the bounds $T(12, 3) \geq 25,340$ and $T(12, 2) \geq 1,086,562$ take several days to calculate. These new permutation arrays can be used for error-correcting codes.

Algorithm 2: Compute T-Distance from Identity to All Permutations

---

Input: Integer n

Output: T-distance $D[\pi]$ from identity to all permutations $\pi \in S_n$

1 Initialize $D[\pi] \leftarrow \infty$ for all π of $\{1, \ldots, n\}$;

2 $\pi_{\text{id}} \leftarrow (1, 2, \ldots, n)$;

3 $r_{id} \leftarrow rank(\pi_{id})$;

4 $D[r_{\text{id}}] \leftarrow 0$;

5 $d \leftarrow 0$;

6 level $\leftarrow \{r_{id}\}$;

7 **while** *level is not empty* **do**

8   nextLevel $\leftarrow \emptyset$;

9   **foreach** r *in level* **do**

10        $\pi = \text{unrank}(r)$;

11   **foreach** *transposition* τ *applicable to* π **do**

12           $\pi' \leftarrow \tau(\pi)$;

13           $r' \leftarrow \text{rank}(\pi')$;

14   **if** $D[r'] = \infty$ **then**

15              $D[r'] \leftarrow d + 1$;

16   nextLevel $\leftarrow$ nextLevel $\cup \{r'\}$;

17     $d \leftarrow d + 1$;

18   level $\leftarrow$ nextLevel;

19 **return** D

Table 2. Lower bounds for $T(n, d)$ found using the greedy algorithm and lexicographic order of permutations.

$n \backslash d$	2	3	4	5	6	7
4	4	2	1	1	1	1
5	15	4	1	1	1	1
6	74	8	2	1	1	1
7	367	29	6	1	1	1
8	2,205	128	14	4	1	1
9	15,285	637	49	9	1	1
10	121,909	3,755	208	22	5	1
11	1,086,562	25,340	1,011	73	11	1

5.3 Space-Efficient Computation of PAs

The space for array D is very large for $n \geq 12$. We design a new approach for computing distances $d_T(\pi, \sigma)$ fast without storing the entire array D. Recall that this problem is NP-hard. Let π be a permutation in S_n. It would be desirable to find a transposition yielding a permutation π' such that $d_T(id, \pi') = d_T(id, \pi) - 1$. However, this problem is NP-hard. We will use the cycle graphs $G(\pi)$ for this.

Let π be a permutation in S_n. We define $S_t(\pi)$ as the set of all transpositions that can be applied to π. A transposition can be represented by a triple (i, j, k). For a triple (i, j, k), let $G(\pi, i, j, k)$ be the cycle graph for a permutation σ which is the permutation in S_n obtained by the (i, j, k)-transposition on permutation π, see Fig. 1 for an example. The edges of the cycle graph $G(\pi, i, j, k)$ are colored gray and black. The graph $G(\pi, i, j, k)$ is the union of alternating cycles. Let $c_{odd}(\pi, i, j, k)$ be the number of odd alternating cycles in $G(\pi, i, j, k)$. Let $c^*(\pi)$ be the maximum of odd alternating cycles

$$c^*(\pi) = \max_{(i,j,k) \in S_t(\pi)} c_{odd}(\pi, i, j, k). \tag{4}$$

Notice that the maximum can be achieved by multiple transpositions. Let σ^* be the permutation in S_n obtained by the (i, j, k)-transposition on permutation π using $c^*(\pi)$ odd alternating cycles such that the triple (i, j, k) is lexico-smallest.

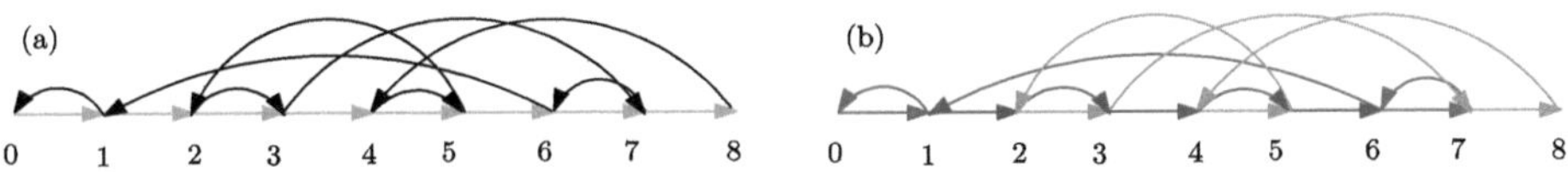

Fig. 6. Let $\pi = (1, 6, 5, 4, 7, 3, 2)$ from Fig. 1 and $\sigma = (1, 6, 7, 3, 2, 5, 4)$ is the permutation obtained by $(3, 5, 7)$-transposition from π. a) The cycle graph $G(\pi, 3, 5, 7)$. (b) The cycle decomposition of $G(\pi, 3, 5, 7)$ into 4 alternating cycles. All cycles are odd. $G(\pi, 3, 5, 7)$ achieves the maximum number of odd cycles, so $c^*(\pi) = 4$.

Definition. We call a permutation π *good* if the corresponding permutation σ^* satisfies $d_T(id, \pi) = d_T(id, \sigma^*) + 1$. We call a permutation π *bad* if it is not the identity permutation and it is not a good permutation. Let B_n be the set of all bad permutations in S_n.

Example. The permutation $\pi = (1, 6, 5, 4, 7, 3, 2)$ is good, as the transposition from Fig. 6 can be used to sort π using minimum number of transpositions.

1) swap blocks $(5, 4)$ and $(7, 3, 2)$: $(1, 6, \boxed{5, 4}\ \boxed{7, 3, 2})$ $\longrightarrow$ $(1, 6, 7, 3, 2, 5, 4)$

2) swap blocks $(6, 7, 3)$ and $(2, 5)$: $(1, \boxed{6, 7, 3}\ \boxed{2, 5}, 4)$ $\longrightarrow$ $(1, 2, 5, 6, 7, 3, 4)$

3) swap blocks $(5, 6, 7)$ and $(3, 4)$: $(1, 2, \boxed{5, 6, 7}\ \boxed{3, 4})$ $\longrightarrow$ $(1, 2, 3, 4, 5, 6, 7)$.

In order to save space, we do not store the array D. Instead, we store an array P_n which is a sorted list of pairs

$$P_n = \{(rank(\pi), d_T(id, \pi)) \mid \pi \in B_n\}.$$

For a permutation π, the corresponding permutation σ^* can be computed in polynomial time using Eq. 4. This subroutine can be used to compute array P_n using the array D. The array P_n can be used as preprocessing step (stored in a file) such that the distance queries can be answered efficiently. The distance between two permutations can be reduced to the distance between a permutation and the identity permutation.

Now, we explain how the distance $d_T(id, \pi)$ can be computed for a permutation π. If $\pi = id$ then $d_T(id, \pi) = 0$. Compute the rank r of π and search for it in P_n. If it is found then π is bad and a pair (r, d) is found, then $d_T(id, \pi) = d$. If π is good (not found in P_n), then we find the first transposition increasing number of odd cycles in the cycle graph. We apply it yielding π'. Then we repeat this algorithm for π'.

This approach is space efficient and the Table 3 shows the number of bad permutations for $n \leq 10$. We are planning to use this approach for computing PAs for $n \geq 12$.

Table 3. Number of bad permutations for $4 \leq n \leq 10$.

n	4	5	6	7	8	9	10
a_6	0	2	29	309	3227	33857	363506

6 Testing PAs

Since computing the T-distance between two permutations in S_n is NP-hard, the problem of verifying a given (n, d)-array A is also NP-hard. Typically, the verification algorithm checks that pairwise distances are at least d. We use preprocessing with BFS algorithm and ranking permutations from Sect. 5. Using array D, the verification algorithm computes $d_T(\pi, \sigma)$ in $O(n)$ time by Equation 2. Therefore the running time for testing a PA is $O(P(n)+nN^2)$ time where $P(n)$ is the preprocessing time. This is very expensive for $d = 2$ as $N = \Omega((n-3)!)$ by the Gilbertâ\u{A}\c{S}Varshamov bound. We show that an array can be verified faster for $d = 2$.

Theorem 6. *An $(n, 2)$-PA of size N under the generalized Kendall-τ distance can be verified in $O(n^4 N \log N)$ time.*

Proof. Let A be a set of N permutations in S_n. First, we sort permutations in A in lexicographic order. It takes $O(nN \log N)$ time since the comparison of two permutations can be done in $O(n)$ time. In $O(nN)$ time we check if two permutations π, σ in A have T-distance 0 (or equivalently $\pi = \sigma$). See lines 2–4 in the pseudocode of Algorithm 3. It remains to check T-distance 1.

For each permutation $\pi \in A$, the algorithm computes $\binom{n+1}{3}$ neighbors of π in G_n by applying transpositions. We search each neighbor τ of π in A using binary search. The total time is $O(n^4 \log N)$ for each permutation. The theorem follows. $\square$

Algorithm 3: Verifying a $(n, 2)$-PA

Input: An array A of N permutations in S_n
Output: Decide whether $d_T(A) \geq 2$

1 Sort A in lexicographical order;
2 **for** $i = 1$ *to* $N - 1$ **do**
3   **if** $A[i] = A[i + 1]$ **then**
4   **return** *False*;

5 **foreach** π *in* A **do**
6   **foreach** *(i,j,k)-transposition of* π *where* $1 \leq i \leq j \leq k \leq n$ **do**
7   τ is transposed from π;
8   **if** *using binary search not found* τ *in* A **then**
9   **return** *False*;

10 **return** *True*;

Disclosure of Interests. The authors have no competing interests to declare that are relevant to the content of this article.

References

1. Abdollahi, A., Bagherian, J., Jafari, F., Khatami, M., Parvaresh, F., Sobhani, R.: New upper bounds on the size of permutation codes under Kendall τ-metric. Cryptogr. Commun. **15**, 891–903 (2023)
2. Alexandrino, A.O., Brito, K.L., Oliveira, A.R., Dias, U., Dias, Z.: A 1.375-approximation algorithm for sorting by transpositions with faster running time. In: Scherer, N.M., de Melo-Minardi, R.C. (eds) Advances in Bioinformatics and Computational Biology, BSB 2022. LNCS, vol. 13523, pp. 147–157. Springer, Cham (2022). https://doi.org/10.1007/978-3-031-21175-1_16
3. Bafna, V., Pevzner, P.A.: Sorting by transpositions. SIAM J. Discret. Math. **11**(2), 224–240 (1998)
4. Barg, A., Mazumdar, A.: Codes in permutations and error correction for rank modulation. IEEE Trans. Inf. Theor. **56**(7), 3158–3165 (2010)
5. Bereg, S., Levy, A., Sudborough, I.H.: Constructing permutation arrays from groups. Des. Codes Crypt. **86**(5), 1095–1111 (2018)
6. Blake, I.F., Cohen, G.D., Deza, M.: Coding with permutations. Inf. Control **43**(1), 1–19 (1979)
7. Bona, M., Flynn, R.: Sorting a permutation by block moves (2008). arXiv e-prints, abs/0806.2787
8. Bulteau, L., Fertin, G., Rusu, I.: Sorting by transpositions is difficult. SIAM J. Discret. Math. **26**(3), 1148–1180 (2012)
9. Buzaglo, S., Etzion, T.: Bounds on the size of permutation codes with the Kendall τ-metric. IEEE Trans. Inf. Theor. **61**(6), 3241–3250 (2015)
10. Cameron, P.J.: Permutation codes. Eur. J. Comb. **31**(2), 482–490 (2010)

11. Chee, Y.M., Jiang, X., Kiah, M., Wang, C., Purkayastha, P., Ling, S.: Breakpoint analysis and permutation codes in generalized Kendall tau and Cayley metrics. In: 2014 IEEE International Symposium on Information Theory, pp. 2959–2963 (2014)
12. Chu, W., Colbourn, C.J., Dukes, P.: Constructions for permutation codes in powerline communications. Des. Codes Cryptogr. **32**(1–3), 51–64 (2004)
13. Cunha, L.F.I., Kowada, L.A.B., de A. Hausen, R., de Figueiredo, C.M.H.: Advancing the transposition distance and diameter through lonely permutations. SIAM J. Discret. Math. **27**(4), 1682–1709 (2013)
14. Eriksson, H., Eriksson, K., Karlander, J., Svensson, L., Wästlund, J.: Sorting a bridge hand. Discret. Math. **241**(1), 289–300 (2001)
15. Farnoud, F., Skachek, V., Milenkovic, O.: Error-correction in flash memories via codes in the Ulam metric. IEEE Trans. Inf. Theor. **59**(5), 3003–3020 (2013)
16. Firoz, J.S., Hasan, M., Khan, A.Z., Rahman, M.S.: The 1.375 approximation algorithm for sorting by transpositions can run in $O(n \log n)$ time. J. Comput. Biol. **18**(8), 1007–1011 (2011)
17. Ganesan, G.: Recursive methods for some problems in coding and random permutations. In: Mudgal, A., Subramanian, C.R. (eds.) CALDAM 2021. LNCS, vol. 12601, pp. 373–384. Springer, Cham (2021). https://doi.org/10.1007/978-3-030-67899-9_30
18. Hartman, T., Shamir, R.: A simpler and faster 1.5-approximation algorithm for sorting by transpositions. Inf. Comput. **204**(2), 275–290 (2006)
19. Jiang, A., Mateescu, R., Schwartz, M., Bruck, J.: Rank modulation for flash memories. IEEE Trans. Inf. Theory **55**(6), 2659–2673 (2009)
20. Jiang, A., Schwartz, M., Bruck, J.: Correcting charge-constrained errors in the rank-modulation scheme. IEEE Trans. Inf. Theory **56**(5), 2112–2120 (2010)
21. Kendall, M.G.: A new measure of rank correlation. Biomet. **30**(1/2), 81–93 (1938)
22. Myrvold, W.J., Ruskey, F.: Ranking and unranking permutations in linear time. Inf. Process. Lett. **79**(6), 281–284 (2001)
23. Parvaresh, F., Sobhani, R., Abdollahi, A., Bagherian, J., Jafari, F., Khatami, M.: Improved bounds on the size of permutation codes under Kendall τ-metric (2024). arXiv e-prints, abs/2406.06029
24. Silva, L.A.G., Kowada, L.A.B., Walter, M.E.M.T.: A barrier for further approximating sorting by transpositions. J. Comput. Biol. **30**(12), 1277–1288 (2023)
25. Silva, L.A.G., Kowada, L.A.B., Rocco, N.R., Walter, M.E.M.T.: A new 1.375-approximation algorithm for sorting by transpositions. Algorithms Mol. Biol. **17**(1), 1 (2022)
26. Slepian, D.: Permutation modulation. Proc. IEEE **53**(3), 228–236 (1965)
27. Vijayakumaran, S.: Largest permutation codes with the Kendall τ-metric in s_5 and s_6. IEEE Commun. Lett. **20**(10), 1912–1915 (2016)
28. Wang, S., Chee, Y.M., Vu, V.K.: Permutation codes in Levenshtein, Ulam and generalized Kendall-tau metrics. In: IEEE International Symposium on Information Theory, pp. 43–48 (2024)
29. Wang, X., Wang, Y., Yin, W., Fu, F.W.: Nonexistence of perfect permutation codes under the Kendall τ-metric. Des. Codes Cryptogr. **89**(11), 2511–2531 (2021)
30. Wang, X., Zhang, Y., Yang, Y., Ge, G.: New bounds of permutation codes under Hamming metric and Kendall's τ-metric. Des. Codes Cryptogr. **85**(3), 533–545 (2017)

On Optimal Achromatic Coloring of 3-Uniform Tight Cycles

Srimanta Bhattacharya, Pawan Kumar, Deepak Rajendraprasad[ID],
and Jishnu Sen[(✉)][ID]

Indian Institute of Technology Palakkad, Palakkad 678623, Kerala, India
{srimanta,deepak}@iitpkd.ac.in, 212214002@smail.iitpkd.ac.in,
senjishnu5@gmail.com

Abstract. For two integers n and k, $n > k \geqslant 2$, an n-vertex k-uniform tight cycle TC_n^k is a hypergraph on n vertices which can be arranged in a cyclic sequence such that every set of k consecutive vertices (and only those) form an edge. A vertex-coloring of a hypergraph H is called achromatic if every edge of H contains vertices of at least two colors and every pair of distinct colors appears together on vertices of some edge. The achromatic number $\psi(H)$ is the maximum number of colors possible in an achromatic coloring of H. In this paper, we study the achromatic number of k-uniform tight cycles and give early results on the 3-uniform case. We find it more natural to study an extremal function $f_k(t)$ which denotes the smallest n for which $\psi(TC_n^k) = t$. We give a general lower bound for $f_k(t)$ and an upper bound matching the same for many values of t when $k = 3$. We can extend this to infinitely many values of t when $k = 3$ if a well-known conjecture on the infinitude of Sophie Germain primes is true. We also introduce a new conjecture and show that, if this conjecture is true, then $f_3(t) = t \left\lceil \frac{t-1}{4} \right\rceil$ for all $t \geqslant 3$ except for $t = 9$. We have computationally verified this conjecture for t up to 101.

Keywords: Achromatic coloring · Tight cycles · Hypergraph · Sophie Germain Primes

1 Introduction

A *proper coloring* of a graph is an assignment of colors to its vertices such that every two adjacent vertices receive different colors. A proper coloring is *complete* if each pair of distinct colors appear on at least one pair of adjacent vertices. For a given graph G, the minimum number of colors in any complete coloring is the *chromatic number* $\chi(G)$, whereas the maximum number is the *achromatic number* $\psi(G)$ *of* G. Complete colorings were first studied by Harary et al. [6] in 1967. For a detailed survey on achromatic coloring of graphs, we refer the readers to Edwards [3].

A hypergraph H is a pair (V, E) where V is the set of *vertices* and E, the set of *hyperedges*, is a family of subsets of V. Two vertices are *adjacent* in H if

© The Author(s), under exclusive license to Springer Nature Switzerland AG 2026
N. Misra and A. Pandey (Eds.): CALDAM 2026, LNCS 16445, pp. 71–83, 2026.
https://doi.org/10.1007/978-3-032-17156-6_6

there exists a hyperedge that contains both vertices. If every hyperedge contains exactly k vertices, then we have a *k-uniform hypergraph*. There are two common ways to generalize proper coloring to hypergraphs. In the first variant, a coloring is said to be *proper* if every edge contains at least two vertices of different colors, and this type was first introduced by Erdős and Hajnal [4] in 1966. This type of coloring in hypergraphs is the most common and is often referred to as *weak coloring*. On the other hand, if no two vertices of an edge share the same color, then we have a *rainbow coloring* (or *strong coloring*) on hypergraphs (see [1]).

Considering properness in the weak sense, Jucovič and Olejník [8] in 1974 generalized the concept of *complete coloring* to uniform hypergraphs, where for every pair of distinct colors there is a hyperedge that contains at least two vertices assigned to those colors. It is evident that if a proper coloring uses the minimum number of colors, then it is complete. For a uniform hypergraph H, the maximum number of colors possible in a complete coloring of H is called the *achromatic number* $\psi(H)$ of H. In fact, Olejník [10] in 1981 extended this notion of achromatic coloring to general hypergraphs. Later in 1983, Nešetřil et al. [9] defined another variant of complete coloring in hypergraphs, in which for every pair of colors, there is a hyperedge whose vertices are exclusively colored with those two colors.

However, in 2017, Dębski et al. [2] defined a complete coloring on uniform hypergraphs of a flavor different from that of [9]. A rainbow coloring on a k-uniform hypergraph is referred to as *achromatic* if for every k-subset of colors, there exists *at least* one hyperedge whose vertices are assigned with those k colors. It is evident that every simple graph has an achromatic coloring, but this fact differs in the case of uniform hypergraphs, as Dębski et al. [2] discovered an infinite family of uniform hypergraphs that do not admit any achromatic coloring.

1.1 Our Contributions

Even though it has been about 50 years since the achromatic number was generalized to hypergraphs, very little is known about the behaviour of these parameters, even on standard hypergraphs.

For two integers n and k, $n > k \geqslant 2$, the *tight cycle* TC_n^k is a k-uniform hypergraph having a cyclically ordered set of n vertices such that any k consecutive vertices form a hyperedge. In this article, we initiate a study of the achromatic number of k-uniform tight cycles TC_n^k. We mainly focus on the earliest generalisation of achromatic coloring to hypergraphs due to Jucovič and Olejník [8], which is made precise in the following definition.

Definition 1. [8,10] *Let H be a hypergraph and c be a positive integer. An achromatic c-coloring on H is a weak coloring on H with c colors such that every pair of distinct colors appears on at least one pair of adjacent vertices. The largest integer c such that H has an achromatic c-coloring is the* achromatic number $\psi(H)$ of H.

By two colors being *adjacent*, we mean that there exists a pair of adjacent vertices assigned with those two colors. Interestingly, for tight cycles, we observe that the achromatic number is not well behaved in the sense that $\psi(TC_n^k)$ may not be a monotone function of n. As preliminary evidence, we list $\psi(TC_n^3)$ for some small values of n in Table 1 to indicate that $\psi(TC_6^3) = 4 < 5 = \psi(TC_5^3)$. However, in Proposition 1, we show that if $k \geqslant 3$ and $m \geqslant k - 1$, then $\psi(TC_n^k) \leqslant \psi(TC_{n+m}^k)$. In particular, for the 3-uniform case, this shows that the monotonicity of $\psi(TC_n^3)$ is restored if the values n differ by at least 2.

Table 1. Achromatic numbers for some 3-uniform tight cycles.

n	3	4	5	6	7
$\psi(TC_n^3)$	3	4	5	4	5

Given the above observation, it is natural to try to find good bounds on $\psi(TC_n^k)$. With this motivation, we introduce and study a new extremal parameter $f_k(t)$ which can be thought of as an "inverse" of the achromatic number $\psi(TC_n^k)$.

Definition 2. *For the integer* $k \geqslant 2$, $f_k(t)$ *is the* least *integer* n *such that* $\psi(TC_n^k) = t$.

Finding $f_k(t)$ is a natural problem, and it is more meaningful to pursue as we show in Proposition 2 that $f_k(t)$ increases monotonically with t. So, finding $f_k(t)$ is indirect, but in our view, a more systematic way of finding $\psi(TC_n^k)$. Moreover, as an extremal problem, finding $f_k(t)$ is quite appealing and, in our opinion, is of independent interest. So, in this work, we exclusively study $f_k(t)$ and most of our results concern the 3-uniform case ($k = 3$). We find that it is already challenging to obtain the exact value of $f_3(t)$. Towards the above goal, our first result is the following lower bound on $f_k(t)$.

Theorem 1. *For any two positive integers* k *and* t, *such that* $k > 1$,

$$f_k(t) \geqslant t \left\lceil \frac{t - 1}{2(k - 1)} \right\rceil .$$

This gives a lower bound of $t \left\lceil \frac{t-1}{4} \right\rceil$ for $f_3(t)$. Our next two theorems show that the above lower bound is nearly tight for many values of t. We define $\mathrm{ord}_p(2)$ as the smallest positive integer k such that $2^k \equiv 1 \pmod{p}$. If $k = p - 1$, then 2 is a *primitive root* modulo p. We call a prime number p a *good prime* if $\mathrm{ord}_p(2) = (p - 1)$ or $\frac{(p-1)}{2}$. One can verify that $3, 5, 7, 11, 13, 19, 23, 29, 37$ and 47 are examples of good primes. A famous conjecture by Artin (see [7]) states that for every integer a which is neither a perfect square nor -1, there are infinitely many primes p such that $\mathrm{ord}_p(a) = p - 1$. If this is true for $a = 2$, then we have an infinite sequence of good primes. A prime p is said to be a *Sophie Germain prime* if $2p + 1$ is also a prime, and the corresponding prime of the type $2p + 1$ is called a *safe prime*. Since $\mathrm{ord}_{2p+1}(2) \in \{2, p, 2p\}$ and $\mathrm{ord}_{2p+1}(2) > 2$ when $2p + 1 > 3$, we can conclude that every safe prime is good. Further, every safe prime greater than 5 is of the form $4k + 3$ for some integer k. Hence, if the well-known conjecture on the infinitude of Sophie Germain primes (see [12]) is true, we have an infinite sequence of good primes of the form $4k + 3$ for some integer k. By the first two cases in Theorem 3, this will in turn give us an infinite sequence of t for which the lower bound in Theorem 1 is tight for 3-uniform tight cycles. If we know that $f_3(t) = n$ for some t, then it follows from the definition of f_k that $\psi(TC_n^3) = t$. Hence, we can determine $\psi(TC_n^3)$ exactly for infinitely many n, if we know the values of $f_3(t)$ at infinitely many values of t.

Theorem 2. *Let p be a good prime such that $p \equiv 1 \pmod 4$. Then*

1. $f_3(p) \leqslant (p + 1) \left\lceil \frac{p-1}{4} \right\rceil + 1.$
2. $f_3(p - 1) \leqslant p \left\lceil \frac{p-1}{4} \right\rceil.$

Theorem 3. *Let p be a good prime such that $p \equiv 3 \pmod 4$. Then*

1. $f_3(p) = p \left\lceil \frac{p-1}{4} \right\rceil.$
2. $f_3(p - 1) = (p - 1) \left\lceil \frac{p-1}{4} \right\rceil.$
3. $f_3(p - 2) \leqslant (p - 2) \left\lceil \frac{p-1}{4} \right\rceil.$
4. $f_3(p - 3) \leqslant (p - 3) \left\lceil \frac{p-1}{4} \right\rceil.$

Since the upper bounds in Theorems 2 and 3 apply only to primes with special properties, we continue our investigation for a tight upper bound which will work for all values t. We believe that the lower bound for $f_3(t)$ given by Theorem 1 is tight for all values of t except $t = 9$. We could prove that it is the case if the following new conjecture is true.

Conjecture 1. Let k be an integer larger than 2 and $t = 4k + 1$. Then there exists a sequence $x_0, x_1, \ldots, x_{k-1} \in \mathbb{Z}_t^*$ such that,

(i) $\mathbb{Z}_t^* = \bigcup_{i=0}^{k-1} \{\pm x_i, \pm(x_i + x_{(i+1 \mod k)})\}$, and
(ii) $\gcd(\sum_{i=0}^{k-1} x_i, t) = 1.$

We discuss our observations and evidence for this conjecture in Sect. 4. However, we note here that the conclusion does not hold for $k = 2$.

Theorem 4. *Let $t \neq 9$ be a positive integer. If Conjecture 1 is true, then*

$$f_3(t) = t \left\lceil \frac{t - 1}{4} \right\rceil.$$

1.2 Proof Technique

In order to show an upper bound of the form $f_3(t) \leqslant n$, it is enough to show an achromatic t-coloring of TC_n^3. This is because $t' = \psi(TC_n^3)$ is at least t and hence $f_3(t) \leqslant f_3(t')$ (Proposition 2) which is at most n.

For the proof of Theorem 2 and Theorem 3, we bank on a common approach of obtaining an achromatic coloring of TC_n^3 by concatenating a bunch of finite sequences of colors in a cyclic order. These sequences are arithmetic progressions on $\mathbb{Z}_p$ whose common differences are chosen carefully. For every good prime p, we find a set $\mathcal{D}$ of integers that, in some sense, "covers" $\mathbb{Z}_p^*$. We will have one sequence for each $d \in \mathcal{D}$; an arithmetic progression with common difference d of certain length, starting with a carefully chosen element. However, in some cases, we add one "special" sequence to obtain the achromaticity.

While we use multiple arithmetic progressions to construct a coloring in Theorem 2 and Theorem 3, we construct the coloring in Theorem 4 from a single multi-difference arithmetic progression. The success of this approach relies on a special property of the set of common differences, which is essentially the statement of Conjecture 1.

1.3 Related Results

Achromatic number of simple cycles and paths was determined by Geller and Kronk [5] in 1974. In our terminology, it follows from Theorem 5 that $f_2(p) = p \left\lceil \frac{p-1}{2} \right\rceil$ for every $p \geqslant 3$.

Theorem 5. *[5] For any natural number $p \geqslant 3$, if $p \left\lceil \frac{p-1}{2} \right\rceil \leqslant n \leqslant (p+1) \left\lceil \frac{p}{2} \right\rceil$, then $\psi(C_n) = p$, unless p is odd and $n = \frac{p(p-1)}{2} + 1$, in that case $\psi(C_n) = p - 1$.*

Theorem 6. *[5] Let $p \geqslant 3$ be any natural number, and $n = p \left\lceil \frac{p-1}{2} \right\rceil$. If p is even, then $\psi(P_{n-1}) < \psi(P_n) = p$; otherwise $\psi(P_n) < \psi(P_{n+1}) = p$.*

A notable work on the variant of the achromatic number of hypergraphs that we study here is a Nordhaus-Gaddum-type bound for uniform hypergraphs by Olejník [10]. Let $H = (V, E)$ be a k-uniform hypergraph. The *complement* $\overline{H} = (V, E')$ of H is a k-uniform hypergraph where E' consists of all the k-element subsets of V that are not in E.

Theorem 7. *[10] For a k-uniform hypergraph H, $k \geqslant 3$, with n vertices,*

$$n \leqslant \psi(H) + \psi(\overline{H}) \leqslant 2n.$$

2 Tight Cycles: Monotonicity and Lowerbound

In this section, we prove a "partial" monotonicity of $\psi(TC_n^k)$ as a function of n, the strict monotonicity of $f_k(t)$ as a function of t and the lower bound of $f_k(t)$.

Proposition 1. *For any three integers n, k and m, $n > k \geqslant 3$ and $m \geqslant k - 1$, $\psi(TC_n^k) \leqslant \psi(TC_{n+m}^k)$.*

Proof. Let $V = (v_1, \ldots, v_n, v_1)$ be the cyclic sequence of $V(TC_n^k)$. Let TC_{n+m}^k be the tight cycle whose vertex set is the cyclic super-sequence of V obtained by adding m vertices, say $u_1, u_2, \ldots, u_m$ between v_n and v_1. Consider any achromatic t-coloring of $V(TC_n^k)$. Now, consider a coloring of $V(TC_{n+m}^k)$ that retains the colors of $V(TC_n^k)$ and the vertex u_i is assigned the same color as v_i for $1 \leqslant i \leqslant k - 1$. The remaining vertices are colored with any of the used colors. In this coloring any two colors that are adjacent in TC_n^k are still adjacent in TC_{n+m}^k. Any monochromatic edge formed in this process can be made non-monochromatic by recoloring one of the intermediate vertices preserving achromaticity. Hence, we get an achromatic t-coloring of TC_{n+m}^k. As we can start with an optimal achromatic coloring of TC_n^k, the result follows.

Remark 1. We cannot extend Proposition 1 to the graph case ($k = 2$) when $m = k - 1 = 1$. For any odd integer $p \geqslant 3$ and $n = \binom{p}{2}$, $\psi(TC_n^2) = p$ while $\psi(TC_{n+1}^2) = p - 1$ (Theorem 5).

Proposition 2. *For any two integers k, t, both greater than 1, $f_k(t) < f_k(t+1)$.*

Proof. Let $f_k(t + 1) = n$ and consider an achromatic $(t + 1)$-coloring $C : V(TC_n^k) \to \{0, 1, \ldots, t\}$. Let V be the cyclic sequence of vertices in TC_n^k, and $TC_{n'}^k$ be the tight cycle whose vertex set is the cyclic subsequence of V obtained by removing all the vertices of the color t and then replacing any sequence of consecutive vertices of the same color with a single vertex of that color. We argue next that the coloring C restricted to $V(TC_{n'}^k)$ is an achromatic t-coloring of $TC_{n'}^k$. Since we have removed consecutive vertices of the same color, there is no monochromatic edge in $TC_{n'}^k$. Any two colors, other than color t, adjacent in TC_n^k are still adjacent in $TC_{n'}^k$. Hence $f_k(t) \leqslant n' < n = f_k(t + 1)$.

Theorem 1. *For any two positive integers k and t, such that $k > 1$,*

$$f_k(t) \geqslant t \left\lceil \frac{t - 1}{2(k - 1)} \right\rceil.$$

Proof. In a k-uniform tight cycle, each vertex has exactly $2(k - 1)$ neighbors. So, in any achromatic t-coloring, each color must be repeated at least $\left\lceil \frac{t-1}{2(k-1)} \right\rceil$ times, since each color must see other $t - 1$ colors. Hence, to have an achromatic t-coloring, a k-uniform tight cycle should have at least $t \left\lceil \frac{t-1}{2(k-1)} \right\rceil$ vertices. Therefore, $f_k(t) \geqslant t \lceil \frac{t-1}{2(k-1)} \rceil$.

Though the above lower bound is a direct counting argument, finding a matching upper bound turns out to be non-trivial. In the following sections, we discuss the tightness of the above lower bound for the case $k = 3$. Even for this case, we could show a matching upper bound only for finitely many cases. We can extend this to infinitely many cases if certain well-known conjectures on prime numbers are true.

3   Arithmetic Progression Based Constructions

In this section, we give upper bounds (Theorems 2 and 3) for $f_3(t)$ when one of the numbers in $\{t, \ldots, t+3\}$ is a good prime. These bounds match or closely match the lower bound $f_3(t) \geqslant t \left\lceil \frac{t-1}{4} \right\rceil$ from Theorem 1. As mentioned in the introduction, we will have infinitely many good primes if either Artin's Conjecture or the conjecture on the infinitude of Sophie Germain primes is true. In particular, if the latter is true, we get $f_3(t) = t \left\lceil \frac{t-1}{4} \right\rceil$ for infinitely many values of t. We start with a lemma that helps in the proof of Theorem 2.

Lemma 1. *If $p \equiv 1 \pmod 4$ is a prime and $\mathrm{ord}_p(2) = \frac{p-1}{2}$, then $\frac{p-1}{4}$ is even.*

Proof. Let $p = 4k+1$ be a prime for some positive integer k. Since $\mathrm{ord}_p(2) = \frac{p-1}{2}$, according to Euler's criterion, 2 is a quadratic residue modulo p. If p is an odd prime, then we know that 2 is a quadratic residue modulo p if and only if $p \equiv \pm 1 \pmod 8$. Therefore, from the two different representations of p, it is evident that $\frac{p-1}{4}$ is even.

Theorem 2. *Let p be a good prime such that $p \equiv 1 \pmod 4$. Then*

1. $f_3(p) \leqslant (p+1) \left\lceil \frac{p-1}{4} \right\rceil + 1$.
2. $f_3(p-1) \leqslant p \left\lceil \frac{p-1}{4} \right\rceil$.

Proof. Let $p = 4k + 1$, $n = (p+1) \left\lceil \frac{p-1}{4} \right\rceil + 1 = k(4k + 2) + 1$ and $V(TC_n^3) = \{v_i : 1 \leqslant i \leqslant n\}$. Given two elements $a, d \in \mathbb{Z}_p$ and $\ell \in \mathbb{N}$, $A(a, d, \ell)$ denotes the ℓ-length sequence $a, a + d, \ldots, a + (\ell - 1)d$. Note that this is an arithmetic progression in $\mathbb{Z}_p$ which may wrap around when $\ell > p$. To prove the upper bound for $f_3(p)$, we first pick a set $\mathcal{D}$ of common differences for the arithmetic progressions we use. We need $|\mathcal{D}|$ to be k and $\bigcup_{d_i \in \mathcal{D}} \{\pm d_i, \pm 2d_i\}$ to cover all the $4k$ elements of $\mathbb{Z}_p^*$. When $\mathrm{ord}_p(2) = p - 1$, we can take

$$\mathcal{D} = \{d_i = 2^{2i} \pmod p : 0 \leqslant i \leqslant k - 1\}.$$

If $\mathrm{ord}_p(2) = \frac{p-1}{2} = 2k$, let $S_1 = \{\pm 2^i \pmod p : 0 \leqslant i \leqslant k - 1\}$. Since $2^k = -1 \pmod p$, S_1 is the cyclic subgroup generated by 2. Let $a \in \mathbb{Z}_p^* \setminus S_1$ and define $S_2 = \{ax_i \pmod p : x_i \in S_1\}$. Then $|S_1| = |S_2| = 2k$ and $S_1 \cap S_2 = \emptyset$, as if $a2^i \equiv \pm 2^j \pmod p$ for some $0 \leqslant i, j \leqslant k - 1$, then $a \equiv \pm 2^{j-i} \pmod p, 1 \leqslant j - i \leqslant k - 1$, a contradiction to $a \notin S_1$. So, $S_1 \cup S_2 = \mathbb{Z}_p^*$. Since k is even, due to Lemma 1, we define

$$\mathcal{D} = \left\{ 2^{2i} \pmod p : 0 \leqslant i \leqslant \frac{k}{2} - 1 \right\} \bigcup \left\{ a2^{2i} \pmod p : 0 \leqslant i \leqslant \frac{k}{2} - 1 \right\},$$

and let

$$d_i = \begin{cases} 2^{2i} \pmod p & \text{if } 0 \leqslant i \leqslant \frac{k}{2} - 1 \\ a2^{2(i - \frac{k}{2})} \pmod p & \text{if } \frac{k}{2} \leqslant i \leqslant k - 1. \end{cases}$$

Hence, in both cases, $\mathbb{Z}_p^* = \bigcup_{d_i \in \mathcal{D}}\{\pm d_i, \pm 2d_i\}$. Now, we define

$$\pi_{d_i} = \begin{cases} A(a_i, d_i, p+1) & \text{if } 0 \leqslant i \leqslant k-2 \\ A(a_i, d_i, p+2) & \text{if } i = k-1, \end{cases}$$

where $a_i = \sum_{j=0}^{i-1} d_j$ (where a_0 is the empty sum and hence 0.) Notice that the first element a_j in the j-th progression, $1 \leqslant j \leqslant k-1$, is same as the second element in $(j-1)$-th progression. By concatenating $\pi_{d_i}, 0 \leqslant i \leqslant k-1$, we get a p-coloring of $V(TC_n^3)$.

Since no two adjacent elements in π_{d_i} are the same, this coloring is proper. Let $j, l \in \mathbb{Z}_p$ be any two distinct colors. As $j - l \in \mathbb{Z}_p^*$, $j - l \in \{\pm d_i, \pm 2d_i\}$ for some i. Then j and l appear within a distance of 2 in the subsequence formed by π_{d_i} and the first term following it. Hence, the coloring is achromatic.

To prove the upper bound for $f_3(p-1)$, we start with a achromatic p-coloring C of TC_n^3, where $n = (p+1)\left\lceil\frac{p-1}{4}\right\rceil + 1$, as we obtained above. Let V be the cyclic sequence of vertices in TC_n^3. Let V' be the cyclic subsequence of V obtained by removing all the vertices of the color 0 and then replacing any sequence of cyclically consecutive vertices of the same color with a single vertex of that color. Let $TC_{n'}^3$ be the 3-uniform tight cycle on the sequence V'. Now, we consider a coloring of $TC_{n'}^3$ which is the restriction of C on $V(TC_{n'}^3)$. Since color 0 appears twice in π_{d_0} and once in each of the other $k-1$ progressions, so $|V'| \leqslant p(\frac{p-1}{4})$. Since we removed consecutive vertices of the same color, there is no monochromatic edge in $TC_{n'}^3$. Any two colors, other than color 0, that were adjacent in TC_n^3 are still adjacent in $TC_{n'}^3$. Hence, the resulting coloring is an achromatic $(p-1)$-coloring of $TC_{n'}^3$ and hence, $f_3(p-1) \leqslant n' = p\left\lceil\frac{p-1}{4}\right\rceil$. Therefore, $f_3(p-1) \leqslant p\left\lceil\frac{p-1}{4}\right\rceil$.

Since any odd prime is of the type $4k+1$ or $4k+3$, we will give an upper bound for $f_3(p)$ when p is of the type $4k+3$. Here we have more freedom as we can write $\left\lceil\frac{p-1}{4}\right\rceil = k+1$ permutations. So we can give an additional permutation compared to the case where $p = 4k+1$ type.

Lemma 2. *Let p be a good prime such that $p \equiv 3 \pmod 4$. Then*

$$f_3(p) \leqslant p\left\lceil\frac{p-1}{4}\right\rceil$$

Proof. Let p be a good prime of type $4k+3$, $n = p\left\lceil\frac{p-1}{4}\right\rceil = (k+1)(4k+3)$ and $V(TC_n^3) = \{v_i : 1 \leqslant i \leqslant n\}$. Define

$$\mathcal{D} = \{d_i = 2^{2i+1} \pmod p : 0 \leqslant i \leqslant k-1\}.$$

If $\text{ord}_p(2) = p-1$, then $\mathbb{Z}_p^* = \bigcup_{d_i \in \mathcal{D}}\{\pm d_i, \pm 2d_i\} \cup \{\pm 1\}$. If the $\text{ord}_p(2) = \frac{p-1}{2}$, then let $S_1 = \{2^j \pmod p : 1 \leqslant j \leqslant 2k\} = \bigcup_{d_i \in \mathcal{D}}\{d_i, 2d_i\}$. Since -1 is not in the cyclic subgroup of $\mathbb{Z}_p^*$ generated by 2, we have $\mathbb{Z}_p^* = S_1 \cup -S_1 \cup \{\pm 1\} = \{\pm 2^j \pmod p : 1 \leqslant j \leqslant 2k\} \cup \{\pm 1\} = \bigcup_{d_i \in \mathcal{D}}\{\pm d_i, \pm 2d_i\} \cup \{\pm 1\}$. Hence, if p is a good prime of the form $4k+3$, then $\mathbb{Z}_p^* = \bigcup_{d_i \in \mathcal{D}}\{\pm d_i, \pm 2d_i\} \cup \{\pm 1\}$. For any given two elements $a, d \in \mathbb{Z}_p$ and $\ell \in \mathbb{N}$, let $A(a, d, \ell)$ denotes the ℓ-length sequence $a, a + d, \ldots, a + (\ell - 1)d$. Now, we define

$$\pi_{d_i} = A(0, d_i, p),\ 0 \leqslant i \leqslant k - 1,\ \text{and}$$

$$\pi = 0, 1, p - 1, 2, p - 2, \ldots, \frac{p-1}{2}, p - \left(\frac{p-1}{2}\right).$$

By concatenating $\pi_{d_i}, 0 \leqslant i \leqslant k - 1$ and then π into a cyclic sequence, we get a p-coloring of $V(TC_n^3)$. Since no two adjacent elements in π_{d_i} and π are the same, this coloring is proper. Let $j, l \in \mathbb{Z}_p$ be any two distinct colors. As $j - l \in \mathbb{Z}_p^*$, $j - l \in \{\pm d_i, \pm 2d_i, \pm 1\}$ for some i. If $j - l = \pm 1$, then j and l appear within a distance of 2 in the subsequence formed by π. If $j - l \in \{\pm d_i, \pm 2d_i\}$, then j and l lie within a distance of 2 in the subsequence formed by π_{d_i} and the first term following it, except for the instance when $\{j, l\} = \{(p - 1)d_i, d_i\}$ for some i. In this case, since $j + l = p$, j and l appear within a distance of 2 in the subsequence formed by π. Hence, the coloring is achromatic. Therefore, $f_3(p) \leqslant p \left\lceil \frac{p-1}{4} \right\rceil$. $\qquad\qquad\square$

Theorem 3. *Let p be a good prime such that $p \equiv 3 \pmod 4$. Then*

1. $f_3(p) = p \left\lceil \frac{p-1}{4} \right\rceil$.
2. $f_3(p - 1) = (p - 1) \left\lceil \frac{p-1}{4} \right\rceil$.
3. $f_3(p - 2) \leqslant (p - 2) \left\lceil \frac{p-1}{4} \right\rceil$.
4. $f_3(p - 3) \leqslant (p - 3) \left\lceil \frac{p-1}{4} \right\rceil$.

Proof. To prove the upper bound in all the four cases, it is sufficient to show that $f_3(p - m) \leqslant (p - m) \left\lceil \frac{p-1}{4} \right\rceil$, where $m \in \{0, 1, 2, 3\}$. When $m = 0$, the upper bound follows from Lemma 2. Now, we give proofs for the other values of m.

Let us consider an achromatic p-coloring for $TC_n^3, n = p \left\lceil \frac{p-1}{4} \right\rceil = p(k + 1)$, as given in the proof of Lemma 2. In this coloring each color is repeated exactly $k + 1$ times. Let V be the cyclic sequence of vertices in TC_n^3, and V' be the cyclic subsequence of V obtained by removing all the vertices of m color classes and replacing any sequence of cyclically consecutive vertices of the same color with a single vertex of that color. Let $TC_{n'}^3$ be a tight cycle with $V(TC_{n'}^3) = V'$. As we removed all the vertices of colors from those m color classes, $n' \leqslant (p - m)(k + 1)$. This coloring is proper, and any two colors, other than those colors from the m color classes, that were adjacent in TC_n^3 are still adjacent in $TC_{n'}^3$. Hence, the resulting coloring is an achromatic $(p - m)$-coloring of $TC_{n'}^3$ and therefore, $f_3(p - m) \leqslant (p - m) \left\lceil \frac{p-1}{4} \right\rceil$.

As the upper bound for $f_3(p - m)$ matches the lower bound in Theorem 1 for $m \in \{0, 1\}$, we have the exact values for $f_3(p)$ and $f_3(p - 1)$.

Example 1. For $p = 19 = 4 \times 4 + 3$, we choose $\mathcal{D} = \{2, 8, 13, 14\}$ which are the first four odd powers of 2 modulo 19. The five permutations are

$$\pi_2 = 0, 2, 4, 6, 8, 10, 12, 14, 16, 18, 1, 3, 5, 7, 9, 11, 13, 15, 17$$
$$\pi_8 = 0, 8, 16, 5, 13, 2, 10, 18, 7, 15, 4, 12, 1, 9, 17, 6, 14, 3, 11$$
$$\pi_{13} = 0, 13, 7, 1, 14, 8, 2, 15, 9, 3, 16, 10, 4, 17, 11, 5, 18, 12, 6$$
$$\pi_{14} = 0, 14, 9, 4, 18, 13, 8, 3, 17, 12, 7, 2, 16, 11, 6, 1, 15, 10, 5$$
$$\pi = 0, 1, 18, 2, 17, 3, 16, 4, 15, 5, 14, 6, 13, 7, 12, 8, 11, 9, 10$$

One can verify that their concatenation $\pi_2 \pi_8 \pi_{13} \pi_{14} \pi$ gives an achromatic 19-coloring of TC_{95}^3. If we remove the color 18 from each of these five permutations, we get an achromatic 18-coloring of TC_{90}^3.

4 Covering Sequence Based Construction

In this section, we obtain $f_3(t)$ for all values of t except $t = 9$. We do this by finding an achromatic t-coloring of TC_{tk}^3, where $k = \lceil \frac{t-1}{4} \rceil$. For the most part (for $t \geqslant 10$) the coloring is based on Conjecture 1, which is restated below for convenience. Based on this conjecture, we first show, in Lemma 3, an achromatic t-coloring of TC_{tk}^3 for $t = 4k + 1$ with $k > 2$, and finally, in Theorem 4, we show that $f_3(t) = t \lceil \frac{t-1}{4} \rceil$ for all $t \neq 9$.

Conjecture 1. Let k be an integer larger than 2 and $t = 4k + 1$. Then there exists a sequence $x_0, x_1, \ldots, x_{k-1} \in \mathbb{Z}_t^*$ such that,

(i) $\mathbb{Z}_t^* = \bigcup_{i=0}^{k-1} \{\pm x_i, \pm(x_i + x_{(i+1 \mod k)})\}$, and
(ii) $\gcd(\sum_{i=0}^{k-1} x_i, t) = 1$.

Remark 2. (a) Henceforth, we will call a sequence that satisfies the requirements of the conjecture as a *covering sequence*.
(b) The claim made in Conjecture 1 is not true for $k = 2$ since the set $\{\pm x_0, \pm x_1, \pm(x_0 + x_1)\}$ has cardinality at most 6 and hence cannot cover $\mathbb{Z}_9^*$.
(c) In Table 2 below, we list a covering sequence each for small values of t. We have verified the conjecture using a dynamic programming approach for all values of k from 3 to 25, that is t up to 101.
(d) Given a covering sequence $x_0, x_1, \ldots, x_{k-1} \in \mathbb{Z}_t^*$, we observe the following.
   (i) Any cyclic shift of the sequence is also a covering sequence.
   (ii) Given $a \in \mathbb{Z}_t$ with $\gcd(a, t) = 1$, the sequence $ax_0, ax_1, \ldots, ax_{k-1}$ is also a covering sequence.
   The above two observations, together with the estimate $\phi(t) > \frac{t}{e^\gamma \log \log t + \frac{3}{\log \log t}}$ (γ is Euler's constant, see e.g. [11]) indicate that there are good number of covering sequences for any t provided there is at least one such.

Table 2. Covering Sequence for $t = 4k + 1$, $3 \leqslant k \leqslant 8$

t	Covering Sequence	t	Covering Sequence
13	1, 2, 4	25	1, 2, 4, 5, 7, 10
17	1, 2, 6, 4	29	1, 2, 4, 5, 8, 7, 10
21	1, 2, 7, 6, 4	33	1, 2, 4, 5, 11, 15, 8, 13

To prove Lemma 3, we need the following elementary number-theoretic fact (see e.g. [12]): For every positive integer t and $a, b \in \mathbb{Z}_t$, if $\gcd(a, t) = 1$, then there exists a unique $z \in \mathbb{Z}_t$ such that $az = b$ (in $\mathbb{Z}_t$).

Lemma 3. *Let $k > 2$ be a natural number and let $t = 4k + 1$. If Conjecture 1 is true, then $f_3(t) \leqslant \frac{t(t-1)}{4}$.*

Proof. Let $k > 2$, $V(TC_{tk}^3) = \{v_i : 0 \leqslant i \leqslant tk - 1\}$. Suppose that Conjecture 1 is true for k, and $x_0, \ldots, x_{k-1} \in \mathbb{Z}_t$ be a covering sequence. Then we define the coloring $C : V(TC_{tk}^3) \to \mathbb{Z}_t$ as follows,

$$C(v_i) = \sum_{j=0}^{i-1} x_j, \quad 0 \leqslant i \leqslant tk - 1,$$

where the subscripts are interpreted modulo k, *i.e.*, $x_i = x_{i \bmod k}$. By definition, C is proper, as adjacent vertices receive different colors. Now, to show that C is achromatic, it is enough to show that any two distinct colors $c_1, c_2 \in \mathbb{Z}_t$, are adjacent in TC_{tk}^3.

Let $c_1, c_2 \in \mathbb{Z}_t$ be any two distinct colors. Then by Conjecture 1, $c_1 - c_2 = \pm x_i$ or $c_1 - c_2 = \pm(x_i + x_{i+1})$ for some $0 \leqslant i \leqslant k - 1$. First, we consider the case where $c_1 - c_2 = x_i$. Let $a = \sum_{r=0}^{k-1} x_r$. Now, since $\gcd(a, t) = 1$ (by Conjecture 1), using the number-theoretic fact we stated before this lemma, there is a unique $d \in \mathbb{Z}_t$ such that $da = c_2 - \sum_{r=0}^{\ell} x_r$, where $\ell = k - 1$ if $i = 0$, otherwise $\ell = i - 1$. So, we have $c_2 = da + \sum_{r=0}^{\ell} x_r = \sum_{r=0}^{dk-1} x_r + \sum_{r=dk}^{dk+\ell} x_r = \sum_{r=0}^{dk+\ell} x_r = C(v_j)$, where $j = dk + \ell$. Next, following $c_1 = c_2 + x_i$, we also have $c_1 = c_2 + x_i = \sum_{r=0}^{dk+l} x_r + x_i = C(v_{j+1})$. So, c_1 and c_2 are adjacent in this case. For the case where $c_1 - c_2 = x_i + x_{i+1}$, following the same argument as above, it follows that $C(v_j) = c_2$ and $C(v_{j+2}) = c_1$ for $j = dk + \ell$. Hence, c_1, c_2 are adjacent in this case also. For the remaining two cases, *i.e.*, for the cases where $c_1 - c_2 \in \{-x_i, -(x_i + x_{i+1})\}$, we follow the steps of the corresponding preceding cases verbatim, while exchanging the roles of c_1 and c_2, to conclude that c_1, c_2 are adjacent for these cases as well. So, C is achromatic, which shows $f_3(t) \leqslant tk = t\frac{(t-1)}{4}$.

Example 2. From Table 2, we get for $t = 13$, the sequence $1, 2, 4$ is a covering sequence, and gives an achromatic 13-coloring of TC_{39}^3 as given below. Note that in the following sequence, the last 7 is adjacent to the first 0, and the last 9 is adjacent to the first 0 and the next 1.

$$0, 1, 3, 7, 8, 10, 1, 2, 4, 8, 9, 11, 2, 3, 5, 9, 10, 12, 3, 4,$$
$$6, 10, 11, 0, 4, 5, 7, 11, 12, 1, 5, 6, 8, 12, 0, 2, 6, 7, 9$$

Theorem 4. *Let $t \neq 9$ be a positive integer. If Conjecture 1 is true, then*

$$f_3(t) = t \left\lceil \frac{t-1}{4} \right\rceil.$$

Proof. The proof is broadly divided into two parts: first, we show the result for $t \geqslant 10$, and then we settle the cases for $t \leqslant 8$. Following the lower bound of Theorem 1, it is enough to show the coloring that meets this lower bound.

Let $t = 4k + 1$, with $k > 2$, and C be *any* (optimal) achromatic t-coloring of $V(TC_{tk}^3)$ (note that the existence of such a coloring is guaranteed by Lemma 3). Then C satisfies the following properties.

(a) Each color class has exactly k vertices.
(b) Any 5 consecutive vertices of TC_{tk}^3 have distinct colors.

Now, for $1 \leqslant i \leqslant 3$, consider the $(t-i)$-coloring of $TC_{(t-i)k}^3$ obtained from the coloring C of TC_{tk}^3 by removing the vertices of any i color classes. So, the vertices of $TC_{(t-i)k}^3$ retain their colors, *i.e.*, they are colored as they were colored under C. It may be observed that under this $(t-i)$-coloring of $TC_{(t-i)k}^3$, any two colors are adjacent (as they were adjacent in TC_{tk}^3 under C). Furthermore, all edges of $TC_{(t-i)k}^3$ remain properly colored as at most 3 vertices from among any 5 consecutive vertices of TC_{tk}^3 were removed (by property (b) above). So, the coloring is an achromatic $(t-i)$-coloring of $TC_{(t-i)k}^3$. Hence, $f_3(t-i) \leqslant (t-i)k$. The conclusion now follows from Lemma 3.

Next, for $t \in \{3, \dots, 8\}$, first note that for $t \in \{1, 2\}$ we cannot have a 3-uniform cycle. For the remaining cases, we have the following.

(i) For $t \in \{3, 4, 5\}$, by giving all the t vertices different colors, we get an achromatic t-coloring of TC_t^3. So, $f_3(t) \leqslant t = t \left\lceil \frac{t-1}{4} \right\rceil$ and by 1 we get $f_3(t) = t \left\lceil \frac{t-1}{4} \right\rceil$.
(ii) For $t \in \{6, 7\}$, we have $f_3(t) = t \left\lceil \frac{t-1}{4} \right\rceil$ directly by Theorem 3.
(iii) For $t = 8$, we have an achromatic 8-coloring of TC_{16}^3 as $1, 2, 3, 6, 5, 2, 7, 8, 4, 1, 6, 7, 8, 3, 5, 4$. So, $f_3(8) \leqslant 16 = 8 \left\lceil \frac{8-1}{4} \right\rceil$ and by using 1 we get $f_3(8) = 16$.

5   Conclusion and Future Work

In this article, we reinitiated the study of the achromatic number of hypergraphs, and our main focus is $\psi(TC_n^k)$ - the achromatic number of k-uniform tight cycles on n vertices. There are multiple notions of achromatic number in the case of hypergraphs; we considered the one put forward by Jucovič and Olejník [8] in 1974. We observed that $\psi(TC_n^k)$ is not well-behaved. To understand the behavior of $\psi(TC_n^k)$, we introduced the extremal parameter $f_k(t)$. We gave a lower bound on $f_k(t)$ using a counting argument. Subsequently, we showed that the lower bound is tight for the 3-uniform case for infinitely many values of t subject to certain well-known number-theoretic results and conjectures. Finally, we showed that we can determine $f_3(t)$ for all $t \neq 9$ based on a (seemingly) new number theoretic conjecture.

A natural extension of this work will be to generalize our results to higher uniformity, $i.e.$, to find $f_k(t)$ for $k > 3$. Also, since all of our upper bounds on $f_3(t)$ depend on number theoretic conjectures, it will be important to have tight/ good upper bounds on $f_k(t)$, even for the case $k = 3$, which are independent of any conjecture. In another direction, it will be interesting to find the achromatic number of other classes of hypergraphs.

Disclosure of Interests. The authors have no competing interests to declare that are relevant to the content of this article.

References

1. Agnarsson, G., Halldórsson, M.M.: Strong colorings of hypergraphs. In: International Workshop on Approximation and Online Algorithms (2004), LNCS 3351, pp. 253–266. Springer (2005)
2. Dębski, M., Lonc, Z., Rzążewski, P.: Harmonious and achromatic colorings of fragmentable hypergraphs. Eur. J. Comb. **66**, 60–80 (2017)
3. Edwards, K.: The harmonious chromatic number and the achromatic number. In: Bailey, R.A. (ed.) Surveys in Combinatorics, London Mathematical Society Lecture Note Series, vol. 241, pp. 13–47, Cambridge University Press (1997)
4. Erdős, P., Hajnal, A.: On chromatic number of graphs and set-systems. Acta Math. Acad. Sci. Hungar **17**(61–99), 1 (1966)
5. Geller, D., Kronk, H.: Further results on the achromatic number. Fund. Math. **85**, 285–290 (1974)
6. Harary, F., Hedetniemi, S., Prins, G.: An interpolation theorem for graphical homomorphisms. Portugal. Math **26**, 453–462 (1967)
7. Ireland, K., Rosen, M.: A classical introduction to modern number theory. Graduate Texts in Mathematics. Springer (1990), ISBN 9780387973296
8. Jucovič, E., Olejník, F.: On chromatic and achromatic numbers of uniform hypergraphs. Časopis pro pěstování matematiky **99**(2), 123–130 (1974)
9. Nešetřil, J., Phelps, K.T., Rödl, V.: On the achromatic number of simple hypergraphs. Ars Combin. **16**, 95–102 (1983)
10. Olejník, F.: Theorems of the Nordhaus-Gaddum type for k-uniform hypergraphs. Math. Slovaca **31**(3), 311–318 (1981)
11. Rosser, J.B., Schoenfeld, L.: Approximate formulas for some functions of prime numbers. Ill. J. Math. **6**(1), 64–94 (1962)
12. Shoup, V.: A computational introduction to number theory and algebra. Cambridge University Press (2009)

A Resonance Neural Network
for the K-Median Problem

Andrew Bloch-Hansen[1,2(✉)], Cody Rossiter[1], and Roberto Solis-Oba[1]

[1] Western University, London, ON, Canada
`ablochha@uwo.ca`
[2] Mount Royal University, Calgary, AB, Canada

Abstract. The k-median problem is a classical clustering problem in which the goal is to select a set $S \subseteq V$ of k nodes in a given weighted graph $G = (V, E)$, so as to minimize the sum of distances from each node in $V \setminus S$ to its nearest node in S. The nodes in S are called facilities, and the remaining nodes are clients; a solution for the problem naturally partitions the nodes of the graph into k clusters, where each cluster has one facility. We assume that the edge lengths satisfy the triangle inequality. We present an asymmetric resonance neural network for solving the k-median problem. The network consists of 2 layers of neurons encoding assignments of clients to clusters and selections of facilities for clusters. We present an efficient implementation of the algorithm that outperforms other known local search and neural network-based algorithms for the k-median problem. Furthermore, we show that a modification of our neural network leads to an approximation algorithm with the same approximation ratio as a local search algorithm by Arya et al. (STOC 2001).

Keywords: Resonance Networks · Hopfield Networks · K-Median Problem · Neural Networks · Local Search · Combinatorial Optimization · Experimental Evaluation · Approximation Algorithms

1 Introduction

Given a graph $G = (V, E)$ with n nodes and non-negative distances $d_{i,j}$ on the edges that satisfy the triangle inequality, in the *metric k-median problem* (KMP) the goal is to select k nodes, called *facilities*, that minimize the sum of distances from every non-selected node, called a *client*, to the facility closest to it. KMP is a classical clustering problem [1] with a large number of applications in network design [2], data mining [3], web access [4], and machine learning [5], among others. KMP is known to be NP-hard [6] and cannot be approximated within a factor strictly less than $1 + \frac{2}{e} \sim 1.736$, unless $\mathsf{NP} \subseteq \mathsf{DTIME}(n^{O(\log \log n)})$ [7].

Research on KMP and its variants has focused on exact algorithms [8,9], approximation algorithms [10,11], and heuristics [12,13]. The best approximation algorithms with provable performance guarantees for KMP are based on

N. Misra and A. Pandey (Eds.): CALDAM 2026, LNCS 16445, pp. 84–95, 2026.
https://doi.org/10.1007/978-3-032-17156-6_7

linear program rounding or local search. On the linear program side, the current best algorithm is the 2.671-approximation algorithm of Cohen-Addad et al. [14], which improves on the 2.675-approximation algorithm of Byrka et al. [15], and the 2.732-approximation algorithm of Li and Svensson [16]. For the Euclidean KMP, Cohen-Addad et al. [5] have improved the approximation ratio to 2.406.

On the local search side, the current best result is the $(2.836 + \epsilon)$-approximation algorithm of Cohen-Addad et al. [10], which improves on the $(3 + \frac{2}{p})$-approximation algorithm of Arya et al. [11], where $p \geq 1$ is a constant. Peng et al. [17] also designed a local search algorithm that achieves the same $(3 + \frac{2}{p})$-approximation ratio. Other local search algorithms have been proposed for KMP [18–20].

Several neural networks have been designed for solving KMP. Merino et al. [21–23] and Mishrra and Barman [24] designed Hopfield networks, and Haralampiev, Rossiter, and Bloch-Hansen designed competition-based neural networks [25–29].

In this paper, we present a resonance network-based algorithm that is faster and more accurate than other local search and neural network algorithms for KMP. The algorithm partitions the nodes of the input graph into k clusters and selects a facility for each cluster that minimizes the sum of distances to the nodes in the cluster. The neural network consists of two layers of neurons. The neurons in one layer determine the selection of a facility for each cluster and the neurons in the second layer partition the nodes into clusters. In this network information flows between the neurons in different layers until a stable state is reached. An interesting feature of our network is that the connections between neurons are not symmetric; what this means is that the weight of the connection from a neuron f in the first layer to a neuron c in the second layer is not the same as the weight of the connection from c to f, different from other resonant networks, BAMs, and Hopfield networks in which the weights are symmetric [23, 26, 27].

We designed our neural network similarly as in [23], so that updates to the states of the neurons always define a feasible solution for the problem without having to add Lagrange multipliers to the objective function. We show that despite the asymmetry in the weights of the neuron connections, our network is guaranteed to reach a stable state. Furthermore, we show that our neural network is equivalent to a local search algorithm that performs swap operations that replace certain cluster facilities with specific cluster nodes and in which each swap operation reduces the cost of the solution. Thus, a modification of our neural network can be shown to be equivalent to a local search algorithm that attempts all possible swap operations, and for such a modified neural network we can show that it has the same approximation ratio as the algorithm in [11]. As there have been only a few notable successes in establishing mathematically provable bounds on the approximation ratios of neural networks, this work is a step towards developing a stronger mathematical understanding of these models.

The rest of the paper is organized as follows. In Sect. 2 we describe the architecture of our neural network, explain how the states of the neurons are updated, and show that the neural network stabilizes in a finite amount of time. In Sect. 3

we compute the approximation ratio of the modified neural network algorithm, in Sect. 4 we present our experimental results, and in Sect. 5 we present our conclusions.

2   The k-Median Problem

As explained above, in the k median problem we must select k facility nodes that minimize the sum of distances from client nodes to the nearest facilities. These k facilities partition the nodes into k clusters, the facilities are the "centers" of the clusters, and a cluster includes all nodes closest to its center. Hence, KMP can be formulated as the following quadratic program [23]:

$$\text{Minimize} : \sum_{i=1}^{n} \sum_{j=1}^{n} \sum_{h=1}^{k} d_{i,j} c_{i,h} f_{j,h} \tag{1a}$$

$$\text{Subject to} : \sum_{h=1}^{k} c_{i,h} = 1, \ \forall i = 1, ...n \tag{1b}$$

$$\sum_{j=1}^{n} f_{j,h} = 1, \ \forall h = 1, ...k \tag{1c}$$

$$c_{i,h}, f_{j,h} \in \{0,1\}, \quad \forall i, j = 1, \dots, n, \ h = 1, \dots, k \tag{1d}$$

where $c_{i,h} = 1$ if client node i is assigned to cluster h, and $f_{j,h} = 1$ if node j is the facility of cluster h. Equation (1b) ensures that each client is assigned to exactly one cluster, Equation (1c) requires that each cluster has exactly one assigned facility, and Equation (1d) ensures that each $c_{i,h}$ and $f_{j,h}$ is a binary variable. In the following, we denote clients with i, facilities with j, and clusters with h.

We use min-max normalization to scale the distances $d_{i,j}$ to the range $[0,1]$ and then define a modified distance function $D'_{i,j} = 1 - d_{i,j}$. The modified distance $D'_{i,j}$ can be interpreted as the value that a facility gains from serving a client. We define the *inner value* of a facility j to be the sum of the modified distances $D'_{i,j}$ to the clients i assigned to j. Note that maximizing the sum of inner values of the facilities is equivalent to minimizing the sum of distances from clients to facilities.

To motivate the use of these modified distances, let us consider an example. Let F_1 be a facility that has a set S_1 of several nearby clients assigned to it and let a second facility F_2 have a set S_2 with only a single client assigned to it and located far from F_2. The contribution, $\sum_{i \in S_1} d_{i,F_1}$, of F_1 to the cost of the solution could be the same as the contribution, $\sum_{i \in S_2} d_{i,F_2}$, of F_2, but the inner value of F_1 would be much higher than the inner value of F_2, which is consistent with the intuition that F_1 is a more desirable facility to add to the solution than F_2. Thus, we expect that our inner value metric will help us identify the most valuable facilities to select as part of the solution.

2.1 Network Architecture

The above quadratic formulation suggests the architecture of a 2-layer asymmetric resonance network, which is similar to the recurrent neural network of [23]. A resonance network, or bidirectional associative memory, has two layers x and y of neurons and connections between the neurons are specified by a weight matrix M. As the network evolves, the states of the neurons in a layer are obtained by multiplying the weight matrix with the vector representing the states of the neurons in the other layer ($y = Mx$, and $x = M^T y$). As suggested in several recent neural network designs for solving combinatorial optimization problems [21, 23, 25, 27], we incorporate the problem constraints into the network architecture instead of in an energy function through Lagrange multipliers.

The network consists of two n by k layers of neurons: neurons $f_{j,h}$ in the first layer represent facility-to-cluster assignments and neurons $c_{i,h}$ in the second layer represent client-to-cluster assignments. For cluster h only one neuron $f_{j,h}$ has value 1 (facility j is assigned to cluster h) and for node i only one neuron $c_{i,h}$ has value 1 (client i is assigned to cluster h), this ensures that constraints (1b) and (1c) are satisfied.

Assume an initial feasible integer assignment for the states of neurons $f_{j,h}$ and $c_{i,h}$. The states of the neurons will change as information is transferred between the layers of the network, as described below.

Let $\delta(x) = 0$ for $x \leq 0$ and $\delta(x) = x$ for $x > 0$. Let q be the index of the cluster whose facility has minimum inner value, and let F_q be the facility currently assigned to cluster q. Let $m_{i,q} = max\{\sum_{j=1}^{n} D'_{i,j} f_{j,h} \mid 1 \leq h \neq q \leq k\}$ for all clients i, $1 \leq i \leq n$, so $m_{i,q}$ is the modified distance $D'_{i,\phi_q(i)}$ from i to the nearest facility $\phi_q(i)$, where $\phi_q(i)$ is not the facility for cluster q. The state of each neuron, $c_{i,h}$, $f_{j,h}$, depends on its activation potential, $\theta(c_{i,h})$, $\theta(f_{j,h})$, which is defined as follows:

$$\theta(c_{i,h}) = \sum_{j=1}^{n} f_{j,h} D'_{i,j}, \ \forall \ 1 \leq i \leq n, 1 \leq h \leq k \tag{2}$$

$$\theta(f_{j,h}) = \begin{cases} \sum_{i=1}^{n} \delta(D'_{i,j} - m_{i,q}) - \sum_{i=1}^{n} \delta(D'_{F_q,i} - m_{i,q}) & \forall \ 1 \leq j \leq n, \text{ if } h = q \\ 0 & \text{if } h \neq q \end{cases} \tag{3}$$

The state of each neuron changes according to the activation potentials. Let $f'_{i,h}$ be the current state of neuron $f_{i,h}$, then the new states of the neurons are computed as follows:

$$c_{i,h} = \begin{cases} 1 & \text{if } \theta(c_{i,h}) = \max \ \{\theta(c_{i,j}) \mid 1 \leq j \leq k\} \\ 0 & \text{otherwise} \end{cases} \tag{4}$$

$$f_{j,h} = \begin{cases} 1 & \text{if } f'_{j,h} = 1 \text{ and } h \neq q, \text{ or } \max \ \{\theta(f_{\ell,q}) \mid 1 \leq \ell \leq n\} = 0 \text{ and } h = q \\ 1 & \text{if } \theta(f_{j,h}) = \max \ \{\theta(f_{\ell,h}) \mid 1 \leq \ell \leq n\} > 0 \text{ and } h = q \\ 0 & \text{otherwise} \end{cases} \tag{5}$$

To understand the activation potentials $\theta(c_{i,h})$ recall that only one neuron $f_{j,h}$ has the value 1 for each cluster h, so only one of the terms in the summation (2) is nonzero and, therefore, $\theta(c_{i,h})$ is the modified distance D' from i to the facility of cluster h. Thus, for each client node i the state value for neuron $c_{i,h}$ is 1 only for the cluster h whose facility is closest to i.

The interpretation of the meaning of the state values for neurons $f_{j,q}$ is a bit more complicated. The value $D'_{i,j} - m_{i,q} = (1-d_{i,j}) - (1-d_{i,\phi_q(i)}) = d_{i,\phi_q(i)} - d_{i,j}$ is positive only if node i is closer to j than $\phi_q(i)$. Recall that q is the cluster whose facility F has the smallest inner value among the cluster facilities. Let S be the set of facilities selected by the algorithm and let $S' = S \setminus F$. Set S', like S, defines a solution in which each client is assigned to the cluster for the closest facility in S'. Hence, $\theta(f_{j,q})$ is the improvement, if any, in the value of the solution S' if j is selected as the facility for the cluster q. Thus, the state value for neuron $f_{j,q}$ is 1 if j is the facility that would improve by the largest amount the value of solution S' if chosen as the facility for cluster q. If no facilities improve the value of the solution S then no neuron $f_{j,h}$ changes its state, and at this point the neural network would have stabilized.

The algorithm for solving KMP using the above resonance neural network is the following.

Algorithm 1. NeuralNetworkKMP(G, k)

1: **Input:** Graph $G = (V, E)$ and number k of facilities.
2: **Output:** A feasible solution for the k-median problem.
3: Choose random feasible values for the neurons $f_{j,h}$ and use (2) and (4) to compute state values for the neurons $c_{i,h}$.
4: **while** the network is not stabilized **do**
5:     Compute the facility F with smallest inner value for which $f_{F,q} > 0$ for some cluster $1 \leq q \leq k$.
6:     Update the state of all neurons $f_{j,h}$ using (3) and (5).
7:     Update the state of all neurons $c_{i,h}$ using (2) and (4).
8: **end while**
9: **Return** the set S of facilities j for which $f_{j,h} = 1$ for some $1 \leq h \leq k$.

Note the order in which the neuron states are updated. Observe that each iteration decreases the value of the solution defined by the facilities j with $f_{j,h} = 1$ for some h, so the algorithm will terminate. The time complexity of the algorithm is not guaranteed to be polynomial, but if we assume that the network stabilizes when the change in the value of the solution is small we can bound the number of iterations by a polynomial function as described in [11].

A very efficient implementation of the algorithm can be made through the use of an $n \times n$ matrix storing the normalized modified distances $D'_{i,j}$ and five $n \times k$ matrices storing the facility inner values, the client $c_{i,j}$ and facility $f_{i,j}$ neuron state values, and the activation potential values $\theta(c_{i,h})$ and $\theta(f_{j,h})$. The algorithm can then perform each step using efficient matrix operations. This allows also the possibility of performing many of these operations in parallel, although in our experiments we did not use a parallel implementation of the algorithm to make a fair comparison with the other algorithms for KMP that cannot be easily parallelized.

3   Approximation Algorithm

Algorithm 1 starts with a random selection of k facilities for the clusters and assigns each client to the cluster whose facility is closest. In each iteration the facility F with the smallest inner value is selected and the activation potentials $\theta(f_{j,h})$ are computed with respect to the cluster q for F. The neuron state update function (5) selects the node j that would improve the most the cost of the solution and sets it as the facility for q. This is equivalent to performing one of the swap operations defined in [11] where facility F is swapped with j. Note that since j is the node with maximum $\theta(f_{j,q})$ then j is the non-facility node with maximum inner value. Also, if swapping F with j does not improve the cost of the solution, then swapping F with any other node would not improve the solution either. Thus, the solution produced by the algorithm is a local optimal solution under the above specific set of swap operations, where a facility with minimum inner value can only be replaced with the non-facility node with maximum inner value.

We can modify the algorithm so that the analysis in [11] can be used to compute its approximation ratio. Let $F_1, F_2, \cdots, F_k$ be the facilities for the k clusters; these are initially selected randomly as specified in Algorithm 1. We modify the definitions of the activation potential $\theta(f_j, h)$ and the neuron state function for neurons $f_{j,h}$ as follows. Let m_i be the k-dimensional vector $\langle m_{i,q_1}, m_{i,q_2}, \cdots, m_{i,q_k} \rangle$, where q_i is the cluster with facility F_i and m_{i,q_j} is the modified distance $D'_{i,\tilde{F}_j}$, where $\tilde{F}_j \neq F_j$ is the nearest facility to i. Then,

$$\theta(f_{j,h}) = \sum_{i=1}^{n} \delta(D'_{i,j} - m_{i,q_h}), \text{ for all } 1 \leq j \leq n, \, 1 \leq h \leq k \qquad (6)$$

Let $M = \max \{\theta(f_{\ell,h}) \mid 1 \leq \ell \leq n, \, 1 \leq h \leq k\}$ and L, μ be such that $M = \theta(f_{L,\mu})$. Let $f'_{j,h}$ be the current state of neuron $f_{j,h}$, then for all $1 \leq j \leq n$ and $1 \leq h \leq k$ define the new states of the neurons as follows:

$$f_{j,h} = \begin{cases} 1 & \text{if } f'_{j,h} = 1, \text{ and } i \text{ is not the facility for cluster } q_\mu \text{ or } M = 0 \\ 1 & \text{if } \theta(f_{j,h}) = M > 0 \\ 0 & \text{otherwise} \end{cases} \qquad (7)$$

Using this new definition of $\theta(f_{j,h})$ and neuron states $f_{j,h}$ note that $f_{j,h} = 1$ for the node j that would cause the largest possible improvement in the cost of the solution if node j were to replace the facility for cluster h. Also, if no node improves the cost of the solution by replacing a cluster facility, then the state of the neurons will not further change and so the network would stabilize.

Corollary 1. *Algorithm 1 with the modified activation potential and the neuron state functions (6), (7) is equivalent to the single-swap local search algorithm in [11], as swapping any of the facilities selected by the algorithm with any other node in the graph does not improve the cost of the solution. Therefore, by Theorem 3.1 in [11], the modified Algorithm 1 has approximation ratio 5 if the edge lengths satisfy the triangle inequality.*

Proof. When Algorithm 1 updates the facilities' states, if the facility for cluster q changes, by (5) the replacement facility is the one which causes the largest improvement in the value of the solution. So, if the replacement facility does not improve the solution, neither of the other potential replacement facilities for the cluster q could improve the solution. The modifications described in (6), (7) simply check whether replacing the facility in any cluster would improve the value of the solution. Therefore, if no facility can be replaced to improve the solution, then there is no single-swap operation that can improve the solution, which allows us to use Theorem 3.1 from [11] to compute the approximation ratio of our algorithm.

We note that it would be possible to modify the definition of the vector m_i, the activation potential functions $\theta(c_{i,h})$ and $\theta(f_{j,h})$ and the neuron state functions $c_{i,h}$ and $f_{j,h}$ so that Algorithm 1 can consider all possible simultaneous swaps of $p \geq 1$ facilities with p non-facility nodes as done by the local search algorithm of [11] to achieve approximation ratio $3 + \frac{2}{p}$. We do not present or implement these modifications, as they would significantly increase the time complexity of the algorithm.

4   Experimental Results

We compared our neural network algorithm with several of the best local search and neural network-based algorithms for KMP: The local search algorithm of Arya et al. [11], the algorithm of Pan and Zhu [30], the local search algorithm of Cohen-Addad et al. [10], the improved fast interchange algorithm of Resende and Werneck's [18], Haralampiev's competition-based neural network [27], and Domínguez and Muñoz's recurrent neural network [23]. We tested all these algorithms using benchmarks from the OR-Library [31] and from TSPLIB [32].

We implemented the algorithms in Python making extensive use of the PyTorch framework [33], which allows the efficient manipulation of matrices as tensors. The experiments were performed on a computer using an Intel Core i7-12700K with 32GB of RAM.

Test cases from the OR-LIB provide weighted graphs and values for n and k, whereas test cases from the TSPLIB provide only weighted graphs and the value n. For each input from TSPLIB we chose multiple values of k and we referred to the paper by Garcia et al. [8] for the optimal values, as they presented an exact algorithm for KMP.

Due to space limitations we have omitted some results that are less interesting[1]: We do not present results for Haralampiev's network as this algorithm is the slowest and it does not compute more accurate solutions than other algorithms; we also do not show results for the algorithm of Pan and Zhu because it did not perform better than the local search algorithms of [10,11].

Since each implemented algorithm was initialized with random feasible solutions, we ran each algorithm ten times per test case to obtain a more reliable evaluation of its performance. For fairness, all algorithms were tested using the same initial feasible solutions. For each test, we report two statistics per algorithm: The average approximation ratio and the average total runtime in seconds. The approximation ratio is calculated as S/S^* where S is the value of the solution computed by an algorithm and S^* is the value of an optimal solution.

Table 1 shows a selection of results across a broad range of input sizes. The results for Algorithm 1 with client and facility activation potentials and neuron state functions (2)–(5) are under the heading ARN, and the results for our algorithm that uses the modified activation potential and neuron state functions (6) and (7) are under the heading MARN. Our implementation of MARN runs the ARN algorithm until a local optimum solution is found, then switches to the modified activation potential and state functions of MARN. Results for the algorithms in [10,11,18], and [23] are under the headings LS, SC, FI, and NA-L, respectively.

Algorithms ARN and NA-L were the fastest and were allowed to run until they stabilized. MARN was allowed to run for the same amount of time that NA-L used and then allowed to continue until it completed computing a feasible solution; since LS, SC, and FI took much longer to stabilize, they were given additional time depending on the size of the graph. The implementations for LS and SC try all possible single-swaps and accept the first improving-swap they find; FI was implemented to check all possible single-swaps and select the one that produces the best improvement.

Observe that even with the additional running time, LS, SC, and FI do not produce the best solutions in any test case. Therefore, we focus our analysis on the performance of ARN, MARN, and NA-L.

While ARN and MARN are similar to the simple local search algorithm of LS, they differ in a significant way: Instead of trying arbitrary swap operations, ARN and MARN replace the facility with the lowest inner value, and this is the reason for the speed increase over LS and SC.

4.1 Algorithm Speed

For each test case shown in Table 1, we show in bold the fastest runtime and the best approximation ratio. Observe that the ARN algorithm holds the majority of the fastest runtimes across the test cases. Intuitively, each of the competing algorithms performs more computations than our algorithm: LS and SC, might

---

[1] The complete results are available at https://www.csd.uwo.ca/~ablochha/KMPResults.pdf.

Table 1. Approximation ratios and runtimes of the algorithms on test cases from the OR-LIB for the k-median problem and test cases from TSPLIB with different values for k. Each algorithm was run 10 times and we report the average approximation ratio and runtime.

Test	n	k	ARN		MARN		LS		SC		FI		NA-L	
			Ratio	Time	Ratio	Time	Ratio	Time	Ratio	Time	Ratio	Time	Ratio	Time
OR-LIB	100	33	1.328	**0.005**	**1.029**	0.137	1.035	2.449	1.049	4.335	1.106	4.389	1.370	0.013
	200	67	1.500	**0.007**	**1.083**	0.204	1.289	5.010	1.355	5.005	1.389	5.007	1.412	0.031
	300	100	1.393	**0.014**	**1.114**	0.204	1.375	5.008	1.410	5.006	1.524	5.008	1.379	0.048
	400	133	1.423	**0.027**	**1.160**	0.203	1.479	5.015	1.502	5.007	1.644	5.009	1.383	0.063
	500	167	1.481	**0.026**	**1.181**	0.202	1.597	5.011	1.620	5.007	1.712	5.012	1.402	0.083
	600	200	1.513	**0.026**	**1.187**	0.203	1.589	5.012	1.615	5.021	1.702	5.013	1.376	0.112
	700	140	1.320	**0.031**	**1.153**	0.204	1.440	5.025	1.461	5.021	1.540	5.014	1.263	0.136
	800	80	1.094	**0.058**	**1.046**	0.204	1.327	5.024	1.340	5.021	1.427	5.017	1.166	0.158
	900	90	1.099	**0.066**	**1.055**	0.204	1.309	5.025	1.340	5.026	1.426	5.016	1.163	0.176
TSPLIB	1304	100	1.307	**0.076**	**1.206**	1.009	1.866	10.041	1.903	10.036	1.965	10.061	1.354	0.933
	1304	300	1.363	**0.222**	**1.236**	1.008	2.070	10.038	2.076	10.043	2.105	10.049	1.423	0.791
	1304	500	1.569	**0.184**	**1.243**	1.008	2.066	10.041	2.070	10.036	2.094	10.041	1.520	0.720
	1400	100	2.453	**0.025**	**1.609**	1.008	3.065	10.042	3.080	10.042	3.260	10.065	2.430	0.854
	1400	300	2.672	**0.049**	**1.914**	1.004	3.725	10.042	3.734	10.038	4.006	10.051	2.993	0.691
	1400	500	2.346	**0.106**	**1.705**	1.006	3.518	10.043	3.529	10.042	3.755	10.043	2.998	0.712
	1432	100	1.317	**0.106**	**1.258**	1.008	1.684	10.039	1.690	10.041	1.712	10.066	1.361	0.863
	1432	300	1.245	**0.298**	**1.194**	1.006	1.646	10.042	1.648	10.042	1.653	10.060	1.339	0.803
	1432	500	1.186	**0.200**	**1.061**	1.009	1.399	10.040	1.401	10.039	1.403	10.056	1.192	0.668
	1748	100	1.372	**0.116**	**1.266**	2.011	1.798	10.042	1.825	10.047	1.983	10.086	1.381	1.750
	1748	300	1.484	**0.272**	**1.369**	2.005	1.938	10.039	1.946	10.044	1.990	10.076	1.489	1.275
	1748	500	1.625	**0.337**	**1.475**	2.011	2.042	10.041	2.064	10.057	2.088	10.072	1.597	1.229
	3038	100	1.366	**0.395**	**1.309**	3.012	1.752	10.079	1.754	10.084	1.760	10.171	1.362	2.820
	3038	300	1.347	**1.134**	**1.307**	3.011	1.750	10.084	1.751	10.084	1.753	10.165	1.383	2.473
	3038	500	1.421	**1.151**	**1.339**	3.015	1.757	10.074	1.758	10.089	1.760	10.156	1.378	2.525
	5934	500	1.367	**5.855**	**1.279**	20.052	1.873	10.164	1.872	10.183	1.889	10.348	1.364	11.025
	5934	1000	1.417	**8.265**	**1.328**	20.068	1.889	15.163	1.889	15.189	1.894	15.346	1.365	13.620
	5934	1500	1.353	17.634	**1.339**	20.085	1.964	15.168	1.964	15.205	1.968	15.334	1.394	**16.357**
	13509	1000	1.611	**39.773**	1.565	75.364	1.988	200.471	1.988	200.554	1.989	201.186	**1.508**	63.430
	13509	3000	1.806	**79.849**	1.706	200.515	2.138	200.489	2.138	200.686	2.140	201.149	**1.660**	156.825
	13509	5000	2.066	**77.550**	1.895	200.555	2.321	200.481	2.321	200.746	2.323	201.153	**1.845**	193.964

make only small improvements and hence need many iterations to get better solutions, FI spends too much time considering all the possible swaps, NA-L holds a competition in all clusters to choose facilities.

In contrast, the ARN algorithm only needs to compute the facility assignment for a single cluster, as the only facility being replaced is the one that had the lowest inner value in each iteration. Therefore, the ARN algorithm only considers swaps between a single facility and a non-facility node. The MARN algorithm performs slightly more work in each iteration: If replacing the facility with the lowest inner value does not result in an improved solution, MARN tries replacing the facility with the second lowest inner value, and so on, accepting the first swap that improves the solution.

We note that the ARN algorithm is significantly faster than the NA-L algorithm for the majority of the test cases. On the largest graphs from TSPLIB with over 13500 vertices, ARN took about half as much time as NA-L, while computing solutions of value within 12% of the solutions of NA-L.

4.2 Algorithm Accuracy

Observe that the MARN algorithm holds the majority of the best approximation ratios across the test cases. Intuitively, by considering a larger number of potential swaps than ARN, MARN can transition out of local minima that ARN might get stuck in. For example, MARN computed solutions significantly closer to an optimal solution for the test cases with 1400 nodes.

The ARN and NA-L algorithms computed solutions of similar quality compared to the optimal values. We note that the NA-L algorithm is the only algorithm in these results that was allowed to perform multi-swaps (this means that in each iteration of this algorithm multiple facilities might be simultaneously swapped with the same number of non-facility nodes). The LS and SC algorithms have been proven to have better theoretical worst-case approximation ratios when multi-swaps are allowed [10, 11], and our full results also indicate that LS and SC compute better solutions within similar time limits when 2-swaps are allowed. The ARN and MARN algorithms behave as if they perform single-swaps.

5 Conclusions

We have presented a resonance neural network based algorithm for the k-median problem that is very fast and produces solutions of value close to the optimum. Our experimental results show that our algorithm outperforms the best existing single-swap local search algorithms and is faster than the best neural network-based algorithms for the k-median problem. Our research shows that it might be possible to compute the approximation ratio of algorithms for neural networks with a simple architecture, so we do not need to see neural networks as mysterious black boxes that work in unfathomable ways and for which it is impossible to predict the quality or guarantee the feasibility of the solutions that they produce.

We attribute the speed of our algorithm to its simplicity. Our algorithm does not consider all possible swaps; instead, it uses the inner value metric to prioritize which facilities to select, reducing the number of computations that it needs to perform. Moreover, the accuracy of our algorithm could be improved if we modify the neural network model to allow multi-swap operations.

Our work introduces the interesting idea of designing shallow neural networks that are mathematically tractable and admit provable performance guarantees for the solutions they produce.

References

1. Hakimi, S.L.: Optimum distribution of switching centers in a communication network and some related graph theoretic problems. Oper. Res. **13**(3), 462–475 (1965)
2. Andrews, M., Zhang, L.: The access network design problem. In: Proceedings 39th Annual Symposium on Foundations of Computer Science, pp. 40–49. IEEE (1998)
3. Bradley, P.S., Fayyad, U.M., Mangasarian, O.L.: Mathematical programming for data mining: formulations and challenges. INFORMS J. Comput. **11**(3), 217–238 (1999)
4. Jamin, S., Jin, C., Jin, Y., Raz, D., Shavitt, Y., Zhang, L.: On the placement of internet instrumentation. In: Proceedings IEEE INFOCOM 2000, vol. 1, pp. 295–304. IEEE (2000)
5. Cohen-Addad, V., Esfandiari, H., Mirrokni, V., Narayanan, S.: Improved approximations for Euclidean k-means and k-median, via nested quasi-independent sets. In: Proceedings of the 54th Annual ACM SIGACT Symposium on Theory of Computing, pp. 1621–1628 (2022)
6. Kariv, O.: An algorithmic approach to network location problems. Part 1: the p-centers. SIAM J. Appl. Math. **37**, 3 (1979)
7. Jain, K., Mahdian, M., Saberi, A.: A new greedy approach for facility location problems. In: Proceedings of the Thiry-Fourth Annual ACM Symposium on Theory of Computing, pp. 731–740 (2002)
8. García, S., Labbé, M., Marín, A.: Solving large p-median problems with a radius formulation. INFORMS J. Comput. **23**(4), 546–556 (2011)
9. Duran-Mateluna, C., Alès, Z., Elloumi, S.: An efficient benders decomposition for the p-median problem. Eur. J. Oper. Res. **308**(1), 84–96 (2023)
10. Cohen-Addad, V., Gupta, A., Hu, L., Oh, H., Saulpic, D.: An improved local search algorithm for k-median. In: Proceedings of the 2022 Annual ACM-SIAM Symposium on Discrete Algorithms, SIAM, 2022, pp. 1556–1612 (2022)
11. Arya, V., Garg, N., Khandekar, R., Meyerson, A., Munagala, K., Pandit, V.: Local search heuristic for k-median and facility location problems. In: Proceedings of the Thirty-Third Annual ACM Symposium on Theory of Computing, pp. 21–29 (2001)
12. Basu, S., Sharma, M., Ghosh, P.S.: Metaheuristic applications on discrete facility location problems: a survey. Opsearch **52**, 530–561 (2015)
13. Irawan, C., Salhi, S.: Aggregation and non aggregation techniques for large facility location problems: a survey. Yugoslav J. Oper. Res. **25**(3), 313–341 (2015)
14. Cohen-Addad Viallat, V., Grandoni, F., Lee, E., Schwiegelshohn, C.: Breaching the 2 lmp approximation barrier for facility location with applications to k-median. In: Proceedings of the 2023 Annual ACM-SIAM Symposium on Discrete Algorithms, SIAM, pp. 940–986 (2023)
15. Byrka, J., Pensyl, T., Rybicki, B., Srinivasan, A., Trinh, K.: An improved approximation for k-median and positive correlation in budgeted optimization. ACM Trans. Algorithms **13**(2), 1–31 (2017)
16. Li, S., Svensson, O.: Approximating k-median via pseudo-approximation. In: Proceedings of the Forty-Fifth Annual ACM Symposium on Theory of Computing, pp. 901–910 (2013)
17. Peng, X., Xia, X., Zhu, R., Lin, L., Gao, H., He, P.: A comparative performance analysis of evolutionary algorithms on k-median and facility location problems. Soft. Comput. **22**(23), 7787–7796 (2018)
18. Resende, M.G., Werneck, R.F.: A fast swap-based local search procedure for location problems. Ann. Oper. Res. **150**, 205–230 (2007)

19. Hansen, P., Mladenović, N.: Variable neighborhood search for the p-median. Locat. Sci. **5**(4), 207–226 (1997)
20. Resende, M.G., Werneck, R.F.: A hybrid heuristic for the p-median problem. J. Heuristics **10**, 59–88 (2004)
21. Dominguez Merino, E., Muñoz Perez, J.: An efficient neural network algorithm for the p-median problem. In: Garijo, F.J., Riquelme, J.C., Toro, M. (eds.) IBERAMIA 2002. LNCS (LNAI), vol. 2527, pp. 460–469. Springer, Heidelberg (2002). https://doi.org/10.1007/3-540-36131-6_47
22. Merino, E.D., Muñoz-Pérez, J., Jerez-Aragonés, J.M.: Neural network algorithms for the p-median problem. In: ESANN, pp. 385–392 (2003)
23. Domínguez, E., Muñoz, J.: A neural model for the p-median problem. Comput. Oper. Res. **35**(2), 404–416 (2008)
24. Mishrra, S., Barman, S.: Cost reduction techniques in p-median method using artificial neural network. JSM Bioinformatics, Genomics and Proteomics (2016)
25. Haralampiev, V.: Neural networks for facility location problems, Annual of Sofia University St. Kliment Ohridski. Faculty Math. nform. **106**, 3–10 (2019)
26. Haralampiev, V.: Theoretical justification of a neural network approach to combinatorial optimization. In: Proceedings of the 21st International Conference on Computer Systems and Technologies, pp. 74–77 (2020)
27. Haralampiev, V.: Neural network approaches for a facility location problem. Math. Model. **4**(1), 3–6 (2020)
28. Rossiter, C.: A modified hopfield network for the k-median problem, Master's thesis, The University of Western Ontario (Canada) (2023)
29. Bloch-Hansen, A.: Approximation algorithms for high multiplicity strip packing, thief orienteering, and k-median, Ph.D. thesis, The University of Western Ontario (Canada) (2024)
30. Pan, R., Daming, Z.: An efficient local search algorithm for k-median problems. In: Proceedings of the 6th WSEAS International Conference on Applied Computer Science, pp. 19–24 (2006)
31. Beasley, J.E.: Or-library: distributing test problems by electronic mail. J. Oper. Res. Soc. **41**(11), 1069–1072 (1990)
32. Reinelt, G.: Tsplib–a traveling salesman problem library. ORSA J. Comput. **3**(4), 376–384 (1991)
33. Paszke, A., et al.: Pytorch: an imperative style, high-performance deep learning library. In: Advances in Neural Information Processing Systems, vol. 32 (2019)

Prime Factorization of Hierarchical Products of Infinite Graphs

Clemens Brand and Wilfried Imrich[(✉)]

Department Mathematics and Information Technology, Montanuniversität Leoben,
A-8700 Leoben, Austria
{brand,imrich}@unileoben.ac.at

Abstract. Connected finite graphs have unique prime factorizations with respect to the hierarchical product, but this property does not extend to connected infinite graphs. Here we show that all connected infinite rayless graphs have unique prime factorization with respect to the hierarchical product, and provide conditions under which this is also true for locally finite infinite graphs. Furthermore, we investigate the factorization properties of special classes of graphs, the structure of the automorphism groups of hierarchical products, and formulate several conjectures.

Keywords: Graph products · Locally finite infinite graphs · Rayless graphs · Automorphisms · Trees

1 Introduction

In 2009 Barrière, Comellas, Dalfó, and Fiol [4] introduced a product of graphs, which they called hierarchical. It is a special case of a product, whose spectral properties were studied in 1978 by Godsil and McKay [6]. They seem to have provided the first formal definition. It is also known as the comb product, which was studied in the context of quantum probability and spectral analysis of graphs in 2007 by Accardi, Lenczewski, and Sałapata [1], and Hora and Obata [8]. Recent papers pertain to prime factorization, automorphisms, domination, the Laplacian spectrum, the distinguishing number, and the complexity of recognizing hierarchical products [2,3,5,9,10].

In [9] it was shown that each finite nontrivial connected graph X is the hierarchical product $G \sqcap H[h]$ of a unique prime graph G by a unique rooted graph $H[h]$. Here we provide conditions under which this also holds for connected locally finite infinite graphs, Theorem 3, and show that it holds for infinite connected rayless graphs under no further conditions, Theorem 4.

Because the hierarchical product is neither commutative nor associative, one cannot arbitrarily rebracket a product presentation of a graph. For finite graphs this was treated in [5]. Here we extend the results to several classes of infinite graphs.

© The Author(s), under exclusive license to Springer Nature Switzerland AG 2026
N. Misra and A. Pandey (Eds.): CALDAM 2026, LNCS 16445, pp. 96–105, 2026.
https://doi.org/10.1007/978-3-032-17156-6_8

We also devote some effort to graphs of bounded degrees with essentially different prime factorizations, and their automorphisms. For locally finite graphs we also provide examples of graphs with infinitely many factors.

2   The Hierarchical Product

We denote the vertex set of a graph G by $V(G)$ and its edge set, which consists of unordered pairs of distinct vertices, by $E(G)$. If $uv \in E(G)$, then u and v are called *adjacent*, in symbols $u \sim v$. Occasionally we also write $[u,v]$ for uv. A *rooted* graph $G[g]$ is a graph G with a distinguished vertex g, called the *root* of G.

Given a graph G and a rooted graphs $H[h]$, the *hierarchical product* $G \sqcap H[h]$ is an unrooted graph with vertex set $V(G) \times V(H)$, where the edges are defined by

$$(u,v)(u',v') \in E(G \sqcap H[h]) \text{ if } \begin{cases} uu' \in E(G) \text{ and } v = v' = h, \text{ or} \\ vv' \in E(H) \text{ and } u = u'. \end{cases}$$

Hence, $G \sqcap H[h]$ is formed from G by attaching to each vertex of G a copy of H via its root.

Figure 1 shows $G \sqcap H[1]$, where $G = K_2$, with $V(K_2) = \{0,1\}$, and $H[h] = K_2[1]$.

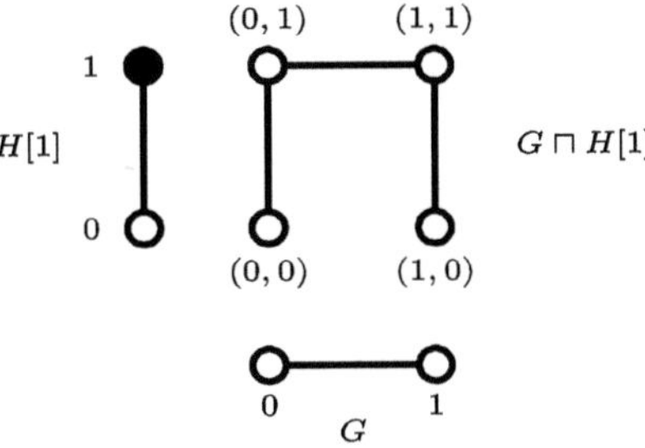

Fig. 1. A hierarchical product.

Hierarchical multiplication is neither commutative nor associative, because in the equations

$$G \sqcap H[h] = H[h] \sqcap G, \text{ and } (A \sqcap B[b]) \sqcap C[c] = A \sqcap (B[b] \sqcap C[c]),$$

the right sides are not defined. However, hierarchical multiplication is semi-associative in the sense that we may always move brackets from left to right:

$$(A \sqcap B[b]) \sqcap C[c] = A \sqcap (B \sqcap C[c])[b,c]. \tag{1}$$

In the other direction,

$$A \sqcap (B \sqcap C[c])[b, c'] = (A \sqcap B[b]) \sqcap C[c] \qquad (2)$$

is clearly is only possible if $c = c'$.

The graphs G and H are called *factors* of $G \sqcap H[h]$. Note that $G \sqcap H[h]$ and $G \sqcap H[h']$ need not be isomorphic for different vertices $h, h' \in V(H)$.

A graph on at least two vertices is called *indecomposable*, or *prime*, with respect to the hierarchical product if it cannot be represented as the hierarchical product of two nontrivial factors.

Clearly, each nontrivial finite graph can be represented as a product of prime graphs. That the factorization need not be unique was shown by Anderson, Guob, Tenney, and Wash [3] for disconnected graphs. In general this follows from Eq. 1 for prime graphs A, B and C.

As observed in [1] the hierarchical product becomes associative in the class of rooted graphs by defining the product $G[g] \sqcap H[h]$ of the rooted graphs $G[g]$ and $H[h]$ as $G \sqcap H[h]$ with root (g, h). We set

$$G[g] \sqcap H[h] = (G \sqcap H[h])[(g, h)].$$

In [9] this product was called the rooted hierarchical product, and it was shown that prime factorization of finite connected rooted graphs with respect to the rooted hierarchical product is unique. Note that each rooted hierarchical product $G[g] \sqcap H[h]$ is the hierarchical product $G \sqcap H[h]$ if one ignores the root (g, h).

The main aim of this paper is the generalization of the following two results to infinite graphs.

Proposition 1 (Imrich, Kalinowski and Pilśniak [9, Corollary 2.2]). *Each finite connected graph X has a unique standard prime factorization as a hierarchical product*

$$X = G_1 \sqcap (G_2 \sqcap (G_3 \sqcap (\cdots \sqcap G_k[v_k])[v_{k-1}]) \cdots [v_3])[v_2] \qquad (3)$$

of uniquely determined prime factors $G_1, \ldots, G_k$ and uniquely determined roots $v_2, \ldots v_k$, where each root v_i, $2 \leq i \leq k$, is a vertex of the product of the last $k - i + 1$ factors, that is,

$$v_i \in V(G_i) \times \cdots \times V(G_k).$$

For the next result we need the concept of layers. Let X be a hierarchical product of graphs $G_1, \ldots, G_k$. Then $V(X) = \prod_{i=1}^{k} V(G_i)$, and the subgraph induced by the set of vertices that differ from vertex $v = (v_1, \ldots, v_i, \ldots, v_k)$ only in v_i is called the *i-layer of G_i through v.*

For example, let X be the graph defined in Eq. 3. Then the subgraph induced by the set of vertices $\{(g, v_2) \mid g \in V(G_1)\}$, where $v_2 \in \prod_{i=2}^{k} V(G_i)$ is the right-most root in Eq. 3, is the unique G_1-layer of X.

If $X = G \sqcap H[h]$ we shall denote the G-layer of X through (g, h), where $g \in V(G)$, by $G \times h$.

Proposition 2 (Brand and Imrich [5, Theorem 2.2]). *Let X be a finite connected graph. Then any representation of X as a product of prime graphs with respect to the hierarchical product can be transformed into the standard prime factorization by rebracketing and the choice of appropriate roots.*

Moreover, for all i, the i-th factors in both representations are isomorphic as unrooted graphs, and the set of G_i-layers for any $i \geq 1$ is invariant under $\mathrm{Aut}(X)$.

This means that finite connected graphs have essentially unique prime factorization with respect to the hierarchical product. We reformulate Proposition 2 as follows.

Theorem 1 (Unique prime factorization for finite connected graphs). *The prime factors of finite connected graphs X with respect to the hierarchical product are unique as unrooted graphs. Their order, up to isomorphisms of the factors as unrooted graphs, is also unique, and* $\mathrm{Aut}(X)$ *preserves the set of layers with respect to any factor.*

We call this the *unique prime factorization property*.

3   Locally Finite Graphs

If G is not locally finite, then it can be the product of infinitely many graphs. An example is the infinite regular tree $T_{\aleph_0}$ of countably infinite degree. It is isomorphic to $T_{\aleph_0}^\infty$, that is

$$\lim_{n \to \infty} (\cdots (T_{\aleph_0} \sqcap T_{\aleph_0}[v_1]) \sqcap \cdots) \sqcap T_{\aleph_0}[v_{n-1}].$$

Note that the product is independent of the choice of the roots, because $T_{\aleph_0}$ is vertex transitive, Moreover, graphs that are not locally finite may have many different, non-isomorphic first factors. For an example, given any finite tree T, it is easy to see that $T \sqcap T_{\aleph_0}[v] \cong T_{\aleph_0}$, and that $\mathrm{Aut}(T_{\aleph_0})$ does not preserve the layer $T \times v$. As there are infinitely many pairwise non-isomorphic finite trees that are prime, neither unique prime factorization nor the invariance of the first prime factor holds for $T_{\aleph_0}$.

But, even connected locally finite infinite graphs may be products of infinitely may prime graphs. For the construction of an example, let $G_i \cong K_2[1]$, for $i \geq 1$, and consider

$$Y_k = (\cdots (G_k \sqcap G_{k-1}) \sqcap \cdots) \sqcap G_2) \sqcap G_1,$$

where $\sqcap$ denotes the rooted hierarchical product. We will embed each Y_k into $Y_{k+1} = G_{k+1} \sqcap Y_k$ such that Y_k is the Y_k-layer of Y_{k+1} through the root of Y_{k+1}. Note that this means that all leaves of Y_k are mapped into leaves of Y_{k+1}, and that the degrees of all other vertices are also preserved in the embedding, except for the root, whose degree increases by 1. Then we form the limit

$$Y = \lim_{k \to \infty} Y_k.$$

We construct Y as follows. Let $V(Y) = \{1, 2, 3, \ldots\}$, and

$$E(Y) = \{[2^{j-1}(2i-1), 2^{j-1}(2i+1)] \mid i, j = 1, 2, \ldots\},$$

which can also be written in the form

$$E(Y) = \{\ [1,2], [3,4], [5,6], \ldots$$
$$[2,4], [6,8], [10,12], \ldots$$
$$[4,8], [12,16], [20,24] \ldots$$
$$\ldots\}$$

Then sets of vertices $\{1, \ldots, 2k\}$ induce subgraphs isomorphic to Y_k with root $2k$, the root of Y_k has degree $k+1$ in Y. By construction Y is locally finite. It is also unrooted, because the root of the partial product Y_k is not the root of the following partial products.

Furthermore, if one relabels each vertex i of Y by $2i$, adds new vertices $\{1, 2, 3 \ldots\}$ and edges $[2i, 2i-1]$, one obtains $Y \sqcap K_2[1]$. Clearly

$$Y \sqcap K_2[1] \cong Y$$

via the mapping $i \mapsto 2i$ for all positive natural numbers.

Because the rooted hierarchical product is associative, the bracketing in the presentation of Y_k is irrelevant. This implies that each finite rooted hierarchical power of $K_2[1]$ is a factor of Y. It is not hard to show that Y has no other factors, and that it cannot be represented in the form

$$G \sqcap Y[r],$$

for any nontrivial G and any $r \in V(Y)$.

By contrast, if we embed each Y_k into Y_{k+1} such that the roots coincide, and take the limit,

$$Z = \lim_{k \to \infty} (G_1 \sqcap (G_2 \sqcap \cdots (G_{k-1} \sqcap G_k) \cdots)),$$

then we obtain a rooted tree, each vertex of which has infinite degree. Hence, $Z = T_{\aleph_0}$. We consider Z as the right limit of the Y_k, and Y as the left limit. Y is locally finite, but not Z, which is also rooted, and neither of them can be represented as the product of finitely many prime graphs.

If the degrees are bounded, then we have the following result.

Lemma 1. *Each connected finite or infinite graph G with finite maximum degree $\Delta(G)$ has a prime factorization*

$$G = G_1 \sqcap (G_2 \sqcap (\ldots \sqcap G_k[g_k]) \cdots)[g_2],$$

for some $k \leq \Delta(G)$.

Proof. A product of k nontrivial connected graphs has at least one vertex of degree k, hence the number of factors in any factorization of a connected graph G is bounded by $\Delta(G)$. Clearly all factors in a factorization of maximal length must be prime.

The factorization need not be unique though. An easy example is the free product $G = K_2 * K_3$ of an edge by a triangle. It is best described as a graph in which each vertex is incident with an edge and a triangle, and where contraction of the triangles to single vertices result in a T_3. In other words, the only cycles of G are the triangles. It is easily seen that G can be represented as a hierarchical product with K_2 as a first factor, and also is a hierarchical product with K_3 as a first factor.

If one chooses an arbitrary vertex v in G and deletes all edges whose endpoints have the same distance from v, then one obtains a tree G^*, which can also be represented in two ways as a hierarchical product, in one case with K_2 as a first factor, and in the other with P_3 as a first factor, see [10]. G^* is neither a free product, nor vertex transitive.

In [10] it was also shown that the regular trees T_k of finite degree k have unique prime factorizations. For the sake of completeness we include a short proof of a slight extension of the result.

Theorem 2. *Each regular tree of finite degree has a unique standard prime factorization with respect to the h-product of graphs. It allows arbitrary bracketings.*

Proof. Clearly all factors of a tree must be trees, and if a regular tree T_k is a product $G \sqcap H[h]$, then G must be a T_i with $1 \leq i < k$. Clearly the second factor, say $H_i[h_i]$, is uniquely determined for each i. It consists of a root vertex h_i of degree $k - i$ and all other vertices have degree k. Note that $T_1 = K_2$ and that H_2 is a ray. Clearly

$$T_k = (\cdots (T_1 \sqcap \ldots \sqcap H_{k-1}[h_{k-1}]) \sqcap H_k[h_k].$$

As T_k cannot have more than k factors by Lemma 1, this is a prime factorization. All partial products $(\cdots (T_1 \sqcap \ldots \sqcap H_{i-1}[h_{i-1}]) \sqcap H_i[h_i]$ are first factors of T_k, and there are no other first factors than $T_1, \ldots, T_{k-1}$. Hence these are the partial products and, by induction the factorization is unique.

Note that by semi-associativity all possible bracketings are allowed in the presentation of T_k.

Given a product $X = G \times H[h]$, the subgraph of X induced by $V(G) \times \{h\}$ is isomorphic to G, and can be considered as a natural embedding of G into the product. It is the G-layer of $G \sqcap H[h]$ through (g, h), where g can be arbitrarily chosen. We denoted it by $G \times h$. By Theorem 2 it is invariant under $\text{Aut}(G \sqcap H[h]$ if G is prime, and if G and H are finite.

For infinite graphs this need not hold. For an infinite locally finite example, observe that K_2 is a prime factor for all T_k, and if $T_k = K_2 \sqcap H[h]$, then $K_2 \times h$ is not invariant under $\text{Aut}(T_k)$.

Given $G \times H[h]$ we are interested in the case, when $G \times h$ is invariant under $\mathrm{Aut}(G \sqcap H[h])$, whether G is prime or not. As a step in this direction we prove the following theorem. It is motivated by the observation that the second factors in the presentations of $G^* = K_2 * K_3$ as hierarchical products have induced subgraphs that are isomorphic to the factors. Thus, if G is such a factor, then it admits an endomorphisms φ with $\varphi(G) \cong G$, where both G and $\varphi(G)$ are considered to be unrooted. Such endomorphisms are called *auto-endomorphisms*.

Theorem 3. *Let X be a locally finite infinite connected graph. If in each factorization $G \sqcap H[h]$ of X, where G is prime, the second factor H has no nontrivial auto-endomorphisms, then G and $H[h]$ are uniquely determined up to isomorphisms, and $G \times h$ is invariant under $\mathrm{Aut}(X)$.*

Proof. Let $G \sqcap H[h] \cong A \sqcap B[b]$, where G and A are prime. We assume that the vertex sets of the two products are the same, which implies that we have two coordinatizations, in other words, each vertex has a pair of coordinates for the first factorization, and another pair for the other one.

We wish to show that $G \times h$ and $A \times b$ are the same, and consider the following possibilities for $G \times h$ and $A \times b$: they are disjoint, their intersection is a proper subgraph of both graphs, or one is contained in the other.

(i) $G \times h \cap A \times b = \emptyset$. Then $A \times b$ is contained in some $g' \times H$. There is an $a' \times B$ that contains $G \times h$. Hence all $a \times B$ with $a' \neq a \in A$ are in $g' \times H$ and all $g \times H$ with $g' \neq g \in V(G)$ are in $a' \times B$. Hence B is isomorphic to a proper induced subgraph of H, and H isomorphic to a proper induced subgraph of B, but B cannot be isomorphic to a proper induced subgraph of itself.

(ii) $G \times h \cap A \times b$ is a proper subgraph of $G \times h$ and $A \times b$. Then there is a $(g', h) \in G \times h \cap A \times b$ with a neighbor $(a', b) \in A \times b \setminus G \times h$. Clearly $a' \times B$ does not meet $G \times h$, but as a neighbor of g' it is in $g' \times H$. Hence B is isomorphic to a proper induced subgraph of H. By symmetry, H is isomorphic to a proper induced subgraph of B. So B is isomorphic to a proper induced subgraph of itself, which we excluded.

(iii) One of the graphs $G \times h$ and $A \times b$ is a proper subgraph of the other. We assume without loss of generality that $A \times b$ is a proper subgraph of $G \times h$. Then there must be an $a \times B$ that contains vertices of $G \times h$ that are different from (a, b), say $(a, b_i), i \in I$. Set $(g_0, h) = (a, b)$, $(g_i, h) = (a, b_i)$, and consider the layers $g_i \times H$ of $G \sqcap H[h]$, where $i \in \{0\} \cup I$.

By definition $g_0 \times H$ contains no vertex of $G \times h$, hence no vertex of $A \times b$, and is completely contained in $a \times B$. If any of the other layers $g_i \times H$, $i \in I$, contained a vertex of $a \times B$, then it would also have to contain (g_0, h), which is not possible. Hence $a \times B$ consists of $(a \times B) \cap (G \times h)$ and the $g_i \times H$, where $i \in \{0\} \cup I$. In other words,

$$a \times B \cong ((a \times B) \cap (G \times h)) \sqcap H[h],$$

which implies that B contains a proper subgraph that is isomorphic to H.

If there is an a' such that

$$(a' \times B) \cap (G \times h) = (a', b),$$

then $a' \times B = g' \times H$, where $(g', h) = (a', b)$. But, this is not possible, because then H is isomorphic to B, and B contains a proper subgraph that is isomorphic to H, which is not possible.

If there is an $a'' \in V(A)$ such that $(a'' \times B) \cap (G \times h) \neq (a'', b)$, then we get, by the previous argument, a nontrivial factorization of B with H as a second factor. If these factorizations are not isomorphic for all $a'' \in V(A)$, then we are back to the case where $G \sqcap H[h] = A \sqcap B[b]$, but now H and B are isomorphic. If $g \times h \neq A \times b$, then the previous argument leads to a non-trivial factorization of H by itself, which is not possible.

Hence, these factorizations must be the same. Then any two intersections of B-layers with $G \times h$ are isomorphic, and G is isomorphic to $A \sqcap B'[(a, b)]$, where $B' = (a \times B) \cap (G \times h)$. This contradicts the assumption that G is prime.

We have thus shown that $G \times h = A \times b$, which implies that $G \times h$ is invariant under automorphisms.

This implies that all graphs satisfying the assumptions of Theorem 3 have a unique standard prime factorization. By semiassociativity and the same arguments as in the finite case we thus obtain the following corollary.

Corollary 1 (Prime factorization of connected locally finite graphs). *Locally finite graphs satisfying the assumptions of Theorem 3 have the unique prime factorization property with respect to the hierarchical product.*

4   Rayless Graphs

Rayless graphs are graphs without rays, that is, without one-sided infinite paths. Each infinite rayless graph contains a vertex of infinite degree, because each locally finite infinite graph has a ray by Kőnig's Theorem. A much more remarkable property is that each rayless graph G contains a uniquely defined finite set of vertices that is invariant under $\mathrm{Aut}(G)$. It is called the *kernel* of G and denoted by $K(G)$. This, and other results about rayless graphs are due to Schmidt [12]. See Halin [7] for an extended exposition in English.

For connected graphs the kernel is not empty. Clearly the kernel $K(T)$ of any rayless tree T induces a finite, nonempty subtree, say S. This implies that the center of S is a uniquely defined subgraph of T. Hence each rayless tree has a uniquely defined center, consisting of a single vertex or an edge. For a direct proof of this result see Polat and Sabidussi [11].

It is an example of properties that rayless graphs share with finite graphs. As a further example, we show in this section that connected rayless graphs have the unique prime factorization property.

Lemma 2. *Let $G[g]$ be a rooted connected graph and φ an auto-endomorphism of G, where $g \notin \varphi(G)$, and where g and $\varphi(g)$ are separated by a cut vertex x of G that is not in $\varphi(G)$. Then G contains a ray.*

Proof. Let C be the connected component of $G - \varphi(x)$ that contains x. Then there is a shortest $x, \varphi(x)$-path P in C, and $P \cap \varphi(P) = \varphi(x)$. Hence $P \cup \varphi(P)$ is a path whose length is twice the length of P. Iterating this process we see that $P \cup \bigcup_{i=1}^{\infty} \varphi^i(P)$ is a ray in G.

Lemma 3. *Each connected rayless graph has a prime factorization.*

Proof. Let $G \sqcap H[h]$ be the product of two non-trivial graphs, and $P = g_0 \cdots g_k$ a path of length k in G, where each inner vertex of P is a cut vertex. Then $P \cup (g_k, h)(g_k, h')$, where $hh' \in E(H)$, is a path P' of length $k + 1$ in $G \sqcap H[h]$ that extends P, and where each inner point is a cut vertex.

Hence, an h-product G of k nontrivial graphs contains a path of length k, where each inner point is a cut vertex. Suppose G can also be represented as a product of $k + 1$ nontrivial factors. Then the coordinates of the endpoints of P can only differ in at most k consecutive coordinates, so it can be extended to a longer path either at the origin or at its terminus. In either case P can be extended to a longer path, each inner point of which is a cut vertex. This means, if G is representable as a product of arbitrarily many factors, then it contains a ray or a double ray.

Thus, there is a bound on the number of factors, and any factorization with the maximum number of factors is a prime factorization.

Theorem 4. *Let X be a connected rayless graph. Then X is a uniquely representable as a product $X = G \sqcap H[h]$, where G is prime, and where $G \times h$ is invariant under* $\mathrm{Aut}(X)$.

Proof. Suppose X has two factorizations $G \sqcap H[h]$ and $A \sqcap B[b]$, where G and A are prime. We assume that the vertex sets of X and the products are the same. It suffices to show that $G \times h = A \times b$.

If this is not the case, then we have the possibilities that the graphs $G \times h$ and $A \times b$ are disjoint, that their intersection is a proper subgraph of both $G \times h$ and $A \times b$, or that one is contained in the other. As in the proof of Theorem 3 we consider them separately.

(i) $G \times h \cap A \times b = \emptyset$. As before we observe that there is an $a' \times B$ that contains $G \times h$. Hence all $a \times B$, where $a' \neq a \in V(A)$ are contained in $g' \times H$, and, by symmetry in A and B, all $g \times H$, where $b' \neq g \in V(G)$ are in $a' \times B$. Hence $g \times H \subset a' \times B$, and $a \times B \subset g' \times H$.

There are isomorphisms $\varphi \in \mathrm{Aut}(X)$ from $a' \times B$ onto $a \times B$, and ψ from $g' \times H$ to $g \times H$. Then

$$\varphi(\psi(g' \times H)) = \varphi(g \times H) \subset \varphi(a' \times B) = a \times B \subset g' \times H.$$

The root (g', h) of $g' \times H$ is in $a' \times B$, and thus not in $a \times B$, which contains $\varphi(\psi(g' \times H))$. Because $a' \times B$ and $a \times B$ are separated by the cut vertex (a, b), the vertices (g', h) and $\varphi(\psi(g' \times H))$ are also separated by (a, b). Hence, by Lemma 2, H contains a ray, contrary to assumption.

(ii) $G \times h \cap A \times b$ is a proper subgraph of $G \times h$ and $A \times b$. By the same arguments as in the first case, we obtain a ray in X.

(iii) One of the graphs $G \times h$ and $A \times b$ is a proper subgraph of the other. A slight extension of the arguments in the proof of Theorem 3 is possible here too, and shows that X has a ray under these conditions, or that G, respectively A, is not prime.

This immediately implies that connected rayless graphs have unique standard prime factorization with respect to the hierarchical product. Again, using semi-associativity, we obtain the following corollary.

Corollary 2 (Unique prime factorization of connected rayless graphs). *Connected rayless graphs have the unique prime factorization property with respect to the hierarchical product.*

Disclosure of Interests. The authors have no competing interests to declare that are relevant to the content of this article.

References

1. Accardi, L., Lenczewski, R., Sałapata, R.: Decompositions of the free product of graphs. Infin. Dimens. Anal. Quantum Probab. Relat. Top. **10**(3), 303–334 (2007). https://doi.org/10.1142/S0219025707002750, https://hdl.handle.net/2108/44028
2. Amouzegar, T.: Distinguishing number of hierarchical products of graphs. Bull. Sci. Math. **168**, 102975 (2021). https://doi.org/10.1016/j.bulsci.2021.102975
3. Anderson, S.E., Guob, Y., Tenney, A., Wash, K.A.: Prime factorization and domination in the hierarchical product of graphs. Discuss. Math., Graph Theory **37**(4), 873–890 (2017). https://doi.org/10.7151/dmgt.1952
4. Barrière, L., Comellas, F., Dalfó, C., Fiol, M.A.: The hierarchical product of graphs. Discrete Appl. Math. **157**(1), 36–48 (2009). https://doi.org/10.1016/j.dam.2008.04.018, https://hdl.handle.net/2117/672
5. Brand, C., Imrich, W.: The laplacian spectrum of hierarchical product graphs and unique prime factorization. Discrete Math. Submitted for publication
6. Godsil, C.D., McKay, B.D.: A new graph product and its spectrum. Bull. Aust. Math. Soc. **18**, 21–28 (1978). https://doi.org/10.1017/S0004972700007760
7. Halin, R.: The structure of rayless graphs. Abh. Math. Semin. Univ. Hamb. **68**, 225–253 (1998). https://doi.org/10.1007/BF02942564
8. Hora, A., Obata, N.: Quantum probability and spectral analysis of graphs. With a foreword by Professor Luigi Accardi. Theor. Math. Phys. (Cham), Berlin: Springer (2007). https://doi.org/10.1007/3-540-48863-4
9. Imrich, W., Kalinowski, R., Pilśniak, M.: Hierarchical product graphs and their prime factorization. Art Discrete Appl. Math. **8**(1), 17 (2025). https://doi.org/10.26493/2590-9770.1701.10e, id/No p1.06
10. Imrich, W., Makar, G., Palmen, J., Zając, P.J.: On the hierarchical product of graphs. Discrete Math. Appl. **9**, 163–171 (2024)
11. Polat, N., Sabidussi, G.: Fixed elements of infinite trees. Discrete Math. **130**(1–3), 97–102 (1994). https://doi.org/10.1016/0012-365X(92)00526-W
12. Schmidt, R.: Ein Ordnungsbegriff für Graphen ohne unendliche Wege mit einer Anwendung auf n-fach zusammenhaengende Graphen. Arch. Math. **40**, 283–288 (1983). https://doi.org/10.1007/BF01192782

Cycle Transit Function, Interval Function and Betweenness

Manoj Changat[1(✉)], Lekshmi Kamal K. Sheela[1,3],
Athulyakrishna Kuruppankandy[2], and T.V. Ramakrishnan[2]

[1] Department of Futures Studies, University of Kerala, Thiruvananthapuram
695581, India
`mchangat@keralauniversity.ac.in`
[2] Department of Mathematical Sciences, Kannur University, Mangattuparamba,
Kannur 670567, India
`{athulyakrishna,ramakrishnantv}@kannuruniv.ac.in`
[3] Department of Mathematics, University of Kerala, Thiruvananthapuram 695581,
India

Abstract. The cycle convexity of a graph, introduced by Norbert Polat
in [13], is a convexity finer than the standard geodesic convexity. It arose
from his investigation of special classes of partial cubes known as netlike
partial cubes.

In this paper, we prove that cycle convexity is associated with a transit function of arity three, which we name the cycle transit function,
denoted as the C_3-function. We investigate some natural extensions of the
betweenness properties of arbitrary 2-ary transit functions as they apply
to the C_3-function. Using a set of first-order axioms, we also characterize
the betweenness properties of its arity-two counterpart, the C_2-function,
for certain special cases. We also study another natural extension of the
geodesic interval function I: the I_3-function and prove that, in any connected graph, some of the natural extensions of the betweenness axioms
for the C_3-function and the I_3-function are equivalent.

Keywords: Cycle Convexity · Arity · Cycle Transit Function ·
Betweenness · Interval Function

1 Introduction

The concepts of betweenness, intervals, and convexity are fundamental notions
that appear in several branches of mathematics. Transit functions, introduced
in [9], are a useful tool for studying these fundamental concepts in graphs and
other discrete structures. Several types of transit functions have been studied
in graphs. Three prominent examples of transit functions and their associated
convexities are those defined by shortest paths, induced paths, and all paths in
a connected graph. Perhaps the most studied of these is the geodesic interval
function, denoted $I(u, v)$, which is defined for any two vertices $u, v \in V$ in a
connected graph $G = (V, E)$ as: $I(u, v) = \{x \in V | x$ lies on some u, v-shortest

N. Misra and A. Pandey (Eds.): CALDAM 2026, LNCS 16445, pp. 106–119, 2026.
https://doi.org/10.1007/978-3-032-17156-6_9

path in G}. Key references for the geodesic transit function include $[6, 8, 11, 12, 14]$. A survey of convexities associated with graph transit functions can be found in $[2]$.

Formally, a *transit function* on a non-empty set V is a function $R{:}V \times V \to 2^V$ satisfying the following axioms (transit axioms), for any $u, v \in V$.

$(t1)$ $u \in R(u, v)$ (law of extension).
$(t2)$ $R(u, v) = R(v, u)$ (law of symmetry).
$(t3)$ $R(u, u) = \{u\}$ (law of idempotence).

The three basic axioms of a transit function correspond to three fundamental properties of betweenness.

Mulder introduced more complex, non-trivial axioms of betweenness in $[9]$ and has been studied by others (for example, in $[3]$). Some of the natural betweenness axioms for a transit function, necessary for geometricity, are as follows:

$(b2)$ $x \in R(u, v) \Rightarrow R(u, x) \subseteq R(u, v)$;
(m) $x, y \in R(u, v) \Rightarrow R(x, y) \subseteq R(u, v)$.

The *underlying graph* G_R of a transit function R is the graph with vertex set V, where two distinct vertices u and v are joined by an edge if and only if $R(u, v) = \{u, v\}$.

A family $\mathcal{C}$ of subsets of a nonempty set V (not necessarily finite) is a *convexity* on V if $\emptyset, V \in \mathcal{C}$ and $\mathcal{C}$ is closed under arbitrary intersections and nested unions. The elements of $\mathcal{C}$ are called *convex sets*. For a subset W of V the smallest convex set containing W is called the *$\mathcal{C}$-convex hull* of W or simply *convex hull* of W denoted as $\langle W \rangle_{\mathcal{C}}$. A convexity $\mathcal{C}$ on V is said to be of arity $\leq n$, if it satisfies the following condition: $\mathcal{C} = \{K \subseteq V \mid F \subseteq K, |F| \leq n \Rightarrow \langle F \rangle_{\mathcal{C}} \subseteq K\}$. For a detailed discussion on n-ary convexities, see van de Vel $[14]$. It can be easily shown that all convexities are n-ary convexities for some $n > 1$. Furthermore, a convexity of arity 1 is simply the trivial convexity consisting of the power set of V.

An interesting question is: *Whether an n-ary convexity ($n > 2$) is the associated convexity of a transit function of arity $n > 2$?* This question is answered in $[4]$ by proving that every n-ary convexity is the associated convexity of some n-ary transit function.

Formally, a function $R : \underbrace{V \times V \times \ldots \times V}_{n \text{ times}} \longrightarrow 2^V$ is a *transit function of arity* n (or n-ary transit function) on V if R satisfies the following axioms.

$(t1')$ $u_i \in R(u_1, u_2, \ldots, u_n)$ for all $u_i \in V$, $i = 1, \ldots, n$,
$(t2')$ $R(u_1, u_2, \ldots, u_n) = R(\pi(u_1, u_2, \ldots, u_n))$ for all $u_i \in V$, where $\pi(u_1, u_2, \ldots, u_n)$ is any permutation of $(u_1, u_2, \ldots, u_n)$,
$(t3')$ $R(u, u, \ldots, u) = \{u\}$ for all $u \in V$.

Let R be an n-ary transit function on a nonempty set V. A set $W \subseteq V$ is called R *-convex* if $R(u_1, u_2, \ldots, u_n) \subseteq W$, for every $u_i \in W$, $i = 1, \ldots, n$. It is easy to verify that the family of all R-convex sets $\mathcal{C}_R$ of V is a convexity in which every singleton set is convex, by the axiom $(t3')$. The convexity in which every singleton is convex is called S_1-convexity [14].

The R-convex hull $\langle A \rangle_R$ of a subset A of V can be generated iteratively from any n-elements of A. That is, the convex hull of A is obtained by first setting $A^1 = \cup\{R(u_1, u_2, \ldots, u_n), u_i \in A\}$ then repeating this operation on subsequent $A^2 = A^1(A^1)$, $A^3 = A^1(A^2)$, etc. until the process stabilizes. When $A^k = A^{k+1}$, for some integer k, the set A^k is the R-convex hull $\langle A \rangle$ of A. The convexity associated with a transit function (2-ary function) R is known as interval convexity or 2-ary convexity. The name reflects that the convex hull of any subset A of V can be generated from pairs of elements of A.

The following betweenness axioms are considered for an n-ary transit function R in [1] as extensions of the axioms $(b2)$ and (m). For any $u_1, \ldots, u_n, x, x_1, \ldots, x_n \in V$,

$(b2)$ $x \in R(u_1, u_2, \ldots, u_n) \implies R(x, u_2, \ldots, u_n) \subseteq R(u_1, u_2, \ldots u_n)$

(m) $\forall x_1, \ldots, x_n \in R(u_1, \ldots u_n) \implies R(x_1, \ldots, x_n) \subseteq R(u_1, \ldots, u_n)$

$(b2(i))$ $x_1, \ldots, x_i \in R(u_1, \ldots, u_n) \implies R(x_1, \ldots, x_i, u_{i+1}, \ldots, u_n) \subseteq R(u_1, \ldots, u_n)$, for $1 \le i \le n$.

The last axiom, for $i = 2$, is introduced in [5]. It may be noted that the axiom $(b2(1)) = (b2)$ and $(b2(n)) = (m)$. Mulder in [8] proved that for any connected graph, the interval function I satisfies the axiom $(b2)$ (for $n = 2$). However, I does not necessarily satisfy the axiom (m) in general, and the characterization of graphs for which I satisfies (m) remains an open problem.

A natural generalization of the geodesic interval $I(u, v)$ to the n-ary case is the n-Steiner interval $S(u_1, u_2, \ldots, u_n)$. To define this, we first need the concept of a *Steiner tree* for a multiset $W \subseteq V(G)$, which is a minimum-order tree in G that contains all vertices of W. The *n-Steiner interval* $S(u_1, u_2, \ldots, u_n)$, then consists of all vertices in G that lie on some Steiner tree with respect to $(u_1, u_2, \ldots, u_n)$. Furthermore, a set $A \subseteq V(G)$ is *n-Steiner convex* if it is closed for n-Steiner intervals, i.e., $S(u_1, u_2, \ldots, u_n) \subseteq A$, for every n-element multiset $\{u_1, u_2, \ldots, u_n\}$ of A.

The betweenness properties $(b2)$ and (m) of the n-Steiner function S have been studied for $n \ge 4$ in [1] and for $n = 3$ in [5]. These studies often characterize the class of graphs for which the Steiner interval satisfies the relation $S(u_1, \ldots, u_n) = \cup_{i \ne j} I(u_i, u_j)$. This is defined as the *union property* of the n-Steiner interval in [1]. Although the union property holds trivially in all graphs for $n = 2$, it turns out that for $n > 3$, the class of graphs satisfying it coincides with the class in which the n-Steiner interval satisfies both the betweenness axiom $(b2)$ and the monotone axiom (m).

The case for $n = 3$ is more nuanced. In [1], it is proved that the class of graphs satisfying the union property is properly contained in the class satisfying the monotone axiom (m), which is in turn properly contained in the class satisfying the betweenness axiom $(b2)$. Furthermore, [5] establishes that the 3-Steiner interval on a connected graph G satisfies the axiom $(b2)$ if and only if each block of G is a geodetic graph of diameter at most 2. The same paper also shows that for the 3-Steiner interval function, the axiom $(b2(2))$ is equivalent to the monotone axiom (m).

In this paper, motivated by studies on the Steiner function (especially the 3-Steiner function), we explore two arity-3 extensions of the interval function in a connected graph G. The first is the function $I_3(u, v, w)$, defined for any triple of vertices u, v, w in G as $I_3(u, v, w) = I(u, v) \cup I(v, w) \cup I(u, w)$. The second function is derived from the I_3-function, and we name it the *cycle transit function* of G. We justify this name by proving that it is the transit function associated with the cycle convexity, a convexity finer than the geodesic convexity introduced by Norbert Polat in [13].

1.1 Preliminaries on Graphs

Throughout this paper, we consider only finite, simple, and connected graphs $G = (V, E)$. A u, v-path $P_{u,v}$ is a sequence of distinct vertices $u = u_1 u_2 \ldots u_k = v$, where u_i is adjacent to u_{i+1},for $i \in \{1, \ldots, k - 1\}$. The *length* of this path is $k - 1$. A *cycle* C is a path that begins and ends at the same vertex and has a length of at least 3. A path of minimum length between two vertices is a u, v-geodesic, and its length is *distance*, denoted $d_G(u, v)$. A subgraph H of G is *isometric* if for every pair of vertices u, v in H, the distance $d_H(u, v) = d_G(u, v)$. A subgraph H is *induced* if for any two vertices u, v in H, the edge uv exists in H if and only if it exists in G. A path that is also an induced subgraph is an *induced path*.

Graph Families and Properties: We will refer to several graph families. The *complete graph* K_n is the graph on the n vertices where every pair of distinct vertices is adjacent. A k-wheel ($k \geq 3$) is a graph formed by a cycle C_k and a central vertex adjacent to all the k vertices of the cycle. An n-fan is a graph formed by a path P_{n+1} and a central vertex adjacent to all vertices of the path. A graph G is *geodetic* if every pair of vertices has a unique shortest path; G is a *block graph* if every block (maximal 2-connected component) is a clique; G is *chordal* if it has no induced cycles of length greater than three; G is *distance-hereditary* if every induced path is a shortest path, and a *Ptolemaic graph* is a chordal and distance-hereditary graph.

Graph Operations: The *Cartesian product* of two graphs G_1 and G_2, denoted $G_1 \square G_2$, is a graph with vertex set $V(G_1) \times V(G_2)$. Two vertices (g_1, g_2) and (h_1, h_2) are adjacent if $g_1 = h_1$ and $g_2 h_2 \in E(G_2)$ or $g_2 = h_2$ and $g_1 h_1 \in E(G_1)$. An $n \times m$ *grid* is the Cartesian product $P_n \square P_m$, where P_n and P_m are paths with n and m vertices, respectively. Let $P_1 = v_1 v_2 \ldots v_n$ and $P_2 = u_1 u_2 \ldots u_m$ be two internally vertex-disjoint paths in a graph such that the last vertex of P_1 coincides with the first vertex of P_2, then their concatenation, denoted $P_1 + P_2$ is the path $P_1 + P_2 = v_1 v_2 \ldots v_n u_2 \ldots u_m$.

The paper is organized as follows. In Sect. 2, we define *cycle convexity* (denoted as $\mathscr{C}$-convexity), which is derived from geodesic convexity. We then prove that $\mathscr{C}$ is a convexity of arity 3 and determine its associated 3-ary transit function, denoted as C_3. In Sect. 3, we study the 2-ary version of this function

(C_2-function) and its underlying graph structure. We also provide an axiomatic characterization for C_2 using a set of first-order axioms that apply to any transit function R. Finally, in Sect. 4, we investigate the betweenness properties $(b1), (b2), (b2(2))$, and (m) for both the I_3 and C_3-functions. We prove a key equivalence: the C_3-function satisfies axioms $(b2), (b2(2))$, and (m) if and only if the I_3-function also satisfies these same axioms.

2   The Cycle Convexity

The idea of the cycle transit function originated from the work of Norbert Polat, who studied a particular type of partial cubes (isometric subgraphs of hypercubes) known as net-like partial cubes. Norbert Polat in [13] introduced the cycle convexity $\mathscr{C}$ as a convexity finer than the geodesic convexity.

To formally define $\mathscr{C}$-convexity, we need a couple of definitions. Let G be a connected graph with the interval function I, define the map $\mathcal{I}_G : 2^V \to 2^V$ as $\mathcal{I}_G(A) = \bigcup_{u,v \in A} I(u,v)$, for each $A \subseteq V$. The set of vertices of G that belong to a cycle of G is denoted by $CV(G)$ and the set of vertices of the subgraph of G induced by $\mathcal{I}_G(A)$ is denoted by $G[\mathcal{I}_G(A)]$. In [13], a set $A \subseteq V(G)$ is defined as $\mathscr{C}$-convex if $CV(G[\mathcal{I}_G(A)]) \subseteq A$. For a subset A of $V(G)$, the smallest $\mathcal{C}$-convex subset containing A is called the $\mathcal{C}$-convex hull of A, denoted as $\langle A \rangle_{\mathscr{C}}$.

Now, the function $I_3 : V \times V \times V :\to 2^V$ is defined as $I_3(u,v,w) = I(u,v) \cup I(v,w) \cup I(u,w)$. We define a new 3-ary-transit function from the 3-ary function I_3-function as $C_3 : V \times V \times V \longrightarrow 2^V$ such that $C_3(u,v,w) = \{u,v,w\} \cup \{w : w \in CV(G[I_3(u,v,w)]) \}$. If $I_3(u,v,w)$ does not contain any cycle, then $C_3(u,v,w) = \{u,v,w\}$. It is interesting to observe that the function $C_3(u,v,w)$ is a well-defined 3-ary transit function in the sense that it satisfies all the axioms $(t1'), (t2')$ and $(t3')$ of a 3-ary function. From the definition of the C_3-function it follows that $C_3(u,v,w) \subseteq I_3(u,v,w)$, for every triple of vertices u,v,w in G.

In addition, $C_3(u,v,v) = C_3(u,u,v)$. It can be easily observed that if the set $A \subseteq V(G)$ is C_3-convex, then $C_3(u,v,w) \subseteq A$, for all $u,v,w \in A$. Now, define $C_3(A) = \bigcup_{u,v,w \in A} C_3(u,v,w)$. The convex hull of A with respect to C_3 is denoted by $\langle A \rangle_{C_3}$ and it can be shown that $\langle A \rangle_{C_3} = \bigcup_{k \in N}(A^k)$, where the sets $A^{k'}$ are defined iteratively as $A^1 = C_3(A), A^2 = C_3(A^1), \ldots A^k = C_3(A^{k-1})$. Let $\langle A \rangle_{\mathscr{C}}$ be the convex hull of A with respect to $\mathscr{C}$. We now show that the convexity induced by the 3-ary transit function C_3 coincides with $\mathcal{C}$.

Theorem 1. *The $\mathcal{C}$-convexity of graph G is of arity 3 and is generated by the 3-ary transit function $C_3(u,v,w)$.*

Proof. Let C be a C_3 convex set of G. We need to prove that C is a $\mathcal{C}$ convex set of G. Suppose, if possible, C is not $\mathcal{C}$ convex. Then there exists a cycle C that belongs to $\mathcal{I}_G(C)$ but not to C. Let $C_1 = v_1 v_2 \ldots v_k v_1$ be a cycle in $\mathcal{I}_G(C)$, not in C with minimum cardinality. Since $C_1 \nsubseteq C$, there exists a $v_i \in C_1 \setminus C$. Since $v_i \in \mathcal{I}_G(C)$, there exist $d_1, d_2 \in C$ such that $v_i \in I(d_1, d_2)$. Now, let P be a d_1, d_2-geodesic that contains v_i. Let v_m and v_n be the first and last vertices in P that intersect C_1. Since P is a geodesic and $v_m, v_n \in V(P)$ implies that the v_m, v_n-path through P is a geodesic, say $P_{v_m v_n}$. Now, $v_i \in V(P_{v_m v_n})$, and every v_m, v_n-geodesic contains v_i. Otherwise, $v_i \in C_3(d_1, d_2, d_1) \subseteq C$, which is not possible. By choosing C_1, the cycle $P_{v_m v_n} + v_n v_{n+1} \ldots v_m$ has a cardinality greater than or equal to $|C_1|$ (See Fig. 1(a)). It follows that the length of $P_{v_m v_n}$ equals the length of $v_m v_{m+1} \ldots v_n$. Thus, $v_m v_{m+1} \ldots v_n$ is a v_m, v_n-geodesic containing v_i. Therefore, we may assume that $P_{v_m v_n} = v_m v_{m+1} \ldots v_n$. Consider v_{n+l} where $l = \lfloor \dfrac{k}{2} \rfloor$ (Fig. 1).

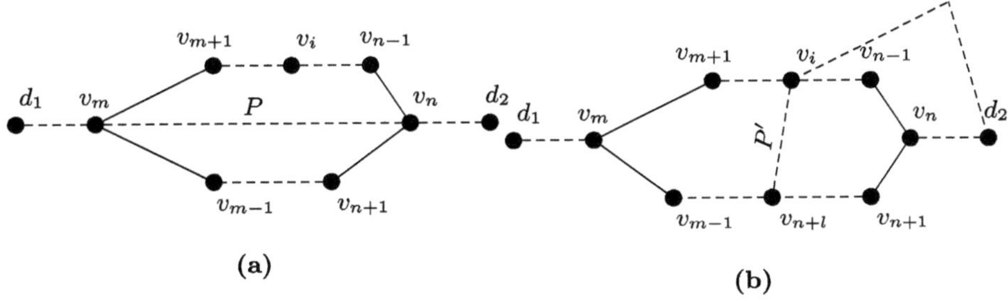

Fig. 1. Illustration of the proof of Theorem 1

Claim 1: $v_{n+l} \notin C$ if $v_i \neq v_n$.

If not, since $|C_1| = k$, we see that the length of $v_{n+l} v_{n+l-1} \ldots v_n \ldots v_i$ is greater than or equal to $v_{n+l} v_{n+l+1} \ldots v_m \ldots v_i$. Now, if $v_i \in I(v_{n+l}, d_2)$, then by the choice of C_1, $v_{n+l}, v_{n+l+1}, \ldots, v_m, \ldots v_i \in I(v_{n+l}, d_2)$. (Otherwise, let P' be the v_{n+l}, d_2-geodesic containing v_i (See Fig. 1(b)). Then, the subgraph induced by $V(P'_{v_{n+l} v_i}) \cup \{v_{n+l}, \ldots, v_n, \ldots, v_i\}$ contains a cycle in $\mathcal{I}_G(C)$ that includes v_i and has a length less than $|C_1|$, a contradiction).

Since P is a d_1, d_2-geodesic, and $v_i \in V(P)$ implies that the v_i, d_2-path through P is a v_i, d_2-geodesic. Thus, $v_{n+l} v_{n+l+1} \ldots v_m \ldots v_i + P_{v_i d_2}$ is a v_{n+l}, d_2-geodesic. Then $v_{n+l} v_{n+l+1} \ldots v_m \ldots v_n$ and $v_{n+l} v_{n+l-1} \ldots v_n$ have the same length. It follows that $v_{n+l} v_{n+l-1} \ldots v_n + P_{v_n d_2}$ is also a v_{n+l}, d_2-geodesic, which implies $C_1 \subseteq I(v_{n+l}, d_2)$, a contradiction.

Thus, $v_i \notin I(v_{n+l}, d_2)$. If $v_i \notin I(v_{n+l}, d_1)$, then $v_i \in C_3(v_{n+l}, d_1, d_2) \subseteq C$, which is not possible. Thus, $v_i \in I(v_{n+l}, d_1)$. Let Q be a v_{n+l}, d_1-geodesic containing v_i. The choice of C_1 and $v_{n+l}, v_i \in C_1$ and $v_{n+l} v_{n+l-1} \ldots v_n \ldots v_i$ has a length greater than l, which implies that $v_{n+l} v_{n+l+1} v_{n+l+2} \ldots v_m \ldots v_i + Q_{v_i d_1}$ is a v_{n+l}, d_1-geodesic containing v_i. Thus, we can assume that $v_{n+l+1}, v_{n+l+2}, \ldots, v_m, \ldots, v_i \in V(Q)$. Since v_m is the first vertex in $V(P)$ that intersects C_1, $d(v_m, d_1) \leq d(v_i, d_1)$, which implies $v_m = v_i$. Since $v_i \notin I(v_{n+l}, d_2)$, we have $v_i = v_m \in C_3(v_{n+l}, d_1, d_2)$, a contradiction. Thus, $v_{n+l} \notin C$. Similarly, if $v_m \neq v_i$, then $v_{m-l} \notin C$.

Claim 2: $v_m = v_i = v_n$

If not, either $v_{n+l} \notin C$ or $v_{m-l} \notin C$. Let $v_{n+l} \notin C$. Since $v_{n+l} \in \mathcal{I}_G(C)$, there exist $d_3, d_4 \in C$ such that $v_{n+l} \in I(d_3, d_4)$. If $I(d_1, d_3)$ and $I(d_1, d_4)$ do not contain v_{n+l}, then $v_{n+l} \in C_3(d_1, d_3, d_4)$, implying $v_{n+l} \in C$, which is not possible. Thus, $I(d_1, d_3)$ or $I(d_1, d_4)$ contains v_{n+l}. Let $I(d_1, d_3)$ contain v_{n+l}, and let P_1 be a d_1, d_3-geodesic containing v_{n+l}. Since $v_{n+l} \notin C_3(d_1, d_2, d_3)$, $v_{n+l} \in I(d_3, d_2)$. Let P_2 be a d_3, d_2-geodesic containing v_{n+l}. Now, either P_1 contains v_i or P_2 contains v_i, since $v_i \notin C_3(d_1, d_2, d_3)$. Let $v_i \in V(P_1)$. Then $v_{n+l}, v_{n+l+1}, \ldots, v_m, \ldots, v_i \in V(P_1)$, by the choice of C_1 implying that $v_i = v_m$. If $v_i \notin V(P_2)$, then $v_i \in C_3(d_1, d_2, d_3)$, a contradiction. If $v_i \in V(P_2)$, then $P_{2d_3 v_i} + P_{v_i d_2}$ is a d_3, d_2-geodesic. Then $P_{2d_3 v_{n+l}} + v_{n+l-1} \ldots v_n + P_{v_n d_2}$ is a d_3, d_2-geodesic by choice of C_1. Then $C_1 \subseteq C_3(d_2, d_3, d_3)$. Thus, $v_i = v_m = v_n$.

If $v_{n-l} \in C$ and $v_{n+l} \in C$, then $v_n \in I(d_1, v_{n-l})$ and $v_i \in I(d_2, v_{n-l})$, since $v_i \notin C_3(v_{n-l}, d_1, d_2)$. Similarly, $v_n \in I(v_{n+l}, d_1)$ and $v_n \in I(v_{n+l}, d_2)$. Then $v_n \in C_3(v_{n+l}, v_{n-l}, d_1)$, a contradiction. Thus, $v_{n+l} \notin C$ or $v_{n-l} \notin C$. Let $v_{n+l} \notin C$ and let $v_{n+l} \in I(d_3, d_4)$. Let Q be the d_3, d_4-path that contains v_{n+l}. Since $v_{n+l} \notin C_3(d_2, d_3, d_4)$, $v_{n+l} \in I(d_2, d_3)$ or $v_{n+l} \in I(d_2, d_4)$. Let $v_{n+l} \in I(d_2, d_3)$. Then $v_{n+l} \in I(d_1, d_3)$ since $v_{n+l} \notin C_3(d_1, d_2, d_3)$. Since $v_n \notin C_3(d_1, d_2, d_3)$, either $v_n \in I(d_1, d_3)$ or $v_n \in I(d_2, d_3)$. Let $v_n \in I(d_2, d_3)$. Then $v_{n+l}, v_{n+l-1}, \ldots, v_n \in I(d_2, d_3))$. Now replace P by the d_3, d_2-geodesic P' that contains the vertices $v_{n+l}, v_{n+l-1}, \ldots, v_n$. Here v_{n+l} and v_n are the first and last vertices of Q that intersect C_1. Then by Claims 1 and 2 to the path P' (with v_{n+l} and v_n in place of v_m and v_n), we have $v_n = v_{n+l}$, a final contradiction. Thus, C is $\mathcal{C}$ convex. $\square$

3   The Cycle Transit Function $C_2(u, v)$

It may be noted that when the cycles are contained in the interval $I(u, v)$ of a graph G, then $C_3(u, v, w)$ will reduce to $C_3(u, v, v) = C_3(u, u, v)$ and consist of the vertices of the cycles C in $I(u, v)$, and hence $C_3(u, v, v)$ will reduce to a transit function of arity two. We first consider the 2-ary version of the C_3-transit function, that is, the function $C_3(u, v, v) = C_3(u, u, v)$, denote it as $C_2(u, v)$. The main aim of this section is to prove that a first-order axiomatic characterization of the function $C_2(u, v)$ of a special case is possible, similar to the characterization of $I(u, v)$ provided by Mulder and Nebeský in [10].

Formally, define the *cycle transit function* $C_2(u, v)$ as $C_2(u, v) = \{u, v\} \cup \{w : w \in I(x, y)$ where $x, y \in I(u, v)$ and $I(x, y)$ is a cycle $\}$.

If $I(u, v)$ does not contain any cycle, then $C_2(u, v) = \{u, v\}$. Clearly, C_2 is a transit function derived from I. The function C_2 is introduced in [7], where some of its properties of betweenness are studied. Note that the convexity induced by C_2 and C_3 is different. For example, in Fig. 2, consider the set $W = \{u, v, x, y\}$. The convex hull of W with respect to the C_2-transit function is $W = \{u, v, x, y\}$ itself, because the intervals $I(a, b), a, b \in W$ do not contain any cycle. Now, from the definition of the C_3-function, we observe that the convex hull of W associated with the C_3-transit function is $V(G)$. That is, W is a convex set with respect to the C_2-transit function, and W is not a convex set associated with the C_3-transit function.

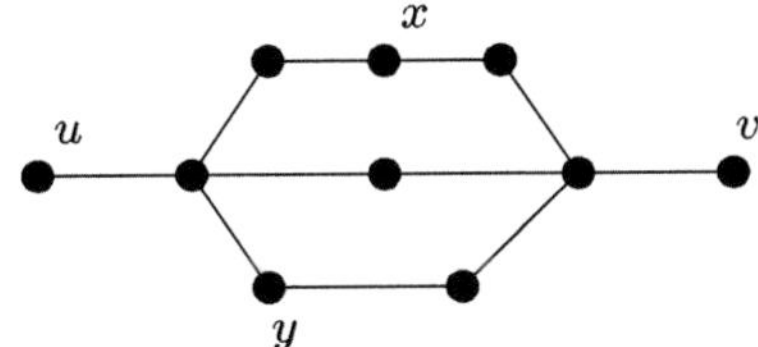

Fig. 2. Example of a graph G showing that the convexity with respect to C_2 and C_3 are not the same.

Mulder and Nebeský in [10] obtained an axiomatic characterization of the interval function $I(u, v)$ in terms of a set of first-order axioms defined on an arbitrary transit function. The following five axioms, namely, the transit axioms $(t1)$, $(t2)$, and $(b2)$, $(b3)$, and $(b4)$ are essential in all these characterizations. The axiom $(b2)$ was already defined in the introductory section. The axioms $(b3)$ and $(b4)$ are defined as

    $(b3)$ If $x \in R(u, v)$ and $y \in R(u, x)$, then $x \in R(y, v)$.

    $(b4)$ If $x \in R(u, v)$, then $R(u, x) \cap R(x, v) = \{x\}$.

We quote two more axioms and the Mulder-Nebeský Theorem from [10] characterizing the interval function of a connected graph.

$(s1)$ If $R(u, \bar{u}) = \{u, \bar{u}\}, R(v, \bar{v}) = \{v, \bar{v}\}, u \in R(\bar{u}, \bar{v})$ and $\bar{u}, \bar{v} \in R(u, v)$, then $v \in R(\bar{u}, \bar{v})$.

$(s2)$ If $R(u, \bar{u}) = \{u, \bar{u}\}, R(v, \bar{v}) = \{v, \bar{v}\}, \bar{u} \in R(u, v), v \notin R(\bar{u}, \bar{v}), \bar{v} \notin R(u, v)$, then $\bar{u} \in R(u, \bar{v})$.

A graph G is called a *thick graph* if, for every u, v with $d(u, v) = 2$, there exist at least two shortest paths between u and v. The following proposition and theorem were proved in [7].

Proposition 1. *[7] Let G be a connected graph. Then the C_2-transit function satisfies the axioms $(b1)$, $(b2)$, $(b4)$ and $(s1)$.*

Theorem 2. *[7] Let G be a connected graph. Then $C_2(u, v) = I(u, v)$ for all $u, v \in V(G)$ if and only if G is a thick graph.*

Consider the axiom (th) for the characterization of the $C_2(u, v)$ of a thick graph. From Theorem 2, it is clear that I and C_2 coincide for a thick graph. The following theorem characterizes the cycle transit function of a thick graph.

(th) If $x \in R(u, v), x \neq u, x \neq v$, then $R(u, x) \cup R(x, v) \subset R(u, v), \forall u, v \in V$.

Theorem 3. *If G is either a thick graph or a geodetic graph, then the cycle transit function C_2 of G satisfies the axiom (th).*

Proof. Let G be a thick graph or a geodetic graph. If G is a geodetic graph, then $C_2(u, v) = \{u, v\}$, for all u, v so that C_2 trivially satisfies the axiom (th) on G. Now assume that G is a thick graph and let u and v be two vertices of G with $d(u, v) = n$. Let $x \in C_2(u, v), x \neq u, x \neq v$. Let

$P : u_0u_1u_2 \ldots u_{k-1}u_ku_{k+1} \ldots u_n$ where $u_0 = u, u_n = v$, and $u_k = x$ be a u, v-shortest path between u and v containing x. Since $d(u_{k-1}, u_{k+1}) = 2$, there exist two shortest paths between them. Let $u_{k-1}wu_{k+1}$ be another shortest path between u_{k-1} and u_{k+1}. Then $w \notin C_2(u, x)$ and $w \notin C_2(x, v)$ but $w \in C_2(u, v)$. That is $C_2(u, x) \cup C_2(x, v) \neq C_2(u, v)$, since C_2 satisfies the axiom $(b2)$, which implies that $C_2(u, x) \cup C_2(x, v) \subseteq C_2(u, v)$, hence C_2 satisfies the axiom (th). $\square$

Theorem 4. *Let G be a graph and I be the interval function on G. Then I satisfies the axiom (th) if and only if G is a thick graph.*

Proof. First, assume that G is a thick graph. We have to show that I satisfies the axiom (th). Suppose $x \in I(u, v)$ with $d(u, v) = n$ and $x \neq u$ and $x \neq v$. Then $I(u, x) \subseteq I(u, v)$ and $I(x, v) \subseteq I(u, v)$, since I satisfies the axiom $(b2)$. Let $P : u_1u_2 \ldots u_{k-1}u_ku_{k+1} \ldots u_n$ where $u_0 = u, u_n = v$, and $u_k = x$ be a u, v-shortest path between u and v. Since $d(u_{k-1}, u_{k+1}) = 2$, there exist two shortest paths between them. Let $u_{k-1}wu_{k+1}$ be another shortest path between u_{k-1} and u_{k+1}. Then $w \notin I(u, x)$ and $w \notin I(x, v)$ but $w \in I(u, v)$. That is $I(u, x) \cup I(x, v) \neq I(u, v)$ and therefore by $(b2)$, $I(u, x) \cup I(x, v) \subset I(u, v)$.

Conversely, suppose that C_2 satisfies the axiom (th). We have to show that G is a thick graph. Assume G is not a thick graph so that $u, v \in V(G)$ are such that $d(u, v) = 2$ and there is only one shortest path between u and v, say $P : uxv$. Then $I(u, x) = \{u, x\}$, $I(x, v) = \{x, v\}$ and $I(u, v) = \{u, x, v\}$. That is $I(u, x) \cup I(x, v) = I(u, v)$ and therefore I does not satisfy axiom (th). $\square$

If a transit function R satisfies the axiom (th), then $R(u, x) \subset R(u, v)$ for $x \in R(u, v)$, $x \neq u$ and $x \neq v$ and $R(u, x) = R(u, v)$ for $x = v$. That is, if R satisfies the axiom (th), then R satisfies the axiom $(b2)$. Then we have the following theorem for characterizing the interval function of thick graphs. The proof is obvious since the axiom (th) implies the axiom $(b2)$. Also, if R satisfies the axioms $(b2)$, $(b3)$, $(s1)$ and $(s2)$ then $R = I$ and by Theorem 4, G_R is a thick graph

Theorem 5. *Let R be a transit function defined on the set V. Then R satisfies the axiom $(b3)$, $(s1)$, $(s2)$ and (th) if and only if $R = I$ and G_R is a thick graph.*

Taking together Theorems 2 and 5, we obtain a first-order axiomatic characterization of the function C_2 of thick graphs.

Theorem 6. *Let R be a transit function defined on V. Then R satisfies the axiom $(b3)$, $(s1)$, $(s2)$ and (th) if and only if $R = C_{2_{G_R}}$ and G_{C_2} is a thick graph.*

The following example establishes the independence of the axioms $(b3)$, $(s1)$, $(s2)$ and (th). In all the examples, we have $R(a, a) = \{a\}$ for all $a \in V$.

Example 1 $((b3)$, $(s1)$, $(s2)$, but not (th) $)$.

Let $V = \{u, v, x, y, z\}$ and define a transit function R on V as follows: $R(u, v) = V$, $R(u, y) = \{u, x, y\}$, $R(u, z) = \{u, x, z\}$, $R(x, v) = R(y, z) = \{x, y, z, v\}$, $R(a, b) = \{a, b\}$ for all other $a, b \in V$. It is straightforward to see that R satisfies axioms $(b3)$, $(s1)$, $(s2)$. In additions $x \in R(u, y), x \neq u, x \neq y$ but $R(u, x) \cup R(x, v) = R(u, v)$ hence R does not satisfies axiom (th).

Example 2 ((th), (s1), (s2), but not (b3)).

Let $V = \{u, v, x, y, w, z\}$ and define a transit function R on V as follows: $R(u, v) = V$, $R(x, v) = \{x, y, w, v\}$, $R(y, w) = \{y, u, x, v, w\}$, $R(y, z) = \{y, u, x, v, z\}$, $R(a, b) = \{a, b\}$ for all other $a, b \in V$. It is straightforward to see that R satisfies axioms (th), $(s1)$, $(s2)$. In additions $x \in R(u, v)$, $y \in R(x, v)$ but $x \notin R(u, y)$ hence R does not satisfies axiom $(b3)$.

Example 3 ((th), (b3), (s2) but not (s1)).

Let $V = \{u, v, x, y, w\}$ and define a transit function R on V as follows: $R(u, v) = R(x, y) = \{u, v, x, y\}$, $R(u, w) = \{u, x, y, w\}$, $R(v, w) = \{v, x, y, w\}$, $R(a, b) = \{a, b\}$ for all other $a, b \in V$. It is straightforward to see that R satisfies axioms (th), $(s2)$, $(b3)$. In additions $R(u, x) = \{u, x\}$, $R(y, w) = \{y, w\}$, $x, y \in R(u, w)$, $u \in R(x, y)$ but $w \notin R(x, y)$ hence R does not satisfies axiom $(s1)$.

Example 4 ((th), (b3), (s1) but not (s2)).

Let $V = \{u, v, x, y, w\}$ and define a transit function R on V as follows: $R(u, v) = R(x, w) = \{u, v, x, w\}$, $R(a, b) = \{a, b\}$ for all other $a, b \in V$. It is straightforward to see that R satisfies axioms (th), $(b3)$, $(s1)$. In additions $R(u, x) = \{u, x\}$, $R(v, y) = \{v, y\}$, $x \in R(u, v)$, $v \notin R(x, y)$, $y \notin R(u, v)$ but $x \notin R(u, y)$ hence R does not satisfies axiom $(s2)$.

4   The C_3-Function and the I_3-Function

This section aims to understand the betweenness properties of the C_3-function and the I_3-function along the same lines as the 3-Steiner function studied in [5]. Unlike the 3-Steiner function, we prove that the betweenness axioms $(b2)$, $(b2(2))$, and (m) are equivalent for the C_3-function and the I_3-function. We have the following interesting theorem.

Theorem 7. *The following statements are equivalent for the functions C_3 and I_3 of a graph G.*

(i) I_3-function satisfies the axiom $(b2)$.
(ii) I_3-function satisfies the axiom $(b2(2))$.
(iii) I_3-function satisfies the axiom (m).
(iv) C_3-function satisfies axiom $(b2)$.
(v) C_3-function satisfies axiom $(b2(2))$.
(vi) C_3-function satisfies axiom (m).

Proof. **(i)** $\Rightarrow$ **(ii)**: Suppose that I_3 satisfies the axiom $(b2)$, we have to show that I_3 satisfies the axiom $(b2(2))$. For this, let $x, y \in I_3(u, v, w)$. Since $x \in I_3(u, v, w)$ implies that $I_3(x, v, w) \subseteq I_3(u, v, w)$, $I_3(u, x, w) \subseteq I_3(u, v, w)$, and $I_3(u, v, x) \subseteq I_3(u, v, w)$ by the axiom $(b2)$. Also, $y \in I_3(u, v, w)$ implies that y is in $I(u, v)$ or $I(u, w)$ or $I(v, w)$ and this means that y is in $I_3(x, v, w)$ or $I_3(u, x, w)$ or $I_3(u, v, x)$. However, we may assume that $y \in I_3(x, v, w) = I_3(v, x, w)$. Then by axiom $(b2)$, $I_3(y, x, w) \subseteq I_3(x, v, w) \subseteq I_3(u, v, w)$, and $I_3(y, x, v) \subseteq I_3(x, v, w) \subseteq I_3(u, v, w)$. Now it remains to show that $I_3(y, x, u) \subseteq I_3(u, v, w)$, for that we need the following. Since $I_3(y, x, w) \subseteq I_3(u, v, w)$, it is clear that $I(x, y) \subseteq I_3(u, v, w)$. Also, $y \in I_3(u, v, w)$ implies that $I_3(y, u, w) \subseteq I_3(u, v, w)$ and that means $I(u, y) \subseteq I_3(u, v, w)$. Now $I_3(x, u, v) \subseteq I_3(u, v, w)$ implies that $I(u, x) \subseteq I_3(u, v, w)$. That is $I(x, y) \cup I(x, u) \cup I(y, u) \subseteq I_3(u, v, w)$ so that $I_3(x, y, u) \subseteq I_3(u, v, w)$. Thus, I_3 satisfies $(b2(2))$.

(ii) $\Rightarrow$ **(iii)**: Suppose that I_3 satisfies the axiom $(b2(2))$, we have to show that I_3 satisfies the axiom (m). For this, let $x, y, z \in I_3(u, v, w)$. Since $x, y, z \in I_3(u, v, w)$ implies by $(b2(2))$ that $I_3(x, y, w) \subseteq I_3(u, v, w)$, $I_3(x, z, w) \subseteq I_3(u, v, w)$ and $I_3(y, z, w) \subseteq I_3(u, v, w)$. Hence, the union, $I_3(x, y, w) \cup I_3(x, z, w) \cup I_3(y, z, w) \subseteq I_3(u, v, w)$. This implies that $I(x, y) \cup I(y, z) \cup I(x, z) \subseteq I_3(u, v, w)$ so that $I_3(x, y, z) \subseteq I_3(u, v, w)$.

(iii) $\Rightarrow$ **(iv)**: Suppose $x \in C_3(u, v, w)$ and $y \in C_3(x, v, w)$. Then $y \in I_3(x, v, w) \subseteq I_3(u, v, w)$. This means that $y \in C_3(u, v, w)$ implies $C_3(x, v, w) \subseteq C_3(u, v, w)$ and, therefore, C_3 satisfies $(b2)$.

(iv) $\Rightarrow$ **(iii)**: Assume that C_3 satisfies $(b2)$. We have to show that I_3 satisfies the axiom $(b2)$, for this, let $x \in I_3(u, v, w)$ be with $x \notin \{u, v, w\}$. To prove that $I_3(x, v, w) \subseteq I_3(u, v, w)$. We may assume that $x \in I(u, v)$, then $I(u, x) \subseteq I(u, v)$ and $I(v, x) \subseteq I(u, v)$ since I always satisfies $(b2)$. To prove $I_3(x, v, w) \subseteq I_3(u, v, w)$, we have to show that $I(x, v) \cup I(x, w) \cup I(v, w) \subseteq I_3(u, v, w)$. Since both $I(x, v)$ and $I(v, w)$ are contained in $I_3(u, v, w)$, we have to prove $I(x, w) \subseteq I_3(u, v, w)$. Assuming $y \in I(x, w)$, we have to prove that $y \in I_3(u, v, w)$. If y is in either $I(u, v)$ or $I(u, w)$ or $I(v, w)$, then $y \in I_3(u, v, w)$. So, assume $y \notin I(u, v) \cup I(u, w) \cup I(v, w)$. Now we have $x \in I(u, v)$, $y \in I(x, w)$ and $y \notin I(u, w)$. That is, we have a u, x-geodesic say P, x, w-geodesic say Q containing y and a u, w-geodesic R does not contain y. Clearly, the path R does not contain both x and y. Then the vertices on the u, x-subpath of P, Q, and R form a cycle, say C, and y belong to the cycle C. That is $y \in C_3(x, v, w)$, then $y \in C_3(x, v, w) \subseteq C_3(u, v, w) \subseteq I_3(u, v, w)$. Similarly, we can prove $I_3(u, x, w) \subseteq I_3(u, v, w)$ and $I_3(u, v, x) \subseteq I_3(u, v, w)$. Hence I_3 satisfies $(b2)$. Then it follows that I_3 satisfies (m) from $(i) \Rightarrow (ii)$ and $(ii) \Rightarrow (iii)$.

(iv) $\Rightarrow$ **(v)**: Suppose that C_3 satisfies $(b2)$. This implies that I_3 satisfies (m) (by case $(iv) \Rightarrow (iii)$, which implies that I_3 satisfies $(b2(2))$. Assume $x, y \in C_3(u, v, w)$ and $z \in C_3(x, y, w) \subseteq I_3(x, y, w) \subseteq I_3(u, v, w)$. This means that $z \in C_3(u, v, w)$. Hence C_3 satisfies $(b2(2))$.

(v) $\Rightarrow$ **(vi)**: Suppose that C_3 satisfies $(b2(2))$, then C_3 satisfies $(b2)$ implies I_3 satisfies (m) (by $(iv) \Rightarrow (iii)$). Assume $x, y, z \in C_3(u, v, w)$ and $t \in C_3(x, y, z) \subseteq I_3(x, y, z) \subseteq I_3(u, v, w)$. This means that $t \in C_3(u, v, w)$. Hence C_3 satisfies (m).

(vi) $\Rightarrow$ **(i)** Suppose C_3 satisfies (m), then C_3 satisfies $(b2)$ and hence I_3 satisfies $(b2)$ (by case $(iv) \Rightarrow (iii)$).

We have proved that all six conditions of the theorem are equivalent, and hence the theorem follows. $\qquad\square$

Now, let us look at some of the classes of graphs that satisfy Theorem 7. We denote the family of graphs that satisfy the conditions of Theorem 7 as $\mathcal{G}_{I_3}$.

Proposition 2. *The following classes of graphs belong to $\mathcal{G}_{I_3}$.*

1. *Ptolemaic graphs*
2. *Grids*
3. *Cycles*
4. *Wheel graphs and fan graphs*

Proof. 1. To prove that the class of Ptolemaic graphs belongs to the family $\mathcal{G}_{I_3}$, we show that the function I_3 satisfies axiom $(b2)$ on a Ptolemaic graph. So, let G be a Ptolemaic graph. For, $u, v, w, x, y \in V(G)$, let $x \in I_3(u, v, w)$ and $y \in I_3(x, v, w)$. Since x is in $I(u, v)$, $I(u, w)$, or $I(v, w)$, we can assume that $x \in I(u, v)$ and hence $I(v, x) \subseteq I_3(u, v, w)$ and $I(v, w) \subseteq I_3(u, v, w)$. So $y \in I(u, v, w)$ whenever $y \in I(v, x)$ or $y \in I(v, w)$. We have to show that $y \in I(u, v, w)$ whenever $x \in I(u, v)$ and $y \in I(x, w)$. If $x = u$ or $x = v$, then $y \in I(u, w)$ or $y \in I(v, w)$. So $x \neq u$ and $x \neq v$. Similarly, if $y = x$ or $y = w$, then $y \in I(u, v)$ or $y \in I(u, w)$ or $y \in I(v, w)$. So $y \neq x$ and $y \neq w$. If possible assume $y \notin I_3(u, v, w)$. That is, $x \in I(u, v)$, $y \in I(x, w)$, $y \notin I(u, v)$, $y \notin I(u, w)$ and $y \notin I(v, w)$. We may choose vertices x, y, u, v, w such that $d(x, w)$ is minimal. Suppose P is a u, v-geodesic containing x and Q, is a x, w-geodesic containing y. Since $y \notin I(u, v)$, y is not on P. Similarly, $y \notin I(u, w)$ and $y \notin I(v, w)$ imply the existence of a u, w-geodesic R which does not contain y and a v, w-geodesic S that also avoids y. If $\ell(P) = 2$, $\ell(R) = 1$ and $\ell(S) = 1$, then the vertices in the paths P, R and S induces a cycle of length four. If $\ell(P) = 2$, $\ell(R) = 1$ and $\ell(S) = 2$ then the vertices in the paths P, R and S forms a cycle of length five. But there may be chords from P to S. Let v' be the vertex between v and w in the path S. The vertices in the paths P, R and S induce a cycle of length five if there are no chords from P to S, other than uw, induces a house if either one of $uv' \in E(G)$ or $xv' \in E(G)$ and induces a fan graph if both $uv' \in E(G)$ and $xv' \in E(G)$. If $\ell(P) \geq 2$, $\ell(R) > 1$, and $\ell(S) \geq 2$, vertices in the paths P, Q, R, and S form a cycle of length at least six and will induce a house, hole, domino, or fan graph. Hence I_3 satisfies axiom $(b2)$ on Ptolemaic graphs.

2. Let G be a grid graph with mn vertices formed by the Cartesian product of the paths $P : u_1 u_2 \ldots u_a$ and $Q : v_1 v_2 \ldots v_b$. Then the vertices of G are (u_i, v_j), where $i \in 1, 2, \ldots, m$ and $j \in 1, 2, \ldots, n$. Suppose that I_3 does not satisfy b_2. Then there exist $u = (u_i, v_j)$, $v = (u_m, v_n)$, $w = (u_r, v_s)$, $x = (u_k, v_l)$, and $y = (u_p, v_q)$ such that $b \in I(u, v)$ and $y \in I(x, w)$, but $y \notin I(u, v, w)$. We have $d(u, v) = d(u_i, u_m) + d(v_j, v_n)$, $d(u, x) = d(u_i, u_k) + d(v_j, v_l)$, and $d(x, v) = d(u_m, u_k) + d(v_n, v_l)$. From these three equalities and since d is a metric, we obtain $d(u_i, u_m) = d(u_i, u_k) + d(u_m, u_k)$ and $d(v_j, v_n) = d(v_j, v_l) + d(v_n, v_l)$. Thus, $u_k \in I(u_i, u_m)$ and $v_l \in I(v_j, v_n)$. Similarly, we obtain $u_p \in I(u_k, u_r)$ and $v_q \in I(v_l, v_s)$. Since $y \notin I(u, w)$, we have $u_p \notin I(u_i, u_r)$ or $v_q \notin I(v_j, v_s)$, using similar argument. If $u_k \in I(u_i, u_r)$, then $I(u_k, u_r) \subseteq I(u_i, u_r)$, which implies $u_p \in I(u_i, u_r)$, a contradiction. Thus, $u_k \in I(u_r, u_m)$, which implies $p \in I(u_k, u_r) \subseteq I(u_r, u_m)$. If $u_r \in I(u_i, u_k)$, then $u_p \in I(u_i, u_m)$. If $i \in I(u_r, u_k)$, then since $p \notin I(u_i, u_r)$, which implies $u_p \in I(u_i, u_k) \subseteq I(u_i, u_m)$. Hence, in both cases, $u_p \in I(u_i, u_m) \cap I(u_r, u_m)$. Now, $v_q \notin I(v_j, v_n)$ since $u_p \in I(u_i, u_m)$ and $y \notin I(u, v)$. It follows that $v_s \notin I(v_j, v_n)$, since $v_q \in$

$I(v_l, v_s)$. Thus, either $v_n \in I(v_l, v_s)$ or $v_j \in I(v_s, v_l)$. If $v_n \in I(v_l, v_s)$, then $v_q \notin I(v_l, v_n)$ since $v_q \notin I(v_j, v_n)$, which implies $v_q \in I(v_n, v_s)$. If $v_j \in I(v_s, v_l)$, then $v_l \in I(v_n, v_s)$ and $v_q \in I(v_j, v_n)$. In both cases, we obtain $y \in I(v, w)$, a contradiction. Thus, I_3 satisfy b_2 on G.

3. Let $x \in I(u, v)$, $y \in I(x, w)$ in a cycle graph G. Assume the u, v-geodesic containing x be P. If y is on the path P, then $y \in I(u, v)$. So we may assume that y is not on the path P. Then the x, w-geodesic containing y contains either u or v. If x, w-geodesic contain u, then $y \in I(u, w)$ and if x, w-geodesic contain v, then $y \in I(v, w)$. So $y \in I_3(u, v, w)$ in either cases.

4. Suppose G is a wheel graph or a fan graph and $x \in I(u, v)$, $y \in I(x, w)$. If x is a central vertex, then y will be either x or w, and in both cases, $y \in I_3(u, v, w)$. So, assume that x is not the central vertex, then $x \in I(u, v)$ means that u and v are at a distance 2 and $y \in I(x, w)$ means that y is the central vertex. Hence $y \in I(u, v)$ and $y \in I_3(u, v, w)$. $\square$

Remark 1. Although odd cycles and block graphs belong to the family $\mathcal{G}_{I_3}$, it is not true that every geodetic graph or chordal graph belongs to $\mathcal{G}_{I_3}$, as shown in Fig. 3. The graph (a) in Fig. 3 is geodetic; however, $x \in I_3(u, v, w)$ and $y \in I_3(x, v, w)$, but $y \notin I_3(u, v, w)$ implies that $(b2)$ does not hold for I_3. The graph (b) in Fig. 3 is chordal; however, $u_3 \in I_3(u_1, u_4, u_6)$ and $u_9 \in I_3(u_3, u_4, u_6)$, but $u_9 \notin I_3(u_1, u_4, u_6)$, implies that $(b2)$ does not hold for I_3.

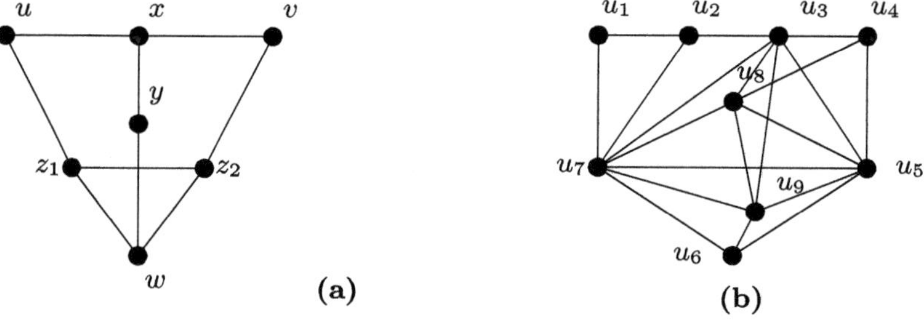

Fig. 3. Example for Remark 1.

Concluding Remarks: Finally, we note that the symmetry required of an n-ary transit function R allows for an alternative and equivalent formulation. Instead of viewing R as acting over an ordered n-tuples, one may regard R as a function from $P(V, n)$ to 2^V where $P(V, n)$ denotes the set of all subsets of V of cardinality at most n. Then, the n-ary transit function R can be defined as a function that meets the following two conditions.

$(t1')$: $U \subseteq R(U)$, for all $U \in P(V, n)$
$(t3')$: $R(\{u\}) = \{u\}$ for all $u \in V$

Equivalently, the axiom $(b2(i))$ can be defined as follows.

$(b2(i))$: For all $U_i \subseteq R(U)$, $|U_i| = i \geq 1$, $R((U \setminus X_i) \cup U_i) \subseteq R(U)$, for every subset $X_i \subseteq U$ with $|X_i| = i$ and $U \in P(V, n)$. (Note that $(b2(1) = b2)$.

It may be noted that with this formulation of the function R, all the results of the Sects. 2 and 4 hold.

Dedication : We dedicate this work to the memory of Norbert Polat, who did pioneering work in metric graph theory (Polat passed away in July 2025), as his work on cycle convexity is the basis of this work.

Acknowledgement. L.K.K.S. acknowledges financial support from the University of Kerala (U.O. No. Ac.EVII/2630/2025/UOK). MC acknowledges DST (ANRF), Government of India (Grant No. DST/INT/DAAD/P-03/2023 (G)). A.K. acknowledges University Grants Commission of India for JRF(NTA.Ref.no 191620026270)

References

1. Brešar, B., et al.: Steiner intervals, geodesic intervals, and betweenness. Discrete Math. **309**, 6114–6125 (2009)
2. Changat, M., Mulder, H.M., Sierksma, G.: Convexities related to path properties on graphs. Discrete Math. **290**, 117–131 (2005)
3. Changat, M., Narasimha-Shenoi, P.G., Pelayo, I.M.: The longest path transit function and betweenness. Util. Math. **82**, 111–127 (2010)
4. Changat, M., Mathews, J., Peterin, I., Narasimha-Shenoi, P.G.: n-ary transit functions in graphs. Discuss. Math. Graph Theory **30**, 671–685 (2010)
5. Changat, M., et al.: A note on 3-Steiner intervals and betweenness. Discrete Math. **311**(22) 2601–2609 (2011)
6. Duchet, P.: Discrete convexity: retractions, morphisms and partition problem. In: Proceedings of the conference on graph connections, India, pp. 10–18. Allied Publishers, New Delhi (1998)
7. Kamal, K.S.L., Changat, M., Paily, A.: Cycle transit function and betweenness. Commun. Combinatorics Optim. **8**(4), 725–735 (2023)
8. Mulder, H.M.: The interval function of a graph. Mathematical Centre Tracts 132, Mathematisch Centrum, Amsterdam (1980)
9. Mulder, H.M.: Transit functions on graphs (and posets). In: Convexity in Discrete Structures (Changat, M., Klavžar, S., Mulder, H.M., Vijayakumar, A., eds.) Lecture Notes Ser. 5, Ramanujan Math. Soc., pp. 117–130 (2008)
10. Mulder, H.M., Nebeský, L.: Axiomatic characterization of the interval function of a graph. Eur. J. Combin. **30**, 1172–1185 (2009)
11. Nebeský, L.: A characterization of the interval function of a connected graph. Czechoslovak Math. J. **44**(119), 173–178 (1994)
12. Nebeský, L.: A Characterization of the interval function of a (finite or infinite) connected graph. Czechoslovak Math. J. **51**(126), 635–642 (2001)
13. Polat, N.: Netlike partial cubes I. General properties. Discrete Math. **307**, 2704–2722 (2007). https://doi.org/10.1016/j.disc.2007.01.018
14. van de Vel, M.L.J.: Theory of Convex Structures. North Holland, Amsterdam (1993)

A New Characterization of VNP via Colored Determinant

Prasad Chaugule$^{(\boxtimes)}$

Indian Institute of Technology Delhi, Delhi, India
`chaugule@cse.iitd.ac.in`

Abstract. The main theme of this work is to provide a new characterization of the algebraic complexity class VNP. We achieve this via following the two main steps -
- We define a combinatorial variant of the determinant polynomial which we call the colored determinant and show it to be VNP-complete under p-projections for all fields. This adds a new non-monotone VNP-complete polynomial sequence to the already known list of non-monotone VNP-complete polynomial sequences [1].
- Using the colored determinant polynomial, we show that a new variant of stack branching program, called the conditional stack branching program, characterizes the class VNP.

Keywords: Colored Determinant · VNP · Algebraic Complexity

1 Introduction

The main goal of algebraic complexity is to separate the complexity classes VP and VNP. Therefore, to understand these classes better, it is important to study them also through a lens different from the classical definitions of these classes. One well-studied and successful way to characterize a given algebraic class is to study a polynomial sequence complete for that class. For example, the determinant sequence is known to almost characterize the class VP[8] and the permanent sequence is known to characterize the class VNP[8]. Moreover, there are also variants of the determinant sequence known to characterize the classes VP and VNP [2].

Inspired by the study of Non-Auxiliary Pushdown automata (NAuxPDA), Mengel initiated the study of an algebraic branching program appended with memory and gave new characterizations of VP and VNP. Mengel proved that the models *stack branching program* and *random access branching program* characterize the classes VP and VNP respectively [5]. Recently, this line of work was further investigated by S. Chillara and N. Raja [3]. In [5], a *stack branching program* is defined as an algebraic branching program with a single stack memory. In [3], a *stack branching program* with two or more stacks is investigated and shown to characterize the class VNP; in other words, it is shown that adding at

N. Misra and A. Pandey (Eds.): CALDAM 2026, LNCS 16445, pp. 120–134, 2026.
https://doi.org/10.1007/978-3-032-17156-6_10

least two stacks to an algebraic branching program is sufficient to blow up its power all up to VNP.

In this work, we study the trade-off question between the power given to the model to access the stack elements to decide the stack operation to be executed along a given edge and the number of stacks appended to the branching program. We study a new model of computation called the *conditional stack branching program* (with one stack) and show that it characterizes VNP. Our work shows that even a single stack is sufficient to blow the power of an algebraic branching program all up to VNP, given access to the topmost element of the stack to decide the stack operation associated with an edge. We believe that the importance of studying such a model is twofold: one, to give a new characterization of VNP , and the other to investigate and get meaningful insights in the context of the said trade-off. To achieve our result, we show that a variant of determinant, called the colored determinant $\mathsf{ColDet}_n(X)$ is VNP-complete under p-projections for all fields. Moreover, we show that the colored determinant polynomial, $\mathsf{ColDet}_n(X)$ can be computed by a conditional stack branching program (with one stack) of size polynomially bounded in n and for any sequence (f_n) where f_n can be computed by a conditional stack branching program (with one stack) of size polynomially bounded in n is in VNP. This establishes a new characterization of the class VNP via colored determinant sequence. Our result is stronger than existing ABP-with-memory models because it captures a strictly richer class of computations under comparable resource bounds. In particular, the conditional stack–based framework allows dependencies and structural constraints that cannot be simulated efficiently in prior memory-augmented ABP models, thereby yielding a more expressive and robust characterization. We refer the readers to Table 1 for more details regarding various algebraic branching program with memory models studied in the literature.

Table 1. Characterizations of VP and VNP. ✓ represents the work in this paper

Class	Characterization	Ref.
VP	Stack Branching Program with one stack (SBP with one stack)	[5]
VNP	Random Access Branching Program (RABP)	[5]
VNP	Stack Branching Program with two stacks (SBP with two stacks)	[3]
VNP	Queue Branching Program (QBP)	[3]
VNP	Conditional Stack Branching Program with one stack (CSBP with one stack)	✓

2 Preliminaries and Related Work

2.1 Algebraic Complexity Classes

In this section, we review the standard definitions of p-bounded polynomial sequences, the computational models studied in algebraic complexity, and the associated complexity classes, as well as the notions of hardness and completeness relevant to our results. We refer the readers to [7] for more details about algebraic complexity.

Definition 1 (p-bounded sequence). *A polynomial sequence (f_n) is called a p-bounded sequence if the number of variables and the degree of f_n are both polynomially bounded in n.*

Definition 2 (Arithmetic circuit and formula). *An arithmetic circuit is a directed acyclic graph (DAG) in which the vertices whose in-degree is 0 are called the input vertices. There is a unique out-degree 0 vertex called the output vertex, and all other vertices are called the internal vertices. The internal vertices are labelled with either $+$ or $\times$. An input gate is labelled with a variable from a set of variables X or a constant from $\mathbb{F}$. We define the polynomial computed by a circuit inductively. An input vertex labelled x_i (or $c \in \mathbb{F}$) is said to compute the polynomial x_i (or c, resp.). Let g be a vertex with inputs g_1, g_2. Let $p_1(X), p_2(X)$ be the polynomials computed by g_1, g_2 respectively. If g is labelled with $\times$ (or $+$) then the polynomial computed by g is simply $p_1(X) \times p_2(X)$ (resp. $p_1(X)+p_2(X)$). The polynomial computed by the circuit is the polynomial computed by the output gate. If the out-degree of every vertex in the circuit is one then such a circuit is called a formula.*

Definition 3 (Algebraic Branching Program (ABP) and layered ABP). *An algebraic branching program (ABP) is a directed acyclic graph $G = (V, E)$ with two special designated nodes, called the source node s and the sink node t. The edges are labelled with variables or field constants. For any s to t path ρ, we use f_ρ to denote the product of the variables or constants labelling the edges in ρ. The polynomial computed by an ABP is $\sum_\rho f_\rho$, where ρ is an s to t path. A layered algebraic branching program is the same as the algebraic branching program except that it is a layered, directed acyclic graph DAG with edges connecting vertices only between adjacent layers.*

Definition 4 (VF, VBP and VP). *A p-bounded polynomial sequence (f_n) is in VP (or VF or VBP, respectively), iff f_n can be computed by an arithmetic circuit (or an arithmetic formula or an algebraic branching program, respectively) of size polynomially bounded in n.*

Definition 5 (The complexity class VNP (or VNP_e, respectively)). *A polynomial sequence (f_n) is said to be in VNP (or, VNP_e, respectively) if there is a polynomial sequence (g_n) in VP (or, (g_n) in VF respectively) such that*

$$f_n(X) = \sum_{y_1 \in \{0,1\},\ldots,y_{m(n)} \in \{0,1\}} g_n(X, Y)$$

Note that $m : \mathbb{N} \to \mathbb{N}$ is polynomially bounded in n.

It is known that $\mathsf{VF} \subseteq \mathsf{VBP} \subseteq \mathsf{VP} \subseteq \mathsf{VNP}$ [6]. But none of the containments are known to be strict. However, it is known that $\mathsf{VNP} = \mathsf{VNP}_e$ [9].

Lemma 1 (Valiant's criterion). *Let $\phi : \{0,1\}^* \to \mathbb{N}$ be a function in* #P/poly, *then the sequence (f_n) defined by*

$$f_n(X) = \sum_{e \in \{0,1\}^n} \phi(n) \times \prod_{i=1}^{n} X_i^{e_i}$$

is in VNP.

Definition 6 (p-projections, Hardness and Completeness). *A polynomial sequence (f_n) is a p-projection of another polynomial sequence (g_n) if there is a polynomially bounded function $m : \mathbb{N} \to \mathbb{N}$ such that for each $n \in \mathbb{N}$, f_n can be obtained from $g_{m(n)}$ by setting its variables to one of the variables of f_n or field constants. It is denoted by $(f_n) \leq_p (g_n)$. A polynomial sequence (g_n) is said to be hard for a class $\mathcal{C}$ if for every sequence $(f_n) \in \mathcal{C}$, $(f_n) \leq_p (g_n)$. A polynomial sequence (g_n) is said to be complete for a class $\mathcal{C}$, if $(g_n) \in \mathcal{C}$ and is hard for $\mathcal{C}$.*

2.2 Related Work

This section provides an overview of the results and models that our work builds upon. We briefly recall the framework of stack-realizable paths and stack branching programs, along with the characterizations of VP and VNP due to S. Mengel [5] and to S. Chillara and N. Raja [3].

Definition 7 (Stack realizable path). *Let Σ be a set of alphabets and k be the number of stacks. Let G be a directed graph with edge label $\phi : E(G) \to \{\mathtt{Push}(\square, j), \mathtt{Pop}(\square, j), \mathtt{No\text{-}op}\}$, where $\square \in \Sigma$ and $j \in [k]$. The label $\mathtt{Push}(\square, j)$ represents the operation of pushing the symbol $\square$ on stack j and the label $\mathtt{Pop}(\square, j)$ represents the operation of popping the symbol $\square$ from the top of the stack. The label $\mathtt{No\text{-}op}$ represents no operation. Let $s, t \in V(G)$. An s to t directed path is a stack-realizable path if and only if the following conditions are satisfied.*

- *We start with an empty stack and perform all the stack operations along the path and end with an empty stack.*
- *There are no faults while executing the stack operations along the path, that is, we do not attempt to pop an element from the stack that is not at the top of the stack.*

We now state the definition of a stack branching program with k stacks.

Definition 8 (Stack Branching Program with k stacks). *Let Σ denote a finite set of symbols. A stack branching program with k stacks is defined as an algebraic branching program $\mathcal{A}$ (having s and t as the source and the sink vertices, respectively) with two kinds of edge labels, the variable label and the stack label, where the variable label is $\phi_v : E(\mathcal{A}) \to \mathbb{F} \cup X$ and the stack label is $\phi_s : E(\mathcal{A}) \to \{\mathtt{Push}(\square, j), \mathtt{Pop}(\square, j), \mathtt{No\text{-}op}\}$, where $\square \in \Sigma$ and $j \in [k]$. The polynomial computed by $\mathcal{A}$ is $\sum_{p \in \mathcal{P}} mon(p)$, where $\mathcal{P}$ denote the set of all stack realizable paths and $mon(p)$ is the monomial formed by mutliplying all the variable labels of the edges in path p. The size of a stack branching program is the number of vertices in it.*

Finally, we restate the characterization results.

Lemma 2 (Characterization of VP and VNP). *[5] [3] A sequence (f_n) is in VP if and only if f_n can be computed by a stack branching program (with one stack) of size polynomially bounded in n. A sequence (f_n) is in VNP if and only if f_n can be computed by a stack branching program (with k stacks) of size polynomially bounded in n, where $k \geq 2$.*

3   Our Work

In this section, we present our contributions. We introduce conditional stack branching programs, define properly colored cycle covers, provide a formal definition of the colored determinant, and then state our main results.

Definition 9 (Conditional Stack Branching Program with one stack). *Let Σ be a sequence of finite set of symbols. A conditional stack branching program is defined as an algebraic branching program $\mathcal{A}$ (with s and t as the source and the sink vertices, respectively) with two kinds of edge labels, the variable label Φ_v and the stack label Φ_s, where the variable label is $\Phi_v : E(\mathcal{A}) \to \mathbb{F} \cup X$ and the stack label Φ_s is of the form*

$$if \quad (top == t) \quad \texttt{Pop}(t) \quad else \quad \texttt{Push}(t)$$

where the top represents the topmost element of the stack. An s to t path is stack realizable if and only if the following condition holds.

- *We start with an empty stack and perform all the stack operations along the path and end with an empty stack.*

The polynomial computed by $\mathcal{A}$ is $\sum_{p \in \mathcal{P}} mon(p)$, where $\mathcal{P}$ denote the set of all stack realizable paths and $mon(p)$ is the monomial formed by mutliplying all the variable labels of the edges in path p.

Definition 10 (cycle cover, colored graph and a properly colored cycle cover). *A cycle cover of a directed graph G is a subgraph of G which is a disjoint union of cycles such that every vertex of G participates in exactly one of the cycles. A graph G is called a colored graph if we associate every edge of G with some color. A properly colored cycle cover of a colored graph G is a cycle cover of G such that each cycle in the cover is monochromatic.*

Definition 11 (Colored Determinant). *Let $\mathcal{H}_n(V, E)$ be a complete directed graph. Let $V(\mathcal{H}_n) = \mathcal{H}_1 \cup \mathcal{H}_2$, where $\mathcal{H}_1 = \{1, 2, \ldots, n\}$ and $\mathcal{H}_2 = \{n + 1, n + 2, \ldots, 2n\}$. Let $E(\mathcal{H}_n) = \{(i, j) | 1 \leq i, j \leq 2n\}$. Let $E(\mathcal{H}_n) = E_1 \cup E_2 \cup E_3$, where E_1 denote the set of edges within the vertices in $\mathcal{H}_1$, the set E_2 denote the set of edges within the vertices in $\mathcal{H}_2$ and the set E_3 denote the set of edges across the vertex sets $\mathcal{H}_1$ and $\mathcal{H}_2$. We color the edges in the sets E_1 and E_2 with color c_r. We color the edges in the set E_3 with c_b. Let $X = \{x_{i,j} | 1 \leq i, j \leq 2n\}$.*

Let $\Phi : E(\mathcal{H}_n) \rightarrow X$, where $\Phi((i,j)) = x_{i,j}$ be the label function. The colored determinant is defined as follows:

$$\mathsf{ColDet}_n = \sum_{c \in \mathcal{C}} (-1)^{n+k} mon(c)$$

where $\mathcal{C}$ is the set of all properly colored cycle covers of G and $mon(c)$ is formed by multiplying the labels on all the edges in c and k is the number of cycles in cycle cover c (see Fig. 1 (left)).

We now state our main results.

Lemma 3. $\mathsf{ColDet}_n(X)$ *is* VNP-*complete for all fields under p-projections.*

Using the Valiant's criterion (Lemma 1), the containment of $\mathsf{ColDet}_n(X)$ in VNP is straightforward. The VNP-hardness proof idea is similar to the classical VNP-hardness proof of permanent by Valiant [8] with minor modifications. The complete proof appears in Sects. 4 and 5.

Lemma 4. *The polynomial sequence* $\mathsf{ColDet}_n(X)$ *can be computed by a conditional stack branching program (with one stack) of size polynomially bounded in n.*

The proof of Lemma 4 is discussed in Sects. 6

Lemma 5. *For a given sequence (f_n), if f_n can be computed by a conditional stack branching programs S_n of size polynomially bounded in n, then $(f_n) \in$* VNP.

The proof of Lemma 5 is presented in Sect. 7

4   Proof of Lemma 3

The proof is a careful modification of the classical VNP-completeness proof of the permanent by Valiant. Section 4.1 discusses the rosette construction, Sect. 4.2 presents the *Determinantal Colored if and only if* gadget, and Sect. 4.3 combines these ingredients to construct a graph T_k. We conclude Sect. 4.3 with a technical lemma (Lemma 7) describing the colored determinant of the graph T_k.

4.1   Rosette Construction $R(n)$ for any odd n[1]

Consider a cycle C_n of length n, where $V(C_n) = \{u_1, u_2, \ldots, u_n\}$ and $E(C_n) = \{e_i = (u_i, u_{i+1}) | 1 \le i \le n-1\} \cup \{e_n = (u_n, u_1)\}$. We call the vertices and the edges in the set $V(C_n)$ and $E(C_n)$ the connector vertices and the connector edges, respectively. For each edge e_k, we add a new vertex α_k. For each $k \in [n-1]$, we add directed edges from u_k to α_k and from α_k to u_{k+1}. We add directed edges

---

[1] In the case of any even n, the above construction works except that the product of the signature and the monomial associated with each cycle cover will be -1 instead of 1. This issue could be handled very easily and therefore we skip the details here.

(u_n, α_n) and (α_n, u_1). We add self-loops on all the vertices. We arbitrarily pick a connector vertex and set the weight of its self-loop to 1. We set the weights of the self-loops on all other vertices to -1. We set the weights of all the remaining edges (other than the self-loops) to 1. Let the color of each edge of $R(n)$ be c_r. This completes the construction of $R(n)$ (see Fig. 1 (middle)). It is trivial to observe the following properties of Rosette $R(n)$.

1. Each cycle cover is a properly colored cycle cover.
2. Let $\phi \neq S \subseteq E(C_n)$, there exists exactly one cycle cover which includes all the edges in S and excludes all the edges in $\bar{S}$.
3. There are exactly two cycle covers which does not use any of the edges in $E(C_n)$. The first one is the cycle cover consisting of a single cycle $(u_1, \alpha_1, u_2, \alpha_2, \ldots, u_n, \alpha_n, u_1)$, and the second one is where every vertex is covered by a self-loop.
4. The total number of properly colored cycle covers is $2^n + 1$.
5. The product of the signature and the weight of the monomial associated with each of the cycle covers in this rosette is the same and is one.

4.2   Determinantal Colored if and only if Gadget $\mathcal{G}$

We now discuss the construction of the determinantal *colored if and only if* gadget. This gadget is in the spirit of the classical Valiant's *if and only if* gadget. Let $\mathcal{G} = (V, E)$ be a colored directed graph with $V(\mathcal{G}) = \{a_1, a_2\}$. Let $E(\mathcal{G}) = \{(a_1, a_2), (a_2, a_1)\}$. We set the weight of (a_1, a_2) as -1 and the weight of (a_2, a_1) as 1. We set the colors of both edges as c_b. This finishes the construction of $\mathcal{G}$.

Placing $\mathcal{G}$ between two edges (u, v) and (u', v'): Let G be a colored directed graph with a pair of distinct edges (u, v) and (u', v') which does not share any common vertex. We assume that all the edges of the graph G are set with a color c_r. We say $\mathcal{G}$ is placed between (u, v) and (u', v'), if we delete edges $(u, v), (u', v')$ in graph G, and add edges in the set $\{(u, a_1), (a_1, v), (u', a_2), (a_2, v')\}$. We set the color of all the newly added edges as c_r. Moreover, we set the weight of edge (u, a_1) as the weight of edge (u, v) and the weight of edge (u', a_2) as the weight of edge (u', v'). We set the weight of all other edges as 1 (see Fig. 1 (right)).

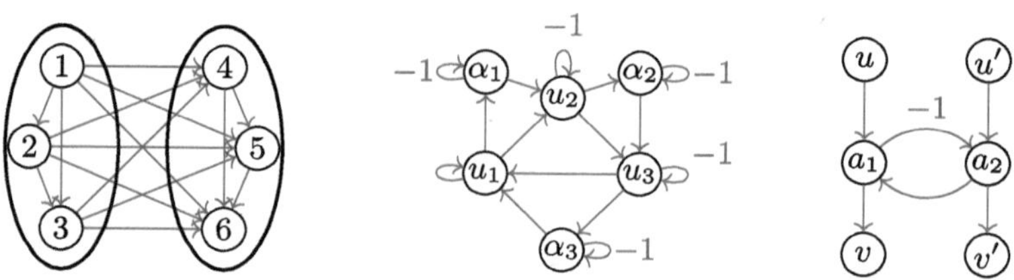

Fig. 1. Graph $\mathcal{H}_3$. To avoid clutter, edge labels are omitted in the figure and only the edges of the form (i, j), where $i < j$ are shown (left), the colored rosette $R(3)$ (middle), the colored determinantal if and only if gadget (right).

We now state our lemma, which describes the properties of the gadget $\mathcal{G}$.

Lemma 6 (Determinantal colored if and only if gadget). *Let G be a directed graph where every edge is colored with c_r. We fix two arbitrary edges (u, v) and (u', v') in graph G that do not share any endpoints. Let $\mathcal{G}$ (with color c_b) is placed between (u, v) and (u', v'). Let us call the new graph G'. Then the following holds.*

- *For any cycle cover C of graph G that uses both the edges (u, v) and (u', v') with signature s and monomial m, there exists a corresponding properly colored cycle cover C' in G' with the signature s' and m' such that $s' = s$ and $m' = m$. In other words, $s \times m = s' \times m'$.*
- *For any cycle cover C of graph that uses none of the edges (u, v) and (u', v') with signature s and monomial m, there exists a corresponding properly colored cycle cover C' in G' with the signature s' and monomial m' such that $s' = -s$ and $m' = -m$. In other words, $s \times m = s' \times m'$.*
- *There are no other properly colored cycle covers in G'.*

The proof of Lemma 6 will appear in an extended version.

4.3   Construction of T_k

1. Let $(f_n) \in$ VNP. Then, there exists a sequence $g_n(X, y_1, y_2, \ldots, y_{p(n)}) \in$ VF, such that $f_n = \sum_{y_1,\ldots,y_{p(n)} \in \{0,1\}} g_n(X, y_1, y_2, \ldots, y_{p(n)})$, where $p(n)$ is polynomially bounded in n. Since $(g_n) \in$ VF, it is known that there exists a layered algebraic branching program $\mathcal{B}_n$ of size polynomially bounded in n that computes g_n. Let s and t be the source and the sink vertices of $\mathcal{B}_n$.
2. We add a new vertex θ and edges (θ, s) and (t, θ) to graph $\mathcal{B}_n$. We add self-loops on all the vertices except θ. We now color all the edges of $\mathcal{B}_n$ with color c_r. We call this new graph $\widehat{\mathcal{B}}_n$. Let $\mathbf{Y}_i$ be the number of edges in $\widehat{\mathcal{B}}_n$ labelled by variable y_i.
3. Consider the graph formed with the disjoint union of $\widehat{\mathcal{B}}_n$ and $R(\mathbf{Y}_i)$ for all $i \in [p(n)]$. Recall that the color of every edge of $\widehat{\mathcal{B}}_n$ and $R(\mathbf{Y}_i)$ is c_r. Let us call it T'_k, where k' is the size of this graph.
4. For each $i \in [p(n)]$, let $e_{y_i}^{(1)}, e_{y_i}^{(2)}, \ldots, e_{y_i}^{(\mathbf{Y}_i)}$ be the edges in $\widehat{\mathcal{B}}_n$ labelled with y_i and let $\mathbf{c}_{y_i}^{(1)}, \mathbf{c}_{y_i}^{(2)}, \ldots, \mathbf{c}_{y_i}^{(\mathbf{Y}_i)}$ be the $\mathbf{Y}_i$ connector edges of rosette $R(\mathbf{Y}_i)$. We finally place a *colored determinantal if and only if gadget* (with color c_b) between e_i and $\mathbf{c}_i$. We now set all the Y variables to 1. We call this final graph T_k, where k is the size of the graph, and it is clearly polynomially bounded in n.

Lemma 7. *The colored determinant of the graph T_k is $f_n(X)$.*

We present the proof details of Lemma 7 in Sect. 5.

5   Proof of Lemma 7

5.1   Cycle Covers of Graph $\widehat{\mathcal{B}}_n$

In this section, we study the cycle covers of the graph $\widehat{\mathcal{B}}_n$.

Lemma 8. *Let $\mathcal{B}_n$ be the layered algebraic branching program of size $s = \text{poly}(n)$ computing $g_n(X,Y)$. Let $\mathcal{P} = \{\mathcal{P}_i | 1 \leq i \leq \mu\}$ be the set of all $s - t$ paths in $\mathcal{B}_n$. Let m_i be the monomial associated with path $\mathcal{P}_i$, formed by multiplying all the edges of $\mathcal{B}_n$ participating in path $\mathcal{P}_i$. Let $\widehat{\mathcal{B}}_n$ be the graph formed from $\mathcal{B}_n$ as described in point 2 of Sect. 4.3. The graph $\widehat{\mathcal{B}}_n$ satisfies the following properties.*

- *Every cycle cover of $\widehat{\mathcal{B}}_n$ is properly colored. For each $i \in [\mu]$, there exists a unique cycle cover C in $\widehat{\mathcal{B}}_n$, such that the monomial associated with C is m_i. Moreover, there are no other cycle covers in $\widehat{\mathcal{B}}_n$.*
- *The signature of each cycle cover of graph $\widehat{\mathcal{B}}_n$ is same. We assume it to be one.*

We skip the proof of Lemma 8, which will appear in the extended version.

5.2   Proof of **VNP** Completeness

In this section, we state our main proof. Let $f_n(X) \in$ VNP. Let $\mathcal{T}_{k'}$ and T_k denote the graphs constructed from $f_n(X)$ as stated in Sect. 4.3. We know that there exists an algebraic branching program $\mathcal{B}_n$ which computes the polynomial $g_n(X,Y)$. Let $\mathcal{P}$ be the set consisting of all the $s-t$ paths in $\mathcal{B}_n$. Let $\mathcal{P} = \{\mathcal{P}_i | 1 \leq i \leq \mu\}$. Let $c_i m_i$ denotes the monomial associated with path $\mathcal{P}_i$ in $\mathcal{B}_n$, where $c_i \in \mathbb{F}$ is the coefficient of m_i. Therefore, $g_n(X,Y) = c_1 m_1 + c_2 m_2 + \ldots + c_\mu m_\mu$. We have, $f_n(X) = \sum_{y_1, y_2, \ldots, y_{p(n)} \in \{0,1\}} c_1 m_1 + c_2 m_2 + \ldots + c_\mu m_\mu$

$= \sum_{y_1, y_2, \ldots, y_{p(n)} \in \{0,1\}} c_1 m_1 + \ldots + \sum_{y_1, y_2, \ldots, y_{p(n)} \in \{0,1\}} c_\mu m_\mu.$

Let $\mathcal{J}_j$ denotes the polynomial $\sum_{y_1, \ldots, y_{p(n)}} c_j m_j$. Therefore, $f_n(X) = \sum_{j=1}^{\mu} \mathcal{J}_j.$

We know that m_j is a monomial over the variable set $X \cup Y$. Therefore, m_j is the product of two monomials $m_j^{(X)}$ and $m_j^{(Y)}$, where $m_j^{(X)}$ is over the variable set X and $m_j^{(Y)}$ is over the variable set Y.

That is,

$$g_n(X,Y) = c_1 m_1^{(X)} m_1^{(Y)} + c_2 m_2^{(X)} m_2^{(Y)} + \ldots + c_\mu m_\mu^{(X)} m_\mu^{(Y)}.$$

Consider a monomial $m_j = c_j m_j^{(X)} m_j^{(Y)}$, let $V_j \subseteq Y$ is the set of variables whose degree is at least one in $m_j^{(Y)}$. Let $V_j = \{y_{t_1}, y_{t_2}, \ldots, y_{t_\ell}\}$ and the corresponding index set $V_j' = \{t_1, t_2, \ldots, t_\ell\}$, where $1 \leq t_1 < t_2 < \ldots < t_\ell \leq p(n)$. An evaluation of monomial $c_j m_j$ at any point where at least one variable from $V_j \subseteq Y$ is set to 0 vanishes. The evaluations where the monomial $c_j m_j$ survives is only at the points of the form $y_{t_1} = 1, \ldots, y_{t_\ell} = 1$ and for all $\Gamma \notin V_j'$, $y_\Gamma = *$, where $* \in \{0, 1\}$. That is, the coefficient of $c_j m_j^{(X)}$ is equal to the $2^{p(n)-|V_j|}$ in $\mathcal{J}_j$. This is true for every monomial $c_j m_j$, for all $1 \leq j \leq \mu$.

Cycle covers of $\mathcal{T}_{k'}$: Consider the graph $\mathcal{T}_{k'}$ as described in point 3 of Sect. 4.3. We analyse the properly colored cycle covers of graph $\mathcal{T}_{k'}$. Observe that any properly colored cycle cover of $\mathcal{T}_{k'}$ is a disjoint union of the properly colored cycle covers of each of its disjoint components. Since the graph $\mathcal{T}_{k'}$ is monochromatic, we hereafter drop the mention of cycle covers as properly colored cycle covers. For the component $\widehat{\mathcal{B}}_n$, any cycle cover must cover the vertex θ by a cycle that uses the edge (θ, s), followed by a directed $s-t$ path in the underlying branching program followed by (t, θ) and all other vertices must get covered by self-loops. Let $\mathcal{C}_j$ be a cycle cover of graph $\widehat{\mathcal{B}}_n$, where the vertex θ is covered by an edge (θ, s), followed by a directed $s-t$ path $\mathcal{P}_j$ followed by an edge (t, θ) and all other vertices are covered with self-loops. We assume the signature of every cycle cover of $\widehat{\mathcal{B}}_n$ to be one. For each $i \in [p(n)]$, we know that the rosette $R(\mathbf{Y}_i)$ can be covered in $2^{\mathbf{Y}_i}+1$ ways. Moreover, the product of the signature and the monomial associated with each rosette is one. Therefore, the colored determinant of graph $\mathcal{T}_{k'}$ is $\sum_{j \in [\mu]} \left(c_j m_j \prod_{t=1}^{p(n)} (2^{\mathbf{Y}_t} + 1) \right)$.

Cycle covers of T_k: Recall that $V_j = \{y_{t_1}, y_{t_2}, \ldots, y_{t_\ell}\}$ is the set of variables whose degree is at least one in m_j. By the property of *colored determinantal if and only if* gadget, the cycle cover of the rosettes $R(\mathbf{Y}_\alpha)$ for $\alpha \in V_j'$ must pick a subset of its connector edges which can be done in only way (by property 2 of rosette) and the cycle cover of the rosettes $R(\mathbf{Y}_\beta)$ for $\beta \in \overline{V_j'}$ must not pick any of the connector edges which can be done in two ways (by property 3 of rosette). Therefore, for all $\alpha \in V_j$, the factor $(2^{\mathbf{Y}_\alpha} + 1)$ drops to one whereas for all $\beta \in \overline{V_j'}$ the factor $(2^{\mathbf{Y}_i} + 1)$ drops to 2. Therefore, the determinant of the graph T_k will be $\sum_{j \in [\mu]} 2^{p(n)-|V_j|} c_j m_j^{(X)}$.

6 Proof of Lemma 4

We prove Lemma 4 in the following three steps.

- We define a variant of colored determinant which is the sum of over all properly colored clow sequences instead of cycle covers (see Subsect. 6.1).
- We then show that such a variant is the same as the colored determinant (see Subsect. 6.2).
- We construct a conditional stack branching program to compute the colored determinant over properly colored clow sequences instead of properly colored cycle covers (See Subsect. 6.3).

6.1 Colored Determinant as a Sum of Clow Sequences

We first recall the definitions of a closed walk and a clow sequence from [4].

Definition 12 (A closed walk, A clow sequence). *[4] A closed walk (clow) over a directed graph $G = (V, E)$, where $V = [n]$ is a walk where the minimum numbered vertex is visited exactly once. The minimum numbered vertex is called the head of a clow. A sequence $C = \langle c_1, c_2, \ldots, c_k \rangle$ is a clow sequence if for all $i \in [k]$, c_k is a clow. Moreover, $Head(c_1) < Head(c_2) < Head(c_3) \ldots < Head(c_k)$. The degree of the clow sequence C is the sum of the total number of edges participating in each clow in c (counted with multiplicity).*

We now extend these definitions to colored graphs and define a properly colored clow and a properly colored clow sequence.

Definition 13 (Properly colored clow). *A clow of a colored graph is called a properly colored clow, if every simple cycle within the clow is colored with the same color (see Fig. 2).*

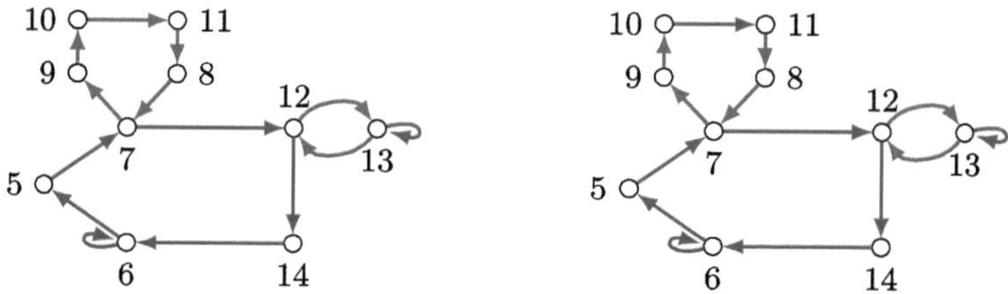

Fig. 2. A properly colored clow $\langle 5, 7, 9, 10, 11, 8, 7, 12, 13, 13, 12, 14, 6, 6, 5 \rangle$ with head 5 and degree 14 (left). A clow with head 5 and degree 14 which is not properly colored (right).

Definition 14 (Properly colored clow sequence). *A sequence $C = \langle c_1, \ldots, c_k \rangle$ over a colored graph is called a properly colored clow sequence if C is a clow sequence and for all $i \in [k]$, c_k is a properly colored clow.*

We define a new variant of Colored Determinant, which, instead of summing over all properly colored cycle covers, it sums over all properly colored clow sequences.

Definition 15. *Let $\mathcal{H}_n(V, E)$ be a complete directed graph. Let $V(\mathcal{H}_n) = \mathcal{H}_1 \cup \mathcal{H}_2$, where $\mathcal{H}_1 = \{1, 2, \ldots, n\}$ and $\mathcal{H}_2 = \{n + 1, n + 2, \ldots, 2n\}$. Let $E(\mathcal{H}_n) = \{(i, j) | 1 \leq i, j \leq 2n\}$. Let $E(\mathcal{H}_n) = E_1 \cup E_2 \cup E_3$, where E_1 denote the set of edges within the vertices in $\mathcal{H}_1$, the set E_2 denote the set of edges within the vertices in $\mathcal{H}_2$ and the set E_3 denote the set of edges across the vertex sets $\mathcal{H}_1$ and $\mathcal{H}_2$. We color the edges in the sets E_1 and E_2 with color c_r. We color the edges in the set E_3 with c_b. Let $X = \{x_{i,j} | 1 \leq i, j \leq 2n\}$. Let $\Phi : E(\mathcal{H}_n) \to X$, where $\Phi((i, j)) = x_{i,j}$ be the label function. The colored determinant (over clow sequences) is defined as follows:*

$$ColDet_n^*(X) = \sum_{c \in \mathcal{C}} (-1)^{n+k} mon(c)$$

where C is the set of all properly colored clow sequences of degree $2n$ of graph $\mathcal{H}_n$ and $mon(c)$ is formed by multiplying the labels on all the edges in c and k is the number of clows in c.

6.2   Involution on the Set of Properly Colored Clow Sequences

Like in the case of the determinant [4], we show that the colored determinant defined as a signed sum of properly colored cycle covers can also be expressed as the signed sum of properly colored clow sequences. Lemma 9 formalizes this.

Lemma 9. $ColDet_n^*(X) = ColDet_n(X)$

The main idea is to show that the properly colored clow sequences that are not cycle covers always appear in pairs with opposite signatures and, therefore, contribute 0 to the overall sum. The formal proof of Lemma 9 will appear in the extended version.

6.3   Construction of a Conditional Stack Branching Program $\mathcal{A}'_{\mathcal{H}_n}$ to Compute $\mathsf{ColDet}_n^*(X)$

We present the construction in three major steps.

– We recall the structure of an $s - t$ path in the algebraic branching program computing the determinant [4] (see Subsect. 6.3.1 for details).
– We modify the above construction to remember the color of the last traversed edge in the clow sequence (see Subsect. 6.3.2).
– We finally use the stack operation labels on the edges of the modified branching program such that it computes $\mathsf{ColDet}_n^*(X)$ (see Subsect. 6.3.3 for more details).

6.3.1   Determinant as a Sum of Signed Clow Sequences

We recall the definition of the determinant, which is defined as the sum of signed clow sequences.

Definition 16. *[4] Consider a directed graph $G = (V, E)$ with $V = [n]$ and $E = \{e_{i,j} = (i,j)|1 \leq i,j \leq n\}$. Let $X = \{x_{i,j}|1 \leq i,j \leq n\}$ be the set of variables. Let $\Phi : E(G) \to X$, where $\phi(e_{i,j}) = x_{i,j}$ The determinant is $Det_n(X) = \sum_{c \in C}(-1)^{n+k}mon(c)$ where the sum is over all clow sequences C of degree n and k is the number of clows in the clow sequence c.*

We know that there exists an algebraic branching program, say $\mathcal{F}_n$ of size $\mathcal{O}(n^3)$ that computes $Det_n(X)$. Moreover, the graph $\mathcal{F}_n$ satisfies the following properties [4].
For every clow sequence $\mathcal{C} = \langle \mathcal{C}_1, \mathcal{C}_2, \ldots, \mathcal{C}_k \rangle$ of degree n and positive signature (or negative signature), let $\mathcal{P}_1, \mathcal{P}_2, \ldots, \mathcal{P}_k$ be the paths formed by unwinding the clows $\mathcal{C}_1$, $\mathcal{C}_2$, ..., $\mathcal{C}_k$, respectively. For every clow sequence $\mathcal{C} = \langle \mathcal{C}_1, \mathcal{C}_2, \ldots, \mathcal{C}_k \rangle$,

there exists a unique $s - t$ path $\mathcal{P}$ in $\mathcal{F}_n$ such that the path $\mathcal{P}$ is an edge (s, s') followed by paths $\mathcal{P}_1, \mathcal{P}_2, \ldots, \mathcal{P}_k$ and then followed by a single edge $\hat{e}$ labelled by $+1$ (-1, respectively). There are no $s - t$ paths in $\mathcal{F}_n$ other than the kind of paths stated above.

6.3.2   Determinant over a Colored Graph G

We present the construction of an algebraic branching program $\mathcal{A}_G$ to compute the determinant polynomial $\mathrm{Det}_n(X)$ over graph G. Our construction is essentially the same as the construction stated in [4], except for a few minor modifications. Let $G = (V, E)$ be a colored directed graph such that $|V(G)| = n$ and a function Φ such that $\Phi((i, j)) \to x_{i,j}$. Let $\Phi' : E(G) \to \{c_b, c_r\}$ be the color function. Let H_n denote the vertex set of $\mathcal{A}_G$ where, $H_n = \{s, t\} \cup H_n'$ and

$$H_n' = \{[p, h, u, i, c] | p \in \{0, 1\}, h \in [n], u \in [n], i \in [n], c \in \{c_r, c_b, \otimes\}\}.$$

The meanings associated with p, h, u, and i in our construction are the same as given in [4]. However, in our construction, since we are working with a colored graph, we also wish to remember the color of the last traversed edge in our traversal; therefore, a new component c is added in the tuple to remember the color of the last traversed edge. We now add the following edges in $\mathcal{A}_G$.

- $(s, [p, h, h, 0, \otimes])$ for $h \in \{1, \ldots, n\}$, where $p = n \bmod 2$.
- $([p, h, u, i, c], [p, h, v, i + 1, c_r])$ if $v > h$ and $i + 1 \le n$; the color of the edge (u, v) is c_r, the weight of this edge is set to x_{uv}.
- $([p, h, u, i, c], [p, h, v, i + 1, c_b])$ if $v > h$ and $i + 1 \le n$; the color of the edge (u, v) is c_b and the weight of this edge is set to x_{uv}.
- $([p, h, u, i, c], [\bar{p}, h', h', i + 1, c_r])$ if $h' > h$ and $i + 1 \le n$; the color of the edge (u, h) is c_r and the weight of this edge is x_{uh}.
- $([p, h, u, i, c], [\bar{p}, h', h', i + 1, c_b])$ if $h' > h$ and $i + 1 \le n$; the color of the edge (u, h) is c_b and the weight of this edge is x_{uh}.
- $([1, h, u, n, c], t)$; the weight of this edge is $+1$
- $([0, h, u, n, c], t)$; the weight of this edge is -1.

This completes the construction of $\mathcal{A}_G$. We now state our lemma formally.

Lemma 10. *The algebraic branching program $\mathcal{A}_{\mathcal{H}_n}$ computes the determinant $Det_n(X)$ over graph $\mathcal{H}_n$.*

The proof of Lemma 10 is straightforward and immediately follows from [4].

6.3.3   Modifying $\mathcal{A}_{\mathcal{H}_n}$ to $\mathcal{A}'_{\mathcal{H}_n}$ Such that $\mathcal{A}'_{\mathcal{H}_n}$ Computes Colored Determinant

In this section, we add the conditional stack operations to the edges of $\mathcal{A}_{\mathcal{H}_n}$ to construct a conditional stack branching program $\mathcal{A}'_{\mathcal{H}_n}$ such that $\mathcal{A}'_{\mathcal{H}_n}$ computes $\mathrm{ColDet}_n^*(X)$. The main idea here is to remember in which clow and at what vertex in the traversal of the clow sequence, there is a color change and record it onto the stack. Therefore, we define an alphabet set sequence $\Sigma_n = \{\triangle_{h,a} | h \in$

$[n], a \in [n]\}$, where h denotes the head of the clow in which the color change occurs and a denotes the vertex at which the color change occurs.

For each edge e in $\mathcal{A}_{\mathcal{H}_n}$ of the form $([p, h, u, \delta, r])$ to $([p, h, v, \delta + 1, b])$ (or $([p, h, u, \delta, b])$ to $([p, h, v, \delta + 1, r]))$, we set the stack label as

$$if(top == \triangle_{h,u}) \ \mathsf{Pop}(\triangle_{h,u}) \quad else \quad \mathsf{Push}(\triangle_{h,u}).$$

A closing edge of a clow is the last edge of the clow. Note that edges directed from $([p, h, u, \delta, r])$ to $([p, h', v, \delta + 1, b])$ (or, $([p, h, u, \delta, b])$ to $([p, h', v, \delta + 1, r]))$ are labelled with $\mathsf{No\text{-}op}$ (no operation) as these edges are closing edges of a clow and a color change here is not a true color change in the clow traversal. We set the stack operation labelings of all other edges to $\mathsf{No\text{-}op}$ (no operation) (see Fig. 3). We state our lemma formally.

Lemma 11. *The conditional algebraic branching program $\mathcal{A}'_{\mathcal{H}_n}$ computes the colored determinant $\mathsf{ColDet}^*_n(X)$ of graph $\mathcal{H}_n$.*

The proof of Lemma 11 will appear in the extended version.

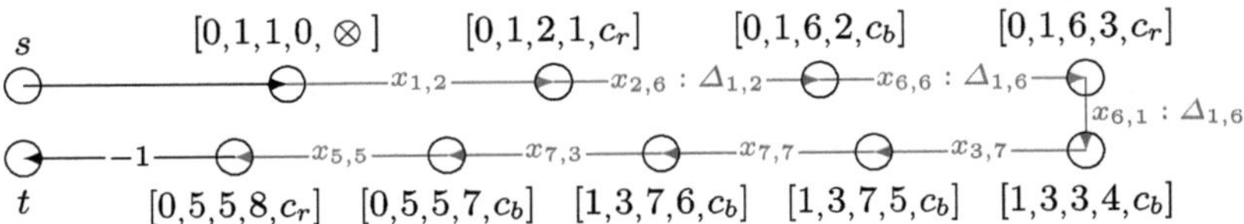

Fig. 3. A $s - t$ path in $\mathcal{A}'_{\mathcal{H}_4}$ corresponding to clow sequence $\langle c_1, c_2, c_3 \rangle$ of degree 8, where $c_1 = (1, 2, 6, 6, 1)$, $c_2 = (3, 7, 7, 3)$, $c_3 = (5, 5)$. Since a conditional stack operation can be uniquely identified by the stack symbol, we write $x : \gamma$ as a shorthand to mean x as the variable label and γ as the conditional stack operation with γ as the stack symbol.

7   Proof of Lemma 5

Proof. Since checking if a path through a conditional stack branching program is realizable or not is easy and can be done in P, (f_n) is in VNP follows from Valiant's criterion (Lemma 1). $\qquad\square$

Acknowledgements. The author would like to thank Nikhil Balaji for helpful discussions on this work and is grateful to the anonymous reviewers for their careful evaluation and valuable feedback.

References

1. Chaugule, P., Limaye, N.: On the closures of monotone algebraic classes and variants of the determinant. Algorithmica **86**(7), 2130–2151 (2024)
2. Chaugule, P., Limaye, N., Pandey, S.: Variants of the determinant polynomial and the VP-completeness. In: The 16th International Computer Science Symposium in Russia. vol. 12730, pp. 31–55. Springer (2021). https://doi.org/10.1007/978-3-030-79416-3_3
3. Chillara, S., Raja, N.: Branching programs with extended memory: new insights. In: International Conference on Algorithms and Complexity, pp. 91–104. Springer (2025)
4. Mahajan, M., et al.: Determinant: Combinatorics, algorithms, and complexity. Chicago J. Theor. Comput. Sci. **1997**(5) (1997)
5. Mengel, S.: Arithmetic branching programs with memory. In: International Symposium on Mathematical Foundations of Computer Science, pp. 667–678. Springer (2013)
6. Saptharishi, R.: A survey of lower bounds in arithmetic circuit complexity. Github Survey **95** (2015)
7. Shpilka, A., Yehudayoff, A.: Arithmetic circuits: A survey of recent results and open questions. Found. Trends® Theor. Compute.Sci. **5**(3–4), 207–388 (2010)
8. Valiant, L.G.: Completeness classes in algebra. In: Proceedings of the Eleventh Annual ACM Symposium on Theory of Computing, pp. 249–261. STOC '79 (1979)
9. Valiant, L.G.: Reducibility by Algebraic Projections. University of Edinburgh, Department of Computer Science (1980)

Lower Bounds of Location Numbers

Rahul Das[1], Soumen Nandi[2], and Ushnish Sarkar[1(✉)]

[1] Department of Mathematics, School of Sciences, Netaji Subhas Open University,
WB, India
`usn.prl@gmail.com,usarkar.sosci@wbnsou.ac.in`
[2] Department of Computer Science, School of Sciences, Netaji Subhas Open
University, WB, India

Abstract. We consider both the backtrack and non-backtrack versions
of the robber locating game on a simple finite connected graph where
a cop wants to locate an invisible mobile and omniscient robber using
distance queries. Also an adaptive sequential version of the metric dimension problem, referred as the sequential locating game, has been studied
where the robber is invisible, but immobile and the cop's objective is to
locate the robber by distance queries.

So far there is no known lower bound for location number, i.e., the
minimum number of rounds taken by the cop to win in the robber locating game, of any locatable graph except trees, irrespective of a backtracking or non-backtracking robber, up to the best of our knowledge. The
same is true for the lower bound of sequential location number which is
the minimum number of rounds taken by the cop to win in the sequential
locating game.

In this article, we show that the lower bound of location number
for any locatable graph G with maximum degree Δ is $\lceil \log_2 \Delta \rceil$, irrespective of a backtracking or a non-backtracking robber. We also show
that there are infinitely many graphs with $\Delta = 3$ for which the lower
bound is tight for both the backtrack and non-backtrack versions. Furthermore, we prove that the lower bound for sequential location number
is $\lceil \log_3(\Delta + 1) \rceil$. This bound is tight for $\Delta = 3^m$, $m \geq 1$ being any
positive integer. Interestingly, this provides us with a difference between
the lower bounds of location numbers (both backtrack as well as non-
backtrack) and sequential location number by a linear multiplicative factor (of $\log_2 3 \approx 1.585$).

Keywords: Robber locating game · Location number · Sequential
location number · Lower bound · Maximum degree

1 Introduction

Introduced by Parsons [14] and Petrov [15], the graph searching game between an
evader and a set of searchers has interesting applications in the field of Artificial
Intelligence, Robotics and Graph Algorithms. The graph searching and graph
locating games are graph theoretic models of this problem. The cop and robber

N. Misra and A. Pandey (Eds.): CALDAM 2026, LNCS 16445, pp. 135–146, 2026.
https://doi.org/10.1007/978-3-032-17156-6_11

game, introduced by Nowakowski and Winkler [12], and Quilliot [16] independently, is an interesting variation of this game which is studied on graphs. Here first the searchers or the cops and then the evader or the robber move in rounds to their neighboring vertices or stay put on a connected graph. Such graph based games are crucial in understanding in problems like surveillance and tracking. The cop number of a graph is the minimum required number of the cops to ensure a winning strategy for capturing the robber. There is an interesting conjecture regarding the cop number of an n-vertex connected graph. This is known as the Meyniel's conjecture, introduced by Frankl [7]. This says that the cop number of a graph is $O(\sqrt{n})$. Bollobás et al. [3] studied the cop-number for sparse random graphs. They showed that the cop number has order of magnitude $n^{1/2+o(1)}$. Łuczak and Prałat [11] proved that the cop number of a random graph $G(n, p)$ is a function of an average degree which forms an intriguing zigzag shape. Alon and Prałat [1] obtained a tight $O(1/r^2)$ upper bound for the number of rounds needed to capture the robber on a random geometric graph $\mathcal{G}_d(n, r)$.

In the graph locating problem, the cop chooses a set of vertices known as probes. The invisible robber hides at an unknown vertex. The cop's objective is to locate the robber. For this, the cop receives the distances from each of these probes to the robber simultaneously. The cop has to determine the location of the robber from these distances.

Slater [20] and Harary and Melter [8] independently investigated a non- adaptive variation of this problem where the cop must choose all the probes at once to locate an invisible but immobile robber hiding at some vertex. So this game concludes in just one round. The minimum number of probes to locate the robber in a graph G is the *metric dimension* of G, denoted by $MD(G)$.

The Robber Locating Game

We assume that the graphs we consider here are finite simple connected. The robber locating game on a graph engages one cop and one robber. As per the practice of the existing literature and also for the ease of discussion, we assume the cop to be a female and the robber to be a male. The game is played in rounds. Unlike the cop robber game, each round consists of the robber's move followed by the cop's move. At any round, the robber hides in a vertex of the graph although, unlike the usual cop robber game, the cop is not in the graph. The robber is mobile and invisible to the cop. Moreover, the robber knows every possible strategy of the cop, i.e., the robber is omniscient. The cop's objective is to locate the robber using distance queries within finite number of rounds and the robber's goal is to avoid being located indefinitely. This game has two versions depending upon the moves of the robber.

The Non-backtrack Version

Seager [17] proposed the model where in the first round, a robber occupies a vertex h_1 of a graph without disclosing its location to the cop. The cop then selects any vertex p_1 as the first probe and obtain the distance d_1 from p_1 to h_1. If the distance suggests that the location of the robber is uniquely determined, then we say the cop wins. Otherwise the game continues to the second round. The robber now has two choices. He may either stay put or move to an adjacent vertex $h_2 \in N[h_1] \setminus \{p_1\}$, $N[h_1]$ being the closed neighbourhood of the vertex h_1. Now the cop chooses a second probe p_2 and obtain the distance d_2 from p_2 to h_2. If the robber is not located, then the game continues. Thus each subsequent round i consists of hiding of the robber at some vertex $h_i \in N[h_{i-1}] \setminus \{p_{i-1}\}$ followed by the distance query made from the i-th probe p_i chosen by the cop, for $i \geq 2$. The cop wins if the robber can be located by obtaining the unique distance d_i at some i-th round, for $i \geq 1$. This game is referred as the *robber locating game with a non-backtracking robber*.

Remark 1. By "the location of the robber is uniquely determined" or "obtaining the unique distance d_i (from p_i) at some i-th round, for $i \geq 1$" we mean that there is no possible location of the robber, other than h_i, at distance d_i from p_i in the i–th round, for $i \geq 1$.

The Backtrack Version

The other version allows the robber to hide at any vertex in the closed neighbourhood of the vertex she was hiding at in the previous round. This alternate version was introduced by Carraher, Choi, Delcourt, Erickson and West in [5]. It is known as the *robber locating game with a backtracking robber*. Note that the robber becomes stronger when it is allowed to backtrack.

An Adaptive Sequential Game Version When the Robber is Invisible but Immobile

In [19], Seager introduced another adaptive sequential game version, referred as sequential locating game, where the invisible robber cannot move to another location throughout the game, once he chooses some vertex to hide at in the beginning. The cop chooses a single vertex as a probe in each round to obtain the distance of the robber from its probe in that round. The minimum number of probes required to locate the robber in any scenario is known as the sequential location number of the graph G (or sequential metric dimension as referred in [2]), denoted by $SL(G)$. In [19], Seager studied the sequential location numbers of paths, cycles, complete graphs, complete bipartite graphs and wheel graphs etc. This problem was further studied by Bensmail et al. in [2] in a multiprobe set up with emphasis on its complexity aspect. In [13], this problem was further studied on Erdős- Rényi graphs.

2   Some Preliminaries

In this article, we deal with both the versions of the robber locating game.

Definition 1. *In both the versions of the robber locating game, if the robber is located, i.e., the cop wins after a finite number of rounds, then we say the graph is backtrack locatable or non-backtrack locatable, if the robber is allowed to or forbidden to backtrack respectively.*

Definition 2. *The minimum number of rounds required to locate the robber is referred as backtrack location number, denoted by $loc_b(G)$, and non-backtrack location number, denoted by $loc(G)$, of the graph G if G is backtrack locatable or non-backtrack locatable respectively.*

In either case, if the robber cannot be located within finite number of rounds, then we say G is not locatable while the robber is backtracking or non-backtracking accordingly.

We recall that each round consists of the robber's move followed by the cop's move. For both variations of the game on G, after the cop makes the distance query from the $i-$th probe, the $i-$th round ends. If the cop does not win at the end of the $i-$th round, then the $(i+1)-$th round begins with the robber's move. Now the robber either stays put or move to one of his neighbours (except the previously probed vertex, if any, in case of the non-backtrack condition). In the i-th round, let the cop find that the robber is hiding at a distance d_i measured from the i-th probe p_i. Here V_{i+1} denotes the set of all possible locations (vertices) in G where the robber can hide at the $(i+1)-$th round, considering the information based on the distance d_i measured from p_i in the previous round, for $i \geq 1$. Also we take the set V_1 of all possible locations of the robber at the first round to be the entire vertex set of G.

3   Previous Works

Carraher et al. [5] has shown that a sufficiently large equal-length subdivision of any graph G is backtrack locatable. Let $G^{\frac{1}{m}}$ be the graph obtained from G by replacing each edge of G by a path of length m, adding $m-1$ new vertices for each such path. Carraher et al. showed that $G^{\frac{1}{m}}$ is also backtrack locatable whenever $m > \min\{|V(G)|-1, \max\{\mu(G) + 2^{\mu(G)}, \Delta(G)\}\}$, within $m(\Delta(G)+1)$ rounds, where $\mu(G)$ is the metric dimension and $\Delta(G)$ is the maximum degree of any vertex of G. They also conjectured that for $n \geq 4$, $K_n^{\frac{1}{m}}$ is backtrack locatable if and only if $m \geq n$. In [9], Haslegrave et al. proved that $K_n^{\frac{1}{m}}$ is backtrack locatable if and only if $m > \frac{n}{2}$, for $n \geq 11$. In [10], Haslegrave et al. studied the problem for any graph and proved that $G^{\frac{1}{m}}$ is backtrack locatable whenever $m \geq \frac{|V(G)|}{2}$. Moreover, they also showed that an unequal subdivision is also backtrack locatable if each edge receives subdivision of at least a certain length.

In [17], Seager has proved that in the non-backtrack condition, a cycle C_n is locatable if and only if $n \neq 5$. She also obtained the non-backtrack location number for cycles of any length except five. Seager has also proved that any tree is non-backtrack locatable. Brandt et al. [4] have provided an upper bound of $loc(T)$ in terms of number of pendant vertices, i.e., vertices of degree one, $\Delta(T)$ and radius of T. In [18], Seager gave the characterization of the backtrack locatable trees. She has shown that with the backtrack condition, any tree T is non locatable if and only if it contains a copy of $T_{3,3}$, where $T_{3,3}$ is the tree with ten vertices such that one vertex v has three neighbours w, x, y and each of w, x, y is adjacent to two leaves.

4   Our Contribution

So far there is no known lower bound for location number of any general finite simple connected graph, irrespective of a backtracking or non-backtracking robber, up to the best of our knowledge. The only existing results were obtained by Seager [17] for any tree. She showed that the lower bound of location number of a tree T for a non-backtracking robber is $\Delta(T) - 1$, $\Delta(T)$ being the maximum degree of T. The tightness of the bound is known to hold for only spiders, up to the best of our knowledge.

In [19], Seager showed the lower bound of the sequential location number of any tree T is $\Delta(T) - 1$. The lower bound was shown to be tight for any tree T, which is not a path, if there exists a path P in T such that all vertices of degree at least 3 lie on P. However, the lower bound for location number of any general finite simple connected graph is still not known, up to the best of our knowledge.

In this article, we obtain the lower bound of backtrack and non-backtrack location numbers of any locatable graph G. Also the lower bound of sequential location number is given. All the lower bounds are given in terms of maximum degree Δ.

We show that the lower bound of location number of any locatable graph G with maximum degree Δ is $\lceil \log_2 \Delta \rceil$, irrespective of a backtracking or non-backtracking robber. For this, first we prove the result for a non-backtracking robber. The lower bound for the backtrack location number follows from the facts that any backtrack locatable graph is non-backtrack locatable and $loc_b(G) \geq loc(G)$, for any backtrack locatable graph G. Also we prove that there are infinitely many graphs with $\Delta = 3$ for which the lower bound is tight for both the backtrack and non-backtrack versions.

For sequential location number, our lower bound $\lceil \log_3(\Delta + 1) \rceil$ is tight for maximum degree $\Delta = 3^m$, $m \geq 1$ being any positive integer. On a concluding note, our lower bounds show a difference between the location numbers (both backtrack as well as non- backtrack) and sequential location number by a linear multiplicative factor (of $\log_2 3 \approx 1.585$).

5 Lower Bound of Location Numbers

The following result from [17] will be used in our article.

Proposition 1. *(Proposition 2.1 of [17]) A graph is locatable with $loc(G) = 1$ if and only if G is a path where the backtracking is not allowed.*

Remark 2. Although Seager proved Proposition 1 for the location game where the backtracking is not allowed, it is quite easy to check that the same is also true if we allow the robber to backtrack.

Now we prove the lower bound of any non-backtrack locatable graph.

Theorem 1. *Let $G = (V, E)$ be a locatable graph such that $\Delta \geq 3$. Then $loc(G) \geq \lceil \log_2 \Delta \rceil$ where backtracking is not allowed.*

Proof. If $\Delta = 3$ or 4, then G is not a path. By Proposition 1, $loc(G) \geq 2 = \lceil \log_2 \Delta \rceil$. We assume $\Delta \geq 5$. Also let $u \in V$ be such that $d(u) = \Delta$.

We will show that whatever strategy the cop adopts, the omniscient invisible robber can avoid being located till $(\lceil \log_2 \Delta \rceil - 1)$-th round by choosing suitable vertices in $N(u) \cup \{u\}$ to hide.

First Round:

Case I: Let $d(p_1, u) > 1$ and u is not at a unique distance from p_1. The omniscient robber can avoid being located in the first round if he hides in u. Since $p_1 \notin N(u)$, therefore for $d_1 = d(p_1, u)$, V_2 contains the vertices of the star $K_{1,\Delta}$ with u as its centre.

Case II: Let $d(p_1, u) > 1$ and u is at a unique distance from p_1. Let $d = d(p_1, u)$. The vertices of $N(u)$ are at distance $(d - 1)$ or $(d + 1)$ from p_1. By pigeon hole principle, $\exists\, \delta \in \{d - 1, d + 1\}$ so that the set $A = \{w \in N(u) : d(p_1, w) = \delta\}$ contains at least $\lceil \frac{\Delta}{2} \rceil$ vertices. As $|N(u)| = \Delta \geq 5$, the omniscient robber hides in any vertex of A to avoid being located in the first round. Hence V_2 contains vertices of a star $K_{1, \lceil \frac{\Delta}{2} \rceil}$ with u as centre.

Case III: Let $d(p_1, u) = 1$ and u is not at a unique distance from p_1. Then the omniscient robber avoids being located in the first round if he hides in u. Now $p_1 \in N(u)$ and backtracking is not allowed. Hence, for $d_1 = 1$, V_2 contains vertices of a star $K_{1,\Delta-1}$ with centre u.

Case IV: Let $d(p_1, u) = 1$ and u is at a unique distance from p_1. If the robber chooses to hide in any vertex of $N(u) \setminus \{p_1\}$, then the cop cannot locate him in the first round. Also for $d_1 = 2$, V_2 contains vertices of a star $K_{1,\Delta-1}$ with centre u.

Case V: Let $d(p_1, u) = 0$, i.e., $p_1 = u$. Then the cop cannot locate the robber hiding in any neighbour of u. Since backtracking is not allowed, for $d_1 = 1$, each vertex of the star $K_{1,\Delta}$, centred at u, except the vertex u itself, belongs to V_2.

Thus, for every choice of p_1 by the cop, the robber has a move to avoid being located at the first round. Moreover, combining all the cases of the first round, we have V_2 contains vertices of a star $K_{1, \lceil \frac{\Delta}{2} \rceil}$ with $u \in V_2$ as its centre (Cases I to IV, i.e., when $p_1 \neq u$) or all Δ vertices of $N(u)$ where $u \notin V_2$ (Case V, i.e., when $p_1 = u$).

Second Round:

Scenario I: First we assume V_2 contains vertices of the star $K_{1,\lceil\frac{\Delta}{2}\rceil}$ with $u \in V_2$ as its centre. Here $d(p_1, u) \neq 0$.

Case I: Let $d(p_2, u) > 1$ and u is not at a unique distance from p_2. The cop cannot locate the omniscient robber if he hides in u in the second round. Since $p_2 \notin N(u)$, therefore V_3 contains the vertices of the star $K_{1,\Delta}$ with u as its centre.

Case II: Let $d(p_2, u) > 1$ and u is at a unique distance from p_2. Let $d = d(p_2, u)$. Then the neighbours of u in V_2 are at distance $(d-1)$ or $(d+1)$ from p_2. By the pigeon hole principle, $\exists\ \delta \in \{d-1, d+1\}$ such that the set $B = \{w \in N(u) \cap V_2 : d(p_2, w) = \delta\}$ contains at least $\lceil\frac{\lceil\frac{\Delta}{2}\rceil}{2}\rceil = \lceil\frac{\Delta}{4}\rceil \geq 2$ vertices. Choosing any vertex of B to hide in, the robber avoids being located in the second round. Since $p_2 \neq u$, therefore V_3 contains the vertices of a star $K_{1,\lceil\frac{\Delta}{4}\rceil}$ with u at its centre.

Case III: Let $d(p_2, u) = 1$ and u is not at a unique distance from p_2. The omniscient robber avoids being located in the second round by choosing to hide in u. Since backtracking is not allowed, $p_2 \notin V_3$. Thus V_3 contains vertices of a star $K_{1,\Delta-1}$ with its centre $u \in V_3$.

Case IV: Let $d(p_2, u) = 1$ and u is at a unique distance from p_2 in V_2. The robber chooses to hide in any vertex of $V_2 \cap N(u) \setminus \{p_2\}$. The cop cannot locate him in the second round. Also for $d_2 = 2$, V_3 contains vertices of a star $K_{1,\lceil\frac{\Delta}{2}\rceil-1}$ with centre u.

Case V: Let $d(p_2, u) = 0$, i.e., $p_2 = u$. If the robber hides in any vertex of $N(u) \cap V_2$, the cop cannot locate him in the second round. As backtracking is not allowed, for $d_2 = 1$, V_3 contains every vertex of a star $K_{1,\lceil\frac{\Delta}{2}\rceil}$ having its center at u, except u itself.

Scenario II: We now consider the other scenario in second round when V_2 contains all Δ vertices of $N(u)$ and $u \notin V_2$ (i.e., $p_1 = u$).

Case A: If $p_2 = u$, then the robber hides in any neighbour of u in the second round and the cop cannot locate him. Note that V_3 contains the vertices $N(u)$ but $u \notin V_3$.

Case B: Let $p_2 \neq u$, i.e., $d(p_2, u) \neq 0$. Note that $N(u) \subset V_2$. Let $d'' = \min\{d(p_2, w) : w \in N(u)\}$. Then $d'' \geq 0$. By the pigeon hole principle, $\exists\ \delta \in \{d'', d''+1, d''+2\}$ such that the set $C = \{w \in V_2 : d(p_2, w) = d_2\}$ contains at least $\lceil\frac{\Delta}{3}\rceil \geq 2$ vertices of $N(u)$. The robber avoids being located in the second round by hiding in any vertex of C. Note that V_3 contains the vertices of a star $K_{1,\lceil\frac{\Delta}{3}\rceil}$ with u at its centre.

Thus for every cop strategy, there are moves of the omniscient robber such that for a choice of d_2, the set $\{w \in V_2 : d(w, p_2) = d_2\}$ contains at least two vertices. Hence the robber has strategy to avoid being located at the second round.

Combining all the scenarios, V_3 contains vertices of the star $K_{1,\lceil\frac{\Delta}{2^2}\rceil}$ with $u \in V_3$ as its centre or at least $\lceil\frac{\Delta}{2}\rceil$ vertices of $N(u)$ where $u \notin V_3$.

Proceeding similarly, we can say that for every cop strategy, there is a sequence of moves of the robber such that for a choice of d_i in the i−th round, the set $\{w \in V_i : d(w, p_i) = d_i\}$ contains at least two vertices, while V_{i+1} contains vertices of a star $K_{1, \lceil \frac{\Delta}{2^i} \rceil}$ with $u \in V_{i+1}$ as its centre or at least $\lceil \frac{\Delta}{2^{i-1}} \rceil$ vertices of $N(u)$ where $u \notin V_{i+1}$, as long as $\lceil \frac{\Delta}{2^{i-1}} \rceil > 2$.

Thus the robber has a strategy to avoid being located till the i−th round as long as $\lceil \frac{\Delta}{2^{i-1}} \rceil > 2$, i.e., for $i \leq (\log_2 \Delta - 1)$, if $\Delta = 2^k$, for some positive integer k, and for $i \leq \lfloor \log_2 \Delta \rfloor$, otherwise.

Hence for every cop strategy, the robber has a sequences of moves such that he will be located at least at $\lceil \log_2 \Delta \rceil$−th round. Hence $loc(G) \geq \lceil \log_2 \Delta \rceil$, where the backtracking is not allowed.

Since any backtrack locatable graph G is non-backtrack locatable and $loc_b(G) \geq loc(G)$, therefore the lower bound of backtrack location number follows using Theorem 1.

Corollary 1. *Let $G = (V, E)$ be a backtrack locatable graph such that $\Delta \geq 3$. Then $loc_b(G) \geq \lceil \log_2 \Delta \rceil$.*

Tightness of lower bound

We show that there are infinitely many graphs with $\Delta = 3$ for which the lower bounds proposed in Theorem 1 and in Corollary 1 are tight. For this, first we consider the following graph operation.

Definition 3. *Let $G_1 = (V_1, E_1)$ and $G_2 = (V_2, E_2)$ be two disjoint graphs with $u_1 \in V_1$ and $u_2 \in V_2$. The vertex sum of G_1 and G_2 over u_1 and u_2, denoted by $G_1 \underset{u_1 \vee u_2 \to u}{+} G_2$ and defined as the graph G obtained by identifying the vertices u_1 and u_2 to a unique vertex u.*

Remark 3. Note that the vertex set of G is $V = (V_1 \setminus \{u_1\}) \cup (V_2 \setminus \{u_2\}) \cup \{u\}$ and the edge set of G is given by $E = J_1 \cup J_2$ where $J_1 = E_1 \setminus \{(x, u_1) : x \in N(u_1) \text{ in } G_1\} \cup \{(x, u) : x \in N(u_1) \text{ in } G_1\}$ and $J_2 = E_2 \setminus \{(x, u_2) : x \in N(u_2) \text{ in } G_2\} \cup \{(x, u) : x \in N(u_2) \text{ in } G_2\}$.

For $m, n \in \mathbb{N}$ with $m \geq 2, n \geq 6$, we consider the graphs $G_{m,n} = P_m \underset{v \vee t \to w}{+} C_{2n}$, where P_m is a path with two end vertices u, v and C_{2n} is a cycle with a vertex t. Also P_m and C_{2n} are two disjoint graphs. Here v, t have been identified to the vertex w in $G_{m,n}$ (as illustrated in Fig. 1). Clearly $\Delta(G_{m,n}) = \Delta = 3$.

By Theorem 1 and Corollary 1, $loc_b(G_{m,n}) \geq loc(G_{m,n}) \geq 2$.

We prove now $loc_b(G_{m,n}) = loc(G_{m,n}) = 2 = \lceil \log_2 \Delta \rceil$. For this, we show that the cop has a strategy to locate the backtracking robber within two rounds. The non-backtrack version follows trivially.

The cop chooses u as the first probe, i.e., $p_1 = u$. Then $0 \leq d_1 \leq m + n - 1$.

If $0 \leq d_1 \leq m - 1$ or $d_1 = m + n - 1$, the cop locates the robber in the first round.

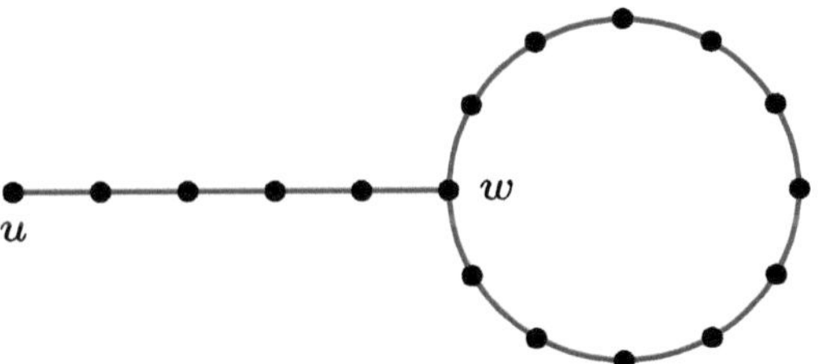

Fig. 1. The Graph $G = P_6 \underset{v \vee t \rightarrow w}{+} C_{12}$ with $\Delta = 3$

If $m \leq d_1 \leq m + n - 2$, then there are exactly two vertices, say u_1, u_2, at distance d_1 from p_1. Let $N[u_1] = \{u_1', u_1, u_1''\}$ and $N[u_2] = \{u_2', u_2, u_2''\}$. Without loss of generality, we assume that u_1' and u_2' are neighbors of u_1 and u_2 respectively such that $d(u_1', u_2') = \min\{d(x, y) : x \in N(u_1),\ y \in N(u_2)\}$. The cop chooses $p_2 = u_1''$ if $d(u_1', u_2') \leq 2$ and $p_2 = u_1'$ if $d(u_1', u_2') > 2$. Then all the possible locations of the robber are at unique distance from p_2. Therefore the robber is located within two rounds. Hence $loc_b(G_{m,n}) = 2, \forall m \geq 2, n \geq 6$. From this, we can also conclude $loc(G_{m,n}) = 2, \forall m \geq 2, n \geq 6$.

Remark 4. In the above discussion, had the cop chosen $p_2 = u_1'$ when $d(u_1', u_2') \leq 2$, then there would have been two possible locations of the robber at distance two from p_2. Hence the game would continue to the third round.

Remark 5. Note that if we add another path at the antipodal vertex of t (i.e., w in $G_{m,n}$) in C_{2n} then the tightness of lower bound still holds. Here we can have more than one vertex of degree 3 in graphs whose location numbers achieve the tightness.

6   Lower Bound of Sequential Location Number

Here we give a lower bound of sequential location number in terms of maximum degree Δ of a graph G.

Theorem 2. *For any graph* $G = (V, E)$, $SL(G) \geq \lceil \log_3(\Delta + 1) \rceil$ *where* Δ *is the maximum degree of* G.

Proof. If $1 \leq \Delta \leq 2$, G is either a path or a cycle. Therefore $SL(G) \geq 1 = \lceil \log_3(\Delta + 1) \rceil$. Also, for $3 \leq \Delta \leq 8$, $SL(G) \geq 2 = \lceil \log_3(\Delta + 1) \rceil$.

We now take $\Delta \geq 9$. Let $u \in V$ such that $d(u) = \Delta$. We will show that whatever strategy the cop adopts, there will be a scenario such that the immobile but invisible robber will not be located till $(\lceil \log_3(\Delta + 1) \rceil - 1)$-th round if he chooses some suitable vertex in $N[u] = N(u) \cup \{u\}$ to hide.

Consider the first probe p_1. Then there exists a $\delta_1 \in \{d(p_1, u), d(p_1, u) - 1, d(p_1, u) + 1\} \cap \mathbb{N}$ such that, by the pigeon hole principle, $A_1 = \{x \in N[u] :$

$d(p_1, x) = \delta_1\}$ contains at least $\lceil \frac{\Delta+1}{3} \rceil$ vertices. Clearly the position of the robber hiding at A_1 is not determined in the first round.

Proceeding similarly, in the i–th round ($2 \le i < SL(G)$), there exists $\delta_i \in \{d(p_i, u), d(p_i, u) - 1, d(p_i, u) + 1\} \cap \mathbb{N}$ such that $A_i = \{x \in A_{i-1} : d(p_i, x) = \delta_i\}$ contains at least $\lceil \frac{\Delta+1}{3^i} \rceil > 1$ vertices. Note that $A_i \supseteq A_{i+1}$ for $1 \le i \le SL(G)-2$. Also the robber hiding at A_i is not located in the i–th round, for $1 \le i \le SL(G) - 1$.

Hence, in the i–th round, the robber is not located as long as $i < \log_3(\Delta+1)$. Therefore $SL(G) \ge \lceil \log_3(\Delta + 1) \rceil$.

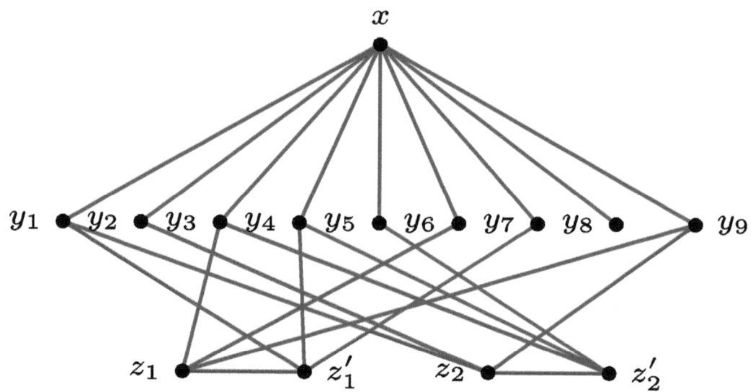

Fig. 2. The Graph $\mathcal{G}_2$ with $\Delta = 3^2$.

Tightness of the Lower Bound

We provide a construction showing the lower bound is tight for the simple finite connected graph $\mathcal{G}_m$ of maximum degree $\Delta = 3^m$, $m \ge 1$ being a positive integer. **The Construction:** The vertex set of $\mathcal{G}_m$ is given by $\{x\} \cup \{y_1, y_2, \ldots, y_\Delta\} \cup \{z_1, z_2, \ldots, z_m\} \cup \{z'_1, z'_2, \ldots, z'_m\}$ where $m = \lceil log_3(\Delta + 1) \rceil - 1$. The edges are assigned as follows: x is adjacent to y_k for all $1 \le k \le \Delta$. Here z_i is adjacent to z'_i for all $1 \le i \le m$. Moreover, z_i is adjacent to y_k if and only if the i–th ternary digit of k (from right to left) is 0 and z'_i is adjacent to y_k if and only if the i–th ternary digit of k is 1.

Now we state the following theorems. These will be required to establish the tightness.

Theorem 3. *[6] For any simple connected finite graph G, $MD(G) \ge \lceil log_3(\Delta + 1) \rceil$.*

Theorem 4. *[19] For any simple connected finite graph G, $SL(G) \le MD(G)$.*

Proof of Tightness: By Theorem 3, $MD(\mathcal{G}_m) \ge m + 1$. The set $\{x, z_1, z_2, \ldots, z_m\}$ is a resolving set of $\mathcal{G}_m$, since any y_k is either of $1, 2$ or 3 distance apart from any z_i, for $1 \le k \le \Delta$ and $1 \le i \le m$. Hence $MD(\mathcal{G}_m) = m + 1$. This implies $SL(\mathcal{G}_m) \le m + 1$, using Theorem 4. But by Theorem 2, $SL(\mathcal{G}_m) \ge m + 1$. Hence $SL(\mathcal{G}_m) = m + 1$.

7 Concluding Remarks

In this article, we have shown that the lower bound for both backtrack as well as non- backtrack location numbers is $\lceil \log_2 \Delta \rceil$ which is tight for $\Delta = 3$. On the other hand, the lower bound for sequential location number is $\lceil log_3(\Delta + 1) \rceil$. This bound is tight for $\Delta = 3^m$, $m \geq 1$ being any positive integer. This gives a difference between the lower bounds of location numbers (both backtrack as well as non- backtrack) and sequential location number by a linear multiplicative factor (of $\log_2 3 \approx 1.585$).

Acknowledgements. The authors thanks the anonymous reviewers for their insightful comments. These comments helped improve the manuscript. The research of the corresponding author has been supported by NSOU Project granted vide Memo. No. Reg/0522 dated 07/06/2024 funded by Netaji Subhas Open University. The research of the second author has been supported by NSOU Project granted vide Memo. No. Reg/0520 dated 07/06/2024 funded by Netaji Subhas Open University.

References

1. Alon, N., Prałat, P.: Chasing robbers on random geometric graphs–an alternative approach. Discret. Appl. Math. **178**, 149–152 (2014)
2. Bensmail, J., Mazauric, D., Inerney, F.M., Nisse, N., Pérennes, S.: Sequential metric dimension. Algorithmica **82**(10), 2867–2901 (2020)
3. Bollobás, B., Kun, G., Leader, I.: Cops and robbers in a random graph. J. Combin. Theory Ser. B **103**(2), 226–236 (2013)
4. Brandt, A., Diemunsch, J., Erbes, C., LeGrand, J., Moffatt, C.: A robber locating strategy for trees. Discret. Appl. Math. **232**, 99–106 (2017)
5. Carraher, J., Choi, I., Delcourt, M., Erickson, L.H., West, D.B.: Locating a robber on a graph via distance queries. Theoret. Comput. Sci. **463**, 54–61 (2012)
6. Chartrand, G., Poisson, C., Zhang, P.: Resolvability and the upper dimension of graphs. Comput. Math. Appl. **39**(12), 19–28 (2000)
7. Frankl, P.: Cops and robbers in graphs with large girth and cayley graphs. Discret. Appl. Math. **17**(3), 301–305 (1987)
8. Harary, F., Melter, R.A.: On the metric dimension of a graph. Ars Combin. **2**(191-195), 1 (1976)
9. Haslegrave, J., Johnson, R.A.B., Koch, S.: The robber locating game. Discrete Math. **339**(1), 109–117 (2016)
10. Haslegrave, J., Johnson, R.A.B., Koch, S.: Subdivisions in the robber locating game. Discret. Math. **339**(11), 2804–2811 (2016)
11. Łuczak, T., Prałat, P.: Chasing robbers on random graphs: zigzag theorem. Random Struct. Algorithms **37**(4), 516–524 (2010)
12. Nowakowski, R., Winkler, P.: Vertex-to-vertex pursuit in a graph. Discret. Math. **43**(2–3), 235–239 (1983)
13. Odor, G., Thiran, P.: Sequential metric dimension for random graphs. J. Appl. Probab. **58**(4), 909–951 (2021)
14. Parsons, T.D.: Pursuit-evasion in a graph. In: Alavi, Y., Lick, D.R. (eds.) Theory and Applications of Graphs, pp. 426–441. Springer, Heidelberg (2006). https://doi.org/10.1007/BFb0070400

15. Petrov, N.N., Petrov, N.N.: The "cossack-robber" differential game. Differentsial'nye Uravneniya **19**(8), 1366–1374 (1983)
16. Quilliot, A.: Jeux et pointes fixes sur les graphes. Ph.D. thesis, Université de Paris VI (1978)
17. Seager, S.: Locating a robber on a graph. Discret. Math. **312**(22), 3265–3269 (2012)
18. Seager, S.: Locating a backtracking robber on a tree. Theoret. Comput. Sci. **539**, 28–37 (2014)
19. Seager, S.M.: A sequential locating game on graphs. Ars Comb. **110**, 45–54 (2013)
20. Slater, P.J.: Leaves of trees. Congr. Numer. **14**(549-559), 37 (1975)

An Update on Pushable Homomorphisms

Tapas Das[1], Pavan P. D[2(✉)], Sagnik Sen[1], and S. Taruni[3]

[1] Indian Institute of Technology Dharwad, Dharwad, India
[2] University of Turku, FI-20014 Turku, Finland
pavanpdevaraj@gmail.com
[3] Centro de Modelamiento Matemático (CNRS IRL2807), Universidad de Chile, Santiago, Chile

Abstract. The notion of pushable homomorphisms of oriented graphs was introduced by Klostermeyer and MacGillivray (Discrete Mathematics 2004) as a modification of homomorphisms of oriented graphs and was further studied in a number of research articles. Our work attempts to fill some gaps in its theory, and also explores the connections of pushable homomorphisms of oriented graphs with homomorphisms of signed graphs and graph coloring.

1 Introduction and Preliminaries

An *oriented graph* $\overrightarrow{G}$ is a directed graph without any loops or parallel arcs in opposite directions. We denote an oriented graph by $\overrightarrow{G}$, while the underlying undirected graph is denoted by G. For an oriented graph $\overrightarrow{G}$, $V(\overrightarrow{G})$ and $A(\overrightarrow{G})$ denote the set of vertices and the set of arcs in $\overrightarrow{G}$. Furthermore, given an arc uv, the vertex u is an *in-neighbor* of v, and the vertex v is an *out-neighbor* of u. The set of all in-neighbors (resp., out-neighbors) of u is denoted by $N^-(u)$ (resp., $N^+(u)$).

A *homomorphism* of an oriented graph $\overrightarrow{G}$ to another oriented graph $\overrightarrow{H}$ is a vertex mapping $f : V(\overrightarrow{G}) \to V(\overrightarrow{H})$ such that for any arc uv of $\overrightarrow{G}$, its image $f(u)f(v)$ is also an arc of $\overrightarrow{H}$. If $\overrightarrow{G}$ admits a homomorphism to $\overrightarrow{H}$, then we say that $\overrightarrow{G}$ is $\overrightarrow{H}$*-colorable*, and denote it by $\overrightarrow{G} \to \overrightarrow{H}$. A bijective homomorphism such that the images of non-adjacent vertices are also non-adjacent is an *isomorphism*. The *oriented chromatic number* of an oriented graph $\overrightarrow{G}$, denoted by $\chi_o(\overrightarrow{G})$, is the minimum $|V(\overrightarrow{H})|$ such that $\overrightarrow{G}$ is $\overrightarrow{H}$-colorable.

Pushable homomorphisms: To *push* a vertex v of $\overrightarrow{G}$ is to reverse the direction of all arcs of $\overrightarrow{G}$ that are incident to v. We denote the so-obtained oriented graph (after pushing v of $\overrightarrow{G}$) by $\overrightarrow{G}^v$. Observe that, if we push multiple vertices of $\overrightarrow{G}$, then the order in which we push the vertices does not affect the final resultant graph. Moreover, pushing one vertex two times is the same as not pushing it at all. Therefore, it makes sense to define the push operation for a set of vertices as well. Given a set $S \subseteq V(\overrightarrow{G})$, to push S is to push each vertex of S exactly once.

N. Misra and A. Pandey (Eds.): CALDAM 2026, LNCS 16445, pp. 147–164, 2026.
https://doi.org/10.1007/978-3-032-17156-6_12

The so-obtained graph is denoted by $\overrightarrow{G}^S$ and we say that $\overrightarrow{G}$ and $\overrightarrow{G}^S$ are *push equivalent*, denoted by $\overrightarrow{G} \equiv_p \overrightarrow{G}^S$. Moreover, $[\overrightarrow{G}]$ denotes the push equivalence class of oriented graphs, where $\overrightarrow{G}$ is any representative of it. That means, given a simple graph G, the set of all its orientations is partitioned into push equivalence classes.

A *pushable homomorphism* of an oriented graph $\overrightarrow{G}$ to another oriented graph $\overrightarrow{H}$ is a vertex mapping $f : V(\overrightarrow{G}) \to V(\overrightarrow{H})$ such that there exists $\overrightarrow{G}' \equiv_p \overrightarrow{G}$ so that f is a homomorphism of $\overrightarrow{G}'$ to $\overrightarrow{H}$. If $\overrightarrow{G}$ admits a pushable homomorphism to $\overrightarrow{H}$, then we say that $\overrightarrow{G}$ is *pushably $\overrightarrow{H}$-colorable*, and we denote it by $\overrightarrow{G} \xrightarrow{p} \overrightarrow{H}$. Moreover, the *pushable chromatic number* of an oriented graph $\overrightarrow{G}$, denoted by $\chi_p(\overrightarrow{G})$ is the smallest $|V(\overrightarrow{H})|$ such that $\overrightarrow{G}$ is pushably $\overrightarrow{H}$-colorable.

Previous works: The concept of pushable homomorphisms was first introduced and studied by Klostermeyer and MacGillivray [19] in 2004. Along with their work, Sen [29] further explored the relation between the pushable chromatic number and the oriented chromatic number. The complexity dichotomy problems regarding finding the pushable chromatic number were studied in [13,19], and the analogue of clique in this set up was studied in [6]. Furthermore, the pushable chromatic number of different graph families, such as, graphs with bounded maximum degree [4], graphs with bounded acyclic chromatic number [29], subcubic graphs [4], planar graphs and planar graphs having girth restrictions [8,10,19,29], sparse graphs defined by maximum average degrees [4,8,10], outerplanar graphs and outerplanar graphs having girth restrictions [19,29], and grids [5] have been explored across several papers and by different researchers.

Apart from the above-mentioned works on pushable homomorphisms of oriented graphs, the push operation on digraphs has been studied in different contexts. A few notable works include the works of Babai and Cameron [1] on automorphism groups of the pushable (switching) classes of tournaments, of Huang, MacGillivray, and Wood [16] on characterizing multipartite tournaments which can be made acyclic through pushing a set of vertices, and of Klostermeyer and Söltés [20] who characterized the multipartite tournaments that can be made Hamiltonian through pushing a set of vertices.

Remark 1. The problem of determining whether an oriented graph $\overrightarrow{G}$ is pushably $\overrightarrow{H}$-colorable or not can be viewed as a graph modification problem [9,12,31] where the actual problem considered here is determining whether $\overrightarrow{G}$ is $\overrightarrow{H}$-colorable, and the modification is pushing a vertex subset of $\overrightarrow{G}$.

Motivation, our contributions, and organization: In the following we list our section-wise contributions, and provide some specific motivation for the same.

- In Sect. 2, we characterize push equivalent orientations of graphs using cycles. This result is an analogue of the well-known lemma due to Zaslavsky [33] from the theory of signed graphs. This lemma is one of the most important structural results in the rich theory of signed graphs and contributes greatly in building the foundation of this theory. Using our main result, we show that the problem of determining whether two given orientations of a graph are push equivalent or not can be solved in polynomial time, to count the number of non-equivalent orientations of a graph, and to provide the canonical definition of pushable homomorphism.

- In Sect. 3, we provide a one-to-one correspondence between oriented and signed graphs which respects the pushable and the switchable homomorphism orders. In terms of category theory, we prove a categorical isomorphism between the categories of bipartite oriented and signed graphs (where morphisms are pushable and switchable homomorphisms, respectively). Moreover, we apply our main theorem to translate a number of important results directly from the theory of signed graphs to oriented graphs. In particular, we show that pushable homomorphisms of bipartite graphs capture the entire theory of graph coloring as a subcase.

- In Sect. 4, we prove a connection between the graph coloring problem and pushable homomorphism to directed odd cycles. Note that, a complete complexity dichotomy characterization of the decision problem of determining whether an input oriented graph $\overrightarrow{G}$ admits a pushable $\overrightarrow{H}$-coloring or not is known due to Klostermeyer and MacGillivray [19]. We apply our main result to show that the pushable $\overrightarrow{H}$-coloring problem, where $\overrightarrow{H}$ is a directed odd cycle, is NP-complete even for sparse graphs. Also we show how the pushable $\overrightarrow{H}$-coloring problem completely encodes graph coloring.

- In Sect. 5, we share our concluding remarks, open problems, and future research directions.

Note: In this article, we follow the book "Introduction to graph theory" by West [32] for the standard graph theoretic notation and terminology.

2   Characterization of Push Equivalent Orientations

Given a graph G, an *ordered closed walk* $C = v_1 v_2 \cdots v_k v_1$ is a closed walk together with the prescribed traversal rule which mandates us to traverse the vertex v_{i+1} after v_i, where $i \in \{1, 2, \cdots, k\}$ and the $+$ operation in the subscript of vertex names is taken modulo k. Thus, it will make sense to speak about *forward arcs* and *backward arcs* of C in this context. Notice that, given a closed walk $C = v_1 v_2 \cdots v_k v_1$, there can be two distinct ordered closed walks on C, that is, (i) the ordered closed walk $v_1 v_2 \cdots v_k v_1$ and (ii) the ordered closed walk $v_1 v_k v_{k-1} \cdots v_2 v_1$. The two ordered closed walks are called each other's *conjugates*. In particular, a cycle is also a closed walk. Thus, the above definitions hold for cycles as well.

An oriented ordered cycle $\overrightarrow{C}$ is *directable* if it is possible to push some vertices of $\overrightarrow{C}$ and make it a directed cycle. In particular, if it is possible to push some vertices of $\overrightarrow{C}$ and make it a directed cycle with all forward (resp., backward) arcs, then $\overrightarrow{C}$ is *forward (resp., backward) directable*. Finally, if $\overrightarrow{C}$ is not directable, then it is *non-directable*. Based on these definitions, oriented ordered cycles, and by extension, oriented ordered closed walks, can be classified into four types.

(i) *Odd forward directable closed walk:* An oriented ordered closed walk of odd length that has an odd number of forward arcs, or equivalently, an even number of backward arcs. In particular, if the oriented ordered closed walk is an oriented ordered cycle, then the cycle is forward directable.

(ii) *Odd backward directable closed walk:* An oriented ordered closed walk of odd length that has an even number of forward arcs, or equivalently, an odd number of backward arcs. In particular, if the oriented ordered closed walk is an oriented ordered cycle, then the cycle is backward directable.

(iii) *Even directable closed walk:* An oriented ordered closed walk of even length that has an even number of forward arcs, or equivalently, an even number of backward arcs. In particular, if the oriented ordered closed walk is an oriented ordered cycle, then the cycle is directable (both forward and backward).

(iv) *Even non-directable closed walk:* An oriented ordered closed walk of even length that has an odd number of forward arcs, or equivalently, an odd number of backward arcs. In particular, if the oriented ordered closed walk is an oriented ordered cycle, then the cycle is non-directable.

If two oriented ordered closed walks are of the same type, then their *directability* is the same, while it is different otherwise.

Theorem 1. *Let $\overrightarrow{G}^1$ and $\overrightarrow{G}^2$ be two orientations of the graph G. The two orientations of G are push equivalent if and only if every ordered cycle of G has the same directability in both $\overrightarrow{G}^1$ and $\overrightarrow{G}^2$.*

Proof. It is enough to prove the statement when G is a connected graph.

Note that, pushing the vertices of an oriented ordered cycle does not change the parity of the number of its forward (resp., backward) arcs. Thus, if $\overrightarrow{G}^1$ and $\overrightarrow{G}^2$ are push equivalent, any ordered cycle of G has the same directability in $\overrightarrow{G}^1$ and $\overrightarrow{G}^2$. This proves the "only if" part of the statement.

To prove the "if" part, let us assume that any ordered cycle of G has the same directability in $\overrightarrow{G}^1$ and $\overrightarrow{G}^2$. Take a spanning tree T of G. Let $\overrightarrow{T}^1$ and $\overrightarrow{T}^2$ be the induced orientations of T under $\overrightarrow{G}^1$ and $\overrightarrow{G}^2$, respectively. Suppose that uv is an arc of $\overrightarrow{T}^1$ while vu is an arc of $\overrightarrow{T}^2$. Notice that, uv is a cut edge of T (since T is a tree). Let A and B be the sets of vertices of the two different connected components of $T - uv$, respectively. Observe that, if we push all the vertices of A in $\overrightarrow{T}^2$, then the arc vu changes its direction while all the other arcs of $\overrightarrow{T}^2$ retain their direction. Thus, we can repeat this process for all edges of T where the direction of their respective arcs in $\overrightarrow{T}^1$ and $\overrightarrow{T}^2$ are different. As a result, we will obtain a push equivalent orientation $\overrightarrow{G}^3$ of $\overrightarrow{G}^2$ in which the tree T has exactly the orientation $\overrightarrow{T}^1$. As the directability of an oriented ordered cycle is invariant under pushing, the directability of all cycles of $\overrightarrow{G}^1$ and $\overrightarrow{G}^3$ must also be the same. As any edge $u'v'$ from $E(G) \setminus E(T)$ is part of a cycle in G (since T is a spanning tree), the arcs corresponding to the edge $u'v'$ in $\vec{G}^1$ and $\vec{G}^3$ must have the same direction. Therefore, $\vec{G}^1$ and $\vec{G}^3$ must be the same orientation of G which implies that $\vec{G}^1$ is push equivalent to $\vec{G}^2$.

Remark 2. In the classification of ordered closed walks presented before the statement of Theorem 1, in each item, the last line is a remark that may be considered as an observation. These observations also follow directly from Theorem 1.

The general notion of homomorphism (not necessarily of graphs) usually preserves some important property of the object for which it is defined. In the case of pushable homomorphisms, it preserves the directability of ordered cycles according to the above theorem. However, from the given definition, it is not so easy to realise which properties get preserved. Hence, the following alternative definition can be regarded as the canonical definition of pushable homomorphism.

Definition 1. (The canonical definition) *A vertex mapping* $f : V(\vec{G}) \rightarrow V(\vec{H})$ *is a pushable homomorphism of* $\vec{G}$ *to* $\vec{H}$ *if* f *preserves the directability of each ordered closed walk of* $\vec{G}$.

Apart from motivating and justifying the canonical definition of a pushable homomorphism, the main result of this section, that is, Theorem 1, has a number of other interesting consequences. An immediate corollary is the following.

Corollary 1. *Any two orientations of a forest F are push equivalent.*

Proof. As a forest F does not have any cycle, according to Theorem 1, any two orientations of F must be push equivalent.

Moreover, Theorem 1 implies a polynomial algorithm for deciding push equivalence between two orientations of a graph.

Theorem 2. *Let* $\vec{G}^1$ *and* $\vec{G}^2$ *be two orientations of the graph G. There is a polynomial time algorithm to determine whether* $\vec{G}^1$ *is push equivalent to* $\vec{G}^2$.

Proof. Note that there are well-known [32] linear time algorithms (for example, Kruskal's algorithm) for finding a spanning tree of a graph. Therefore, we can apply the procedure used in the proof of Theorem 1 to devise a polynomial algorithm.

Furthermore, we can count the number of distinct orientations, up to push equivalence, of a graph.

Corollary 2. *Let G be a graph with n vertices, m edges, and c components. Then the number of orientations of G, that are not push equivalent to each other, is 2^{m-n+c}.*

Proof. Let F be a spanning forest of G. As any two orientations of F are push equivalent due to Corollary 1, the distinct orientations of G will be determined by the orientations of the edges that do not belong to F. We know that a forest on n vertices having c components has exactly $(n - c)$ edges. Thus, there will be 2^{m-n+c} distinct choices for orienting the edges of G that do not belong to F.

Notice that the above count is for a labeled graph G and does not take isomorphism into account. In contrast, next, we list the oriented graphs that remain invariant (up to isomorphism) under the push operation, with the proof of the result deferred to the appendix due to space constraints.

Theorem 3. *Let $\overrightarrow{G}$ be a connected oriented graph that is isomorphic to each of its push equivalent oriented graphs. Then $\overrightarrow{G}$ is either the one vertex (oriented) graph $\overrightarrow{K}_1$, the orientation $\overrightarrow{K}_2$ of a complete graph on two vertices, or the non-directable 4-cycle.*

3   Connection with Signed Graphs

In the following, we establish a connection between pushable homomorphisms of oriented graphs and switchable homomorphisms of signed graphs. The theory of signed graphs has been well studied [2, 14, 22, 24, 25, 33] and it is deemed important due to its strong connections with graph coloring, graph minor theory, Ramsey Theory, flows, etc. However, let us recall some basics of signed graphs to keep our paper self contained.

Preliminaries of signed graphs: A *signed graph* (G, σ) is a graph G along with a *signature function* $\sigma : E(G) \to \{+, -\}$ that assigns a positive or a negative sign to the edges of G. A *sign-preserving homomorphism* of a signed graph (G, σ) to another signed graph (H, π) is a vertex mapping $f : V(G) \to V(H)$ such that for any edge uv of (G, σ), its image $f(u)f(v)$ is also an edge of (H, π) satisfying $\sigma(uv) = \pi(f(u)f(v))$.

To *switch* a vertex v of (G, σ) is to change the sign of all the edges incident to v. Given a set $S \subseteq V(G)$, to switch S is to push each vertex of S exactly once. The so-obtained signed graph is denoted by (G, σ^S) and we say that (G, σ) and (G, σ^S) are *switch equivalent*, denoted by $(G, \sigma) \equiv_s (G, \sigma^S)$. Moreover, $[(G, \sigma)]$ denotes the switch equivalent class of signed graphs, where (G, σ) is any representative of it. That means, given a simple graph G, the set of all of its signatures is partitioned into switch equivalent classes.

A *switchable homomorphism* of a signed graph (G, σ) to another signed graph (H, π) is a vertex mapping $f : V(G) \to V(H)$ such that there exists a switch equivalent graph (G, σ^S) of (G, σ) so that f is a sign-preserving homomorphism of (G, σ^S) to (H, π). If (G, σ) admits a switchable homomorphism to (H, π), then we denote it by $(G, \sigma) \xrightarrow{s} (H, \pi)$. Moreover, the *switchable chromatic number* of a signed graph (G, σ), denoted by $\chi_s((G, \sigma))$ is the smallest $|V(H)|$ such that $(G, \sigma) \xrightarrow{s} (H, \pi)$.

Let $\mathcal{BC}_p$ (resp., $\mathcal{BC}_s$) denote the set of all push (resp., switch) equivalence classes of oriented (resp., signed) bipartite graphs. We say $f : V(G) \to V(H)$ is a *morphism* of $[\overrightarrow{G}]$ to $[\overrightarrow{H}]$ (resp., of $[(G, \sigma)]$ to $[(H, \pi)]$) if $f : \overrightarrow{G} \xrightarrow{p} \overrightarrow{H}$ (resp., $f : (G, \sigma) \xrightarrow{s} (H, \pi)$) is a pushable (resp., switchable) homomorphism.

Theorem 4. *There exists a bijection* $\Phi : \mathcal{BC}_p \to \mathcal{BC}_s$ *such that f is a morphism of $[\overrightarrow{G}]$ to $[\overrightarrow{H}]$ if and only if f is a morphism of $\Phi([\overrightarrow{G}]) = [(G, \sigma)]$ to $\Phi([\overrightarrow{H}]) = [(H, \pi)]$.*

Proof. Firstly, we describe the function Φ through a construction. Let $\overrightarrow{G}$ be a bipartite oriented graph with partite sets A and B (in this order). Thus, all arcs of $\overrightarrow{G}$ have exactly one endpoint in A and the other in B. Now we construct a signed graph $(G, \sigma_{(\overrightarrow{G}, A, B)})$ in the following manner: we replace each arc from A to B with a positive edge and each arc from B to A with a negative edge. Now we set $\Phi([\overrightarrow{G}]) = [(G, \sigma_{(\overrightarrow{G}, A, B)})]$. Notice that the choice of A and B is not unique unless the graph G is connected. Also, even if G is connected, swapping the order of the partite sets A and B would produce a different signed graph $(G, \sigma_{(\overrightarrow{G}, B, A)})$. Therefore, it is important to prove that Φ is well-defined.

Let G be a bipartite graph with partite sets A, B (resp., A', B'). Note that, for showing that Φ is well-defined, it is enough to prove $(G, \sigma_{(\overrightarrow{G}, A, B)})$ and $(G, \sigma_{(\overrightarrow{G}^S, A', B')})$ are switch equivalent, where S is a vertex subset of G. To do so, we will find a way to switch a set of vertices of $(G, \sigma_{(\overrightarrow{G}, A, B)})$ to obtain $(G, \sigma_{(\overrightarrow{G}^S, A', B')})$.

As a connected bipartite graph has a unique bipartition, for any connected component of G, either $A \cap V(C) = A' \cap V(C)$ and $B \cap V(C) = B' \cap V(C)$ or $A \cap V(C) = B' \cap V(C)$ and $B \cap V(C) = A' \cap V(C)$. Let $A \cap A' = A_1$, $B \cap B' = B_1$, $A \cap B' = A_2$, and $B \cap A' = B_2$. Thus, it is possible to express G as a disjoint union of G_1 and G_2 such that A_1, B_1 are partite sets of G_1 and A_2, B_2 are partite sets of G_2. Therefore, if we switch the vertices of A_2 in $(G, \sigma_{(\overrightarrow{G}, A, B)})$, we will obtain the graph $(G, \sigma_{(\overrightarrow{G}, A', B')})$. Next, observe that it is possible to obtain the graph $(G, \sigma_{(\overrightarrow{G}^S, A', B')})$ from $(G, \sigma_{(\overrightarrow{G}, A', B')})$ by simply switching the set S of vertices. This proves that Φ is well-defined.

Next, we will show that the function Φ is injective. Suppose that $\Phi([\overrightarrow{G}]) = \Phi([\overrightarrow{G'}])$. To show Φ is injective, we need to show that $\overrightarrow{G'}$ is push equivalent to $\overrightarrow{G}$. First of all, note that $\overrightarrow{G}$ and $\overrightarrow{G'}$ must have the same underlying graph G for $\Phi([\overrightarrow{G}]) = \Phi([\overrightarrow{G'}])$ to hold. Let A, B be partite sets of G. Thus, in particular, we know that $(G, \sigma_{(\overrightarrow{G}, A, B)})$ and $(G, \sigma_{(\overrightarrow{G'}, A, B)})$ are switch equivalent. Suppose that $(G, \sigma_{(\overrightarrow{G'}, A, B)})$ can be obtained from $(G, \sigma_{(\overrightarrow{G}, A, B)})$ by switching the set S of vertices. Thus, if we switch the vertices of S in $\overrightarrow{G}$, we will obtain the oriented graph $\overrightarrow{G'}$, proving $\overrightarrow{G} \equiv_p \overrightarrow{G'}$. Hence, Φ is indeed injective.

To prove that Φ is a bijection, we are left to show that Φ is a surjection. Let $[(G, \sigma)]$ be a switch equivalence class of signed bipartite graphs. Moreover, let A, B be partite sets of G. Let us describe a construction of an oriented graph based on (G, σ). The oriented graph $\overrightarrow{G}$ is obtained by replacing all the positive edges with arcs from A to B and all the negative edges with arcs from B to A. Notice that, we will have $(G, \sigma_{(\overrightarrow{G}, A, B)}) = (G, \sigma)$. That means, we have found an oriented graph $\overrightarrow{G}$ satisfying $\Phi([\overrightarrow{G}]) = [(G, \sigma)]$. Hence, the function Φ is surjective, and thus, bijective.

Finally, suppose that $f : V(G) \to V(H)$ is a morphism of $[\overrightarrow{G}]$ to $[\overrightarrow{H}]$, This essentially means that there exists a homomorphism of an element of $[\overrightarrow{G}]$ to an element of $[\overrightarrow{H}]$. Without loss of generality, we may assume that there exists a homomorphism of $\overrightarrow{G}$ to $\overrightarrow{H}$. Next, let us fix a bipartition (A, B) of G. Furthermore, we will fix a bipartition (A', B') of H satisfying the properties $f(A) \subseteq A'$ and $f(B) \subseteq B'$. It is possible to find such a bipartition (A', B') of H as both G, H are bipartite graphs. Now, note that f is a sign-preserving homomorphism of $(G, \sigma_{(\overrightarrow{G}, A, B)})$ to $(H, \sigma_{(\overrightarrow{H}, A', B')})$. That implies, f is a morphism of $[\Phi(\overrightarrow{G})]$ to $[\Phi(\overrightarrow{H})]$. This proves the "only if" part of the statement. The "if" part of the statement can be proved similarly.

Remark 3. If we consider $\mathcal{BC}_p$ and $\mathcal{BC}_s$ (along with their morphisms) as categories, then the above mentioned one-to-one correspondence given by Φ is an isomorphic covariant functor which shows that the two categories are isomorphic. This is probably the underlying reason why we find similarities between the theory of pushable homomorphisms of oriented graphs and switchable homomorphisms of signed graphs.

3.1   Applications

We begin by showing how the entire theory of graph coloring can be viewed as a special case of pushable homomorphisms of oriented bipartite graphs.

Given a graph G, we construct an oriented bipartite graph $S(\overrightarrow{G})$ as follows: we replace each edge uv by a 4-cycle of the type uw_1vw_2u and orient the cycle to form a non-directable 4-cycle.

Let A, B be the two partite sets of the complete bipartite graph $K_{n,n}$. Let $\overrightarrow{K}^*_{n,n}$ be the orientation of $K_{n,n}$ in which a perfect matching is oriented from A to B and the rest of the edges are oriented from B to A.

Then, the following result is an analogue of one by Naserasr, Rollova, and Sopena [24].

Theorem 5. *Let G be a graph. Then, $\chi(G) \leq k$ if and only if $S(\overrightarrow{G}) \xrightarrow{p} \overrightarrow{K}^*_{k,k}$.*

Proof. Let $\hat{S}(G)$ be the signed graph obtained by replacing each edge uv of G by a (signed) 4-cycle uw_1vw_2u having three positive edges and one negative edge. Let $(K_{k,k}, \tau)$ be the signed graph whose underlying graph is $K_{k,k}$, and the set of negative edges is a perfect matching.

Notice that, $\hat{S}(G) \in \Phi([S(\overrightarrow{G})])$ and $(K_{k,k}, \tau) \in \Phi([\overrightarrow{K}^*_{n,n}])$. We know due to Naserasr, Rollova, and Sopena (see Theorem 6.2 in [24]) that $\chi(G) \leq k$ if and only if $\hat{S}(G) \xrightarrow{s} (K_{k,k}, \tau)$. Thus we are done using Theorem 4.

Observe that using Theorem 5, the equivalent formulation of the Four-Color Theorem will be: all $S(\overrightarrow{G})$ admits a pushable homomorphism to $\overrightarrow{K}^*_{4,4}$, where G is a planar graph. Note that the oriented graph $S(\overrightarrow{G})$ is thus an oriented planar bipartite graph, and furthermore, not all oriented planar bipartite graphs can be expressed as $S(\overrightarrow{G})$ for some planar G. Therefore, the next result is stronger than the Four-Color Theorem. We will prove it using Theorem 9.5 from [24], whose proof uses the Four-Color Theorem.

Theorem 6. *Every oriented planar bipartite graph admits a pushable homomorphism to $\overrightarrow{K}^*_{4,4}$.*

Proof. The signed graph $(K_{4,4}, \tau)$ is obtained by assigning negative signs to the edges of a perfect matching in $K_{4,4}$ and by assigning positive signs to the rest of the edges. We know that planar signed bipartite graphs admit a homomorphism to $(K_{4,4}, \tau)$ due to Naserasr, Rollová and Sopena [24]. The proof is completed by using Theorem 4 as $\overrightarrow{K}^*_{4,4} \in \Phi^{-1}([(K_{4,4}, \tau)])$.

The Folding Lemma by Klostermeyer and Zhang [18] is very useful to study the homomorphism properties of a planar graph. This lemma claims that for each planar graph G of shortest odd cycle length $2r+1$, and for each $k \leq r$, there is a planar homomorphic image H of G where every face of H is of length $2k+1$ and the shortest odd-length cycle of H is also of length $2k+1$. By considering negative cycles instead of odd-length cycles Naserasr, Rollova and Sopena [23], proved a similar result for the class of planar signed bipartite graphs. The analogous result for oriented graphs uses the notion of the "balance of an even cycle" for oriented graphs.

An oriented even ordered cycle $\overrightarrow{C}_{2n}$ on $2n$ vertices is *balanced* if there exists $S \subseteq V(\overrightarrow{C}_{2n})$ such that $\overrightarrow{C}^S_{2n}$ has exactly n forward and n backward arcs. Notice that in the definition of balanced, whether $\overrightarrow{C}_{2n}$ is ordered or not is redundant. Furthermore, an oriented even cycle is *unbalanced* if it is not balanced. By Theorem 1, one can observe that $\overrightarrow{C}_{2n}$ is balanced if and only if either n is even and $\overrightarrow{C}_{2n}$ is directable or n is odd and $\overrightarrow{C}_{2n}$ is non-directable. The length of a shortest balanced (resp., unbalanced) cycle of an oriented bipartite graph is its *balanced* (resp., *unbalanced*) girth.

Theorem 7. *Let $\overrightarrow{G}$ be a planar oriented bipartite graph with unbalanced girth g. If $C = v_0 \cdots v_{r-1} v_0$ is a balanced facial cycle of $\overrightarrow{G}$, or an unbalanced facial cycle of $\overrightarrow{G}$ with $r > g$, then there is an integer $i \in \{0, \cdots, r-1\}$ such that the oriented graph $\overrightarrow{G}'$ obtained from $\overrightarrow{G}$ by identifying v_{i-1} and v_{i+1} (subscripts are taken modulo r) is a pushable homomorphic image of $\overrightarrow{G}$ of unbalanced girth g.*

156   T. Das et al.

Proof. A signed cycle is positive (resp., negative) if it has even (resp., odd) number of negative edges. Observe that, if $\vec{C}$ is an even balanced (resp., unbalanced) oriented cycle, then any $(C, \sigma) \in \Phi([\vec{C}])$ is a positive (resp., negative) signed cycle. It is known (see Lemma 3.1 in [23]) that the exact statement is true if we replace oriented by signed, balanced by positive, and unbalanced by negative. Thus, the proof follows directly applying Theorem 4.

The repeated application of this result, gives us the following.

Theorem 8. *Given a planar oriented bipartite graph $\vec{G}$ of unbalanced girth g, there is a pushable homomorphic image $\vec{G}'$ of $\vec{G}$ such that:*

(i) $\vec{G}'$ is planar,
(ii) $\vec{G}'$ is an oriented bipartite graph,
(iii) $\vec{G}'$ is of unbalanced girth g,
(iv) every face of $\vec{G}'$ is an unbalanced cycle of length g.

Proof. In [23], the equivalent result for signed graphs is proved (see Corollary 3.2 in [23]). Therefore, the proof follows by applying Theorem 4.

The next result is the oriented graph analogue of one due to Naserasr, Pham and Wang [22], which was proved using the potential method [21]. An oriented graph $\vec{G}$ is *pushably $\vec{H}$-critical* if there does not exist a pushable homomorphism of $\vec{G}$ to $\vec{H}$ but every proper subgraph of $\vec{G}$ admits a pushable homomorphism to $\vec{H}$.

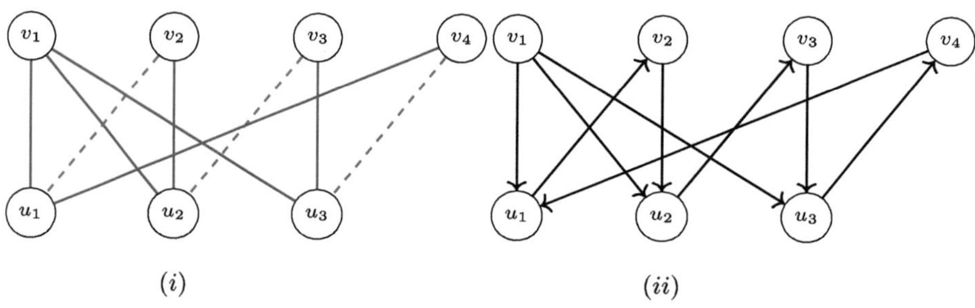

Fig. 1. (*i*) The signed graph (W, τ). (*ii*) The oriented graph $\vec{W}$.

Theorem 9. *Let $\vec{C}_4$ be the unbalanced directed 4-cycle. If $\vec{G}$ is a pushably $\vec{C}_4$-critical oriented graph which is not isomorphic to $\vec{W}$ depicted in Fig. 1(ii), then*

$$|A(\vec{G})| \geq \frac{4 |V(\vec{G})|}{3}.$$

Proof. Let (C_4, ϵ) denote the signed graph whose underlying graph is a 4-cycle, and it has exactly one negative edge. Notice that $\Phi([\overrightarrow{C}_4]) = [(C_4, \epsilon)]$ and $\Phi([\overrightarrow{W}]) = [(W, \tau)]$ (see Fig. 1).

Let (G, σ) be a signed graph that does not admit a switchable homomorphism to (C_4, ϵ), but any of its proper subgraph does. It is known (see Theorem 3.8 in [22]) that $|E(G)| \geq \frac{4|V(G)|}{3}$ unless (G, σ) is switchably isomorphic to (W, τ). Notice that the only graphs that can admit a homomorphism to (C_4, ϵ) (resp., $\overrightarrow{C}_4$) are bipartite graphs. Therefore, the proof follows using Theorem 4.

The equivalent statement of the above theorem for signed graphs is important for multiple reasons [22]. Interested readers are suggested to consult [22] for further details.

Remark 4. The pushable chromatic number of oriented hexagonal grids and square grids are studied (see Theorems 3.1, 3.4, 4.1, 4.2 in [5]). On the other hand, the switchable chromatic number of signed hexagonal grids and square grids are studied (see Theorem 7 in [17] and Theorems 10, 12 in [11]). One can note, both hexagonal and square grids are bipartite graphs. Thus, due to Theorem 4 one can conclude that Theorems 3.1, 3.4 in [5] (resp., Theorems 4.1, 4.2 in [5]) are equivalent to Theorems 7 in [17] (resp., Theorems 10, 12 in [11]).

4 Pushable Homomorphisms to Directed Odd Cycles

In this section, we analyse the complexity of the problem of determining whether an input oriented graph $\overrightarrow{G}$ admits a pushable homomorphism to an oriented odd cycle. Notice that, the complete complexity dichotomy of this problem is known due to Klostermeyer and MacGillivray [19] for general (unrestricted families) oriented graphs. However, in our result, the input graph is from restricted (sparse) graph families, and hence in the cases of pushable homomorphisms to oriented odd cycles, our result is a (partial) refinement to the findings of Klostermeyer and MacGillivray [19]. We state the decision problem as follows.

PUSHABLE $\overrightarrow{H}$-COLORING

Instance: A graph $\overrightarrow{G}$.

Question: Does there exist a pushable homomorphism of $\overrightarrow{G}$ to $\overrightarrow{H}$?

To prove our complexity result, we make use of a general correspondence between the pushable homomorphism to directed odd cycles and graph coloring. A simple graph G is *k-colorable* if it is possible to label the vertices of G using a set of k colors in such a way that adjacent vertices of G receive distinct labels. The decision version of the problem is as follows.

---

k-COLORING

Instance: A graph G.

Question: Is G k-colorable?

It is well-known that the k-COLORING problem is NP-complete for $k \geq 3$ [15]. We make use of this to show that the PUSHABLE $\overrightarrow{C}_{2k+1}$-COLORING is an NP-complete problem even for sparse graphs. The required construction and intermediate results have been deferred to the appendix due to space constraints.

Theorem 10. *The* PUSHABLE $\overrightarrow{C}_{2k+1}$-*COLORING problem is NP-complete for the family of bipartite graphs having girth at least $8k$, where $\overrightarrow{C}_{2k+1}$ is any oriented cycle on $2k + 1$ vertices.*

5   Conclusions

(1) While the lemma due to Zaslavsky [33] acts as the backbone for the theory of signed graphs, one of the major reasons why signed graphs have a deep connection to graph minor theory is due to its definition of minor [24] that preserves the signs of the cycles. Since we already have an oriented analogue of the lemma due to Zaslavsky [33] (Theorem 1), is it possible to have an analogue of "signed minor" in oriented graphs that interacts "nicely" with the push operation? The term "nicely" will vaguely refer to preserving the directability of ordered cycles.
(2) There are two "consistent" types of signed graphs that play important roles in the theory of homomorphisms of signed graphs [24]: odd signed graphs (possible to switch some vertices and make all edges negative) and bipartite signed graphs (underlying graphs are bipartite) [24]. Due to Theorem 4 proved in this article, we have seen that the exact oriented analogues of the latter are bipartite oriented graphs. However, what a natural oriented analogue of the former could be, remains an open question.
(3) There are several notions that are (or could be) defined for oriented graphs, but not for signed graphs such as oriented arc coloring and oriented chromatic index, oriented total coloring and oriented total chromatic number, deeply critical oriented graphs [3,7,27], etc. On the other hand, many signed graph notions like 0-free-coloring, circular chromatic number, flows [26,28], etc. are yet to be translated in the theory of oriented graphs. It seems to be a challenging task to, firstly, find analogues of these notions, and secondly, justify why the notion is a natural analogue. However, Theorem 4 addresses the second issue partially. We know that any analogous notions in the theory of oriented graphs and signed graphs must behave exactly the same way when restricted to the class of bipartite graphs. Therefore, it will be an interesting task to look for suitable analogues of notions to enrich both the theories.
(4) It will be interesting to study oriented graphs with respect to pushable homomorphisms as a graph category. To be precise, let $\mathcal{C}_p$ be the category in which oriented graphs play the role of objects and pushable homomorphisms play the role of morphisms. As a special case of the works done in [30], we know that categorical product and co-product exist for $\mathcal{C}_p$. Restricting two general open questions asked in [30], we would like to enquire that: (i) does there exist an analog of the notion of exponential graphs [15] in $\mathcal{C}_p$, (ii) does there exist an

analog of the density theorem [15] in $\mathcal{C}_p$. Both these questions are also open in the context of signed graphs.

(5) Since the complete dichotomy characterization for the PUSHABLE $\overrightarrow{H}$-COLORING problem is known due to Klostermeyer and MacGillivray [19], it is interesting to study the complexity of the problem for restricted classes of graphs. We did this for sparse graphs where our $\overrightarrow{H}$ was directed odd cycles. However, one may choose a different $\overrightarrow{H}$ and study the complexity of the problem for various graph classes. In particular, it will be interesting to have polynomial time solution for the problem of determining the pushable chromatic number for some important graph families such as graphs having low maximum degree, maximum average degree, edge density, treewidth, etc.

Acknowledgments. Pavan P D was supported by Research Council of Finland grants 338797 and 358718. Sagnik Sen was supported by SERB-MATRICS "Oriented chromatic and clique number of planar graphs" (MTR/2021/000858). S Taruni was supported by Centro de Modelamiento Matemático (CMM) BASAL fund FB210005 for center of excellence from ANID-Chile.

Appendix

Theorem 3. *Let $\overrightarrow{G}$ be a connected oriented graph that is isomorphic to each of its push equivalent oriented graphs. Then $\overrightarrow{G}$ is either the one vertex (oriented) graph $\overrightarrow{K}_1$, the orientation $\overrightarrow{K}_2$ of a complete graph on two vertices, or the non-directable 4-cycle.*

Proof. Let $\overrightarrow{G}$ be an oriented graph such that $|V(\overrightarrow{G})| \leq 3$. It is trivial to see that $\overrightarrow{K}_1$ and $\overrightarrow{K}_2$ are isomorphic to their push equivalent oriented graphs. Furthermore, there are no such oriented graphs on three vertices since the transitive triangle and directed triangle are push equivalent but not isomorphic.

Observe that the non-directable 4-cycle is isomorphic to all its push equivalent oriented graphs. Let $\overrightarrow{G}$ be a connected oriented graph with $|V(\overrightarrow{G})| \geq 4$ which is not the non-directable 4-cycle. It is enough to find two non-isomorphic push equivalent orientations of $\overrightarrow{G}$.

If G has diameter 1, that is, if G is a complete graph, then push all in-neighbors of a particular vertex v of $\overrightarrow{G}$ to obtain its push equivalent oriented graph $\overrightarrow{G}^1$. Notice that $\overrightarrow{G}^1$ has a source, and no sink. Next, push all out-neighbors of v of $\overrightarrow{G}$ to obtain its push equivalent oriented graph $\overrightarrow{G}^2$. Note that $\overrightarrow{G}^2$ has a sink, and no source. Hence, $\overrightarrow{G}^1$ and $\overrightarrow{G}^2$ are two non-isomorphic push equivalent orientations of $\overrightarrow{G}$.

If G has diameter at least 3, then we can take a spanning tree T of G and push some vertices of $\overrightarrow{G}$ so that the orientation of T has a single source v (say). This is possible due to Corollary 1. Let us denote this push equivalent orientation of $\overrightarrow{G}$ by $\overrightarrow{G}^1$. Observe that $\overrightarrow{G}^1$ has at most one source in it. On the other hand, choose two vertices v_1 and v_2 of G that are at a distance equal to 3. Then push the in-neighbors of v_1 and v_2 to obtain the push equivalent oriented graph $\vec{G}^2$ of $\vec{G}$. Notice that, both v_1 and v_2 are sources in $\vec{G}^2$. In particular, that means $\vec{G}^2$ has at least two sources, and thus cannot be isomorphic to $\vec{G}^1$.

If G has diameter 2, then it can have at most one cut vertex. If G has a cut vertex v, then we can push all in-neighbors of v in $\vec{G}$ to obtain its push equivalent oriented graph $\vec{G}^1$. On the other hand, we can push all out-neighbors of v in $\vec{G}$ to obtain its push equivalent oriented graph $\vec{G}^2$. Note that the cut vertex v is a source in $\vec{G}^1$ and a sink in $\vec{G}^2$. Thus, $\vec{G}^1$ cannot be isomorphic to $\vec{G}^2$ as v must map to v under any isomorphism.

Hence, suppose that G has diameter 2 and is 2-connected. Further, if G has a directable cycle C, then consider the graph $G' = G[V(G) \setminus V(C)]$ induced by the vertices which are not part of C. Take a spanning forest F of G', and assume that the components of F are $T_1, T_2, \cdots, T_r$. As G is connected and has diameter 2, there is a vertex v_i in T_i which is adjacent to a vertex of C, for each $i \in \{1, 2, \cdots, r\}$. Now push the vertices of $\vec{G}$ to obtain an orientation $\vec{G}^1$ where C is oriented as a directed cycle, T_i is oriented as a directed tree with a single source v_i (source inside the tree) for all $i \in \{1, 2, \cdots, r\}$, and v_i has an in-neighbor from $V(C)$. Note that, $\vec{G}^1$ does not have any source. On the other hand, pick any vertex v of $\vec{G}$ and push all its in-neighbors to obtain the push equivalent orientation $\vec{G}^2$ with v as a source. As $\vec{G}^1$ does not have any source, and $\vec{G}^2$ has at least one source v, they cannot be isomorphic.

Therefore, as all odd cycles are directable (forward or backward), the only case that we are yet to consider is when G is a 2-connected bipartite graph of diameter 2 whose cycles are all non-directable. Therefore, G must be a complete bipartite graph $K_{m,n}$, where $n \geq 3$ and $m \geq 2$. However, it is not possible to obtain an orientation of $K_{2,3}$ so that all its cycles are non-directable. Thus, such a graph $\vec{G}$ cannot exist. This completes the proof.

Construction A1. *Given a simple graph G, and any integer $k \geq 1$, we construct an oriented graph as follows. Each edge uv in G is replaced by two internally disjoint oriented paths $\vec{P}$ and $\vec{P}'$ of length $4k$ that connect u and v. The parities of forward arcs (with respect to traversal from u to v) in $\vec{P}$ and $\vec{P}'$ are even and odd, respectively. Let the so-obtained oriented graph be $\vec{G}^{(k)}$. Notice that there are several ways to obtain $\vec{G}^{(k)}$ from G. Irrespective of that, we get the same $\vec{G}^{(k)}$ up to push equivalence. Thus, in the context of pushable homomorphism, the construction of $\vec{G}^{(k)}$ from G is well-defined.*

Theorem A2. *For any $k \geq 1$, a simple graph G is $(2k + 1)$-colorable if and only if $\vec{G}^{(k)}$ admits a pushable homomorphism to the directed cycle $\vec{C}_{2k+1}$.*

Proof. Notice that the portion corresponding to an edge uv of G in $\vec{G}^{(k)}$ is two internally disjoint oriented paths $\vec{P}$ and $\vec{P}'$ connecting the vertices u and v. Moreover, $\vec{P}$ (resp., $\vec{P}'$) has even (resp., odd) number of forward and backward arcs, with respect to traversal from u to v. For convenience, let us denote this particular induced subgraph of $\overrightarrow{G}^{(k)}$ by $\overrightarrow{X}$.

Notice that, if we push some of the internal vertices of $\overrightarrow{P}$ (resp., $\overrightarrow{P'}$), the parity of the number of forward arcs remains unchanged. Since there are exactly $4k - 1$ internal vertices in $\overrightarrow{P}$ (resp., $\overrightarrow{P'}$), it is possible to obtain 2^{4k-1} different orientations of P having even (resp., odd) number of forward and backward arcs. On the other hand, as there are exactly $2^{4k} = 2 \cdot 2^{4k-1}$ different orientations of P (since it has $4k$ edges), we can say that it is possible to obtain any orientation of P having even (resp., odd) number of forward and backward arcs by pushing only the internal vertices of $\overrightarrow{P}$ (resp., $\overrightarrow{P'}$).

If we push one of the vertices among u and v in $\overrightarrow{X}$, the oriented path $\overrightarrow{P}$ (resp., $\overrightarrow{P'}$) gets modified to an oriented path having odd (resp., even) number of forward and backward arcs. That means, pushing exactly one of u and v exchanges the role of $\overrightarrow{P}$ and $\overrightarrow{P'}$ in $\overrightarrow{X}$.

Let $\overrightarrow{C}_{2k+1} = w_0 w_1 w_2 \cdots w_{2k} w_0$ be the directed cycle on $2k + 1$ vertices. Next, we want to show that it is possible to find a pushable homomorphism $f : \overrightarrow{X} \overset{p}{\to} \overrightarrow{C}_{2k+1}$ with $f(u) = w_i$ and $f(v) = w_j$ if and only if $i \neq j$, where $i, j \in \{0, 1, 2, \cdots, 2k\}$. Notice that this local information actually can be implemented globally. That is, if the above is correct, then we can say that $\overrightarrow{G}^{(k)}$ admits a pushable homomorphism to $\overrightarrow{C}_{2k+1}$ if and only if it is possible to label the vertices of G using the vertices of $\overrightarrow{C}_{2k+1}$ in such a way that adjacent vertices obtain different labels. This is essentially equivalent to the statement of the theorem we are trying to prove. Therefore, it is enough to show that given a partial function $f : V(\overrightarrow{X}) \to V(\overrightarrow{C}_{2k+1})$ defined by $f(u) = w_i$ and $f(v) = w_j$, it is possible to extend f to a pushable homomorphism of $\overrightarrow{X}$ to $\overrightarrow{C}_{2k+1}$ if and only if $i \neq j$. Notice that it is the same as obtaining a push equivalent orientation $\overrightarrow{X'}$ that admits a homomorphism to $\overrightarrow{C}_{2k+1}$, which is an extension of f if and only if $i \neq j$. Moreover, since pushing the vertex u or v only toggles the roles of $\overrightarrow{P}$ and $\overrightarrow{P'}$, if such a $\overrightarrow{X'}$ exist, without loss of generality we may assume that it is possible to obtain $\overrightarrow{X'}$ from $\overrightarrow{X}$ without pushing the vertices u and v. We will now formulate this problem algebraically.

Suppose that the oriented paths $\overrightarrow{P}$ and $\overrightarrow{P'}$ got modified to $\overrightarrow{P}_1$ and $\overrightarrow{P'}_1$, respectively. Moreover, suppose that there are exactly b (a positive even number less than or equal to $4k$) backward arcs in $\overrightarrow{P}_1$ and b' (a positive odd number less than $4k$) backward arcs in $\overrightarrow{P'}_1$. Therefore, it is possible to extend f to a homomorphism of $\overrightarrow{X'}$ to $\overrightarrow{C}_{2k+1}$ if and only if the following equations are satisfied:

$$i + (4k - b) - b \equiv j \pmod{2k + 1}, \tag{1}$$

$$i + (4k - b') - b' \equiv j \pmod{2k + 1}. \tag{2}$$

Since b (resp., b') is an even (resp., odd) number less than or equal to $4k$, we may assume that $b = 2a$ and $b' = 2a' + 1$ where $a \in \{0, 1, \cdots, 2k\}$ and

$a' \in \{0, 1, \cdots, 2k - 1\}$. Moreover, we may also assume that $(j - i) = \ell$ is some integer between 0 to $2k$. Thus, the modified Eqs. (5) and (5) will be as follows.

$$4k - 4a = 4(k - a) \equiv \ell \pmod{2k + 1}, \tag{3}$$

$$4k - 4a' - 2 = 4(k - a') - 2 \equiv \ell \pmod{2k + 1}. \tag{4}$$

As 4 and $2k + 1$ are co-primes and a varies among all integers from 0 to $2k$, the numbers of the form $4(k - a) \pmod{2k + 1}$ gives us $2k + 1$ distinct numbers modulo $2k + 1$. To be precise, Eq. (5) has a solution for all $\ell \in \{0, 1, \cdots, 2k\}$.

Similarly, as 2 and $2k + 1$ are co-primes and a varies among all integers from 1 to $2k - 1$, the numbers of the form $4(k - a') - 2 \pmod{2k + 1}$ gives us $2k$ distinct numbers modulo $2k + 1$. To be precise, Eq. (5) has a solution for all $\ell \in \{0, 1, \cdots, 2k\}$ except one value. Notice that, for $\ell = 0$, Eq. (5) will have any solution if and only if there is a solution for one of the following equations:

$$4k = 4a' + 2. \tag{5}$$

$$4k - 4a' - 2 = \pm(2k + 1). \tag{6}$$

However, Eq. (5) doesn't have a solution as the left hand side is divisible by 4, whereas the right hand side of it is not. Furthermore, Eq. (6) does not have a solution as the left hand side is even while the right hand side is odd. Thus, the proof is completed.

Corollary A3. *Let G^* be a graph obtained by adding a new vertex z to G which is adjacent to all vertices of G and let $\overrightarrow{G}^{(k)}$ be the oriented graph due to Construction A1. Then the following are equivalent for all $k \geq 3$:*

(i)   *The graph G is $(2k + 1)$-colorable.*
(ii) *The graph G^* is $(2k + 2)$-colorable.*
(iii) *The oriented graph $\overrightarrow{G}^{(k)}$ admits a pushable homomorphism to the directed cycle $\overrightarrow{C}_{2k+1}$ of order $2k + 1$.*

Proof. The equivalence of (i) and (ii) is trivial while the equivalence of (i) and (iii) follows directly from Theorem A2.

Theorem 10. *The PUSHABLE $\overrightarrow{C}_{2k+1}$-COLORING problem is NP-complete for the family of bipartite graphs having girth at least $8k$, where $\overrightarrow{C}_{2k+1}$ is any oriented cycle on $2k + 1$ vertices.*

Proof. Let G be a simple graph with n vertices and m edges. Then notice that $\overrightarrow{G}^{(k)}$ has $n + 2m(4k - 1) = n + 8mk - 2m$ vertices and $8km$ arcs. That means, the number of vertices and arcs of $\overrightarrow{G}^{(k)}$ is $O(n + m)$. Moreover, observe that the underlying graph of $\overrightarrow{G}^{(k)}$ is a bipartite graph with girth at least $8k$. It is well-known that the k-COLORING problem is NP-complete for all $k \geq 3$. Thus, using the correspondence established in Theorem A2, we can conclude this proof.

References

1. Babai, L., Cameron, P.J.: Automorphisms and enumeration of switching classes of tournaments. Electron. J. Comb. **7**(1), R38 (2000)
2. Beck, M., Zaslavsky, T.: The number of nowhere-zero flows on graphs and signed graphs. J. Comb. Theor. Series B **96**(6), 901–918 (2006)
3. Bensmail, J., et al.: Oriented total-coloring of oriented graphs. Discret. Math. **347**(11), 114174 (2024)
4. Bensmail, J., et al.: Pushable chromatic number of graphs with degree constraints. Discret. Math. **344**(1), 112151 (2021)
5. Bensmail, J., Das, T., Lajou, D., Nandi, S., Sen, S.: On the pushable chromatic number of various types of grids. Discret. Appl. Math. **329**, 140–154 (2023)
6. Bensmail, J., Nandi, S., Sen, S.: On oriented cliques with respect to push operation. Discret. Appl. Math. **232**, 50–63 (2017)
7. Borodin, O.V., Fon-Der-Flaass, D., Kostochka, A.V., Raspaud, A., Sopena, É.: On deeply critical oriented graphs. J. Comb. Theor. Series B **81**(1), 150–155 (2001)
8. Borodin, O.V., Kostochka, A.V., Nešetřil, J., Raspaud, A., Sopena, É.: On universal graphs for planar oriented graphs of a given girth. Discret. Math. **188**(1–3), 73–85 (1998)
9. Chung, F.R., Mumford, D.: Chordal completions of planar graphs. J. Comb. Theor. Series B **62**(1), 96–106 (1994)
10. Das, T., Sen, S.: Pushable chromatic number of graphs with maximum average degree at most $\frac{14}{5}$. Discret. Appl. Math. **334**, 163–171 (2023)
11. Dybizbański, J., Nenca, A., Szepietowski, A.: Signed coloring of 2-dimensional grids. Inf. Process. Lett. **156**, 105918 (2020)
12. Fomin, F.V., Villanger, Y.: Subexponential parameterized algorithm for minimum fill-in. SIAM J. Comput. **42**(6), 2197–2216 (2013)
13. Guégan, G., Ochem, P.: Complexity dichotomy for oriented homomorphism of planar graphs with large girth. Theoret. Comput. Sci. **596**, 142–148 (2015)
14. Harary, F.: On the notion of balance of a signed graph. Michigan Math. J. **2**(2), 143–146 (1953)
15. Hell, P., Nesetril, J.: Graphs and homomorphisms, vol. 28. OUP Oxford (2004)
16. Huang, J., MacGillivray, G., Wood, K.L.: Pushing the cycles out of multipartite tournaments. Discret. Math. **231**(1–3), 279–287 (2001)
17. Jacques, F.: On the chromatic numbers of signed triangular and hexagonal grids. Inf. Process. Lett. **172**, 106156 (2021)
18. Klostermeyer, W., Zhang, C.Q.: $(2+\epsilon)$-coloring of planar graphs with large odd-girth. J. Graph Theor. **33**(2), 109–119 (2000)
19. Klostermeyer, W.F., MacGillivray, G.: Homomorphisms and oriented colorings of equivalence classes of oriented graphs. Discret. Math. **274**(1–3), 161–172 (2004)
20. Klostermeyer, W.F., SÇŠltés, L.: Hamiltonicity and reversing arcs in digraphs. J. Graph Theory **28**(1), 13–30 (1998)
21. Kostochka, A., Yancey, M.: Ore's conjecture on color-critical graphs is almost true. J. Comb. Theor. Series B **109**, 73–101 (2014)
22. Naserasr, R., Pham, L.A., Wang, Z.: Density of C_4-critical signed graphs. J. Comb. Theor. Series B **153**, 81–104 (2022)
23. Naserasr, R., Rollová, E., Sopena, E.: Homomorphisms of planar signed graphs to signed projective cubes. Discrete Math. Theor. Comput. Sci. **15** (2013)
24. Naserasr, R., Rollová, E., Sopena, É.: Homomorphisms of signed graphs. J. Graph Theor. **79**(3), 178–212 (2015)

25. Naserasr, R., Sopena, É., Zaslavsky, T.: Homomorphisms of signed graphs: an update. Eur. J. Comb. **91**, 103222 (2021)
26. Naserasr, R., Wang, Z., Zhu, X.: Circular chromatic number of signed graphs. Electron. J. Comb. **28**(2), 2 (2021)
27. Ochem, P., Pinlou, A., Sopena, E.: On the oriented chromatic index of oriented graphs. J. Graph Theor. **57**(4), 313–332 (2008)
28. Raspaud, A., Zhu, X.: Circular flow on signed graphs. J. Comb. Theor. Series B **101**(6), 464–479 (2011)
29. Sen, S.: On homomorphisms of oriented graphs with respect to the push operation. Discret. Math. **340**(8), 1986–1995 (2017)
30. Sen, S., Sopena, É., Taruni, S.: Homomorphisms of (n, m)-graphs with respect to generalised switch. arXiv preprint arXiv:2204.01258 (2022)
31. Villanger, Y., Heggernes, P., Paul, C., Telle, J.A.: Interval completion is fixed parameter tractable. SIAM J. Comput. **38**(5), 2007–2020 (2009)
32. West, D.B.: Introduction to graph theory. Prentice Hall, 2 edn. (2000)
33. Zaslavsky, T.: Signed graphs. Discret. Appl. Math. **4**(1), 47–74 (1982)

Semi-online Scheduling with Look-Ahead

Debasis Dwibedy$^{(\boxtimes)}$ and Rakesh Mohanty$^{(\boxtimes)}$

Veer Surendra Sai University of Technology, Burla, Odisha 768018, India
{debasis.dwibedy,rakesh.iitmphd}@gmail.com

Abstract. The availability of partial future information can significantly improve the design of online algorithms for scheduling. In this work, we study the role of lookahead as an extra piece of information (EPI) in semi-online scheduling of jobs on identical machines with the makespan minimization objective. While competitive analysis remains the standard tool to evaluate online algorithms, incorporating lookahead raises fundamental questions about achievable bounds on the competitive ratio.

We introduce the k-lookahead semi-online scheduling model, where the processing times of the next k jobs are known when scheduling the current one. For the two-machine setting, we prove a lower bound of 4/3 on the competitive ratio and design a deterministic semi-online algorithm with 1-lookahead that matches this bound, showing that increasing the lookahead size beyond one does not yield further improvement. For the three-machine setting, we establish a lower bound of 15/11 and propose a 1-lookahead algorithm achieving an upper bound of 16/11, thus improving upon the best-known competitive ratio of 5/3 for classical online scheduling. Our results provide the first tight characterization of lookahead-based semi-online scheduling on two machines and significantly advance the known bounds for three machines, demonstrating the practical and theoretical impact of the lookahead model.

Keywords: Online Algorithm · Competitive Analysis · Semi-online Algorithm · Multi-processor Scheduling · Look-Ahead

1 Introduction

The m-machine scheduling is a well-studied NP-Complete problem with immense practical significance in diversified areas of computing, such as multiprocessor scheduling, network routing, transportation, task delegation, project management, and manufacturing, to name a few [14,20]. In this problem, we have to schedule the processing of a list of n jobs on a set of m machines with well-defined constraints and objectives, where $n > m$. Several practically-significant variants of the m-machine scheduling problem have been defined and studied in the literature based on machine models, availability of jobs, constraints, and objective functions. Offline and online scheduling are well-known variants of the m-machine scheduling problem.

N. Misra and A. Pandey (Eds.): CALDAM 2026, LNCS 16445, pp. 165–178, 2026.
https://doi.org/10.1007/978-3-032-17156-6_13

Offline and Online Scheduling. In offline scheduling, the whole list of jobs is available at the outset, while in online scheduling, jobs are given one by one in order [7,11]. Each received one must be scheduled immediately and irrevocably without knowledge of successive jobs with an objective to minimize the completion time of the job schedule, i.e., *makespan*.

Lookahead Model and its Significance. Lookahead is an impactful realistic concept that helps improve the performance of online scheduling algorithms. When we make a scheduling decision without the knowledge of future inputs, foreseeing a part of the future can help us to make better decisions with additional information. Lookahead can be considered as an *extra piece of information (EPI)* and constitutes a practically-significant model for defining semi-online scheduling problems.

Semi-online Scheduling. In practice, neither all jobs are available in advance as in the offline case, nor do they appear in an online fashion. The occurrence of jobs follows an intermediate framework known as semi-online [13,16]. Semi-online scheduling is a variant of online scheduling, where in addition to the current job, some EPI about the future ones is given. Semi-online scheduling has practical significance in resource allocation, network bandwidth utilization, packet routing, distributed data management, client request processing, manufacturing and production planning.

Competitive Analysis. The competitive analysis provides a mathematical framework to measure the performance of online and semi-online scheduling algorithms [22]. Let σ be a job sequence, $C_{ALG}(\sigma)$ be the makespan obtained by the semi-online algorithm ALG for σ, and $C_{OPT}(\sigma)$ be the optimum makespan for σ. The algorithm ALG is c-competitive, if there exists a positive number c and a non-negative constant b such that $C_{ALG}(\sigma) \leq c \cdot C_{OPT}(\sigma) + b$ for all σ. Here, c is the *upper bound (UB)* on the *competitive ratio (CR)* and $c \geq 1$. The smaller value of c indicates a better performance of ALG. The *lower bound (LB)* on the CR for a semi-online scheduling problem specifies an instance of the problem for which no semi-online algorithm obtains a better CR than the LB. If g is the *LB* of a semi-online scheduling problem, then there must be an instance σ such that no semi-online algorithm achieves a CR less than g. The common practice is to maximize the lower bound and minimize the upper bound on the CR. A semi-online algorithm is optimal if it achieves a CR matching the LB of the problem.

Related Work. The semi-online scheduling problem has been extensively investigated in the literature [3,6,9]. In semi-online scheduling, the objective is to improve the best-known bounds on the CR for a specific semi-online setting or to explore practically-significant new EPI to define innovative semi-online models. Here, we highlight only the most relevant semi-online settings and well-known competitive analysis results. Kellerer et al. [13] introduced the buffer concept into semi-online scheduling. A buffer is a storage structure that allows an online algorithm to defer the scheduling decision on the current job by storing the incoming jobs until there is a space in the buffer. When the buffer is full, and a job arrives, an algorithm selects a job from the available ones and schedules it on a machine. Different ways of selecting and assigning jobs lead to various semi-online algorithms. A buffer capable of storing k jobs allows an online algorithm to see $k+1$ jobs before making a scheduling decision. By considering a buffer of size $k(\geq 1)$ Kellerer et al. [13] proposed an optimal algorithm H_1 and achieved a CR of $\frac{4}{3}$. Zhang [25] proved a upper bound of $\frac{4}{3}$ on the CR by considering $k = 1$. The recent works on deterministic non-preemptive semi-online scheduling with a buffer are [5, 8, 15, 21].

Semi-online Scheduling with Lookahead. The semi-online scheduling with k-lookahead model enables an online algorithm to foresee the processing times of k future jobs on receiving an incoming one, where $k \geq 1$. The model does not require an explicit structure like a buffer to store and access future jobs. The additional information provided by the lookahead model on future jobs helps an online algorithm minimize the makespan of the job schedule. On the other hand, the unbounded size of the lookahead increases the execution time of an algorithm due to the overheads of extra lookups. We can carefully reduce the time by specifying a bound on the lookahead size.

Comparison with the Buffer Model. The buffer model with a k-sized buffer allows an algorithm to store and reorder the next k jobs before scheduling them. In contrast, the k-lookahead model reveals the processing times of the next k jobs as extra piece of information upon the arrival of a new job and requires each received job to be scheduled immediately, without any additional storage or reordering capability. Thus, the k-lookahead model avoids the storage overhead of the buffer model while still achieving improved competitive ratios through its extra piece of information.

Practical Significance. In a dynamic project scheduling scenario, relevant jobs arrive on the fly. A project manager allocates resources to each incoming job without knowledge of the future ones. However, a manager often foresees and estimates the processing times of future incoming jobs based on intuition, experience, or good connections with the concerned client to meet the requirements. Nevertheless, efficient resource allocation is a non-trivial challenge in such a scenario, which motivates our study on semi-online scheduling with a lookahead.

Research Motivation. The best-known CR for non-preemptive online scheduling with 2 and 3 identical machines settings is $\frac{3}{2}$ and $\frac{5}{3}$ respectively [11]. Faigle et al. [10] proved that no online algorithm achieves a better CR than the best-known ones. An important research challenge is: how much improvement in the CR can be achieved if an online algorithm foresees some portion of future jobs. Several semi-online scheduling models and algorithms answer the above question by considering various new EPI. However, less attention has been paid to online scheduling with lookahead [4, 18, 19]. According to our knowledge, a maiden work [26] has been reported in the literature about semi-online scheduling with lookahead in identical parallel machine settings for the makespan minimization objective. Here, the authors considered that an online algorithm knows the total processing time of first k jobs a priori. By considering $\sum_{i=1}^{k} p_i \geq \alpha\delta,$

the authors achieved a UB of $\frac{\alpha+2}{\alpha+1}$ on the CR, where p_i is the processing time of a job J_i, $\alpha \geq 2$, and δ is the largest processing time. In this paper, we introduce a new realistic lookahead model and address the following non-trivial research challenge: how much lookahead is sufficient for a semi-online algorithm to beat the best-known CR in 2 and 3 identical machine settings?

Our Contribution. First, we define the k-lookahead model in the context of semi-online scheduling. In addition, we prove a lower bound of $\frac{4}{3}$ on the CR for a 2-identical machine setting with $k \geq 1$. Subsequently, we design an optimal deterministic semi-online algorithm named 2-LA_1 and achieve a matching UB on the CR with $k = 1$. Furthermore, we highlight interesting remarks and critical observations on the 1-lookahead in the context of semi-online scheduling on a 2-identical machine setting. Next, we consider the 3-identical machine setup and design a $\frac{16}{11}$-competitive deterministic semi-online algorithm named 3-LA_1. We conclude by exploring several research challenges for prospective future work.

2   Preliminaries

In this paper, we use the standard terminologies and notations defined in a recent survey article on semi-online scheduling [6]. To denote a semi-online scheduling problem setting, we follow the 3-field $(\alpha|\beta|\gamma)$ notation framework of Graham et al. [12].

2.1 m-Machine Scheduling Problem ($P_m|C_{max}$)

We define the m-machine offline scheduling problem as follows.
Given a set $M = \{M_1, M_2, \ldots, M_m\}$ of $m(\geq 2)$ identical parallel machines and a list $J = \langle J_1, J_2, \ldots, J_{n-1}, J_n \rangle$ of n independent jobs, where each job J_i is characterized by its processing time p_i. We have to schedule n jobs on m machines such that the makespan C_{max}, or the maximum load on any machine is minimized, where load l_j of a machine M_j is the total sum of processing times of the jobs assigned to the machine M_j . We denote the problem as $P_m \mid C_{max}$.
Optimal Offline Algorithm assigns a given list of n jobs on m machines such that $\max\{l_j \mid 1 \leq j \leq m\}$ is the minimal.
In the context of the m-machine scheduling problem, it is an open problem to formally define the generic behavior of the optimal offline algorithm. However, the following lower bounds of the optimal makespan C_{OPT} have been used in the literature while computing the CR of online and semi-online scheduling algorithms.

Lemma 1. *[11]. Let $p_{max} = \max\{p_i \mid 1 \leq i \leq n\}$. The optimal makespan C_{OPT} for the problem $P_m|C_{max}$ is such that*

$$C_{OPT} \geq \begin{cases} p_{max} \\ \frac{1}{m} \cdot \sum_{i=1}^{n} p_i \end{cases}$$

Corollary 1.1. *For any instance σ of $P_m|C_{max}$, we have $C_{OPT} \geq \max\{p_{max}, \frac{1}{m} \cdot \sum_{i=1}^{n} p_i\}$.*

Corollary 1.2. *Let l_j be the load of machine M_j and $C_{ALG}(\sigma)$ be the makespan incurred by any algorithm ALG for an instance σ. Then $C_{ALG}(\sigma) = \max\{l_j \mid 1 \leq j \leq m\}$.*

Next, we will formally define the lookahead model in the context of semi-online scheduling.

2.2   Our Proposed Semi-online Scheduling Model with k-Lookahead

A semi-online algorithm receives n independent jobs one by one over a list. Upon receiving a job J_i, the algorithm in addition to p_i knows $p_{i+1}, p_{i+2}, \ldots, p_{i+k}$, where $1 \leq k < n$ and $p_i > 0$, $\forall i$. The received one must be scheduled non-preemptively and irrevocably without a further clue about $p_{i+(k+1)}, p_{i+(k+2)}, \ldots, p_n$. The objective is to minimize the makespan, denoted by C_{max}.

3   Our Competitive Analysis Results on 2 Identical Machines with k-Lookahead ($P_2|LA_k|C_{max}$)

3.1   Lower Bound Result

Theorem 1. *Let ALG be a deterministic semi-online algorithm for the problem $P_2|LA_k|C_{max}$. Then there exists an instance σ such that $\frac{C_{ALG}(\sigma)}{C_{OPT}(\sigma)} \geq \frac{4}{3}$.*

Proof. We construct an instance $\sigma = \langle J_1, J_2, \ldots, J_n \rangle$, where $p_i = x$ for $1 \leq i \leq n - k - 1$, $p_i = 1$, for $n - k \leq i \leq n - 1$, and $p_n = y$.

Given the instance σ, we compute $C_{OPT}(\sigma)$ and $C_{ALG}(\sigma)$. When the $(n-k)^{th}$ job arrives, the jobs $J_1, J_2, \ldots, J_{n-k-1}$ have been scheduled, and the processing time $p_{n-k}, p_{n-k+1}, \ldots, p_{n-1}, p_n$ are known. At this time, let the loads of machines M_1 and M_2 be l_1 and l_2. We consider the follwing two cases by assuming $l_1 \geq l_2$.

Case 1: $l_1 \geq 2l_2$. Consider $p_n = y = k$. We have $C_{ALG}(\sigma) \geq l_1$, while
$$C_{OPT}(\sigma) = \frac{l_1 + l_2 + 2k}{2} \leq \frac{l_1 + (l_1/2) + k}{2} \leq \frac{3l_1 + 4k}{4}. \text{ Implies,}$$
$$\frac{C_{ALG}(\sigma)}{C_{OPT}(\sigma)} \geq \frac{4l_1}{3l_1 + 4k} \to \frac{4}{3}, \text{ for large } n.$$

Case 2: $l_1 < 2l_2$. Consider $p_n = y = 2l_1 - l_2$. We have $C_{ALG}(\sigma) \geq l_2 + p_n = 2l_1$, while
$$C_{OPT}(\sigma) = \frac{l_1 + l_2 + k + 2l_1 - l_2}{2} \leq \frac{3l_1 + k}{2}. \text{ Implies,}$$
$$\frac{C_{ALG}(\sigma)}{C_{OPT}(\sigma)} \geq \frac{4l_1}{3l_1 + k} \to \frac{4}{3}, \text{ for large } n. \qquad \square$$

Research Question: *How much lookahead is sufficient to achieve a upper bound of $\frac{4}{3}$ on the CR for $P_2|LA_k|C_{max}$?*

In the following section, we will address the above research question.

3.2   An Optimal Semi-online Algorithm with 1-Lookahead : 2-LA_1

We design a deterministic semi-online algorithm named 2-LA_1 for 2 identical parallel machines setting by considering a lookahead of size 1. We prove that algorithm 2-LA_1 is optimal for the problem $P_2|LA_1|C_{max}$ and has a upper bound of $\frac{4}{3}$ on the CR.

The algorithm works as follows: upon receiving a new job J_i, it knows the processing times p_i and p_{i+1} in addition to the current loads of machines M_1 and M_2. The algorithm assigns J_i to machine M_1 if the updated load satisfies

$$l_1 + p_i \leq \frac{2}{3}(l_1 + l_2 + p_i + p_{i+1}).$$

Otherwise, J_i is scheduled on machine M_2. After scheduling the first $n - 1$ jobs, the last job J_n is assigned to the currently least-loaded machine. The threshold is chosen carefully so that the final maximum load on either machine never exceeds $\frac{4}{3}$ times the optimal makespan. The pseudocode of 2-LA1 is presented as Algorithm 1.

Algorithm 1. 2-LA_1

```
Initially, l₁ = l₂ = 0
When a new job Jᵢ arrives, pᵢ and pᵢ₊₁ are known.
FOR i = 1 to n − 1
  BEGIN
     IF (l₁ + pᵢ) ≤ ⅔ · (l₁ + l₂ + pᵢ + pᵢ₊₁)
        THEN assign job Jᵢ to machine M₁
        UPDATE l₁ = l₁ + pᵢ
     ELSE
        Assign job Jᵢ to machine M₂
        UPDATE l₂ = l₂ + pᵢ
  END
lₘᵢₙ ← {l₁, l₂}
Assign job Jₙ to machine Mⱼ for which lⱼ = lₘᵢₙ, where j = {1, 2}
UPDATE lⱼ = lⱼ + pᵢ
UPDATE l₁, l₂
Return      C₂₋LA₁ = max{l₁, l₂}
```

Algorithm 2-LA_1 considers the EPI p_{i+1} for efficient scheduling of each incoming job J_i. The objective is to show that $k = 1$ is sufficient to achieve an improved CR over the best-known upper bound of $\frac{3}{2}$ on the CR. Algorithm 2-LA_1 imbalances the current total load between machines M_1 and M_2 such that $l_1 \leq 2 \cdot l_2$ and the final maximum load of any machine is not more than $\frac{2}{3} \cdot \sum_{i=1}^{n} p_i$, or p_{max}.

In Theorem 2, we prove that algorithm 2-LA_1 has a CR of at least $\frac{4}{3}$ for $P_2|LA_1|C_{max}$.

Theorem 2. *There exists an instance σ of the problem $P_2|LA_1|C_{max}$ such that* $\frac{C_{2-LA_1}(\sigma)}{C_{OPT}(\sigma)} \geq \frac{4}{3}.$

Proof. Consider an instance σ with a sequence of n independent jobs, where $p_i = 1$, for $1 \leq i \leq n - 3$, $p_{n-2} = n$, $p_{n-1} = 2n + 3$, and $p_n = 2n$. Initially, the load $l_1 = l_2 = 0$. In σ, jobs are given one by one in order, and each job J_i from $i = 1$ to $n - 1$ satisfies the condition $l_1 + p_i \leq \frac{2}{3} \cdot (l_1 + l_2 + p_i + p_{i+1})$. Therefore, algorithm 2-LA_1 schedules each incoming job J_i to machine M_1, where $1 \leq i \leq n - 1$.

We now have $l_1 = 4n$ and $l_2 = 0$. Algorithm 2-LA_1 assigns the last job J_n to the current least loaded machine M_2. Hence, $C_{2-LA_1}(\sigma) \geq 4n$, while $C_{OPT} = 3n$. Therefore, $\frac{C_{2-LA_1}(\sigma)}{C_{OPT}(\sigma)} \geq \frac{4}{3}$. $\qquad\square$

3.3 Upper Bound Result

Before proving the upper bound on the CR of algorithm 2-LA_1, we consider some practically-significant problem instances. We show by following lemmas that algorithm 2-LA_1 is at most $\frac{4}{3}$-competitive for the considered instances of $P_2|LA_1|C_{max}$.

Lemma 2. *Let σ_1 be an instance of $P_2|LA_1|C_{OPT}$, where $p_i = x$, $\forall J_i$, and $x \geq 1$. Algorithm 2-LA_1 is such that $\frac{C_{2-LA_1}(\sigma_1)}{C_{max}(\sigma_1)} \leq \frac{4}{3}$.*

Proof. Let us consider n jobs in σ_1. Given that $\sum_{i=1}^{n} p_i = n \cdot x$ as $p_i = x$, $\forall J_i$. Irrespective of n is even or odd, algorithm 2-LA_1 assigns $\lfloor \frac{2n}{3} \rfloor$ jobs to machine M_1 and $\lceil \frac{n}{3} \rceil$ jobs to machine M_2. Therefore, $C_{2-LA_1}(\sigma_1) \leq \frac{2n}{3} \cdot x$. If n is even, algorithm OPT schedules $\frac{n}{2}$ jobs to each of the machines, and incurs $C_{OPT}(\sigma_1) = \frac{n}{2} \cdot x$. Therefore, $\frac{C_{2-LA_1}(\sigma_1)}{C_{OPT}(\sigma_1)} \leq \frac{4}{3}$.

If n is odd, algorithm OPT assigns $\lceil \frac{n}{2} \rceil$ jobs to one machine and the remaining jobs to the other, incurring $C_{OPT}(\sigma_1) = \lceil \frac{n}{2} \rceil \cdot x$, while $C_{2-LA_1}(\sigma_1) \leq \frac{2n}{3} \cdot x$. Therefore, $\frac{C_{2-LA_1}(\sigma_1)}{C_{OPT}(\sigma_1)} \leq \frac{4}{3}$. $\qquad\square$

Corollary 2.1. *Let σ_1 consists of n jobs, where $n = 6x$, $x \geq 1$, and $p_i = 1$, $\forall i$. Then $\frac{C_{2-LA_1}(\sigma_1)}{C_{OPT}(\sigma_1)} \leq \frac{4}{3}$.*

Remark 1: No algorithm can outperform Graham's LS strategy for scheduling a sequence of equal length jobs in 2-identical machine setting. Therefore, LS algorithm is optimal for σ_1.

Theorem 3. *Let σ be an instance of $P_2|LA_1|C_{max}$. Algorithm 2-LA_1 is such that $\frac{C_{2-LA_1}(\sigma)}{C_{OPT}(\sigma)} \leq \frac{4}{3}$ for all σ.*

Proof. We use the method of contradiction to prove the theorem.

Assume that there exists an instance σ with minimum number of jobs, which contradicts the theorem, where $\sigma = \langle J_1, J_2, \ldots, J_t \rangle$. Implies, $\frac{C_{2-LA_1}(\sigma)}{C_{OPT}(\sigma)} > \frac{4}{3}$.

Suppose, in the schedule generated by algorithm 2-LA_1 for σ, job J_x is the last job that completes its execution and $x < t$. Therefore, the instance $\sigma_1 = \langle J_1, J_2, \ldots, J_x \rangle$ makes $C_{2-LA_1}(\sigma_1) = C_{2-LA_1}(\sigma)$, while $C_{OPT}(\sigma_1) < C_{OPT}(\sigma)$. Implies, $\frac{C_{2-LA_1}(\sigma_1)}{C_{OPT}(\sigma_1)} > \frac{C_{2-LA_1}(\sigma)}{C_{OPT}(\sigma)} > \frac{4}{3}$.

This conveys, σ_1 is the smallest counterexample with minimum number of jobs, which contradicts our assumption on minimality of σ with t jobs. Hence, job J_x does not exist and the instance σ is the smallest counterexample in terms of number of jobs such that $\frac{C_{2-LA_1}(\sigma)}{C_{OPT}(\sigma)} > \frac{4}{3}$.

Therefore, for all other instances σ' with $t - 1$ jobs, $\frac{C_{2-LA_1}(\sigma')}{C_{OPT}(\sigma')} \leq \frac{4}{3}$.

Algorithm 2-LA_1 imbalances the total load on machines M_1 and M_2 such that $l_1 \geq l_2$ and $l_1 \leq \frac{2}{3} \cdot \sum_{i=1}^{t-1} p_i$. Therefore, we have $\frac{C_{2-LA_1}(\sigma')}{C_{OPT}(\sigma')} \leq \frac{4}{3}$. Clearly, the inclusion and scheduling of job J_t makes $\frac{C_{2-LA_1}(\sigma)}{C_{OPT}(\sigma)} > \frac{4}{3}$.

Prior to scheduling J_t, we have $l_{max} = \frac{2}{3} \cdot \sum_{i=1}^{t-1} p_i$ and $l_{min} = \frac{1}{3} \cdot \sum_{i=1}^{t-1} p_i$. Algorithm 2-$LA_1$ schedules J_t on M_j with $l_j = l_{min}$. Implies,

$$l_{min} + p_t > \frac{2}{3} \cdot \sum_{i=1}^{t} p_i$$
$$\implies \frac{1}{3} \cdot \sum_{i=1}^{t-1} p_i + p_t > \frac{2}{3} \cdot \sum_{i=1}^{t-1} p_i + \frac{2}{3} \cdot p_t$$

Without loss of generality, we assume that $l_2 < l_1$, implies, $l_2 + p_t > l_1 + \frac{2}{3} \cdot p_t$

$$\implies l_2 + p_t > 2 \cdot l_2 + \frac{2}{3} \cdot p_t$$
$$\implies p_t > l_1 + l_2, \text{ implies, } p_t > \sum_{i=1}^{t-1} p_i.$$

In such a case algorithm 2-LA_1 schedules each incoming job J_i from $i = 1$ to $t - 1$ on machine M_1 and assigns job J_t to machine M_2 to incur $C_{2-LA_1}(\sigma) = p_t$, while $C_{OPT}(\sigma) = p_t$. Implies, $\frac{C_{2-LA_1}(\sigma)}{C_{OPT}(\sigma)} = 1 \leq \frac{4}{3}$, which is a contradiction on our assumption on σ.

Therefore, there does not exist a counterexample to the theorem. We can now conclude that the theorem holds for all instances. $\qquad\square$

Observations

- Increasing the value of lookahead from 1 to any k would not lead to a better CR than $\frac{4}{3}$.
- The proposed 1-lookahead model and algorithm consider the arrival and scheduling of one job at a time with knowledge of non-zero positive processing time of the current and the next job. The model can not achieve a better CR than $\frac{3}{2}$ if it considers $p_i = 0$ for any J_i, and scheduling of J_{i+1} without knowledge of p_{i+2}. For example, consider an instance $\sigma = \langle J_1, J_2, J_3 \rangle$ with $p_1 = p_2 = 1$, and $p_3 = y$. Let ALG be a semi-online algorithm. If ALG assigns J_1 and J_2 to the same machine, then consider $p_3 = y = 0$. Implies, $C_{ALG}(\sigma) = 2$, while $C_{OPT}(\sigma) = 1$. ALG schedules J_1 and J_2 on different machines, then consider $p_3 = y = 2$. Implies, $C_{ALG}(\sigma) = 3$, while $C_{OPT}(\sigma) = 2$. Therefore, $\frac{C_{ALG}(\sigma)}{C_{OPT}(\sigma)} \geq \frac{3}{2}$., our model has significance over the traditional lookahead models concerning performance improvement.

4   Our Competitive Analysis Results on 3 Identical Machines with 1-Lookahead ($P_3|LA_1|C_{max}$)

Graham [11] explored the following multiprocessing timing anomaly. An increase in the number of machines does not necessarily guarantee the minimization of makespan. Graham proved that the LS algorithm is the optimal one with CR of $\frac{5}{3} \approx 1.66$ for 3-identical machine setting. However, Graham's optimal result does not hold in the semi-online setting with a bounded lookahead. In this section, we study the problem $P_3|LA_1|C_{max}$, design the algorithm $3\text{-}LA_1$, and show that the incorporation of a 1-lookahead allows $3\text{-}LA_1$ to achieve a strictly improved competitive ratio of $\frac{16}{11} \approx 1.45$, implying that the LS algorithm is no longer optimal when such an extra piece of information is available.

4.1   Lower Bound Result

Theorem 4. *Let ALG be a deterministic semi-online algorithm for the problem $P_3|LA_1|C_{max}$. Then there exists an instance σ such that $\frac{C_{ALG}(\sigma)}{C_{OPT}(\sigma)} \geq \frac{15}{11}$.*

Proof. We consider the adverserial instances of the problem to prove the lower bound. Let us assume that the adversary knows the nature of ALG and dynamically generates an instance σ such that $\frac{C_{ALG}(\sigma)}{C_{OPT}(\sigma)} \geq \frac{15}{11}$. We examine through the following cases, various strategies of scheduling σ that ALG could employ. We use the notation J_i/p_i to denote a job J_i and its corresponding processing time p_i. Without loss of generality, let us consider an instance $\sigma = \{J_1/7, J_2/4, J_3/p_3, \ldots, J_n/p_n\}$, where $n \geq 3$. Intially, the loads $l_1 = l_2 = l_3 = 0$.

Case 1: If ALG assigns J_1, J_2 and J_3 to three different machines (say M_1, M_2 and M_3 respectively), or to the same machine. Consider $p_3 = 4$, $p_4 = 7$ and $p_5 = 11$. We have $C_{ALG}(\sigma) \geq 15$, while $C_{OPT}(\sigma) = 11$. Therefore, $\frac{C_{ALG}(\sigma)}{C_{OPT}(\sigma)} \geq \frac{15}{11}$.

Case 2: If ALG assigns J_1, J_2 to the same machine (say M_1) and J_3 to a different machine (say M_2). Consider $p_3 = 4$, now $l_1 = 11$, $l_2 = 4$ and $l_3 = 0$.
    Sub-case 2.1: If J_4 is assigned to M_3. Consider $p_4 = 7$ and $p_5 = 11$. We have $C_{ALG}(\sigma) \geq 15$, while $C_{OPT}(\sigma) = 11$. Therefore, $\frac{C_{ALG}(\sigma)}{C_{OPT}(\sigma)} \geq \frac{15}{11}$.
    Sub-case 2.2: If J_4 is assigned to M_2. Consider $p_4 = 11$ and $p_5 = 7$. Implies, $C_{ALG}(\sigma) \geq 15$, while $C_{OPT}(\sigma) = 11$. Therefore, $\frac{C_{ALG}(\sigma)}{C_{OPT}(\sigma)} \geq \frac{15}{11}$.
    Sub-case 2.3: If J_4 is assigned to M_1. Consider $p_4 = 4$ and $p_5 = 11$. Clearly, $C_{ALG}(\sigma) \geq 15$, while $C_{OPT}(\sigma) = 11$. Therefore, $\frac{C_{ALG}(\sigma)}{C_{OPT}(\sigma)} \geq \frac{15}{11}$.

Case 3: If ALG assigns J_1, J_2 to different machines (say M_1 and M_2 respectively).
    Sub-case 3(a): Job J_3 is assigned to M_1. Consider $p_3 = 4$, now $l_1 = 11$, $l_2 = 4$ and $l_3 = 0$.
    Sub-case 3(a).1: If J_4 is assigned to M_3. Consider $p_4 = 7$ and $p_5 = 11$. We have $C_{ALG}(\sigma) \geq 15$, while $C_{OPT}(\sigma) = 11$. Therefore, $\frac{C_{ALG}(\sigma)}{C_{OPT}(\sigma)} \geq \frac{15}{11}$.

Sub-case 3(a).2: If J_4 is assigned to M_1. Consider $p_4 = 4$ and $p_5 = 11$. Clearly, $C_{ALG}(\sigma) \geq 15$, while $C_{OPT}(\sigma) = 11$. Therefore, $\frac{C_{ALG}(\sigma)}{C_{OPT}(\sigma)} \geq \frac{15}{11}$.

Sub-case 3(a).3: If J_4 is assigned to M_2. Consider $p_4 = 11$ and $p_5 = 7$. Implies, $C_{ALG}(\sigma) \geq 15$, while $C_{OPT}(\sigma) = 11$. Therefore, $\frac{C_{ALG}(\sigma)}{C_{OPT}(\sigma)} \geq \frac{15}{11}$.

Sub-case 3(b): Job J_3 is assigned to M_2. Consider $p_3 = 4$, now $l_1 = 7$, $l_2 = 8$ and $l_3 = 0$.

Sub-case 3(b).1: If J_4 is assigned to M_2 or M_3. Consider $p_4 = 7$ and $p_5 = 8$. We have $C_{ALG}(\sigma) \geq 15$, while $C_{OPT}(\sigma) = 11$.

Sub-case 3(b).2: If J_4 is assigned to M_1. Consider $p_4 = 8$ and $p_5 = 7$. Implies, $C_{ALG}(\sigma) \geq 15$, while $C_{OPT}(\sigma) = 11$. Therefore, $\frac{C_{ALG}(\sigma)}{C_{OPT}(\sigma)} \geq \frac{15}{11}$.

Therefore, we conclude that there exists an instance σ of the problem $P_3|LA_1|C_{max}$ such that $\frac{C_{ALG}(\sigma)}{C_{OPT}(\sigma)} \geq \frac{15}{11}$. $\qquad\square$

4.2 An Improved Semi-online Algorithm with 1-Lookahead : $3\text{-}LA_1$

We design a deterministic semi-online algorithm named $3\text{-}LA_1$ for 3 identical parallel machines setting by considering a lookahead of size 1. We prove that algorithm $3\text{-}LA_1$ has a CR of at most $\frac{16}{11} \approx 1.45$.

The algorithm works as follows: upon receiving a new job J_i $(1 \leq i \leq n-1)$, it knows the processing times p_i and p_{i+1} in addition to the current loads of machines M_1, M_2, and M_3. The algorithm assigns J_i to M_1 if the updated load of M_1 after the assignment satisfies the following condition

$$l_1 + p_i \leq \frac{16}{33}(l_1 + l_2 + l_3 + p_i + p_{i+1}).$$

If the condition fails, the algorithm assigns J_i to M_2 provided that the updated load of machine M_2 after the assignment satisfies the following condition

$$l_2 + p_i \leq \frac{15}{33}(l_1 + l_2 + l_3 + p_i + p_{i+1}).$$

If both conditions fail, then J_i is scheduled on machine M_3. After scheduling the first $n-1$ jobs, the final job J_n is assigned to the currently least-loaded machine. The algorithm then returns the makespan, which is the largest final load across the three machines. We present the pseudo code of $3\text{-}LA_1$ in *Algorithm 2*.

Algorithm $3\text{-}LA_1$ considers the EPI p_{i+1} for efficient scheduling of each incoming job J_i. Our goal is to show that $k = 1$ is sufficient to achieve an improved CR over the best-known upper bound of $\frac{5}{3} \approx 1.66$ on the CR. The objective of algorithm $3\text{-}LA_1$ is to keep an imbalance in the loads of machines M_1, M_2 and M_3 such that the final loads $l_1 \leq \frac{16}{33} \cdot \sum_{i=1}^{n} p_i$, $l_2 \leq \frac{15}{33} \cdot \sum_{i=1}^{n} p_i$ and $l_3 \leq C_{OPT}$. Our goal is to show that $C_{3-LA_1} \leq \frac{16}{33} \cdot \sum_{i=1}^{n} p_i$, while $C_{OPT} \geq \frac{1}{3} \cdot \sum_{i=1}^{n} p_i$, or $C_{OPT} \geq p_{max}$ such that $\frac{C_{3-LA_1}}{C_{OPT}} \leq \frac{16}{11} \approx 1.45$ for all instances of $P_3|LA_1|C_{max}$.

Algorithm 2. 3-LA_1

```
Initially, l₁ = l₂ = l₃ = 0
When a new job Jᵢ arrives, pᵢ and pᵢ₊₁ are known.
FOR i = 1 to n − 1
   BEGIN
      IF (l₁ + pᵢ) ≤ 16/33 · (l₁ + l₂ + l₃ + pᵢ + pᵢ₊₁)
         THEN assign job Jᵢ to machine M₁
         UPDATE l₁ = l₁ + pᵢ
      ELSE IF (l₂ + pᵢ) ≤ 15/33 · (l₁ + l₂ + l₃ + pᵢ + pᵢ₊₁)
         Assign job Jᵢ to machine M₂
         UPDATE l₂ = l₂ + pᵢ
      ELSE
         Assign job Jᵢ to machine M₃
   END
lₘᵢₙ ← min{l₁, l₂, l₃}
Assign job Jₙ to machine Mⱼ for which lⱼ = lₘᵢₙ, where j = {1, 2, 3}
UPDATE lⱼ = lⱼ + pᵢ
UPDATE l₁, l₂, l₃
Return     C₃₋ₗₐ₁ = max{l₁, l₂, l₃}
```

4.3   Upper Bound Result

We prove the competitiveness of algorithm 3-LA_1 by considering several critical cases as Lemmas to establish the upper bound on the CR.

Theorem 5. *Let σ be an instance of $P_3|LA_1|C_{max}$. Algorithm 3-LA_1 is such that $\frac{C_{3-LA_1}(\sigma)}{C_{OPT}(\sigma)} \leq \frac{15}{11}$ for all σ.*

Proof. We prove Theorem 5 by Lemma 3–6 as follows.

Lemma 3. *Let $T = \sum_{i=1}^{n} p_i$ for any instance σ of the problem $P_3|LA_1|C_{max}$. If the final load of machine M_3 is such that $l_3 \leq \frac{2}{33} \cdot T$, then $\frac{C_{3-LA_1}(\sigma)}{C_{OPT}(\sigma)} \leq \frac{16}{11}$.*

Proof. This is the obvious case, no matter the order of arrival of jobs in σ and their processing times, we have $C_{3-LA_1}(\sigma) = l_{max} = l_1 \leq \frac{16}{33} \cdot T$, while $C_{OPT}(\sigma) \geq \frac{1}{3} \cdot T$. Therefore, $\frac{C_{3-LA_1}(\sigma)}{C_{OPT}(\sigma)} \leq \frac{16}{11}$.   □

The following lemma put some insights into the maximum value of p_{max} and its impact on the performance of algorithm 3-LA_1.

Lemma 4. *If $p_{max} \geq \frac{16}{33} \cdot T$, then $C_{3-LA_1} = C_{OPT} = p_{max}$.*

Proof. If $p_{max} \geq \frac{16}{33} \cdot T$, then $\sum p_i \leq \frac{17}{33} \cdot T$ for the rest $n-1$ jobs. As we consider $n \geq 3$, w.o.l.g, let us normalize the processing times of the jobs such that $T = 33$. Clearly, $p_{max} \leq \frac{31}{33} \cdot T$. Consider the sequence $\sigma = \langle J_1/16, J_2/16, J_3/1 \rangle$ with least number of jobs. We have, $3 - LA_1(\sigma) = C_{OPT}(\sigma) = p_{max}$.
□

Lemma 5. *If $p_{max} = p_1$ or p_n, and $p_{max} \geq \sum_i p_i$ for rest $n-1$ jobs, then $C_{3-LA_1} = C_{OPT} = p_{max}$.*

Proof. Following the proof of the previous lemma, let us consider the sequences $\sigma_2 = \langle J_1/17, J_2/14, J_3/1, J_2/1 \rangle$ and $\sigma_3 = \langle J_1/1, J_2/1, J_3/14, J_2/17 \rangle$, where $p_{max} = 17$. For both sequences, we have, $3 - LA_1(\sigma) = C_{OPT}(\sigma) = p_{max}$.
□

Lemma 6. *Let σ be an instance with n jobs, where $p_i = 1$, $\forall J_i$, then* $\frac{C_{3-LA_1}(\sigma)}{C_{OPT}(\sigma)} \leq \frac{16}{11}$.

Proof. We explore and consider a worst job sequence σ to establish our claim. Let $n = 33x$, where $x \geq 1$. Here, $T = n$. Consider an instance σ, where $x = 1$, implies $T = 33$. We have $C_{3-LA_1} \leq 16$, while $C_{OPT}(\sigma) = 11$. Therefore, $\frac{C_{3-LA_1}(\sigma)}{C_{OPT}(\sigma)} \leq \frac{16}{11}$.
□

5   Research Challenges

Based on the related works from the literature till date, we explore the following non-trivial research challenges and present then as follows.

- Can the gap between our achieved UB of $\frac{16}{11}$ and LB of $\frac{15}{11}$ on the CR for the problem $P_3|LA_1|C_{max}$ be minimized?
- How much lookahead is sufficient to achieve a tight bound on the CR for the 3-identical machine setup?
- How much effective is the k-lookahead model for semi-online scheduling on a m-identical machine setting, where $m > 3$?
- Can the k-lookahead model be further characterized for special job sequences based on real-world applications?

6   Conclusion and Future Work

We introduced the k-lookahead semi-online scheduling model for minimizing makespan on identical machines and analyzed its impact on competitiveness. For two identical machines, we proved a tight lower bound of 4/3 and presented an optimal deterministic algorithm with 1-lookahead that achieves the bound, establishing that larger lookahead does not yield additional improvement. For three machines, we derived a lower bound of 15/11 and designed a 16/11-competitive semi-online algorithm with 1-lookahead, improving upon the classical bound of 5/3 for online scheduling.

These results provide the first tight characterization of lookahead-based semi-online scheduling on two machines and significantly improve known bounds for three machines. Our study demonstrates that limited lookahead can be both theoretically powerful and practically meaningful in online decision-making. This work opens several research directions as mentioned below:

- **Tight Bounds for Three Machines:** Narrowing the gap between the lower bound (15/11) and our achieved upper bound (16/11) remains an intriguing open problem.

- **Beyond Three Machines:** Extending the k-lookahead model to $m > 3$ identical machines and understanding how competitiveness scales with machine count.
- **Heterogeneous and Related Machines:** Investigating whether lookahead yields stronger improvements in non-identical machine models.
- **Randomization and Predictions:** Studying the benefits of randomized algorithms or machine learning-based predictions in combination with lookahead.
- **Practical Job Sequences:** Characterizing performance on special classes of job sequences arising in real-world applications (e.g., cloud scheduling, distributed data processing).

We believe that combining theoretical insights with practical constraints will further establish the k-lookahead model as a powerful paradigm in semi-online scheduling and resource allocation.

References

1. Ben-David, S., Borodin, A.: A new measure for the study of online algorithms. Algorithmica **11**(1), 73–91 (1994)
2. Bohm, M., Chrobak, M., Jez, L., Li, F., Sgall, J., Vesely, P.: Online packet scheduling with bounded delay and lookahead. In: Proc. 27th International Symposium on Algorithms and Computations, vol. 21, pp. 1–13 (2016)
3. Boyar, J., Favrholdt, L.M., Kudahl, C., Larsen, K.S., Mikkelsen, J.W.: Online algorithms with advice: a survey. ACM Comput. Surv., **50**(2), 1–34 (2017)
4. Coleman, B.J.: Lookahead scheduling in a real-time context: models, algorithms, and analysis. Dissertations, Theses, and Masters Projects, W & M ScholarWorks. Paper 1539623445
5. Ding, N., Lan, Y., Chen, X., Dosa, G., Guo, H., Han, X.: Online minimum makespan scheduling with a buffer. Int. J. Found. Comput. Sci. **25**(5), 525–536 (2014)
6. Dwibedy, D., Mohanty, R.: Semi-online scheduling: a survey. Comput. Operations Res., **139**, 105646 (2022)
7. Dwibedy, D., Mohanty, R.: Online list scheduling for makespan minimization: a review of the state-of-the-art results, research challenges, and open problems. SIGACT News. ACM **53**(2), 84–105 (2022)
8. Englert, M., Ozmen, D., Westermann, M.: The power of re-ordering for online minimum makespan scheduling. In: Proc. of the 49th Annual IEEE symposium on Foundations of Computer Science (2008)
9. Epstein, L.: A survey on makespan minimization in semi-online environments. J. Sched. **21**(3), 269–284 (2018). https://doi.org/10.1007/s10951-018-0567-z
10. Faigle, U., Kern, W., Turan, G.: On the performance of online algorithms for partition problems. Acta Cybernet. **9**(2), 107–119 (1989)
11. Graham, R.L.: Bounds for certain multiprocessor anomalies. Bell Syst. Tech. J. **45**(1), 1563–1581 (1966)
12. Graham, R.L., Lawer, E.L., Lenstra, J.K., Rinnooy kan, A.H.: Optimization and approximation in deterministic sequencing and scheduling: a survey. Ann. Discrete Math., **5**(1), 287–326 (1979)

13. Kellerer, H., Kotov, V., Speranza, M.G., Tuza, T.: Semi-online algorithms for the partition problem. Oper. Res. Lett. **21**, 235–242 (1997)
14. Kress, D., Meiswinkel, S., Pesch, E.: Mechanism design for machine scheduling problems: classification, and literature review. OR Spectrum **40**(3), 583–611 (2018)
15. Lan, Y., Chen, X., Ding, N., Dosa, G., Han, X.: Online makespan scheduling with a buffer. Front. Algorithms Aspects Inf. Manag., 161–171 (2012)
16. Liu, W.P., Sidney, J.B., Vliet, A.: Ordinal algorithms for parallel machine scheduling. Oper. Res. Lett. **18**, 223–232 (1996)
17. Ma, R., Guo, S., Miao, C.: A semi-online algorithm and its competitive analysis for parallel machine scheduling problem with rejection. Appl. Math. Comput. **392**, 125670 (2021)
18. Mao, W., Kincaid, R.K.: A lookahead heuristic for scheduling jobs with release dates on a single machine. Comput. Operations Res., **10**, 1041–1050 (1994)
19. Motwani, R., Saraswat, V., Torng, E.: Online scheduling with lookahead: multipass assembly lines. INFORMS J. Comput. **10**(3), 331–340 (1998)
20. Pinedo, M.L.: Scheduling, Theory, Algorithms, and Systems. 5th Edn, Springer Cham (2016). https://doi.org/10.1007/978-3-319-26580-3
21. Sun, H., Fan, R.: Improved semi-online makespan scheduling with a reordering buffer. Inf. Process. Lett. **113**(12), 434–439 (2013)
22. Tarjan, R.E., Sleator, D.D.: Amortized computational complexity. SIAM J. Algebraic Discrete Methods **6**(2), 306–318 (1985)
23. Torng, E.: A unified analysis of paging and caching. In: Proc. 36th Annual IEEE Symposium on Foundations of Computer Science, pp. 194–203 (1995)
24. Yong, H.: Semi-online scheduling problems for maximizing the minimum machine completion time. Acta Mathematicae Applicate sinica **17**(1), 107–113 (2001)
25. Zhang, G.: A simple semi-online algorithm for $P_2//C_{max}$ with a buffer. Inf. Process. Lett. **61**, 145–148 (1997)
26. Zheng, F., Chen, Y., Liu, M., Xu, Y.: Semi-online scheduling on two identical parallel machines with initial lookahead information. Asia Pacific J. Oper. Res. (2023). https://doi.org/10.1142/S0217595923500033

Algorithms and Hardness for Geodetic Set on Tree-Like Digraphs

Florent Foucaud[1], Narges Ghareghani[2], Lucas Lorieau[1]($\boxtimes$),
Morteza Mohammad-Noori[2], Rasa Parvini Oskuei[2], and Prafullkumar Tale[3]

[1] Université Clermont Auvergne, CNRS, Clermont Auvergne INP, Mines
Saint-Étienne, LIMOS, Clermont-Ferrand 63000, France
`florent.foucaud@uca.fr, lucas.lorieau@limos.fr`
[2] School of Mathematics, Statistics, and Computer Science, University of Tehran,
Tehran, Iran
`{ghareghani,mmnoori}@ut.ac.ir`
[3] Indian Institute of Science Education and Research Pune, Pune, India
`prafullkumar@iiserpune.ac.in`

Abstract. In the GEODETIC SET problem, an input is a digraph G and integer k, and the objective is to decide whether there exists a vertex subset S of size k such that any vertex in $V(G) \setminus S$ lies on a shortest path between two vertices in S. The problem has been studied on undirected and directed graphs from both algorithmic and graph-theoretical perspectives.

We focus on directed graphs and prove that GEODETIC SET admits a polynomial-time algorithm on ditrees, that is, digraphs *with* possible 2-cycles when the underlying undirected graph is a tree (after deleting possible parallel edges). This positive result naturally leads us to investigate cases where the underlying undirected graph is 'close to a tree'.

Towards this, we show that GEODETIC SET on digraphs *without* 2-cycles and whose underlying undirected graph has feedback edge set number fen, can be solved in time $2^{\mathcal{O}(\mathsf{fen})} \cdot n^{\mathcal{O}(1)}$, where n is the number of vertices. To complement this, we prove that the problem remains NP-hard on DAGs (which do not contain 2-cycles) even when the underlying undirected graph has constant feedback vertex set number. Our last result significantly strengthens the result of Araújo and Arraes [Discrete Applied Mathematics, 2022] that the problem is NP-hard on DAGs when the underlying undirected graph is either bipartite, cobipartite or split.

F. Foucaud—This author was supported by the French government IDEX-ISITE initiative 16- IDEX-0001 (CAP 20–25), the International Research Center "Innovation Transportation and Production Systems" of the I-SITE CAP 20–25, and by the ANR project GRALMECO (ANR-21-CE48-0004).

N. Ghareghani—The work of this author is based upon research funded by Iran National Science Fundation (INSF) hander project No. 4032600.

L. Lorieau—This author has received financial support from the CNRS through the MITI interdisciplinary programs and the IRL ReLaX.

P. Tale—Work supported by INSPIRE Faculty Fellowship offered by DST, GoI, India.

N. Misra and A. Pandey (Eds.): CALDAM 2026, LNCS 16445, pp. 179–193, 2026.
https://doi.org/10.1007/978-3-032-17156-6_14

Keywords: Geodetic Set · Directed Trees · NP-hardness · Parameterized Complexity · Feedback Edge Set Number · Feedback Vertex Set Number

1   Introduction

Harary, Loukakos, and Tsuros [25] introduced the concept of a *geodetic set*, defined as a set S of vertices of an undirected graph G such that every vertex of G lies on some geodesic (shortest path) between two vertices of S. Since then, the problem of computing a geodetic set has been extensively studied, both from the structural and algorithmic perspectives. It has become a central topic in *geodesic convexity* in graphs [18,29], and has found applications in diverse settings. We refer the reader to [17] and the references therein for a representative list of applications. For example, computing a minimum-size geodetic set can be seen as a network design problem, where one seeks to determine optimal locations of public transportation hubs in a road network [5].

Let GEODETIC SET be the corresponding algorithmic problem of determining a smallest possible geodetic set. Harary, Loukakos, and Tsuros [25] proved that the problem is NP-hard. See [16] for the earliest rigorous proof. Later works established that the problem remains NP-hard even for restricted graph classes [4–6,15,17]. This has motivated the study of algorithms for structured graph classes [1,4,6,15–17,28]. To cope with this hardness, the problem has also been investigated through the lenses of parameterized complexity [4,20,21,26,30] and approximation algorithms [5,12]. More recently, interest in this problem has been rejuvenated, as its 'metric-based nature' has led to interesting conditional lower bounds in parameterized complexity [20,22,30] and enumeration complexity [3], together with related problems of a similar 'metric-based' nature. Here, we use the term 'metric-based graph problem' as an umbrella notion for graph problems whose solutions are defined using a graph metric, for example, shortest distance between two vertices in the case of GEODETIC SET.

While the majority of studies on the GEODETIC SET problem focuses on undirected graphs, the problem has also been investigated for directed graphs (digraphs). Most of the existing work on geodetic sets in digraphs has concentrated on non-algorithmic questions like determining the minimum and maximum values of these sets, as well as the range of possible values. See [7–9,14,19,27] and the book [29, Chapter 6]. More recently, Araújo and Arraes [2] initiated the algorithmic study of the GEODETIC SET problem for digraphs. Before presenting their results, we formally define the problem addressed in this article and compare the results for undirected and directed graphs.

GEODETIC SET

Input: A directed graph D and an integer k.

Task: Determine whether there exists a subset $S \subseteq V(D)$ of size k such that every vertex in $V(D)$ lies on some directed shortest path between two vertices in S.

We define a 2-*cycle* of D as a directed cycle of length 2: a pair of vertices u, v such that both arcs (u, v) and (v, u) are present in D. Note that directed graphs with 2-cycles inherit the difficulties of designing algorithms for the undirected case (since the problem is equivalent on an undirected graph G and on the digraph obtained from G by replacing each edge by a directed 2-cycle). Hence, the presence of 2-cycles plays a critical role when stating results for digraphs.

As noted by Araújo and Arraes [2, Sec. 6], the unique minimum-sized geodetic set of an undirected tree T is equal to the set of its leaves. Define an *extremal vertex* of a digraph as a vertex that has either no incoming or no outgoing arcs, and denote by $\mathrm{Ext}(D)$ the set of extremal vertices of D (in an oriented graph, this set contains all leaves). The authors remark that in oriented trees, *i.e.* digraphs that have no 2-cycles and whose underlying undirected graph is a tree, the similar following property holds.

Proposition 1. ([2], Proposition 6.1). *Let D be an oriented tree. Then, $\mathrm{Ext}(D)$ is a minimum geodetic set of D.*

Using a non-trivial set of ideas and careful case analysis, the authors of [2] generalized this algorithm to digraphs without 2-cycles whose underlying undirected graph is a cactus. (A graph is called a *cactus* if each block is either an edge or a cycle, and hence every tree is a cactus graph.) We note that their case analysis holds only when the input digraph does not contain a 2-cycle. A digraph whose underlying graph is a tree is called a *ditree* [11,23]. Note that a ditree, contrary to oriented trees, might contain 2-cycles. As our first result, we present a polynomial-time algorithm for ditrees.

Theorem 2. GEODETIC SET *on ditrees admits a linear-time algorithm.*

This naturally leads us to investigate the problem for cases where the underlying undirected graph of the input digraph is 'close to a tree'. We consider the following two definitions of 'closeness to trees' for an undirected graph G: the minimum number of edges (respectively, vertices) that must be deleted from G to obtain a forest. The minimum size of a set of such edges (respectively, vertices) is called the *feedback edge set number* (respectively, *feedback vertex set number*) of the graph, and is denoted by $\mathrm{fen}(G)$ and $\mathrm{fvn}(G)$, respectively.

Kellerhals and Koana [26] proved that the GEODETIC SET problem admits an algorithm running in time $2^{\mathcal{O}(\mathrm{fen}(G)^2)} \cdot n^{\mathcal{O}(1)}$ when the input is an undirected graph. In the authors' own words, *"It turns out to be quite effortful to obtain fixed-parameter tractability, requiring the design and analysis of polynomial-time data reduction rules and branching before employing the main technical trick: Integer*

Linear Programming (ILP) with a bounded number of variables." Improving the running time of this algorithm has remained a challenging open problem. Note that obtaining a fixed-parameter tractable algorithm parameterized by fen for GEODETIC SET when the input digraph is allowed to have 2-cycles inherently encodes the difficulties encountered by the authors of [26]. As our next result, we show that a significantly faster algorithm (that does not need to use any ILP) can be obtained when considering the case in which 2-cycles are not allowed.

Theorem 3. GEODETIC SET *on digraphs without 2-cycles whose underlying undirected graph has feedback edge set number* fen, *admits an algorithm running in time* $2^{\mathcal{O}(\mathsf{fen})} \cdot n^{\mathcal{O}(1)}$, *where n is the number of vertices in the input digraph.*

Finally, we turn our attention to the feedback vertex set number and show that a fixed-parameter tractable algorithm for this parameter is not possible. Recently, Tale [30] proved that GEODETIC SET, restricted to undirected graphs, remains NP-hard even when the feedback vertex set fvn of the input graph is bounded. This implies that GEODETIC SET, restricted to directed graphs with possible 2-cycles, remains NP-hard even when the underlying undirected graph has constant fvn. We prove that a similar result holds even when 2-cycles (which help channel the hardness from the undirected case into the directed case), are absent. In fact, we prove the result not only for digraphs without 2-cycles but also for directed acyclic graphs (DAGs), which do not have directed cycles of any length. This significantly strengthens the results of Araújo and Arraes [2], which state that GEODETIC SET is NP-hard on DAGs even when the underlying undirected graph is bipartite, co-bipartite, or a split graph.

Theorem 4. GEODETIC SET *on DAGs (which do not have 2-cycles) whose underlying undirected graph has feedback vertex set number* 12, *is NP-hard.*

We remark that, although our reduction is inspired by the ideas in [30], it is significantly simpler than the one presented there.

Outline. Due to space constraints, we omit proofs of the results marked with ($\star$) and refer to the upcoming full version of the paper. We start with a linear-time algorithm on ditrees in Sect. 2 proving Theorem 2, followed by describing the fixed-parameter tractable algorithm mentioned in Theorem 3 in Sect. 3. We prove Theorem 4 in Sect. 4 and conclude with some open problems in Sect. 5.

Preliminaries. We present general definitions and results that will be used throughout the paper. We refer to the book [10] for terminology and details on parameterized complexity.

A *directed graph* (digraph for short) D consists of vertex set $V(D)$ and arc set $A(D)$, where each arc is an ordered pair of vertices. An arc from vertex v to vertex u is denoted by vu, and v is its *tail* and u is its *head*. A digraph is called an *oriented graph* if it does not contain a directed 2-cycle. The *underlying undirected graph* (or simply *underlying graph*) of some digraph D is the graph obtained by removing the orientation of each arc of D. An *oriented path* of a digraph D is a subgraph of D whose underlying graph is a path. A *directed path* (or *dipath*) is an oriented path for which all arcs are oriented in the same direction. A digraph is called *strongly connected* if every pair of vertices are connected by a directed path. The *in-neighborhood* of vertex u is denoted by $N^-(u)$ and its *out-neighborhood* is denoted by $N^+(u)$. The in-neighborhood of a subset S of $V(D)$ is $N^-(S) = (\cup_{u \in S} N^-(u)) \setminus S$, and similarly the out-neighborhood of S is defined. A vertex x is called a *source*, if $N^-(x) = \emptyset$, and it is called a *sink* if $N^+(x) = \emptyset$. A vertex that is a source or a sink is called *extremal*. For two vertices u and v, the set of vertices that lie in some shortest path from u to v is denoted by $I(u, v)$ and for a subset S of V, the *geodetic closure* of S, denoted by $I(S)$, is the set of all vertices which lie in some shortest path between two vertices of S. In other words, $I(S) = \cup_{u,v \in S} (I(u,v) \cup I(v,u))$. We also say that a vertex v *is covered* by two vertices u and w if $v \in I(\{u, w\})$. In a directed path P from u to v, vertices u and v are called the *tail* and the *head* of P, respectively. The vertices of P which are neither the tail nor the head of P are called its *inner vertices*. A *directed acyclic graph* (DAG for short) is a digraph which does not contain any directed cycle.

A vertex v is *transitive* if, for every in-neighbor u_1 and out-neighbor u_2 of v, either the arc $u_1 u_2$ exists, or $u_1 = u_2$.[1] A vertex of a digraph is called a *leaf* if, in the underlying undirected graph, it has degree 1. Note that any leaf of a digraph is either a sink, a source, or transitive.

We conclude this section with the following lemma.

Lemma 5. ([2]). *In any digraph D, every source, sink and transitive vertex of D (in particular, every leaf of D) belongs to every geodetic set of D.*

2   Linear-Time Algorithm for Ditrees

In this section, we consider the problem of finding the geodetic number of a ditree and prove the following theorem.

Theorem 2. GEODETIC SET *on ditrees admits a linear-time algorithm.*

In the context of digraphs admitting 2-cycles, we say that a vertex of a digraph D is a *leaf* of D if it is a leaf of the underlying graph of D. When T is an oriented tree, a minimum-size geodetic set (mgs for sort) may contain some non-leaf vertices: as observed in Proposition 1, in this case, an optimal mgs always consists of all sources and sinks of the tree. To prove Theorem 2, we reduce the problem to finding a mgs in directed trees where the only 2-cycles present are adjacent to some leaf of the graph. We argue then that taking all extremal vertices of the graph, in addition to the leaves contained in a 2-cycle, yields a mgs. Intuitively, the graph obtained behaves almost exactly as an oriented tree, which enables us to extend Proposition 1 naturally.

Let S be a maximal strongly connected component of D. Then we call S a *source set* if $N^-(S) \setminus S = \emptyset$. Similarly, we call S a *sink set* if $N^+(S) \setminus S = \emptyset$. We can state a simple observation about sink and source sets.

---

[1] Note that the latter condition was not present in the definition from [2], since the authors only considered digraphs without 2-cycles.

Observation 1. *Let S be a source set or a sink set. Then, any two adjacent vertices of S form a 2-cycle.*

We define now a new graph, whose structure is very similar to an oriented tree.

Definition 6. *Let T be a ditree. We define the* contracted *ditree T^c of T by iteratively contracting every 2-cycle of T that does not contain any leaf vertex into a single vertex.*

In particular, in T^c, the only 2-cycles are adjacent to some leaf of the graph. T^c is thus an oriented tree to which some leaves are contained in 2-cycles. Lemma 7 shows that in contrast to oriented trees, an mgs in a general ditree may contain vertices which are neither sources, sinks nor leaves. Then, Lemma 8 states that if all source and sink sets are either composed of a unique vertex or contain a leaf, then we can easily find a mgs in the considered ditree. In particular, this can be applied on T^c. Finally, Lemmas 9,10 justify that we can extend any mgs of T^c to T.

Lemma 7 ($\star$). *Let T be a ditree and let S be a source set or sink set of T. Then, every geodetic set of T includes at least one vertex from S.*

Lemma 8 ($\star$). *Let T be ditree for which any source (resp. sink) set of size at least 2 contains a leaf. Let $S \subseteq V(T)$ be the set of all sink (resp. source) vertices and all leaves of T. Then, S is a geodetic set of T of minimum size.*

Lemma 9. *Let T be a ditree and S be a source set or sink set of size at least 2 that contains no leaves. Suppose M is an mgs for T that includes exactly one vertex from S. Define $N \subseteq V(T)$ as the set obtained from M by replacing the unique vertex of S with any other vertex of S. Then, N is also an mgs for T.*

Proof. Without loss of generality, assume that S is a source set. Arguments are similar for sink sets. Let $u \in S \cap M$, we first show that the lemma holds when u is replaced by one of its neighbours in S. Let $P_1, \ldots, P_m$ be all maximal directed paths starting from u and ending at some vertex of M. Choose a vertex $v \in S$ such that v is an out-neighbour of u. Since S is a source set, v is also an in-neighbour of u. Because M is an mgs and S is a source set, some paths among $P_1, \ldots, P_m$ must contain v. let $P_1, \ldots, P_i$ denote exactly those paths that contain v. Since T is a ditree, each such path necessarily begins with the arc uv. Moreover, as u is not a leaf, there should also exist paths among $P_1, \ldots, P_m$ that do not contain v. Assume without loss of generality, that $i < m$ and $P_{i+1}, \ldots, P_m$ are all these paths.

Now, for each $1 \leq j \leq i$, let P_j' be the path obtained from P_j by deleting the vertex u and the arc uv. For each j with $i + 1 \leq j \leq m$, let P_j' be the directed path obtained from P_j by prepending the arc uv. It follows that the set N covers exactly the same set of vertices as M, and hence N also is an mgs.

Next, we show that the lemma remains valid if u is replaced by some vertex $w \in S$ that is not a neighbour of u. Since S is a source set, there is a directed path from u to w all whose internal vertices are in S. Moreover, each arc in this path is part of a 2-cycle in T. Therefore, starting from u, we may successively replace the current vertex by its neighbour along the path, until reaching w. At each step, the resulting set is an mgs and thus, after the final replacement, the set obtained by substituting u with w is also an mgs, as required. $\square$

Lemma 10. *Let T be a ditree and $\mathcal{S} = \{S_1, S_2, \ldots, S_t\}$ be the set of all source sets and sink sets of T which do not contain any leaf. Let $\mathcal{L}$ be the set consisting of all leaves of T and M be a subset of $V(T)$ that contains $\mathcal{L}$ and exactly one vertex from each S_i. Then, M is an mgs for T.*

Proof. We prove the theorem using induction on the number of vertices of T. If T has 2 vertices, then every vertex of T is a leaf. Using Lemma 5, $\mathcal{L}$ is an mgs for T. By the induction hypothesis, suppose that the theorem is true for any ditree of size less than n. Let T be a ditree with n vertices. If each source (resp. sink) set of T of size at least 2 contains a leaf, then each source (resp. sink) set which does not contain a leaf is a source (resp. sink) vertex. So in this case, the theorem holds, using Lemma 8.

Now, without loss of generality, suppose that $S_t \in \mathcal{S}$ is a source set and it has at least two vertices. Let u, v be two vertices in S_t which are adjacent and T' be a ditree obtained from T by contracting the edge between u and v. Let w be the new vertex in T' that replaces two vertices u and v of T. Note that T' is a ditree with the same set of leaves as T, and with the same set of source (resp. sink) sets as T, except that S_t is replaced by S_t', where $S_t' = (S_t \setminus \{u, v\}) \cup \{w\}$. Hence, by the induction hypothesis, there is an mgs of T' that contains all leaves and a vertex from each of the set $S_1, S_2, \ldots, S_t'$.

By Lemma 9, we conclude that a subset of vertices of T' that contains all leaves of T' and an arbitrary vertex from each of the set $S_1, S_2, \ldots, S_t'$ is an mgs.

Let $P_1', \ldots, P_r'$ be the set of all maximal directed paths in T' starting from w. Since w is not a leaf, $r \geq 2$. One can see that the end-vertex of each of these paths is either a leaf or belongs to a sink set. Moreover, no two of these paths terminate at the same sink set; Otherwise the underlying undirected graph would contain a cycle, which is impossible. Since T' has an mgs which contains all leaves and exactly one vertex from each source (resp. sink) set which does not contain any leaf, using Lemma 9, T' has an mgs, M', which contains w and the end-vertices of each path $P_1', \ldots, P_r'$. Define $M = M' \setminus \{w\} \cup \{u\}$. We claim that M is an mgs for T. Since T and T' have the same set of leaves and the same number of sink (resp. source) sets, Lemma 8 implies that every mgs of T contains at least $|M'|$ vertices. Therefore, it remains to show that M is a geodetic set for T.

Since u and v are in a same source set, T contains both arcs uv and vu. Since M' is an mgs for T', each vertex of T other than v is covered by a path with both ends in M. Now, it suffices to show that v is covered by some path with ends in M (note that T is a ditree, so each path is a shortest path between its end vertices), Now, we claim that v belongs to some P_i, $1 \leq i \leq r$. If not, $T \setminus \{v\}$ is connected and this happens only if v is a leaf, which is not the case, a contradiction. $\square$

Proof. (of Theorem 2). By Lemma 10, every mgs of T contains all leaves and an arbitrary vertex of any source (sink) set that does not contain any leaf. Since any leaf in T is also a leaf in T^c and vice-versa, the algorithm proceeds as follows. First, we construct the contracted ditree T^c from T in linear time by contracting every 2-cycle that is not incident with a leaf. Then, we determine all source (resp. sink) vertices (and the corresponding source (sink) sets in T), and leaves of T^c, which can also be done in linear time by examining the in-degree and out-degree of each vertex. Since each step requires only linear time in the number of vertices and edges, the overall complexity is $O(|V(T)| + |E(T)|)$.   □

3   Algorithm Parameterized by Feedback Edge Set Number

In this section, we present an algorithm solving GEODETIC SET parameterized by the feedback edge set number of the input graph and prove Theorem 3.

Theorem 3. GEODETIC SET *on digraphs without 2-cycles whose underlying undirected graph has feedback edge set number fen, admits an algorithm running in time* $2^{\mathcal{O}(\mathsf{fen})} \cdot n^{\mathcal{O}(1)}$*, where n is the number of vertices in the input digraph.*

We introduce useful notions from [13]. A *core vertex* is a vertex of degree at least 3. A *core path* of digraph D is a path in the underlying graph of D between two core vertices with only degree 2 internal vertices. Note that both endpoints are allowed to be the same core vertex: in this case, we call the core path a *core cycle*. We call *proper core path* a core path whose endpoints are two different core vertices. A *leg* of the underlying graph of D is a (non-empty) path between a core vertex and a leaf in said graph. The *base graph* of some undirected graph G is the graph obtained by removing iteratively leaves from G until no leaf is present. We say that the base graph of a digraph D is the base graph of its underlying undirected graph. Note that a vertex of a base graph is either a core vertex, or an inner vertex of some core path. The following observation comes from [26, Observation 5].

Observation 2. *The base graph of any undirected graph G has at most $2\mathsf{fen}(G) - 2$ core vertices and at most $3\mathsf{fen}(G) - 3$ core paths.*

We say that the *base digraph D_b* of D is the subgraph of D such that its underlying graph is the base graph of D. A *hanging tree* of the underlying graph of D is the union of some legs removed to form the base graph of D so that the union of those legs forms a connected component. The *root* of a hanging tree of D is the vertex of the base graph that was linked to the last removed leg of the hanging tree considered. It is easily seen that the underlying undirected graph of D can be decomposed into its base graph and a set of maximal hanging trees. A *hanging ditree* of D is a subgraph of D such that its underlying graph is a hanging tree of D, and its root is the root of the associated hanging tree. Similarly, D can be decomposed into its base digraph and a collection of maximal hanging ditrees. We call an *oriented core path* any core path of the base graph of

D whose edges are oriented based on the orientations in D. If an oriented core path forms a directed path in D, we call it a *core dipath*.

First, we will argue that deciding which vertices to take in a solution in the hanging ditrees of D is not difficult, using the following observation.

Observation 3 ($\star$). *Let v be a vertex of a hanging ditree T of D rooted in r.*

- *If v has an outgoing arc, v can either reach r or a sink $w \in V(T)$.*
- *If v has an incoming arc, v can be reached by either r or a source $u \in V(T)$.*

Claim 1 ($\star$). *Let D be a digraph with S^0 its set of extremal vertices. Suppose S is a geodetic set of D. Then, $S' = (S \cap V(D_b)) \cup S^0$ is a geodetic set of D.*

Claim 1 identifies which vertices of hanging ditrees are part of minimum geodetic sets, so we next focus on vertices of oriented core paths. We identify three different cases, depending on the number of extremal vertices present in the considered oriented core path. Note that when there is no extremal vertex among the inner vertices of an oriented core path, it is in fact a core dipath.

Claim 2. *Let D be a digraph and S^0 the set of extremal vertices of D. Consider an oriented core path P of D and denote by V_P its inner vertices. Number them as $V_P = \{v_1, \ldots, v_l\}$ so that two neighbors in P have consecutive indexes. Suppose that $|S^0 \cap V_P| \geq 2$, and denote by v_i (respectively v_j) the vertex with minimum index (respectively with maximum index) of $S^0 \cap V_P$. We have $\{v_i, \ldots, v_j\} \subseteq I(S^0)$ and there exists an* **mgs** *S of D so that $V_P \cap S = V_P \cap S^0$.*

Proof. We argue that for any $k \in [i, j]$, v_k is covered by two vertices of S^0. Indeed, either v_i and v_j are the only two extremal vertices of V_P, and then there exists a dipath between the two covering all considered vertices, or there exist other extremal vertices between v_i and v_j. Consider v_m the vertex with minimum index among those vertices. If v_i is a sink, then v_m is a source (and vice versa), so if $k \in [i, m]$, $v_k \in I(v_i, v_m)$, otherwise one can apply recursively this argument with vertices v_m and v_j.

Suppose now that S is an **mgs** of D, and that some vertex v_p belongs to $(V_P \cap S) \setminus S^0$. Since $\{v_i, \ldots, v_j\} \subseteq I(S^0)$, if $i < p < j$, $S \setminus \{v_p\}$ is still a geodetic set, which is in contradiction with the minimality of S. Suppose then without loss of generality that $p < i$. Since S is a geodetic set, the core vertex $v^{\leftarrow}$ neighboring v_1 is either in S or covered by a shortest path starting at some vertex $u \in S$ and ending at some vertex $w \in S$. In the former case, vertices $v_1, \ldots, v_i$ are covered by the unique shortest path between $v^{\leftarrow}$ and v_i and v_p can be removed from S as above. In the latter case, consider $S' = (S \setminus \{v_p\}) \cup \{v^{\leftarrow}\}$. Any shortest path for which v_p is an endpoint can either be extended to a shortest path with $v^{\leftarrow}$ as an endpoint, or goes through $v^{\leftarrow}$. Thus, S' is also an **mgs** of D. The same reasoning applies if $j < p$ by considering $v^{\rightarrow}$, the core vertex neighboring v_l. $\square$

Claim 3 ($\star$). *Let P be an oriented core path of some digraph D and denote by V_P the inner vertices of P. Number vertices of P such that $V_P = \{v_1, \ldots, v_l\}$. Suppose that $|V_P \cap S^0| = 1$ and denote the vertex of $V_P \cap S^0$ by v_p. There exists an* **mgs** *S of D such that $|V_P \cap S| \leq 3$ and $V_P \cap S \subseteq \{v_1, v_p, v_l\}$. In particular, $V_P \cap S$ contains at most two non-extremal vertices.*

Claim 4 ($\star$). *Let P be a core dipath of some digraph D. Denote by V_P the inner vertices of P and number them in the order induced by the arcs of the dipath so that $V_P = \{v_1, \dots, v_l\}$. There exists an **mgs** S of D such that $V_P \cap S \subseteq \{v_1\}$.*

Claims 2, 3, 4 imply an upper bound of two non-extremal vertices that can be part of an **mgs** in any oriented core path of D. Define V_C as the set of core vertices of D and V_I as the set of inner vertices of oriented core paths that are neighbors of vertices of V_C.

Lemma 11. *Let D be a digraph, with S^0 its set of extremal vertices. There exists an **mgs** S of D such that $S \subseteq V_C \cup V_I \cup S^0$, and $|S \setminus S^0| \leq 8\mathsf{fen}(D) - 8$.*

Proof. By Claims 2,3,4, we can suppose that vertices of S' belonging to the base digraph of D are in $V_C \cup V_I \cup S^0$. We can then apply Claim 1 on S' to obtain the geodetic set $S = (S' \cap V(D_b)) \cup S_0$ where D_b is the base digraph of D. It follows that $S \subseteq V_C \cup V_I \cup S^0$. Let P^c the set of core paths of D_b. Again by Claims 2,3,4, we have $|V_C \cup V_I| \leq |V_C| + 2|P^c|$, and by Observation 2, we obtain $|S \setminus S^0| \leq 8\mathsf{fen}(D) - 8$. $\square$

We can now describe the algorithm solving GEODETIC SET parameterized by the feedback edge set number of the underlying undirected graph of the input digraph. This algorithm guesses the vertices to add in a solution among vertices in $V_C \cup V_I$. The algorithm can be stated as follows:

- For all subsets S^1 of $V_C \cup V_I$, if $S = S^0 \cup S^1$ is a geodetic set of D, mark S
- Return the marked set S of minimum size.

Proof. (of Theorem 3). The correctness of the algorithm is clear from Lemma 11. The running time follows, since by Lemma 11, we have $|V_C \cup V_I| \leq 8\mathsf{fen}(D) - 8$. Thus, the algorithm checks $2^{\mathcal{O}(\mathsf{fen}(D))}$ different vertex sets. Each check is polynomial-time in n since GEODETIC SET belongs to NP. $\square$

4   NP-Hardness on Restricted DAGs

In this section, we sketch the proof of Theorem 4. We present a reduction from the classic NP-complete problem 3-DIMENSIONAL MATCHING [24]. In this problem, an input is a ground set U partitioned in three sets X^α, X^β and X^γ such that $|X^\alpha| = |X^\beta| = |X^\gamma| = n$ and a collection of 3D edges $E \subseteq X^\alpha \times X^\beta \times X^\gamma$. The objective is to decide if there exists a set S of n edges of E so that any element of U is covered by an edge of S.

Reduction. Consider an instance of 3-DIMENSIONAL MATCHING with a ground set U and its partition in three sets X^α, X^β and X^γ, and an edge set E. In the following, the notation δ will designate any of the letters α, β or γ. Number the vertices of X^δ so that $X^\delta = \{x_1^\delta, \dots, x_n^\delta\}$ and denote by m the cardinality of E. We construct a digraph D as follows:

Edge vertices. Add n sets $M_1, \ldots, M_n$ of m vertices, each corresponding to some edge of E. Denote by u_i^e the vertex in M_i associated with edge e. Add also for $1 \leq i \leq n$ a vertex d_i and an arc from any vertex in M_i to d_i.

Ensuring edge vertices are covered. Add three vertices a, b and c and connect them as follows:

- Add arcs from a to each vertex in $\bigcup_{1 \leq i \leq n} M_i$.
- Add arcs from each vertex in $\bigcup_{1 \leq i \leq n} M_i$ to b.
- Add arcs from each vertex in $\{d_i \mid 1 \leq i \leq n\}$ to c.
- Add an arc from a to c.

Element vertices. Add vertices v_i^δ, w_i^δ and t_i^δ associated with the element x_i^δ in X^δ. Add arcs from w_i^δ to v_i^δ and from t_i^δ to v_i^δ.

Encoding adjacency. Add nine vertices α_1, α_2, α_3, β_1, β_2, β_3, γ_1, γ_2, γ_3 and add an outgoing pendant vertex to each of them. For $i \in \{1, 2, 3\}$, denote by δ_i' the pendant vertex associated with δ_i. Denote by e some edge of E such that $e = (x_i^\alpha, x_j^\beta, x_k^\gamma)$ and call m any of its associated vertices in $\bigcup_{1 \leq i \leq n} M_i$. Define $\lambda = m^2$. We add some paths of defined length as follows:

- Add a path of length $\lambda^2 - i\lambda$ from m to α_1, a path of length λ^2 from m to α_2, and a path of length $\lambda^2 + i\lambda$ from m to α_3;
- Add a path of length $\lambda^2 - j\lambda$ from m to β_1, a path of length λ^2 from m to β_2, and a path of length $\lambda^2 + j\lambda$ from m to β_3
- Add a path of length $\lambda^2 - k\lambda$ from m to γ_1, a path of length λ^2 from m to γ_2, and a path of length $\lambda^2 + k\lambda$ from m to γ_3;
- For any $\delta \in \{\alpha, \beta, \gamma\}$, add a path of length $\lambda^2 + l\lambda$ from δ_1 to w_l^δ, a path of length λ^2 from δ_2 to w_l^δ, a path of length λ^2 from δ_2 to t_l^δ and a path of length $\lambda^2 - l\lambda$ from δ_3 to w_l^δ.

Ensuring the edge paths are covered. For each path from an edge vertex of $\bigcup_{1 \leq i \leq n} M_i$ to a vertex δ_i for $i \in \{1, 2, 3\}$ a pendant outgoing vertex to the vertex right before δ_i in said path.

Shortcutting the vertex a. Add all arcs from a to vertices v_i^δ associated with elements of U.

Figure 1 illustrates the main component of the obtained graph, namely the edge vertices and the encoding of adjacencies described above. We argue that there exists a solution to 3-DIMENSIONAL MATCHING for the original instance if and only if there exists a geodetic set of size $6nm + 4n + 10$ in D. The proof follows from the correctness of the reduction (which is presented in the full version) and the fact that deleting vertices α_1, α_2, α_3, β_1, β_2, β_3, γ_1, γ_2, γ_3, a, b and c removes any cycle in the underlying graph of D.

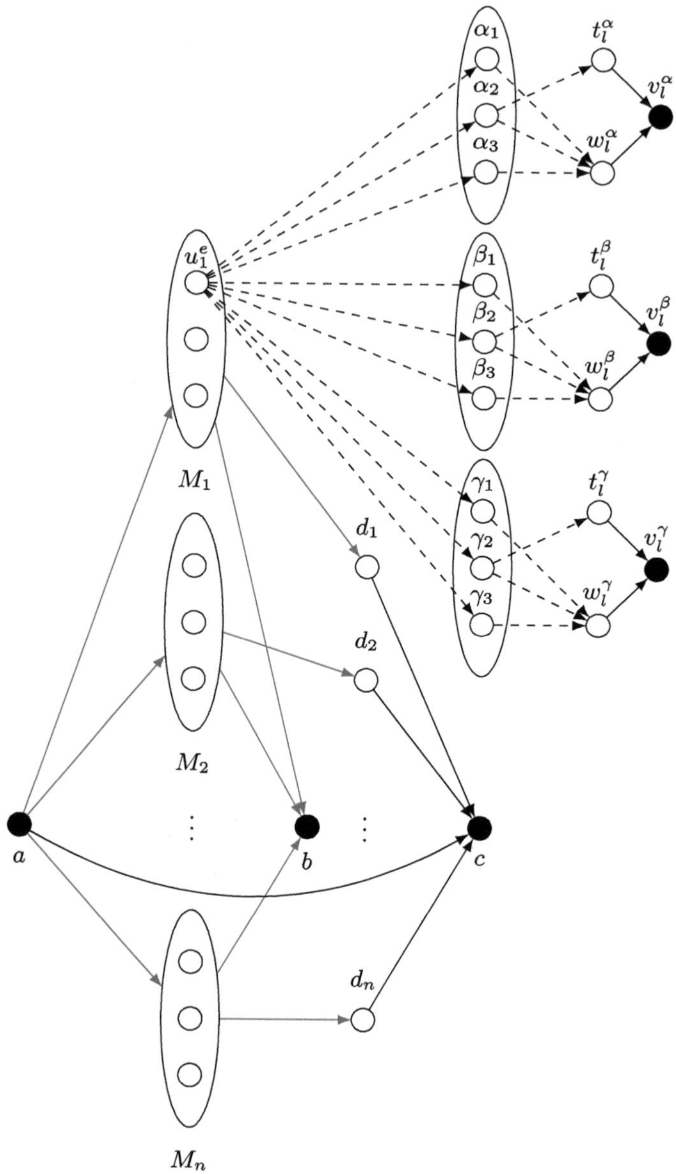

Fig. 1. A partial representation of the digraph D constructed during the reduction. Dashed arcs represent paths of length greater than one. Red arcs between a vertex and a vertex set mean that there exists such an arc for all vertices in the vertex set. Filled vertices are extremal vertices of D, that belong to any geodetic set. Arcs adjacent to vertices δ_i are represented for only one edge vertex and vertices associated with three different elements, each of them belonging to a different partition set of U. Pending vertices to δ_i and paths from edge vertices to δ_i are not represented. (Color figure online)

5   Conclusion

We continued the study of GEODETIC SET for digraphs. As directions for further research, it would be interesting to extend our algorithm for ditrees to more general classes of digraphs. For instance, what can be said about directed cactus graphs (dicactii)? An algorithm for directed cactus graphs without 2-cycles (oriented cactii) was given in [2]. More broadly, one could consider directed outerplanar graphs. Note that an algorithm for undirected outerplanar graphs appears in [28]. Regarding parameterized complexity, it is natural to ask whether a polynomial kernel exists with respect to the feedback edge set number of the underlying undirected graph. This remains open even for undirected graphs [26].

References

1. Ahn, J., Jaffke, L., Kwon, O., Lima, P.T.: Well-partitioned chordal graphs. Discret. Math. **345**(10), 112985 (2022)
2. Araújo, J., Arraes, P.S.M.: Hull and geodetic numbers for some classes of oriented graphs. Discret. Appl. Math. **323**, 14–27 (2022)
3. Bergougnoux, B., Defrain, O., Mc Inerney, F.: Enumerating minimal solution sets for metric graph problems. In: Proc. of the 50th International Workshop on Graph-Theoretic Concepts in Computer Science (WG 2024). Lecture Notes in Computer Science, Springer (2024)
4. Chakraborty, D., Das, S., Foucaud, F., Gahlawat, H., Lajou, D., Roy, B.: Algorithms and complexity for geodetic Sets on planar and chordal graphs. In: 31st International Symposium on Algorithms and Computation (ISAAC 2020). Leibniz International Proceedings in Informatics (LIPIcs), vol. 181, pp. 7:1–7:15. Schloss Dagstuhl–Leibniz-Zentrum für Informatik, Dagstuhl, Germany (2020)
5. Chakraborty, D., Foucaud, F., Gahlawat, H., Ghosh, S.K., Roy, B.: Hardness and approximation for the geodetic set problem in some graph classes. In: Proceedings of the 6th International Conference on Algorithms and Discrete Applied Mathematics (CALDAM 2020). Lecture Notes in Computer Science, vol. 12016, pp. 102–115. Springer International Publishing, Cham (2020)
6. Chakraborty, D., Gahlawat, H., Roy, B.: Algorithms and complexity for geodetic sets on partial grids. Theoretical Comput. Sci. **979**, 114217 (2023)
7. Chang, G.J., Tong, L.D., Wang, H.T.: Geodetic spectra of graphs. Eur. J. Comb. **25**(3), 383–391 (2004)
8. Chartrand, G., Fink, J.F., Zhang, P.: The hull number of an oriented graph. Int. J. Math. Math. Sci. **2003**(36), 2265–2275 (2003)
9. Chartrand, G., Zhang, P.: The geodetic number of an oriented graph. Eur. J. Comb. **21**(2), 181–189 (2000)
10. Cygan, M., et al.: Parameterized Algorithms. Springer (2015)
11. Dailly, A., Foucaud, F., Hakanen, A.: Algorithms and hardness for metric dimension on digraphs. In: Paulusma, D., Ries, B. (eds.) Graph-Theoretic Concepts in Computer Science - 49th International Workshop, WG 2023, Fribourg, Switzerland, June 28-30, 2023. Lecture Notes in Computer Science, vol. 14093, pp. 232–245. Springer (2023)

12. Davot, T., Isenmann, L., Thiebaut, J.: On the approximation hardness of geodetic set and its variants. In: Chen, C.-Y., Hon, W.-K., Hung, L.-J., Lee, C.-W. (eds.) COCOON 2021. LNCS, vol. 13025, pp. 76–88. Springer, Cham (2021). https://doi.org/10.1007/978-3-030-89543-3_7

13. Dev, S.R., Dey, S., Foucaud, F., Narayanan, K., Sulochana, L.R.: Monitoring edge-geodetic sets in graphs. Discret. Appl. Math. **377**, 598–610 (2025)

14. Dong, L., Lu, C., Wang, X.: The upper and lower geodetic numbers of graphs. Ars Combin. **91**, 401–409 (2009)

15. Dourado, M.C., Protti, F., Rautenbach, D., Szwarcfiter, J.L.: Some remarks on the geodetic number of a graph. Discret. Math. **310**(4), 832–837 (2010)

16. Douthat, A.L., Kong, M.C.: Computing geodetic bases of chordal and split graph. J. Comb. Math. Comb. Comput. **22**, 67–77 (1996)

17. Ekim, T., Erey, A., Heggernes, P., van 't Hof, P., Meister, D.: Computing minimum geodetic sets of proper interval graphs. In: Fernández-Baca, D. (ed.) LATIN 2012. LNCS, vol. 7256, pp. 279–290. Springer, Heidelberg (2012). https://doi.org/10.1007/978-3-642-29344-3_24

18. Farber, M., Jamison, R.E.: Convexity in graphs and hypergraphs. SIAM J. Algebraic Discrete Methods **7**(3), 433–444 (1986)

19. Farrugia, A.: Orientable convexity, geodetic and hull numbers in graphs. Discret. Appl. Math. **148**(3), 256–262 (2005)

20. Foucaud, F., Galby, E., Khazaliya, L., Li, S., Inerney, F.M., Sharma, R., Tale, P.: Problems in NP can admit double-exponential lower bounds when parameterized by treewidth or vertex cover. In: Bringmann, K., Grohe, M., Puppis, G., Svensson, O. (eds.) 51st International Colloquium on Automata, Languages, and Programming, ICALP 2024, July 8-12, 2024, Tallinn, Estonia. LIPIcs, vol. 297, pp. 66:1–66:19. Schloss Dagstuhl - Leibniz-Zentrum für Informatik (2024)

21. Foucaud, F., Galby, E., Khazaliya, L., Li, S., Inerney, F.M., Sharma, R., Tale, P.: Metric dimension and geodetic set parameterized by vertex cover. In: Beyersdorff, O., Pilipczuk, M., Pimentel, E., Nguyen, K.T. (eds.) 42nd International Symposium on Theoretical Aspects of Computer Science, STACS 2025, March 4-7, 2025, Jena, Germany. LIPIcs, vol. 327, pp. 33:1–33:20. Schloss Dagstuhl - Leibniz-Zentrum für Informatik (2025)

22. Foucaud, F., et al.: Metric dimension and geodetic set parameterized by vertex cover. In: Beyersdorff, O., Pilipczuk, M., Pimentel, E., Nguyen, K.T. (eds.) 42nd International Symposium on Theoretical Aspects of Computer Science, STACS 2025, March 4-7, 2025, Jena, Germany. LIPIcs, vol. 327, pp. 33:1–33:20. Schloss Dagstuhl - Leibniz-Zentrum für Informatik (2025)

23. Foucaud, F., Ghareghani, N., Sharifani, P.: Extremal digraphs for open neighbourhood location-domination and identifying codes. Discret. Appl. Math. **347**, 62–74 (2024)

24. Garey, M.R., Johnson, D.S.: Computers and Intractability: A Guide to the Theory of NP-Completeness. W. H. Freeman (1979)

25. Harary, F., Loukakis, E., Tsouros, C.: The geodetic number of a graph. Math. Comput. Model. **17**(11), 89–95 (1993)

26. Kellerhals, L., Koana, T.: Parameterized complexity of geodetic set. J. Graph Algorithms Appl. **26**(4), 401–419 (2022)

27. Lu, C.h.: The geodetic numbers of graphs and digraphs. Sci. China Seri. A Math. **50**(8), 1163–1172 (2007)

28. Mezzini, M.: Polynomial time algorithm for computing a minimum geodetic set in outerplanar graphs. Theoret. Comput. Sci. **745**, 63–74 (2018)
29. Pelayo, I.M.: Geodesic Convexity in Graphs. Springer (2013)
30. Tale, P.: Geodetic set on graphs of constant pathwidth and feedback vertex set number. In: Agrawal, A., van Leeuwen, E.J. (eds.) 20th International Symposium on Parameterized and Exact Computation, IPEC 2025, September 17-19, 2025, University of Warsaw, Poland. LIPIcs, vol. to appear. Schloss Dagstuhl - Leibniz-Zentrum für Informatik (2025)

On the Word-Representability of 5-Regular Circulant Graphs

Suchanda Roy and Ramesh Hariharasubramanian$^{(\boxtimes)}$

Department of Mathematics, Indian Institute of Technology Guwahati,
Guwahati 781039, Assam, India
`{r.suchanda,ramesh_h}@iitg.ac.in`

Abstract. A graph $G = (V, E)$ is *word-representable* if there exists a word w over the alphabet V such that, for any two distinct vertices $x, y \in V$, $xy \in E$ if and only if x and y alternate in w. Two letters x and y are said to *alternate* in w if, after removing all other letters from w, the resulting word is of the form $xyxy\ldots$ or $yxyx\ldots$ (of even or odd length). For a given set $R = \{r_1, r_2, \ldots, r_k\}$ of jump elements, an undirected *circulant graph* $C_n(R)$ on n vertices has vertex set $\{0, 1, \ldots, n-1\}$ and edge set $E = \{\{i, j\} \mid |i - j| \bmod n \in \{r_1, r_2, \ldots, r_k\}\}$, where $0 < r_1 < r_2 < \cdots < r_k < \frac{n}{2}$. Recently, Kitaev and Pyatkin showed that every 4-regular circulant graph is word-representable. Also, Srinivasan and Hariharasubramanian studied word-representability of circulant graphs and obtained some bounds on the representation number for k-regular circulant graphs with $2 \leq k \leq 4$. In addition to these positive results, their work also presents examples of non-word-representable circulant graphs. In this work, we extend these investigations to 5-regular circulant graphs. We study word-representability and the representation number of 5-regular circulant graphs via techniques from elementary number theory and group theory, as well as graph coloring, graph factorization, and morphisms.

Keywords: word-representable graph · circulant graph · semi-transitive orientation · representation number

1   Introduction

The theory of *word-representable graphs* is a rich and promising area of research, with strong connections to algebra, graph theory, combinatorics on words, formal languages, and scheduling problems. This concept was first introduced by Kitaev and Pyatkin in [8], based on the study of Kitaev and Seif on the celebrated *Perkins semigroup* [10]. Word-representable graphs generalize several fundamental graph classes, including *circle graphs, comparability graphs, and 3-colorable graphs*. We refer the reader to [6,7] for motivation and a detailed discussion of the connections between word-representable graphs and various fields. Notably, determining whether a graph is word-representable or not is a *NP-complete* problem.

N. Misra and A. Pandey (Eds.): CALDAM 2026, LNCS 16445, pp. 194–208, 2026.
https://doi.org/10.1007/978-3-032-17156-6_15

The notion of *semi-transitive orientation* (Definition 10), introduced by Halldórsson in [3], plays a central role in characterizing *word-representable graphs*. In fact, a graph is word-representable if and only if it admits such an orientation (Theorem 3).

Circulant graph is an important family in graph theory due to their high degree of symmetry, as they are regular and vertex-transitive. These properties, together with their connections to number theory and applications in areas such as network design, coding theory, and cryptography, make them a natural object of study. These graphs have been extensively investigated with respect to their structural properties, including connectivity, planarity, and factorization. In recent years, word-representability of circulant graphs has attracted attention. In [9], Kitaev and Pyatkin showed that every 4-regular circulant graph is word-representable, while Srinivasan and Hariharasubramanian studied the word-representability of circulant graphs and established bounds on the representation number for k-regular circulant graphs with $2 \leq k \leq 4$ in [12]. However, the case of 5-regular circulant graphs has not been thoroughly investigated, and this motivates our present work.

This paper presents several contributions to the study of word-representability of 5-regular circulant graphs. We begin by dividing this class into subclasses according to the parity of the jump set elements and analyze their word-representability (Theorem 17, Remark 3). Also, we give particular attention to the subclass of 5-regular circulant graphs in which at least one jump set element is relatively prime to the number of vertices. Their word-representability is examined in Theorems 21 and Corollary 1, along with upper bounds on their representation number in Theorems 24 and 25. We then extend the analysis to the general case of 5-regular circulant graphs (Theorems 20, 19, 22 and 23). The paper ends with several open problems and directions for future research.

This paper is organized as follows. Section 2 reviews basic results from elementary number theory and group theory, provides the necessary background on word-representability in 2.1, and introduces circulant graphs along with their key properties in 2.2. In Sect. 3, we present our results on the word-representability of 5-regular circulant graphs and give upper bounds on their representation number. Finally, Sect. 4 concludes the paper with open problems and directions for future research.

2 Preliminaries

Before presenting the necessary background on word-representability, we first state some basic results and definitions from elementary number theory and group theory that will be used in this work.

Theorem 1. *A linear congruence of the form* $bx \equiv a \pmod{n}$ *has a unique solution modulo* n *if and only if* $\gcd(b, n) = 1$.

Definition 1. *Let H be a group and $a \in H$. The* order *of a, denoted $o(a)$, is the smallest positive integer m such that $a^m = e$, where e is the identity element of H. If no such m exists, a is said to have infinite order.*

Proposition 1. *Let $a \in H$ be an element of finite order $o(a)$, and let x be a positive integer. Then the order of a^x is given by $o(a^x) = \frac{o(a)}{\gcd(o(a), x)}$.*

Definition 2. *A group H is called* cyclic *if there exists an element $g \in H$ such that every element of H can be written as a power of g; that is, $H = \{g^k \mid k \in \mathbb{Z}\}$. In this case, g is called a* generator *of H.*

Proposition 2. *Let $H = \langle g \rangle$ be a cyclic group of order n. An element $g^k \in H$ is a generator of H if and only if $\gcd(k, n) = 1$.*

We have applied these results to the cyclic group $(\mathbb{Z}_{2n}, +)$ in Theorems 18 and 19. As $(\mathbb{Z}_{2n}, +)$ is an additive group, instead of g^k we write $k.g$, where g is a generator of $(\mathbb{Z}_{2n}, +)$. In the following, we present a brief overview of word-representability.

2.1   Word-Representable Graphs

Throughout this paper, we assume that all graphs are finite, simple, and undirected. For a graph G, we denote its vertex set by $V(G)$ and its edge set by $E(G)$. For vertices $x, y \in V(G)$, we write $x \sim y$ if there is an edge between x and y, and $x \nsim y$ if there is no edge between them. We begin this section with the definition of word-representable graphs.

Definition 3 ([7]). *A simple graph $G = (V, E)$ is word-representable if there exists a word w over the alphabet V such that letters x and y, $\{x, y\} \in V$ alternate in w if and only if $xy \in E$, i.e., x and y are adjacent for each $x \neq y$. If a word w represents G, then w contains each letter of $V(G)$ at least once.*

Example 1. For instance, the word 1234 represents the complete graph K_4 on four vertices. More generally, any permutation of n distinct symbols represents the complete graph K_n on n vertices. Also, 162132435465 represents C_6, the cycle graph on 6 vertices.

Definition 4 ([7]). *A word w contains a word u as a factor if $w = xuy$ where x and y can be empty words.*

Definition 5 ([7]). *A subword of a word w is a word obtained by removing certain letters from w. In a word w, if x and y alternate, then w contains $xyxy \cdots$ or $yxyx \cdots$ (odd or even length) as a subword.*

Example 2. The word 1241154325 contains the words 154 and 24 as factors, while 24142 is a subword of it.

Definition 6 ([7]). *k-uniform word A word w is said to be a k-uniform word if every letter occurs exactly k times in it.*

Definition 7 ([7]). *A graph is said to be k-word-representable if there exists a k-uniform word representing it.*

The following proposition shows that a cyclic shift of a k-uniform representing word does not affect the represented graph.

Proposition 3 ([7]). *Let $w = uv$ be a k-uniform word representing a graph G, where u and v are two, possibly empty, words. Then, the word $w' = vu$ also represents G.*

Definition 8 ([6]). *For a word-representable graph G, the representation number is the least k such that G is k-representable and it is denoted by $R(G)$.*

The following theorem gives the representation number of a cycle graph.

Theorem 2 ([7]). *Let G be a cycle graph. Then the representation number of G is 2.*

In the following, we state one of the key developments in the study of word-representable graphs, namely its characterization in terms of semi-transitive orientations, which is defined based on shortcuts in [3].

Definition 9 ([7]). *A* semi-cycle *is the directed acyclic graph obtained by reversing the direction of one edge of a directed cycle.*

Definition 10 ([3]). *An acyclic digraph is called a* shortcut *if it is induced by the vertices of a semi-cycle and contains a pair of non-adjacent vertices. In particular, any shortcut*

- *is acyclic (that is, it has no directed cycles);*
- *contains at least four vertices;*
- *has exactly one source (a vertex with no incoming edges) and exactly one sink (a vertex with no outgoing edges), with a directed path from the source to the sink passing through every vertex;*
- *includes an edge connecting the source to the sink, called the* shortcutting *edge;*
- *is not transitive (that is, there exist vertices u, v, and z such that $u \to v$ and $v \to z$ are edges, but $u \to z$ is not).*

Definition 11 ([3]). *An orientation of a graph is* semi-transitive *if it is acyclic and shortcut-free.*

Theorem 3 ([3]). *A graph G is word-representable if and only if it admits a semi-transitive orientation.*

Remark 1. The notion of semi-transitive orientation generalizes the classical concept of transitive orientation, where an orientation is *transitive* if the presence of $u \to v$ and $v \to z$ implies $u \to z$. Graphs admitting such an orientation are called *comparability graphs*. In particular, path graphs, bipartite graphs and complete graphs admit transitive orientations, and therefore they are word-representable.

Theorem 4 ([7]). *If a graph G is 3-colorable, then it is word-representable.*

The above theorem establishes a connection between colorability and word-representability, which plays an important role in the proofs of Theorems 20 and 19. Furthermore, the following theorem allows us to restrict the study of word-representable graphs to connected graphs.

Theorem 5 ([7]). *Let G be a graph with connected components $G_1, G_2, \ldots, G_k$. Then G is word- representable if and only if each G_i is word-representable. Moreover, the representation number of G satisfies $R(G) = \max\{R(G_1), R(G_2), \ldots, R(G_k)\}$.*

In the following, we present results on the word-representability of Cartesian products of graphs. We begin with the definition of the Cartesian product.

Definition 12. *Let $G_1 = (V(G_1), E(G_1))$ and $G_2 = (V(G_2), E(G_2))$ be two graphs. The Cartesian product $G_1 \square G_2$ is the graph defined as follows:*

- *The vertex set is $V(G_1 \square G_2) = V(G_1) \times V(G_2)$.*
- *Two vertices (u_1, v_1) and (u_2, v_2) are adjacent in $G_1 \square G_2$ if and only if either*
    1. *$u_1 = u_2$ and v_1 is adjacent to v_2 in G_2, or*
    2. *$v_1 = v_2$ and u_1 is adjacent to u_2 in G_1.*

Theorem 6 ([7]). *If G_1 and G_2 are word-representable graphs, then their Cartesian product $G_1 \square G_2$ is also word-representable.*

Theorem 7 ([11]). *Let G_1 and G_2 be two word-representable graphs with their representation numbers r_1 and r_2, respectively. Then the Cartesian product $G_1 \square G_2$ is $(r_1 + r_2 + \min\{|G_1|, |G_2|\})$-representable, i.e., $R(G) \leq (r_1 + r_2 + \min\{|G_1|, |G_2|\})$*

2.2   Circulant Graph

A circulant graph $C_n(R)$ for a set $R = \{r_1, r_2, \ldots, r_k\}$ is defined as the graph with vertex set $V(G) = \{0, 1, \ldots, n-1\}$ and edge set $E(G) = \{ij \mid |i - j| \pmod{n} \in \{r_1, r_2, \ldots, r_k\}\}$, where $0 < r_1 < r_2 < \cdots < r_k < \frac{n+1}{2}$. R is known as the jump set and $r_1, r_2, \cdots r_k$ are known as jump elements of the circulant graph.

The graph $C_n(r_1, r_2, \ldots, r_k)$ is regular of degree $d = \begin{cases} 2k, & \text{if } r_k \neq \frac{n}{2}, \\ 2k - 1, & \text{if } r_k = \frac{n}{2}. \end{cases}$

In the following, we will discuss some fundamental properties of circulant graphs.

Theorem 8 ([14]). *Let $C_n(R)$ be a circulant graph and let $r \in R$. Then a cycle of period r in $C_n(R)$ has length $\frac{n}{\gcd(n,r)}$, and there are exactly $\gcd(n, r)$ vertex-disjoint periodic cycles of period r.*

Theorem 9 ([1,13]). *A circulant graph $C_n(r_1, r_2, \ldots, r_k)$ is connected if and only if $\gcd(n, r_1, r_2, \ldots, r_k) = 1$. Moreover, if $d = \gcd(n, r_1, r_2, \ldots, r_k)$, then $C_n(r_1, r_2, \ldots, r_k) \cong d \cdot C_{\frac{n}{d}}\left(\frac{r_1}{d}, \frac{r_2}{d}, \ldots, \frac{r_k}{d}\right)$, i.e., $C(n; r_1, \ldots, r_k)$ is isomorphic to the disjoint union of d copies of $C\left(\frac{n}{d}; \frac{r_1}{d}, \ldots, \frac{r_k}{d}\right)$.*

Theorem 10 ([4]). *Let $G := C_n(r_1, r_2, \ldots, r_k)$ be a connected circulant graph. Then G is bipartite if and only if n is even and each r_i is odd for $1 \leq i \leq k$.*

The following two theorems establish that circulant graphs can be factorized uniquely into a Cartesian product of prime graphs.

Theorem 11 ([14]). *For $n \in \mathbb{N}$ and a set $R = \{r_1, r_2, \ldots, r_k\}$, the Cartesian product $P_2 \,\square\, C(2n+1; R) \cong C\big(2(2n+1); 2R \cup \{2n+1\}\big) \cong C\big(2(2n+1); 2dR \cup \{2n+1\}\big)$, where $\gcd\big(2(2n+1), d\big) = 1$.*

Theorem 12 ([14] **Factorization Theorem of Circulant Graphs**). *Let p and q be relatively prime integers. If $R \subseteq [1, p/2]$, $S \subseteq [1, q/2]$, and $T \subseteq [1, pq/2]$ with $T = dqR \cup dpS$ for some d such that $\gcd(pq, d) = 1$, then $C_{pq}(T) \cong C_p(R) \,\square\, C_q(S)$.*

We now present the known results regarding the word-representability and non-word-representability of circulant graphs (Figs. 1 and 2).

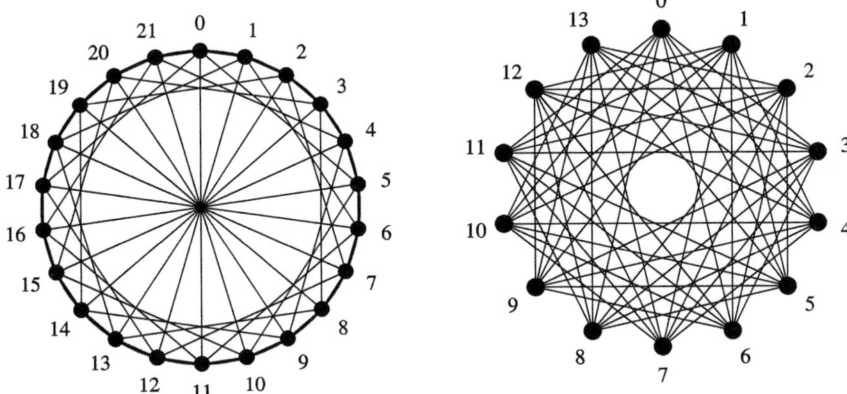

Fig. 1. An Example of Word-Representable Circulant Graph $C_{22}(1, 5, 11)$.

Fig. 2. An Example of Non-Word-Representable Circulant Graph $C_{14}(3, 4, 5, 6)$.

Theorem 13 ([12]). *Let $G \cong C_{2n}(a, n)$ be a 3-regular circulant graph. Then it is word-representable and $R(G) \leq 3$.*

Theorem 14 ([9]). *Let $G \cong C_n(a, b)$ be a 4-regular circulant graph. Then G is word-representable.*

Theorem 15 ([12]). *Let $G \cong C_n(1, a)$ be a connected 4-regular circulant graph with $\frac{n}{3} \leq a < \frac{n}{2}$ and $n > 6$. Then G is word-representable and $R(G) \leq 4$.*

Remark 2. As we know that any 2-regular connected graph is a cycle, then any 2-regular connected circulant graph is also a cycle. Hence, by Theorem 2, G is word-representable and $R(G) = 2$ where G is any 2-regular circulant graph.

The following theorem will provide examples of non-word-representable circulant graphs.

Theorem 16 ([12]). *Let $G = C_n(r, r + 1, \ldots, 2r)$. Then G is not word-representable whenever $\frac{n+1}{5} < r < \frac{n-1}{4}$.*

Here, we presented some results concerning the word-representability of k-regular circulant graphs for $k \leq 4$, and provided explicit examples of circulant graphs that are not word-representable. These observations naturally motivate the investigation of 5-regular circulant graphs, which we pursue in the following section.

3   Word-Representability of 5-Regular Circulant Graph

In this section, we first divide the class of 5-regular circulant graphs into subclasses based on the parity of the jump set elements and analyze their word-representability (Theorem 17, Remark 3). We then study word-representability more generally and, in certain cases, establish upper bounds on the representation number by employing techniques such as colorability (3.1), morphisms (3.3) and factorization into smaller circulant graphs (3.2).

Theorem 17. *Let $G = C_{2n}(a, b, n)$ be a 5-regular circulant graph, where n is odd. If a and b have the same parity, i.e., both are even or both are odd, then G is word-representable.*

Proof. Let $G = C_{2n}(a, b, n)$ be a 5-regular circulant graph, where n is odd. We consider two cases as below.

- *Both a and b are even. In this case, using Theorem 11 we get $G = C_{2n}(a, b, n) \cong P_2 \square C_n(\frac{a}{2}, \frac{b}{2})$, where P_2 is the path of 2 vertices. Now, combining Remark 1 and Theorem 6, we conclude that G is word-representable.*
- *Both a and b are odd. In this case, using Theorem 10, we conclude that G is bipartite. Therefore, G is word-representable by Remark 1.*

Remark 3. Let $G = C_{2n}(a, b, n)$ be a 5-regular circulant graph where n, a, and b are even. As $gcd(a, b, n) \geq 2$, then by Theorem 9, G is not connected. Hence, G is word-representable if and only if all of its connected components are word-representable (by Theorem 5). From this point onward, we restrict our attention to connected 5-regular circulant graphs.

Theorem 18. *Let $G = C_{2n}(a, b, n)$ be a circulant graph with $gcd(b, 2n) = 1$. Then $G \cong C_{2n}(x, 1, n)$, where $bx \equiv a \pmod{2n}$.*

Proof. Since $\gcd(b, 2n) = 1$, the element b is a generator of the cyclic group $\mathbb{Z}_{2n}$. Hence, the vertices of the circulant graph can be arranged on a cycle as $\{0, 1 \cdot b, 2 \cdot b, \ldots, (n-1) \cdot b\}$. Consider the isomorphism $f : V\big(C_{2n}(a, b, n)\big) \longrightarrow V\big(C_{2n}(x, 1, n)\big)\}$ defined by $f(i \cdot b) = i, \forall i \in \{0, 1, 2, \ldots, n-1\}$.

- If $|i - j| = 1$, then $i \cdot b \sim j \cdot b \iff i \sim j$.
- If $|i - j| > 1$, then $i \cdot b \sim j \cdot b \implies i \cdot b - j \cdot b \equiv \pm a \pmod{2n}$ or $i \cdot b - j \cdot b \equiv n \pmod{2n}$.

  Now, $i \cdot b - j \cdot b \equiv \pm a \pmod{2n} \iff i - j \equiv \pm x \pmod{2n}$ by *Theorem 1*. Hence $i \cdot b \sim j \cdot b \iff i \sim j$.

  Since $2n \equiv 0 \pmod{2n}$, $o(n) = 2$. Let $n = k.b$. Then $o(k.b) = \frac{o(b)}{\gcd(k, o(b))}$, which implies $\frac{2n}{\gcd(k, 2n)} = 2$, i.e., $k = n$. Hence, $n.b = n$. Then, $i \cdot b - j \cdot b \equiv n \pmod{2n} \implies (i - j) \equiv n \pmod{2n}$ by *Theorem 1*. Again, $i - j \equiv n \pmod{2n} \implies i.b - j.b \equiv n.b \equiv n \pmod{2n}$. Hence, $i \cdot b \sim j \cdot b \iff i \sim j$.

Hence, $C_{2n}(a, b, n) \cong C_{2n}(x, 1, n)$.

In the following sections, for the class of 5-regular circulant graphs with either $\gcd(a, 2n) = 1$ or $\gcd(b, 2n) = 1$, we restrict our attention to the graphs of the form $C_{2n}(x, 1, n)$.

3.1 Word-Representability and Colorability of 5-Regular Circulant Graphs

In this section, we establish the word-representability of 5-regular circulant graphs by relating it to their colorability. Our approach relies on Theorem 4 together with the embedding technique described in [5]. Basically, for a given graph G, we find a word-representable graph H with a homomorphism $f : G \to H$, orient G according to a fixed acyclic orientation of H, and then check H for shortcutting edges; if none exist, we conclude that the induced orientation of G is semi-transitive and if there exists a shortcutting edge, we check whether it is inducing a shortcut in G or not. If no shortcut is induced, then we conclude that the induced orientation in G is semitransitive.

Theorem 19. *Let $G = C_{2n}(a, b, n)$ be a 5-regular circulant graph. G is word-represent- able if there exists a generator g of the underlined cyclic group $(Z_{2n}, +)$ such that $r = min\{p, q, 2n - p, 2n - q\} \geq \lceil \frac{2n}{3} \rceil$, where $a = p.g$ and $b = q.g$.*

Proof. We claim that $r = min\{p, q, 2n - p, 2n - q\} < n$. To prove this we first prove that for any $j \in (Z_{2n}, +)$, $o(j) = 2 \iff j = n.g$. Let $o(j) = 2$ and $j = x.g$ for some $x \in \{0, 1, 2 \cdots 2n - 1\}$. Then $o(j) = o(x.g) \implies 2 = \frac{o(g)}{\gcd(x, o(g))} \implies \gcd(x, 2n) = n \implies x = n$. Again, let $j = n.g$. Then $o(j) = o(n.g) \implies o(j) = \frac{o(g)}{\gcd(n, o(g))} \implies o(j) = 2$.

   Moreover, since $o(n) = 2$, the element $n = n.g$ is the unique element of order 2 in $(\mathbb{Z}_{2n}, +)$. As $a \neq n$, it follows that p is either greater than n or less than*

n. *Without loss of generality, assume $p > n$. Then $2n - p < n$, which in turn implies $r < n$.*

Next, we define $X_0 = \{j \in V(G) \mid j = k.g \text{ where }, 0 \leq k \leq r - 1\}$, $X_1 = \{j \in V(G) \mid j = k.g \text{ where }, r \leq k \leq 2r - 1\}$ and $X_2 = V(G) \setminus (X_0 \cup X_1)$. Clearly, X_0, X_1, X_2 form a partition of $V(G)$. We assign color c_i to every vertex in X_i for $i \in \{0, 1, 2\}$.

Let $j_1 \in X_i$ and $j_1 = k_1.g$. Suppose, for contradiction, $j_1 \sim j_2(= k_2.g)$ and $j_2 \in X_i$. Then $|k_1 - k_2| \in \{p, q, 2n - p, 2n - q, n\}$, but $|k_1 - k_2| \leq r - 1$ where $r = min\{p, q, 2n - p, 2n - q\}$. Hence, we get a contradiction. Thus, G is 3-colorable and word-representable by Theorem 4.

In Theorems 20 and 21, we color the vertices of the circulant graph according to their congruence classes modulo 3. Since the vertices are indexed modulo $2n$, the modulo-3 class of any neighbour of a vertex j is obtained by first computing its adjacency modulo $2n$ (yielding a value in $\{0, \ldots, 2n - 1\}$) and then reducing this value modulo 3. For example, the neighbour $(j - x)_{2n}$ has modulo-$2n$ representative $j - x$ when $j \geq x$, and $j + 2n - x$ otherwise. In either case, its color is obtained simply by reducing this representative modulo 3.

Theorem 20. *Let $G = C_{2n}(a, b, n)$ be a 5-regular circulant graph. If 3 does not divide any of $a, b, 2n - a, 2n - b$, or n, then G is word-representable.*

Proof. We define $X_i = \{j \in V(G) \mid j \equiv i \pmod{3}\}$, $i = 0, 1, 2$. Clearly, X_0, X_1, X_2 form a partition of $V(G)$. We assign color c_i to every vertex in X_i.

Let $j \in X_i$. Suppose, for contradiction, $(j + a)_{2n} \in X_i$. Then $j \equiv (j + a)_{2n} \equiv i \pmod{3} \implies a \equiv 0 \pmod{3}$, Which contradicts the assumption $3 \nmid a$. The same argument can be applied to $b, 2n - a, 2n - b$, and n, ensuring that no two vertices within X_i are adjacent. Therefore, G is 3-colorable and by Theorem 4, it follows that G is word-representable.

Corollary 1. *Let $G = C_{2n}(x, 1, n)$ be a 5-regular circulant graph. If $n \equiv x \equiv 1 \pmod{3}$, then G is word-representable.*

Theorem 21. *Let $G = C_{2n}(x, 1, n)$ be a 5-regular circulant graph. If $n \equiv x \equiv 2 \pmod{3}$, then G is word-representable.*

Proof. We define $X_i = \{j \in V(G) \setminus \{2n - 1\} \mid j \equiv i \pmod{3}\}$, $i = 0, 1, 2$. Thus, X_0, X_1, X_2 form a partition of $V(G) \setminus \{2n - 1\}$. We assign color c_i to the vertices of X_i and color the remaining vertex $2n - 1$ with c_3. Note that in Figs. 3 and 4, the colors are fixed as follows: c_0 is red, c_1 is blue, c_2 is yellow, and c_3 is green.

Let $j \neq 0$ be any vertex in X_i. Without loss of generality, let $j \in X_0$. Now, since $j - 1 \equiv 2 \pmod{3}$, the vertices j and $j - 1$ necessarily receive different colors. Since 3 does not divide $1, x, 2n - x$, or n, similar arguments can be applied to all other adjacencies, as in Theorem 20. If $j = 0$, then its neighbours are $1, 2n - 1, x$, and $2n - x$. Here $1 \in X_1$ has color c_1, while x and $2n - x \in X_2$ have color c_2; and $2n - 1$ has color c_3. Thus, none of the neighbours of 0 shares its color. Since $2n - 1$ is the unique vertex colored c_3, none of its neighbours has color c_3. Therefore, G is 4-colorable, with $2n - 1$ being the only vertex assigned color c_3.

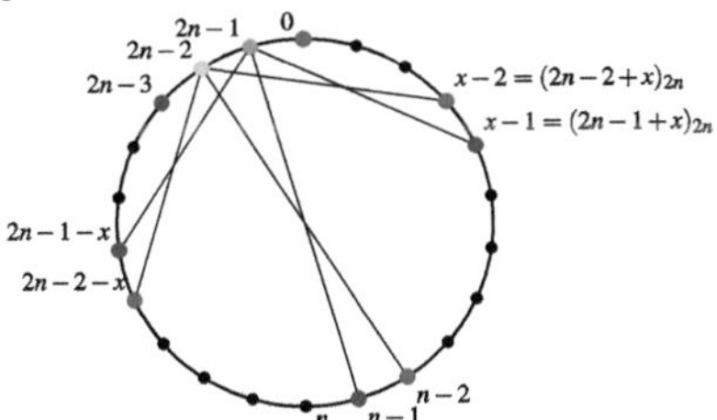

Fig. 3. A circulant graph showing only the neighbours of the vertices $2n - 1$ and $2n - 2$ together with their assigned colors; the coloring of all other vertices is omitted.

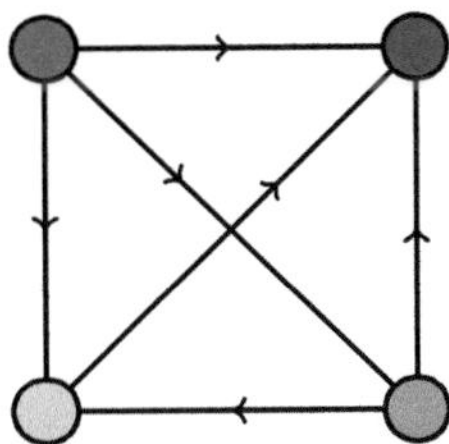

Fig. 4. A semi-transitive orientation of K_4.

We embed the given graph into K_4, the complete graph on four vertices. Let $f : G \to K_4$ be a homomorphism defined as $f(i) = $ color of the vertex i, $\forall i \in \{0, 1, \ldots, 2n - 1\}$. We consider an acyclic orientation of K_4 as shown in Fig. 4. Clearly, the induced orientation of G through f is acyclic. We claim that this orientation is also semi-transitive. Note that the orientation of K_4 contains only one shortcutting path: $c_0 \to c_3 \to c_2 \to c_1$ with the shortcutting edge: $c_0 \to c_1$. To prove our claim, it is sufficient to prove that this shortcutting path and shortcutting edge do not induce any shortcut in G. Now, from the vertices $2n - 1$ and $2n - 2$, together with their adjacencies and assigned colors in Fig. 3, it follows that the shortcutting path $c_0 \to c_3 \to c_2 \to c_1$ corresponds uniquely to the path $0 \to 2n - 1 \to 2n - 2 \to 2n - 3$ in G. But $0 \not\sim 2n - 3$ as $|2n - 3 - 0| \notin \{1, 2n - 1, x, 2n - x, n\}$. Hence $c_0 \to c_3 \to c_2 \to c_1$ with the shortcutting edge: $c_0 \to c_1$ does not induce any shortcut in G. Thus, G is semi-transitive as well as word-representable.

3.2 Word-Representability of 5-Regular Circulant Graph Through Factorization

In this section, we factorize 5-regular circulant graphs into smaller regular circulant graphs. Using this factorization, we study their word-representability (Theorem 22 and Theorem 23) and the representation number (Remark 4). The factorization step primarily relies on Theorem 12.

Theorem 22. *Let $G = C_{2n}(a, b, n)$ be a 5-regular connected circulant graph, where n, b are odd and a is even. Then $C_{2n}(a, b, n)$ can be factorized as a Cartesian product of a 3-regular and a 2-regular circulant graph using Theorem 12 iff there exist positive integers $p(> 2), q(1 < q < n)$ such that $p \mid a$, $q \mid b$, $pq = 2n$ and $\gcd(p, q) = 1$.*

Proof. $(\Rightarrow)$ *Let $G = C_{2n}(a, b, n)$ be a 5-regular connected circulant graph, where n, b are odd and a is even. We assume that it can be factorized as a Cartesian product of a 3-regular and a 2-regular circulant graph using Theorem 12. Then* *by Theorem 12 there exist $p(> 1), q(> 1)$ with $\gcd(p, q) = 1$ and $pq = 2n$ such that $C_{pq}(T) = C_p(R) \square C_q(S)$. Here, $R \subseteq [1, \frac{p}{2}]$, $S \subseteq [1, \frac{q}{2}]$ and $T = qR \cup pS = \{a, b, n\}$.*

If possible, let $p = 2$ and $q = n$. Then $R = \{1\}$, i.e., $qR = \{n\}$. Since $p = 2$, all the elements of pS are even, which contradicts the fact that $b \in pS$. Hence, we must have $p > 2$ and $q < n$.

Now, since n is odd, it follows that $p = 2n_1$ and $q = n_2$, where both n_1 and n_2 are odd integers strictly greater than 1.

Next, we claim that $R = \{\frac{p}{2}, \frac{b}{q}\}$ and $S = \{\frac{a}{p}\}$. As p is even, neither b nor n can belong to pS. Thus, $pS = \{a\}$ and $qR = \{b, n\}$ and our claim is true.

Since every element of R and S is an integer, it follows that p divides a and q divides b.

($\Leftarrow$) Let us assume there exist positive integers $p(> 2), q(1 < q < n)$ such that $p \mid a$, $q \mid b$, $pq = 2n$ and $\gcd(p, q) = 1$. We take $R = \{\frac{p}{2}, \frac{b}{q}\}$ and $S = \{\frac{a}{p}\}$. Then, $T = qR \cup pS = \{a, b, n\}$. Also, we claim that, $R \subseteq [1, \frac{p}{2}]$, $S \subseteq [1, \frac{q}{2}]$. If possible, suppose $\frac{b}{q} \geq \frac{p}{2}$ and $\frac{a}{p} \geq \frac{q}{2}$, which is equivalent to $b \geq n$ and $a \geq n$. This contradicts the fact that $a < n$ and $b < n$. Hence, our claim holds. By Theorem 12, G can be factorized as the Cartesian product of $C_p(R)$ and $C_q(S)$.

Theorem 23. *Let $G = C_{2n}(a, b, n)$ be a 5-regular circulant graph, where n, a are even and b is odd. Then $C_{2n}(a, b, n)$ can be factorized as a Cartesian product of a 3-regular and a 2-regular circulant graph using Theorem 12 iff there exist positive integers $p(> 2), q(1 < q < n)$ such that $p \mid a$, $q \mid b$, $pq = 2n$ and $\gcd(p, q) = 1$.*

Proof. ($\Rightarrow$) Let $G = C_{2n}(a, b, n)$ be a 5-regular connected circulant graph, where n, b are odd and a is even. We assume that it can be factorized as a Cartesian product of a 3-regular and a 2-regular circulant graph using Theorem 12. Then by Theorem 12 there exist $p(> 1), q(> 1)$ with $\gcd(p, q) = 1$ and $pq = 2n$ such that $C_{pq}(T) = C_p(R) \square C_q(S)$. Here, $R \subseteq [1, \frac{p}{2}]$, $S \subseteq [1, \frac{q}{2}]$ and $T = qR \cup pS = \{a, b, n\}$.

If possible, let $p = 2$ and $q = n$. Then $R = \{1\}$, i.e., $qR = \{n\}$. Since $p = 2$, all the elements of pS are even, which contradicts the fact that $b \in pS$. Hence, we must have $p > 2$ and $q < n$.

Now, since $\gcd(p, q) = 1$, it follows that $p = 2^k n_1$ and $q = n_2$, where both n_1 and $n_2(> 1)$ are odd integers.

Next, we claim that $R = \left\{\frac{p}{2}, \frac{b}{q}\right\}, S = \left\{\frac{a}{p}\right\}$.

Since p is even, we have $b \notin pS$. Now, if $n \in pS$ then $\frac{q}{2}$ must be in S. But $q = n_2$ is odd, hence $\frac{q}{2}$ is not an integer. So, n can not be in pS. Therefore our claim is true.

Since every element of R and S is an integer, it follows that p divides a and q divides b.

($\Leftarrow$) Let us assume there exist positive integers $p(> 2), q(1 < q < n)$ such that $p \mid a$, $q \mid b$, $pq = 2n$ and $\gcd(p, q) = 1$. We take $R = \{\frac{p}{2}, \frac{b}{q}\}$ and $S = \{\frac{a}{p}\}$. Then, $T = qR \cup pS = \{a, b, n\}$. Also, we claim that, $R \subseteq [1, \frac{p}{2}]$, $S \subseteq [1, \frac{q}{2}]$. If

possible, suppose $\frac{b}{q} \geq \frac{p}{2}$ and $\frac{a}{p} \geq \frac{q}{2}$, which is equivalent to $b \geq n$ and $a \geq n$. This contradicts the fact that $a < n$ and $b < n$. Hence, our claim holds. By Theorem 12, G can be factorized as the Cartesian product of $C_p(R)$ and $C_q(S)$.

Remark 4. It is well established that all 3-regular (Theorem 13) and 2-regular (Remark 2) circulant graphs are word-representable. Therefore, using Theorem 6, we conclude that all the 5-regular circulant graphs in Theorems 22 and 23 are word-representable.

Therefore, combining the above results, we deduce from Theorem 7 that $R(G) \leq 5 + \min\{p, q\}$ for all graphs G mentioned in Theorems 22 and 23.

3.3 Word-Representability and the Representation Number of 5-Regular Circulant Graph Using Morphisms

This section addresses the word-representability and the representation number of certain subclasses of 5-regular circulant graphs. The proofs primarily follow the approach via morphisms as presented in [12]. For any two words w and u, we write $\underbrace{w}_{u}$ to indicate that u is a subword of w. Moreover, for two letters y, z in a word w, we write $y \prec_w z$ if y occurs to the left of z in w.

Theorem 24. *Let $G = C_{2n}(x, 1, n)$ be a 5-regular circulant graph. Then G is word-representable for $\frac{n}{2} < x \leq \frac{2n}{3}$ and $R(G) \leq 5$.*

Proof. Consider the connected 5-regular circulant graph $G \cong C(x, 1, n)$ with $\frac{n}{2} < x < \frac{2n}{3}$. Since G is a 5-regular circulant graph, then $1 < x < n$ and hence $n \geq 3$. The vertex set of G is $V(G) = \{0, 1, 2, \ldots, n-1\}$. By the definition of a circulant graph, each vertex $i \in V(G)$ is adjacent to each of $(i+1)_{2n}, (i-1)_{2n}, (i+x)_{2n}, (i-x)_{2n}$ and $(i+n)_{2n}$. Define a morphism $f : V(G)^* \to V(G)^*$ by

$$f(i) = (i)_{2n}(i-1)_{2n}(i-x)_{2n}(i+n)_{2n}(i+x)_{2n}, \forall i \in V(G).$$

, where We show that the word $f(w)$ represents G, where $w = 0\,1\,2\,\ldots\,(n-1)$. For each $0 \leq i \leq n-1$, let $w_i = i\,(i+1)\,\ldots\,(n-1)\,0\,1\,\ldots\,(i-1)$. By Proposition 3, $f(w_i)$ represents G for all $0 \leq i \leq n-1$. Consequently, two vertices i and j alternate in $f(w)$ if and only if they alternate in $f(w_i)$ for every i. From the definition of the morphism f, each vertex i occurs exactly once in each of the words $f(i)$, $f((i+1)_{2n})$, $f((i+x)_{2n})$, $f((i+n)_{2n})$, and $f((i-x)_{2n})$. Hence, $f(w)$ is a 5-uniform word. Moreover,

$$f(w_i) = (i)\,(i-1)_{2n}\,(i-x)_{2n}\,(i+n)_{2n}\,(i+x)_{2n}\,(i+1)_{2n}\,(i)\,\ldots\,(i-1)_{2n}\,(i-2)_{2n}$$

$$(i-1-x)_{2n}\,(i-1+n)_{2n}\,(i-1+x)_{2n}.$$

For any vertex $i \in V(G)$, the occurrence of the factor $f(i)\,f(i+1)$ in $f(w_i)$ guarantees that i does not alternate with any vertex $j \in V(G) \setminus \{(i+1)_{2n}, (i-1)_{2n}, (i+x)_{2n}, (i-x)_{2n}, (i+n)_{2n}\}$. Therefore, it is sufficient to verify that i alternates with each vertex $j \in \{(i+1)_{2n}, (i-1)_{2n}, (i+x)_{2n}, (i-x)_{2n}, (i+n)_{2n}\}$ within $f(w_i)$, for all $0 \leq i \leq n-1$.

Case(1): *Here we will show the alternation between i and $(i+1)_{2n}$ in $f(w_i)$.*

Since $\frac{n}{2} < x \le \frac{2n}{3}$ and $1 < x < n$, we have $(i+x+1)_{2n} \preceq_{w_i} (i+n)_{2n}$ and $(i+n+1)_{2n} \preceq_{w_i} (i-x)_{2n}$.

If possible let, $(i+n)_{2n} \prec_{w_i} (i+x+1)_{2n}$ and $(i-x)_{2n} \prec_{w_i} (i+n+1)_{2n}$, i.e., $i+n < i+x+1$ and $2n+i-x < i+n+1$. Both of these lead to $n < x+1$, which is a contradiction as $x < n$. Hence, for every $0 \le i \le 2n-1$, the following ordering holds in w_i:

$$i \prec_{w_i} (i+1)_{2n} \prec_{w_i} (i+2)_{2n} \prec_{w_i} (i+x)_{2n} \prec_{w_i} (i+x+1)_{2n} \preceq_{w_i} (i+n)_{2n} \prec_{w_i} (i+n+1)_{2n} \preceq_{w_i}$$

$$(i-x)_{2n} \prec_{w_i} (i-x+1)_{2n}.$$

By the definition of the morphism f, we have four subcases,

- $(i+x+1)_{2n} = (i+n)_{2n}$ and $(i+n+1)_{2n} = (i-x)_{2n}$

$$f(w_i) = \underbrace{f(i)}_{i}\ \underbrace{f(i+1)}_{(i+1)_{2n},i}\ \underbrace{f(i+2)\cdots f(i+x)}_{(i+1)_{2n}}\ \underbrace{f(i+x+1)}_{i}\ \underbrace{f(i+n+1)\cdots f(i-x+1)}_{(i+1)_{2n},i}\ \underbrace{}_{(i+1)_{2n},i}\ \underbrace{}_{(i+1)_{2n}}.$$

- $(i+x+1)_{2n} \ne (i+n)_{2n}$ and $(i+n+1)_{2n} = (i-x)_{2n}$

$$f(w_i) = \underbrace{f(i)}_{i}\ \underbrace{f(i+1)}_{(i+1)_{2n},i}\ \underbrace{f(i+2)\cdots f(i+x)}_{(i+1)_{2n}}\ \underbrace{f(i+x+1)\cdots f(i+n)}_{i}\ \underbrace{f(i+n+1)\cdots}_{(i+1)_{2n}}\ \underbrace{}_{i}\ \underbrace{}_{(i+1)_{2n},i}$$

$$\underbrace{f(i-x+1)}_{(i+1)_{2n}}.$$

- $(i+x+1)_{2n} = (i+n)_{2n}$ and $(i+n+1)_{2n} \ne (i-x)_{2n}$

$$f(w_i) = \underbrace{f(i)}_{i}\ \underbrace{f(i+1)}_{(i+1)_{2n},i}\ \underbrace{f(i+2)\cdots f(i+x)}_{(i+1)_{2n}}\ \underbrace{f(i+x+1)}_{i}\ \underbrace{f(i+n+1)\cdots\cdots f(i-x)}_{(i+1)_{2n},i}\ \underbrace{}_{(i+1)_{2n}}\ \underbrace{}_{i}$$

$$\underbrace{f(i-x+1)}_{(i+1)_{2n}}.$$

- $(i+x+1)_{2n} \ne (i+n)_{2n}$ and $(i+n+1)_{2n} \ne (i-x)_{2n}$

$$f(w_i) = \underbrace{f(i)}_{i}\ \underbrace{f(i+1)}_{(i+1)_{2n},i}\ \underbrace{f(i+2)\cdots f(i+x)}_{(i+1)_{2n}}\ \underbrace{f(i+x+1)\cdots f(i+n)}_{i}\ \underbrace{f(i+n+1)\cdots}_{(i+1)_{2n}}\ \underbrace{}_{i}\ \underbrace{}_{(i+1)_{2n}}$$

$$\underbrace{f(i-x)}_{i}\ \underbrace{f(i-x+1)}_{(i+1)_{2n}}.$$

Hence, i and $(i+1)_{2n}$ alternate in $f(w_i)$ for all $0 \le i \le 2n-1$

The alternation between i and $(i-1)_{2n}$, i and $(i+x)_{2n}$, i and $(i-x)_{2n}$, as well as i and $(i+n)_{2n}$ can be proved in a similar manner. Hence, in all cases, the required alterations hold. Thus, the word $f(w)$ represents the graph G and consequently $R(G) \le 5$.

Theorem 25. *Let $G = C_{2n}(x, 1, n)$ be a 5-regular circulant graph and $x = \frac{n}{2}$. Then G is word-representable and $R(G) \leq 5$.*

Proof. Consider the connected 5-regular circulant graph $G \cong C(x, 1, n)$ with $x = \frac{n}{2}$. Since G is a 5-regular circulant graph, then $1 < x < n$, i.e., $x \geq 2$ and $n \geq 4$. The vertex set of G is $V(G) = \{0, 1, 2, \ldots, n-1\}$. By the definition of a circulant graph, each vertex $i \in V(G)$ is adjacent to each of $(i+1)_{2n}, (i-1)_{2n}, (i+x)_{2n}, (i-x)_{2n}$ and $(i+n)_{2n}$. Define a morphism $f : V(G)^* \to V(G)^*$ by

$$f(i) = (i)_{2n}(i-1)_{2n}(i+x)_{2n}(i+n)_{2n}(i-x)_{2n}, \forall i \in V(G).$$

We shall show that the word $f(w)$ represents G, where $w = 0\,1\,2\,\ldots\,(n-1)$. For each $0 \leq i \leq n-1$, define $w_i = i\,(i+1)\,\ldots\,(n-1)\,0\,1\,\ldots\,(i-1)$. By Proposition 2.6, the word $f(w_i)$ represents G for all $0 \leq i \leq n-1$. Consequently, two vertices i and j alternate in $f(w)$ if and only if they alternate in $f(w_i)$ for every i. From the definition of the morphism f, each vertex i occurs exactly once in each of the words $f(i)$, $f((i+1)_{2n})$, $f((i+x)_{2n})$, $f((i+n)_{2n})$, and $f((i-x)_{2n})$. Hence, $f(w)$ is a 5-uniform word. Moreover,

$$f(w_i) = (i)\,(i-1)_{2n}\,(i+x)_{2n}\,(i+n)_{2n}\,(i-x)_{2n}\,(i+1)_{2n}\,(i)\,\ldots\,(i-1)_{2n}\,(i-2)_{2n}\,(i-1+x)_{2n}$$

$$(i-1+n)_{2n}\,(i-1-x)_{2n}.$$

For any vertex $i \in V(G)$, the presence of the factor $f(i)\,f(i+1)$ in $f(w_i)$ guarantees that i does not alternate with any vertex $j \in V(G) \setminus \{(i+1)_{2n}, (i-1)_{2n}, (i+x)_{2n}, (i-x)_{2n}, (i+n)_{2n}\}$.

Therefore, it suffices to verify that i alternates with each vertex $j \in \{(i+1)_{2n}, (i-1)_{2n}, (i+x)_{2n}, (i-x)_{2n}, (i+n)_{2n}\}$ within $f(w_i)$, for all $0 \leq i \leq n-1$.

The alternation between i and $(i+1)_{2n}$, i and $(i-1)_{2n}$, i and $(i+x)_{2n}$, i and $(i-x)_{2n}$, as well as i and $(i+n)_{2n}$ can be proved in a similar process using the morphism f, as shown in Theorem 24. Hence, in all cases, the required alterations hold. Thus, the word $f(w)$ represents the graph G and consequently $R(G) \leq 5$.

4   Conclusion and Future Work

We have studied 5-regular circulant graphs with respect to word-representability and established upper bounds for certain subclasses. We now present a conjecture and pose several open problems that provide possible ways to address it.

Conjecture 1. All 5-regular circulant graphs are word-representable.

Problem 1. In Theorem 21 and Corollary 1, we established the word-representability of $C_{2n}(x, 1, n)$ for the case $n \equiv x \equiv 1 \pmod 3$ and $n \equiv x \equiv 2 \pmod 3$. A natural extension is to examine the remaining cases, such as $n \equiv x \equiv 0 \pmod 3$, $n \equiv 1; x \equiv 2 \pmod 3$, and $n \equiv 2; x \equiv 0 \pmod 3$, etc. Addressing these cases would lead to a complete characterization of the class $C_{2n}(x, 1, n)$. Upper bounds on the representation number can be obtained using morphisms, particularly for the intervals $1 < x < \frac{n}{2}$ and $\frac{2n}{3} < x < n$.

Problem 2. We have studied word-representability of 5-regular circulant graphs through Cartesian products. An alternative direction can be the factorization of this class via rooted products. For a detailed study of rooted products and their connection with word-representability, we refer the reader to [7,11], and [2].

References

1. Boesch, F., Tindell, R.: Circulants and their connectivities. J. Graph Theor. **8**(4), 487–499 (1984)
2. Broere, B.: Word-representable graphs. Master thesis at Radboud University, Nijmegen (2018)
3. Halldórsson, M.M., Kitaev, S., Pyatkin, A.: Semi-transitive orientations and word-representable graphs. Discret. Appl. Math. **201**, 164–171 (2016)
4. Heuberger, C.: On planarity and colorability of circulant graphs. Discret. Math. **268**(1–3), 153–169 (2003)
5. Huang, S., Kitaev, S., Pyatkin, A.: An embedding technique in the study of word-representability of graphs. Discret. Appl. Math. **346**, 170–182 (2024)
6. Kitaev, S.: A comprehensive introduction to the theory of word-representable graphs. In: International Conference on Developments in Language Theory, pp. 36–67. Springer (2017)
7. Kitaev, S., Lozin, V.: Words and graphs. Springer (2015)
8. Kitaev, S., Pyatkin, A.: On representable graphs. J. Autom. Lang. Comb. **13**(1), 45–54 (2008)
9. Kitaev, S., Pyatkin, A.: On semi-transitive orientability of triangle-free graphs. arXiv preprint arXiv:2003.06204 (2020)
10. Kitaev, S., Seif, S.: Word problem of the Perkins semigroup via directed acyclic graphs. Order **25**(3), 177–194 (2008)
11. Srinivasan, E., Hariharasubramanian, R.: Minimum length word-representants of graph products. Discret. Appl. Math. **358**, 91–104 (2024)
12. Srinivasan, E., Hariharasubramanian, R.: On semi-transitive orientability of circulant graphs. Discret. Appl. Math. **377**, 498–509 (2025)
13. Vilfred, V.: σ-labelled graph and circulant graphs. University of Kerala, India (1994)
14. Vilfred, V.: A theory of cartesian product and factorization of circulant graphs. Hindawi Pub. Corp. J. Discrete Math **2013** (2013)

Matrix Multiplication in the MPC Model

Lakshya Joshi[1]([✉]) [iD], Arya Deshmukh[2]([✉]) [iD], Atharv Chhabra[1]([✉]) [iD],
and Chetan Gupta[3]([✉]) [iD]

[1] Squarepoint Technologies, Bengaluru, India
[2] Microsoft, Aurangabad, India
[3] Indian Institute of Technology, Roorkee, India
`chetan.gupta@cs.iitr.ac.in`

Abstract. In this paper, we present algorithms to solve matrix multiplication problems in the MPC model. In particular, we consider the problem under various processor/memory constraints in the MPC model and prove the following results.

1. Multiplication of two rectangular matrices of size $d \times n$ and $n \times d$ (where $d \leq n$) respectively can be done in,
    (a) $O(\sqrt{d} + \log_d n)$ rounds with n processors and $\Theta(d)$ memory per processor
    (b) $O\left(\frac{d}{\sqrt{n}}\right)$ rounds with d processors and $\Theta(n)$ memory per processor.
2. Multiplication of two rectangular matrices of size $n \times d$ and $d \times n$ (where $d \leq n$) respectively, with n processors of $\Theta(n)$ memory per processor can be done in $O\left(\frac{d}{\sqrt{n}}\right)$ rounds.
3. The multiplication of two d-sparse matrices (matrices that contain at most d nonzero elements in each row and in each column) with n processors and $\Theta(d)$ memory per processor can be done in $O(d^{0.9})$ rounds.

Keywords: Distributed and Parallel Algorithms · Matrix Multiplication · MPC model

1 Introduction

We are given two matrices A and B, and our task is to compute $C = A \times B$. Matrix multiplication has been extensively studied in the centralized setting, leading to a long line of work on improving the asymptotic complexity of sequential algorithms (see, e.g., [1,5,6,11–13]). However, these algorithms are not efficiently parallelizable due to the inherent dense communication pattern in them. This limitation has motivated the investigation of matrix multiplication in distributed and parallel computational models, including the clique model, low-bandwidth models, and related frameworks [2–4,7,8,10]. Building on this line of research, in this paper we explore matrix multiplication in the Massively Parallel Computation (MPC) model, a framework that captures large-scale parallel

N. Misra and A. Pandey (Eds.): CALDAM 2026, LNCS 16445, pp. 209–220, 2026.
https://doi.org/10.1007/978-3-032-17156-6_16

computation with restricted memory per machine. We consider the problem $C = A \times B$ in the following three settings: when matrices A and B are (i) rectangular matrices of dimension $d \times n$ and $n \times d$, respectively, (ii) rectangular matrices of dimension $n \times d$ and $d \times n$, respectively and, (iii) d-sparse matrices i.e. both are $n \times n$ size matrices such that each matrix contains at most d non-zero elements in each row and each column. For completeness, we provide a brief overview of the MPC model in Sect. 1.1, followed by a precise description of the computational settings in which we address these three variants of the matrix multiplication problem.

1.1 MPC Model

In this paper, we work with the *massively parallel computation* model (MPC) [9]. Given any problem of size (input + output) m words, such that m is too big to store on a single computer, in the MPC model, we assume that the problem is distributed over a fully connected network of computers (or processors). The computation proceeds in synchronous rounds. Processors perform computation locally, and at the end of a round, they exchange their messages. However, unlike some other models of distributed and parallel computation, in the MPC model, we assume that each computer has a limited memory. In particular, we assume that each computer has $\Theta(m^\delta)$ memory and the number of computers in the network is $\Theta(m^{1-\delta})$, where $0 < \delta < 1$. Therefore, the total memory of the system is $\Theta(m)$ sufficient to store the given problem. Since each computer has $\Theta(m^\delta)$ memory, it is allowed to send and receive $O(m^\delta)$ words in each round. If any processor receives more than $\Theta(m^\delta)$ words during the course of an algorithm, the algorithm is considered failed.

1.2 Setup

We assume that initially, matrices A and B are distributed evenly among all the processors. The structure of the distribution does not matter because we can reach any desired even distribution in one round (even if we start with an uneven distribution). At the end of the algorithm the elements of the output matrix C should also be evenly distributed. We analyse the problem in the following cases.

(i) (d, n, d): when A and B are rectangular matrices size $d \times n$ and $n \times d$ respectively, where $d \leq n$

(ii) (n, d, n) : A and B are of size $n \times d$ and $d \times n$ respectively

(iii) d-sparse: A and B are d-sparse $n \times n$ matrices, i.e. each row and column of matrices A and B contains at most d non-zero elements. Also we are interested in at most d nonzero entries in each row and each column in C.

1.3   Our Results

(d, n, d): In this case, both the input and output matrices have size $O(dn)$. Accordingly, it suffices to assume that the total memory available in the system is $\Theta(dn)$. We analyze this instance under two different processor–memory configurations:

(i) For n processor each with $\Theta(d)$ memory we design an algorithm with round complexity $O(\sqrt{d} + \log_d n)$

(ii) For d processor each with $\Theta(n)$ memory we design an algorithm with round complexity $(\frac{d}{\sqrt{n}})$ rounds.

(n, d, n): Here, the input matrices have size $O(nd)$, while the output matrix may be as large as $O(n^2)$, corresponding to an $n \times n$ matrix. Therefore, we analyse this case under the configuration of n processors, each with $\Theta(n)$ memory, and present an algorithm with round complexity $O(\frac{d}{\sqrt{n}})$.

d-Sparse Case: Finally, we consider the case where both A and B are $n \times n$ matrices, each with at most d nonzero entries per row and per column. Since we are interested in d elements in the output matrix, the size of both input and output matrices is bounded by $O(nd)$. Therefore, we analyse this case with n processors each with $\Theta(d)$ memory. The state-of-the-art for this is a trivial algorithm that takes $O(d)$ rounds. We improve that by giving an $O(d^{0.9})$ round algorithm.

1.4   High-level Idea of Our Technique

In rectangular cases, we observed that in order to reduce inter-processor communication, the best case is when all processors try to compute a square submatrix of C. Thus, we divide the matrices A and B into smaller square sub-matrices of equal size depending on the local memory of the processor, and assign each submatrix to one of the processors. At the end, each processor needs to produce the corresponding submatrix of C. For the d-sparse case, we adapt the idea of Gupta et al. [7]. We first refine the analysis given in their paper and improve upon their results. Here, we would like to highlight that we improve the bound present in their paper from $O(d^{1.927})$ to $O(d^{1.925})$ for solving d-sparse matrix multiplication in the low bandwidth model. We do not explicitly mention that result here because that result is about the low-bandwidth model, and in this paper, we stick to the MPC model. However, the improved bound can be directly derived from the Lemma 3 that we prove in this paper (by appropriately setting the value of ϵ for the low bandwidth model). Then we use Lemma 3 to prove our result for d-sparse matrix multiplication in the MPC model. In their paper, they show that d-sparse matrix multiplication is equivalent to processing triangles in a d-degree graph. In other words, processing triangles is equivalent to a product $a_{ij}b_{jk}$, where a_{ij} and b_{jk} are some elements of matrices A and B, respectively. They show that given any d-degree graph, say T, we can divide the graph into two parts such that one part, say T', is *clustered* - it contains layers of subgraphs of the original graph such that in each layer we have disjoint clusters of triangles which are small and dense, and the other part, say T', contains a low number of triangles. Now, in order to process all the triangles, we can first process all the triangles in the clustered part layer-by-layer using square matrix multiplication, then compute all the triangles T' using brute force. We noticed that there was some suboptimality in bring down the unprocessed triangles from T to T'. In this paper, we remove that suboptimality and then use that result as a black box to solve d-sparse matrix multiplication in the MPC model.

2   Rectangular Matrix Multiplication

In this section, we consider the case when matrices A and B are rectangular matrices. There can be many types of instances in rectangular matrix multiplication. In this paper, we focus on two primary cases of type (d, n, d) and (n, d, n) and analyse them separately in subsequent sections.

2.1   (d, n, d) Matrix Multiplication

In the (d, n, d) case, the size of all three matrices is bounded by nd; thus, it is sufficient to have a total memory of size $O(nd)$. We analyse this case under two different memory constraints, two subcases: (i) when there are n processors each with $O(d)$ memory, and (ii) when there are d processors each with $O(n)$, where $d \leq n$. First, we prove the following two lemmas that will help us to analyse both these cases.

Lemma 1. *If there are t processors each containing matrices of size $\sqrt{k} \times \sqrt{k}$, and the memory of each processor is $\Theta(k)$, then the sum of these matrices can be computed in $O(\log_k t)$ rounds in the MPC model.*

Proof. Suppose $M_1, M_2, \ldots, M_t$ are the t matrices and $X = \sum M_l$. Let m_{ij}^l denote the element at i^{th} row and j^{th} column of M_l. Therefore $x_{ij} = \sum m_{ij}^l$. We divide the t processors into k groups of $(\frac{t}{k})$ processors such that each group is responsible for computing one x_{ij}. Let's just focus on the first group, the same algorithm will be followed in the other groups. Let $p_1, p_2, \ldots, p_{t/k}$ be the processors in the first group. In the first round, all t processor exchange messages such that p_1 receives elements $m_{11}^1, m_{11}^2, \ldots m_{11}^k$; p_2 receives elements $m_{11}^{k+1}, m_{11}^{k+2}, \ldots m_{11}^{2k}$ and so on (see Fig. 1). Now, each p_i computes the sum of all the elements it receives. Notice that all the values that are required to compute x_{11} are present in t/k processors, such that each processor contains only one value (the sum it has calculated). Now, in the next round, processor $p_2, p_3, \ldots p_k$ will send all the intermediate values they have calculated in the previous round to p_1. Similarly, $p_{k+2}, p_{k+3}, \ldots p_{2k}$ will send all the intermediate values they have calculated in the previous round to p_{k+1} and so on (see Fig. 2 for reference). Now, notice that all the values that are required to compute x_{11} are now present in the $(\frac{n}{k^2})$ processors, and each contains only one value. Repeat

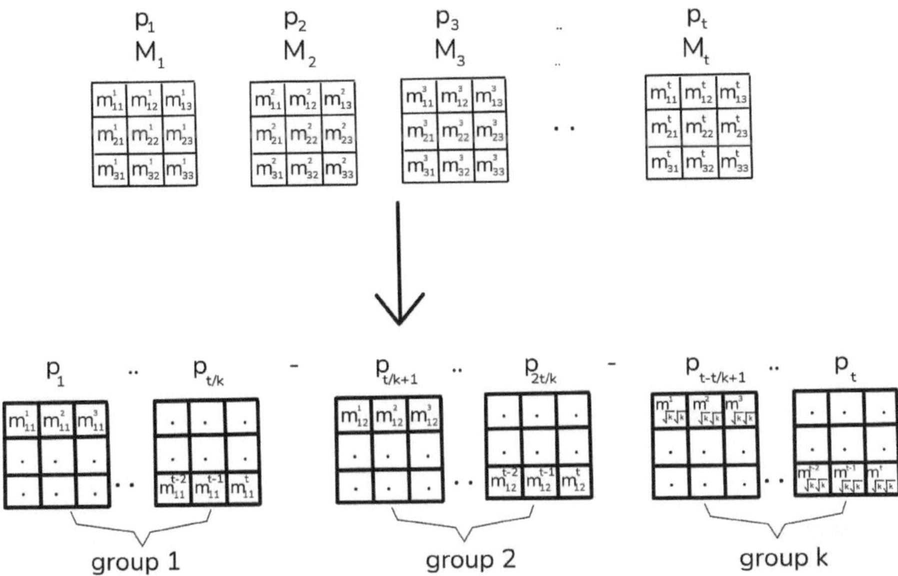

Fig. 1. Upper layer represents the initial distribution of elements, and lower layer represents the distribution after the first round.

the previous step for $O(\log_k t)$ rounds. After $O(\log_k t)$, the final value of x_{11} will be present in one processor. The same thing will happen in the other groups. Therefore, after $O(\log_k t)$ rounds processors will calculate X. This finishes the proof of the lemma. Notice that in the case where $t \leq k$, the algorithm will terminate in merely a single round. Therefore, in the above proof, we can assume $t > k$.

Now we are all set to analyse the (d, n, d) case in the aforementioned settings. First, we discuss the case with n processors with $O(d)$ memory.

n Processors with $O(d)$ Memory: Since C is of size $d \times d$, let us divide C into d sub-matrices of size $\sqrt{d} \times \sqrt{d}$ each, denoted by C^{ij}, where $i, j \in [\sqrt{d}]$. We assign $(\frac{n}{d})$ processors to evaluate each sub-matrix. Also divide the matrices A and B into sub-matrices of size $\sqrt{d} \times \sqrt{d}$, each being denoted as A^{iq} and B^{qj} respectively, where $i, j \in [\sqrt{d}]$ and $q \in [\frac{n}{\sqrt{d}}]$. We know that,

$$C^{ij} = \sum_q A^{iq} * B^{qj}$$

Observe that there would be a total of n submatrices of the forms A^{iq} and B^{qj}. In one round of communication, let's distribute the input among the n processors such that each processor contains exactly two sub-matrices, namely, A^{iq} and B^{qj} for some values of i, j, q such that the sub-matrices stored by any two processors are distinct. This is possible since storing A^{iq} and B^{qj} requires

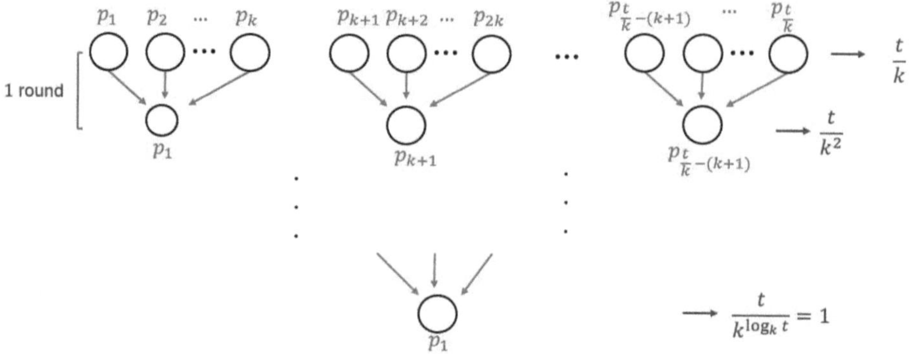

Fig. 2. Processors in the first group computing the value of x_{11}.

only $O(d)$ memory. Let us focus on computation on C^{11}, other submatrices are computed in the same way. We have assigned $\frac{n}{d}$ processors to compute it. Let $P^{11} = \{p_1, p_2, \ldots p_{\frac{n}{d}}\}$ be the set of processor assigned to C^{11}. We set the communication pattern as follows. In round k, processor p_i receives submatrices corresponding to $q = (i-1)\sqrt{d}+k$. In other words, in round 1, it receives matrices $A^{1((i-1)\sqrt{d}+1)}$ and $B^{((i-1)\sqrt{d}+1)1}$, in round 2, it receives $A^{1((i-1)\sqrt{d}+2)}$ and $B^{((i-1)\sqrt{d}+2)1}$ and so on. In each round, p_i computes the product of summatrices received in the current round, and adds the newly computed product to the existing matrix it computed till the previous round using simple matrix addition. It is easy to observe that after $\sqrt{d}$ rounds, processor p_i will contain matrix $\sum_{k=1}^{\sqrt{d}} A^{1((i-1)\sqrt{d}+k)} * B^{((i-1)\sqrt{d}+k)1}$. Now we have $(\frac{n}{d})$ submatrices distributed over the processors in P^{11}, each containing a matrix of size $\sqrt{d} \times \sqrt{d}$ such that the sum of those matrices will produce C^{11}. According to Lemma 1, C^{11} can be computed in $O(\log_d n)$ rounds. Similarly, the other C^{ij} are also computed in parallel. Thus, matrix C can be computed in $O(\sqrt{d} + \log_d n)$ rounds. The only thing missing in the proof is to show that each processor is sending messages of $O(d)$ size per round. Notice that each element of matrices A and B needs to be sent to at most $\sqrt{d}$ processors. Each processor initially holds d elements of both A and B. Thus, it needs to send $d\sqrt{d}$ elements to other processors, which can easily be done in $O(\sqrt{d})$ rounds, such that each processor sends $O(d)$ elements in each round. From this we get the following theorem.

Theorem 1. *Multiplication of two rectangular matrices of size $d \times n$ and $n \times d$ respectively, with n processors of $O(d)$ memory can be done $O(\sqrt{d} + \log_d n)$ rounds in the MPC model, where $d \leq n$.*

d Processors with O(n) Memory: We handle this case also in a similar way to the previous one, except the division of processors and matrices is slightly different. Without loss of generality, assume that we scale the value of d such that $d := (\lceil \frac{d}{\sqrt{n}} \rceil \sqrt{n})$. Let us divide the matrix C into $(\frac{d^2}{n})$ sub-matrices of size

$\sqrt{n} \times \sqrt{n}$ each, denoted by C^{ij}, where $i, j \in [\frac{d}{\sqrt{n}}]$. We assign $(\frac{n}{d})$ processors to evaluate each sub-matrix. Also divide the matrices A and B into sub-matrices of size $\sqrt{n} \times \sqrt{n}$, each being denoted as A^{iq} and B^{qj} respectively, where $i, j \in [\frac{d}{\sqrt{n}}]$ and $q \in [\sqrt{n}]$. We know that,

$$C^{ij} = \sum_q A^{iq} * B^{qj}$$

Observe that there would be a total of d submatrices of the form A^{iq} and B^{qj}. In one round of communication, let's distribute the input among the d processors such that each processor contains exactly two sub-matrices, namely, A^{iq} and B^{qj} for some values of i, j, q such that the sub-matrices stored by any two processors are distinct. This is possible since storing A^{iq} and B^{qj} requires only $O(n)$ memory. Let us focus on computation on C^{11}, other submatrices are computed in the same way. We have assigned $\frac{n}{d}$ processors to compute it. Let $P^{11} = \{p_1, p_2, \ldots p_{\frac{n}{d}}\}$. In round k, processor p_i receives submatrices corresponding to $q = (i - 1)\frac{d}{\sqrt{n}} + k$ $i.e.$ in round 1, it receives $A^{1((i-1)\frac{d}{\sqrt{n}}+1)}$ and $B^{((i-1)\frac{d}{\sqrt{n}}+1)1}$, in round 2, it receives $A^{1((i-1)\frac{d}{\sqrt{n}}+2)}$ and $B^{((i-1)\frac{d}{\sqrt{n}}+2)1}$ and so on. It is easy to observe that p_i would receive these sub-matrices from a single other processor because of the way in which the initial distribution of A^{iq} and B^{qj} was done among the processors. Thus, the amount of information that a processor ends up sending in any communication round would be $O(n)$. Processor p_i then computes the product of these submatrices locally after each round, adding the newly computed product to the existing product that it computed in the previous round using simple matrix addition. It's easy to observe that after $\frac{d}{\sqrt{n}}$ rounds, processor p_i will contain $\sum_{k=1}^{\frac{d}{\sqrt{n}}} A^{1((i-1)\frac{d}{\sqrt{n}}+k)} * B^{((i-1)\frac{d}{\sqrt{n}}+k)1}$. Now we have $(\frac{n}{d})$ submatrices distributed over the processors in P^{ij}, each containing a matrix of size $\sqrt{n} \times \sqrt{n}$ such that the sum of those matrices will produce C^{11}. According to Lemma 1, C^{11} can be computed in $O(\log_n \frac{n}{d})$ rounds. But since $\frac{n}{d} \leq n$, it can be computed in a constant number of rounds. Similarly, the other C^{ij} are also computed in parallel. Thus, matrix C can be computed in $O(\frac{d}{\sqrt{n}})$ rounds.

Theorem 2. *Multiplication of two rectangular matrices of size $d \times n$ and $n \times d$ respectively, with d processors of $O(n)$ memory can be done in $O(\frac{d}{\sqrt{n}})$ rounds ounds in the MPC model, where $d \leq n$.*

2.2   (n, d, n) Matrix Multiplicaton

In this case, we assume that we have n processors with $O(n)$ memory each. Notice that in this case the matrices A and B can be stored in $O(nd)$ memory; however, the resulting C matrix is a square matrix of size $n \times n$. Thus, we limit the total memory of the system to $O(n^2)$. We divide the matrix C into n sub-matrices of size $\sqrt{n} \times \sqrt{n}$ each, denoted by C^{ij} (similar to Sect. 2.1) where

$i, j \in [\sqrt{n}]$. Assign the computation of C^{ij} to processor p^{ij}. Without loss of generality, assume that we scale the value of d such that $d := (\lceil \frac{d}{\sqrt{n}} \rceil \sqrt{n})$. Divide the matrices A and B into sub-matrices of size $\sqrt{n} \times \sqrt{n}$, each being denoted as A^{iq} and B^{qj} respectively, where $i,\ j \in [\sqrt{n}]$ and $q \in [\frac{d}{\sqrt{n}}]$. Observe that

$$C^{ij} = \sum_{q=1}^{\frac{d}{\sqrt{n}}} A^{iq} * B^{qj}, \text{where } * \text{ denotes matrix multiplication}$$

We can assume that each processor p_{ij} contains one row and one column of A and B respectively. Note that an element of A needs to be sent to at most $\sqrt{n}$ processors. Thus, p_{ij} needs to send $d\sqrt{n}$ elements. In one round, it can send n elements and thus, in $\frac{d}{\sqrt{n}}$ rounds, it can send all the elements to the destination processors. Now let us analyse how many elements of A and B are received by each processor. A^{iq} and B^{qj} are $\sqrt{n} \times \sqrt{n}$ matrices containing a total of n elements. Therefore, p_{ij} needs a total of $(\frac{d}{\sqrt{n}} \cdot n) = d\sqrt{n}$ elements of A and B in order to compute C^{ij}. In one round, p_{ij} can receive at most $O(n)$ elements and thus, in $\frac{d\sqrt{n}}{n} = \frac{d}{\sqrt{n}}$ rounds, it can receive all the elements that it requires for the computation of C^{ij}.

Theorem 3. *Multiplication of two rectangular matrices of size $n \times d$ and $d \times n$ respectively, with n processors of $O(n)$ memory requires $\Theta(\frac{d}{\sqrt{n}})$ rounds, where $d \leq n$.*

3   Sparse Matrix Multiplication

In this section, we will discuss the case when A and B are d-sparse. Notice that in this case, the resulting matrix $C = A * B$ can be d^2-sparse. But in the same spirit as discussed in Gupta et al. [7], in order to keep the model more meaningful, we assume that we are interested in only at most d elements of each row and column of C. Therefore, both the input and output can be stored on $O(nd)$ memory. We analyse the case when we have n processors, each with $O(d)$ memory. We assume that initially each processor holds one row and one column (i.e. $O(d)$ elements in total) of A and B, and at the end each processor outputs d elements of one of the rows of C. We assume that the position of these d elements of C that a processor needs to output is already known to the processor.

A Trivial Algorithm: Notice each element of C is the sum of d pairwise products of elements of A and B. Thus, there are at most nd^2 products that need to be calculated. In one round, a processor can compute any d products of C that are required to compute an element of C. It can do this by obtaining the d elements of matrices A and B, then performing d multiplications locally. Thus, in one round of local computation, the n processors can compute any nd terms of C. Hence, to compute the nd^2 of C, we only require $O(d)$ communication rounds. Now, we give a better algorithm than this as follows.

We try to find a more efficient method to compute the terms of C. For that, we adapt the idea of Gupta et al. [7]. In particular, we use Lemma 4.2 of their paper. As we discussed earlier, in their paper, they translated the problem of matrix multiplication to triangle processing. So first, we translate their results according to the terminology we used in this. We call each product of type $a_{ij}b_{jk}$ a *term*, where a_{ij} and b_{jk} are elements of A and B respectively. As we mentioned, there are at most nd^2 terms that need to be computed inorder to compute C, Gupta et al. [7] computed them in a sequence (or layer) of rounds as follows.

Lemma 2 ([7]). *Let $0 \leq \varepsilon_1 < \varepsilon_2$, and assume d is sufficiently large. Let T denote the remaining number of terms of C that need to be computed such that*

$$|T| = O(d^{2-\varepsilon_1}n).$$

Then, we can compute these terms in l sequential iterations where

$$l = O(d^{5\varepsilon_2 - \varepsilon_1})$$

and after these iterations only T' terms remain to be computed, such that

$$|T'| = O(d^{2-\varepsilon_2}n)$$

Let us try to explain the lemma in simple words. The lemma basically states that the multiplication of two $n \times n$, d-sparse matrices ($C = A * B$) that contain $O(d^{2-\epsilon_1}n)$ *terms* to be computed can be divided into two parts C' and C'' such that

1. $C' = A_1 \times B_1 + A_2 \times B_2 + \ldots + A_l \times B_l$ (l is same as mentioned in the lemma)
   where
   (i) A_i and B_i are submatrices of A and B respectively.
   (ii) A_i and B_i consist of $(\frac{n}{d})$ submatrices A_i^j and B_i^j, for $j \in [\frac{n}{d}]$, respectively. Each A_i^j and B_i^j is of size $d \times d$ such that $A_i \times B_i = \sum A_i^j * B_i^j$.

2. $C'' = A' \times B'$ such that, it requires at most $O(d^{2-\epsilon_2}n)$ terms to compute C''
3. $C = C' + C''$

Thus, the total time required to compute C is the time to compute C' and C'' one after another. The idea is to use square matrix multiplication as a black box algorithm to compute C' (because computing A_i and B_i requires multiplication of many $d \times d$ square matrices) and use a brute force algorithm to compute C''.

To compute C', we can run l sequential iterations, each for one $A_i \times B_i$. In one iteration, we can assign d processors for the computation of $A_i^j * B_i^j$ for each j. These $(\frac{n}{d})$ computations will happen in parallel in one iteration. In this paper, we improve the bound on the number of iterations (l) required to go from T to T'. Notice that, in the MPC model, the time required for one iteration is $O(\sqrt{d})$. Because this involves $\frac{n}{d}$ instance matrix multiplications of $d \times d$ matrices happening in parallel, each with the help of d computers. Thus, if we plug $n = d$ in Theorem 3, we get an $O(\sqrt{d})$ rounds algorithm for this task. Now we improve the bounds on the number of iterations required to bring down the number of terms from $O(d^{2-\epsilon_1})$ to $O(d^{2-\epsilon_2})$ mentioned in Lemma 2

3.1   Tight Bound for Number of Iterations

Lemma 3. *Number of rounds required to bring down the number of uncomputed terms from $|T| = O(d^{2-\varepsilon_1}n)$ to $|T'| = O(d^{2-\varepsilon_2}n)$ is $l = O(d^{4\varepsilon_2})$, where $0 \le \varepsilon_1 < \varepsilon_2$.*

Proof. According to Lemma 2 it will require $O(d^{5\varepsilon_2-\varepsilon_1})$ iteration to go from T to T'. Let us see what happens when we introduce an intermediate step in the reduction of terms from T to T', i.e. instead of directly going from $O(d^{2-\varepsilon_1}n)$ to $O(d^{2-\varepsilon_2}n)$. We first go from $O(d^{2-\varepsilon_1}n)$ to $O(d^{2-\varepsilon_3}n)$ and then go from $O(d^{2-\varepsilon_3}n)$ to $O(d^{2-\varepsilon_2}n)$ for some $\varepsilon_1 < \varepsilon_3 < \varepsilon_2$, using Lemma 2. And calculate the number of iterations consumed in this process. Let $\varepsilon_3 = \frac{\varepsilon_1+\varepsilon_2}{2}$ such that

$$|T| = O(d^{2-\varepsilon_1}n) \quad , \quad |T''| = O(d^{2-\varepsilon_3}n) \quad , \text{ and } \quad |T'| = O(d^{2-\varepsilon_2}n).$$

Let l_1 denote the number of iterations required to reduce the number of uncomputed terms from T to T'', and similarly, l_2 denote the number of iterations required to reduce the number of uncomputed terms from T'' to T'. By Lemma 2, we have $l_1 = O(d^{5\varepsilon_3-\varepsilon_1})$ and $l_2 = O(d^{5\varepsilon_2-\varepsilon_3})$. Thus, the total number of iterations required to reduce the number of uncomputed terms from T to T' is $l = l_1 + l_2 = O(d^{5\varepsilon_3-\varepsilon_1}) + O(d^{5\varepsilon_2-\varepsilon_3}) = O(d^{5\varepsilon_2-\varepsilon_3})$. Since $5\varepsilon_2 - \varepsilon_3 > 5\varepsilon_3 - \varepsilon_1$, we get a strictly better bound than presented in Lemma 2.

Now the idea is that we recursively insert intermediate steps in this process i.e. insert ε_4 between ε_3 and ε_2 (such that ε_4 is the arithmetic mean of ε_3 and ε_2) and similarly ε_5 between ε_4 and ε_2 and do this x number of times (where x is big constant), then, we can see that number of iterations required to reduce the number of *terms* step by step would be $l = l_1 + l_2 + .. + l_{x+1}$, where l_1 is the number of iterations to go from ε_1 to ε_3, l_2 from ε_3 to ε_4, .., l_{x+1} from ε_{x+2} to ε_2. Observe that the order of l is dominated by the term l_{x+1}. Thus, $l = O(d^{5\varepsilon_2-\varepsilon_{x+2}})$. Therefore, by inserting enough intermediate steps (bounded by some constant) we can make ε_{x+2}, infinitesimally close to ε_2. Hence l will converge to

$$O(d^{5\varepsilon_2-\varepsilon_{x+2}}) \approx O(d^{4\varepsilon_2})$$

Final Computation of Terms: As we mentioned, the multiplication of two d-sparse matrices can be done in two phases. In one phase, we iteratively use square matrix multiplication, and in the other phase, we use the brute-force algorithm. Our task is to compute $O(d^2n)$ terms. We will first reduce the number of uncomputed terms from $O(nd^2)$ to $O(nd^{2-\varepsilon})$ $(\varepsilon \ge 0)$ by iteratively using square matrix multiplication, and then compute the rest of the terms using trivial computation. By Lemma 3, the number of iterations to reduce the number of uncomputed terms from $O(nd^2)$ to $O(nd^{2-\varepsilon})$ is $O(d^{4\varepsilon})$, where each iteration takes $O(\sqrt{d})$ communication rounds. Thus the total time for phase one is $O(d^{4\varepsilon+\frac{1}{2}})$. After phase one, we are left with $O(nd^{2-\varepsilon})$ terms that we compute using a trivial brute force algorithm. We know that in one round, one processor can compute $O(d)$ terms (because it has $O(d)$ memory), therefore total rounds required to compute $O(nd^{2-\varepsilon})$ temrs will be $O(\frac{nd^{2-\varepsilon}}{nd}) = O(d^{1-\varepsilon})$. There is a tradeoff between the round complexity of these two phases, i.e. if we spend less time in phase one, it will take more time (polynomially) in phase two. Therefore, in order to optimise the total number of communication rounds needed, which is $O(d^{4\varepsilon+1/2}) + O(d^{1-\varepsilon})$, should be minimised. Which happen for $\varepsilon = 0.1$, taking $O(d^{0.9})$ communication rounds. Thus, we prove the following theorem.

Theorem 4. *Multiplication of two d-sparse matrices in the MPC model with n processors, each with $\Theta(d)$ memory, can be performed in $O(d^{0.9})$ rounds.*

Acknowledgments. This work was supported by the ANRF India Prime Minister Early Career Research Grant: ANRF/ECRG/2024/004480/ENS.

References

1. Alman, J., Williams, V.V.: A refined laser method and faster matrix multiplication. In: Proceedings of the ACM-SIAM Symposium on Discrete Algorithms (SODA 2021), pp. 522–539 (2021). https://doi.org/10.1137/1.9781611976465.32
2. Censor-Hillel, K., Dory, M., Korhonen, J.H., Leitersdorf, D.: Fast approximate shortest paths in the congested clique. Distrib. Comput. **34**(6), 463–487 (2020). https://doi.org/10.1007/s00446-020-00380-5
3. Censor-Hillel, K., Kaski, P., Korhonen, J.H., Lenzen, C., Paz, A., Suomela, J.: Algebraic methods in the congested clique. Distrib. Comput. **32**(6), 461–478 (2016). https://doi.org/10.1007/s00446-016-0270-2
4. Censor-Hillel, K., Leitersdorf, D., Turner, E.: Sparse matrix multiplication and triangle listing in the congested clique model. In: Proceedings of the OPODIS 2018 (2018). https://doi.org/10.4230/LIPIcs.OPODIS.2018.4
5. Coppersmith, D., Winograd, S.: Matrix multiplication via arithmetic progressions. In: Proceedings of the 19th Annual ACM Symposium on Theory of Computing (STOC), pp. 1–6 (1987)
6. Coppersmith, D., Winograd, S.: Matrix multiplication via arithmetic progressions. J. Symb. Comput. **9**(3), 251–280 (1990)
7. Gupta, C., Hirvonen, J., Korhonen, J.H., Studený, J., Suomela, J.: Sparse matrix multiplication in the low-bandwidth model. In: Agrawal, K., Lee, I.T.A. (eds.) SPAA '22: 34th ACM Symposium on Parallelism in Algorithms and Architectures, Philadelphia, PA, USA, July 11–14, 2022. pp. 435–444. ACM (2022). https://doi.org/10.1145/3490148.3538575
8. Gupta, C., Korhonen, J.H., Studený, J., Suomela, J., Vahidi, H.: Brief announcement: low-bandwidth matrix multiplication: Faster algorithms and more general forms of sparsity. In: Agrawal, K., Petrank, E. (eds.) Proceedings of the 36th ACM Symposium on Parallelism in Algorithms and Architectures, SPAA 2024, Nantes, France, June 17–21, 2024. pp. 305–307. ACM (2024). https://doi.org/10.1145/3626183.3660270, https://doi.org/10.1145/3626183.3660270
9. Karloff, H.J., Suri, S., Vassilvitskii, S.: A model of computation for mapreduce. In: Charikar, M. (ed.) Proceedings of the Twenty-First Annual ACM-SIAM Symposium on Discrete Algorithms, SODA 2010, Austin, Texas, USA, January 17–19, 2010. pp. 938–948. SIAM (2010). https://doi.org/10.1137/1.9781611973075.76

10. Le Gall, F.: Further algebraic algorithms in the congested clique model and applications to graph-theoretic problems. In: Gavoille, C., Ilcinkas, D. (eds.) DISC 2016. LNCS, vol. 9888, pp. 57–70. Springer, Heidelberg (2016). https://doi.org/10.1007/978-3-662-53426-7_5
11. Le Gall, F.: Powers of tensors and fast matrix multiplication. In: Proceedings of the International Symposium on Symbolic and Algebraic Computation (ISSAC), pp. 296–303. ACM (2014)
12. Pan, C.: Strassen's algorithm is not optimal. Trilinear technique of aggregating, uniting and canceling for constructing fast algorithms for matrix operations. In: 19th Annual Symposium on Foundations of Computer Science (FOCS), pp. 166–174. IEEE (1978)
13. Vassilevska Williams, V.: Multiplying matrices faster than coppersmith–winograd. In: Proceedings of the 44th Annual ACM Symposium on Theory of Computing (STOC), pp. 887–898 (2012)

Complexity and Approximation Algorithm of 3-Component Domination in Graphs

Subhasmita Joshi[(⊠)] and Bhawani Sankar Panda

Department of Mathematics, Indian Institute of Technology Delhi,
Hauz Khas, New Delhi 110016, India
{maz238454,bspanda}@maths.iitd.ac.in

Abstract. Let s be an integer and $G = (V, E)$ be a graph. A subset $D \subseteq V$ is called an *s-component dominating set* of $G = (V, E)$ if every vertex $v \in V \setminus D$ is adjacent to a vertex in D and each component of $G[D]$ has at least s vertices. The *s-component domination number* of $G = (V, E)$, denoted as $\gamma_s(G)$, is the minimum cardinality among all s-component dominating sets of G. Note that an s-component dominating set is just a dominating set for $s = 1$ and a total dominating set for $s = 2$. Given a graph G and a positive integer k, the DECIDE 3-COMP DOM is the problem to decide whether G admits a 3-component dominating set of cardinality at most k, and MIN 3-COMP DOM is the problem of computing $\gamma_3(G)$. In this paper, we first show the complexity difference between domination and 3-component domination. We, then, show that DECIDE 3-COMP DOM is NP-complete for bipartite graphs as well as for chordal graphs. We also show that MIN 3-COMP DOM is $\frac{3}{2}(1 + \ln \Delta)$-approximable for general graphs, where Δ is the maximum degree of G. Next, we show that MIN 3-COMP DOM cannot be approximated within $(1 - \epsilon) \ln |V|$ for any $\epsilon > 0$ unless P = NP even when $G = (V, E)$ is bipartite and even when $G = (V, E)$ is chordal. We prove that MIN 3-COMP DOM for bounded degree graphs admits a constant approximation algorithm. Finally, we show that MIN 3-COMP DOM is APX-complete for bipartite graphs with maximum degree 4.

Keywords: Domination · Total domination · 3-component domination · NP-complete · APX-complete · Approximation algorithm

1 Introduction

Let $G = (V, E)$ be a graph. A set $D \subseteq V$ of a graph $G = (V, E)$ is called a *dominating set* of G if every vertex $v \in V \setminus D$ is adjacent to a vertex in D. Over the years, several variations of domination have been introduced to address different application-based requirements. One notable variation is total domination. A set $D \subseteq V$ of a graph $G = (V, E)$ is called a *total dominating set* of G if every vertex $v \in V$ is adjacent to a vertex in D. A natural extension of domination and

N. Misra and A. Pandey (Eds.): CALDAM 2026, LNCS 16445, pp. 221–235, 2026.
https://doi.org/10.1007/978-3-032-17156-6_17

total domination is the *s-component domination*, which generalizes domination by incorporating connectivity constraints.

Let s be a positive integer. An (s-cd set in short) of a graph $G = (V, E)$ is a set $D_s \subseteq V$ such that every $v \in V \setminus D_s$ is adjacent to a vertex in D_s and each component of $G[D_s]$ has at least s vertices. The *s-component domination number* of $G = (V, E)$ is the minimum cardinality among all *s-cd* sets of G and it is denoted as $\gamma_s(G)$. The concept of *s-cd* set was introduced by Alvarado et al. [2]. An *s-cd* set reduces to a dominating set when $s = 1$ and reduces to a total dominating set when $s = 2$. Gao et al. [5] initiated the study of *s-cd* set for $s = 3$. They established several bounds for $\gamma_3(G)$. Bounds for $\gamma_3(G)$ have been obtained in [6], when G is a tree and when G is a generalized Petersen graph. The MIN *s*-COMP DOM problem is to compute the *s*-component domination number, $\gamma_s(G)$ and DECIDE *s*-COMP DOM is its decision version as stated below.

<u>DECIDE *s*-COMP DOM</u>

Instance: A graph G, a positive integer k, and a fixed positive integer s.

Question: Does there exist an *s-cd* set of G of cardinality at most k?

In this paper, we start the algorithmic and complexity study of the 3-component domination problem in graphs. The main contributions of this paper are as follows.

1. We establish the complexity difference between MIN DOM and MIN 3-COMP DOM.
2. We show that DECIDE 3-COMP DOM is NP-complete for bipartite graphs and for chordal graphs.
3. We prove that the MIN 3-COMP DOM is $\frac{3}{2}(1 + \ln \Delta)$-approximable for general graphs with maximum degree Δ.
4. We show that MIN 3-COMP DOM cannot be approximated within $(1 - \epsilon) \ln |V|$ for any $\epsilon > 0$ unless $P = NP$, even for bipartite graphs and even for chordal graphs.
5. We prove that MIN 3-COMP DOM for bounded degree graphs admits a constant approximation algorithm.
6. We show that MIN 3-COMP DOM is APX-complete for bipartite graphs with maximum degree 4.

2   Preliminaries

In this section, we give some pertinent definitions. Let $G = (V, E)$ be a finite, simple, and undirected graph of order $n = |V|$ and size $m = |E|$. The *open neighborhood* of a vertex v is defined as $N(v) = \{u \in V | uv \in E\}$ and the *closed neighborhood* is defined as, $N[v] = \{v\} \cup N(v)$. The degree of a vertex v is denoted by $d(v)$ and defined as $d(v) = |N(v)|$. The maximum degree of the graph G is denoted by Δ. For $S \subseteq V$, $G[S] = (S, E_S)$, where $E_S = \{uv \in E : u, v \in S\}$, denotes the *subgraph induced by* S. A set S of vertices in which every pair of distinct vertices is non-adjacent is called an *independent set* and if every pair of distinct vertices in S is adjacent, then S is called a *clique*. We use the notation $[k] = \{1, 2, \ldots, k\}$.

A graph $G = (V, E)$ is called a *bipartite graph* if the vertex set V can be partitioned into two disjoint sets X and Y such that every edge in E has one end vertex in X and the other in Y. A bipartite graph with bipartition (X, Y) is denoted as $G = (X \cup Y, E)$. A graph $G = (V, E)$ is said to be a *chordal graph* if every cycle in G of length at least four has a chord, that is, an edge joining two non-consecutive vertices of the cycle. A graph $G = (V, E)$ is said to be a *split graph* if the vertex set V can be partitioned into two sets, say C and I, such that C is a clique and I is an independent set of G. A split graph with split partition (C, I) is denoted as $G = (C \cup I, E)$.

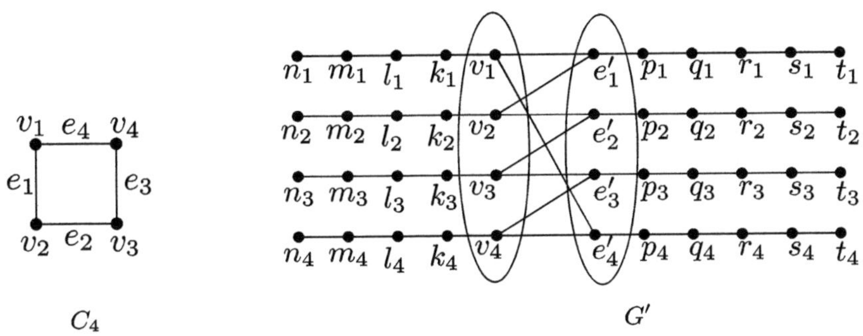

Fig. 1. Construction of the graph G' from $G = C_4$ as described in Definition 1.

3   Complexity Difference

In this section, we show that 3-component domination problem is computationally different from the domination problem. To achieve this, we construct special graph classes where the DECIDE DOM problem is NP-complete while the MIN 3-COMP DOM problem is solvable in polynomial time for one graph class and for the other graph class, MIN DOM problem is solvable in polynomial time while DECIDE 3-COMP DOM problem is NP-complete. Firstly, we define a family of graphs, denoted as $\mathcal{F}$, such that for every graph in $\mathcal{F}$, DECIDE 3-COMP DOM is NP-complete but MIN DOM is solvable in polynomial time.

Definition 1. *Let* $\mathcal{F} = \{G' = (V', E') : G = (V, E) \text{ is a graph}\}$, *where* $G' = (V', E')$ *is obtained from* $G = (V, E)$ *with* $V = \{v_1, v_2, \ldots, v_n\}$ *and* $E = \{e_1, e_2, \ldots, e_m\}$ *as follows:*

- *First, we take the subdivision of the graph G, that is by inserting a vertex e'_j on the edge e_j of G, for $j \in [m]$.*
- *For each $v_i \in V$, $i \in [n]$, we attach the path $P_4 = k_i, l_i, m_i, n_i$ to v_i by adding the edge $v_i k_i$.*

- *For each e'_j, $j \in [m]$, we attach the path $P_5 = p_j, q_j, r_j, s_j, t_j$ to e'_j by adding the edge $e'_j p_j$.*

The construction of G' from G is illustrated in Fig. 1. Now, we prove the following theorem.

Theorem 1. *Let $G' = (V', E')$ be the graph in $\mathcal{F}$ obtained from a graph $G = (V, E)$ of order n and size m, as defined in Definition 1, then $\gamma(G') = 2n + 2m$.*

Proof. Note that in G', for each $i \in [n]$, the subgraph $G'[\{n_i, m_i, l_i, k_i, v_i\}]$ and for each $j \in [m]$, the subgraph $G'[\{p_j, q_j, r_j, s_j, t_j\}]$ are paths of length 5. Since each of these paths needs at least two vertices in its minimum dominating set, the cardinality of any dominating set of G' is at least $2n + 2m$. Let $D = \{m_i, k_i : i \in [n]\} \cup \{s_j, p_j : j \in [m]\}$. Then it is easy to see that D is a dominating set of G' and hence $\gamma(G') = 2n + 2m$. $\qquad\square$

Now, we prove the following theorem.

Theorem 2. DECIDE 3-COMP DOM *is* NP-*complete for the family $\mathcal{F}$.*

Proof. It is easy to see that DECIDE 3-COMP DOM is in NP for the graphs in $\mathcal{F}$. To show the hardness, we give a polynomial reduction from the VERTEX COVER problem, which is known to be NP-complete [3]. The VERTEX COVER is defined below.

VERTEX COVER

Instance: A graph G and a positive integer k.
Question: Does there exist a vertex cover in G of cardinality at most k?

Let (G, k) be an instance of VERTEX COVER, where G is of order n and size m. Let $G' \in \mathcal{F}$ be the graph constructed from G as described in Definition 1. Let $k' = k + 3n + 3m$. Now (G', k') be the instance of DECIDE 3-COMP DOM. We prove the following claim.

Claim 1. G has a vertex cover of size at most k if and only if G' has a 3-*cd* set of size at most $k' = k + 3n + 3m$.

Proof. Let $G = (V, E)$ has a vertex cover, say V_c, such that $|V_c| \le k$. We define the set $D_3 = V_c \cup \{m_i, l_i, k_i : i \in [n]\} \cup \{q_j, r_j, s_j : j \in [m]\}$. Clearly, D_3 is a 3-*cd* set of G' with cardinality at most $k + 3n + 3m$.

Conversely, assume that D'_3 is a 3-*cd* set of G' of cardinality at most $k + 3n + 3m$. Now we will construct a 3-*cd* set D_3 of G' from D'_3 such that $|D_3| \le |D'_3|$ and D_3 does not contain any e'_j, $j \in [m]$. This is achieved through the following replacements.

- For each $j \in [m]$, either $\{r_j, s_j, t_j\} \subset D'_3$ or $\{q_j, r_j, s_j\} \subset D'_3$. If $\{r_j, s_j, t_j\} \subset D'_3$ then replace it with $\{q_j, r_j, s_j\}$.
- For each $j \in [m]$, if $e'_j \in D'_3$, then replace it with one of its neighbors in V.
- For each $i \in [n]$, either $\{n_i, m_i, l_i\} \subset D'_3$ or $\{m_i, l_i, k_i\} \subset D'_3$. If $\{n_i, m_i, l_i\} \subset D'_3$ then replace it with $\{m_i, l_i, k_i\}$.

Clearly, D_3 is a 3-*cd* set and $|D_3| \leq |D_3'|$. Also, in D_3, for each $j \in [m]$, e_j' is dominated exclusively by some $v_i \in V$. Hence, $D_3 \cap V$ is a vertex cover of G. Notice that $|D_3 \cap V| = |D_3 \setminus (\{m_i, l_i, k_i : i \in [n]\} \cup \{q_j, r_j, s_j : j \in [m]\})| = |D_3| - 3n - 3m \leq |D_3'| - 3n - 3m \leq k$. This completes the proof of the claim.□

Therefore, DECIDE 3-COMP DOM is NP-complete for the family $\mathcal{F}$.   □

Hence, from Theorem 1 and Theorem 2, we have the following observation.

Corollary 1. *For the family of graphs* $\mathcal{F}$, DECIDE 3-COMP DOM *is* NP-*complete, whereas* MIN DOM *is solvable in polynomial time.*

On the other hand, we construct a graph class, say $\mathcal{F}'$, for which MIN 3-COMP DOM is solvable in polynomial time and DECIDE DOM is NP-complete.

Definition 2. *Let* $\mathcal{F}' = \{G' = (V', E') : G = (V, E)$ *is a graph* $\}$, *where* $G' = (V', E')$ *is obtained from* $G = (V, E)$ *with* $V = \{v_1, v_2, \ldots, v_n\}$ *and* $E = \{e_1, e_2, \ldots, e_m\}$ *as follows:*

- *For each* $v_i \in V$, $i \in [n]$, *we attach a path* $P_3 = a_i, b_i, c_i$ *to* v_i *by adding the edge* $v_i a_i$.

Figure 2 illustrates the construction of G' from G. We then prove the following theorem.

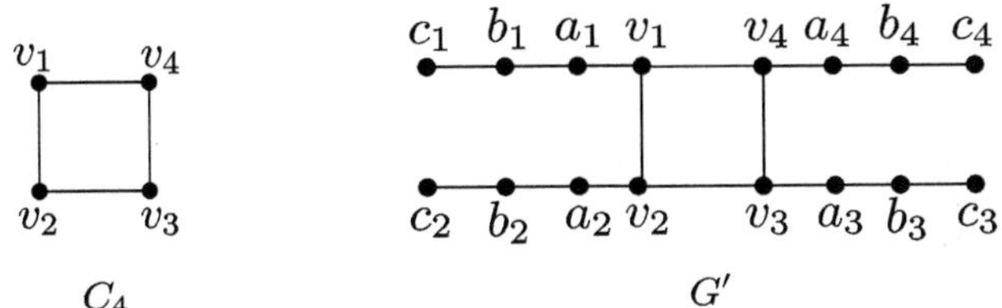

Fig. 2. Construction of the graph G' from C_4 as descibed in Definition 2.

Theorem 3. *Let* G' *be a graph in the family* $\mathcal{F}'$. *Then,* $\gamma_3(G') = 3n$.

Proof. Consider the subgraph, $G'[\{v_i, a_i, b_i, c_i\}]$, for some $i \in [n]$, which is a path of length 4. Note that any minimum 3-*cd* set contains either $\{a_i, b_i, c_i\}$ or $\{v_i, a_i, b_i\}$, for each $i \in [n]$. Hence, $\gamma_3(G') \geq 3n$. Consider the set $D_3 = \{v_i, a_i, b_i : i \in [n]\}$. Clearly, D_3 is a 3-*cd* set of G'. Thus, we have $\gamma_3(G') = 3n$.□

Next, in the following theorem, we show that DECIDE DOM is NP-complete for graphs in the family $\mathcal{F}'$.

Theorem 4. DECIDE DOM *for the graphs in the family* $\mathcal{F}'$ *is* NP-*complete.*

Proof. Clearly, DECIDE DOM is in NP. We show the hardness by giving a polynomial reduction from the DECIDE DOM for general graphs. For a given instance of DECIDE DOM for general graphs, say (G, k), where G is of order n, let G' be the graph in $\mathcal{F}'$ constructed from G according to the Definition 2. Let $k' = k+n$. Now (G', k') is an instance of DECIDE DOM. We then prove the following claim.

Claim 2. G has a dominating set of cardinality at most k if and only if G' has a dominating set of cardinality at most $k + n$.

Proof. Suppose that D is a dominating set of G of cardinality at most k. Then, the set $D \cup \{b_i : i \in [n]\}$ is a dominating set of G' of cardinality at most $k' = k+n$.

Conversely, assume that G' has a dominating set, say D' of cardinality at most $k' = k + n$. For each $i \in [n]$, if $c_i \in D'$, then replace it with b_i and if $a_i \in D'$, then replace it with v_i. Let the resultant set be D''. Clearly, $D'' \cap V$ is a dominating set of G such that $|D''| \leq k$. This completes the proof of the claim. $\square$

Therefore, DECIDE DOM is NP-complete for the graphs in the family $\mathcal{F}'$. $\square$

Now, we get the following corollary using Theorem 3 and Theorem 4.

Corollary 2. *For the graphs in $\mathcal{F}'$,* MIN 3-COMP DOM *is solvable in polynomial time, whereas* DECIDE DOM *is* NP-*complete.*

4 NP-Completeness Results

In this section, we show that the DECIDE 3-COMP DOM is NP-complete for bipartite graphs as well as for chordal graphs.

4.1 NP-complete for Bipartite Graphs

In Theorem 2, we establish that DECIDE 3-COMP DOM is NP-complete for the class $\mathcal{F}$. Since the family $\mathcal{F}$ forms a subclass of bipartite graphs, we have the following result.

Theorem 5. DECIDE 3-COMP DOM *is* NP-*complete for bipartite graphs.*

4.2 NP-complete for Chordal Graphs

In this subsection, we show that DECIDE 3-COMP DOM for chordal graphs is NP-complete by showing the NP-completeness for split graphs, a subclass of chordal graphs. To show the hardness, we shall use the following well-known NP-complete problem [3].

SET COVER

Instance: A set system $(U, \mathcal{C})$, where U is the ground set, $\mathcal{C}$ is a collection of subsets of U, and a positive integer k.

Question: Does there exist a set cover $\mathcal{S} = \{S_1, S_2, \ldots, S_k\} \subseteq \mathcal{C}$ of U such that $|\mathcal{S}| \leq k$?

Theorem 6. DECIDE 3-COMP DOM *is* NP-*complete for split graphs.*

Proof. Clearly, the DECIDE 3-COMP DOM for split graphs is in NP. To demonstrate the hardness result, we present a polynomial reduction from SET COVER. Given an instance of SET COVER, say $(U, \mathcal{C})$, we construct an instance of DECIDE 3-COMP DOM for split graphs as (G, k), where $G = (C \cup I, E)$ is a split graph. The construction is done in polynomial time as follows (Fig. 3):

- For each element $u_i \in U$, we include a vertex y_i in the set I of G.
- For each S_j in the collection $\mathcal{C}$, we add a vertex x_j in the set C of G.
- Add an edge between y_i and x_j if the element u_i belongs to set S_j.
- For each pair of vertices x_i, x_j in C, add an edge.

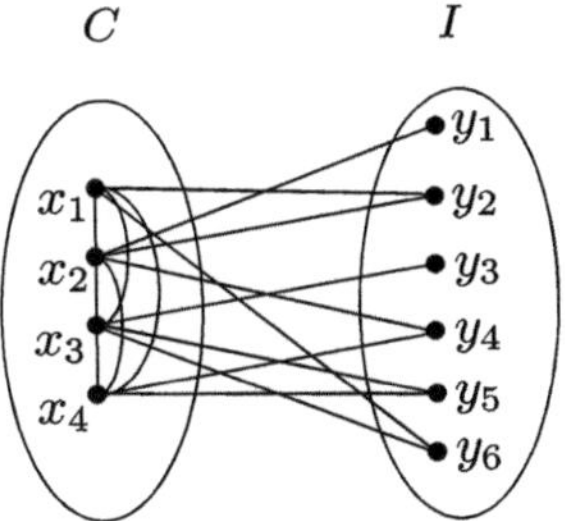

Fig. 3. Construction of the split graph $G = (C \cup I, E)$ from set system $(U, \mathcal{C})$ with $U = \{u_1, u_2, u_3, u_4, u_5, u_6\}$ and $\mathcal{C} = \{S_1 = \{u_2, u_6\}, S_2 = \{u_1, u_2, u_4\}, S_3 = \{u_3, u_5, u_6\}, S_4 = \{u_4, u_5\}\}$ for Theorem 6.

Clearly, the constructed graph G is a split graph with split partitions: (C, I), where C is a clique and I is an independent set. We now prove the following claim.

Claim 3. The set system $(U, \mathcal{C})$ has a set cover of cardinality at most k if and only if the graph G has a 3-*cd* set of cardinality at most k.

Proof. Let $\mathcal{S}$ be a set cover of $(U, \mathcal{C})$ such that $|\mathcal{S}| \leq k$. Define a set $D_3 = \{x_j : S_j \in \mathcal{S}\}$. D_3 is clearly a 3-*cd* set of G with cardinality at most k. Conversely, suppose that G has a 3-*cd* set of cardinality at most k, say D_3. If $y_i \in D_3$, then replace it with one of its neighbors in C and let D_3' be the resultant set. Clearly, $\mathcal{S} = \{S_j : x_j \in D_3'\}$ is a set cover of the system $(U, \mathcal{C})$ of cardinality at most k.□
    Thus, DECIDE 3-COMP DOM is NP-complete for split graphs. □

5   Approximation Results

In this section, we design an approximation algorithm for MIN 3-COMP DOM for general graphs and also for bounded-degree graphs.

Algorithm 1: APPROX-3CD

1 **Input:** A graph $G = (V, E)$.
2 **Output:** A 3-*cd* set, D_3 of G.
3 **begin**
4   Compute a total dominating set D_t of G using the algorithm APPROX-TD;
5   Compute the components of $G[D_t]$. Let it be $C_1, C_2, \ldots C_p$;
6   **for** $i = 1$ *to* p **do**
7          $T = T \cup \{z_i\}$, where z_i is the vertex as described above;
8   **end**
9      $D_3 = D_t \cup T$;
10   **return** D_3 ;
11 **end**

5.1   General Graphs

Before describing the approximation algorithm for MIN 3-COMP DOM for general graphs, we first recall the following result on total domination.

Theorem 7. *[7]* MIN TOTAL DOM *in any graph* $G = (V, E)$ *with maximum degree* Δ *can be approximated with an approximation ratio of* $1 + \ln \Delta$.

We now prove the following theorem.

Theorem 8. MIN 3-COMP DOM *in a graph* G *with maximum degree* Δ *can be approximated with an approximation ratio of* $\frac{3}{2}(1 + \ln \Delta)$.

Proof. We propose an algorithm, APPROX-3CD to compute an approximate solution of MIN 3-COMP DOM of a graph $G = (V, E)$. From Theorem 7, we infer that there exists a polynomial time algorithm, say APPROX-TD, which takes a graph G as input and outputs a total dominating set, say D_t, of G such that $|D_t| \leq (1 + \Delta)\gamma_t(G)$, where $\gamma_t(G)$ denotes the *total domination number* of G. The algorithm APPROX-3CD works in two stages: in the first stage, we use the algorithm APPROX-TD to get a total dominating set, D_t of G. In the second stage, we find an additional set $T \subset V$ such that $D_t \cup T$ is a 3-*cd* set of G. Now, we discuss the construction of the set T.

Let $C_1, C_2, \ldots, C_p$ be the components of $G[D_t]$ such that $|V(C_i)| = 2$, for each $i \in [p]$. Since each $v \in D_t$ has at least one neighbor in D_t, the number of connected components in $G[D_t]$ is at most $|D_t|/2$. Thus, $p \leq |D_t|/2$. Let $V(C_i) = \{x_i, y_i\}$, for each $i \in [p]$. As G is connected, either $d(x_i) \geq 2$ or $d(y_i) \geq 2$. Without loss of generality, assume that $d(x_i) \geq 2$. Then, clearly $|N(x_i) \cap D_t| = 1$ and $|N(x_i) \cap \overline{D_t}| \geq 1$, where $\overline{D_t} = V \setminus D_t$. Let $z_i \in N(x_i) \cap \overline{D_t}$. Define T to be the collection of such z_i, for each $1 \leq i \leq p$. Clearly, $D_3 = D_t \cup T$ is a 3-*cd* set of G. From the construction of the set T, we have the following inequality:

$$|T| = p \leq |D_t|/2.$$

Next, we use the fact that $\gamma_t(G) \leq \gamma_3(G)$ and do the following computations to get the approximation factor.

$$|D_3| = |D_t| + |T| \leq |D_t| + \frac{|D_t|}{2} = \frac{3}{2}|D_t|$$
$$\leq \frac{3}{2}(1 + \ln \Delta)\gamma_t \leq \frac{3}{2}(1 + \ln \Delta)\gamma_3.$$

Therefore, the 3-*cd* set D_3 returned by the Algorithm 1 is an approximate solution of MIN 3-COMP DOM with an approximation ratio of $\frac{3}{2}(1 + \ln \Delta)$. $\square$

5.2 Bounded Degree Graphs

In this subsection, we show that MIN 3-COMP DOM for any graph G with maximum degree Δ can be approximated within an approximation ratio of $1 + \Delta$.

Let D_3 be a 3-*cd* set of the graph $G(V, E)$. Then, we have

$$\bigcup_{v \in D_3} N[v] = V \implies |V| \leq \sum_{v \in D_3} |N[v]| \implies |V| \leq (1 + \Delta)\gamma_3(G).$$

Since V is a 3-*cd* set of a graph G, we have the following result.

Theorem 9. MIN 3-COMP DOM *for any graph G with maximum degree Δ can be approximated within an approximation ratio of $1 + \Delta$.*

6 Lower-Bound of Approximation

In this section, we provide a lower bound on the approximation ratio of MIN 3-COMP DOM for bipartite graphs and for chordal graphs.

6.1 Bipartite Graphs

To get the lower bound on the approximation ratio of MIN 3-COMP DOM for bipartite graphs, we give an approximation-preserving reduction from MIN SET COVER. For this, we need the following theorem.

Theorem 10. *[4]* MIN SET COVER *for set system $(U, \mathcal{C})$ cannot be approximated within $(1 - \epsilon) \ln |U|$ for any $\epsilon > 0$ unless* P=NP.

Now, we prove the following theorem.

Theorem 11. MIN 3-COMP DOM *for bipartite graph $G = (V, E)$ cannot be approximated within $(1 - \epsilon) \ln |V|$ for any $\epsilon > 0$ unless* P=NP.

Proof. Given an instance $(U, \mathcal{C})$ of MIN SET COVER, where $U = \{u_1, u_2, \ldots, u_p\}$ and $\mathcal{C} = \{S_1, S_2, \ldots, S_q\}$. We construct a bipartite graph $G = (X \cup Y, E)$ in polynomial time as follows:

- For each S_j, $j \in [q]$ in the collection $\mathcal{C}$, we include a vertex x_j in the partite set X of G.
- For each element $u_i \in U$, $i \in [p]$, we add a vertex y_i in the partite set Y of G.
- Add an edge between x_j and y_i if the element u_i belongs to set S_j.
- Further, add a vertex x_s in X and add a vertex y_t in Y. Add the edge $x_s y_t$ and add the edges $x_j y_t$, for each $j \in [q]$.

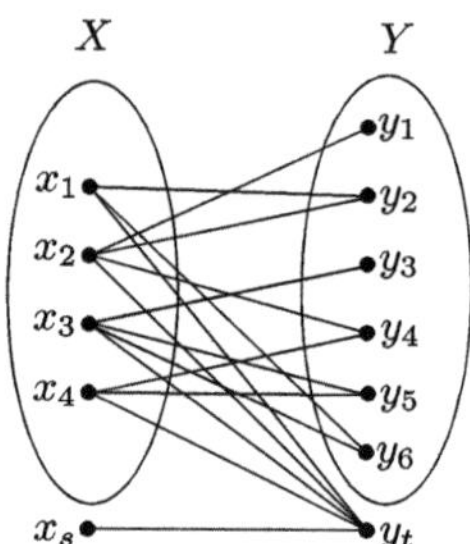

Fig. 4. Construction of bipartite graph $G = (X \cup Y, E)$ from set system $(U, \mathcal{C})$, where $U = \{u_1, u_2, u_3, u_4, u_5, u_6\}$ and $\mathcal{C} = \{S_1 = \{u_2, u_6\}, S_2 = \{u_1, u_2, u_4\}, S_3 = \{u_3, u_5, u_6\}, S_4 = \{u_4, u_5\}\}$ for Theorem 11.

Figure 4 illustrates the above construction. Now, we prove the following claim.

Claim 4. The set system $(U, \mathcal{C})$ has a set cover of cardinality at most k if and only if G has a 3-*cd* set of cardinality at most $k + 1$.

Proof. Suppose that $\mathcal{S}$ is a set cover of $(U, \mathcal{C})$ such that $|\mathcal{S}| \leq k$. Define the set $D_3 = \{x_j : S_j \in \mathcal{S}\} \cup \{y_t\}$. Clearly, D_3 is a 3-*cd* set of G and $|D_3| \leq k + 1$.

Conversely, let D_3 be a 3-*cd* set of G with cardinality at most $k+1$. If $x_s \in D_3$, then replace it with y_t. Also, if $y_i \in D_3$ for some $i \in [p]$, replace y_i with one of its neighbors in X. Let D_3' be the newly constructed set. Clearly, D_3' is a 3-*cd* set of cardinality at most $k + 1$. Note that the set $\mathcal{S} = \{S_j : x_j \in D_3'\}$ is a set cover of cardinality at most k. $\qquad\square$

From the Claim 4, it follows that if D_3^* is any minimum 3-*cd* set of G and $\mathcal{S}^*$ is any optimal set cover of $(U, \mathcal{C})$, then $|D_3^*| = |\mathcal{S}^*| + 1$.

Now, to the contrary, suppose that MIN 3-COMP DOM can be approximated within a ratio of α, where $\alpha = (1 - \epsilon) \ln |V|$ for some (fixed) $\epsilon > 0$, by some polynomial-time approximation algorithm, say Algorithm A. By using Algorithm A, we propose an algorithm A' to compute a set cover of the given system $(U, \mathcal{C})$ in polynomial time.

Note that since l is a constant, line 4 of the Algorithm A' can be executed in polynomial time. Also, as Algorithm A runs in polynomial time, Algorithm A' is a polynomial-time algorithm. Now, if $\mathcal{S}$ is computed in step 1, then $\mathcal{S}$ is optimal. Thus, we restrict our attention to the case where $|\mathcal{S}| \geq l$.

Algorithm 2: A'

1 **Input:** A set system $(U, \mathcal{C})$ and a constant integer l.
2 **Output:** A set cover $\mathcal{S}$ of $(U, \mathcal{C})$.
3 **begin**
4   **if** *there exists a minimum set cover $\mathcal{S}$ of $(U, \mathcal{C})$ such that $|\mathcal{S}| < l$* **then**
5 | **return** $\mathcal{S}$;
6   **end**
7   **else**
8   Construct a bipartite graph G as described at the beginning of the proof;
9   Compute a 3-*cd* set D_3 of G using Algorithm A;
10   Modify D_3 to get the set D_3' as described in the proof of the Claim 4 of Theorem 11;
11       $\mathcal{S} = \{S_j : x_j \in D_3'\}$;
12   **end**
13   **return** $\mathcal{S}$;
14 **end**

Let D_3^* be any minimum 3-*cd* set of G and $\mathcal{S}^*$ be an optimal set cover in $(U, \mathcal{C})$. Then, $|\mathcal{S}^*| \geq l$ and $|D_3^*| = |\mathcal{S}^*| + 1$. Let $\mathcal{S}$ be the set cover computed by Algorithm 2. Then

$$|\mathcal{S}| = |D_3| - 1 \leq |D_3| \leq \alpha |D_3^*|$$
$$= \alpha(|\mathcal{S}^*| + 1)$$
$$= \alpha \left(1 + \frac{1}{|\mathcal{S}^*|}\right) |\mathcal{S}^*|$$
$$\leq \alpha \left(1 + \frac{1}{l}\right) |\mathcal{S}^*|$$

Hence, Algorithm 2 approximates MIN SET COVER for a given set system $(U, \mathcal{C})$ within ratio $\alpha(1 + \frac{1}{l})$.

Let l be a positive integer such that $\frac{1}{l} < \epsilon$. Then

$$\alpha \left(1 + \frac{1}{l}\right) \leq (1 - \epsilon) \ln\left(|U| + |\mathcal{C}| + 2\right)(1 + \epsilon)$$
$$\leq (1 - \epsilon^2) \ln |U| \ \text{(as } \ln\left(|U| + |\mathcal{C}| + 2\right) \approx \ln |U| \text{ for sufficiently large}$$
$$\text{value of } |U|)$$
$$\leq (1 - \epsilon') \ln |U| \ \text{(assuming } \epsilon' = \epsilon^2)$$

Therefore, Algorithm 2 approximates MIN SET COVER within ratio $(1 - \epsilon) \ln |U|$ for some $\epsilon > 0$. Then by Theorem 10, P $=$ NP. Hence, our assumption that MIN 3-COMP DOM can be approximated within a ratio of $(1 - \epsilon) \ln |U|$ for some $\epsilon > 0$ is wrong. This proves that MIN 3-COMP DOM for bipartite graphs cannot be approximated $(1 - \epsilon) \ln |V|$ for some $\epsilon > 0$ unless P $=$ NP. $\qquad\qquad\square$

Algorithm 3: B'

1 Input: A set system $(U, \mathcal{C})$.
2 Output: A minimum set cover $\mathcal{S}$ of $(U, \mathcal{C})$.
3 begin
4   Construct a split graph G as described at the beginning of the proof;
5   Compute a 3-*cd* set D_3 of G using Algorithm B;
6   Modify D_3 to get the set D_3' as described in the proof of Claim 3;
7 $\mathcal{S} = \{S_j : y_j \in D_3'\}$;
8 end
9 return S;

6.2 Chordal Graphs

We now establish the inapproximability of MIN 3-COMP DOM through the following theorem.

Theorem 12. MIN 3-COMP DOM *for a chordal graph* $G = (V, E)$ *cannot be approximated within* $(1 - \epsilon) \ln |V|$ *for any* $\epsilon > 0$ *unless* P=NP.

Proof. Let $(U, \mathcal{C})$ be an instance of MIN SET COVER. From this instance, we construct a split graph $G = (V, E)$ with split partition (C, I), an instance of MIN 3-COMP DOM for chordal graphs as described below.

- For each element $u_i \in U$, we include a vertex y_i in the set I of G.
- For each S_j in the collection $\mathcal{C}$, we add a vertex x_j in the set C of G.
- Add an edge between y_i and x_j if the element u_i belongs to set S_j.
- For each pair of distinct vertices x_i, x_j in C, add the edge $x_i x_j$.

Note that the above construction is the same as in Theorem 6. Moreover, by Claim 3 of Theorem 6, we have $\gamma_3(G) = |\mathcal{S}^*|$, where $\mathcal{S}^*$ is a minimum set cover of the system $(U, \mathcal{C})$.

To the contrary, suppose that MIN 3-COMP DOM can be approximated within a ratio of α, where $\alpha = (1 - \epsilon) \ln |V|$ for some fixed $\epsilon > 0$, by some polynomial-time approximation algorithm, say algorithm B. Next, we design a polynomial-time algorithm B' to compute a set cover of the given system $(U, \mathcal{C})$ as given in Algorithm 3. Clearly, Algorithm 3 is a polynomial-time algorithm as Algorithm B runs in polynomial time. Let D_3^* be any minimum 3-*cd* set of G and $\mathcal{S}^*$ be an optimal set cover in $(U, \mathcal{C})$. Then, $|D_3^*| = |\mathcal{S}^*|$. Let $\mathcal{S}$ be the set cover computed by Algorithm 3 and D_3 be the 3-*cd* set computed by Algorithm B. Then,

$$|\mathcal{S}| = |D_3| \leq \alpha |D_3^*| = (1 - \epsilon) \ln (|U| + |\mathcal{C}|) |\mathcal{S}^*| \approx (1 - \epsilon) \ln (|U|) |\mathcal{S}^*|$$

Hence, Algorithm B' approximates MIN SET COVER within a ratio $(1 - \epsilon) \ln |U|$ for some $\epsilon > 0$. Then by Theorem 10, P=NP. Hence, our assumption that MIN 3-COMP DOM can be approximated within a ratio of $(1 - \epsilon) \ln |U|$ for some $\epsilon > 0$ is wrong. This shows that MIN 3-COMP DOM for a chordal graph $G = (V, E)$ cannot be approximated within $(1 - \epsilon) \ln |V|$ for some $\epsilon > 0$ unless P = NP. $\square$

7   APX-complete Result

By Theorem 9, MIN 3-COMP DOM for graphs with maximum degree 4 is in APX. Now, we show that MIN 3-COMP DOM is APX-complete for bipartite graphs with maximum degree 4. For this purpose, we first recall the concept of L-reduction.

Definition 3. *Given two* NP *optimization problems* π_1 *and* π_2 *and a polynomial time transformation* f *from instances of* π_1 *to instances of* π_2, *we say that* f *is an L-reduction if there are positive constants* α *and* β *such that for every instance* x *of* π_1:

- $opt_{\pi_2}(f(x)) \leq \alpha opt_{\pi_1}(x)$.
- *for every feasible solution* y *of* $f(x)$ *with objective value* $m_{\pi_2}(f(x), y) = c_2$, *we can find a solution* y' *of* x *in polynomial time with* $m_{\pi_1}(f(x), y') = c_1$ *such that* $|opt_{\pi_1}(x) - c_1| \leq \beta |opt_{\pi_2}(f(x)) - c_2|$.

To show the APX-completeness of a problem $\pi \in$ APX, it suffices to show that there is an L-reduction from some APX-complete problem to π. Here, we give an L-reduction from MIN VERTEX COVER. For this, we first recall the following result.

Theorem 13. *[1]* MIN VERTEX COVER *is* APX-*complete for graphs with degree at most* 3.

Now we prove the following theorem.

Theorem 14. MIN 3-COMP DOM *is* APX-*complete for bipartite graphs with maximum degree* 4.

Proof. Since by Theorem 13, MIN VERTEX COVER is APX-complete for graphs with maximum degree 3, it is sufficient to establish an L-reduction from the instances of MIN VERTEX COVER for graphs with maximum degree 3 to the instances of MIN 3-COMP DOM for bipartite graphs with maximum degree 4. Given a graph $G = (V, E)$, where $V = \{v_1, v_2, \ldots, v_n\}$ and $E = \{e_1, e_2, \ldots, e_m\}$, an instance of minimum vertex cover, we construct a bipartite graph $G' = (X' \cup Y', E')$ as follows:

- First, we obtain a subdivision graph of G by inserting a vertex e'_j on the edge e_j of G, for $j \in [m]$.
- For each $v_i \in V$, $i \in [n]$, we attach the path $P_4 = a_i, b_i, c_i, d_i$ to v_i by adding the edge $v_i d_i$.

Figure 5 illustrates the construction described above. Note that the graph G' is clearly a bipartite graph with maximum degree 4. Now, we first prove the following claim.

Claim 5. $\gamma_3(G') = |VC^*| + 3n$, where VC^* is a minimum cardinality vertex cover of G.

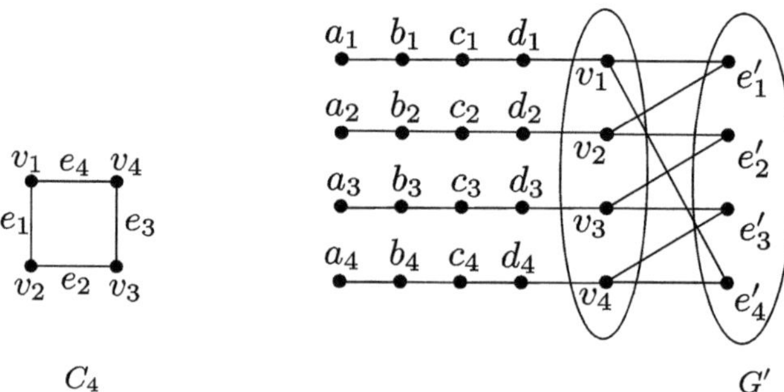

Fig. 5. Construction of the graph G' from $G = C_4$ as described in the proof of Theorem 14.

Proof. Let VC^* be a minimum cardinality vertex cover of G. Consider the set $D_3 = VC^* \cup \{b_i, c_i, d_i | i \in [n]\}$. Clearly, D_3 is a 3-*cd* set of G'. Hence, $\gamma_3(G') \leq |D_3| = |VC^*| + 3n$.

Conversely, let D_3^* be a minimum 3-*cd* set of G'. If $e_j' \in D_3^*$ for some $j \in [m]$, then replace it with one of its neighbors in V. Note that for each $i \in [n]$, a_i can be dominated by either a_i or b_i. If $a_i \in D_3^*$, then replace it with d_i. Let D_3 be the resultant set. Clearly, $|D_3| \leq |D_3^*|$. Note that for each $i \in [n]$, as $a_i \notin D_3$, then clearly, $\{b_i, c_i, d_i\} \subset D_3$ (as D_3^* is a 3-*cd* set and so as now D_3). It is easy to see that $D_3 \cap V$ is a vertex cover of G with cardinality at most $|D_3| - 3n \leq |D_3^*| - 3n$. Hence, $|VC^*| \leq |D_3^*| - 3n$ implies $|VC^*| + 3n \leq \gamma_3(G')$. This completes the proof of the Claim 5. $\qquad\square$

Now, we return to the proof of Theorem 14. Note that if VC^* is a vertex cover of G, then

$$\sum_{v \in VC^*} |N[v]| \geq n \implies |VC^*|(1 + \Delta(G)) \geq n.$$

Since the instance G is a graph with maximum degree 3, we have $n \leq 4|VC^*|$. Now, by Claim 5,

$$\gamma_3(G') = |VC^*| + 3n \leq |VC^*| + 3 \cdot 4|VC^*| = 13|VC^*|.$$

Now, consider a 3-*cd* set, say D_3' of G' with cardinality k. We can convert it into a 3-*cd* set D_3 with $|D_3| \leq k$, such that $e_j' \notin D_3$ for each $j \in [m]$ and $a_i \notin D_3$ for each $i \in [n]$. Then, say $VC = D_3 \cap V$ is a vertex cover of G such that $|VC| \leq k - 3n$ (as explained in the proof of Claim 5). Analogously the set VC is a vertex cover of G and $|VC| \leq |D_3| - 3n$. Hence, $||VC| - |VC^*|| \leq ||D_3| - 3n - |D_3^*| + 3n| = ||D_3| - |D_3^*||$.

From these two inequalities, it is clear that the above reduction is an L-reduction with $\alpha = 13$ and $\beta = 1$. Therefore, MIN 3-COMP DOM is APX-complete for bipartite graphs with maximum degree 4. $\qquad\square$

8 Conclusion

In this paper, we have initiated the computational complexity of 3-component domination problem. We proved that the decision version of this problem is NP-complete. However, on the positive side, we showed that this problem admits an $O(\ln n)$ approximation algorithm. It would be interesting to find some subclasses of bipartite graphs and chordal graphs that admit polynomial-time algorithms.

References

1. Alimonti, P., Kann, V.: Some APX-completeness results for cubic graphs. Theoret. Comput. Sci. **237**(1–2), 123–134 (2000)
2. Alvarado, J.D., Dantas, S., Rautenbach, D.: Dominating sets inducing large components. Discret. Math. **339**(11), 2715–2720 (2016)
3. Thomas, H.C., Leiserson, C.E., Rivest, R.L., Stein, C.: Introduction to Algorithms. MIT Press (2022)
4. Dinur, I., Steurer, D.: Analytical approach to parallel repetition. In: Proceedings of the Forty-Sixth Annual ACM Symposium on Theory of Computing, pp. 624–633 (2014)
5. Gao, Z., Lang, R., Xi, C., Yue, J.: 3-component domination numbers in graphs. Discret. Math. **347**(4), 113859 (2024)
6. Gao, Z., Lang, R., Xi, C., Yue, J.: On 3-component domination numbers in graphs. Discret. Appl. Math. **366**, 53–62 (2025)
7. Pradhan, D.: Algorithmic aspects of k-tuple total domination in graphs. Inf. Process. Lett. **112**(21), 816–822 (2012)

Directed Weakly Modular Graphs and Their Directed Interval Functions

Lekshmi Kamal K. Sheela[1], Manoj Changat[2(✉)], M. R. Chithra[1],
and Peter F. Stadler[3,4,5,6,7]

[1] Department of Mathematics, University of Kerala, Karyavattom Campus,
Thiruvananthapuram 695581, India
lekshmikamal@keralauniversity.ac.in, chithra@keralauniversity.ac.in
[2] Department of Futures Studies, University of Kerala, Karyavattom Campus,
Thiruvananthapuram 695 581, India
mchangat@keralauniversity.ac.in
[3] Department Computer Science, and Interdisciplinary Center for Bioinformatics,
University Leipzig, Härtelstr. 16–18, Leipzig, Germany
studla@bioinf.uni-leipzig.de
[4] Department Theoretical Chemistry, University Vienna, Währingerstr. 17, Wien,
Austria
[5] Facultad de Ciencias, Universidad Nacional de Colombia, Sede Bogotá, Ciudad
Universitaria, COL-111321 Bogotá, D.C., Colombia
[6] MPI Mathematics in the Sciences, Inselstr 22, Leipzig, Germany
[7] Santa Fe Institute, 1399 Hyde Park Road, Santa Fe, USA

Abstract. Weakly modular graphs are a well-studied graph class in metric graph theory. Here we extend the concept to directed graphs in such a way that symmetric (undirected) case is included and key properties, in particular closure under Cartesian products and gated amalgams continue to be satisfied. We characterize directed weakly modular graphs in terms of their directed interval functions. We then consider extensions to the directed setting for two important subclasses: modular and median graphs.

Keywords: directed graphs · weakly modular graphs · modular graphs · median graphs · interval function · transit function

1 Introduction

Interval functions of undirected simple graphs are a well-studied concept in metric graph theory. It is an important tool for studying *metric betweenness* in graphs. The first systematic study on interval functions was due to H. M. Mulder in 1980 [1]. Since then, the notion has received considerable attention in axiomatic studies [2,3] as well as in characterizing several graph families with strong metric properties.

Our focus will be on finite directed simple graphs (digraphs), which will denote by $D = (V, E)$. Where undirected graphs appear, they will be denoted by

N. Misra and A. Pandey (Eds.): CALDAM 2026, LNCS 16445, pp. 236–249, 2026.
https://doi.org/10.1007/978-3-032-17156-6_18

$G = (V, E)$. The *interval function* I_G of a connected graph G is the function $I_G : V \times V \longrightarrow 2^V$ defined with respect to the standard shortest path distance d in G as $I_G(u, v) := \{w \in V(G) : d(u, w) + d(w, v) = d(u, v)\}$. That is, $w \in I_G(u, v)$ if and only if w lies on some u, v-geodesic in G. Analogously, a *directed interval functions* I_D can be defined for every directed graph D using the same expression in [4]. While d is a metric for undirected graphs, it is only a quasi-metric lacking symmetry for directed graphs. Hence, the directed interval function I_D of a digraph D is the function $I_D : V \times V \longrightarrow 2^V$ defined with respect to the quasi-metric d in D as

$$
I_D(u, v) := \begin{cases} \{w \in V(D) | d(u, w) + d(w, v) = d(u, v)\} & \text{if } d(u, v) < \infty \\ \emptyset & \text{if } d(u, v) = \infty \end{cases} \tag{1}
$$

For a non-empty finite set X, a function $R_D : X \times X \to 2^X$ is a *directed transit function* [5] if it satisfies the axioms:

(t0) If $R_D(u, w) \neq \emptyset$ and $R_D(w, v) \neq \emptyset$, then $R_D(u, v) \neq \emptyset$.
(t1) If $R_D(u, v) \neq \emptyset$, then $\{u, v\} \subseteq R_D(u, v)$.
(t3) $R_D(u, u) = \{u\}$ for all $u \in X$.

The *underlying digraph* D_R of a directed transit function R_D has vertex set $V(D_R) = X$ and edges $(u, v) \in E(D_R)$ if and only if $R_D(u, v) = \{u, v\}$. A directed transit function R_D is called *geometric* if it satisfies (tr2), (b2), and (b3)$_{1,2}$.

(tr2) If $R_D(u, w) = \emptyset$ or $R_D(w, v) = \emptyset$, then $w \notin R_D(u, v)$ for all $u, v \in X$ with $u \neq v$.
 (b2) If $w \in R_D(u, v)$, then $R_D(u, w) \cup R_D(w, v) \subseteq R_D(u, v)$.
(b3)$_1$ If $x \in R_D(u, v)$ and $y \in R_D(u, x)$, then $x \in R_D(y, v)$.
(b3)$_2$ If $x \in R_D(u, v)$ and $y \in R_D(x, v)$, then $x \in R_D(u, y)$.

It is *weakly geometric* if it satisfies (tr2), (b2), and (b1)$_{1,2}$.

(b1)$_1$ If $x \in R_D(u, v)$ and $x \neq v$, then $v \notin R_D(u, x)$.
(b1)$_2$ If $x \in R_D(u, v)$ and $x \neq u$, then $u \notin R_D(x, v)$.

A key observation is that the directed interval function I_D of a digraph is always a geometric directed transit function [5].

Weakly modular graphs were introduced in [6] as a common generalization of many well-studied graph classes that are of particular interest in metric graph theory, such as median graphs, modular graphs, quasi-modular graphs, weakly median graphs, Helly graphs, and bridged graphs. For a detailed survey, see [7]. A graph G is weakly modular if it satisfies the following triangle and quadrangle conditions:

(TC) for any three vertices u, v, w with $1 = d(v, w) < d(v, u) = d(u, w)$, there exists a common neighbor z of v and w such that $d(u, z) = d(u, v) - 1$.

(QC) for any four vertices u, v, w, x with $d(v, x) = d(w, x) = 1$ and $2 = d(v, w) \leq d(u, v) = d(u, w) = d(u, x) - 1$, there exists a common neighbor z of v and w such that $d(u, z) = d(u, v) - 1$.

Bandelt and Chepoi in [8] introduced axiom

(ta) If $I_G(u, v) \cap I_G(u, w) = \{u\}$, $I_G(u, v) \cap I_G(v, w) = \{v\}$, $I_G(u, w) \cap I_G(v, w) = \{w\}$ and $I_G(u, v) = \{u, v\}$, then $I_G(u, w) = \{u, w\}$ and $I_G(v, w) = \{v, w\}$, for all $u, v, w \in V(G)$.

and showed that the interval function of weakly modular graphs satisfies condition (ta). In this contribution, we define a directed analogue of weakly modular graphs, along with some interesting subclasses (directed median graphs and directed modular graphs), and investigate some of their properties. Moreover, we provide a characterization of the directed interval functions of directed weakly modular graphs.

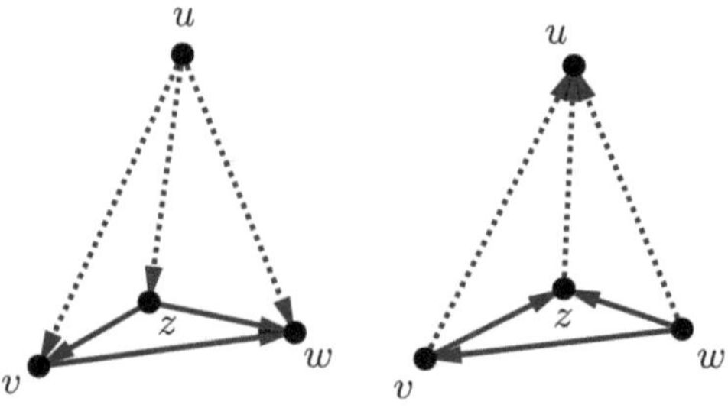

Fig. 1. Directed triangle conditions (DTC$_1$) and (DTC$_2$).

2   Directed Weakly Modular Graphs

The premise of condition (TC) is that vw is an edge, and there exists a shortest path between u and v (or v and u) that does not pass through the vertex w, as well as a shortest path between u and w (or w and u) that does not pass through the vertex v. This situation can be re-interpreted as: for the vertices u and w, there are two u, w-paths (or w, u-paths), one is the u, w-shortest path, say P, and the second is the u, w-path containing the vertex v say Q. Although, the path Q is not a geodesic between u and w, its sub-path from u to v is a geodesic between u and v. Since digraphs are not symmetric, i.e., shortest u, v-paths are not shortest v, u-paths in general, the directed analogues of (TC) covers two distinct cases, shown in Fig. 1.

(DTC$_1$) *Directed triangle condition 1*: for any three vertices u, v, w with $I_D(u, v) \neq \emptyset$, $I_D(v, w) = \{v, w\}$, $v \notin I_D(u, w)$ and $w \notin I_D(u, v)$, there exist $z \in I_D(u, v) \cap I_D(u, w)$ such that $d(z, v) = d(z, w) = 1$.

(DTC$_2$) *Directed triangle condition 2*: for any three vertices u, v, w with $I_D(v,u) \neq \emptyset$, $I_D(w,v) = \{w,v\}$, $v \notin I_D(w,u)$ and $w \notin I_D(v,u)$, there exist $z \in I_D(v,u) \cap I_D(w,u)$ such that $d(v,z) = d(w,z) = 1$.

In an undirected graph, the condition (TC) needs to be verified only when $d(u,v) = d(u,w)$, otherwise, either $v \in I_G(u,w)$ or $w \in I_G(v,w)$ and in such cases the condition (TC) holds trivially. However, in the premises of (DTC$_1$) and (DTC$_2$), the distances $d(u,v)$ and $d(u,w)$ as well as $d(v,u)$ and $d(w,u)$ do not need to be equal. Condition (DTC$_1$) must be verified whenever $I_D(u,v) \neq \emptyset$, $I_D(v,w) = \{v,w\}$, $v \notin I_D(u,w)$ and $w \notin I_D(u,v)$. This situation arises when $d(u,v) = d(u,w)$ or when $d(u,v) + d(v,w) > d(u,w)$ in the digraph. Similarly, the condition (DTC$_2$) must be verified when $d(v,u) = d(w,u)$ or when $d(w,v) + d(v,u) > d(w,u)$.

A digraph D is *symmetric* if $uv \in E(D)$ implies $vu \in E(D)$. In particular, the distance function of symmetric digraphs is symmetric. Symmetric digraphs can also be re-interpreted as undirected graph.

Lemma 1. *Conditions (DTC$_1$) and (DTC$_2$) reduce to (TC) on symmetric digraphs.*

Proof. In a symmetric digraph D, $d(x,y) = d(y,x)$ for all $x,y \in V(D)$. This together with the assumption that $v \notin I_D(u,w)$ and $w \notin I_D(u,v)$ implies that $d(u,v) = d(u,w)$ and $d(v,u) = d(w,u)$ under the premise of conditions (DTC$_1$) and (DTC$_2$) respectively. Then there exist a vertex z such that $z \in I_D(u,v) \cap I_D(u,w)$ and both $zv, zw \in E(D)$ by (DTC$_1$). Since the diagraph is symmetric, the same z is enough for satisfying (DTC$_2$), that is $z \in I_D(v,u) \cap I_D(w,u)$ and $vz, wz \in E(D)$. Hence in a symmetric digraph both (DTC$_1$) and (DTC$_2$) reduce to (TC). $\square$

Let us now turn to the quadrangle condition (QC). Its precondition asks for the existence of two different shortest paths between the two vertices u and x at distance $m > 2$, in which v and w are neighbors of x on these paths, respectively. Let P and Q be two shortest paths from u to x (or x to u) such that the vertex v is the neighbor of x in the path P and w is the neighbor of x in the path Q. A directed version of (QC) again has to consider the two alternative orientations of the paths, shown in Fig. 2:

(DQC$_1$) *Directed quadrangle condition 1*: if there exist two shortest paths from the vertex u to the vertex x at distance $m > 2$, and v and w be the neighbors of x on these paths respectively, then there exist a vertex z such that $z \in I_D(u,v) \cap I_D(u,w)$ and $d(z,v) = d(z,w) = 1$.

(DCQ$_2$) *Directed quadrangle condition 2*: if there exist two shortest paths from the vertex x to u at distance $m > 2$ and v and w be the neighbors of x on these paths respectively, then there exist a vertex z such that $z \in I_D(v,u) \cap I_D(w,u)$ such that $d(v,z) = d(w,z) = 1$.

The directed quadrangle conditions (DQC$_1$), and (DQC$_2$) assumes $d(u,v) = d(u,w)$ and $d(v,u) = d(w,u)$, respectively. Hence it is straightforward to verify:

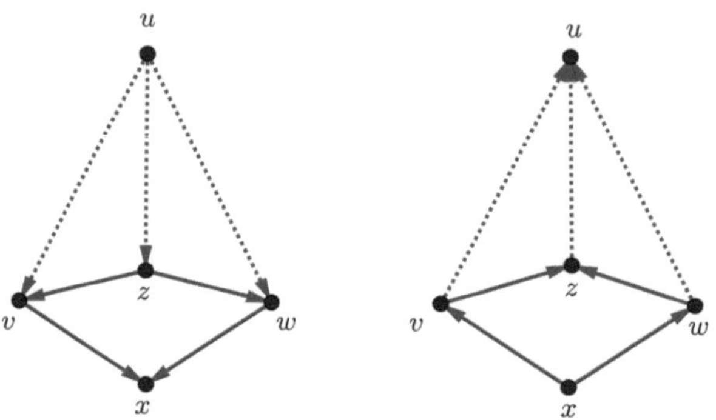

Fig. 2. Directed quadrangle conditions (DCQ$_1$) and (DCQ$_2$).

Remark 1. In a symmetric digraph, conditions (DQC$_1$) and (DQC$_2$) reduce to (QC).

In summary, a symmetric digraph D satisfies (DTC$_1$), (DTC$_2$) (DQC$_1$), and (DQC$_2$) if and only if the corresponding undirected graph G_D satisfied (TC) and (QC), i.e., if and only if G_D is weakly modular. We therefore say that D is a *directed weakly modular graph* if D satisfies (DTC$_1$), (DTC$_2$), (DQC$_1$), and (DQC$_2$).

In addition to the symmetric digraphs corresponding to weakly modular undirected graphs, one easily checks that directed trees and directed cycles belong to the class of directed weakly modular graphs. The underlying undirected graph G_D of a directed weakly modular graph, however, is not weakly modular in general. For example, consider an odd directed cycle of length greater than three. Such a digraph is a weakly modular directed graph, whereas its underlying undirected graph, an odd cycle of lengths at least 5, is not weakly modular, since it does not satisfy the triangle condition in the undirected setting.

On the other hand, any weakly modular graph that contains a cycle as a subgraph can be oriented to produce a digraph that violates the directed triangle condition. In other words, an undirected weakly modular graph may lose its weak modularity in a directed setting under certain orientations. As an example, consider the wheel graphs W_5 and W_6 given in Fig. 3. Both are weakly modular in the undirected setting, but under the given orientations in the Fig. 3, they are not weakly modular as digraphs.

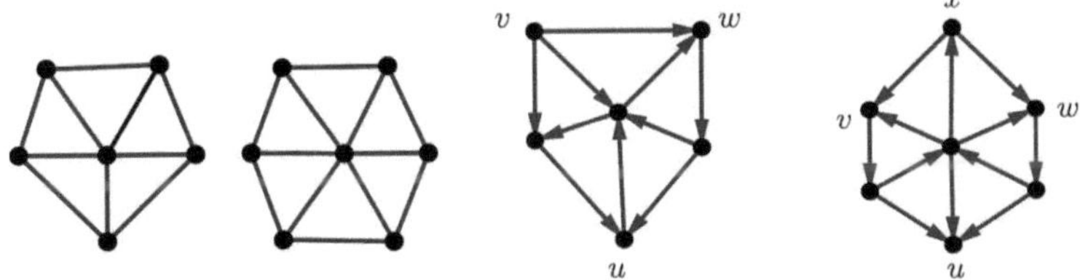

Fig. 3. Orientation of weakly modular wheel graphs W_5 and W_6 so that the resulting digraphs are not directed weakly modular.

Cartesian Products

The *Cartesian product* $D_1 \square D_2$, of two digraphs D_1 and D_2 is a graph with vertex set $V(D_1) \times V(D_2)$. Two vertices (g_1, h_1) and (g_2, h_2) are precisely adjacent if $g_1 = g_2$ and $h_1 h_2 \in E(D_1)$ or $g_1 g_2 \in E(D_2)$ and $h_1 = h_2$.

Lemma 2. *Let D_1 and D_2 be directed graphs that satisfy the conditions (DTC_1) and (DTC_2). Then the Cartesian product $D = D_1 \square D_2$ also satisfies the conditions (DTC_1) and (DTC_2).*

Proof. Assume D be the Cartesian product of two digraphs D_1 and D_2 each satisfying conditions (DTC_1) and (DTC_2). Let $a_i, i \in \{1, 2, \ldots, n\}$ and $b_i, i \in \{1, 2, \ldots, m\}$ denote the vertices of the digraphs D_1 and D_2 respectively. To prove that D satisfies the directed triangle condition (DTC_1), assume $u(a_i, b_j)$, $v(a_k, b_l)$ and $w(a_p, b_q)$ be the vertices of D such that $I_D(u, v) \neq \emptyset$, $I_D(v, w) = \{v, w\}$, $v \notin I_D(u, w)$ and $w \notin I_D(u, v)$. Since $vw \in E_D$, either $a_k = a_p$ or $b_l = b_q$. Without loss of generality, we may assume that $a_k = a_p$ which implies $b_l b_q \in E(D_2)$. Now, there are two cases:

Case (i): $a_i = a_k = a_p$. In this case $d(u, v) = d(b_j, b_l)$, $d(u, w) = d(b_j, b_q)$ and $b_l b_q \in E(D_2)$. Since the graph D_2 satisfied (DTC_1), there exist a vertex z in $I_D(b_j, b_l) \cap I_D(b_j, b_q)$ and both zb_l and zb_k are edges in D_2. Consider the vertex (a_k, z), it satisfies that $(a_k, z)(a_k, b_l)$ and $(a_k, z)(a_k, b_q)$ are edges and, moreover $(a_k, z) \in I_D((a_k, b_j), (a_l, b_l)) \cap I_D((a_k, b_j), (a_l, b_q))$ in the digraph D. Hence the digraph D satisfies (DTC_1) in this case.

Case (ii): $a_i \neq a_k = a_p$. Since D is the Cartesian product of D_1 and D_2, it follows that $d(u, v) = d((a_i, b_j), (a_k, b_l)) = d(a_i, a_k) + d(b_j, b_l)$ and $d(u, w) = d((a_i, b_j), (a_p, b_q)) = d(a_i, a_p) + d(b_j, b_q)$. This implies that $d(b_j, b_l) = d(b_j, b_q)$ since $a_k = a_p$. The vertices in the path $u = (a_i, b_j)(a_{i+1}, b_j) \ldots (a_k, b_j) = u'$ are common to both u, v-geodesic and u, w-geodesic. Moreover, the vertices u', v, w correspond to the situation described in Case (i). So by Case (i) (a_k, z) is the required vertex which fulfill the condition (DTC_1) in D for the vertices u', v, w and hence also for u, v, w.

The proof of (DTC_2) follows in a similar manner. $\qquad\square$

Lemma 3. *Let D_1 and D_2 be directed graphs that satisfy the conditions (DQC_1) and (DQC_2). Then the Cartesian product $D = D_1 \square D_2$ also satisfies the conditions (DQC_1) and (DQC_2).*

Proof. To prove the statement for conditions (DQC_1), assume that the vertices $u(a_r, b_s)$, $v(a_k, b_l)$, $w(a_p, b_q)$ and $x(a_m, b_n)$ of D satisfy the premise of (DQC_1). That is $vx \in E(D)$ and $wx \in E(D)$ and $d(u, v) = d(u, w) = m \geq 2$. Since vx and wx are edges, there are two possible cases for the first coordinate a; the corresponding cases for the second coordinate b are analogous. There are two cases:

Case (i): $a_k = a_p$. Let $a_k = a_p = a$ then $a_k = a_m = a$ follows. If $a_r \neq a$, then the path from $u(a_r, b_s)$ to $u'(a, b_s)$ is common to both u, v-geodesic and u, w-geodesic. Thus, the vertices $u'(a, b_s), v(a, b_l), w(a, b_q)$ satisfy the premise of condition (DQC_1), where the vertices b_l, b_s, b_q are pairwise distinct and different from b_n in D_2. Since the graph D_2 satisfy (DQC_1), there exist a vertex $z \in V(D_2)$ such that $z \in I_D(b_s, b_l) \cap I_D(b_s, b_q)$ and both zb_l and zb_q are edges in D_2. Then the vertex $z'(a, z)$ is the required vertex in D to satisfy condition (DQC_1) for the vertices u', v, w, x and hence also for u, v, w, x.

Case (ii): $a_k \neq a_p$. In this case if $b_l = b_q$, the argument is similar to Case (i). So assume that $b_l \neq b_q$. Since D is the Cartesian product of D_1 and D_2, it follows that $m = d(u, v) = d((a_r, b_s), (a_k, b_l)) = d(a_r, a_k) + d(b_s, b_l)$ and $m = d(u, w) = d((a_r, b_s), (a_p, b_q)) = d(a_r, a_p) + d(b_s, b_q)$. If $a_k \neq a_m$, implies that $b_l = b_n$ and $a_k a_m \in E(D_1)$ because $vx \in E(D)$. Similarly, if $a_p \neq a_m$, implies that $b_q = b_n$ and $a_k a_m \in E(D_1)$. Therefore the assumption $b_l \neq b_q$ implies that it is not possible that $a_k \neq a_m$ and $a_p \neq a_m$ simultaneously. With out loss of generality, we may assume that $a_k \neq a_m$ and $a_p = a_m$ and which implies that $b_l = b_n$ and $b_q \neq b_n$. So $d(a_r, a_k) = d(a_r, a_p) - 1$ in D_1 and $d(b_s, b_l) = d(b_s, b_q) + 1$ in D_2. Then the vertex $z' = (a_k, b_q)$ is the required vertex for satisfying condition (DQC_1) for the vertices u, v, w.

The proof of (DQC_2) follows in a similar manner. $\square$

Taken together Lemmata 2 and 3, we have the following theorem.

Theorem 1. *The class of directed weakly modular graphs is closed w.r.t. to forming Cartesian product.*

Gated Amalgamation

The notions of gates and gated set were introduced by Dress and collaborators in the setting of metric spaces [9]. These concepts are explored further for undirected graphs in [10]. Here, define the notion of a gated set for directed graphs as follows. Let K be a subset of the vertex set in a digraph D, and let $u \in V \setminus K$. An *in-gate* in K for u is a vertex $x \in K$ such that $x \in I_D(u, w)$ for all vertex $w \in K$ with $I_D(u, w) \neq \emptyset$. Clearly, if u has an in-gate in K, then it is unique. Similarly, an *out-gate* for u in K is a vertex y such that $y \in I_D(w, u)$ for every vertex $w \in K$ with $I_D(w, u) \neq \emptyset$. Again, if u has an out-gate in K, then it is unique. A subset K of $V(D)$ is called gated, if every vertex v of D has an in-gate and out-gate in K. In Fig. 4, the vertices marked in blue color forms a gated set of the digraph in the figure.

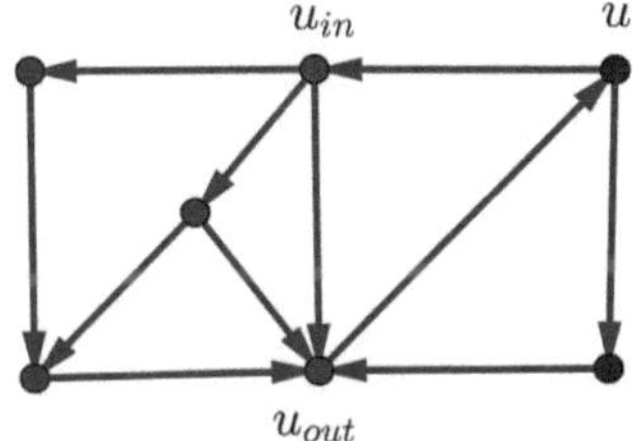

Fig. 4. A gated set is marked in blue. The in-gate of u is u_{in} and out-gate of u is u_{out}.

For symmetric digraphs, the notion of in-gates and out-gates coincides by symmetry of I_D. Moreover, it is easy to see that the notions of gated sets and gates reduce to the corresonding constructions for undirected graphs.

Gated amalgamation is a well-known operation in metric graph theory [10]. The definition of gated amalgamation for undirected graphs extends naturally to directed graphs. A subgraph of D' of D is gated if it is induced by a gated set. The graph D is the gated amalgam of two gated subgraphs D_1 and D_2 if (i) $V(D_1) \cup V(D_2) = V(D)$, (ii) $V(D_1) \cap V(D_2)$ is a gated subset of D_1 and of D_2, and (iii) there are no edges between $V(D_1) \setminus V(D_2)$ and $V(D_2) \setminus V(D_1)$. Equivalently, let H_1 be a gated subgraph of a digraph D_1 and H_2 a gated subgraph of D_2 such that H_1 and H_2 are isomorphic graphs. Then the gated amalgam of D_1 and D_2 is obtained from D_1 and D_2 by identifying their subgraphs H_1 and H_2.

Theorem 2. *The class of directed weakly modular graphs is closed under the gated amalgamation.*

Proof. Let D be the digraph obtained by the gated amalgamation of two directed weakly modular graphs D_1 and D_2. To prove that D is directed weakly modular, it is sufficient to show that D satisfies both the directed triangle and quadrangle conditions. To prove D satisfy condition (DTC$_1$), assume u, v, w be vertices in D such that $I_D(u, v) \neq \emptyset$, $I_D(v, w) = \{v, w\}$, $v \notin I_D(u, w)$ and $w \notin I_D(u, v)$. Since $vw \in E(D)$, it follows from the definition of gated amalgamation that both v and w lie in either D_1 or D_2. Without loss of generality we may assume that both v and w lies in D_1. If u also belong to D_1, then the vertices satisfy condition (DTC$_1$). Therefore, assume $u \in D_2 \setminus D_1$. Since $u \in D_2 \setminus D_1$, $d(u, v) \geq 2$ and $d(u, w) \geq 2$. By the definition of gated amalgamation, D_1 is gated in D, so u has an in-gate in D_1 and let it be u'. Clearly $u' \in D_1$, and if u' is adjacent to both v and w, then u' serves as the required vertex z satisfying condition (DTC$_1$). If u' is not adjacent to v or w or both, then the vertices fulfills the premise of condition (DTC$_1$) in D_1. That is $I_D(u', v) \neq \emptyset$, $I_D(v, w) = \{v, w\}$, $v \notin I_D(u', w)$ and $w \notin I_D(u', v)$, and since D_1 is directed weakly modular, there exists a common neighbor z of v and w such that $d(u', z) = d(u', v) - 1 = d(u', w) - 1$. Hence $d(u, z) = d(u, v) - 1 = d(u, w) - 1$. Therefore the vertices u, v and w satisfy condition (DTC$_1$).

To prove D satisfies condition (DTC$_2$), assume $I_D(v, u) \neq \emptyset$, $I_D(w, v) \neq \emptyset$, $v \notin I_D(w, u)$ and $w \notin I_D(v, u)$. The proof is similar to that of condition (DTC$_1$), the only difference is that, since D_1 is gated in D, the vertex u has an out-gate in D_1 say u'.

Next, to prove D satisfies (DQC$_1$), assume P and Q be two u, x-shortest paths and v and w be the neighbors of x on these paths, respectively. Since xv and xw are edges, it follows that the three vertices x, v and w lies either in D_1 or in D_2. We may assume that x, v and w lies in the digraph D_1. If u is in D_1, the graph D satisfy condition (DQC$_1$) for the vertices u, v, w and x, so we may assume that $u \in D_2 \setminus D_1$. Since D_1 is gated, u has an in-gate say u' in D_1. Since D_1 is directed weakly modular, there is a common neighbor say z of both the vertices v, w and $z \in I_{D_2}(u', v) \cap I_{D_2}(u', w)$. That is $z \in I_D(u, v) \cap I_D(u, w)$ hence D satisfies (DQC$_1$). The proof of condition (DQC$_2$) similar.   □

Characterization in Terms of Directed Interval Functions

In this section, we show that directed weakly modular graphs can be characterized in terms of the directed interval transit function. To see this, we first extending the (ta) axiom of Bandelt and Chepoi [8] to the directed case. Not surprisingly, the lack of symmetry again leads us to two variants:

(ta)$_1$ If $I_D(u, v) \cap I_D(u, w) = \{u\}$, $I_D(u, v) \cap I_D(v, w) = \{v\}$, $I_D(u, w) \cap I_D(v, w) = \{w\}$, and $I_D(v, w) = \{v, w\}$, then $I_D(u, v) = \{u, v\}$ and $I_D(u, w) = \{u, w\}$.

(ta)$_2$ If $I_D(v, u) \cap I_D(w, u) = \{u\}$, $I_D(v, u) \cap I_D(w, v) = \{v\}$, $I_D(w, u) \cap I_D(w, v) = \{w\}$, and $I_D(w, v) = \{w, v\}$, then $I_D(v, u) = \{v, u\}$ and $I_D(w, u) = \{w, u\}$.

To establish the equivalence of (ta)$_1$ with (DTC$_1$) and of (ta)$_2$ with (DTC$_2$) we will make use of the following lemma from [4].

Lemma 4. *[4] A weakly geometric directed transit function R_D satisfies the following properties:*

1. *If $R_D(v, u) \neq \emptyset$ and $R_D(u, w) \neq \emptyset$, then there exists an $x \in R_D(v, u) \cap R_D(u, w)$ such that $R_D(v, x) \cap R_D(x, w) = \{x\}$;*
2. *If $R_D(u, v) \neq \emptyset$ and $R_D(u, w) \neq \emptyset$, then there exists an $x \in R_D(u, v) \cap R_D(u, w)$ such that $R_D(x, v) \cap R_D(x, w) = \{x\}$;*
3. *If $R_D(v, u) \neq \emptyset$ and $R_D(w, u) \neq \emptyset$, then there exists an $x \in R_D(v, u) \cap R_D(w, u)$ such that $R_D(v, x) \cap R_D(w, x) = \{x\}$.*

Since the directed interval function I_D is a geometric directed transit function, Lemma 4 also holds for I_D.

Lemma 5. *The directed interval function I_D of a digraph D satisfies the following two properties;*

1. *The function I_D satisfies (ta)$_1$ if and only if D satisfies (DTC$_1$).*
2. *The function I_D satisfies (ta)$_2$ if and only if D satisfies (DTC$_2$).*

Proof. To prove the first statement, suppose that the directed interval function I_D on a digraph D does not hold (DTC$_1$). That is, there exist three vertices u, v, w with $I_D(u, v) \neq \emptyset$, $I_D(v, w) = \{v, w\}$, $v \notin I_D(u, w)$ and $w \notin I_D(u, v)$, but there does not exist a common neighbor z of v and w such that $d(u, z) = d(u, v) - 1 = d(u, w) - 1$. Since $I_D(u, v) \cap I_D(u, w) \neq \emptyset$, by Lemma 4, there exist x such that $x \in I_D(u, v) \cap I_D(u, w)$ with $I_D(x, v) \cap I_D(x, w) = \{x\}$. But both xv and xw are not edges, as D does not satisfy condition (DTC$_1$). Then $I_D(x, v) \cap I_D(x, w) = \{x\}$, $I_D(x, v) \cap I_D(v, w) = \{v\}$, $I_D(x, w) \cap I_D(v, w) = \{w\}$ and $I_D(v, w) = \{v, w\}$ but both xv and xw are not edges. So the directed interval function I_D does not satisfies axiom (ta)$_1$.

To prove the converse, assume D satisfies (DTC$_1$). That is, for every three $u, v,\ w$ with $I_D(u, v) \neq \emptyset$, $I_D(v, w) = \{v, w\}$, $v \notin I_D(u, w)$ and $w \notin I_D(u, v)$, there exists a common neighbor z of v and w such that $d(u, z) = d(u, v) - 1$. That is $z \in I_D(u, v) \cap I_D(u, w)$. Clearly $z \neq u$ if either $d(u, v) > 1$ or $d(u, w) > 1$ and hence I_D satisfies axiom (ta)$_1$ trivially on D because premise of axiom (ta)$_1$ does not hold. If $d(u, v) = 1$ and $d(u, w) = 1$, then I_D satisfies axiom (ta)$_1$ non-trivially. If possible, assume the vertices does not hold the premise of (DTC$_1$), then either $I_D(u, v) \cap I_D(v, w) \neq \{v\}$ or $I_D(u, w) \cap I_D(v, w) \neq \{w\}$ and hence I_D satisfies axiom (ta)$_1$ trivially on D.

In a similar way we can prove that if I_D satisfies (ta)$_2$ if and only if D satisfies (DTC$_2$). $\square$

The quadrangle conditions (DQC$_1$) and (DQC$_2$) also can be translated to the language of directed interval functions.

(qc)$_1$ If $v, w \in I_D(u, x)$, $I_D(v, x) = \{v, x\}$, $I_D(w, x) = \{w, x\}$, then there exist $z \in I_D(u, v) \cap I_D(u, w)$ such that $I_D(z, v) = \{z, v\}$ and $I_D(z, w) = \{z, w\}$.

(qc)$_2$ If $v, w \in I_D(x, u)$, $I_D(x, v) = \{x, v\}$, $I_D(x, w) = \{x, w\}$, then there exist $z \in I_D(v, u) \cap I_D(w, u)$ such that $I_D(v, z) = \{v, z\}$ and $I_D(w, z) = \{w, z\}$.

Lemma 6. *Let I_D be the directed interval function on a digraph D. Then I_D satisfies (qc)$_1$ if and only if D satisfies (DQC$_1$) and I_D satisfies (qc)$_2$ if and only if D satisfies (DQC$_2$).*

Proof. Suppose D does not satisfy (DQC$_1$). Then there exist two shortest paths from the vertex u to the vertex x at distance $m > 2$, and v and w are the neighbors of x on these paths respectively. However, there does not exists a common neighbor z of v and w such that $d(u, v) = d(u, z) + d(z, v)$, $d(u, w) = d(u, z) + d(z, w)$. By Lemma 4, there exist $y \in I_D(u, v) \cap I_D(u, w)$ such that $I_D(y, v) \cap I_D(y, w) = \{y\}$, so the premise of axiom (qc)$_1$ holds, but there does not exist a vertex z such that $z \in I_D(u, v) \cap I_D(u, w)$, $I_D(z, v) = \{z, v\}$ and $I_D(z, w) = \{z, w\}$. Hence I_D does not satisfy axiom (qc)$_1$. Conversely, suppose that I_D does not satisfy the axiom (qc)$_1$. Then we have $v, w \in I_D(u, x)$, $I_D(v, x) = \{v, x\}$, $I_D(w, x) = \{w, x\}$, but there does not exist an z such that $z \in I_D(u, v) \cap I_D(u, w)$, $I_D(z, v) = \{z, v\}$ and $I_D(z, w) = \{z, w\}$ for some $u, v, w \in V(D)$. Hence D does not satisfy condition (DQC$_1$). Similar arguments pertain to the second statement. $\square$

Taken together Lemmata 5 and 6, we obtain the following theorem, which characterizes directed weakly modular graphs in terms of axioms on the directed interval function.

Theorem 3. *The directed interval function I_D of a directed graph D satisfies axioms (ta)$_1$, (ta)$_2$, (qc)$_1$ and (qc)$_2$ if and only if D is a directed weakly modular graph.*

3   Subclasses of Directed Weakly Modular Graphs

Important subclasses of weakly modular graphs are modular graphs and median graphs. An undirected graph G is modular if for every triple of vertices $u, v, w \in V(G)$, $I_G(u,v) \cap I_G(v,w) \cap I_G(w,u) \neq \emptyset$ and median if $|I_G(u,v) \cap I_G(v,w) \cap I_G(w,u)| = 1$. In [4], it was proposed to define directed modular graphs in terms of the following axiom:

(mod)   If $I_D(u,v) \neq \emptyset$, $I_D(v,w) \neq \emptyset$, then $I_D(u,v) \cap I_D(v,w) \cap I_D(u,w) \neq \emptyset$, for any three vertices u, v, w in a directed graph D.

However, this leads to the situation that directed modular graphs need not be directed weakly modular. As an example, consider the digraph in Fig. 5, the graph is directed modular but not weakly modular as the digraph fails to satisfy condition (DQC$_2$) for the vertices x, u, v, w.

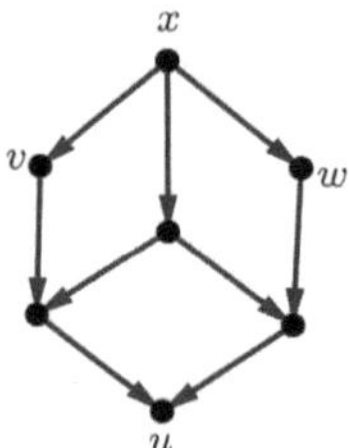

Fig. 5. The digraph satisfies (mod). That is for any three vertices a, b, c in the digraph if $I_D(a,b) \neq \emptyset$ and $I_D(b,c) \neq \emptyset$, then $I_D(a,b) \cap I_D(b,c) \cap I_D(a,c) \neq \emptyset$. However the digraph is not directed weakly modular as the digraph fails to satisfy condition (DQC$_2$) for the vertices x, u, v, w.

As a possible remedy we define a stronger version of the modular axiom as follows:

(s-mod) $I_D(u,v) \cap I_D(v,w) \cap I_D(u,w) \neq \emptyset$, for any three vertices u, v, w.

Thus, in particular $I_D(u,v) \neq \emptyset$ for any two vertices u and v, and hence (s-mod) implies that D is strongly connected. Moreover, for symmetric graphs, (s-mod) reduces to definition of undirected modular graphs. Since the preconditions in (mod) are satisfied exactly for the strongly connected graphs, the following remark is straight forward.

Remark 2. A digraph D satisfies (s-mod) if and only if it is strongly connected and satisfies (mod).

So termed these strongly connected directed modular graphs as directed *s-modular* graphs. The following property of strongly connected directed modular graphs will be useful in the following:

Lemma 7. *If $I_D(u, v) \cap I_D(u, w) = \{u\}$ or $I_D(v, u) \cap I_D(w, u) = \{u\}$ in a strongly connected directed modular graph D, then $u \in I_D(v, w) \cap I_D(w, v)$.*

Proof. First assume, $I_D(u, v) \cap I_D(u, w) = \{u\}$ in a directed s-modular graph D. Since D is directed s-modular, the intersections $I_D(u, v) \cap I_D(v, w) \cap I_D(u, w)$ and $I_D(u, w) \cap I_D(w, v) \cap I_D(u, v)$ are non-empty. As $I_D(u, v) \cap I_D(u, w) = \{u\}$, these non-empty intersections implies that $u \in I_D(v, w) \cap I_D(w, v)$. Similarly, assume $I_D(v, u) \cap I_D(w, u) = \{u\}$. Then the non-empty intersections $I_D(w, v) \cap I_D(v, u) \cap I_D(w, u)$ and $I_D(v, w) \cap I_D(w, u) \cap I_D(v, u)$ implies that $u \in I_D(v, w) \cap I_D(w, v)$. $\square$

Lemma 8. *Let D be a strongly connected directed modular graph, $u \in V(D)$, and $vw \in E(D)$, then*

1. *either $v \in I_D(u, w)$ or $w \in I_D(u, v)$*
2. *either $v \in I_D(w, u)$ or $w \in I_D(v, u)$.*

Proof. To prove the first statement, assume u, v, w be three vertices in a directed s-modular graph D such that vw is an edge. We have to prove that either the vertex v is on u, w-geodesic or w is on u, v-geodesic. Since D is a directed s-modular graph, $I_D(u, v) \cap I_D(v, w) \cap I_D(u, w) \neq \emptyset$. Since $I_D(v, w) = \{v, w\}$, the intersection is non-empty only if either $v \in I_D(u, w)$ or $w \in I_D(u, v)$. The second statement follows in a similar way as $I_D(w, v) \cap I_D(v, u) \cap I_D(w, u) \neq \emptyset$. $\square$

Theorem 4. *Strongly connected directed modular graphs are directed weakly modular graphs.*

Proof. Let D be a directed s-modular graph. To prove D is directed weakly modular graph, it is enough to show that D satisfies both the directed triangle and quadrangle conditions. Lemma 8, establishes that directed s-modular graphs satisfy the directed triangle conditions trivially. It remains to show that directed modular graphs satisfy the directed quadrangle conditions. First, assume that D does not satisfy condition (DQC_1). For that, assume there exist two shortest paths from the vertex u to the vertex x, further more v and w be the neighbors of x on these paths respectively. Suppose that vertex z belongs to both u, v-geodesic and u, w-geodesic such that $I_D(z, v) \cap I_D(z, w) = \{z\}$. Hence from Lemma 7, $z \in I_D(v, w) \cap I_D(w, v)$. Since $d(u, v) = d(u, w)$, it follows that $d(z, v) = d(z, w)$. Moreover, the fact that D does not satisfy condition (DQC_1) implies that neither zv nor zw is an edge and hence $d(z, v) = d(z, w) = n > 1$. Let $d(v, z) = n_1$ and $d(w, z) = n_2$ then $d(v, w) = n + n_1$ and $d(w, v) = n + n_2$ since $z \in I_D(v, w) \cap I_D(w, v)$. Now, since vx and wx are edges, $I_D(v, x) \cap I_D(w, x) = \{x\}$, then $x \in I_D(v, w) \cap I_D(w, v)$ by Lemma 7. As $x \in I_D(v, w)$, we have $d(v, w) = d(v, x) + d(x, w)$. This implies that $d(x, w) = d(v, w) - d(v, x) = n_1 + n - 1$. Similarly $d(x, w) = n_2 + n - 1$.

Now, let v_1 and w_1 be the neighbors of z in z, v-geodesic and z, w-geodesic respectively. Consider the vertices v_1, w_1, x, since D is directed modular graph, $I_D(w_1, v_1) \cap I_D(v_1, x) \cap I_D(w_1, x) \neq \emptyset$. So let $z_1 \in I_D(w_1, v_1) \cap I_D(v_1, x) \cap I_D(w_1, x)$. It follows that $v_1 \notin I_D(w_1, x)$ and $w_1 \notin I_D(v_1, x)$, since $I_D(z, v) \cap I_D(z, w) = \{z\}$. Also, since $d(w_1, v_1) \leq 2n + n_1 + n_2 - 2$, $x \notin I_D(w_1, v_1)$. That is $x \neq z_1$.

Consider the vertices v, x, z_1. As vx is an edge the only possibility to have the intersection $I_D(z_1, v) \cap I_D(v, x) \cap I_D(z_1, x)$ non-empty is either $x \in I_D(z_1, v)$ or $v \in I_D(z_1, x)$. But $x \notin I_D(z_1, v)$ (if $x \in I_D(z_1, v)$ implies $z_1 \in I_D(w, x)$ by axiom $(b3_1)$ as $z_1 \in I_D(u, x)$, not possible as wx is an edge). So the only possibility is $v \in I_D(z_1, x)$. If $v \in I_D(z_1, x)$ together with $z_1 \in I_D(u, x)$ implies by $(b3_1)$ that $z_1 \in I_D(u, v)$. Similarly, if we consider the vertices w, x, z_1, we get $z_1 \in I_D(u, w)$. That is $z_1 \in I_D(u, v) \cap I_D(u, w)$ and also $z_1 \in I_D(z, v) \cap I_D(z, w)$, a contradiction to the assumption that $I_D(z, v) \cap I_D(z, w) = \{z\}$.　　□

Chalopine and Chepoi in [11] defined directed median graph in the context of studying directed version of non-positively curved complexes, as a pair $D = (G, o)$, where G is a median graph and o is an orientation of the edges of G in such a way that opposite edges of squares of G have the same direction. A key property of this class of directed graphs is

Lemma 9. ([12]). *Any directed path of a directed median graph D is a shortest path of the median graph G.*

A natural generalization of undirected median axiom is;

(med)　If $I_D(u, v) \neq \emptyset$, $I_D(v, w) \neq \emptyset$, then $|I_D(u, v) \cap I_D(v, w) \cap I_D(u, w)| = 1$, for any three vertices u, v, w in a directed graph D.

One may note that the axiom (med) is a special case of axiom (mod).

Lemma 10. *The directed median graphs defined in [11, 12] satisfy axiom (med).*

Proof. Let $D = (G, o)$ be a directed median graph. Assume $I_D(u, v) \neq \emptyset$, $I_D(v, w) \neq \emptyset$. Then it is natural that $I_D(u, w) \neq \emptyset$. Let P be the direct path obtained by concatenating u, v-geodesic and v, w- geodesic. Since P is a directed path by Lemma 9 the path corresponding to P is a shortest path in G. So P is a directed shortest path and $v \in I_D(u, w)$. Hence $v \in I_D(u, v) \cap I_D(v, w) \cap I_D(u, w)$. Now we have to show that $I_D(u, v) \cap I_D(v, w) \cap I_D(u, w) = \{v\}$. If possible assume that $z \neq v$ such that $z \in I_D(u, v) \cap I_D(v, w) \cap I_D(u, w)$. Then by Lemma 9 $z \in I_G(u, v) \cap I_G(v, w) \cap I_G(u, w)$, not possible since G is a median graph. Hence $I_D(u, v) \cap I_D(v, w) \cap I_D(u, w) = \{v\}$ and $D = (G, o)$ satisfies axiom (med).　　□

We note, finally that the Cartesian product preserves (med).

4　Concluding Remarks

We have proposed here a generalization of weakly modular graphs to the directed case that includes symmetric digraphs corresponding to the undirected weakly modular graphs. Our construction, moreover, preserve key properties of undirected weakly modular graphs, in particular closure under forming Cartesian products and gated amalgamation. Turning to interesting subclasses, we introduced s-modular graphs, as the strongly connected graphs satisfying (mod). Regarding a directed version of median graphs, one might consider strongly connected graphs satisfying (med), i.e., a proper subclass of directed s-modular graphs. These, however, form a class that is disjoint from the "directed median graphs" introduced in [11]. Hence it would be of interest to find sub-classes of directed weakly modular graphs that share properties with modular and median graphs and include both alternatives. Results in this contribution characterize graph classes in terms of additional properties of their directed interval function. It remains an open problem in which cases simple (first order) axioms exist that identify directed interval functions of given class among all (geometric) directed transit functions.

Acknowledgement. L.K.K.S. acknowledges financial support from the University of Kerala (U.O. No. Ac.EVII/2630/2025/UOK). MC and CMR acknowledge DST (ANRF), Government of India (Grant No. DST/INT/DAAD/P-03/2023 (G)). PFS acknowledges the support by the BMBF (Germany) through DAAD project 57616814 (SECAI, School of Embedded Composite AI), and jointly with SMWK (Saxony) through the *Center for Scalable Data Analytics and Artificial Intelligence Dresden/Leipzig* (SCADS24B).

References

1. Mulder, H.M.: The Interval Function of a Graph, Math. Centre Tracts, vol. 132. Mathematisch Centrum, Amsterdam, NL (1980)
2. Nebeský, L.: The interval function of a connected graph and a characterization of geodetic graphs. Math. Bohem. **126**(1), 247–254 (2001)
3. Mulder, H.M., Nebeský, L.: Axiomatic characterization of the interval function of a graph. Eur. J. Comb. **30**(5), 1172–1185 (2009)
4. Anil, A., et al.: Directed interval transit functions, submitted (2025)
5. Anil, A., et al.: Directed transit functions. RM **80**(2), 45 (2025)
6. Chepoı, V.: Classification of graphs by means of metric triangles. Metody Diskret Analiz **49**, 75–93 (1989)
7. Chalopin, J., Chepoi, V., Hirai, H., Osajda, D.: Weakly modular graphs and non-positive curvature, vol. 268. American Mathematical Society (2020)
8. Bandelt, H.J., Chepoi, V.: A Helly theorem in weakly modular space. Discret. Math. **160**, 25–39 (1996). https://doi.org/10.1016/0012-365X(95)00217-K
9. Dress, A.W., Scharlau, R.: Gated sets in metric spaces. Aequationes Math. **34**, 112–120 (1987)
10. Bandelt, H.J., Mulder, H.M., Wilkeit, E.: Quasi-median graphs and algebras. J. Graph Theor. **18**(7), 681–703 (1994)
11. Chalopin, J., Chepoi, V.: A counterexample to thiagarajan's conjecture on regular event structures. J. Comput. Syst. Sci. **113**, 76–100 (2020)
12. Chalopin, J., Chepoi, V.: 1-safe petri nets and special cube complexes: equivalence and applications. ACM Trans. Comput. Logic (TOCL) **20**(3), 1–49 (2019)

On Word-Representability of Minimal Non-comparability Graphs

Benny George Kenkireth, Gopalan Sajith, and Sreyas Sasidharan[(✉)]

Department of Computer Science and Engineering, IIT Guwahati, Guwahati, India
`{ben,sajith,sreyas.s}@iitg.ac.in`

Abstract. Word-representable graphs, characterized by the existence of a semi-transitive orientation, form a well-studied class in graph theory. Comparability graphs form a subclass of word-representable graphs. Both comparability graphs and word-representable graphs are hereditary, meaning they can be characterized by their forbidden induced subgraphs. The minimal forbidden induced subgraphs of comparability graphs and word-representable graphs are referred to as minimal non-comparability graphs and minimal non-word-representable graphs, respectively. While the set of all minimal non-comparability graphs is known, a complete description of minimal non-word-representable graphs remains open.

In this paper, we classify all minimal non-comparability graphs into those that are word-representable and those that are not, thereby identifying precisely which minimal non-comparability graphs are also minimal non-word-representable. This classification further allows us to describe minimal non-word-representable graphs containing an all-adjacent vertex, obtained by adding such a vertex to each word-representable minimal non-comparability graph. As a result, we identify several infinite families of minimal non-word-representable graphs, thereby advancing the structural understanding of this class.

Keywords: comparability graph · minimal non-comparability graph · word-representable graph · minimal non-word-representable graph · semi-transitivity

1 Introduction

Comparability graphs form a classical family of graphs, characterized by the existence of a transitive orientation. They are hereditary: every induced subgraph of a comparability graph is itself a comparability graph. In general, every hereditary graph class can be described by its (unique) set of minimal forbidden induced subgraphs [5]. In particular, Gallai [2] determined the minimal forbidden induced subgraphs of comparability graphs; these graphs are commonly referred to as *minimal non-comparability graphs*. Each such graph fails to admit a transitive orientation, while all its proper induced subgraphs do admit one.

Word-representable graphs were introduced by Kitaev [7]. A graph is *word-representable* if there exists a word over its vertex set such that two vertices are

N. Misra and A. Pandey (Eds.): CALDAM 2026, LNCS 16445, pp. 250–263, 2026.
https://doi.org/10.1007/978-3-032-17156-6_19

adjacent exactly when their letters alternate in the word. This well-studied class includes, for example, 3-colorable graphs, sub-cubic graphs, and comparability graphs [5], and is hereditary as well. Equivalently, a graph is word-representable if and only if it admits a semi-transitive orientation [5,8]. Despite substantial progress on word-representable graphs (see, e.g., [1,4,6]), a complete description of the minimal forbidden induced subgraphs for this class—called *minimal non-word-representable graphs*—remains open.

The study of minimal non-word-representable graphs is mathematically significant because hereditary classes are determined by their forbidden induced subgraphs, and these minimal forbidden subgraphs capture the essential structural obstructions. Since comparability graphs form a subclass of the word-representable graphs and their minimal forbidden induced subgraphs are known, it is natural to ask which minimal non-comparability graphs are also minimal non-word-representable. By classifying minimal non-comparability graphs into word-representable and non-word-representable cases, we determine exactly which of them constitute minimal non-word-representable graphs. Moreover, this analysis allows us to obtain the complete list of minimal non-word-representable graphs containing an all-adjacent vertex and to identify infinite families of minimal non-word-representable graphs.

Problem. Which minimal non-comparability graphs are also minimal non-word-representable graphs?

Our Contributions. We settle the problem stated above in its entirety. Concretely, our main contributions are:

1. A complete classification of all minimal non-comparability graphs into those that are word-representable (i.e., semi-transitive) and those that are not.
2. A complete list of minimal non-word-representable graphs containing an all-adjacent vertex. Specifically, for every word-representable minimal non-comparability graph, adjoining an all-adjacent vertex produces a minimal non-word-representable graph, giving the full set of such graphs.

The remainder of the paper is organized as follows. Section 2 collects definitions and preliminary results used throughout the paper. In Sect. 3, we present the classification and proofs. Section 4 concludes with a summary of the results and final remarks.

2   Preliminaries

All graphs considered are simple. For a graph G, let $V(G)$ and $E(G)$ denote its vertex and edge sets

Definition 1. *Let w be a word over an alphabet Σ, and let $a, b \in \Sigma$. Define $w|_{ab}$ as the subsequence of w obtained by retaining only the occurrences of a and b, in their original order. The letters a and b alternate in w if $w|_{ab}$ contains neither the substring aa nor bb.*

Example 1. For $w = 312314$, the letters 1 and 2 alternate, since $w|_{12} = 121$ contains no substring of the form 11 or 22. In contrast, 1 and 4 do not alternate, as $w|_{14} = 114$ contains a repetition. (Other pairs can be checked similarly; these two illustrate the concept.)

Definition 2. *A graph G is* word-representable *if there exists a word w over $V(G)$ such that for all distinct $x, y \in V(G)$:*

$$(x, y) \in E(G) \iff x \text{ and } y \text{ alternate in } w.$$

Such a word w is called a word-representant *of G.*

Fig. 1. Graph G_1 is word-representable, while G_2 is not.

The undirected graph G_1 in Fig. 1 is word-representable; a justification for the displayed orientation will be given later. In contrast, G_2 in Fig. 1 is not word-representable.

Definition 3. *A class X of graphs is called* hereditary *if for every $G \in X$, every induced subgraph of G also belongs to X.*

Examples of hereditary classes include planar graphs, comparability graphs, and word-representable graphs [5]. Every hereditary class can be described in terms of its minimal forbidden induced subgraphs, which succinctly capture the structural constraints of the class.

Definition 4. *A graph G is a* minimal forbidden induced subgraph *for a hereditary class X if $G \notin X$, but every proper induced subgraph of G belongs to X.*

The wheel graph G_2 in Fig. 1 is a minimal forbidden induced subgraph for word-representable graphs. Identifying the complete set of minimal non-word-representable graphs remains open.

Definition 5. (Semi-transitive Orientation). *A graph G is semi-transitive if it admits an acyclic orientation such that for every directed path $v_1 \to \cdots \to v_k$, either $v_1 \not\to v_k$, or $v_i \to v_j$ exists for all $1 \leq i < j \leq k$. If $v_1 \to v_k$ is present while some $v_i \to v_j$ with $i < j$ is missing, we call this a* shortcut *[5].*

The orientation of G_1 shown in Fig. 1 is semi-transitive.

Theorem 1 ([5,8]). *G is word-representable iff it admits a semi-transitive orientation. Moreover, any vertex can serve as a source in such an orientation.*

Fact 1 ([3]). *Any 3-colorable graph is semi-transitive.*

Definition 6. *A graph G is a* comparability graph *if its edges admit a transitive orientation: $a \rightarrow b$ and $b \rightarrow c$ imply $a \rightarrow c$. Otherwise, G is a non-comparability graph.*

Fig. 2. Comparability graph and a non-comparability graph.

Graph G_1 in Fig. 2 is a comparability graph, whereas G_2 is a non-comparability graph. Comparability graphs form a hereditary class, and their minimal forbidden induced subgraphs (the *minimal non-comparability graphs*) were identified by Gallai [2]. These graphs are illustrated in Fig. 3.

Remark 1. In Fig. 3, the graph classes G_n^5 and G_n^6 are depicted schematically by indicating only their missing edges, shown as dashed lines. Similarly, the graph classes G_n^7, G_n^8, and G_n^9 are represented in a simplified manner due to their more complex structure: their vertices are arranged in three levels. In these three classes, dashed edges indicate the missing edges within each level, while all other edges within that level are present. Finally, the graphs H_1 through H_{11} in Fig. 3 are individual graphs, whereas the remaining ones form infinite families parameterized by n.

3   Classification of Minimal Non-comparability Graphs by Word-Representability

Figure 3 presents the complete set of minimal non-comparability graphs. This collection consists of two types: individual graphs H_i and infinite families of graphs G_n^i parameterized by n. For each family G_n^i, the parameter n is at least the lower bound indicated in the figure, which corresponds to the smallest instance in the family that is non-comparability.

In this section, we classify minimal non-comparability graphs according to their word-representability. The results are summarized in Theorems 2 and 3.

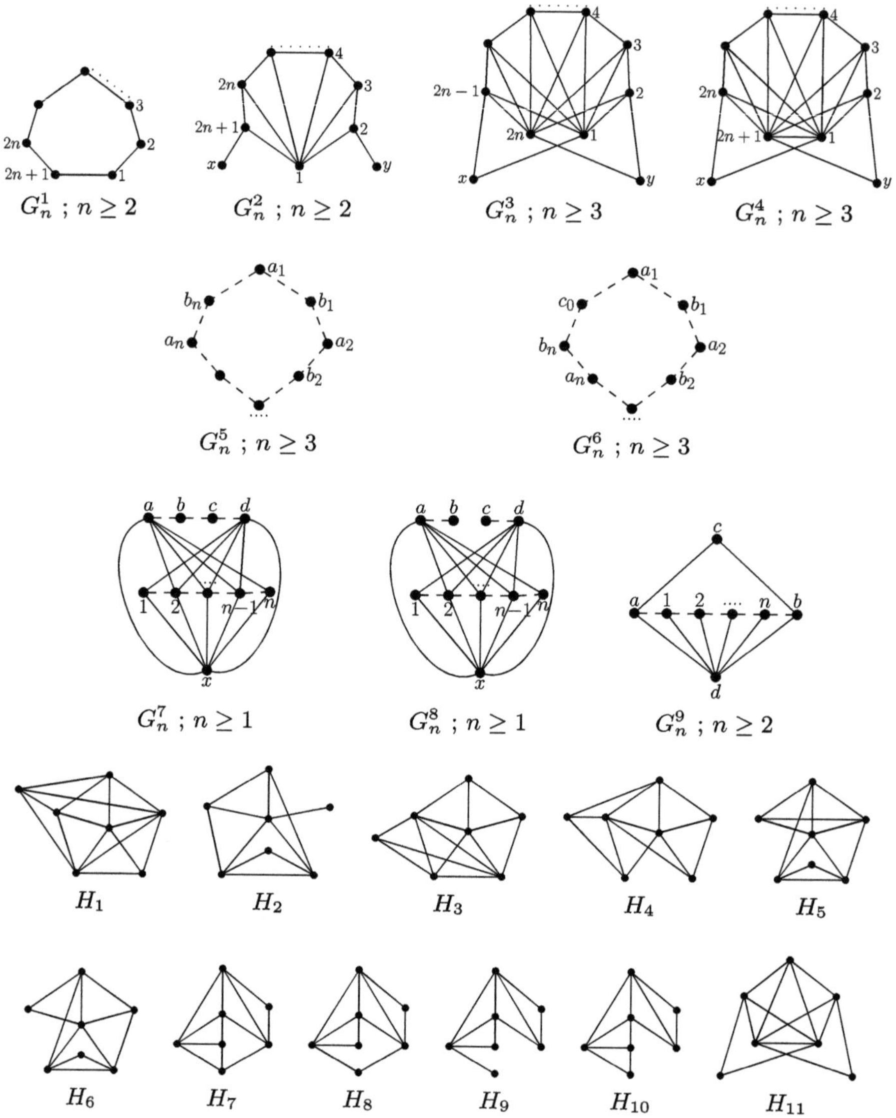

$G_n^1 \; ; \; n \geq 2$

$G_n^2 \; ; \; n \geq 2$

$G_n^3 \; ; \; n \geq 3$

$G_n^4 \; ; \; n \geq 3$

$G_n^5 \; ; \; n \geq 3$

$G_n^6 \; ; \; n \geq 3$

$G_n^7 \; ; \; n \geq 1$

$G_n^8 \; ; \; n \geq 1$

$G_n^9 \; ; \; n \geq 2$

H_1

H_2

H_3

H_4

H_5

H_6

H_7

H_8

H_9

H_{10}

H_{11}

Fig. 3. List of all minimal non-comparability graphs.

Theorem 2 identifies the word-representable graphs among the collection, while Theorem 3 identifies the non-word-representable ones, which form the intersection of minimal non-comparability graphs with minimal non-word-representable graphs.

Theorem 2. *All minimal non-comparability graphs shown in Fig. 3 that are word-representable are precisely the following:*

1. *All graphs in the families $G_n^1, G_n^2, G_n^3, G_n^5, G_n^6, G_n^7, G_n^8$.*
2. *The graphs $H_2, H_3, \ldots, H_{11}$.*

The proof of Theorem 2, establishing the word-representability of the listed graphs and families, is presented in Subsect. 3.1 through a series of claims.

Theorem 3. *All minimal non-comparability graphs shown in Fig. 3 that are non-word-representable are precisely the following:*

1. *All graphs in the families G_n^4 and G_n^9.*
2. *The graph H_1.*

These are exactly the minimal non-comparability graphs that are also minimal non-word-representable graphs.

The proof of Theorem 3 is presented in Subsect. 3.2 through a series of claims.

3.1 Minimal Non-comparability Graphs that Are Word-Representable

In this subsection, we present the proof of Theorem 2 by establishing the word-representability of the listed minimal non-comparability graph families through a series of claims.

Claim 1. The graph families G_n^1, G_n^2, and G_n^3, as depicted in Fig. 3, are word-representable.

Proof. All graphs in these families are 3-colorable. For any graph $G \in G_n^1$, where $n \geq 2$, the vertex set can be partitioned into three independent sets: A, B, and C, where $A = \{1, 3, \ldots, 2n - 1\}$, $B = \{2, 4, \ldots, 2n\}$, and $C = \{2n + 1\}$. Similarly, any graph $G \in G_n^2$, where $n \geq 2$, can be vertex partitioned into three independent sets: A, B, and C, where $A = \{1, x, y\}$, $B = \{2, 4, 6, \ldots, 2n\}$, and $C = \{3, 5, 7, \ldots, 2n + 1\}$. Likewise, for any graph $G \in G_n^3$, where $n \geq 3$, the vertex set can be partitioned into three independent sets: A, B, and C, with $A = \{1, 2n\}$, $B = \{2, 4, 6, \ldots, 2n - 2, x\}$, and $C = \{3, 5, 7, \ldots, 2n - 1, y\}$. Figure 4 demonstrates the 3-colorability of these graph classes (Fig. 4).

By Fact 1, every 3-colorable graph is semi-transitive. Hence, all graphs in these families are word-representable. $\square$

Claim 2. The graph family G_n^5, as shown in Fig. 3, is word-representable.

Proof. Consider an arbitrary graph $G \in G_n^5$, with $n \geq 3$ and vertex set partition $V(G) = A \cup B$, where $A = \{a_1, \ldots, a_n\}$ and $B = \{b_1, \ldots, b_n\}$. Both A and B induce cliques. The missing $2n$ edges form a cycle.

To prove word-representability, we exhibit a semi-transitive orientation (Theorem 1). Orient each edge from the vertex with smaller index to the vertex with larger index. This orientation is acyclic.

Suppose for a contradiction that a directed path $P = u_1 \to \cdots \to u_k$ violates semi-transitivity. Then $u_1 \to u_k$ is present, and some $\{u_i, u_j\} \notin E(G)$ for $1 \leq i < j \leq k$. Missing edges in G are of two types:

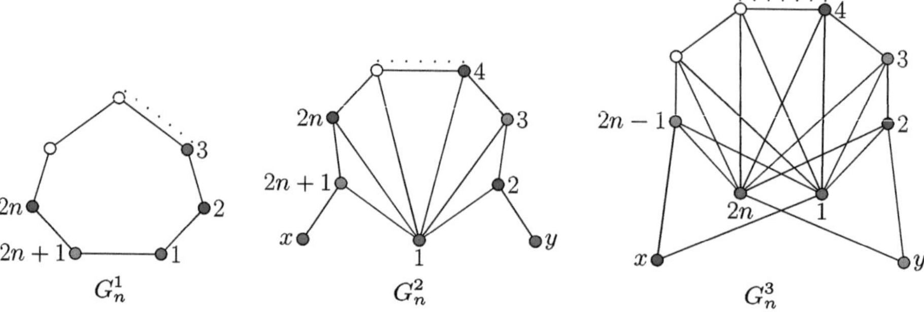

Fig. 4. 3-colorable minimal non-comparability graph classes.

- Type 1: $\{a_i, b_j\}$ with $|i - j| \le 1$,
- Type 2: $\{a_1, b_n\}$.

As all edges are oriented from lower to higher index, vertices in any directed path appear in increasing index order. Thus, a Type 1 edge could only connect adjacent vertices in P, contradicting the assumption that the missing edge lies strictly inside the path. The only other missing edge is $\{a_1, b_n\}$. A path containing both must start at a_1 and end at b_n. Such a path cannot violate semi-transitivity, contradicting the assumption. Hence, no such path P exists.

Therefore, the orientation is semi-transitive, and G is word-representable. $\square$

Claim 3. The graph family G_n^6, as shown in Fig. 3, is word-representable.

Proof. Consider an arbitrary graph $G \in G_n^6$, with $n \ge 3$, and vertex set $V(G) = A \cup B \cup \{c_0\}$, where $A = \{a_1, \ldots, a_n\}$ and $B = \{b_1, \ldots, b_n\}$. The sets A and B each induce a clique in G, and the $2n + 1$ missing edges form a cycle involving all vertices.

To prove word-representability, we construct a semi-transitive orientation (Theorem 1). Orient every edge (u, v) with $u < v$ by index. This orientation is clearly acyclic.

Suppose for a contradiction that a directed path $P = u_1 \to \cdots \to u_k$ violates semi-transitivity. Then $u_1 \to u_k$ is present, and some $\{u_i, u_j\} \notin E(G)$ for $1 \le i < j \le k$. Missing edges in G are of two types:

- Type 1: $\{a_i, b_j\}$ with $|i - j| \le 1$,
- Type 2: $\{c_0, b_n\}$.

As all edges are oriented from lower to higher index, vertices in any directed path appear in increasing index order. Thus, a Type 1 edge could only connect adjacent vertices in P, contradicting the assumption that the missing edge lies strictly inside the path.

The only other missing edge is $\{c_0, b_n\}$. A path containing both must start at c_0 and end at b_n. Such a path cannot violate semi-transitivity, contradicting the assumption. Hence, no such path P exists.

Therefore, the orientation is semi-transitive, and G is word-representable. $\square$

Claim 4. The graph family G_n^7, as shown in Fig. 3, is word-representable.

Proof. Consider an arbitrary $G \in G_n^7$, $n \geq 1$, with vertex partitioning as follows:

- Level 1: $\{a, b, c, d\}$,
- Level 2: $\{1, 2, \ldots, n\}$,
- Level 3: $\{x\}$.

To prove G is word-representable, we show that G is semi-transitive (Theorem 1). Orient the edges of G as follows:

- Within level 1: $d \to a$, $a \to c$, $b \to d$,
- Within level 2: $i \to j$ if $j > i$,
- Between levels: edges oriented from lower to higher level.

Observation 4.1 (Level 2 paths). Any path starting and ending at level 2 vertices preserves semi-transitivity. *Proof.* Missing edges in G are only consecutive pairs $\{i, i+1\}$, so no path can violate semi-transitivity. $\square$

Observation 4.2 (Level 1 paths). Any path starting and ending at level 1 vertices preserves semi-transitivity. *Proof.* The longest path is $b \to d \to a \to c$, and the edge $b \to c$ is absent. $\square$

Observation 4.3 (Level 1 to Level 2 paths). Any path starting at level 1 and ending at level 2 preserves semi-transitivity. *Proof.* Only a and d connect to level 2. Using Observations 4.1 and 4.2, any such path adheres to semi-transitivity. $\square$

Observation 4.4 (Paths ending at Level 3). Any path terminating at x preserves semi-transitivity. *Proof.* Vertex x connects to all except b and c, which do not connect to other levels. Using previous observations(4.1, 4.2, 4.3), semi-transitivity holds. $\square$

The orientation is semi-transitive, as we verified it for all paths in G. Hence, G_n^7 is word-representable. $\square$

Claim 5. The graph family G_n^8, as shown in Fig. 3, is word-representable.

Proof. The proof is analogous to Claim 4. Within level 1, orient edges as $d \to a$, $c \to a$, $c \to b$, $d \to b$. All other edges follow the same orientation rules as in Claim 4. Semi-transitivity verification is identical. $\square$

Remark 2. Except for H_1, all other individual graphs $H_2, \ldots, H_{11}$ in Fig. 3 are semi-transitive. Their semi-transitive orientations are provided in Fig. 5. Since they are smaller graphs (7-vertex graphs), we omit formal proofs of their semi-transitivity.

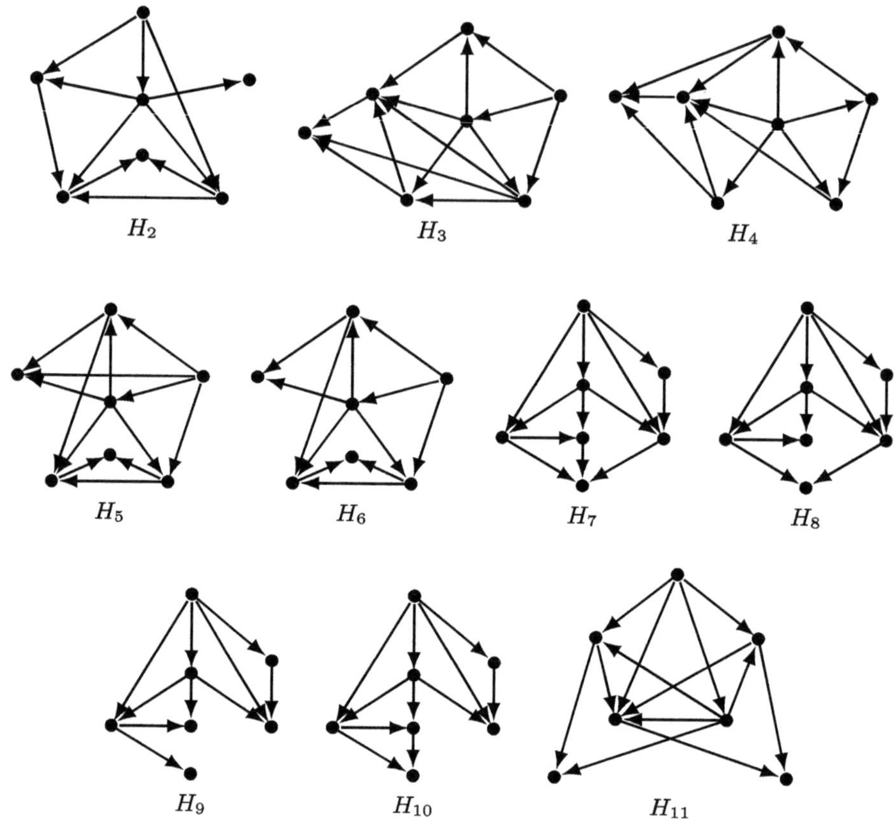

Fig. 5. Semi-transitive orientations of individual graphs in Fig. 3 (H_1 is not semi-transitive).

Minimal Non-Word-Representable Graphs with an All-Adjacent Vertex In this subsubsection, we characterize minimal non-word-representable graphs that contain an all-adjacent vertex.

Lemma 1. *[6] Let G be a graph with n vertices, and let $x \in V(G)$ be a vertex of degree $n - 1$ (that is, x is adjacent to all other vertices in G). Let H be the induced subgraph of G with $V(H) = V(G) \setminus \{x\}$. Then, G is word-representable if and only if H is a comparability graph.*

The following theorem extends Lemma 1. While Lemma 1 establishes a direct correspondence between the word-representability of G and the comparability of H, our classification of minimal non-comparability graphs into semi-transitive and non-semi-transitive cases provides a precise characterization of, and identifies, the complete set of minimal non-word-representable graphs containing an all-adjacent vertex.

Theorem 4. *Let G be a graph with n vertices, where a vertex $x \in V(G)$ has degree $n-1$. Let H be the induced subgraph of G with $V(H) = V(G) \setminus \{x\}$. Then, G is minimal non-word-representable if and only if the following hold:*

1. *H is a minimal non-comparability graph, and*
2. *H is semi-transitive.*

Proof. Suppose H is a minimal non-comparability graph and is semi-transitive. Since H is non-comparability, Lemma 1 implies that G is non-word-representable.

We now verify minimality. Consider any proper induced subgraph $H' \subset G$:

- If $x \in V(H')$, then $H' \setminus \{x\} \subset H$. By minimality of H, $H' \setminus \{x\}$ is comparability, and hence Lemma 1 implies H' is word-representable.
- If $x \notin V(H')$, then $H' \subset H$ is a proper induced subgraph of a semi-transitive graph. Since word-representable graphs are hereditary, H' is word-representable.

Thus, G is minimal non-word-representable.

Conversely, suppose G is minimal non-word-representable with an all-adjacent vertex x. Let $H = G \setminus \{x\}$. By minimality of G, every proper induced subgraph is word-representable. Then, Theorem 1 implies H must be semi-transitive.

If H were comparability, Lemma 1 would imply G is word-representable, a contradiction. Therefore, H is non-comparability. Finally, if H were not minimal non-comparability, there exists a proper induced subgraph $H_2 \subset H$ that is also non-comparability. Then the induced subgraph $H_2 \cup \{x\} \subset G$ would be non-word-representable (Lemma 1), contradicting minimality of G. Hence, H is minimal non-comparability. $\square$

We have identified all minimal non-comparability graphs that are semi-transitive. By Theorem 4, adding an all-adjacent vertex to each of these graphs yields the complete set of minimal non-word-representable graphs containing an all-adjacent vertex.

3.2 Minimal Non-comparability Graphs that Are Non-Word-Representable

In this subsection, we present the proof of Theorem 3.

Claim 6. The graph family G_n^9, as shown in Fig. 3, is minimal non-word-representable.

Proof. Let $G \in G_n^9$ with $n \geq 2$. Suppose, for contradiction, that G is word-representable. Then, by Theorem 1, G admits a semi-transitive orientation. Without loss of generality, assume vertex d is the source. By symmetry between a and b, we may further assume that the edge ab is oriented as $a \to b$. The orientations $a \to c$ and $c \to b$ are not possible, since the path $d \to a \to c \to b$ would violate semi-transitivity. Similarly, the orientations $b \to c$ and $c \to a$ are not possible, as the path $d \to b \to c \to a$ would violate semi-transitivity. Thus, the only possibilities are: $a \to c$, $b \to c$ or $c \to a$, $c \to b$. We analyze both cases. In either case, note that the edge $1b$ must be oriented as $1 \to b$, since $b \to 1$ would create a forbidden path $d \to a \to b \to 1$.

Case 1. $a \to c$, $b \to c$.

Observation 1. *In any semi-transitive orientation of G with d as a source and with $a \to c$, $b \to c$, and $a \to b$, the orientation of edges from each vertex i, $2 \le i \le n-1$, to a and b must be: $i \to a$, $i \to b$.*

Proof. If we orient $a \to i$ and $i \to b$, the path $a \to i \to b \to c$ forms a shortcut. If $b \to i$ and $i \to a$, then the path $b \to i \to a \to c$ violates semi-transitivity. Thus, only two possibilities remain: (i) $a \to i, b \to i$, or (ii) $i \to a, i \to b$. For $i = 2$, option (i) is invalid: if $b \to 2$, then $d \to 1 \to b \to 2$ violates semi-transitivity. Hence, $2 \to a$ and $2 \to b$. Now assume there exists some i with $a \to i$ and $b \to i$. Let i be the smallest such index. Then, the path $d \to (i-1) \to b \to i$ violates semi-transitivity, a contradiction. Hence, for all i, we must have $i \to a$ and $i \to b$. $\square$

Finally, consider the edge an. If $a \to n$, then $d \to (n-1) \to a \to n$ violates semi-transitivity. If $n \to a$, then $d \to n \to a \to b$ violates semi-transitivity. In both cases we reach a contradiction. Thus, no semi-transitive orientation exists in Case 1.

Case 2. $c \to a$, $c \to b$.

Observation 2. *In any semi-transitive orientation of G with d as a source and with $c \to a$, $c \to b$, and $a \to b$, the orientation of edges from each vertex i, $2 \le i \le n-1$, to a and b must be: $i \to a$, $i \to b$.*

The proof of Observation 2 follows exactly the same reasoning as Observation 1.

For the edge an, both orientations $a \to n$ and $n \to a$ yield semi-transitivity violations:$d \to (n-1) \to a \to n$, $d \to n \to a \to b$. Thus, Case 2 also leads to a contradiction. Since no semi-transitive orientation exists in either case, G is non-word-representable. Moreover, as G is minimal non-comparability, every proper induced subgraph of G is word-representable. Therefore, G is minimal non-word-representable. $\square$

Claim 7. The graph family G_n^4, as shown in Fig. 3, is minimal non-word-representable.

Proof. Let $G \in G_n^4$ be arbitrary, where $n \ge 3$. Suppose, for contradiction, G is word-representable. Then, by Theorem 1, it admits a semi-transitive orientation in which vertex 1 is chosen as a source. Denote such an orientation by G'. Let G'' be the subgraph of G' obtained by deleting the vertices x and y from G'. If G'' does not admit a semi-transitive orientation, then neither does G', contradicting our assumption. Therefore, G'' must also be semi-transitively orientable. Let G''_U denote the underlying undirected graph of G''. We now consider the possible semi-transitive orientations of the graph G''_U.

Step 1. Orientations of the Internal Vertices. Consider the edge $\{2, 3\}$, which can be oriented either as $2 \to 3$ or $3 \to 2$.

Observation 3. *If G_U'' is oriented with vertex 1 as the source and $2 \to 3$ is chosen, then for every even vertex i with $4 \leq i \leq 2n - 2$, the orientations $i \to (i - 1)$ and $i \to (i + 1)$ are forced to maintain semi-transitivity.*

Proof. We proceed by induction on even i. **Base case ($i = 4$):** If $3 \to 4$, then the path $1 \to 2 \to 3 \to 4$ violates semi-transitivity. Hence, we must have $4 \to 3$. Similarly, if $5 \to 4$, then $1 \to 5 \to 4 \to 3$ also violates semi-transitivity, so the correct orientation is $4 \to 5$. Thus, the claim holds for $i = 4$.
Inductive step: Assume the observation holds for some even j with $4 \leq j \leq 2n - 4$, i.e., $j \to (j - 1)$ and $j \to (j + 1)$. Consider $i = j + 2$. If $(j + 1) \to (j + 2)$, then the path $1 \to j \to (j + 1) \to (j + 2)$ violates semi-transitivity. Thus, $(j + 2) \to (j + 1)$. Similarly, if $(j + 3) \to (j + 2)$, then $1 \to (j + 2) \to (j + 3) \to (j + 2)$ violates semi-transitivity, hence $(j + 2) \to (j + 3)$.
  By induction, the observation holds for all even i with $4 \leq i \leq 2n - 2$.

Observation 4. *If G_U'' is oriented with vertex 1 as the source and $3 \to 2$ is chosen, then for every even vertex i with $4 \leq i \leq 2n - 2$, the orientations $(i - 1) \to i$ and $(i + 1) \to i$ are forced to maintain semi-transitivity..*

The proof of Observation 4 is identical in structure to that of Observation 3, and follows by applying similar inductive reasoning.

Step 2. Orientation Near $2n$. From Step 1, the orientation at the far end of the chain $2, \ldots, 2n$ is fixed. Concretely, if $2 \to 3$, then $(2n - 1) \to 2n$ would make the path $1 \to (2n - 2) \to (2n - 1) \to 2n$ violate semi-transitivity; hence the required orientation is $2n \to (2n - 1)$. By symmetry, if $3 \to 2$, the required orientation is $(2n - 1) \to 2n$.

Step 3. Orientation at $2n + 1$. The orientation of the edge $\{2n, 2n+1\}$ uniquely determines all edges incident to $2n + 1$. If $(2n + 1) \to 2n$, then for each i with $2 \leq i \leq 2n - 1$ we must also have $(2n + 1) \to i$, otherwise the path $1 \to i \to (2n + 1) \to 2n$ violates semi-transitivity; the base case $i = 2$ holds and induction covers all i. Symmetrically, if $2n \to (2n + 1)$, then for each i with $2 \leq i \leq 2n - 1$, we must have $i \to (2n + 1)$.

Step 4. Classification of Possible Orientations of G_U''. Combining steps 1, 2 and 3, exactly four semi-transitive orientations of G_U'' with vertex 1 as the source: the two choices for $\{2, 3\}$ times the two choices for $\{2n, 2n + 1\}$. These four orientations are depicted in Fig. 6.

Step 5. Extension to G'. We verify that none of the four orientations of G_U'' shown in Fig. 6 can extend to G' once vertices x and y are restored:

(a) Fig. 6 a. If $2n \to x$, then $1 \to (2n+1) \to 2n \to x$ forms a shortcut. If $x \to 2n$, then $1 \to x \to 2n \to (2n - 1)$ violates semi-transitivity.

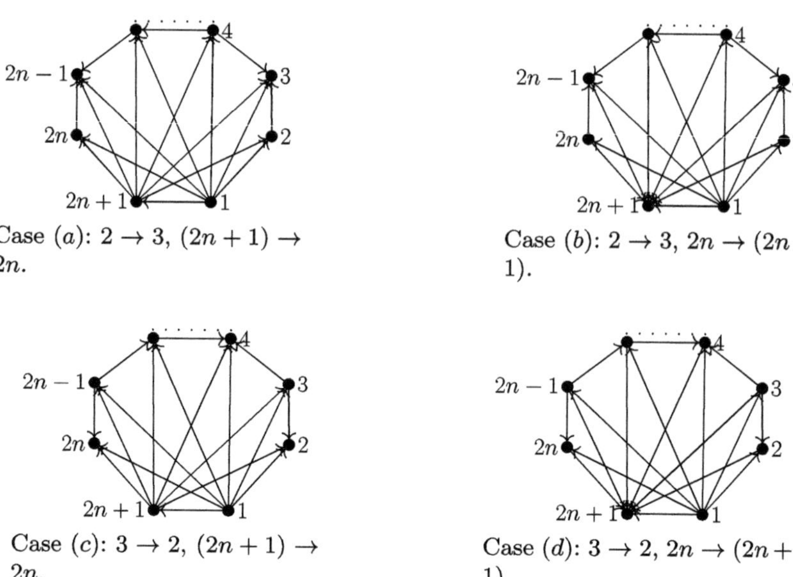

Case (a): $2 \to 3$, $(2n + 1) \to 2n$.

Case (b): $2 \to 3$, $2n \to (2n + 1)$.

Case (c): $3 \to 2$, $(2n + 1) \to 2n$.

Case (d): $3 \to 2$, $2n \to (2n + 1)$.

Fig. 6. Four semi-transitive orientations of G_U'', with 1 as a source.

(b) Fig. 6 b. - If $2 \to y$ and $y \to (2n + 1)$, then $1 \to 2 \to y \to (2n + 1)$, forms a shortcut. - If $2 \to y$ and $(2n+1) \to y$, then $2 \to 3 \to (2n+1) \to y$, is violating semi-transitivity. - If $y \to 2$ and $y \to (2n + 1)$, then $y \to 2 \to 3 \to (2n + 1)$, forms a shortcut. - If $y \to 2$ and $(2n + 1) \to y$, then $1 \to (2n + 1) \to y \to 2$, violates semi-transitivity.

(c) Fig. 6 c. The reasoning parallels case (b), except in the following cases: - $2 \to y$ with $(2n + 1) \to y$ gives $(2n + 1) \to 3 \to 2 \to y$, forming a shortcut. - $y \to 2$ with $y \to (2n + 1)$ yields $y \to (2n + 1) \to 3 \to 2$, violating semi-transitivity.

(d) Fig. 6 d. If $2n \to x$, then $1 \to (2n-1) \to 2n \to x$ forms a shortcut. If $x \to 2n$, then $1 \to x \to 2n \to (2n + 1)$ violates semi-transitivity.

Conclusion. In all four cases, a contradiction arises; thus, G admits no semi-transitive orientation and is not word-representable. Since every proper induced subgraph of G_n^4 is word-representable, it follows that G_n^4 is minimal non-word-representable. □

Note that the graph H_1 shown in Fig. 3 is a known minimal non-word-representable graph [5].

4 Concluding Remarks

In this work, we have determined the set of all graphs that lie in the intersection of minimal non-comparability graphs and minimal non-word-representable graphs. This intersection consists of two infinite families of graphs, namely G_n^9 and G_n^4, as well as the graph H_1, all illustrated in Fig. 3. Our results are obtained by classifying minimal non-comparability graphs into those that are word-representable and those that are not. Building on this classification, we fully identify and characterize the set of all minimal non-word-representable graphs containing an all-adjacent vertex. As a direction for future research, we pose the following open question:

Question. What is the minimum number of comparability graphs required to cover the edge set of any word-representable graph?

References

1. Collins, A., Kitaev, S., Lozin, V.: New results on word-representable graphs. Discrete Appl. Math. **216** (2014). https://doi.org/10.1016/j.dam.2014.10.024
2. Gallai, T.: Transitiv orientierbare graphen. Acta Math. Academiae Scientiarum Hungarica **18**, 25–66 (1967). https://api.semanticscholar.org/CorpusID:119485995
3. Halldórsson, M.M., Kitaev, S., Pyatkin, A.: Alternation graphs. In: Kolman, P., Kratochvíl, J. (eds.) Graph-Theoretic Concepts in Computer Science, pp. 191–202. Springer, Lecture Notes in Computer Science (2011)
4. Kitaev, S.: A comprehensive introduction to the theory of word-representable graphs. In: Charlier, É., Leroy, J., Rigo, M. (eds.) DLT 2017. LNCS, vol. 10396, pp. 36–67. Springer, Cham (2017). https://doi.org/10.1007/978-3-319-62809-7_2
5. Kitaev, S., Lozin, V.: Words and Graphs. Springer, Cham (2015). https://doi.org/10.1007/978-3-319-25859-1
6. Kitaev, S., Pyatkin, A.: On representable graphs. J. Automata Lang. Comb. **13**, 45–54 (01 2008)
7. Kitaev, S., Seif, S.: Word problem of the perkins semigroup via directed acyclic graphs. Order **25**, 177–194 (08 2008). https://doi.org/10.1007/s11083-008-9083-7
8. Kitaev, S., Sun, H.: Human-verifiable proofs in the theory of word-representable graphs. RAIRO - Theoretical Inf. Appl. **58**, 9 (2024). https://doi.org/10.1051/ita/2024004

Exact Recovery of Planted Cliques
in Semi-random Graphs

Yash Khanna[(✉)]

Indian Institute of Science, Bangalore, India
`yashkhanna@alum.iisc.ac.in`

Abstract. In this paper, we study the PLANTED CLIQUE problem in a semi-random model. Our model is inspired from the Feige-Kilian model [15] which has been studied in many other works [8,11,16,25,34,37] for a variety of graph problems. Our algorithm and analysis is on similar lines to the one studied for the DENSEST k-SUBGRAPH problem in the work of Khanna and Louis [24].

As a by-product of our main result, we give an alternate SDP-based rounding algorithm (with similar guarantees) for solving the PLANTED CLIQUE problem in a random graph.

Keywords: Planted cliques · Semi-random models · Beyond worst-case analysis

1   Introduction

Given an undirected graph, the decision problem of checking whether it contains a k-clique, i.e., a subgraph of size k which contains all the possible edges is a famous NP-hard problem and appears in the list of 21 NP-complete problems in the early work of Karp [22]. The best known approximation algorithm by the work of Boppana and Halldórsson [10] has an approximation factor of $\mathcal{O}\left(n/\left(\log n\right)^2\right)$. The results by Håstad and Zuckerman [19,38] shows that no polynomial time algorithm can approximate this to a factor better than $n^{1-\epsilon}$ for every $\epsilon > 0$, unless $P = NP$. This was improved by Khot et al. [26], who showed that there is no algorithm which approximates the maximum clique problem (in the general case) to a factor better than $n/2^{(\log n)^{3/4+\epsilon}}$ for any constant $\epsilon > 0$ assuming $NP \subsetneq BPTIME\left(2^{(\log n)^{\mathcal{O}(1)}}\right)$.

These results led to studying this problem in the average-case, i.e., we plant a clique of size k in a Erdős-Rényi random graph ($G(n,p)$), and study the ranges of parameters of k and p for which this problem can be solved. We give a brief survey in Sect. 1.4.

Another direction is to consider the problem in a restricted family of graphs or "easier" instances. This allows us to design new and interesting algorithms with much better guarantees (as compared to the worst-case models) and might

N. Misra and A. Pandey (Eds.): CALDAM 2026, LNCS 16445, pp. 264–277, 2026.
https://doi.org/10.1007/978-3-032-17156-6_20

possibly help us get away from the adversarial examples which cause the problem to be hard in the first place. This way of studying hard problems falls under the area of "Beyond worst-case analysis". We take this approach and in this work, we study the PLANTED CLIQUE problem in a semi-random model. This is a model generated in multiple stages via a combination of adversarial and random steps. Such generative models have been studied in the early works of [9,13,15,16] in the context of algorithms. We refer the reader to [24] and the references therein for a survey of variety of graph problems which have been subjected to such a study.

We start by establishing some notation used throughout the paper.

1.1   Notation (from [24])

Let $\bar{A}$ denote the adjacency matrix of our input graph $\mathcal{G} = (\mathcal{V}, \mathcal{E})$ whose construction is defined in Sect. 1.2. We use $n \overset{\text{def}}{=} |\mathcal{V}|$, and use $\mathcal{V}$ and $[n] \overset{\text{def}}{=} \{1, 2, \cdots, n\}$ interchangeably. We assume, w.l.o.g., that $\mathcal{G}$ is a complete graph: if $\{i, j\} \notin \mathcal{E}$, we add $\{i, j\}$ to $\mathcal{E}$ and set $\bar{A}_{ij} = \bar{A}_{ji} = 0$.

For $\mathcal{V}' \subseteq \mathcal{V}$, we use $\mathcal{G}[\mathcal{V}']$ to denote the subgraph induced on $\mathcal{V}'$. For a vector v, we use $\|v\|$ to denote $\|v\|_2$. For a matrix M, we use $\|M\|$ to denote the spectral norm, $\|M\| \overset{\text{def}}{=} \max\limits_{x \neq 0} \dfrac{\|Mx\|}{\|x\|}$.

We define probability distributions μ over finite sets Ω. For a random variable (r.v.) $X : \Omega \to \mathbb{R}$, its expectation is denoted by $\mathbb{E}_{x \sim \mu}[X]$. In particular, we define the distribution which we use next. For a vertex set $\mathcal{V}' \subseteq \mathcal{V}$, we define a probability (uniform) distribution $(f_{\mathcal{V}'})$ on the vertex set $\mathcal{V}'$ as follows. For a vertex $i \in \mathcal{V}'$, $f_{\mathcal{V}'}(i) = \dfrac{1}{|\mathcal{V}'|}$. We use $i \sim \mathcal{V}'$ to denote $i \sim f_{\mathcal{V}'}$ for clarity.

Definition 1 (Restatement of Definition 1.10 from [24]).   A graph $\mathcal{H} = (\mathcal{V}_{\mathcal{H}}, \mathcal{E}_{\mathcal{H}})$ is said to be a (s, d, λ)-expander if $|\mathcal{V}_{\mathcal{H}}| = s$, $\mathcal{H}$ is d-regular, and $|\lambda_i| \leq \lambda$, $\forall i \in [s] \setminus \{1\}$, where $\lambda_1 \geq \lambda_2 \cdots \geq \lambda_s$ are the eigenvalues of the adjacency matrix of $\mathcal{H}$.

1.2   Model

In this section, we describe our semi-random model. We first describe it informally. We start with an empty graph on n vertices and partition it arbitrarily into sets $\mathcal{S}$ and $\mathcal{V} \setminus \mathcal{S}$ of sizes k and $n - k$ respectively. We plant a clique onto the subgraph induced on $\mathcal{S}$ (denoted by $\mathcal{G}[\mathcal{S}]$). The bipartite subgraph $\mathcal{G}[\mathcal{S} \times \mathcal{V} \setminus \mathcal{S}]$ is a random subgraph with parameter p, i.e., each edge is added independently with probability p. And finally the subgraph $\mathcal{G}[\mathcal{V} \setminus \mathcal{S}]$ is composed of multiple small subgraphs each of which is far from containing a clique of size k, and these subgraphs are connected by random independent edges again with parameter p. There are three kinds of subgraphs in $\mathcal{G}[\mathcal{V} \setminus \mathcal{S}]$, first we have r disjoint (s, d, λ)-expander graphs (see Definition 1) and, second we have t disjoint subgraphs having the property that any induced subgraph of it has an average degree of at most γk, and third we have a random graph of size w and parameter p. A formal definition is presented below.

Definition 2. An instance of our input graph $\mathcal{G} = (\mathcal{V}, \mathcal{E}) \sim \mathsf{Clique}(n, k, p, r, s, t, d, w, \gamma, \lambda)$ is generated as follows,

1. We divide the vertex set $\mathcal{V}$ ($|\mathcal{V}| = n$) into two sets, $\mathcal{S}$ and $\mathcal{V} \setminus \mathcal{S}$ with $|\mathcal{S}| = k$. We further divide $\mathcal{V} \setminus \mathcal{S}$ into sets Λ, Π, Γ such that
   - The set Λ is arbitrarily divided into disjoint subsets $\Lambda_1, \Lambda_2, \cdots, \Lambda_r$ such that for all $\ell \in [r]$, $|\Lambda_\ell| = s$,
   - the set Π is arbitrarily divided into disjoint subsets $\Pi_1, \Pi_2, \cdots, \Pi_t$ such that for all $\ell \in [t]$, $|\Pi_\ell| > 0$, and
   - the set Γ is such that it has size $|w| > 0$.
2. (*Adding random edges*) We add edges between the following sets of pairs
   - $\mathcal{S} \times \mathcal{V} \setminus \mathcal{S}$,
   - $\Lambda_i \times \Lambda_j$ for $i, j \in [r], i \neq j$,
   - $\Pi_i \times \Pi_j$ for $i, j \in [t], i \neq j$,
   - $\Lambda_i \times \Pi_j$ for $i \in [r], j \in [t]$,
   - $\Lambda_i \times \Gamma$ for $i \in [r]$,
   - $\Gamma \times \Pi_j$ for $j \in [t]$

   independently with probability p. The edges between pairs of vertices in Γ are also added with probability p.
3. (*Adding a clique on $\mathcal{S}$*) We add edges between pairs of vertices in $\mathcal{S}$ such that the graph induced on $\mathcal{S}$ is a *clique*. For the sake of brevity, we also add a self loop on each vertex of $\mathcal{V}$, this will make the arithmetic cleaner (like the average degree of $\mathcal{G}[\mathcal{S}]$ is now k instead of $k - 1$) and has no severe consequences.
4. (*Adding edges in Λ_i's*) For each $i \in [r]$, we add edges between arbitrary pairs of vertices in Λ_i, such that the graph induced on Λ_i is a (s, d, λ)-expander graph.
5. (*Adding edges in Π_i's*) For each $i \in [t]$, we add edges between arbitrary pairs of vertices in Π_i, such that the graph induced on Π_i has the following property,
$$\max_{\mathcal{V}' \subseteq \Pi_i} \left\{ \frac{\sum\limits_{i,j \in \mathcal{V}'} \bar{A}_{ij}}{2 |\mathcal{V}'|} \right\} \leq \gamma k.$$ Or, in other words, for each $i \in [t]$ and

$\mathcal{V}' \subseteq \Pi_i$, the maximum average degree of the subgraph $\mathcal{G}[\mathcal{V}']$ is at most γk for some $\gamma \in (0, 1)$.
6. (*Monotone adversary step*) Arbitrarily delete any of the edges added in Steps 2, 4, or 5.
7. Output the resulting graph.

Note that in our model (Definition 2), the three kinds of subgraphs $\mathcal{G}[\Lambda_i], \mathcal{G}[\Pi_j]$, and $\mathcal{G}[\Gamma]$ which constitute $\mathcal{G}[\mathcal{V} \setminus \mathcal{S}]$ have sparse induced subgraphs by definition (at least in the range of parameters where we study them). It is interesting to see that the first two of them are pairwise exclusive in the sense that a $\mathcal{G}[\Lambda_i]$ graph need not qualify to be $\mathcal{G}[\Pi_j]$ and vice versa. It is an easy exercise to show this.

In this paper, the problem which we study is as follows: Given a graph generated from the above described model, the goal is to recover the planted clique ($\mathcal{G}[\mathcal{S}]$) with high probability. We show that for a "large" range of the input parameters, we can indeed solve this problem.

The key ingredient of our algorithm is the following semidefinite program (SDP 1) which is a standard relaxation of the K-CLIQUE problem (We define the K-CLIQUE problem as the problem of finding a clique of size k, given a large graph as input), however we state it in full for completeness.

SDP 1.

$$\max_{\{\{\bar{X}_i\}_{i=1}^n, \bar{I}\}} \quad \sum_{i,j=1}^n \bar{A}_{ij} \langle \bar{X}_i, \bar{X}_j \rangle \tag{1}$$

$$\text{subject to} \quad \sum_{i=1}^n \langle \bar{X}_i, \bar{X}_i \rangle = k \tag{2}$$

$$\sum_{j=1}^n \langle \bar{X}_i, \bar{X}_j \rangle \leq k \langle \bar{X}_i, \bar{X}_i \rangle \qquad \forall i \in [n] \tag{3}$$

$$\langle \bar{X}_i, \bar{X}_j \rangle = 0 \qquad \forall (i,j) \notin \mathcal{E} \tag{4}$$

$$0 \leq \langle \bar{X}_i, \bar{X}_j \rangle \leq \langle \bar{X}_i, \bar{X}_i \rangle \qquad \forall i, j \in [n],\ (i \neq j) \tag{5}$$

$$\langle \bar{X}_i, \bar{X}_i \rangle \leq 1 \qquad \forall i \in [n] \tag{6}$$

$$\langle \bar{X}_i, \bar{I} \rangle = \langle \bar{X}_i, \bar{X}_i \rangle \qquad \forall i \in [n] \tag{7}$$

$$\langle \bar{I}, \bar{I} \rangle = 1 \tag{8}$$

$$\bar{X}_i \in \mathbb{R}^{n+1} \qquad \forall i \in [n] \tag{9}$$

$$\bar{I} \in \mathbb{R}^{n+1} \tag{10}$$

The above SDP can be solved upto arbitrary precision in polynomial time using the Ellipsoid algorithm to fetch the solution set $\{\{X_i\}_{i=1}^n, I\}$. It is easy to see that since $\mathcal{G}[\mathcal{S}]$ is a clique, the integral solution corresponding to $\mathcal{S}$ does satisfy the above constraints. We state it now,

$$\forall i \in [n], \bar{X}_i = \begin{cases} \hat{v} & i \in \mathcal{S} \\ \hat{0} & i \in \mathcal{V} \setminus \mathcal{S} \end{cases} \qquad \text{and} \qquad \bar{I} = \hat{v}$$

where $\hat{v}$ is any unit vector, $\hat{0}$ is the all zeroes vector, and this feasible solution gives an objective value of k^2.

Note that this semidefinite programming relaxation is quite similar to that of DENSEST k-SUBGRAPH problem from [24] but we also add the following set of constraints to it,

$$\langle X_i, X_j \rangle = 0 \quad \forall (i,j) \notin \mathcal{E}. \tag{11}$$

This is a key difference as compared to the DENSEST k-SUBGRAPH problem and we will use the above set of constraints crucially in our analysis, much of which is inspired from [24]. We will describe this in more detail in Sect. 2.

1.3 Main Result

We propose an algorithm which is based on rounding the above described SDP 1. The algorithm and the analysis uses tools from the recent literature. Roughly speaking, the ranges of parameters where our algorithm works is when the subgraph $\mathcal{G}[\mathcal{S}]$ is a clique while any other k-sized induced subgraph is "far" from containing a clique. An advantage of using SDP-based algorithms is that they are robust against a monotone adversary (Step 6 of the model construction). This is an important point because many of the algorithms based on spectral or combinatorial methods are not always robust and may not work effectively with the presence of such adversaries.

Theorem 2. *There exist universal constants $\kappa, \xi \in \mathrm{I\!R}^+$ and a deterministic polynomial time algorithm, which takes an instance of* $\mathsf{Clique}(n, k, p, r, s, t, d, w, \gamma, \lambda)$ *where*

$$\nu = \frac{36\xi^2 (np)(r + t + 2)}{k^2 \left(1 - 6p - 2\gamma - \dfrac{d}{s} - \dfrac{\lambda}{k}\right)^2},$$

satisfying $\nu \in (0, 1)$, and $p \in [\kappa \log n / n, 1)$, and recovers the planted clique $\mathcal{S}$ with high probability (over the randomness of the input).

Note that our result does not depend on the size of the subgraphs $\mathcal{G}[\Pi_\ell]$'s but only on their counts, i.e., parameter t. Even our model is not parameterized by the sizes of Π_ℓ's. In other words, all the Π_ℓ's can be of different sizes but as long as the average degree requirement of subgraphs $\mathcal{G}[\Pi_\ell]$'s (the one stated in Step 5) is met, our result holds.

We see some interesting observations from Theorem 2. Firstly, there are a few conditions for the algorithm to work,

1. $p = \Omega\left(\dfrac{\log n}{n}\right)$, or to be verbose, p should be "large".
2. The function ν (which is dependent on the input parameters) should lie in the range $(0, 1)$, or stated in other words, ν should be "small".

A setting of input parameters when the value of ν is "small" is as follows:

$$k = \Omega\left(\max\left(\sqrt{np(r + t + 2)}, \lambda\right)\right), \gamma = \mathcal{O}(1), s = \Omega(d).$$

And also, $6p + 2\gamma + \frac{d}{s} + \frac{\lambda}{k} < 1$. The above values of different input parameters suggest that the algorithm will work only when any subgraph of size k will be far from a dense set (or a clique) inside $\mathcal{G}[\mathcal{V} \setminus \mathcal{S}]$. Also since the subgraph $\mathcal{G}[\mathcal{S} \times \mathcal{V} \setminus \mathcal{S}]$ is a random graph, it will not have dense sets either, thus naturally we can think of that the SDP 1 should put most of the its mass on the vertices of $\mathcal{S}$.

1.4   Related Work

Random Models for the Clique Problem. For the Erdős-Rényi random graph: $G(n, 1/2)$, it is known that the largest clique has a size approximately $2 \log_2 n$ [33]. There are several poly-time algorithms which find a clique of size $\log_2 n$, i.e., with an approximation factor roughly $1/2$ [17]. It is a long standing open problem to give an algorithm which finds a clique of size $(1 + \epsilon) \log_2 n$ for any fixed $\epsilon > 0$. This conjecture has a few interesting cryptographic consequences as well [20].

Planted Models for the Clique Problem. In the PLANTED CLIQUE problem, we plant a clique of size k in $G(n, 1/2)$ and study the ranges of k for which this problem can be solved. The work by [29] shows that if $k = \Omega(\sqrt{n \log n})$, then the planted clique essentially comprises of the vertices of the largest degree. Alon, Krivelevich, and Sudakov [1] give a spectral algorithm to find the clique when $k = \Omega(\sqrt{n})$. Feige and Krauthgamer [16] gave a SDP-based algorithm (different than ours) based on the Lovász theta function that works for $k = \Omega(\sqrt{n})$ in the presence of a monotone adversary, which can remove the random edges but not the edges of the planted clique. There is also a nearly linear time algorithm which succeeds w.h.p. when $k \geq (1 + \epsilon)\sqrt{n/e}$ for any $\epsilon > 0$ [14]. When $k = o(\sqrt{n})$, the work by Barak et al. [4] rules out the possibility for a sum of squares algorithm to work.

For $r = t = 0$, i.e., the case when there are no such Λ_ℓ's, and Π_ℓ's. The case when $\mathcal{G}[\mathcal{V} \setminus \mathcal{S}]$ is nothing but a random graph on $n - k$ vertices and probability parameter p, the lower bound on k translates to $\Omega(\sqrt{np})$. Thus in this case, our problem reduces to recovering the planted clique in a random graph and we get a similar threshold value of k to the one already studied in literature [16,24].

We compare our work with that of recent work on PLANTED CLIQUE problem by Błasiok et al. [8], and Buhai et al. [11] in the table below.

Comparison with recent work on the PLANTED CLIQUE problem for $p = 1/2$.			
	This work	Błasiok et al. [8]	Buhai et al. [11]
1) Size of k	$\boldsymbol{\Omega(\sqrt{n})}$	$\Omega(\sqrt{n} \log^2 n)$	$\Omega(n^{1/2+\epsilon})$
2) Structure of $\mathcal{V} \backslash \mathcal{S}$	Union of disjoint sparse graphs	**Arbitrary graph**	**Arbitrary graph**
3) Monotone deletions	**Allowed**	Not Allowed	**Allowed**
4) Recovery (w.h.p.)	**Exact recovery of $\mathcal{S}$**	List-decoding	List-decoding
5) Running Time	$\mathbf{poly}(n)$	$\mathbf{poly}(n)$	$n^{\mathcal{O}(1/\epsilon)}$

There are many applications of the PLANTED CLIQUE problem (and its variants), here is a partial list of works which talk about this problem: [2,3,5,7,18,21,27,35].

Semi-random Models for Related Problems. The semi-random model studied in this paper is inspired from a combination of two works. First is the very generic Feige-Kilian model [15]. In this model, we plant an independent set on $\mathcal{S}$ ($|\mathcal{S}| = k$), the subgraph $\mathcal{G}[\mathcal{S} \times \mathcal{V} \setminus \mathcal{S}]$ is a random graph with parameter p, while the subgraph $G[\mathcal{V} \setminus \mathcal{S}]$ can be an arbitrary graph. Then an adversary is allowed to add edges anywhere without disturbing the planted independent set. McKenzie, Mehta, and Trevisan [34] show that for $k = \Omega\left(n^{2/3}/p^{1/3}\right)$, their algorithm finds a "large" independent set. And for the range $k = \Omega\left(n^{2/3}/p\right)$, their algorithm outputs a list of independent sets (this type of algorithms' output is called the list-decoding variant), one of which is $\mathcal{S}$ with high probability. In the hypergraph case, the work by Khanna et al. [25] generalises their results to r-uniform hypergraphs for an analogous family of instances. Restrictions of this model has also been studied in the works of [12, 37].

It is important to note that while the above model is a pretty generic model and also solves the semi-random model which we study in our paper, however there are some key differences. Firstly, ours is an exact deterministic algorithm based on the SDP relaxation of the K-CLIQUE problem while they use a "crude" SDP (this idea was introduced in [34]) which is not a relaxation of the independent set (or the complementary clique problem). But both the SDPs "clusters" the vectors corresponding to the planted set. Secondly the algorithmic guarantee of the work by [25, 34] is of a different nature where they output a list of independent sets one of which is the planted set, as described above.

The second relevant model is studied by Khanna and Louis [24] for the DENSEST k-SUBGRAPH problem. They plant an arbitrary dense subgraph on $\mathcal{G}[\mathcal{S}]$, the subgraph $\mathcal{G}[\mathcal{S} \times \mathcal{V} \setminus \mathcal{S}]$ is a random subgraph, and the subgraph $\mathcal{G}[\mathcal{V} \setminus \mathcal{S}]$ has a property (Step 3 of model construction) like the one of Λ_ℓ's or Π_ℓ's of this paper. A monotone adversary can delete edges outside $\mathcal{G}[\mathcal{S}]$. Our algorithm, model, and the analysis is inspired from their work. We study the problem in the case when $\mathcal{G}[\mathcal{S}]$ is a clique on k vertices instead of an arbitrary d-regular graph. We get a full recovery of the clique in this paper instead of a "large" recovery of the planted set, for a "wide" range of input parameters.

We now compare our results to the models of Khanna and Louis [24].

- Recall the model, $\mathrm{D}k\mathrm{SReg}(n, k, d, \delta, \gamma)$ introduced in their work. In this model, the subgraph $\mathcal{G}[\mathcal{S}]$ is an arbitrary d-regular graph of size k, $\mathcal{G}[\mathcal{S} \times \mathcal{V} \setminus \mathcal{S}]$ is a random graph with parameter p, and the subgraph $\mathcal{G}[\mathcal{V} \setminus \mathcal{S}]$, has the following property, $\displaystyle\max_{\mathcal{V}' \subseteq \mathcal{V} \setminus \mathcal{S}} \left\{ \frac{\sum_{i,j \in \mathcal{V}'} \bar{A}_{ij}}{2\,|\mathcal{V}'|} \right\} \leq \gamma d$. Clearly, this is analogous to the case when we only have one such Π_1 comprising the whole of $\mathcal{G}[\mathcal{V} \setminus \mathcal{S}]$ such that the maximum average degree of any subgraph of $\mathcal{G}[\Pi_1]$ is at most γk. Now when $r = 0$ and $t = 1$, our model reduces to the case when $\mathrm{D}k\mathrm{SReg}(n, k, d, \delta, \gamma)$ has a clique on $\mathcal{S}$ (instead of a d-regular subgraph). Note that this case can be solved using our algorithm efficiently and we can recover the planted clique, i.e., $\mathcal{S}$ w.h.p. This is a much stronger guarantee as compared to the one in [24] where they output a vertex set with a large intersection with the planted set (but not completely), with the same threshold on k, i.e., $k = \Omega\left(\sqrt{np}\right)$.

– Similarly, in the model, $\mathrm{D}k\mathrm{SExpReg}(n, k, d, \delta, d', \lambda)$ introduced in the work of [24]. In this model, the subgraph $\mathcal{G}[\mathcal{S}]$ is an arbitrary d-regular graph of size k, $\mathcal{G}[\mathcal{S} \times \mathcal{V} \setminus \mathcal{S}]$ is a random graph with parameter p, and the subgraph $\mathcal{G}[\mathcal{V} \setminus \mathcal{S}]$, is a $(n - k, d', \lambda)$-expander graph. This is analogous to the case when we have only one such Λ_1 comprising the whole of $\mathcal{G}[\mathcal{V} \setminus \mathcal{S}]$. Now when $t = 0$ and $r = 1$, our model reduces to the case when $\mathrm{D}k\mathrm{SExpReg}(n, k, d, \delta, d', \lambda)$ has a clique on $\mathcal{S}$. And similar to the previous point, this case can also be solved using our algorithm efficiently and we can recover the planted clique, i.e., $\mathcal{S}$ w.h.p.

Remark 1. It is important to note that it has been pointed to us by anonymous reviewers that the partial recovery in the work of Khanna and Louis [24] as described above can be translated to the full recovery (in the clique case) easily by combining some results from Buhai et al. [11].

The idea of using SDP-based algorithms for solving semi-random models of instances has been explored in multiple works for a variety of graph problems, some of which are [6, 24, 25, 28, 30–32, 34, 36].

1.5   Proof Idea

Our algorithm is based on rounding a SDP solution. The basic idea is to show that the vectors corresponding to the planted set $\mathcal{S}$ are "long". In the integral solution we have exactly k long vectors which correspond to the set $\mathcal{S}$, in our solution we show that the vertices corresponding to the "long" vectors form a subset of $\mathcal{S}$ (the planted clique). This is shown by bounding the contribution of the vectors towards the SDP mass from the rest of the graph (i.e., everything except $\mathcal{G}[\mathcal{S}]$). The decomposable nature of the SDP objective (Eq. 1) into multiple sums corresponding to different subgraphs is leveraged here. This allows us to exploit the geometry of vectors to recover a part of the planted clique. This is possible only because the subgraph $\mathcal{G}[\mathcal{V} \setminus \mathcal{S}]$, which is a combination of expanders, low-degree graphs etc. and thus is a sparse graph by construction (Steps 4 or 5) and the random bipartite subgraph $\mathcal{G}[\mathcal{S} \times \mathcal{V} \setminus \mathcal{S}]$ (Step 2 of the model construction) will not have any dense sets either. Thus qualitatively the SDP should put most of the mass on the vertices of $\mathcal{S}$. We study the range of input parameters when this happens.

Once we have recovered a subset of $\mathcal{S}$, the rest of the vertices can be recovered using a greedy algorithm. Let $\mathcal{T}$ denote the set of long vectors obtained by rounding the SDP 1 such that $\mathcal{T} \subseteq \mathcal{S}$, the remaining vertices of $\mathcal{S}$ can be obtained by iterating over vertices in $\mathcal{V} \setminus \mathcal{T}$ and checking if it has an edge with all vertices of $\mathcal{T}$ and completing the clique this way. Note that for this to work, we crucially use the orthogonality constraints added for each non-edge pair (Eq. 11) and this additional recovery step works only because the planted set is a clique and not an arbitrary dense subgraph. The recovery procedure is explained in Sect. 2 of the paper.

1.6 Action of Monotone Adversary

A monotonicity argument can be used to ignore the action of the adversary as stated below.

Lemma 1. *In the upcoming discussion and analysis, w.l.o.g., we can ignore the adversarial action (Step 6 of the model construction) to have taken place.*

Proof. Let us assume the monotone adversary removes edges arbitrarily from the subgraphs $\mathcal{G}[\mathcal{V} \setminus \mathcal{S}]$, $\mathcal{G}[\mathcal{S}, \mathcal{V} \setminus \mathcal{S}]$ and the new resulting adjacency matrix is $\bar{A}$. Then for any feasible solution $\{\{Y_i\}_{i=1}^{n}, I_Y\}$ of the SDP 1, we have $\sum_{i \in P, j \in Q} \bar{A}_{ij} \langle Y_i, Y_j \rangle \leq \sum_{i \in P, j \in Q} A_{ij} \langle Y_i, Y_j \rangle$ for $\forall P, Q \subseteq \mathcal{V}$. This holds because of the non-negativity SDP constraint 5. Thus the upper bounds on SDP contribution by vectors in $\mathcal{G}[\mathcal{S}, \mathcal{V} \setminus \mathcal{S}]$ and $\mathcal{G}[\mathcal{V} \setminus \mathcal{S}]$ are intact and the rest of the proof follows exactly. Hence we can ignore this step in the analysis of our algorithm. $\square$

Let A denote the adjacency matrix of the input graph before the action of the adversary (before Step 6) and $\bar{A}$ denote the same after the action of the adversary. Due to the Lemma 1, we can work with the adjacency matrix A in the rest of the paper.

1.7 Organization

We present the introduction with all the relevant related work in Sect. 1, we shift the main analysis (due to page limit and also because it has a high overlap with the work in Khanna and Louis [24]) to full version of the paper [23] while we show the recovery part of the clique to Sect. 2 and the conclusion (Sect. 3) next.

2 Recovering the Planted Clique

In the full version of the paper [23], we show that under some mild conditions over the input parameters (namely when, p is "large" and ψ is "small") with high probability (over the randomness of the input), we have,

$$\mathbb{E}_{i \sim \mathcal{S}} \|X_i\|^2 \geq 1 - \psi. \tag{12}$$

ψ is defined in Definition 3. We define a vertex set $\mathcal{T} \stackrel{\text{def}}{=} \{i \in \mathcal{V} : \|X_i\|^2 \geq 1 - \alpha\psi\}$ where $1 < \alpha < 1/\psi$ is a parameter to be chosen later.

We will next show that for a cleverly chosen value of α, we can show that $\mathcal{T}$ is also a clique, and using the facts that $|\mathcal{T} \cap \mathcal{S}| > 0$ and that the boundary of the subgraph $\mathcal{G}[\mathcal{S}]$ is random, we further show that $\mathcal{T} \subseteq \mathcal{S}$. Once we have established this, it is easy to recover the rest of the vertices of $\mathcal{S} \setminus \mathcal{T}$ using a simple greedy heuristic. Before that, we recall two important technical results from [24].

Lemma 2 (Restatement of Lemma 3.5 from [24]). *With high probability (over the randomness of the input),* $|\mathcal{T} \cap \mathcal{S}| \geq \left(1 - \dfrac{1}{\alpha}\right) k.$

Lemma 3 (Corollary of Lemma 2.3 from [24]). *Let* $\{\{Y_i\}_{i=1}^n, I_Y\}$ *be any feasible solution of SDP 1 and* $\mathcal{V}' \subseteq \mathcal{V}$ *such that,*

1. *If* $\|Y_i\|^2 \geq 1 - \epsilon$ *for all* $i \in \mathcal{V}'$ *where* $0 \leq \epsilon \leq 1$, *then* $\langle Y_i, Y_j \rangle \geq 1 - 3\epsilon$ *for all* $i, j \in \mathcal{V}'$.
2. *If* $\mathbb{E}_{i \sim \mathcal{V}'} \|Y_i\|^2 \geq 1 - \epsilon$ *where* $0 \leq \epsilon \leq 1$, *then* $\mathbb{E}_{i,j \sim \mathcal{V}'} \langle Y_i, Y_j \rangle \geq 1 - 4\epsilon$.

Definition 3. Let ψ be the function over the input parameters defined as,

$$\psi \overset{\text{def}}{=} \frac{4\xi^2 (np)(r + t + 2)}{k^2 \left(1 - 6p - 2\gamma - \dfrac{d}{s} - \dfrac{\lambda}{k}\right)^2} \text{ for the sake of brevity.}$$

The next Lemma 4 is perhaps the most important technical result of this paper.

Lemma 4. *For* $\alpha = 1/(3\sqrt{\psi})$ *and* $\psi \in (0, 1/9)$. *With high probability (over the randomness of the input), the subgraph* $\mathcal{G}[\mathcal{T}]$ *is a clique and moreover,* $\mathcal{T} \subseteq \mathcal{S}$.

Proof. By applying Lemma 3 (Part 1) to the set $\mathcal{T}$, we get for all $i, j \in \mathcal{T}$: $\langle X_i, X_j \rangle \geq 1 - 3\alpha\psi$. We set α such that $1 - 3\alpha\psi > 0 \iff \alpha < 1/(3\psi)$. Thus we can set $\alpha = 1/(3\sqrt{\psi})$. It does satisfy the bounds on α, namely $\alpha \in (1, 1/\psi)$ when $\psi \in (0, 1/9)$. By the SDP constraints, $\langle X_i, X_j \rangle = 0 \; \forall \, (i, j) \notin E$ (the extra added constraint, or, Eq. 11), we have that the subgraph induced on $\mathcal{T}$ is a clique. This is easy to see. Consider any two vertices $u, v \in \mathcal{T}$ such that there is no edge between u and v, then by the above SDP constraint, $\langle X_u, X_v \rangle = 0$, however by the definition of set $\mathcal{T}$, $\langle X_u, X_v \rangle > 0$. This is a contradiction and thus $\mathcal{T}$ is indeed a clique.

Next we prove that w.h.p. $\mathcal{T} \subseteq \mathcal{S}$. By Lemma 2, $|\mathcal{T} \cap \mathcal{S}| \geq \left(1 - \dfrac{1}{\alpha}\right) k = \left(1 - 3\sqrt{\psi}\right) k > 0$ when $\psi \in (0, 1/9)$.

$$\therefore \mathbb{P}[\mathcal{T} \subsetneq \mathcal{S}] \leq \mathbb{P}[\exists v \in \mathcal{V} \setminus \mathcal{S} \text{ which has an edge with all the vertices of } \mathcal{T} \cap \mathcal{S}]$$

$$\leq np^{|\mathcal{T} \cap \mathcal{S}|} \leq np^{\left(1 - 3\sqrt{\psi}\right)k} = o(1).$$

where we used the union bound in step 2 and the lower bound on $|\mathcal{T} \cap \mathcal{S}|$ in step 3. The step 4 holds when p, k is "large" and ψ is "small".

We now have all the ingredients to prove our main result.

Proof (Proof of Theorem 2). By Lemma 4 we showed that $\mathcal{T} \subseteq \mathcal{S}$, now we can use a greedy strategy to recover the rest of $\mathcal{S}$. We iterate over all vertices in $\mathcal{V} \setminus \mathcal{T}$ and add them to our set if it has edges to all of $\mathcal{T}$. A calculation similar to the one shown above can be used to ensure that no vertex of $\mathcal{V} \setminus \mathcal{S}$ enters in this greedy step. Also note that $\alpha\psi = \dfrac{\psi}{3\sqrt{\psi}} = \dfrac{\sqrt{\psi}}{3}$. We define $\nu \overset{\text{def}}{=} 9\psi$. Here ν is nothing but a normalization of ψ for a cleaner representation. We summarize this in Algorithm 1 below. It is easy to see that the output of this algorithm, the set $\mathcal{Q}$ is essentially the planted clique $\mathcal{S}$ itself.

Algorithm 1. Algorithm to recover $\mathcal{S}$.

Require: An Instance of $\mathsf{Clique}(n, k, p, r, s, t, d, w, \gamma, \lambda)$.
Ensure: A vertex set $\mathcal{Q}$.
1: Solve SDP 1 to get the vectors $\left\{ \{X_i\}_{i=1}^n, I \right\}$.
2: Let $\mathcal{T} = \left\{ i \in \mathcal{V} : \|X_i\|^2 \geq 1 - (\sqrt{\nu}/9) \right\}$.
3: Initialize $\mathcal{Q} = \mathcal{T}$.
4: **for** vertex $v \in \mathcal{V} \setminus \mathcal{T}$, **do**
5:     If v shares an edge with all the vertices in $\mathcal{Q}$, then update $\mathcal{Q} = \mathcal{Q} \cup \{v\}$.
6:     Else discard v.
7: **end for**
8: Return $\mathcal{Q}$.

---

3   Conclusion and Future Work

In this paper, we studied the PLANTED CLIQUE problem in a semi-random model and presented an SDP-based algorithm to recover the clique exactly.

A powerful semi-random model would have any k sized induced subgraph in $\mathcal{G}[\mathcal{V} \setminus \mathcal{S}]$ have an average degree of γk, and the goal would be to give an efficient algorithm to be still able to recover the planted clique when $k = \Omega(\sqrt{n})$ while tolerating monotone deletions. To the best of our knowledge, this problem hasn't been studied in the literature, so we pose it as an interesting open question.

Acknowledgements. YK thanks Akash Kumar, Anand Louis, and Rameesh Paul for helpful discussions. He also thanks the anonymous reviewers for their useful comments on earlier versions of the paper. He was supported by the Ministry of Education, Government of India during his stay at IISc.

References

1. Alon, N., Krivelevich, M., Sudakov, B.: Finding a large hidden clique in a random graph. In: Proceedings of the Ninth Annual ACM-SIAM Symposium on Discrete Algorithms, SODA 1998, pp. 594–598. Society for Industrial and Applied Mathematics, USA (1998)
2. Arora, S., Barak, B., Brunnermeier, M., Ge, R.: Computational complexity and information asymmetry in financial products. Commun. ACM **54**(5), 101–107 (2011). https://doi.org/10.1145/1941487.1941511
3. Austrin, P., Braverman, M., Chlamtac, E.: Inapproximability of np-complete variants of nash equilibrium (2011). https://arxiv.org/abs/1104.3760

4. Barak, B., Hopkins, S., Kelner, J., Kothari, P.K., Moitra, A., Potechin, A.: A nearly tight sum-of-squares lower bound for the planted clique problem. SIAM J. Comput. **48**(2), 687–735 (2019). https://doi.org/10.1137/17M1138236

5. Berthet, Q., Rigollet, P.: Optimal detection of sparse principal components in high dimension. Ann. Stat. **41**(4), 1780–1815 (2013). https://doi.org/10.1214/13-AOS1127

6. Bhaskara, A., Charikar, M., Chlamtac, E., Feige, U., Vijayaraghavan, A.: Detecting high log-densities: an $o(n^{\frac{1}{4}})$ approximation for densest k-subgraph. In: Proceedings of the Forty-Second ACM Symposium on Theory of Computing, STOC 2010, pp. 201–210. ACM, New York, NY, USA (2010). https://doi.org/10.1145/1806689.1806719

7. Bhaskara, A., Jha, A.V., Kapralov, M., Manoj, N.S., Mazzali, D., Wrzos-Kaminska, W.: On the robustness of spectral algorithms for semirandom stochastic block models (2024). https://arxiv.org/abs/2412.14315

8. Blasiok, J., Buhai, R.D., Kothari, P.K., Steurer, D.: Semirandom planted clique and the restricted isometry property . In: 2024 IEEE 65th Annual Symposium on Foundations of Computer Science (FOCS), pp. 959–969. IEEE Computer Society, Los Alamitos, CA, USA, October 2024. https://doi.org/10.1109/FOCS61266.2024.00064

9. Blum, A., Spencer, J.: Coloring random and semi-random k-colorable graphs. J. Algorithms **19**(2), 204–234 (1995). https://doi.org/10.1006/jagm.1995.1034

10. Boppana, R., Halldórsson, M.M.: Approximating maximum independent sets by excluding subgraphs. In: Gilbert, J.R., Karlsson, R. (eds.) SWAT 1990. LNCS, vol. 447, pp. 13–25. Springer, Heidelberg (1990). https://doi.org/10.1007/3-540-52846-6_74

11. Buhai, R.D., Kothari, P.K., Steurer, D.: Algorithms approaching the threshold for semi-random planted clique. In: Proceedings of the 55th Annual ACM Symposium on Theory of Computing, STOC 2023, pp. 1918–1926. ACM, New York, NY, USA (2023). https://doi.org/10.1145/3564246.3585184

12. Charikar, M., Steinhardt, J., Valiant, G.: Learning from untrusted data. In: Proceedings of the 49th Annual ACM SIGACT Symposium on Theory of Computing, STOC 2017, pp. 47–60. ACM, New York, NY, USA (2017). https://doi.org/10.1145/3055399.3055491

13. Coja-Oghlan, A.: Solving np-hard semirandom graph problems in polynomial expected time. J. Algorithms **62**(1), 19–46 (2007). https://doi.org/10.1016/j.jalgor.2004.07.003

14. Deshpande, Y., Montanari, A.: Finding hidden cliques of size $\sqrt{N/e}$ in nearly linear time. Found. Comput. Math. **15**(4), 1069–1128 (2015). https://doi.org/10.1007/s10208-014-9215-y

15. Feige, U., Kilian, J.: Heuristics for semirandom graph problems. J. Comput. Syst. Sci. **63**(4), 639–671 (2001). https://doi.org/10.1006/jcss.2001.1773

16. Feige, U., Krauthgamer, R.: Finding and certifying a large hidden clique in a semirandom graph. Random Struct. Algorithms **16**, 195–208 (2000). https://doi.org/10.1002/(SICI)1098-2418(200003)16:23.3.CO;2-1

17. Grimmett, G.R., McDiarmid, C.J.H.: On colouring random graphs. Math. Proc. Cambridge Philos. Soc. **77**(2), 313–324 (1975). https://doi.org/10.1017/S0305004100051124

18. Guruswami, V.,Wang, H.P.: Semirandom planted clique via 1-norm isometry property. In: Megow, N., Basu, A. (eds.) Integer Programming and Combinatorial Optimization, pp. 270–282. Springer Nature Switzerland, Cham (2025)

19. Hastad, J.: Clique is hard to approximate within n/sup 1-/spl epsiv//. In: Proceedings of 37th Conference on Foundations of Computer Science, pp. 627–636 (1996). https://doi.org/10.1109/SFCS.1996.548522

20. Juels, A., Peinado, M.: Hiding cliques for cryptographic security. In: Proceedings of the Ninth Annual ACM-SIAM Symposium on Discrete Algorithms, SODA 1998, pp. 678–684. Society for Industrial and Applied Mathematics, USA (1998)

21. Juels, A., Peinado, M.: Hiding cliques for cryptographic security. Des. Codes Cryptography **20**(3), 269–280 (2000). https://doi.org/10.1023/A:1008374125234

22. Karp, R.M.: Reducibility among combinatorial problems. In: Miller, R.E., Thatcher, J.W. (eds.) Proceedings of a symposium on the Complexity of Computer Computations, held March 20-22, 1972, at the IBM Thomas J. Watson Research Center, Yorktown Heights, New York, USA, pp. 85–103. The IBM Research Symposia Series, Plenum Press, New York (1972). https://doi.org/10.1007/978-1-4684-2001-2_9

23. Khanna, Y.: Exact recovery of planted cliques in semi-random graphs. CoRR **abs/2011.08447**, https://arxiv.org/abs/2011.08447 (2020)

24. Khanna, Y., Louis, A.: Planted models for the densest k-subgraph problem. CoRR **abs/2004.13978**, https://arxiv.org/abs/2004.13978 (2020)

25. Khanna, Y., Louis, A., Paul, R.: Independent sets in semi-random hypergraphs. In: Lubiw, A., Salavatipour, M., He, M. (eds.) Algorithms and Data Structures, pp. 528–542. Springer International Publishing, Cham (2021)

26. Khot, S., Ponnuswami, A.K.: Better inapproximability results for MaxClique, chromatic number and Min-3Lin-deletion. In: Bugliesi, M., Preneel, B., Sassone, V., Wegener, I. (eds.) ICALP 2006. LNCS, vol. 4051, pp. 226–237. Springer, Heidelberg (2006). https://doi.org/10.1007/11786986_21

27. Koiran, P., Zouzias, A.: On the certification of the restricted isometry property. ArXiv **abs/1103.4984**, https://api.semanticscholar.org/CorpusID:16886247 (2011)

28. Kolla, A., Makarychev, K., Makarychev, Y.: How to play unique games against a semi-random adversary: Study of semi-random models of unique games. In: 2011 IEEE 52nd Annual Symposium on Foundations of Computer Science, pp. 443–452 (2011). https://doi.org/10.1109/FOCS.2011.78

29. Kučera, L.: Expected complexity of graph partitioning problems. Discrete Appl. Math. **57**(2), 193–212 (1995). https://doi.org/10.1016/0166-218X(94)00103-K, https://www.sciencedirect.com/science/article/pii/0166218X9400103K, combinatorial optimization 1992

30. Louis, A., Venkat, R.: Semi-random graphs with planted sparse vertex cuts: algorithms for exact and approximate recovery. In: Chatzigiannakis, I., Kaklamanis, C., Marx, D., Sannella, D. (eds.) 45th International Colloquium on Automata, Languages, and Programming (ICALP 2018). Leibniz International Proceedings in Informatics (LIPIcs), vol. 107, pp. 101:1–101:15. Schloss Dagstuhl – Leibniz-Zentrum für Informatik, Dagstuhl, Germany (2018). https://doi.org/10.4230/LIPIcs.ICALP.2018.101

31. Louis, A., Venkat, R.: Planted models for k-way edge and vertex expansion. In: Chattopadhyay, A., Gastin, P. (eds.) 39th IARCS Annual Conference on Foundations of Software Technology and Theoretical Computer Science (FSTTCS 2019). Leibniz International Proceedings in Informatics (LIPIcs), vol. 150, pp. 23:1–23:15. Schloss Dagstuhl – Leibniz-Zentrum für Informatik, Dagstuhl, Germany (2019). https://doi.org/10.4230/LIPIcs.FSTTCS.2019.23

32. Makarychev, K., Makarychev, Y., Vijayaraghavan, A.: Constant factor approximation for balanced cut in the pie model. In: Proceedings of the Forty-Sixth Annual ACM Symposium on Theory of Computing, STOC 2014, pp. 41–49. ACM, New York, NY, USA (2014). https://doi.org/10.1145/2591796.2591841

33. Matula, D.: The largest clique in a random graph. Technical Report, Department of Computer Science, Southern Methodist University (1976). https://s2.smu.edu/matula/Tech-Report76.pdf

34. McKenzie, T., Mehta, H., Trevisan, L.: A new algorithm for the robust semi-random independent set problem. In: Proceedings of the Thirty-First Annual ACM-SIAM Symposium on Discrete Algorithms, SODA 2020, pp. 738–746. Society for Industrial and Applied Mathematics, USA (2020)

35. Milo, R., Shen-Orr, S., Itzkovitz, S., Kashtan, N., Chklovskii, D., Alon, U.: Network motifs: simple building blocks of complex networks. Science **298**(5594), 824–827 (2002). https://doi.org/10.1126/science.298.5594.824

36. Mossel, E., Neeman, J., Sly, A.: Consistency thresholds for the planted bisection model. In: Proceedings of the Forty-Seventh Annual ACM Symposium on Theory of Computing, STOC 2015, pp. 69–75. ACM, New York, NY, USA (2015). https://doi.org/10.1145/2746539.2746603

37. Steinhardt, J.: Does robustness imply tractability? A lower bound for planted clique in the semi-random model. Electron. Colloquium Comput. Complex. **24**, 69 (2017). https://eccc.weizmann.ac.il/report/2017/069

38. Zuckerman, D.: Linear degree extractors and the inapproximability of max clique and chromatic number. In: Proceedings of the Thirty-Eighth Annual ACM Symposium on Theory of Computing, STOC 2006, pp. 681–690. ACM, New York, NY, USA (2006). https://doi.org/10.1145/1132516.1132612

Algebraic Expressions of Grid Graphs
with Diagonal Edges

Mark Korenblit$^{(\boxtimes)}$

Holon Institute of Technology, Holon, Israel
`korenblit@hit.ac.il`

Abstract. The paper investigates the relationship between algebraic expressions and labeled graphs. We consider directed graphs based on grid structure, which have m rows and n columns. Our goal is to simplify the expressions associated with these graphs. To this end, we apply two methods that generate expressions for these graphs and analyze the lengths of these expressions as functions of n.

1 Introduction

A two-terminal directed acyclic graph (*st-dag* in [2]) has only one source and only one target. We consider a *labeled graph* in which each edge has a unique label. Each path between the source and the target (a *spanning path*) in an st-dag can be represented as the product of all edge labels along the path. We define the sum of edge-label products corresponding to all possible spanning paths of an st-dag G as the *canonical expression* of G. The order of labels in every product (from the left to the right) is identical to the order of corresponding edges in the path (from source to target). An algebraic expression is called an *st-dag expression* (a *factoring of an st-dag* in [2]) if it is algebraically equivalent to the canonical expression of an st-dag. An st-dag expression consists of edge labels, and the operators + (disjoint union) and · (concatenation, also denoted by juxtaposition). We denote an expression of an st-dag G by $Ex(G)$.

Two algebraic expressions F_1 and F_2 are *algebraically equivalent* if there is a series of algebraic transformations arriving from one of them to another. We consider F_1 as a *representation* of F_2 and F_2 as a *representation* of F_1. For instance, $a(b + c) + ad$ is a representation of $a(b + d) + ac$ and vice versa. We define the total number of labels in an st-dag expression as its *length*. An *optimal representation of the expression* F is an expression of minimum length algebraically equivalent to F. Our goal is to simplify an st-dag expression to its optimal representation or, at least, to an expression of polynomial length in relation to the *st-dag's size* (this may be the number of vertices or another suitable parameter).

A *series-parallel* (*SP*) *graph* is defined recursively: a single edge is an SP graph and a graph obtained by a parallel or a series composition of SP graphs is SP [2]. Its expression has a representation in which each label appears only

N. Misra and A. Pandey (Eds.): CALDAM 2026, LNCS 16445, pp. 278–292, 2026.
https://doi.org/10.1007/978-3-032-17156-6_21

once (a *read-once formula* [3]), being an optimal representation of the SP graph expression.

An st-dag is SP if and only if it does not contain a subgraph which is a homeomorph of the *forbidden subgraph* [2] consisting of edges a_{11}, b_{12}, a_{22}, c_{11}, c_{12} of the graph illustrated in Fig. 1 (a). Possible optimal representations of its expression are $a_{11}(b_{12}a_{22} + c_{12}) + c_{11}a_{22}$ or $(a_{11}b_{12} + c_{11})a_{22} + a_{11}c_{12}$. For this reason, an expression of a non-SP st-dag cannot be represented as a read-once formula. However, generating an optimum factored form for an arbitrary expression is NP-hard [17].

Problems related to computations on graphs whose edges or vertices are associated with additional data have applications in various areas. Specifically, algorithms developed to obtain good factored forms for expressions related to graphs are described in articles devoted to problems of logic synthesis [3,17]. As shown in [12], using max-plus algebra tools [1], an st-dag expression can be interpreted as a special shortest-path algorithm on the st-dag, characterized by improved stability and faster adaptation to data updates for a real-time system represented by this st-dag. As noted in many works, a number of network problems which are generally intractable or involve complex solutions can often be efficiently solved on SP graphs.

Several of our previous works are dedicated to the analysis of algebraic expressions for certain types of non-SP st-dags. For example, in [11] we presented an algorithm that constructs an expression of length $O(n^2)$ for an n-vertex *Fibonacci graph* [4], which serves as a generic example of a non-SP graph. More complex *rhomboidal* graphs were considered in [9]. For these n-vertex graphs, the lengths of the derived expressions are $O(n^{\log_2 6})$.

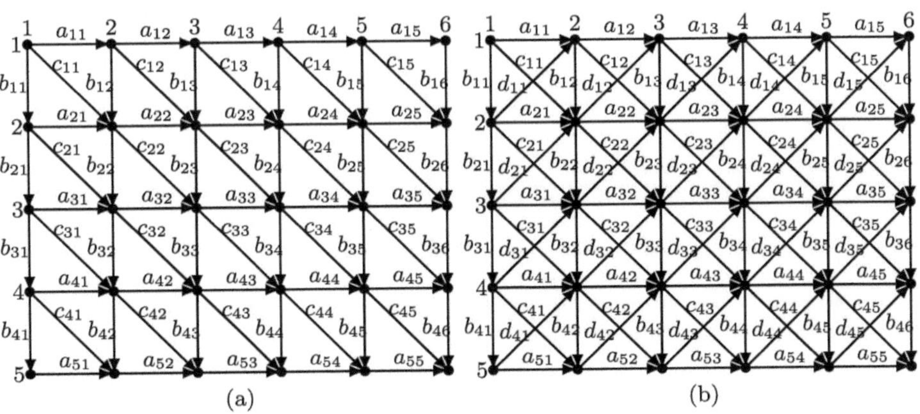

Fig. 1. Directed grid graphs with diagonal edges.

In [10] we investigated a directed *grid graph* $G_{m,n}$ [19] having $m \times n$ vertices and $m(n-1)+n(m-1)$ edges. Each vertex in this graph corresponds to a unique pair of integers (i,j) $(1 \le i \le m,\ 1 \le j \le n)$ which are vertex coordinates. It has edges $\{((i,j),(i+1,j)) \mid 1 \leq i < m, 1 \leq j \leq n\} \cup \{((i,j),(i,j+1)) \mid 1 \leq i \leq m, 1 \leq j < n\}$. Each edge $((i,j),(i,j+1))$ is labeled a_{ij}. Each edge $((i,j),(i+1,j))$ is labeled b_{ij}. We consider m as a constant which determines the *depth* of a grid graph, while n characterizes the *size* of the graph.

This paper focuses on grid graphs augmented with diagonal edges. Such graphs arise naturally in the study of maze-solving problems, in particular reachability problems [6]. One of these graphs is a directed *triangulated grid graph* (*TGG*) $T_{m,n}$ [14,20]. This graph has the same vertices and edges as $G_{m,n}$ and additionally includes edges $\{((i,j),(i+1,j+1)) \mid 1 \leq i < m, 1 \leq j < n\}$ labeled c_{ij}. The second one is a directed *king graph* $K_{m,n}$ [16,21] that in addition to all vertices and edges of $T_{m,n}$, also includes edges $\{((i,j),(i-1,j+1)) \mid 2 \leq i < m, 1 \leq j < n\}$ labeled $d_{i-1,j}$. For example, graphs $T_{5,6}$ and $K_{5,6}$ are presented in Fig. 1 ((a) and (b), respectively).

Our goal is to generate and to simplify the expressions of both TGGs and king graphs. To that end, we present two methods, one of which we call a *backtracking method* and the second one is named a *decomposition method*.

2   Generating Expressions for Directed Triangulated Grid Graphs by A Backtracking Method

The method is universal and suitable for generating expressions of any st-dag. An expression is derived by using intermediate subexpressions which are accumulated in the graph's vertices. Specifically, in our case, a subexpression accumulated at vertex (i,j) of $T_{m,n}$ corresponds to its subgraph located between vertices (i,j) and (m,n), and is denoted by $F_{(i,j)}$. The following recursive procedure is used (the expression of a single-vertex graph in Step 1 is accepted to be 1):

1. $F_{(m,n)} \leftarrow 1$
2. $F_{(i,n)} \leftarrow b_{i,n}b_{i+1,n}...b_{m-1,n}$ $(i < m)$
3. $F_{(m,j)} \leftarrow a_{m,j}a_{m,j+1}...a_{m,n-1}$ $(j < n)$
4. $F_{(i,j)} \leftarrow a_{i,j}F_{(i,j+1)} + b_{i,j}F_{(i+1,j)} + c_{i,j}F_{(i+1,j+1)}$ $(i < m, j < n)$

The subexpression accumulated in vertex $(1,1)$ is the resulting expression.

Proposition 1. *The total number of labels $L_m(n)$ in the expression $Ex(T_{m,n})$ derived by the backtracking method is defined recursively as follows:*

$$L_m(1) = m - 1, \quad L_1(n) = n - 1 \tag{1}$$
$$L_m(n) = L_m(n-1) + L_{m-1}(n) + L_{m-1}(n-1) + 3 \quad (m > 1, n > 1).$$

Proof. Initial statements (1) follow directly from Steps 1, 2, 3 of the recursive procedure. The resulting expression $Ex(T_{m,n})$ is equal to $a_{1,1}F_{(1,2)} + b_{1,1}F_{(2,1)} + c_{1,1}F_{(2,2)}$. $F_{(1,2)}$ is $Ex(T_{m,n-1})$, $F_{(2,1)}$ is $Ex(T_{m-1,n})$ and $F_{(2,2)}$ is $Ex(T_{m-1,n-1})$. Labels a_{11}, b_{11} and c_{11} are three additional labels in $Ex(T_{m,n})$.   □

Theorem 1. *The total number of labels $L_m(n)$ in the expression $Ex(T_{m,n})$ (m is considered as a constant) derived by the backtracking method is $O\left(n^m\right)$.*

Proof. The proof is obtained by induction on m. $L_1(n) = n - 1 = O(n)$. Therefore, the theorem holds for $m = 1$. We will prove it for any $m > 1$ on condition that it is correct for $m - 1$. According to Proposition 1,

$$L_m(n) = L_m(n-1) + L_{m-1}(n) + L_{m-1}(n-1) + 3$$
$$= L_m(n-2) + L_{m-1}(n-1) + L_{m-1}(n-2) + 3 + L_{m-1}(n) + L_{m-1}(n-1) + 3$$
$$= L_m(n-2) + L_{m-1}(n-2) + 2L_{m-1}(n-1) + L_{m-1}(n) + 2 \cdot 3$$
$$= L_m(n-3) + L_{m-1}(n-2) + L_{m-1}(n-3) + 3 + L_{m-1}(n-2) + 2L_{m-1}(n-1)$$
$$+ L_{m-1}(n) + 2 \cdot 3$$
$$= L_m(n-3) + L_{m-1}(n-3) + 2\left(L_{m-1}(n-2) + L_{m-1}(n-1)\right) + L_{m-1}(n) + 3 \cdot 3$$
$$= \ldots = L_m(1) + \sum_{k=1}^{n} L_{m-1}(k) + \sum_{k=2}^{n-1} L_{m-1}(k) + 3(n-1)$$
$$= \sum_{k=1}^{n} L_{m-1}(k) + \sum_{k=2}^{n-1} L_{m-1}(k) + 3(n-1) + m - 1.$$

Hence, by the induction hypothesis, there exists a positive constant c such that

$$L_m(n) \leq c\left(\sum_{k=1}^{n} k^{m-1} + \sum_{k=2}^{n-1} k^{m-1}\right) + 3(n-1) + m - 1.$$

In accordance to *Faulhaber's formula* for the sum of the p-th powers of the first n positive integers

$$\sum_{k=1}^{n} k^p = \frac{1}{p+1} \sum_{k=0}^{p} (-1)^k \binom{p+1}{k} B_k n^{p+1-k},$$

where B_k are *Bernoulli numbers* $\left(B_1 = -\frac{1}{2}\right)$. That is, $\sum_{k=1}^{n} k^p = O\left(n^{p+1}\right)$ and, therefore, $L_m(n) = O\left(n^m\right) + 3(n-1) + m - 1 = O\left(n^m\right)$. $\square$

Specifically,

$$L_2(n) = n^2 + n - 1, \quad L_3(n) = \frac{2}{3}n^3 + n^2 + \frac{4}{3}n - 1.$$

For example, $Ex(T_{3,4})$ derived by the backtracking method is

$a_{11}(a_{12}(a_{13}b_{14}b_{24} + b_{13}(a_{23}b_{24} + b_{23}a_{33} + c_{23}) + c_{13}b_{24}) +$
$b_{12}(a_{22}(a_{23}b_{24} + b_{23}a_{33} + c_{23}) + b_{22}a_{32}a_{33} + c_{22}a_{33}) + c_{12}(a_{23}b_{24} + b_{23}a_{33} + c_{23})) +$
$b_{11}(a_{21}(a_{22}(a_{23}b_{24} + b_{23}a_{33} + c_{23}) + b_{22}a_{32}a_{33} + c_{22}a_{33}) + b_{21}a_{31}a_{32}a_{33} + c_{21}a_{32}a_{33}) +$
$c_{11}(a_{22}(a_{23}b_{24} + b_{23}a_{33} + c_{23}) + b_{22}a_{32}a_{33} + c_{22}a_{33}).$

It contains 63 labels.

Thus, the proposed algorithm optimizes the prefix parts of all subexpressions. The length of $Ex(T_{m,n})$ obtained by the backtracking method grows polynomially with n. Nevertheless, for sufficiently large constant m, problems based on $Ex(T_{m,n})$ may become intractable.

3 Generating Expressions for Directed Triangulated Grid Graphs by a Decomposition Method

The method is known to be highly efficient and is based on recursively revealing subgraphs within the graph of a regular structure. The resulting expression is obtained by a special composition of subexpressions describing these subgraphs. The existence of a decomposition method for a graph G is a sufficient condition for the existence of a polynomial-length expression for G, provided that in each recursive step the graph is split into a constant number of subgraphs whose sizes decrease proportionally.

The graph $T_{m,n}$ is conditionally split into left and right parts connected by *connecting edges* $a_{i,j}$ $(i = 1, 2, ..., m)$ and $c_{i,j}$ $(i = 1, 2, ..., m - 1)$ so that j is chosen in the middle of a row (edges $a_{13}, ..., a_{53}, c_{13}, ..., c_{43}$ in Fig. 1(a)). Each such edge leaves a target of a subgraph belonging to the left part (its source is a source of the graph) and enters a source of a subgraph of the right part (its target is a target of the graph). Any path from the source to the target of the graph passes through one of the connecting edges. This decomposition is applied recursively to each revealed subgraph.

We denote by $F((p_1, p_2), (q_1, q_2))$ a subexpression of a subgraph with source (p_1, p_2) and target (q_1, q_2) (it has depth $q_1 - p_1 + 1$ and size $q_2 - p_2 + 1$). Therefore,

1. $F((p_1, p_2), (p_1, p_2)) \leftarrow 1$
2. $F((p_1, p_2), (q_1, q_2)) \leftarrow b_{p_1,p_2} b_{p_1+1,p_2} ... b_{q_1-1,p_2}$ $(q_1 > p_1, q_2 = p_2)$
3. $F((p_1, p_2), (q_1, q_2)) \leftarrow F((p_1, p_2), (p_1, j)) a_{p_1,j} F((p_1, j + 1), (q_1, q_2)) +$
$$F((p_1, p_2), (p_1 + 1, j)) a_{p_1+1,j} F((p_1 + 1, j + 1), (q_1, q_2)) + ... +$$
$$F((p_1, p_2), (q_1, j)) a_{q_1,j} F((q_1, j + 1), (q_1, q_2)) +$$
$$F((p_1, p_2), (p_1, j)) c_{p_1,j} F((p_1 + 1, j + 1), (q_1, q_2)) +$$
$$F((p_1, p_2), (p_1 + 1, j)) c_{p_1+1,j} F((p_1 + 2, j + 1), (q_1, q_2)) + ... +$$
$$F((p_1, p_2), (q_1 - 1, j)) c_{q_1-1,j} F((q_1, j + 1), (q_1, q_2)) \ (q_2 > p_2)$$
$$j = \left\lceil \tfrac{q_2+p_2-1}{2} \right\rceil \text{ or } \left\lfloor \tfrac{q_2+p_2-1}{2} \right\rfloor$$

The decomposition procedure is initially invoked by substituting $(1, 1)$ and (m, n) for (p_1, p_2) and (q_1, q_2), respectively.

Thus, in the general case $Ex(T_{m,n})$ includes $4m - 2$ subexpressions related to its subgraphs of size $n' \approx \frac{n}{2}$ and $2m - 1$ additional labels corresponding to connecting edges. This expression can be simplified by factoring out subexpressions which appear twice (e.g., from the left). After the simplification, Step 3 of the decomposition procedure takes the form:

$$F((p_1, p_2), (q_1, q_2)) \leftarrow$$
$$F((p_1, p_2), (p_1, j)) \left(a_{p_1,j} F((p_1, j+1), (q_1, q_2)) + c_{p_1,j} F((p_1+1, j+1), (q_1, q_2))\right)$$
$$+ F((p_1, p_2), (p_1+1, j)) \left(a_{p_1+1,j} F((p_1+1, j+1), (q_1, q_2))\right.$$
$$\left. + c_{p_1+1,j} F((p_1+2, j+1), (q_1, q_2))\right) \tag{2}$$
$$+ \ldots + F((p_1, p_2), (q_1-1, j)) \left(a_{q_1-1,j} F((q_1-1, j+1), (q_1, q_2))\right.$$
$$\left. + c_{q_1-1,j} F((q_1, j+1), (q_1, q_2))\right)$$
$$+ F((p_1, p_2), (q_1, j)) a_{q_1,j} F((q_1, j+1), (q_1, q_2)).$$

The number of its subexpressions related to subgraphs of size $n' \approx \frac{n}{2}$ is thereby reduced to $3m - 1$. To make the factored subexpression longer, we choose $j = \left\lceil \frac{q_2 + p_2 - 1}{2} \right\rceil$.

For $m \geq 3$, the expressions $Ex(T_{m,2})$ and $Ex(T_{m,3})$ can be further simplified. Specifically, when decomposing $T_{m,2}$ or $T_{m,3}$, subgraphs of size 1 are revealed on both sides or on one side, respectively, of the decomposed graph. Expressions of these subgraphs are label products (see Step 2 of the decomposition procedure), and their common labels can also be factored out. In particular, the number of labels in $Ex(T_{m,2})$ may decrease by $2(1 + 2 + \ldots + m - 2) = (m-1)(m-2)$, and the number of labels in $Ex(T_{m,3})$ may decrease by $\frac{(m-1)(m-2)}{2}$.

Using the above reasoning, specifically statement (2), we obtain the following proposition.

Proposition 2. *The total number of labels $L_m(n)$ in the expression $Ex(T_{m,n})$ derived by the decomposition method is defined recursively as follows:*

$$L_m(1) = m - 1, \quad L_m(2) = \frac{1}{2}m^2 + \frac{5}{2}m - 2, \quad L_m(3) = \frac{1}{6}m^3 + 2m^2 + \frac{5}{6}m - 1$$

$$L_m(n) = 2 \sum_{k=1}^{m-1} L_k\left(\left\lfloor \frac{n}{2} \right\rfloor\right) + L_m\left(\left\lfloor \frac{n}{2} \right\rfloor\right) + \sum_{k=1}^{m} L_k\left(\left\lceil \frac{n}{2} \right\rceil\right) + 2m - 1 \quad (n > 3).$$

Theorem 2. *The total number of labels $L_m(n)$ in the expression $Ex(T_{m,n})$ (m is considered as a constant) derived by the decomposition method is $O\left(n \log^{m-1} n\right)$.*

Proof. The proof is obtained by induction on m. As follows from Proposition 2, $L_1(n) = L_1\left(\left\lceil \frac{n}{2} \right\rceil\right) + L_1\left(\left\lfloor \frac{n}{2} \right\rfloor\right) + 1 = O(n)$. Therefore, the theorem holds for $m = 1$. We will prove it for any $m > 1$ on condition that it is correct for $1, 2, \ldots, m - 1$. According to Proposition 2,

$$L_m(n) \leq 2L_m\left(\left\lceil \frac{n}{2} \right\rceil\right) + 3 \sum_{k=1}^{m-1} L_k\left(\left\lceil \frac{n}{2} \right\rceil\right) + 2m - 1.$$

By the induction hypothesis, for $k = 1, \ldots, m - 1$ there exists a positive constant c_k such that $L_k\left(\left\lceil \frac{n}{2} \right\rceil\right) \leq c_k \left\lceil \frac{n}{2} \right\rceil \log_2^{k-1} \left\lceil \frac{n}{2} \right\rceil \leq c_k n \log_2^{k-1} n = O\left(n \log^{k-1} n\right).$

Therefore,

$$L_m(n) \leq 2L_m\left(\left\lceil\frac{n}{2}\right\rceil\right) + 3\sum_{k=1}^{m-1} O\left(n\log^{k-1} n\right) + O(1)$$

$$= 2L_m\left(\left\lceil\frac{n}{2}\right\rceil\right) + O\left(n\log^{m-2} n\right).$$

In accordance with the *master theorem*, given constants $\alpha \geq 1$, $\beta > 1$, $k \geq 0$ and the recurrence $\Phi(n) = \alpha\Phi\left(\frac{n}{\beta}\right) + f(n)$ ($\frac{n}{\beta}$ is interpreted as $\left\lfloor\frac{n}{\beta}\right\rfloor$ or $\left\lceil\frac{n}{\beta}\right\rceil$), if $f(n) = O\left(n^{\log_\beta \alpha}\log^k n\right)$ then $\Phi(n) = O\left(n^{\log_\beta \alpha}\log^{k+1} n\right)$. For this reason,

$$L_m(n) \leq 2L_m\left(\left\lceil\frac{n}{2}\right\rceil\right) + O\left(n\log^{m-2} n\right) = O\left(n\log^{m-1} n\right).$$

$$\square$$

Specifically, for n that is a power of two ($n = 2^p$ for some positive integer $p \geq 1$) we obtain the following explicit formulae:

$$L_2(n) = \frac{3}{2}n\log_2 n + n, \quad L_3(n) = \frac{9}{8}n\log_2^2 n + \frac{15}{8}n\log_2 n + 3n - 2.$$

For example, $Ex(T_{3,4})$ derived by the decomposition method is

$$a_{11}(a_{12}((a_{13}b_{14} + c_{13})b_{24} + b_{13}(a_{23}b_{24} + c_{23} + b_{23}a_{33})) + c_{12}(a_{23}b_{24} + c_{23} + b_{23}a_{33})) +$$

$$(a_{11}b_{12} + c_{11} + b_{11}a_{21})(a_{22}(a_{23}b_{24} + c_{23} + b_{23}a_{33}) + c_{22}a_{33}) +$$

$$(a_{11}b_{12}b_{22} + b_{11}a_{21}b_{22} + b_{11}b_{21}a_{31})a_{32}a_{33}.$$

It contains 43 labels.

Thus, given a graph $T_{m,n}$ of any depth m, the total number of labels in $Ex(T_{m,n})$ derived by the decomposition method grows quasi-linearly with the size n of the graph.

As shown in [10], the lengths of expressions $Ex(G_{m,n})$ derived by the backtracking and decomposition methods are also $O\left(n^m\right)$ and $O\left(n\log^{m-1} n\right)$, respectively. In other words, adding diagonal edges labeled c to a directed grid graph only increases lengths of corresponding expressions numerically but does not change their asymptotic order.

4   Generating Expressions for Directed King Graphs by a Backtracking Method

A subgraph enclosed between vertices (i,j) and (m,n) of $G_{m,n}$ or $T_{m,n}$ is also a grid graph or a TGG, respectively. The addition of diagonal edges labeled c results in new paths, which essentially duplicate existing ones in $G_{m,n}$, but provide shorter routes to the right and downward. However, edges labeled d, although they ultimately lead to the target, also lead upwards in certain areas. For this reason, a subgraph of $K_{m,n}$ positioned between source (i, j) and target (m, n) is generally a pentagon (in special cases, this subgraph reduces to a trapezoid or a triangle). This subgraph (see Fig. 2) is a king graph with a cut upper-left corner and will be called an *upper-cut king graph*. We denote such a subgraph by $K_{m,n}^h$, where m and n are the depth and the size of the initial rectangular king graph (with the same meaning for $K_{m,n}^h$), and h is the number of vertices in each side of the cut corner. For example, the graph in Fig. 2 will be denoted by $K_{5,6}^2$. A regular grid graph of depth m and size n may be denoted by $K_{m,n}^0$.

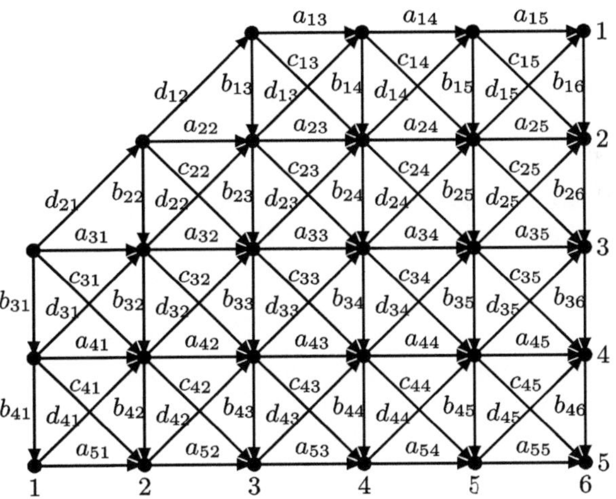

Fig. 2. An upper-cut king graph.

In order to apply a backtracking method to a king graph, we denote by $F_{(i,j)}$ a subexpression accumulated in vertex (i, j) of $K_{m,n}$, corresponds to its subgraph which is positioned between vertices (i, j) and (m, n). We use the following recursive procedure, assuming that $m > 1$ when $n > 1$ (the subexpression accumulated in vertex $(1, 1)$ is the resulting expression):

1. $F_{(m,n)} \leftarrow 1$
2. $F_{(i,n)} \leftarrow b_{i,n}b_{i+1,n}...b_{m-1,n}$ $(i < m)$
3. $F_{(m,j)} \leftarrow a_{m,j}F_{(m,j+1)} + d_{m-1,j}F_{(m-1,j+1)}$ $(j < n)$
4. $F_{(1,j)} \leftarrow a_{1,j}F_{(1,j+1)} + b_{1,j}F_{(2,j)} + c_{1,j}F_{(2,j+1)}$ $(j < n)$
5. $F_{(i,j)} \leftarrow a_{i,j}F_{(i,j+1)} + b_{i,j}F_{(i+1,j)} + c_{i,j}F_{(i+1,j+1)} + d_{i-1,j}F_{(i-1,j+1)}$
   $(1 < i < m, j < n)$

If $m = 1$ and $n > 1$ than $F_{(1,1)} \leftarrow a_{11}a_{1,2}...a_{1,n-1}$.

Proposition 3. *Denote by $L_m^h(n)$ the total number of labels in the expression $Ex(K_{m,n}^h)$ derived by the backtracking method.*

$$L_2^1(1) = 0, \ L_2^0(1) = 1 \tag{3}$$
$$L_2^1(n) = L_2^1(n-1) + L_2^0(n-1) + 2 \quad (n > 1) \tag{4}$$
$$L_2^0(n) = 2\left(L_2^0(n-1) + L_2^1(n-1)\right) + 5 \quad (n > 1) \tag{5}$$

Proof. Initial statements (3) follow directly from Steps 1 and 2 of the recursive procedure. Statement (4) follows from Step 3.

In accordance with Step 4,

$$\begin{aligned}
L_2^0(n) &= L_2^0(n-1) + L_2^1(n) + L_2^1(n-1) + 3 \\
&= L_2^0(n-1) + L_2^1(n-1) + L_2^0(n-1) + L_2^1(n-1) + 2 + 3 \\
&= 2\left(L_2^0(n-1) + L_2^1(n-1)\right) + 5.
\end{aligned}$$

$\square$

Theorem 3. *Denote by $L_m^h(n)$ the total number of labels in the expression $Ex(K_{m,n}^h)$ derived by the backtracking method.*

$$L_2^0(n) = 3^n - 2, \ L_2^1(n) = \frac{3^n}{2} - \frac{3}{2}.$$

Proof. The proof is obtained by induction on n. $L_2^0(1) = 3 - 2 = 1$. $L_2^1(1) = \frac{3}{2} - \frac{3}{2} = 0$. Therefore, the theorem holds for $n = 1$. We will prove it for any $n > 1$ on condition that it is correct for $n - 1$.

By statement (5) of Proposition 3,

$$\begin{aligned}
L_2^0(n) &= 2\left(L_2^0(n-1) + L_2^1(n-1)\right) + 5 \\
&= 2\left(3^{n-1} - 2 + \frac{3^{n-1}}{2} - \frac{3}{2}\right) + 5 = 3 \cdot 3^{n-1} - 4 - 3 + 5 = 3^n - 2.
\end{aligned}$$

By statement (4) of Proposition 3,

$$L_2^1(n) = L_2^1(n-1) + L_2^0(n-1) + 2 = \frac{3^{n-1}}{2} - \frac{3}{2} + 3^{n-1} - 2 + 2 = \frac{3^n}{2} - \frac{3}{2}.$$

$\square$

For example, $Ex(K_{2,3})$ derived by the backtracking method is

$$a_{11}(a_{12}b_{13} + b_{12}(a_{22} + d_{12}b_{13}) + c_{12}) +$$
$$b_{11}(a_{21}(a_{22} + d_{12}b_{13}) + d_{11}(a_{12}b_{13} + b_{12}(a_{22} + d_{12}b_{13})) + c_{12}) + c_{11}(a_{22} + d_{12}b_{13}).$$

It contains 25 labels.

Thus, the backtracking method applied to a king graph of size n produces expressions of exponential length in n, already for graphs of depth 2. For graphs with greater depths, the order of derived st-dag expressions is correspondingly even higher.

5 Generating Expressions for Directed King Graphs by a Decomposition Method

We apply the decomposition method to a king graph in a similar manner as for a TGG in Sect. 3. $K_{m,n}$ is conditionally split into left and right parts connected by *connecting edges* $a_{i,j}$ $(i = 1, 2, ..., m)$, $c_{i,j}$ $(i = 1, 2, ..., m-1)$ and $d_{i,j}$ $(i = 1, 2, ..., m-1)$ so that j is chosen in the middle of a row (edges $a_{13}, ..., a_{53}$, $c_{13}, ..., c_{43}$, $d_{13}, ..., d_{43}$ in Fig. 1(b)). The graph is decomposed into $2m$ subgraphs of size $n' \approx \frac{n}{2}$. The left upper subgraph is connected with two right subgraphs (via edges a_{13} and c_{13} in Fig. 1(b)), the left lower subgraph is connected with two right subgraphs as well (via edges a_{53} and d_{43} in Fig. 1(b)), and each of other left subgraphs is connected with three right subgraphs by edges labeled a, c, d.

In a king graph, targets of left subgraphs can be reached not only from above, but also from below. For this reason, subgraphs revealed from the left except for the lower one are generally pentagons. Each such subgraph (see Fig. 3(a)) is a king graph with a cut lower-right corner and will be called an *under-cut king graph*. We denote such a subgraph by $K_{m,n}^{-h}$, where m and n are the depth and the size of the initial rectangular king graph (with the same meaning for $K_{m,n}^{-h}$) and h is the number of vertices in each side of the cut corner. For example, the graph in Fig. 3(a) will be denoted by $K_{5,4}^{-2}$. It should also be noted that there are paths leading upward from sources of right subgraphs. Thus, these subgraphs, except for the upper one, are upper-cut king graphs (Fig 2).

Upper-cut and under-cut subgraphs are decomposed in the same way. New king subgraphs and upper-cut or under-cut subgraphs are revealed in the course of decomposition. In addition, king subgraphs with both the upper-left and the lower-right corners cut off may appear (see Fig. 3(b)). We call such hexagonal (in the general case) subgraphs *two-cut king graphs* and denote them by $K_{m,n}^{h_1,-h_2}$, where m and n are the depth and the size of the initial rectangular king graph (with the same meaning for $K_{m,n}^{h_1,-h_2}$) and h_1 and h_2 are the numbers of vertices in each side of the upper and the lower cut corners, respectively. For example, the graph in Fig. 3(b) will be denoted by $K_{5,4}^{1,-2}$.

Two-cut subgraphs are decomposed in the same manner. New upper-cut, under-cut and two-cut subgraphs are revealed during decomposition.

Thus, in the general case (when the size of a revealed subgraph is not less than its depth m, which is a constant) each subgraph is decomposed into $2m$ subgraphs with the same depth m. An expression of each decomposed subgraph of size n consists of $6m-4$ subexpressions related to revealed subgraphs ($m-2$ triads of pairs of subexpressions and four additional pairs referred to upper and lower subgraphs) of size $n' \approx \frac{n}{2}$ and $3m-2$ additional labels corresponding to connecting edges. As in Sect. 3, the expression can be simplified by factoring out subexpressions which appear more than once (three times or twice in our case). Depending on where the longer repeating expression is located, bracketing is done on the left or on the right. After the simplification, the number of subexpressions is reduced to $4m-2$.

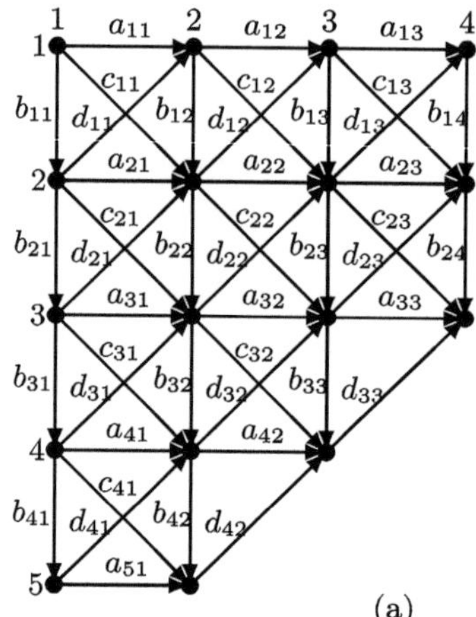

(a)

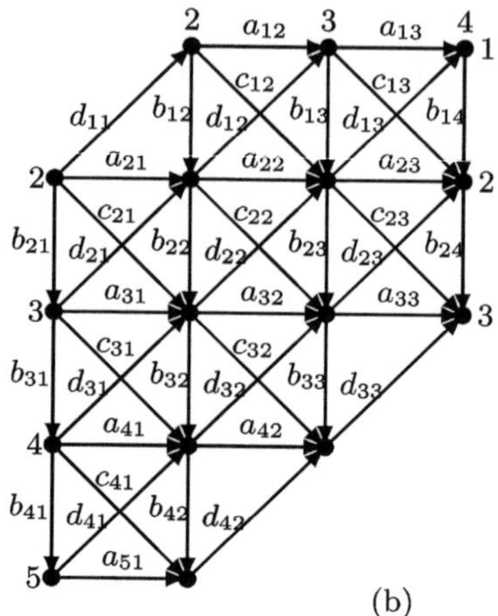

(b)

Fig. 3. Newly formed graphs: (a) an under-cut king graph, (b) a two-cut king graph.

According to the master theorem, given constants $\alpha \geq 1$, $\beta > 1$ and the recurrence $\Phi(n) = \alpha\Phi\left(\frac{n}{\beta}\right) + f(n)$ ($\frac{n}{\beta}$ is interpreted as $\left\lfloor\frac{n}{\beta}\right\rfloor$ or $\left\lceil\frac{n}{\beta}\right\rceil$), if $f(n) = O\left(n^{\log_\beta \alpha - \epsilon}\right)$ for some constant $\epsilon > 0$, then $\Phi(n) = O\left(n^{\log_\beta \alpha}\right)$. Therefore, we conclude the following result.

Theorem 4. *The total number of labels in the expression $Ex(K_{m,n})$ (m is considered as a constant) derived by the decomposition method is $O\left(n^{\log_2(4\,m-2)}\right)$.*

We denote by $F((p_1,p_2),(q_1,q_2))$, $F^h((p_1,p_2),(q_1,q_2))$, $F^{-h}((p_1, q_2))$, and $F^{h_1,-h_2}((p_1,p_2),(q_1,q_2))$ subexpressions related to king subgraphs, upper-cut subgraphs, under-cut subgraphs, and two-cut subgraphs, respectively, with a source (p_1,p_2) and a target (q_1,q_2). The superscripts have the same meaning as in designations of subgraphs themselves.

Below we present the decomposition procedure for generating an expression for a king graph of depth 2.

1. $F((p_1,p_2),(p_1+1,p_2)) \leftarrow b_{p_1,p_2}$,
   $F^1((p_1,p_2),(p_1,p_2)) \qquad \leftarrow \qquad 1, \qquad F^{-1}((p_1,p_2),(p_1,p_2)) \qquad \leftarrow \qquad 1,$
   $F^{1,-1}((p_1,p_2),(p_1,p_2)) \leftarrow 1$

2. $F^1((p_1,p_2),(p_1,p_2+1)) \leftarrow a_{p_1,p_2} + d_{p_1-1,p_2}b_{p_1-1,p_2+1}$,
   $F^{-1}((p_1,p_2),(p_1,p_2+1)) \leftarrow a_{p_1,p_2} + b_{p_1,p_2}d_{p_1,p_2}$,
   $F^{1,-1}((p_1,p_2),(p_1-1,p_2+1)) \leftarrow d_{p_1-1,p_2}$

3. $q_2 > p_2 : F((p_1,p_2),(q_1,q_2)) \leftarrow$
   $F^{-1}((p_1,p_2),(p_1,j))\left(a_{p_1,j}F((p_1,j+1),(q_1,q_2)) + c_{p_1,j}F^1((p_1+1,j+1),(q_1,q_2)) + \right.$
   $F((p_1,p_2),(p_1 \qquad\qquad\qquad\qquad\qquad\qquad\qquad\qquad\qquad\qquad\qquad\qquad +$
   $\left. 1,j))\left(a_{p_1+1,j}F^1((p_1+1,j+1),(q_1,q_2)) + d_{p_1,j}F((p_1,j+1),(q_1,q_2))\right.\right.$

4. $q_2 > p_2 + 1 : F^1((p_1,p_2),(q_1,q_2)) \leftarrow$
   $\left(F^{1,-1}((p_1,p_2),(p_1,j))a_{p_1,j} + F^1((p_1,p_2),(p_1+1,j))d_{p_1,j}\right)F((p_1,j \qquad\qquad +$

$$1), (q_1, q_2))+$$
$$\left(F^{1,-1}((p_1, p_2), (p_1, j))c_{p_1,j} + F^1((p_1, p_2), (p_1 + 1, j))a_{p_1+1,j}\right) F^1((p_1 + 1, j + 1), (q_1, q_2))$$

5. $q_2 > p_2 + 1 : F^{-1}((p_1, p_2), (q_1, q_2)) \leftarrow$
$$F^{-1}((p_1, p_2), (p_1, j)) \left(a_{p_1,j} F^{-1}((p_1, j + 1), (q_1, q_2))+\right.$$
$$c_{p_1,j} F^{1,-1}((p_1 + 1, j + 1), (q_1, q_2))+$$
$$F((p_1, p_2), (p_1 + 1, j)) \left(a_{p_1+1,j} F^{1,-1}((p_1 + 1, j + 1), (q_1, q_2))+\right.$$
$$d_{p_1,j} F^{-1}((p_1, j + 1), (q_1, q_2))$$

6. $q_2 > p_2 + 1 : F^{1,-1}((p_1, p_2), (q_1, q_2)) \leftarrow$
$$F^{1,-1}((p_1, p_2), (p_1, j)) \left(a_{p_1,j} F^{-1}((p_1, j + 1), (q_1, q_2))+\right.$$
$$c_{p_1,j} F^{1,-1}((p_1 + 1, j + 1), (q_1, q_2))+$$
$$F^1((p_1, p_2), (p_1 + 1, j)) \left(a_{p_1+1,j} F^{1,-1}((p_1 + 1, j + 1), (q_1, q_2))+\right.$$
$$d_{p_1,j} F^{-1}((p_1, j + 1), (q_1, q_2))$$

While Steps 1, 2 describe the initial subexpressions, the general cases are processed in Steps 3 - 6. In Steps 3, 5, 6 bracketing is done on the left with $j = \left\lceil \frac{q_2+p_2-1}{2} \right\rceil$. In Step 4 bracketing is done on the right with $j = \left\lfloor \frac{q_2+p_2-1}{2} \right\rfloor$.

Denote by $L_m(n)$, $L_m^h(n)$, $L_m^{h_1,h_2}(n)$ the total numbers of labels in the expressions $Ex(K_{m,n})$, $Ex(K_{m,n}^h)$ or $Ex(K_{m,n}^{-h})$, $Ex(K_{m,n}^{h_1,-h_2})$ derived by the decomposition method.

As follows from the decomposition procedure, for n that is a power of two ($n = 2^p$ for some positive integer $p \geq 2$)

$$\begin{cases} L_2(n) = 3L_2\left(\frac{n}{2}\right) + 3L_2^1\left(\frac{n}{2}\right) + 4 \\ L_2^1(n) = L_2\left(\frac{n}{2}\right) + 3L_2^1\left(\frac{n}{2}\right) + 2L_m^{1,1}\left(\frac{n}{2}\right) + 4 \\ L_2^{1,1}(n) = 3L_2^1\left(\frac{n}{2}\right) + 3L_m^{1,1}\left(\frac{n}{2}\right) + 4, \end{cases} \tag{6}$$

where $L_2(2) = 7$, $L_2^1(2) = 3$, $L_2^1(2) = 1$.

The following explicit formulae ($n = 2^p$, $p \geq 1$) for (6) are obtained by the method for simultaneous linear recurrences solving [13]:

$$L_2(n) = \frac{19}{30}n^{\log_2 6} + \frac{4}{3}n^{\log_2 3} - \frac{4}{5}, \quad L_2^1(n) = \frac{19}{30}n^{\log_2 6} - \frac{4}{5}, \quad L_2^{1,1}(n) = \frac{19}{30}n^{\log_2 6} - \frac{2}{3}n^{\log_2 3} - \frac{4}{5}.$$

For example, $Ex(K_{2,4})$ derived by the decomposition method is

$$(a_{11} + b_{11}d_{11})(a_{12}(a_{13}b_{14} + c_{13} + b_{13}(a_{23} + d_{13}b_{14})) + c_{12}(a_{23} + d_{13}b_{14})) +$$
$$(a_{11}b_{12} + c_{11} + b_{11}(a_{21} + d_{11}b_{12}))(a_{22}(a_{23} + d_{13}b_{14}) + d_{12}(a_{13}b_{14} + c_{13} +$$
$$b_{13}(a_{23} + d_{13}b_{14}))).$$

It contains 34 labels. For comparison, $Ex(K_{2,4})$ derived by the backtracking method contains 79 labels.

Thus, the length of $Ex(K_{m,n})$ derived by the decomposition method grows polynomially with n. Moreover, even for large constant m, the exponent of n remains low.

6   Applications of Grid-Like Graphs

Grid-like graphs, which model systems with a structured, often rectangular, layout, are applied in diverse areas, including routing [8], coloring [5], circuit design [7], chemistry [15], text analysis [18], and others. Their regular structure makes them suitable for representing spatial or logical relationships and for applying efficient algorithms for tasks such as shortest-path finding and data processing.

Specifically, we aim to use grid-structured graphs to model a task-scheduling system, namely, a multi-priority system of robotic lines in which priorities change dynamically. Each vertex of an $m \times n$ graph represents a machine, and each edge (i, j) denotes a link for moving a manufactured product from machine i to machine j. In this context, edge labels serve as weights, each representing a generalized cost computed as a complex function of various parameters.

Row i $(i = 1, ..., m)$ corresponds to a line of priority i. A product may be transferred from machine j to machine $j + 1$ on the same line or, if necessary, to any lower line. Grid graphs allow a transition only to the machine of the same number in the new line, followed by a transfer to the next machine on that line. By contrast, TGGs and king graphs permit a direct passage to the next machine in the new line. Any transition to a lower line decreases the task's priority, since it reduces the number of lines that remain available for future transitions. In systems modeled by grid graphs and TGGs, returning to a higher line is impossible. In systems represented by king graphs, priority can be increased, but only by one level per transition.

In the systems described above, shortest-path problems arise naturally. These problems can be solved symbolically: according to this approach, we use max-plus algebra tools [1] (operators $+, \cdot$ are replaced by $\min, +$, respectively) to interpret an st-dag expression as a shortest-path algorithm. This method allows rapid adaptation to modifications in the robotic lines without recomputing shortest paths from scratch after each change. As a result, it ensures stability and fast responsiveness to data updates, both of which are essential for dynamic real-time environments. Thus, a compact st-dag expression provides an efficient shortest-path algorithm on the underlying st-dag.

In [12], we employed this symbolic technique to schedule a single robotic line modeled by a Fibonacci graph. We plan to extend this work to the models schematically described above, which are based on grid-like graphs.

7   Conclusion and Future Work

Two methods for generating algebraic expressions of triangulated grid graphs and king graphs have been presented. Both types of graphs have m rows and n columns, where m is treated as a constant. The first approach, the backtracking method, produces expressions of length $O(n^m)$ for TGGs and expressions of exponential length in n for king graphs. The second approach, the decomposition method, provides expressions of quasi-linear length, $O(n \log^{m-1} n)$, for TGGs and expressions of length $O(n^{\log_2(4m-2)})$ for king graphs.

Both methods rely on recursive algorithms. The expressions they generate can be implemented using linked lists in which each element represents a symbol of the corresponding expression. Each recursive call produces a list implementing one of the subexpressions, which is concatenated to the unified list in $O(1)$ time. Consequently, the running-time complexities of the algorithms coincide with the complexities of the lengths of the derived expressions themselves for both types of graphs. The amount of memory required by the algorithms is determined solely by the sizes of the derived expressions, and therefore its asymptotic order is equal to the order of the expression length as well.

We plan to develop an improved version of the decomposition method which is expected to yield more compact expressions. In particular, we intend to avoid connecting edges and to identify subgraphs (of, apparently, fairly complex structure) that adjoin one another at the corners. Ultimately, our goal is to find in this way expressions of minimum lengths for both TGGs and king graphs.

Acknowledgements. The author thanks Vadim E. Levit for helpful advice. The author is also grateful to the anonymous reviewers for their careful reading of the manuscript, as well as for their constructive, detailed comments and suggestions, which improved the paper overall.

References

1. Akian, M., Babat, R., Gaubert, S.: Max-Plus Algebra. In: Hogben, L. et al. (eds.) Handbook of Linear Algebra. Discrete Mathematics and Its Applications, vol. 39, chapter 25, Chapman & Hall/CRC, Boca Raton (2007)
2. Bein, W.W., Kamburowski, J., Stallmann, M.F.M.: Optimal reduction of two-terminal directed acyclic graphs. SIAM J. Comput. **21**(6), 1112–1129 (1992)
3. Golumbic, M.C., Mintz, A., Rotics, U.: Factoring and recognition of read-once functions using cographs and normality and the readability of functions associated with partial k-trees. Discrete Appl. Math. **154**(10), 1465–1477 (2006)
4. Golumbic, M.C., Perl, Y.: Generalized Fibonacci maximum path graphs. Discrete Math. **28**, 237–245 (1979)
5. Herva, P., Karl, J.: On forced periodicity of perfect colorings. Theor. Comput. Syst. **67**, 732–759 (2023)
6. Holzer, M., Jacobi, S.: Grid graphs with diagonal edges and the complexity of xmas mazes. In: Kranakis, E., Krizanc, D., Luccio, F. (eds.) FUN 2012. LNCS, vol. 7288, pp. 223–234. Springer, Heidelberg (2012)
7. Kaufmann, M., Gao, S., Thulasiraman, K.: On Steiner minimal trees in grid graphs and its application to VLSI routing. In: Ding-Zhu, D., Zhang, X.-S. (eds.) ISAAC 1994. LNCS, vol. 834, pp. 351–359. Springer, Heidelberg (1994)
8. Kayaturan, G.C., Vernitski, A.: Encoding Paths with Binary Arrays in a King's Graph for Error-Free Data Transmission. Annals of Mathematics and Artificial Intelligence (2025). https://link.springer.com/article/10.1007/s10472-025-09985-7
9. Korenblit, M.: Decomposition methods for generating algebraic expressions of full square rhomboids and other graphs. Discrete Appl. Math. **228**, 60–72 (2017)
10. Korenblit, M.: Generating algebraic expressions for labeled grid graphs. In: Kim, D., Zelikovsky, A. (eds.) COCOA 2018. LNCS, vol. 11346, pp. 138–153. Springer, Heidelberg (2018)

11. Korenblit, M., Levit, V.E.: On algebraic expressions of series-parallel and Fibonacci graphs. In: Calude, C.S., Dinneen, M.J., Vajnovzki, V. (eds.) DMTCS 2003. LNCS, vol. 2731, pp. 215–224. Springer, Heidelberg (2003)
12. Korenblit, M., Levit, V.E: Symbolic solutions of shortest-path problems and their applications. In: Tuba, M., Akashe, S., Joshi, A. (eds.) ICT System and Sustainability, Proc. ICT4SD 2019, Vol. 1, Advances in Intelligent Systems and Computing, vol. 1077, pp. 299–307. Springer, Heidelberg (2019)
13. Korenblit, M., Levit, V.E.: Some systems of simultaneous linear recurrences and their applications to computing of graph expression lengths. JCMCC **114**, 193–223 (2020)
14. Law, H.F., Leung, S.L., Ostrovskii, M.I.: Spanning tree congestion of planar graphs. Involve J. Math. **7**(2), 205–226 (2014)
15. Liu, S., Ou, J., Lin, Y.: On K-resonance of grid graphs on the plane, torus and cylinder. J. Math. Chem. **52**(7), 1807–1816 (2014)
16. Mertens, S.: Domination Polynomials of the Grid, the Cylinder, the Torus, and the King Graph. J. Integer Sequences **27** (2024)
17. Mintz, A., Golumbic, M.C.: Factoring Boolean functions using graph partitioning. Discrete Appl. Math. **149**(1–3), 131–153 (2005)
18. Schmidt, J.P.: All highest scoring paths in weighted grid graphs and their application to finding all approximate repeats in strings. SIAM J. Comput. **27**(4), 972–992 (1998)
19. Weisstein, E.W.: Grid Graph. From MathWorld – A Wolfram Web Resource. http://mathworld.wolfram.com/GridGraph.html
20. Weisstein, E.W.: Triangular Grid. From MathWorld – A Wolfram Web Resource. https://mathworld.wolfram.com/TriangularGrid.html
21. Weisstein, E.W.: King Graph. From MathWorld – A Wolfram Web Resource. https://mathworld.wolfram.com/KingGraph.html

Parameterized Hardness Results for the Restricted Santa Claus Problem

S. Anil Kumar[(✉)] and N. S. Narayanaswamy

Indian Institute of Technology Madras, IIT Campus P.O., Chennai, India 600036
anilgecp@gmail.com, swamy@cse.iitm.ac.in

Abstract. The Restricted Santa Claus Problem [1] is a classic NP-hard fair allocation problem, where the goal is to assign a set of indivisible gifts to children in a way that maximizes the minimum total happiness across all children. Each gift provides zero happiness value to some children and a fixed positive happiness value to the remaining children.

In this paper, we study the parameterized complexity of the problem with respect to the minimum happiness threshold τ and some structural parameters associated with the incidence graph of the input instance, which include treewidth ω, clique-width λ, and diameter δ. The incidence graph is a bipartite graph with one partition of vertices for the children and another for the gifts, where an edge connects a child and a gift if the gift contributes a non-zero happiness value to that child. We show that the problem is W[1]-hard when parameterized by $(\omega, \lambda, \delta)$, even when the incidence graph is planar. We also establish W[1]-hardness under the combined parameters (τ, λ, δ). These results extend to the Restricted Makespan Minimization Problem, highlighting the inherent difficulty of achieving fair allocations even under strong structural restrictions.

Keywords: Restricted Santa Claus Problem · Incidence Graph · Parameterized Complexity

1   Introduction

The Restricted Santa Claus Problem [1] is a cornerstone NP-hard problem in resource allocation problems. In this problem, the input consists of a set of children and a set of gifts such that gift j has a happiness value of $p_{ij} \in \{0, p_j\}$ for child i where $p_j > 0$. If R represents the set of gifts allocated to the child i in some gift allocation, then $\sum_{j \in R} p_{ij}$ represents the total happiness value of the child i in that allocation. The goal of the problem is to find an allocation of all the gifts to the children so that the minimum of the total happiness values of all the children is maximized. While several constant-factor approximation algorithms are known for this problem [2–5], its parameterized complexity landscape remains largely unexplored.

A closely related problem to the Restricted Santa Claus Problem is the Restricted Makespan Minimization Problem. It can be viewed as the

© The Author(s), under exclusive license to Springer Nature Switzerland AG 2026
N. Misra and A. Pandey (Eds.): CALDAM 2026, LNCS 16445, pp. 293–306, 2026.
https://doi.org/10.1007/978-3-032-17156-6_22

min-max analogue to the RESTRICTED SANTA CLAUS PROBLEM. This problem is the special case of the general MAKESPAN MINIMIZATION PROBLEM in which the input is a set of machines and a set of jobs to be allocated to the machines. Each job j has a processing time p_{ij} on machine i. The goal of the problem is to find an allocation of the jobs to the machines in such a way that the maximum total processing time among the machines (known as the makespan) is minimized. In the RESTRICTED MAKESPAN MINIMIZATION PROBLEM, each job j can be allocated to only a specific subset of machines and has a fixed processing time of p_j on all of these machines (the processing time of j is taken to be ∞ on all other machines).

It follows from the hardness result of Jansen *et al.* [6] on the UNARY BIN PACKING PROBLEM parameterized by the number of bins that the MAKESPAN MINIMIZATION PROBLEM is W[1]-hard when parameterized by the number of machines. Mnich and Wiese [7] presented an FPTalgorithm for this problem under the combined parameter (number of distinct processing times, number of machines). They also established that the problem, when restricted to identical machines, can be solved in FPT time when parameterized by the maximum job processing time $p_{\max}$. Chen *et al.* [8] studied the parameterized complexity of the MAKESPAN MINIMIZATION PROBLEM by taking the rank of the processing time matrix and $p_{\max}$ as parameters. The processing time matrix of a MAKESPAN MINIMIZATION instance having m machines and n jobs is an $m \times n$ matrix, where the rows correspond to the machines, the columns correspond to the jobs, and the (i, j)-th entry represents the processing time of job j on machine i. They showed that the problem is fixed-parameter tractable when parameterized by the combined parameter $(p_{\max}, r)$, where r is the rank of the processing time matrix [8]. Koutecký and Zink [9] showed that the MAKESPAN MINIMIZATION PROBLEM, when parameterized by the number of different job types (t), is fixed-parameter tractable on identical machines, but W[1]-hard on unrelated machines. They also showed that the problem can be solved in time $p_{\max}^{\mathcal{O}(t^2)}$. However, it is not clear whether the problem is fixed-parameter tractable when parameterized by the number of different types of machines and the number of distinct processing times.

Jansen *et al.* [10] showed that both the general SANTA CLAUS PROBLEM, where p_{ij} can take arbitrary non-negative real values (unlike the restricted case where $p_{ij} \in \{0, p_j\}$), and the MAKESPAN MINIMIZATION PROBLEM can be solved in time $3^n \cdot \text{poly}(\text{input size})$. Here n represents the number of gifts and the number of jobs for the SANTA CLAUS PROBLEM and the MAKESPAN MINIMIZATION PROBLEM, respectively. They proved that there is no exact algorithm for both the problems with a runtime of $2^{o(n)}$ unless ETH fails. Subsequently, Annamalai and Narayanaswamy [11] designed algorithms for both the problems with a running time of $2^n \cdot \text{poly}(\text{input size})$.

INTEGER LINEAR PROGRAMMING (ILP) serves as a fundamental framework for solving the SANTA CLAUS PROBLEM. The earlier approximation algorithms for the RESTRICTED SANTA CLAUS PROBLEM were based on a linear program, namely the Assignment LP [1]. The Assignment LP is the LP relaxation of an integer programming formulation of the SANTA CLAUS PROBLEM. Ganian *et al.* [12] showed that ILP is W[1]-hard when parameterized by the incidence treewidth. However, no result is currently known in the literature that sheds light on the parameterized complexity of the SANTA CLAUS PROBLEM based on this hardness result.

Our Results: Motivated by the fact that the RESTRICTED SANTA CLAUS PROBLEM has received little attention in the parameterized complexity domain, we study the parameterized complexity of the following decision version of the RESTRICTED SANTA CLAUS PROBLEM.

RESTRICTED SANTA CLAUS PROBLEM (Decision Version)

Input: A pair $(\mathcal{I}, \tau)$, where τ is a non-negative integer and $\mathcal{I}$ is an instance consisting of a set of children $\mathcal{C}$, and a set of gifts $\mathcal{G}$ such that gift j has a rational non-negative happiness value $p_{ij} \in \{0, p_j\}$ for child i.

Question: Does there exist an allocation of the gifts in $\mathcal{G}$ to the children in $\mathcal{C}$ such that the total happiness value of each child is at least τ?

Our parameterized complexity analysis focus on the structure of the incidence graph of the input instance (see Sect. 2 for the definition) because it captures the interactions between children and gifts that directly influences algorithmic behavior. Parameters based solely on the children and the gifts do not reflect these structural dependencies and provide less insight into the feasibility of fixed-parameter algorithms. We choose *treewidth* (ω), *clique-width* (λ), and *diameter* (δ) of the incidence graph, as well as *the minimum total happiness value of the children* (τ), as the parameters. The choice is motivated by the following facts.

1. ω, λ, and δ characterize the structural complexity of a RESTRICTED SANTA CLAUS instance. Specifically, ω captures the local sparsity and tree-likeness of the incidence graph, λ quantifies the connectedness of its subgraphs, and δ reflects the overall spread of the incidence graph. Studying the problem under these parameters clarifies how structural restrictions influence the feasibility of fixed-parameter algorithms.
2. The objective function value τ is a natural and widely used parameter in the study of parameterized complexity for optimization problems [13]. In our setting, τ corresponds to the minimum total happiness value of the children.

We show that the problem is unlikely to have FPT algorithms under the combined parameters $(\omega, \lambda, \delta)$ and (τ, λ, δ) as captured in the following theorems.

Theorem 1. *The* RESTRICTED SANTA CLAUS PROBLEM *is W[1]-hard when parameterized by the combined parameter* $(\omega, \lambda, \delta)$*, where* ω*,* λ*, and* δ *are the treewidth, clique-width, and diameter of the incidence graph of the input instance, respectively, even when the incidence graph is planar.*

Theorem 2. *The* RESTRICTED SANTA CLAUS PROBLEM *is W[1]-hard when parameterized by the combined parameter* (τ, λ, δ)*, where* τ *is the minimum total happiness value, and* λ *and* δ *are the clique-width and diameter of the incidence graph of the input instance, respectively.*

We prove the two theorems by giving two novel polynomial time reductions from the k-SUM PROBLEM (which is W[1]-hard when parameterized by k [14]) to the RESTRICTED SANTA CLAUS PROBLEM. To the best of our knowledge, these are the first parameterized hardness results for the RESTRICTED SANTA CLAUS PROBLEM under combined parameterizations involving the structural parameters ω, λ, and δ. More importantly, our reductions are sufficiently robust to establish analogous hardness results for the RESTRICTED MAKESPAN MINIMIZATION PROBLEM, leading to the following corollaries. Previously, these hardness results were not known for the RESTRICTED MAKESPAN MINIMIZATION PROBLEM.

Corollary 1. *The* RESTRICTED MAKESPAN MINIMIZATION PROBLEM *is W[1]-hard when parameterized by the combined parameter* $(\omega, \lambda, \delta)$*, where* ω*,* λ*, and* δ *are the treewidth, clique-width, and diameter of the incidence graph of the input instance, respectively, even when the incidence graph is planar.*

Corollary 2. *The* RESTRICTED MAKESPAN MINIMIZATION PROBLEM *is W[1]-hard when parameterized by the combined parameter* (τ, λ, δ)*, where* τ *is the makespan, and* λ *and* δ *are the clique-width and diameter of the incidence graph of the input instance, respectively.*

The rest of the paper is organized as follows. Section 2 presents the preliminaries for presenting the results. In Sect. 3, we prove the hardness results. We present our conclusions in Sect. 4.

2   Preliminaries

We use the notation $[m \mathinner{.\,.} n]$ to denote the integer interval $\{m, m+1, \ldots, n\}$ for integers $m \leq n$. An instance of a parameterized problem Q is a tuple (x, k), where x encodes the input and k is a tuple of one or more parameters [13]. If k consists of multiple parameters, we refer to it as a combined parameter. Parameterized problems are classified into a hierarchy of complexity classes known as the *Weft hierarchy*, given by $\mathsf{FPT} \subseteq \mathsf{W}[1] \subseteq \mathsf{W}[2] \subseteq \cdots \subseteq \mathsf{W}[P]$. A problem Q is *fixed-parameter tractable* (in class FPT) with respect to parameter k if it can be solved by an algorithm running in time $f(k) \cdot n^c$, where n is the input size, f is a computable function, and c is a constant.

An FPT reduction from a parameterized problem Q to Q' is an algorithm that, given an instance (x, k) of Q, outputs an instance (x', k') of Q' such that (x, k) is a YES-INSTANCE of Q if and only if (x', k') is a YES-INSTANCE of Q', each parameter in k' is bounded by a computable function of k, and the reduction runs in time $f(k) \cdot |x|^{\mathcal{O}(1)}$ for some computable function f. If a problem is W[t]-hard for some $t \geq 1$, then it is unlikely to be fixed-parameter tractable. If there exists a FPT reduction from a known W[t]-hard problem to Q, then Q is also W[t]-hard.

For a graph G, the distance between vertices u and v, denoted $\mathrm{dist}(u, v)$, is the length of the shortest path between them. The *diameter* of G, denoted $\mathrm{diam}(G)$, is the maximum distance over all pairs of distinct vertices in G. The *clique-width* of a graph is the minimum number of labels needed to construct the graph using a sequence of four fundamental operations. These fundamental operations are introducing a new vertex with a specified label, taking the disjoint union of two labeled graphs, connecting all vertices with one label to all vertices with another label, and relabeling all vertices of one label to another.

A *tree decomposition* of a graph $G = (V, E)$ is a pair (T, χ), where T is a tree and χ assigns to each node $z \in V(T)$ a subset $\chi(z) \subseteq V$, called a bag, such that every vertex of G appears in at least one bag, every edge of G is contained in some bag, and for each vertex $v \in V$, the set of nodes z for which $v \in \chi(z)$ forms a connected subtree of T [13]. The width of a tree decomposition is defined as $\max_{z \in V(T)} |\chi(z)| - 1$, and the treewidth of G is the minimum width over all tree decompositions of G. A *nice tree decomposition* of G is a tree decomposition with the following additional structural properties. T is a rooted binary tree where the root and leaves have empty bags, and every internal node is one of three types—a join node with two children having the same bag as the node itself, an introduce node which adds exactly one vertex to its child's bag, or a forget node which removes exactly one vertex from its child's bag [13].

The incidence graph of a SANTA CLAUS instance $\mathcal{I}$ is a bipartite graph $G_{inc}(\mathcal{I})$ whose vertices correspond to the children on one side and the gifts on the other, with an edge between a child and a gift if the gift has a non-zero happiness value for that child.

3   Hardness Results

We prove the two hardness results by giving two FPT reductions from the k-SUM PROBLEM to the RESTRICTED SANTA CLAUS PROBLEM. The k-SUM PROBLEM parameterized by k is known to be W[1]-hard [14]. The formal statement of the k-SUM PROBLEM is as follows.

---

k-SUM PROBLEM [14]
Input: A pair (S, k), where k is a positive integer and S is a set of n integers in $[-n^{2k} .. n^{2k}]$.
Parameter: k
Question: Does there exist $S' \subseteq S$ such that $|S'| = k$ and $\sum_{j \in S'} j = 0$?

We choose $d = \sum_{j \in S} j$ and $\Delta = \sum_{j \in S} |j|$ throughout this paper. Since each integer in S lies in $[-n^{2k} .. n^{2k}]$, it follows that $-n^{2k+1} \leq d \leq n^{2k+1}$, $0 \leq \Delta \leq n^{2k+1}$, and $-1 \leq \frac{j}{\Delta} \leq 1$ for each $j \in S$.

In both the reductions, the RESTRICTED SANTA CLAUS instance is constructed in such a way that each happiness value is of the form $q \pm \frac{l}{r}$, where q, l and r are integers satisfying the conditions $-n^{2k+1} \leq l \leq n^{2k+1}$, $q \leq k$, and $1 \leq r \leq 3n^{2k+1} + 1$. Each such happiness value can be encoded in binary as a triplet (q, l, r). It follows that q requires $\mathcal{O}(\log k)$ bits and each of l and r requires $\mathcal{O}(k \log n)$ bits. Furthermore, this encoding runs in time $\mathcal{O}(k \log n)$. This leads to the following observation.

Observation 1. *Each rational number of the form $q \pm \frac{l}{r}$ where $q \leq k, -n^{2k+1} \leq l \leq n^{2k+1}, 1 \leq r \leq 3n^{2k+1} + 1$ and $k \leq n$ can be encoded as a triplet (q, l, r) in FPT time parameterized by k.*

It is easy to see that the exact computation of each such happiness value as a rational number $\frac{qr \pm l}{r}$ from the triplet encoding (q, l, r) runs in FPT time parameterized by k.

Theorem 1. *The* RESTRICTED SANTA CLAUS PROBLEM *is W[1]-hard when parameterized by the combined parameter $(\omega, \lambda, \delta)$, where ω, λ, and δ are the treewidth, clique-width, and diameter of the incidence graph of the input instance, respectively, even when the incidence graph is planar.*

Proof. Let (S, k) be an instance of the k-SUM PROBLEM. Construct the pair $(\mathcal{I}, \tau)$ as follows. $\mathcal{I}$ is a RESTRICTED SANTA CLAUS instance in which the treewidth (ω), clique-width (λ), and diameter (δ) of the incidence graph are bounded by a function of k, and $\tau = n$. There are two children (c_1 and c_2) and $2n - k + 1$ gifts in the instance $\mathcal{I}$. These gifts are categorized into two types, *normal gifts* and *special gifts*. There are n normal gifts, each corresponding to an integer in S, and $n - k + 1$ special gifts. The normal gifts can be allocated arbitrarily to both c_1 and c_2. For each $i \in [1 \mathinner{.\,.} n - k]$, i^{th} special gift has non-zero happiness value for only c_1, whereas $(n - k + 1)^{\text{th}}$ special gift has non-zero happiness value for only c_2. The happiness values of these gifts are chosen in such a way that the following two conditions are satisfied.

1. The sum of the happiness values of all the gifts is 2τ. Therefore, $(\mathcal{I}, \tau)$ is a YES-INSTANCE if and only if there is a gift allocation in which each child has a total happiness value of exactly τ.
2. The total happiness value of c_1 from the special gifts is $n - k$. Hence, it follows that in any gift allocation in which c_1 has a total happiness value of exactly τ, the sum of the happiness values of all the normal gifts allocated to c_1 must be exactly k. Moreover, such a gift allocation is possible if and only if exactly k normal gifts are allocated to c_1, and the subset S' of S corresponding to these k normal gifts satisfies the condition $\sum_{j \in S'} j = 0$.

Formally the instance $\mathcal{I}$ is defined as follows.

1. The set of children in the instance $\mathcal{I}$ is $\{c_1, c_2\}$.
2. Each integer $j \in S$ corresponds to a normal gift α_j. The $n - k + 1$ special gifts are denoted by $\beta_1, \ldots, \beta_{n-k+1}$. Hence, the set of gifts in the instance $\mathcal{I}$ is $\{\alpha_j \mid j \in S\} \cup \{\beta_i \mid i \in [1 \mathinner{.\,.} n - k + 1]\}$.
3. For each $j \in S$, α_j has a happiness value of $1 - \frac{j}{3\Delta+1}$ for both c_1 and c_2. For each $i \in [1 \mathinner{.\,.} n - k]$, β_i has happiness values 1 and 0 for c_1 and c_2, respectively. The special gift β_{n-k+1} has happiness values 0 and $k + \frac{d}{3\Delta+1}$ for c_1 and c_2, respectively.

The reduction is illustrated in Fig. 1, which presents the incidence graph of the instance $\mathcal{I}$.

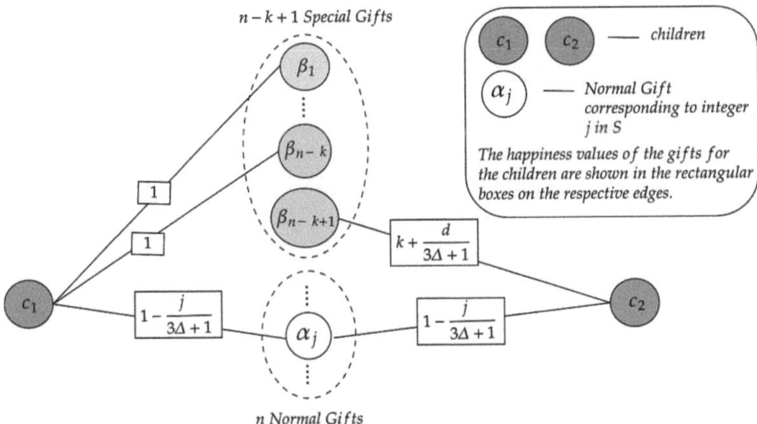

Fig. 1. The incidence graph of the instance $\mathcal{I}$ created in the proof of Theorem 1.

By the construction of the instance $\mathcal{I}$, the following facts are evident.

1. The incidence graph $G_{inc}(\mathcal{I})$ of the instance $\mathcal{I}$ is planar. This follows from the fact that a planar embedding of $G_{inc}(\mathcal{I})$ can be constructed by placing all gift vertices at the center of the plane, positioning the child vertices on either side, and drawing the edges between them. This embedding ensures that no edges cross, as each edge in $G_{inc}(\mathcal{I})$ connects a gift vertex to a child vertex. An illustration of this construction is given in Fig. 1.
2. A tree decomposition (T, χ) of $G_{inc}(\mathcal{I})$ can be constructed as follows. Set T to be a simple path having $2n - k + 1$ nodes, each corresponding to a gift in the instance $\mathcal{I}$. For each z in $V(T)$, set $\chi(z)$ to be the set consisting of c_1, c_2 and the gift corresponding to z. It is not hard to see that (T, χ) satisfies all the conditions of a tree decomposition, and that the treewidth of $G_{inc}(\mathcal{I})$ is 2. Hence, the treewidth of $G_{inc}(\mathcal{I})$ is $2 \leq k + 4$. That is, $\omega \leq k + 4$.
3. $G_{inc}(\mathcal{I})$ is a bipartite graph with one of the partitions containing only two vertices so that the distance between any two vertices of $G_{inc}(\mathcal{I})$ is at most 3. Hence, the diameter of $G_{inc}(\mathcal{I})$ is at most 3. That is, $\delta \leq 3 \leq k + 4$.
4. We can construct the incidence graph $G_{inc}(\mathcal{I})$ using at most four labels, as follows. We begin by creating the two gift vertices c_1 and c_2, assigning them labels 1 and 2, respectively. For each vertex β_i with $i \in [1 .. n - k]$, we create a new vertex with label 3, connect it to c_1 (using the join operation between labels 1 and 3), and then relabel it to 4 to prevent further edge additions. Similarly, we create the vertex β_{n-k+1} with label 3, connect it only to c_2 (via label 2), and relabel it to 4. For each α_j with $j \in S$, we again create a vertex with label 3, connect it to both c_1 and c_2, and relabel it to 4. At every step, we use at most four labels (label 3 for newly introduced vertices, labels 1 and 2 for the fixed vertices c_1 and c_2, and label 4 for processed vertices). This construction shows that the clique-width of $G_{inc}(\mathcal{I})$ is at most 4. That is $\lambda \leq 4 \leq k + 4$.

Thus it follows that the combined parameter $(\omega, \lambda, \delta)$ is bounded by $k + 4$ componentwise. The happiness values associated with the gifts are of the form $q \pm \frac{l}{r}$, where $q \le k, -n^{2k+1} \le l \le n^{2k+1}, 1 \le r \le 3n^{2k+1} + 1$, and $k \le n$. Hence, by Observation 1, the encoding of the happiness value of each gift can be computed in FPT time parameterized by k. Since there are only $2n - k + 1$ gifts in the instance $\mathcal{I}$, it follows that the instance $\mathcal{I}$ can be encoded in FPT time parameterized by k. Hence, the reduction is an FPT reduction.

The sum of the happiness values of all the gifts is

$$\sum_{j \in S} \left(1 - \frac{j}{3\Delta + 1}\right) + k + \frac{d}{3\Delta + 1} + n - k$$

$$= n - \frac{d}{3\Delta + 1} + k + \frac{d}{3\Delta + 1} + n - k$$

$$= 2n = 2\tau \tag{1}$$

Let (S, k) be a YES-INSTANCE of the k-SUM PROBLEM. Therefore, there exists a subset S' of S such that $|S'| = k$ and $\sum_{j \in S'} j = 0$. Construct a solution for the instance $\mathcal{I}$ as follows. Allocate α_j for each $j \in S'$ and the special gifts $\beta_1, \beta_2, \beta_3, \ldots, \beta_{n-k}$ to c_1, and all the remaining gifts to c_2. The total happiness value of c_1 is $\sum_{j \in S'} \left(1 - \frac{j}{3\Delta+1}\right) + n - k = n$. Hence, the total happiness value of c_2 is also exactly n, as evident from Equation (1). Hence, $(\mathcal{I}, \tau)$ is a YES-INSTANCE of the RESTRICTED SANTA CLAUS PROBLEM.

Conversely, let $(\mathcal{I}, \tau)$ be a YES-INSTANCE of the RESTRICTED SANTA CLAUS PROBLEM. Therefore, there exists a gift allocation in which the total happiness values of both c_1 and c_2 are at least n. Therefore, it follows from Equation (1) that both c_1 and c_2 have a total happiness value of exactly n. Hence, this gift allocation will be an optimal allocation for the instance $\mathcal{I}$. Therefore, the special gifts $\beta_1, \ldots, \beta_{n-k}$ are allocated to c_1 and β_{n-k+1} is allocated to c_2 in this allocation, since this is the only allocation in which each of the special gifts contributes a non-zero happiness value to the child to whom it is allocated. Let S' be the subset of S corresponding to the normal gifts allocated to c_1.

If $|S'| < k$, the total happiness value of c_1 is

$$\sum_{j \in S'} \left(1 - \frac{j}{3\Delta + 1}\right) + n - k \le k - 1 + \sum_{j \in S'} \left(\frac{|j|}{3\Delta + 1}\right) + n - k < n.$$

If $|S'| > k$, the total happiness value of c_2 is

$$\sum_{j \in S \setminus S'} \left(1 - \frac{j}{3\Delta + 1}\right) + k + \frac{d}{3\Delta + 1}$$

$$\le n - k - 1 + \sum_{j \in S'} \left(\frac{|j|}{3\Delta + 1}\right) + k + \frac{d}{3\Delta + 1}$$

$$\le n - 1 + \frac{\Delta + d}{3\Delta + 1} < n.$$

Hence, in either case, the total happiness value of one of the children is less than n, which is a contradiction. Therefore, it follows that $|S'| = k$. Since the total happiness value of c_1 is exactly τ, $\sum_{j \in S'} \left(1 - \frac{j}{3\Delta+1}\right) + n - k = n$. That is, $k - \frac{1}{3\Delta+1}\left(\sum_{j \in S'} j\right) - k = 0$. Therefore, $\sum_{j \in S'} j = 0$. Hence, (S, k) is a YES-INSTANCE of the k-SUM PROBLEM. $\qquad\square$

Theorem 2. *The* RESTRICTED SANTA CLAUS PROBLEM *is W[1]-hard when parameterized by the combined parameter (τ, λ, δ), where τ is the minimum total happiness value, and λ and δ are the clique-width and diameter of the incidence graph of the input instance, respectively.*

Proof. Let (S, k) be an instance of the k-SUM PROBLEM. Construct the pair $(\mathcal{I}, \tau)$ as follows. $\mathcal{I}$ is a RESTRICTED SANTA CLAUS instance in which the clique-width (λ), and the diameter (δ) of the incidence graph are bounded by $k+6$, and $\tau = k+2$. There are $n-k+2$ children, labeled $c_1, \ldots, c_{n-k+2}$ and $3n-k+2$ gifts in the instance $\mathcal{I}$. These gifts are categorized into three types, *normal gifts*, *special gifts*, and *dummy gifts*. There are n normal gifts and n special gifts corresponding to the n integers in S. The instance $\mathcal{I}$ has $n - k + 2$ dummy gifts. The normal gifts have non-zero happiness values for all the children except c_{n-k+2}. The special gifts have non-zero happiness values for all the children except c_{n-k+1}. For $i \in [1 .. n - k + 2]$, i^{th} dummy gift has non-zero happiness value for only c_i. The happiness values of these gifts are chosen in such a way that the following two conditions are satisfied.

1. The sum of the happiness values of all the gifts is $(n - k + 2)\tau$. Therefore, $(\mathcal{I}, \tau)$ is a YES-INSTANCE if and only if there is a gift allocation in which each child has a total happiness value of exactly τ.
2. In any gift allocation in which all the children have a total happiness value of exactly τ, all of $c_1, \ldots, c_{n-k}$ must be allocated exactly one normal gift and c_{n-k+1} must be allocated exactly k normal gifts. Moreover, such a gift allocation is possible if and only if the subset S' of S corresponding to the k normal gifts allocated to c_{n-k+1} satisfies the condition $\sum_{j \in S'} j = 0$.

Formally the instance $\mathcal{I}$ is defined as follows.

1. The set of children in the SANTA CLAUS instance $\mathcal{I}$ is $\{c_1, \ldots, c_{n-k+2}\}$.
2. Each integer $j \in S$ corresponds to a normal gift α_j and a special gift β_j. The $n-k+2$ dummy gifts are denoted by $\gamma_1, \ldots, \gamma_{n-k+2}$. Hence, the set of gifts in the instance $\mathcal{I}$ is given by $\{\alpha_j \mid j \in S\} \cup \{\beta_j \mid j \in S\} \cup \{\gamma_i \mid i \in [1 .. n-k+2]\}$.
3. For $j \in S$, α_j has zero happiness value for c_{n-k+2}, and a happiness value of $1 - \frac{j}{3\Delta+1}$ for all other children. For $j \in S$, β_j has zero happiness value for c_{n-k+1}, and a happiness value of $1 + \frac{j}{3\Delta+1}$ for all other children. For $i \in [1 .. n - k]$, γ_i has a happiness value of k for c_i, and zero happiness value for all other children. γ_{n-k+1} has a happiness value of 2 to c_{n-k+1}, and zero happiness value for all other children. γ_{n-k+2} has a happiness value of 2 to c_{n-k+2}, and zero happiness value for all other children.

The reduction is illustrated in Fig. 2 by giving the incidence graph of the instance $\mathcal{I}$.

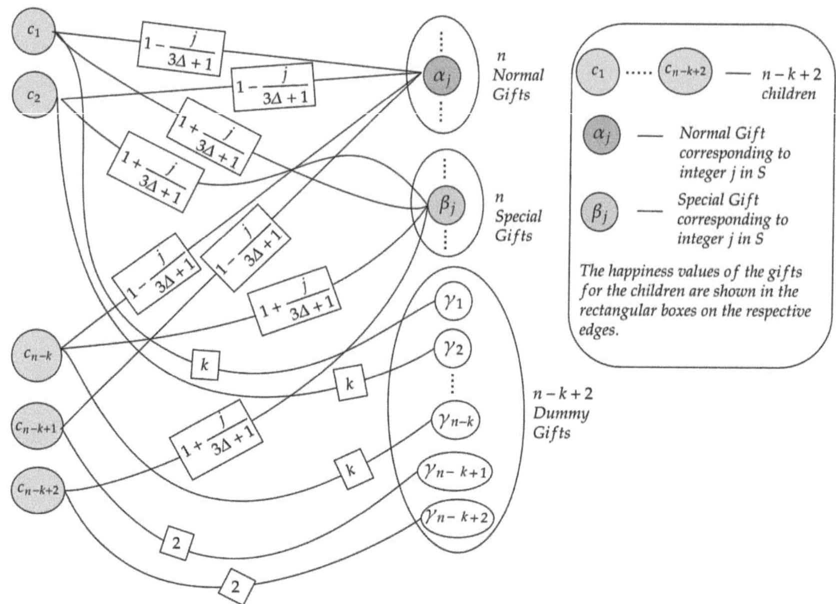

Fig. 2. The incidence graph of the instance $\mathcal{I}$ constructed in the proof of Theorem 2.

By the construction of the instance $\mathcal{I}$, it follows that for each $i \in [1 \mathinner{.\,.} n - k]$, child c_i requires at least two gifts other than γ_i to be allocated to it in order to achieve a total happiness of at least τ. This is because allocating a single normal gift along with γ_i, or a single special gift along with γ_i, results in a total happiness strictly less than $k + 2$. Hence, we state the following observation.

Observation 2. *In any gift allocation, if all of $c_1, \ldots, c_{n-k}$ has a total happiness value of at least τ, then the total number of normal gifts and special gifts allocated among these $n - k$ children will be at least $2(n - k)$.*

By the construction of the instance $\mathcal{I}$, the following facts are evident.

1. For $i \in [1 \mathinner{.\,.} n - k]$ and $i' \in [1 \mathinner{.\,.} n - k + 2]$ such that $i \neq i'$, $\mathrm{dist}(c_i, c_{i'}) = 2$. Further, $\mathrm{dist}(c_{n-k+1}, c_{n-k+2}) = 4$ (path is $c_{n-k+1} - \alpha_j - c_1 - \beta_j - c_{n-k+2}$). For $j, j' \in S$ such that $j \neq j'$, $\mathrm{dist}(\alpha_j, \alpha_{j'}) = \mathrm{dist}(\beta_j, \beta_{j'}) = 2$. For $j, j' \in S$, $\mathrm{dist}(\alpha_j, \beta_{j'}) = 2$. For $j \in S$ and $i \in [1 \mathinner{.\,.} n - k]$, $\mathrm{dist}(\alpha_j, \gamma_i) = \mathrm{dist}(\alpha_j, \gamma_{n-k+1}) = \mathrm{dist}(\beta_j, \gamma_i) = \mathrm{dist}(\beta_j, \gamma_{n-k+2}) = 2$, $\mathrm{dist}(\alpha_j, \gamma_{n-k+2}) = 4$ (path is $\alpha_j - c_1 - \beta_j - c_{n-k+2} - \gamma_{n-k+2}$), and $\mathrm{dist}(\beta_j, \gamma_{n-k+1}) = 4$ (path is $\beta_j - c_1 - \alpha_j - c_{n-k+1} - \gamma_{n-k+1}$). For $i \in [1 \mathinner{.\,.} n - k]$ and $i' \in [1 \mathinner{.\,.} n - k + 2]$ $i \neq i'$, $\mathrm{dist}(\gamma_i, \gamma_{i'}) = 4$ and $\mathrm{dist}(\gamma_{n-k+1}, \gamma_{n-k+2}) = 6$ (path is $\gamma_{n-k+1} - c_{n-k+1} - \alpha_j - c_1 - \beta_j - c_{n-k+2} - \gamma_{n-k+2}$). Hence, the distance between any two vertices in G is at most 6. Therefore, the diameter of $G_{inc}(\mathcal{I})$ is at most 6. That is, $\delta \leq 6 \leq k + 6$.

2. We construct the incidence graph $G_{\mathrm{inc}}(\mathcal{I})$ using at most 6 labels, as follows. The vertices $c_1, \ldots, c_{n-k}$ and $\gamma_1, \ldots, \gamma_{n-k}$, along with the edges between each

c_i and γ_i, are created in $n - k$ iterations. For each $i \in [1..n - k]$, the ith iteration consists of creating the vertices c_i with label 1 and γ_i with label 2, adding them to the current graph, immediately applying the join operation between labels 1 and 2 (which adds exactly the edge connecting c_i and γ_i), and then relabeling c_i to label 3 and γ_i to label 4 to free labels 1 and 2 for the next iteration. After completing the $n-k$ iterations, we create vertices c_{n-k+1} with label 1, c_{n-k+2} with label 2, γ_{n-k+1} with label 5, and γ_{n-k+2} with label 6. We then take their disjoint union with the current graph and add an edge between vertices with label 1 and label 5, and another between vertices with label 2 and label 6. These steps connect c_{n-k+1} to γ_{n-k+1}, and c_{n-k+2} to γ_{n-k+2}. We then relabel both γ_{n-k+1} and γ_{n-k+2} to label 4, ensuring that labels 5 and 6 are again available. Next, we create n vertices corresponding to the normal gifts $\alpha_1, \ldots, \alpha_n$, all with label 5, n vertices corresponding to the special gifts $\beta_1, \ldots, \beta_n$, all with label 6, and take their disjoint union with the current graph. We connect all vertices with label 3 (*i.e.*, $c_1, \ldots, c_{n-k}$) to all vertices with labels 5 and 6. This adds edges between each c_i and every α_j and β_j for $i \in [1..n - k]$ and $j \in S$. Finally, we add edges between vertices with label 1 (*i.e.*, c_{n-k+1}) and those with label 5 ($\alpha_1, \ldots, \alpha_n$), and edges between vertices with label 2 (*i.e.*, c_{n-k+2}) and those with label 6 ($\beta_1, \ldots, \beta_n$). These steps complete the construction by adding edges between c_{n-k+1} and each α_j and the edges between c_{n-k+2} and each β_j, for all $j \in S$. This construction shows that the clique-width of $G_{inc}(\mathcal{I})$ is at most 6. That is $\lambda \leq 6 \leq k + 6$.

Since $\tau = k + 2$, it further follows that the combined parameter (τ, λ, δ) is bounded by $k+6$ componentwise. The happiness values associated with the gifts are of the form $q \pm \frac{l}{r}$, where $q \leq k, -n^{2k+1} \leq l \leq n^{2k+1}, 1 \leq r \leq 3n^{2k+1} + 1$, and $k \leq n$. Hence, by Observation 1, the encoding of the happiness value of each gift can be computed in **FPT** time parameterized by k. Since there are only $3n - k + 2$ gifts in the instance $\mathcal{I}$, it follows that the instance $\mathcal{I}$ can be encoded in **FPT** time parameterized by k. Hence, the reduction is an **FPT** reduction. The sum of the happiness values of all the gifts is

$$= \sum_{j \in S}\left(1 - \frac{j}{3\Delta + 1}\right) + \sum_{j \in S}\left(1 + \frac{j}{3\Delta + 1}\right) + (n - k)k + 2 + 2$$
$$= (n - k + 2)(k + 2) = (n - k + 2)\tau \tag{2}$$

Let (S, k) be a YES-INSTANCE of the k-SUM PROBLEM. Therefore, there exists a subset S' of S such that $|S'| = k$ and $\sum_{j \in S'} j = 0$. Construct a solution for the instance $\mathcal{I}$ as follows. Let $h \colon [1..n - k] \to S \setminus S'$ be an arbitrary bijection from $[1..n - k]$ to $S \setminus S'$. For $i \in [1..n - k]$, allocate $\alpha_{h(i)}, \beta_{h(i)}$, and γ_i to c_i. Hence, for $i \in [1..n - k]$, the total happiness value of c_i is

$$\left(1 - \frac{h(i)}{3\Delta + 1}\right) + \left(1 + \frac{h(i)}{3\Delta + 1}\right) + k = k + 2 = \tau.$$

For each $j \in S'$, allocate α_j and γ_{n-k+1} to c_{n-k+1}. Therefore, the total happiness value of c_{n-k+1} is

$$\sum\nolimits_{j \in S'} \left(1 - \frac{j}{3\Delta + 1}\right) + 2 = k + 2 = \tau.$$

Allocate the remaining gifts (β_j for each $j \in S'$ and γ_{n-k+2}) to c_{n-k+2}. Therefore, it follows from Equation (2) that the total happiness value of c_{n-k+2} is also τ. Hence, $(\mathcal{I}, \tau)$ is a YES-INSTANCE of the RESTRICTED SANTA CLAUS PROBLEM.

Conversely, let $(\mathcal{I}, \tau)$ be a YES-INSTANCE of the RESTRICTED SANTA CLAUS PROBLEM. Therefore, there exists a gift allocation in which the total happiness value of each child is at least $k + 2$. By Equation (2), it follows that each child has a total happiness value of exactly $k + 2$. Moreover, this gift allocation is an optimal gift allocation for the instance $\mathcal{I}$. Hence, it follows that the dummy gift γ_i is allocated to c_i for each $i \in [1..n-k+2]$ in this gift allocation, since this is the only allocation in which each of the dummy gifts contributes a non-zero happiness value to the child to whom it is allocated. Let S' be the subset of S corresponding to the normal gifts allocated to c_{n-k+1}. Let S'' be the subset of S corresponding to the special gifts allocated to c_{n-k+2}.

If $|S''| < k$, then the total happiness value of c_{n-k+2} is

$$\sum\nolimits_{j \in S''} \left(1 + \frac{j}{3\Delta + 1}\right) + 2 \le k - 1 + \sum\nolimits_{j \in S''} \left(\frac{|j|}{3\Delta + 1}\right) + 2 < k + 2.$$

Hence, it follows that $|S''| \ge k$. If $|S'| < k$, the total happiness value of c_{n-k+1} is

$$\sum\nolimits_{j \in S'} \left(1 - \frac{j}{3\Delta + 1}\right) + 2 \le k - 1 + \sum\nolimits_{j \in S'} \left(\frac{|j|}{3\Delta + 1}\right) + 2 < k + 2.$$

If $|S'| > k$, then the total number of normal and special gifts among the children $c_1, \ldots, c_{n-k}$ is $|S \setminus S'| + |S \setminus S''| \le n - k - 1 + n - k < 2(n - k)$. Hence, by Observation 2, the total happiness value of at least one of $c_1, \ldots, c_{n-k}$ is strictly less than $k + 2$. Therefore, in all cases, the total happiness value of at least one of the children is less than $k + 2$, which is a contradiction. Hence, it follows that $|S'| = k$.

Since the total happiness value of c_{n-k+1} is exactly $k + 2$, it follows that $\sum_{j \in S'} \left(1 - \frac{j}{3\Delta+1}\right) + 2 = k + 2$. That is, $k - \frac{1}{3\Delta+1}\left(\sum_{j \in S'} j\right) = k$. Therefore, $\sum_{j \in S'} j = 0$. Hence, (S, k) is a YES-INSTANCE of the k-SUM PROBLEM. $\square$

With small changes, the RESTRICTED SANTA CLAUS instance $(\mathcal{I}, \tau)$ constructed in the proofs of each of Theorem 1 and Theorem 2 can be considered as RESTRICTED MAKESPAN MINIMIZATION instances $(\mathcal{I}', \tau)$. The machines, jobs, and the processing times in the instance $\mathcal{I}'$ correspond to the children, gifts, and the happiness values in the instance $\mathcal{I}$, respectively. The only difference between the instances $\mathcal{I}$ and $\mathcal{I}'$ is that, if any gift j has zero happiness value for a child i in the instance $\mathcal{I}$, then job j has a processing time of ∞ on machine i in the instance $\mathcal{I}'$. Moreover τ represents the makespan. It follows that, in either case, the only job allocation in which all the machines in the instance $\mathcal{I}'$ have a makespan of at most τ is the one where the total processing time of each machine is exactly τ. Moreover, such a job allocation is possible if and only if there exists a subset S' of S such that $|S'| = k$ and $\sum_{j \in S'} j = 0$. Hence, we have the following two corollaries.

Corollary 1. *The* RESTRICTED MAKESPAN MINIMIZATION PROBLEM *is W[1]-hard when parameterized by the combined parameter $(\omega, \lambda, \delta)$, where ω, λ, and δ are the treewidth, clique-width, and diameter of the incidence graph of the input instance, respectively, even when the incidence graph is planar.*

Corollary 2. *The* RESTRICTED MAKESPAN MINIMIZATION PROBLEM *is W[1]-hard when parameterized by the combined parameter (τ, λ, δ), where τ is the makespan, and λ and δ are the clique-width and diameter of the incidence graph of the input instance, respectively.*

4 Concluding Remarks

We showed that the RESTRICTED SANTA CLAUS PROBLEM is W[1]-hard under the combined parameters $(\omega, \lambda, \delta)$ and (τ, λ, δ) for arbitrary non-negative happiness values. We also proved analogous hardness results for the RESTRICTED MAKESPAN MINIMIZATION PROBLEM. These results show that both problems remain difficult even on highly restricted and sparse structures, indicating that the combinatorial nature of the allocation constraints dominates the structural simplicity of the incidence graph. It remains open whether the RESTRICTED SANTA CLAUS PROBLEM is fixed-parameter tractable under alternative parameterizations involving ω, λ, and δ together with parameters such as the number of children or the number of distinct happiness values.

References

1. Bansal, N., Sviridenko, M.: The Santa Claus Problem. In: Proceedings of the Thirty-eighth Annual ACM Symposium on Theory of Computing, pages 31–40. Association for Computing Machinery, Inc, (2006). https://doi.org/10.1145/1132516.1132522
2. Asadpour, A., Feige, U., Saberi, A.: Santa Claus Meets Hypergraph Matchings. ACM Trans. Al. (TALG) **8**(3), 1–9 (2012). https://doi.org/10.1145/2229163.2229168
3. Bamas, É., Lindermayr, A., Megow, N., Rohwedder, L., Schlöter, J.: Santa Claus meets Makespan and Matroids: algorithms and Reductions. In: Proceedings of the 2024 Annual ACM-SIAM Symposium on Discrete Algorithms (SODA), pages 2829–2860. SIAM, (2024). https://doi.org/10.1137/1.9781611977912.100
4. Davies, S., Rothvoss, T., Zhang, Y.: A Tale of Santa Claus, Hypergraphs and Matroids. In: Proceedings of the Fourteenth Annual ACM-SIAM Symposium on Discrete Algorithms, pages 2748–2757. SIAM (2020). https://doi.org/10.1137/1.9781611975994.167

5. Annamalai, C., Kalaitzis, C., Svensson, O.: Combinatorial Algorithm for Restricted Max-Min Fair Allocation. ACM Trans. Al. (TALG) **13**(3), 1–28 (2017). https://doi.org/10.1145/3070694
6. Jansen, K., Kratsch, S., Marx, D., Schlotter, I.: Bin packing with fixed number of bins revisited. J. Comput. Syst. Sci. **79**(1), 39–49 (2013). https://doi.org/10.1016/j.jcss.2012.04.004
7. Mnich, M., Wiese, A.: Scheduling and fixed-parameter tractability. Math. Program. **154**(1), 533–562 (2015). https://doi.org/10.1007/s10107-014-0830-9
8. Chen, L., Marx, D., Ye, D., Zhang, G.: Parameterized and approximation results for scheduling with a low rank processing time matrix. In: 34th Symposium on Theoretical Aspects of Computer Science (STACS 2017), pages 1–14. Schloss Dagstuhl – Leibniz-Zentrum fúr Informatik, (2017). https://doi.org/10.4230/LIPIcs.STACS.2017.22
9. Koutecký, M., Zink, J.: Complexity of scheduling few types of jobs on related and unrelated machines. J. Schedul. **28**(1), 1–18 (2025). https://doi.org/10.1007/s10951-024-00827-8
10. Jansen, K., Land, F., Land, K.: Bounding the running time of algorithms for scheduling and packing problems. SIAM J. Disc. Math. **30**(1), 343–366 (2016). https://doi.org/10.1137/140952636
11. Annamalai, S., Narayanaswamy, N.S.: Exact algorithms for allocation problems. In: Chen, J., Lu, P. (eds.) FAW 2018. LNCS, vol. 10823, pp. 251–262. Springer, Cham (2018). https://doi.org/10.1007/978-3-319-78455-7_19
12. Ganian, R., Ordyniak, S., Ramanujan, M.S.: Going Beyond Primal Treewidth for (M)ILP. In: Proceedings of the AAAI Conference on Artificial Intelligence **31**(1) (2017). https://doi.org/10.1609/aaai.v31i1.10644
13. Cygan, M., et al.: Parameterized algorithms, vol. 4. Springer (2015). https://doi.org/10.1007/978-3-319-21275-3
14. Abboud, A., Lewi, K., Williams, R.: Losing weight by gaining edges. In: Schulz, A.S., Wagner, D. (eds.) ESA 2014. LNCS, vol. 8737, pp. 1–12. Springer, Heidelberg (2014). https://doi.org/10.1007/978-3-662-44777-2_1

Minimum Selective Subset on Unit Disk Graphs and Circle Graphs

Bubai Manna[(✉)]

Indian Institute of Technology Kharagpur, Kharagpur, India
`bubaimanna11@gmail.com`

Abstract. In a connected simple graph $G = (V(G), E(G))$, each vertex is assigned one of c colors, where $V(G) = \bigcup_{\ell=1}^{c} V_\ell$ and V_ℓ denotes the set of vertices of color ℓ. A subset $S \subseteq V(G)$ is called a *selective subset* if, for every ℓ, $1 \leq \ell \leq c$, every vertex $v \in V_\ell$ has at least one nearest neighbor in $S \cup (V(G) \setminus V_\ell)$ that also lies in V_ℓ. The *Minimum Selective Subset* (MSS) problem asks for a selective subset of minimum size.

We show that the MSS problem is log-APX-hard on general graphs, even when $c = 2$. As a consequence, the problem does not admit a polynomial-time approximation scheme (PTAS) unless P = NP. On the positive side, we present a PTAS for unit disk graphs that does not require a geometric representation and applies for arbitrary c. We further prove that MSS remains NP-complete in unit disk graphs for arbitrary c. In addition, we show that the MSS problem is APX-hard on circle graphs, even when $c = 2$. The full version of this work can be found in [1].

Keywords: Nearest-Neighbor Classification · Minimum Consistent Subset · Minimum Selective Subset · Unit Disk Graphs · Circle Graphs · NP-complete · log-APX-hard · Polynomial-time Approximation Scheme

1 Introduction

Many computational tools have been developed for supervised learning methods on a labeled training set T embedded in a metric space (X, d). Each data point $t \in T$ is associated with a label (also called a *character* or *color*), chosen from a set $C = \{1, 2, \ldots, c\}$. The objective is to extract a smallest possible subset $S \subseteq T$ such that every point in T either belongs to S or has at least one nearest neighbor (with respect to the metric d) within S that shares the same character. This optimization problem, called the *Minimum Consistent Subset* (MCS), was originally formulated by Hart [2] in 1968, which has received thousands of citations, highlighting its significant impact in the field. However, the paper [2] did not establish any complexity results or algorithms.

Later in 1991, Wilfong [3] defined two problems MCS and MSS together and proved that the MCS and MSS problems are NP-complete in $\mathbb{R}^2$ for $c \geq 3$ and $c \geq 2$, respectively. It also proposed a polynomial-time algorithm when

© The Author(s), under exclusive license to Springer Nature Switzerland AG 2026
N. Misra and A. Pandey (Eds.): CALDAM 2026, LNCS 16445, pp. 307–320, 2026.
https://doi.org/10.1007/978-3-032-17156-6_23

there is only one red point and all other points are blue in $\mathbb{R}^2$. Later, in 2018, it was proved that MCS remains NP-complete when $c = 2$ in $\mathbb{R}^2$ [4]. Recently, Banerjee et al. [5] showed that MCS is W[2]-hard (for arbitrary c) and MSS is W[1] (for $c = 2$), both parameterized by the solution size. Various algorithms, including those for many restricted inputs for the MCS problem in $\mathbb{R}^2$ have been proposed [5–7], highlighting its significance in machine learning and computational geometry. The only algorithm for the MSS problem is a PTAS, which was established when $c = 2$ [5].

The Minimum Selective Subset (MSS) problem plays a crucial role in optimizing data selection by identifying the smallest subset which preserves essential information. It can be viewed both as a clustering and a proximity problem. This is particularly useful in applications such as fingerprint recognition, character recognition, and pattern recognition, where it helps reduce redundancy and improve decision-making in classification and feature selection tasks. MSS was introduced because the standard MCS methods, like Hart's [2], could not guarantee that the resulting subset was the smallest possible size. MSS applies a stronger consistency condition on the subset to create a mathematical framework that is amenable to an exact minimization procedure, thereby fulfilling the optimization goal that MCS heuristics failed to satisfy. So far, we have discussed MCS and MSS problems along with their published results in $\mathbb{R}^2$. We now turn to these problems in the context of graph algorithms.

Banerjee et al. [5] proved that MCS is W[2]-hard [8] when parameterized by the solution size, even with only two colors on general graphs. Dey et al. [9,10] provided polynomial-time algorithms for MCS on some simple graph classes including paths, spiders, caterpillars, combs, and trees (for trees, $c = 2$). XP, NP-complete, and FPT (when c is a parameter) results on trees, can be found in [11,12]. The MCS problem is also NP-complete on interval graphs [12] and APX-hard on circle graphs [13]. Variants, such as the *Minimum Consistent Spanning Subset* (MCSS) and the *Minimum Strict Consistent Subset* (MSCS) of MCS, have been studied on trees [15–17]. However, the algorithmic results for MSS have not been extensively studied to date. Banerjee et al. [5] only showed that MSS is NP-complete on general graphs. Very recently, the MSS problem has been studied in various settings, including $\mathcal{O}(\log n)$-approximation algorithms for general graphs, NP-complete results for planar graphs, and linear-time algorithms for trees and unit interval graphs [18] (published in CCCG 2025).

Our Contributions. The MSS problem admits an $\mathcal{O}(\log n)$-approximation on general graphs, which raises the question of whether better approximations exist. We show in Sect 3 that MSS is log-APX-hard even when $c = 2$. Hence, the problem is also APX-hard and does not admit a PTAS on general graphs. This leads to the natural question of whether some graph classes allow a PTAS. To date, none are known. We answer this by proving in Sect 5 that MSS admits a PTAS on unit disk graphs for arbitrary c, without requiring a geometric representation. Unit disk graphs are also fundamental in wireless networks, robotics, and computational geometry, where efficient approximation algorithms are highly relevant [20]. Before presenting our **PTAS**, we establish in Sect 4 that MSS remains NP-complete on unit disk graphs when c is arbitrary. We also investigate whether MSS is APX-hard in other graph classes. In Sect 6, we prove that MSS is APX-hard on circle graphs even when $c = 2$. Circle graphs, which model intersecting chords, have applications in VLSI design, scheduling, and bioinformatics [21,22]. All proofs of results marked with (*) can be found in [1].

2   Preliminaries

Let $G = (V(G), E(G))$ be a graph, where $V(G)$ is the vertex set and $E(G)$ is the edge set. For any $U \subseteq V(G)$, $G[U]$ denotes the subgraph of G induced on U, and $|U|$ is the cardinality of U. We denote $[n]$ as the set of integers $\{1, \ldots, n\}$. We use an arbitrary vertex color function $C : V(G) \rightarrow [c]$, which assigns each vertex exactly one color from the set $[c]$. For a subset of vertices $U \subseteq V(G)$, let $C(U)$ represent the set of colors of the vertices in U, formally defined as $C(U) = \{C(u) \mid u \in U\}$. The shortest path distance (i.e., *hop-distance*) between two vertices u and v in G is denoted by $\mathrm{d}(u, v)$. Distance between $v \in V(G)$ and the set $U \subseteq V(G)$ is given by $\mathrm{d}(v, U) = \min_{u \in U} \mathrm{d}(v, u)$. Similarly, the distance between two subgraphs G_1 and G_2 in G is defined as $\mathrm{d}(G_1, G_2) = \min\{\mathrm{d}(v_1, v_2) \mid v_1 \in V(G_1), v_2 \in V(G_2)\}$. The set of nearest neighbors of v in the set U is denoted as $\hat{\mathrm{N}}(v, U)$, formally defined as $\hat{\mathrm{N}}(v, U) = \{u \in U \mid \mathrm{d}(v, u) = \mathrm{d}(v, U)\}$. Therefore, if $v \in U$, then $\hat{\mathrm{N}}(v, U) = \{v\}$. The set of vertices in U adjacent to v is given by $\mathrm{N}(v, U) = \{u \in U \mid (u, v) \in E(G)\}$. We also define $\mathrm{N}[v, U] = \{v\} \cup \mathrm{N}(v, U)$. For any two subsets $U_1, U_2 \subseteq V(G)$, we define $\mathrm{N}(U_1, U_2) = \bigcup_{v \in U_1} \mathrm{N}(v, U_2)$, and $\mathrm{N}[U_1, U_2] = \bigcup_{v \in U_1} \mathrm{N}[v, U_2]$. Most symbols and notations follow standard conventions from [23]. Suppose $G = (V(G), E(G))$ is a given simple connected undirected graph where $\bigcup_{i=1}^{c} V_i = V(G)$ and $V_i \cap V_j = \emptyset$ for $i \neq j$ and each vertex in V_i is assigned color i. A *Minimum Consistent Subset* (MCS) is a subset $S \subseteq V(G)$ of minimum cardinality such that for every vertex $v \in V(G)$, if $v \in V_i$, then $\hat{\mathrm{N}}(v, S) \cap V_i$ is non-empty.

Definition 1 (Selective Subset). *A subset $S \subseteq V(G)$ is called a* Selective Subset (MSS) *if, for each vertex $v \in V(G)$, if $v \in V_i$, the set of nearest neighbors of v in $S \cup (V(G) \setminus V_i)$, denoted as $\hat{\mathrm{N}}(v, S \cup (V(G) \setminus V_i))$, contains at least one vertex u such that $C(v) = C(u)$. An MSS is a selective subset of minimum cardinality. The decision version of the MSS problem is as follows:*

—— DECISION VERSION OF SELECTIVE SUBSET PROBLEM ON GRAPHS ——

__Input:__ A graph $G = (V(G), E(G))$, a coloring function $C : V(G) \rightarrow [c]$, and an integer s.
__Question:__ Does there exist a selective subset of size at most s for (G, C)?

In other words, we seek a vertex set $S \subseteq V(G)$ of minimum cardinality such that every vertex v has at least one nearest neighbor of the same color in the graph, excluding vertices of the same color as v that are not in S. If all the *vertices of a graph G are of the same color (i.e., G is monochromatic), then any vertex in the graph forms a valid MSS. Figure 1 illustrates an example of MSS.*

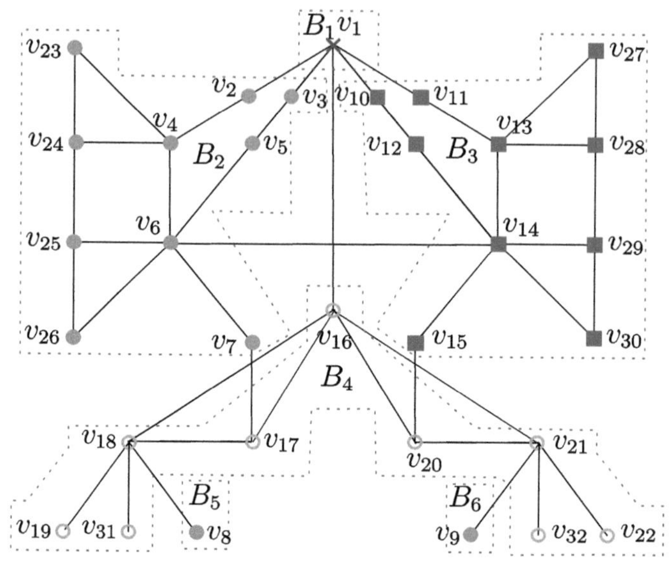

Fig. 1. Colors: *blue $\equiv$ cross, green $\equiv$ disk, red $\equiv$ square,* and *orange $\equiv$ circle*. $V(G) = V_{blue} \cup V_{green} \cup V_{red} \cup V_{orange}$, where $V_{blue} = \{v_1\}$, $V_{green} = \{v_2, \ldots, v_9, v_{23}, \ldots, v_{26}\}$, $V_{red} = \{v_{10}, \ldots, v_{15}, v_{27}, \ldots, v_{30}\}$, and $V_{orange} = \{v_{16}, \ldots, v_{22}, v_{31}, v_{32}\}$. $S = \{v_1, v_2, v_3, v_7, v_8, v_9, v_{10}, v_{11}, v_{15}, v_{16}\}$ is an MSS, and $S = \{v_1, v_4, v_5, v_7, v_8, v_9, v_{12}, v_{13}, v_{15}, v_{16}\}$ is also an MSS. Brown-dotted regions indicate the blocks. The complete list of blocks is $B_1 = \{v_1\}$, $B_2 = \{v_2, \ldots, v_7, v_{23}, \ldots, v_{26}\}$, $B_3 = \{v_{10}, \ldots, v_{15}, v_{27}, \ldots, v_{30}\}$, $B_4 = \{v_{16}, \ldots, v_{22}, v_{31}, v_{32}\}$, $B_5 = \{v_8\}$, $B_6 = \{v_9\}$. $B_{2,1} = \{v_2, v_3, v_6, v_7\}$, $B_{2,2} = \{v_4, v_5, v_{25}, v_{26}\}$. $\{\{v_2\}, \{v_3\}, \{v_7\}\}$ is a collection of 2-distance sets in $B_{2,1}$.(Color figure online)

Definition 2 (Block). *A block is a maximal connected subgraph whose vertices all have the same color (i.e., a maximal connected monochromatic subgraph).*

Figure 1 illustrates an example of the blocks. Suppose $B_1, \ldots, B_k$ is the complete list of blocks in G. We assume that $|V(G)| = n$, so that $k \leq n$. We form the sets B_i^1, B_i^2, $B_{i,3}$ for each $i = 1, \ldots, k$ as follows (see Figure 1):

- Initially, $B_{i,1} := \emptyset$, $B_{i,2} := \emptyset$.
- For each vertex $v \in B_i$, if there exists a vertex $u \in \mathrm{N}(v, V(G))$ such that $C(u) \neq C(v)$, then $v \in B_{i,1}$.
- For any vertex $v \in B_i \setminus B_{i,1}$ if $\mathrm{d}(v, B_{i,1}) = 1$, then $v \in B_{i,2}$.
- We denote $B_{i,3} = B_{i,1} \cup B_{i,2}$.

Lemma 1. * *For any vertex $v \in B_{i,1}$ and a selective subset S, we have $\mathrm{N}[v, B_{i,3}] \cap S \neq \emptyset$ for $1 \leq i \leq k$.*

3   log-APX-hardness of MSS on General Graphs

We establish a reduction from the MINIMUM DOMINATING SET (MDS) problem to the MSS problem. In the MINIMUM DOMINATING SET problem, the input is a graph G together with an integer s. The task is to decide whether there exists a subset $D \subseteq V(G)$ of size at most s such that every vertex $u \in V(G)$ satisfies $N[u, V(G)] \cap D \neq \emptyset$. It is well known that the MINIMUM SET COVER problem is log-APX-hard (see [24] for complexity class definitions), and moreover, it is NP-hard to approximate it within a factor of $\delta \cdot \log n$ for some positive constant δ [25]. Since there exists an L-reduction from the MINIMUM SET COVER problem to the MINIMUM DOMINATING SET problem, the MINIMUM DOMINATING SET problem is also log-APX-hard.

Let (G, s) be an arbitrary instance of the MINIMUM DOMINATING SET problem. We construct an instance $(G', C, s + 1)$ for the MSS problem as follows (see Figure 2(a)). Define the new graph G' with $V(G') = V(G) \cup \{z\}$ and $E(G') = E(G) \cup \{(z, u) \mid u \in V(G)\}$. The color function C assigns color 1 to all vertices $u \in V(G)$, and color 2 to the additional vertex z.

Lemma 2. * *G has a dominating set of size at most s if and only if G' has a selective subset of size at most $s + 1$.*

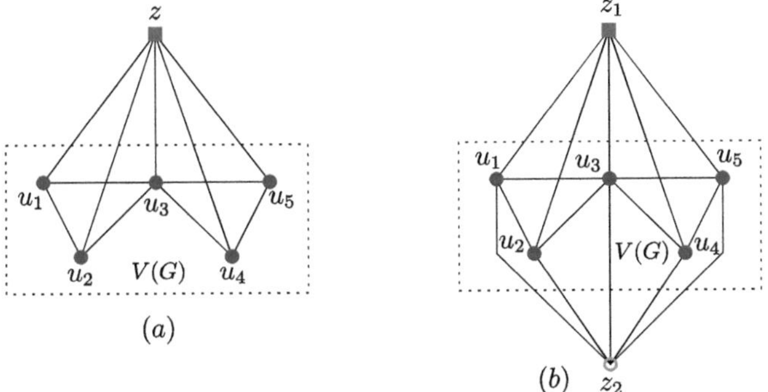

Fig. 2. Colors: *blue $\equiv$ disk*, *red $\equiv$ square*, and *green $\equiv$ circle*. (a) Reduction from an instance of MINIMUM DOMINATING SET problem to an instance of MSS problem when $c = 2$. (b) Example of the reduction when $c = 3$.(Color figure online)

Theorem 1. * *There exists a constant $\delta > 0$ such that it is NP-hard to approximate the MSS problem within a factor of $\delta \cdot \log n$, where n is the number of vertices in the graph.*

Remark 1. Note that the above reduction remains valid even if we add any number of new vertices (each adjacent to all of $V(G)$) and assign each a distinct color. The correctness of Lemma 2 and the resulting hardness theorem continue to hold under this extended construction (see Figure 2(b)).

4   NP-completeness of MSS on Unit Disk Graphs

A graph $U = (V(U), E(U))$ is called a *unit disk graph* (UDG) if its vertices can be represented as points in the Euclidean plane such that an edge exists between two vertices if and only if their Euclidean distance is at most 2.

Formally, U is a UDG if there exists a mapping $f : V \to \mathbb{R}^2$ such that:

$$(u, v) \in E(U) \iff \|f(u) - f(v)\| \leq 2 \tag{1}$$

where $\| \cdot \|$ denotes the Euclidean norm.

Clark et al. [26] showed that MINIMUM DOMINATING SET (MDS) is NP-complete in UDG. We reduce from an instance U of the UDG with $|V(U)| = n$ to an instance U' as follows:

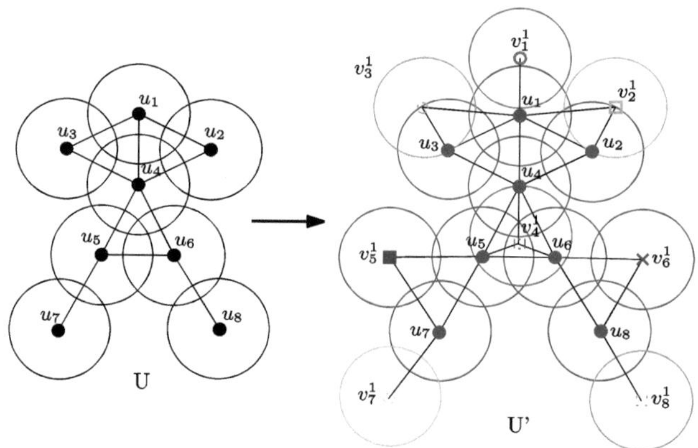

Fig. 3. Colors: *blue $\equiv$ disk, red $\equiv$ circle, orange $\equiv$ fsquare, green $\equiv$ dash dotted circle, brown $\equiv$ dash dotted fsquare, darkred $\equiv$ square, darkgreen $\equiv$ cross, lightgreen $\equiv$ dotted circle,* and *violet $\equiv$ dotted square.* An example of the reduction when $m = 1$. Each *blue* disk in U' is adjacent to a disk of a distinct color, different from all other colors in U'.(Color figure online)

Reduction. Define $V(U') = V(U) \cup V(X)$ and $E(U') = E(U) \cup E(X)$, where we introduce a subgraph X with vertex set $V(X)$ and edge set $E(X)$ as follows (see Figure 3). Initially, set $V(X) := \emptyset$ and $E(X) := \emptyset$. Assign color 0 to all unit disks in $V(U)$, i.e., $C(V(U)) = \{0\}$. For each $u_i \in V(U)$, introduce a total of m (where $m \in \mathbb{N}$) unit disks $v_i^1, \ldots, v_i^m \in V(X)$. Additionally, place $v_i^1, \ldots, v_i^m$ in such a way that u_i and v_i^l are adjacent for $1 \leq l \leq m$ (where v_i^l may also be adjacent to other unit disks). These new edges are added to $E(X)$. All unit disks in $V(X)$ have distinct colors, none of which is 0. Thus, we obtain $|V(X)| = mn$, $|V(U')| = nm + n$, and $|C(V(U'))| = nm + 1$.

Lemma 3. * U *has a dominating set of size* t *if and only if* U' *has a selective subset of size* $nm + t$.

Theorem 2. * MSS *is NP-complete on unit disk graphs.*

5   PTAS of MSS on Unit Disk Graphs

A unit disk graph may admit multiple geometric representations. In this work, we assume that the geometric representation f is either unknown or not explicitly given. We first describe our approach for a general graph before restricting to unit disk graphs. If G is not a connected graph, we apply the algorithm independently on each component; hence, we may assume that G is connected.

The key idea is that blocks are independent within any selective subset solution. Exploiting this property, we compute *local* selective subsets within each block independently. For the set $B_{i,1}$ corresponding to a given block B_i, we define a family of sets $D_{i,1}^1, \ldots, D_{i,1}^{t_i}$ such that $\mathrm{d}(D_{i,1}^j, D_{i,1}^\ell) > 2$ for all $j \neq \ell$. By Lemmas 1 and 4, this ensures that no two sets $D_{i,1}^j$ and $D_{i,1}^\ell$ share a common vertex in optimal solution. Consequently, the solutions for $D_{i,1}^1, \ldots, D_{i,1}^{t_i}$ provide a lower bound on the size of the optimal solution. However, the union of these solutions does not necessarily form a valid solution for the block B_i. To address this, we enlarge each set $D_{i,1}^j$ into a corresponding set $E_{i,1}^j$, ensuring that the union of the solutions for $E_{i,1}^1, \ldots, E_{i,1}^{t_i}$ yields a valid solution for B_i. We refer to the solutions for $D_{i,1}^j$ and $E_{i,1}^j$ as *local solutions*. By combining these local solutions, we obtain a blockwise selective subset, and taking the union over all blocks yields a global solution used in our PTAS.

For any subset $U \subseteq V(G)$, we denote its minimum selective subset (*local*) by $S^{\min}(U)$ and a selective subset (*local*) of U by $S(U)$. The global optimum is denoted by $S^{\min}$.

Lemma 4. * For any minimum selective subset S^{min} of G, we have*

- *No vertex $v \in B_i \setminus B_{i,3}$ belongs to S^{min}.*
- *$S^{min} \subseteq \bigcup_{i=1}^k B_{i,3}$.*

Lemma 1 ensures that for each vertex $v \in B_{i,1}$, either $v \in S^{min}$ or at least one of its adjacent vertices in $B_{i,3}$ belongs to S^{min}. Importantly, the choice of including v itself or one of its adjacent vertices from $B_{i,3}$ in S^{min} does not affect the selection of vertices in other blocks. Combined with Lemma 4, which establishes that $S^{\min} \subseteq \bigcup_{i=1}^k B_{i,3}$, we obtain the following remark:

Remark 2. The blocks are independent in constructing a selective subset; that is, the selection of vertices in one block does not constrain the selection in other blocks.

We now apply an appropriate algorithm to each block separately due to Remark 2. To do this, we first define a selective subset for each block as follows.

Definition 3 (elective Subset of $B_{i,1}$). *A selective subset of $B_{i,1}$, denoted by $S(B_{i,1})$, is a subset of $B_{i,3}$ such that for every vertex $v \in B_{i,1}$, either $v \in S(B_{i,1})$ or $\mathrm{N}(v, B_{i,3}) \cap S(B_{i,1}) \neq \emptyset$. By Lemmas 1 and 4, it follows that $S(B_{i,1}) \subseteq \mathrm{N}[B_{i,1}, B_{i,3}] \subseteq B_{i,3}$.*

Theorem 3. * *Let G be a connected graph with blocks $B_1, \ldots, B_k$, and for each block B_i let $S(B_{i,1})$ be any selective subset of $B_{i,1}$. Let $S = \bigcup_{i=1}^{k} S(B_{i,1})$, then S is a selective subset of G. Moreover, if each $S(B_{i,1})$ satisfies $|S(B_{i,1})| \leq (1 + \epsilon)|S^{\min}(B_{i,1})|$, then $|S| \leq (1+\epsilon)|S^{\min}|$, where $S^{\min}$ denotes a minimum selective subset of G.*

5.1   Finding Local Selective Subsets

We now establish a bound on a local solution using *2-distance subsets* for our problem and then merge all local solutions to obtain the desired solution.

Definition 4 (2-distance Subsets). *A collection of subsets of the vertices in $B_{i,1}$, denoted as $D_i = \{D_{i,1}^1, \ldots, D_{i,1}^{t_i}\}$, is called a collection of 2-distance subsets if the following properties hold (see example in Figure 1):*

- *$D_{i,1}^j \subseteq B_{i,1}$ for all $1 \leq j \leq t_i$.*
- *The subgraph $G[D_{i,1}^j]$ is connected in the induced subgraph $G[B_{i,3}]$, i.e., the induced subgraph $G[D_{i,1}^j]$ may not be connected itself, but any two vertices in $D_{i,1}^j$ must have a path between them in $G[B_{i,3}]$.*
- *The subsets are pairwise at a distance greater than two in $G[B_{i,3}]$, i.e., $\mathrm{d}(D_{i,1}^j, D_{i,1}^l) > 2$ in the subgraph $G[B_{i,3}]$ when $j \neq l$.*

Definition 5 (Local Selective Subset). *A local selective subset of $D_{i,1}^j$, denoted by $S(D_{i,1}^j)$, is a subset of $B_{i,3}$ such that for every vertex $v \in D_{i,1}^j$, either $v \in S(D_{i,1}^j)$ or $\mathrm{N}(v, B_{i,3}) \cap S(D_{i,1}^j) \neq \emptyset$. By Lemmas 1 and 4, it follows that $S(D_{i,1}^j) \subseteq \mathrm{N}[D_{i,1}^j, B_{i,3}] \subseteq B_{i,3}$.*

Lemma 5. * *For any $j \neq l$, the following holds:*

- *$\mathrm{N}[D_{i,1}^j, B_{i,3}] \cap \mathrm{N}[D_{i,1}^l, B_{i,3}] = \emptyset$.*
- *$S^{min}(D_{i,1}^j) \cap S^{min}(D_{i,1}^l) = \emptyset$.*
- *$\left(S^{min} \cap S^{min}(D_{i,1}^j)\right) \cap \left(S^{min} \cap S^{min}(D_{i,1}^l)\right) = \emptyset$.*

Lemma 5 implies that the solutions $S^{min}(D_{i,1}^j)$ and $S^{min}(D_{i,1}^l)$ do not share a common vertex in S^{min} for $j \neq l$.

Lemma 6. * *$S^{min} \cap \mathrm{N}[D_{i,1}^j, B_{i,3}]$ is a local selective subset of $D_{i,1}^j$.*

Lemma 7. * *For any collection of 2-distance subsets $D_i = \{D_{i,1}^1, \ldots, D_{i,1}^{t_i}\}$, where $1 \leq i \leq k$ in the graph G; we have: $\sum_{i=1}^{k} \sum_{j=1}^{t_i} |S^{min}(D_{i,1}^j)| \leq |S^{min}|$.*

Lemma 7 shows that 2-distance subsets yield a lower bound on the size of a minimum selective subset. However, the set $\sum_{i=1}^{k} \sum_{j=1}^{t_i} S^{min}(D_{i,1}^j)$ need not form a selective subset of the entire graph G. To construct a selective subset for G, we enlarge each $D_{i,1}^j$ to a corresponding set $E_{i,1}^j$ that remains locally bounded while still providing a valid local solution. This enlargement allows us to approximate a selective subset of G.

Theorem 4. * *Let $D_i = \{D_{i,1}^1, \ldots, D_{i,1}^{t_i}\}$ be a collection of 2-distance subsets, and $\{E_{i,1}^1, \ldots, E_{i,1}^{t_i}\}$ be the corresponding collection of subsets of $B_{i,1}$ such that $D_{i,1}^j \subseteq E_{i,1}^j$ for all $1 \leq i \leq k$ and $1 \leq j \leq t_i$.*

If there exists a bound $\delta \geq 1$ such that $|S^{min}(E_{i,1}^j)| \leq \delta \cdot |S^{min}(D_{i,1}^j)|$ for all $1 \leq i \leq k$ and $1 \leq j \leq t_i$, and if $\bigcup_{i=1}^{k} \bigcup_{j=1}^{t_i} S^{min}(E_{i,1}^j)$ forms a selective subset of G, then $\bigcup_{i=1}^{k} \bigcup_{j=1}^{t_i} S^{min}(E_{i,1}^j)$ is a δ-approximation of a minimum selective subset of G.

5.2 Finding a Global Selective Subset

For $r = 0, 1, \ldots$, we recursively define the r-th neighborhood of any vertex $v \in B_{i,1}$ in $B_{i,3}$ by

$$N_i^r[v, B_{i,3}] = N[N_i^{r-1}[v, B_{i,3}], B_{i,3}],$$

with

$$N_i^0[v, B_{i,3}] = \{v\}, \quad N_i^1[v, B_{i,3}] = N[v, B_{i,3}].$$

Since $N_i^r[v, B_{i,3}] \subseteq B_{i,3}$, we partition $N_i^r[v, B_{i,3}]$ into $X_i^r \subseteq B_{i,1}$ and $Y_i^r \subseteq B_{i,2}$. We will later use X_i^r and Y_i^r in our algorithm.

As $\delta \geq 1$, we assume that $\delta := (1 + \epsilon)$. The key idea is to determine the neighborhood of a vertex in $B_{i,3}$ and then progressively expand this neighborhood until we obtain sets $D_{i,1}^j$ and $E_{i,1}^j$ (where $E_{i,1}^j \supseteq D_{i,1}^j$) that satisfy Theorem 4. Once this is achieved, we remove the current neighborhood and repeat the process for the remaining graph. Note that $E_{i,1}^j$ is not a 2-distance subset, but $D_{i,1}^j$ is. The complete procedure is described below (see Algorithm 1 in [1]):

- Initially, set $i \leftarrow 1$.
- **Stage 1:** Initialize $j \leftarrow 1$, and set $B_{i,1}^j \leftarrow B_{i,1}$, $B_{i,2}^j \leftarrow B_{i,2}$, and $B_{i,3}^j \leftarrow B_{i,3}$.
- **Stage 2:** Choose an arbitrary vertex v_i^j from $B_{i,1}^j$.
- For $r = 0, 1, \ldots$, consider the r-th neighborhood $N_{i,j}^r[v_i^j, B_{i,3}^j]$. Starting with $N_{i,j}^0[v_i^j, B_{i,3}^j]$ and compute the minimum selective subset while inequality (2) holds.

$$|S^{min}(X_{i,j}^{r+2})| > \delta \cdot |S^{min}(X_{i,j}^r)| \tag{2}$$

  Here $N_{i,j}^r$ (rather than N_i^r) denotes the r-th neighborhood used to compute $D_{i,1}^j$ and $E_{i,1}^j$ from $B_{i,3}$. The same convention applies to $X_{i,j}^r$ and $Y_{i,j}^r$.
- Let $\overline{r_{i,j}}$ be the smallest r for which inequality (2) is violated, i.e.,

$$|S^{min}(X_{i,j}^{\overline{r_{i,j}}+2})| \leq \delta \cdot |S^{min}(X_{i,j}^{\overline{r_{i,j}}})|.$$

– Update the sets as follows:

$$D_{i,1}^{j} \leftarrow X_{i,j}^{\overline{r_{i,j}}},$$
$$E_{i,1}^{j} \leftarrow X_{i,j}^{\overline{r_{i,j}}+2},$$
$$B_{i,3}^{j+1} \leftarrow B_{i,3}^{j} \setminus N_{i,j}^{\overline{r_{i,j}}+2}[v_i^j, B_{i,3}^j],$$
$$B_{i,1}^{j+1} \leftarrow B_{i,1}^{j} \setminus X_{i,j}^{\overline{r_{i,j}}+2},$$
$$B_{i,2}^{j+1} \leftarrow B_{i,2}^{j} \setminus Y_{i,j}^{\overline{r_{i,j}}+2},$$
$$j \leftarrow j + 1.$$

– Repeat the process from **Stage 2** until $B_{i,3}^{j} = \emptyset$.
– Once $B_{i,3}^{j}$ becomes empty, set $i \leftarrow i + 1$ and repeat the process from **Stage 1** until $i = k + 1$.

Suppose the sets $D_{i,1}^{1}, D_{i,1}^{2}, \ldots, D_{i,1}^{t_i}$ and $E_{i,1}^{1}, E_{i,1}^{2}, \ldots, E_{i,1}^{t_i}$ are returned from the above algorithm for $1 \leq i \leq k$. We establish the following lemmas.

Lemma 8. * *The sets* $\{D_{i,1}^{1}, D_{i,1}^{2}, \ldots, D_{i,1}^{t_i}\}$, *where* $1 \leq i \leq k$, *obtained from the above algorithm, form a collection of 2-distance subsets.*

Lemma 9. * *For the collection of sets* $\{E_{i,1}^{1}, \ldots, E_{i,1}^{t_i}\}$ *obtained from the above algorithm, the union* $S = \bigcup_{i=1}^{k} \bigcup_{j=1}^{t_i} S^{min}(E_{i,1}^{j})$ *forms a selective subset of* G.

Combining Lemmas 8 and 9 with Theorem 4, we obtain the following theorem directly:

Theorem 5. *The above algorithm produces a selective subset* $\bigcup_{i=1}^{k} \bigcup_{j=1}^{t_i} S^{min}(E_{i,1}^{j})$ *of size at most* $(1 + \epsilon)$ *times the size of the minimum selective subset of* G.

Until now, we have considered general graphs rather than unit disk graphs. Thus, Theorem 5 holds for general graphs.

5.3   Finding $S^{min}(E_{i,1}^{j})$ on Unit Disk Graphs

The only remaining task is to compute $S^{min}(E_{i,1}^{j})$ in time $n^{f(\epsilon)}$ on unit disk graphs. We assume that $F_{i,j} = N_{i,j}^{\overline{r_{i,j}}+2}[v_i^j, B_{i,3}^j]$. According to the above algorithm,

$$F_{i,j} = X_{i,j}^{\overline{r_{i,j}}+2} \cup Y_{i,j}^{\overline{r_{i,j}}+2} = E_{i,1}^{j} \cup Y_{i,j}^{\overline{r_{i,j}}+2}.$$

We first show that the size of $S^{min}(E_{i,1}^{j})$ is at most the size of the MAXIMUM INDEPENDENT SET of $F_{i,j}$. This provides a bound on $E_{i,1}^{j}$. The MAXIMUM INDEPENDENT SET of a graph G is the largest set of vertices such that no two vertices in the set are adjacent. One might think that $S^{min}(E_{i,1}^{j})$ is the same as a MINIMUM DOMINATING SET of $F_{i,j}$. However, by Theorem 3, some vertices in $B_{i,3} \setminus B_{i,1}$ may have no adjacent vertex (including themselves) in $S(B_{i,1})$. Therefore, with this bound, it suffices to compute each $S^{min}(E_{i,1}^{j})$.

Lemma 10. * *The size of $S^{min}(E_{i,1}^j)$ is at most the size of the maximum independent set of $F_{i,j}$.*

Now, we apply the method described in [27] to find a MAXIMUM INDEPENDENT SET in $F_{i,j}$ for unit disk graphs.

Lemma 11. * *For any unit disk graph U and an independent set $I^r \subseteq N_{i,j}^r[v_i^j, B_{i,3}^j]$, we have $|I^r| = (2r+1)^2 = \mathcal{O}(r^2)$.*

Using Lemmas 10 and 11, we derive the following theorem directly:

Theorem 6. *The size of the minimum selective subset (local) of $E_{i,1}^j$ satisfies*

$$|S^{\min}(E_{i,1}^j)| = \mathcal{O}(r^2).$$

Lemma 12. * *There exists a constant $d(\delta)$, depending on $\delta = (1+\epsilon)$, such that $\overline{r_{i,j}} \leq d(\delta)$. The running time of our algorithm to compute a PTAS is $\mathcal{O}(n^{d^2})$, where $|V(U)| = n$, $d(\epsilon) = \mathcal{O}\left(\frac{1}{\epsilon^2} \log \frac{1}{\epsilon}\right)$, and $0 < \epsilon < \frac{1}{10}$.*

6   APX-Hardness of MSS on Circle Graphs

The vertex set of a circle graph is a set of chords of a given circle, and if two chords intersect, the corresponding vertices share an edge. We obtain a "gap-preserving" reduction from the MAX-3SAT(8) problem to a circle graph using the MINIMUM DOMINATING SET (MDS) problem on circle graphs. The MAX-3SAT(8) problem is as follows [24]:

We are given a set of n variables $\mathcal{X} = \{x_1, \ldots, x_n\}$ and m clauses $\mathcal{C} = \{c_1, \ldots, c_m\}$ such that each clause has at most 3 literals and each variable occurs in at most 8 clauses. The objective is to find a truth-assignment of the variables in $\mathcal{X}$ that maximizes the number of clauses in $\mathcal{C}$ satisfied. MAX-3SAT(8) is APX-hard, and the MDS problem on circle graphs is also APX-hard [28].

Consider an instance ϕ of MAX-3SAT(8). Let $SAT(\phi)$ represent the maximum fraction of clauses in ϕ that can be satisfied. For a given graph G, let $\gamma(G)$ denote the cardinality of its MINIMUM DOMINATING SET. The paper [28] reduces an instance of MAX-3SAT(8) to a circle graph G such that $|V(G)| = m+56n+4$ and states the following theorem:

Theorem 7. *A polynomial-time reduction transforms an instance ϕ of MAX-3SAT(8), consisting of n variables and m clauses, into a circle graph G such that*

$$SAT(\phi) = 1 \implies \gamma(G) \leq 16n + 2,$$

$$SAT(\phi) < \alpha \implies \gamma(G) > 16n + 2 + \frac{(1-\alpha)m}{8}, \quad \text{for any } 0 < \alpha < 1.$$

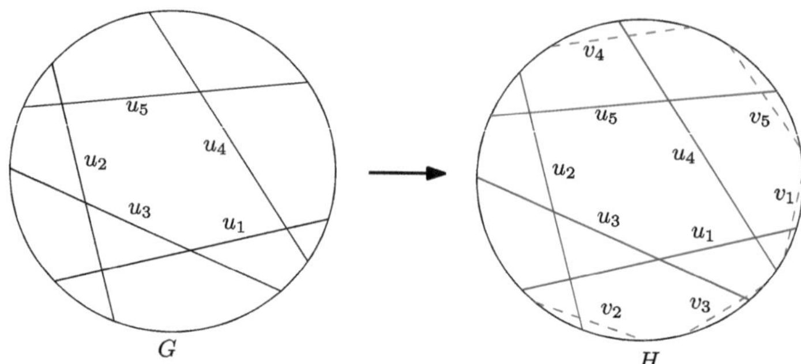

Fig. 4. Colors: *blue* ≡ *normal chord*, and *red* ≡ *dotted chord*. An example of the reduction: each chord of G is colored *blue* in H. Each *red* chord in H is adjacent only to a chord of *blue* color.(Color figure online)

We now reduce the graph G into a graph H as follows:

Reduction. Define $V(H) = V(G) \cup V(X)$ and $E(H) = E(G) \cup E(X)$, where X is a subgraph with vertex set $V(X)$ and edge set $E(X)$ are constructed as follows (see Figure 4). Initially, set $V(X) := \emptyset$ and $E(X) := \emptyset$. Assign color 0 to all chords in $V(G)$, i.e., $C(V(G)) = \{0\}$. For each chord $u_i \in V(G)$, introduce a corresponding chord $v_i \in V(X)$ such that $C(v_i) = 1$. Position v_i so that it intersects only u_i (meaning v_i has degree 1 in H). These newly created edges are then included in $E(X)$. As a result, we obtain $|V(H)| = 2m + 112n + 8$, $|V(X)| = m + 56n + 4$ and $|C(V(H))| = 2$.

Lemma 13. * *G has a dominating set of size t if and only if H has a selective subset of size $m + 56n + 4 + t$.*

Let $|S^{min}(H)|$ denote the cardinality of the minimum selective subset of H. We now show the "gap-preserving" reduction from MAX-3SAT(8) to the graph H in the following theorem:

Theorem 8. * *The MSS problem is APX-hard in circle graphs.*

Remark 3. Our APX-hard result holds not only for circle graphs with two colors; indeed, any color (other than blue) can replace the red chords adjacent to the blue ones.

7   Conclusion

It is still open regarding whether MSS problem is NP-complete on unit disk graphs when c is constant. This raises the open question of whether FPT is possible when the number of colors c is treated as a parameter. Another direction to explore is whether FPT is possible when the number of blocks k is treated as a parameter. Additionally, since MSS is APX-hard on circle graphs, it is natural to ask whether a constant-factor approximation, or a $(2 + \epsilon)$-approximation is possible for circle graphs, in contrast to a $(1 + \epsilon)$-approximation. FPT for MSS problem with respect to parameters c or k also remains an open question in the context of circle graphs.

Moreover, exploring the complexity of MSS on additional graph classes—such as circular-arc graphs, chordal graphs, and permutation graphs—could uncover new tractable cases or reveal deeper structural insights. These graph families often arise in scheduling, bioinformatics, and network analysis, and understanding MSS within these domains may have practical implications as well.

Acknowledgments. I sincerely thank Bodhayan Roy and Aritra Banik for their support and guidance. Also, the motivation for this work arose from discussions on the MCS paper with Aritra Banik, facilitated by the ACM India Anveshan Setu Fellowship program in 2024. I also thank the anonymous reviewers for their thoughtful comments and suggestions, which helped enhance the quality and presentation of the paper.

References

1. Manna, B.: Minimum selective subset on unit disk graphs and circle graphs. arXiv preprint arXiv:2510.01931 (2025)
2. Hart, P.: The condensed nearest neighbor rule (corresp.). IEEE Trans. Inf. Theory **14**(3), 515–516 (1968)
3. Wilfong, G.: Nearest neighbor problems. In: Proceedings of the Seventh Annual Symposium on Computational Geometry, SCG '91, pages 224—-233, (1991)
4. Khodamoradi, K., Krishnamurti, R., Roy, B.: Consistent subset problem with two labels. In: Panda, B.S., Goswami, P.P., eds., Algorithms and Discrete Applied Mathematics, pages 131–142 (2018)
5. Banerjee, S., Bhore, S., Chitnis, R.: Algorithms and hardness results for nearest neighbor problems in bicolored point sets. In: Bender, M.A., Farach-Colton, M., Mosteiro, M.A. (eds.) LATIN 2018. LNCS, vol. 10807, pp. 80–93. Springer, Cham (2018). https://doi.org/10.1007/978-3-319-77404-6_7
6. Biniaz, A., Cabello, S., Carmi, P., De Carufel, J.-L., Maheshwari, A., Mehrabi, S., Smid, M.: On the minimum consistent subset problem. Algorithmica **83**(7), 2273–2302 (2021)
7. Chitnis, R.: Refined lower bounds for nearest neighbor condensation. In: Dasgupta, S., Haghtalab, N., eds., International Conference on Algorithmic Learning Theory, 29 March - 1 April 2022, Paris, France, volume 167 of Proceedings of Machine Learning Research, pages 262–281. PMLR (2022)
8. Cygan, M., et al.: Parameterized Algorithms, 1st edn. Lecture Notes in Computer Science, Springer, Cham (2015). https://doi.org/10.1007/978-3-319-21275-3_15
9. Dey, S., Maheshwari, A., Nandy, S.C.: Minimum consistent subset problem for trees. In: Bampis, E., Pagourtzis, A. (eds.) FCT 2021. LNCS, vol. 12867, pp. 204–216. Springer, Cham (2021). https://doi.org/10.1007/978-3-030-86593-1_14
10. Dey, S., Maheshwari, A., Nandy, S.C.: Minimum consistent subset of simple graph classes. Discret. Appl. Math. **338**, 255–277 (2023)
11. Arimura, H., Gima, T., Kobayashi, Y., Nochide, H., Otachi, Y.: Minimum consistent subset for trees revisited. CoRR, abs/2305.07259 (2023)

12. Banik, A.,et al.: Minimum consistent subset in trees and interval graphs. In: Barman, S., Lasota, S., eds., 44th IARCS Annual Conference on Foundations of Software Technology and Theoretical Computer Science (FSTTCS 2024), volume 323 of Leibniz International Proceedings in Informatics (LIPIcs), pages 7:1–7:15, Dagstuhl, Germany, 2024. Schloss Dagstuhl – Leibniz-Zentrum für Informatik
13. Manna, B., Roy, B.: Some results on minimum consistent subsets of trees. arXiv preprint arXiv:2303.02337 (2023)
14. Manna, B.: Minimum consistent subset in interval graphs and circle graphs. arXiv preprint arXiv:2405.14493 (2024)
15. Banik, A., et al. Minimum consistent subset in trees and interval graphs. arXiv preprint arXiv:2404.15487 (2024)
16. Biniaz, A., Khamsepour, P.: The minimum consistent spanning subset problem on trees. J. Graph Algorithms Appl. **28**(1), 81–93 (2024)
17. Manna, B.: Minimum strict consistent subset in paths, spiders, combs and trees. arXiv preprint arXiv:2405.18569 (2024)
18. Manna, B.: Minimum selective subset on some graph classes. arXiv preprint arXiv:2507.00235 (2025)
19. Manna, B.: Minimum selective subset on some graph classes. In: Kamali, S. (ed.) Proceedings of the 37th Canadian Conference on Computational Geometry, CCCG 2025, York University, Toronto, Canada, 13–15 August 2025, pp. 299–305 (2025)
20. Goldin, D., Attia, S.A.: Unit disk graph based modelling of a network of mobile agents. IFAC Proc. Volumes, **42**(20), 234–239 (2009). 1st IFAC Workshop on Estimation and Control of Networked Systems
21. Ward, M., Gozzard, A., Wise, M., Datta, A.: A faster algorithm for maximum induced matchings on circle graphs. J. Graph Algorithms Appl. **22**, 389–396 (2018)
22. Sherwani, N.A.: Algorithms for VLSI Physical Design Automation. Springer, US (2012)
23. Diestel, R.: Graph theory, volume 173 of. Graduate texts in mathematics, page 7 (2012)
24. Vijay, V.: Vazirani. Approximation Algorithms. Springer Publishing Company, Incorporated (2010)
25. Raz, R., Safra, S.: A sub-constant error-probability low-degree test, and a sub-constant error-probability PCP characterization of np. In: Proceedings of the Twenty-Ninth Annual ACM Symposium on Theory of Computing, STOC '97, page 475–484, New York, NY, USA (1997). Association for Computing Machinery
26. Clark, B.N., Colbourn, C.J., Johnson, D.S.: Unit disk graphs. Discret. Math. **86**(1–3), 165–177 (1990)
27. Nieberg, T., Hurink, J., Kern, W.: A Robust PTAS for maximum weight independent sets in unit disk graphs. In: Hromkovič, J., Nagl, M., Westfechtel, B. (eds.) WG 2004. LNCS, vol. 3353, pp. 214–221. Springer, Heidelberg (2004). https://doi.org/10.1007/978-3-540-30559-0_18
28. Damian, M., Pemmaraju, S.V.: Apx-hardness of domination problems in circle graphs. Inf. Process. Lett. **97**(6), 231–237 (2006)

Boxicity of a Class of Split Graphs Using Words

Khyodeno Mozhui$^{(\boxtimes)}$ and Tithi Dwary

Department of Mathematics, Indian Institute of Technology Guwahati, Guwahati,
Assam, India
{k.mozhui,tithi.dwary}@iitg.ac.in

Abstract. A k-dimensional box is defined as a Cartesian product of k
closed intervals on the real line $\mathbb{R}$. The boxicity of a graph G, denoted by
$\mathrm{box}(G)$, is the smallest natural number k such that G can be represented
as the intersection graph of a set of k-dimensional boxes. A split graph
is a graph whose vertex set can be partitioned into a clique and an
independent set. It is known that the boxicity of a split graph is at most
$\min\{\frac{m}{2}, \frac{n}{2}\}$, where m and n are the sizes of a clique and an independent
set splitting the graph. Moreover, determining whether a split graph has
boxicity at most three is NP-complete.

In this work, we use words as a tool to determine the boxicity of a
class of split graphs. We focus on graphs that are characterized by words
over vertices, in which the adjacency between vertices is determined by
an alternating property of letters in the word. These graphs are referred
to as word-representable split graphs, a class of graphs that includes the
well-known split comparability graphs. We establish that the boxicity of
a word-representable split graph is at most four. In addition, we show
that the boxicity of a split comparability graph is at most three.

Keywords: Boxicity · Split graph · Word-representable graph ·
Comparability graph

1 Introduction

A k-dimensional box is a Cartesian product of k closed intervals on the real line
$\mathbb{R}$, i.e., $[a_1, b_1] \times \cdots \times [a_k, b_k]$, for $a_i, b_i \in \mathbb{R}$ $(1 \leq i \leq k)$. A k-box representation
of a graph $G = (V, E)$ is a function f that maps each vertex of G to a k-
dimensional box such that, for all $a, b \in V$, a and b are adjacent in G if and
only if the k-dimensional boxes $f(a)$ and $f(b)$ intersect. The boxicity of a graph
G, denoted by $\mathrm{box}(G)$, is the smallest integer k such that G admits a k-box
representation. The notion of boxicity was introduced by Roberts in [17], and
it has applications which include problems of niche overlap (competition) in
ecology and fleet maintenance problems in operations research (cf. [6]).

The class of complete graphs corresponds to the graphs with boxicity 0, while
the interval graphs are precisely those graphs with boxicity at most one. In [6],

© The Author(s), under exclusive license to Springer Nature Switzerland AG 2026
N. Misra and A. Pandey (Eds.): CALDAM 2026, LNCS 16445, pp. 321–335, 2026.
https://doi.org/10.1007/978-3-032-17156-6_24

Cozzens proved that computing the boxicity of a graph is NP-hard. Further, Yannakakis [19] established that determining whether a graph has boxicity at most three is NP-complete, and Kratochvíl [15] subsequently proved that the problem remains NP-complete even when restricted to boxicity at most two. Nonetheless, in the literature, there have been many attempts to compute or bound the boxicity of graph classes with special structure such as planar graphs, line graphs, comparability graphs, etc. (cf. [2,4,5] and the references therein).

A simple graph G is said to be a word-representable graph if there exists a word w over its vertex set such that two vertices are adjacent in G if and only if they alternate in the word w. We say that the word w represents the graph G, and it is called a word-representant for the graph G. The class of word-representable graphs includes several important graph classes such as comparability graphs, circle graphs, 3-colorable graphs, cover graphs and parity graphs.

A split graph is a graph whose vertex set can be partitioned into a clique and an independent set. If such a graph is word-representable, then it is referred to as a word-representable split graph. Moreover, if a split graph is also a comparability graph, a graph which admits a transitive orientation, then it is called a split comparability graph. In [11,14], word-representable split graphs were characterized in terms of semi-transitive orientation, and it was shown that deciding whether a given split graph is word-representable can be done in polynomial time. Further, it was proved in [10] that for every word-representable split graph there exists a word-representant in which every letter appears exactly three times. In the context of boxicity, it was established in [7] that the boxicity of a split graph is at most $\min\{\frac{m}{2}, \frac{n}{2}\}$, where m and n are the sizes of a clique and an independent set splitting the graph. Later, in [1], it was proved that determining whether a split graph has boxicity three is NP-complete.

In this paper, we focus on determining the boxicity of word-representable split graphs and, in particular, split comparability graphs. In Sect. 2, we study the structure of the word-representant of a word-representable split graph produced by the algorithm given in [10]. Using the structure of the word-representant, in Sect. 3, we show that the boxicity of a word-representable split graph is at most four. Further, in Sect. 4, we prove that the boxicity of a split comparability is at most three.

2   Preliminaries

In this section, we provide the necessary background material on boxicity, and word-representable graphs; and fix the notations. For the concepts that are not presented here, one may refer to [12,18].

2.1   Boxicity

Let $G = (V, E)$ be a simple, finite and undirected graph. The boxicity of a graph G is the smallest integer k such that G has a k-box representation. If $k = 1$, then

G is called an interval graph, i.e., for all $a, b \in V$, a and b are adjacent in G if and only if the corresponding intervals assigned to the vertices a and b intersect.

Let $G' = (V, E')$ be a graph. If $E \subseteq E'$, then G' is called a super graph of G. For $1 \leq i \leq t$, let $G_i = (V, E_i)$ be a super graph of the graph G. We say that G is the intersection of G_i's, denoted by $G = \bigcap_{i=1}^{t} G_i$, if $E = \bigcap_{i=1}^{t} E_i$. The boxicity of a graph can be characterized in terms of intersection of graphs as follows:

Lemma 1 ([17]). *The boxicity of a non-complete graph G is the smallest positive integer k such that G can be represented as the intersection of k interval graphs.*

If all the closed intervals in a k-dimensional box are of unit length, then it is called a k-cube. A k-cube representation of a graph G is a function that maps each vertex of G to a k-cube such that two vertices are adjacent if and only if the corresponding k-cubes intersect. The smallest value of k such that a graph G has a k-cube representation is called the cubicity of G, denoted by $\mathrm{cub}(G)$. From the definition of cubicity, it is clear that the boxicity of a graph is at most its cubicity. Further, in [3], the following upper bound for the cubicity in terms of boxicity was established.

Theorem 1 ([3]). *For any graph G, $\mathrm{cub}(G) \leq \lceil \log_2 \alpha(G) \rceil \mathrm{box}(G)$, where $\alpha(G)$ is the size of the maximum independent set in G.*

2.2   Word-Representable Graphs

A word over a finite set of letters is a finite sequence which is written by juxtaposing the letters of the sequence. For a word w, $(w)^k = ww \cdots w$ (k times). A subword u of a word w, denoted by $u \ll w$, is defined as a subsequence of the sequence w. For example, $aabb \ll aabbccb$. A subword u of a word w is a factor of w if $w = xuy$, for some words x and y. Let w be a word over a set X, and $A \subseteq X$. Then, $w|_A$ denotes the subword of w that precisely consists of all occurrences of the letters of A. For example, if $w = acabbccb$, then $w|_{\{a,b\}} = aabbb$. For a word w, if $w|_{\{a,b\}}$ is of the form $(ab)^k$ or $(ba)^k$ or $(ab)^k a$ or $(ba)^k b$, for some k, we say the letters a and b alternate in w; otherwise a and b do not alternate in w. A k-uniform word is a word in which every letter occurs exactly k times. Note that a 1-uniform word w is a permutation on the letters of w. For a word w, we write w^R to denote its reversal. If a word is of the form uv, for some words u and v, then the word vu is a cyclic shift of the word uv.

In this paper, we consider only simple, undirected and connected graphs. To distinguish between two-letter words and edges of graphs, we write $\overline{ab}$ to denote an undirected edge between the vertices a and b. The neighborhood of a vertex a in a graph, denoted by $N(a)$, is the set of all vertices adjacent to a in the graph.

A graph $G = (V, E)$ is said to be a word-representable graph if there exists a word w over its vertices such that for all $a, b \in V$, $\overline{ab} \in E$ if and only if a and b alternate in w. A word-representable graph is called k-word-representable if it is represented by a k-uniform word. In [13], it was proved that every word-representable graph is k-word-representable, for some k.

Let $G = (I \cup C, E)$ be a split graph, where I and C induce an independent set and a clique in G, respectively. In this paper, we consider C to be a maximum clique in the graph G. The following theorem gives a necessary and sufficient condition for a split graph to be word-representable. In what follows, for integers $m, n \in \mathbb{Z}$, the set of integers from m to n is denoted in interval format by $[m, n]_{\mathbb{z}}$, i.e., $[m, n] \cap \mathbb{Z}$.

Theorem 2 ([11,14]). *Let $G = (I \cup C, E)$ be a split graph. Then, G is word-representable if and only if the vertices of C can be labeled from 1 to $k = |C|$ in such a way that for each $a, b \in I$, the following holds:*

(i) Either $N(a) = [1, m]_{\mathbb{z}} \cup [n, k]_{\mathbb{z}}$, for $m < n$, or $N(a) = [l, r]_{\mathbb{z}}$, for $l \leq r$.

(ii) If $N(a) = [1, m]_{\mathbb{z}} \cup [n, k]_{\mathbb{z}}$ and $N(b) = [l, r]_{\mathbb{z}}$, for $m < n$ and $l \leq r$, then $l > m$ or $r < n$.

(iii) If $N(a) = [1, m]_{\mathbb{z}} \cup [n, k]_{\mathbb{z}}$ and $N(b) = [1, m']_{\mathbb{z}} \cup [n', k]_{\mathbb{z}}$, for $m < n$ and $m' < n'$, then $m' < n$ and $m < n'$.

In what follows, unless stated otherwise, we consider $G = (I \cup C, E)$ to be a word-representable split graph, where the vertices of C are labeled from 1 to $k = |C|$ such that it satisfies the three properties of Theorem 2. Note that such labeling of the vertices of C can be found in polynomial time [14]. We now consider the following sets:

$$A = \{a \in I \mid N(a) = [1, m]_{\mathbb{z}} \cup [n, k]_{\mathbb{z}}, \text{ for some } m < n\}$$
$$B = \{a \in I \mid N(a) = [l, r]_{\mathbb{z}}, \text{ for some } l \leq r\}$$

We have $A \cap B = \varnothing$ and $I = A \cup B$. For $a \in A$, let $N(a) = [1, m_a]_{\mathbb{z}} \cup [n_a, k]_{\mathbb{z}}$ and for $b \in B$, let $N(b) = [l_b, r_b]_{\mathbb{z}}$. Note that from Theorem 2 (iii), for any $a_1, a_2 \in A$, $m_{a_1} < n_{a_2}$, and hence $\max\{m_a \mid a \in A\} < \min\{n_a \mid a \in A\}$.

Based on Theorem 2, a polynomial-time algorithm was provided in [10] that produces a 3-uniform word-representant for a given word-representable split graph $G = (I \cup C, E)$. In this work, we make use of the structure of the word produced by the algorithm. Therefore, we recall the algorithm from [10] and present it in Algorithm 1, which is helpful in understanding the structure of the word produced by the algorithm. Let w be the 3-uniform word-representant for the graph G obtained by Algorithm 1. It can be observed that the word w can be decomposed into six factors, i.e., $w = u_1 v_1 u_2 u_3 v_2 u_4$, where u_1 is a permutation on the set $I \cup C$, v_1 is a permutation on the set B, u_2 is a permutation on the set $I \cup C$, $u_3 = 1 \cdots d$ and $u_4 = (d+1) \cdots k$, where $d = \max\{m_a \mid a \in A\}$, and v_2 is a permutation on the set A. An example of a split word-representable graph and its 3-uniform word-representant using Algorithm 1 is depicted in Fig. 1.

Notation 3 *Let u be a k-uniform word over a set X. For $x \in X$ and $i \in \{1, \ldots, k\}$, $(x, u)^i$ denotes the position of the ith occurrence of x in the word u. Note that the occurrence of any letter in a word is considered from left to right.*

Remark 1. Based on the construction of the word w given in Algorithm 1, we have the following properties of the vertices in the factors u_1, u_2, u_3 and u_4.

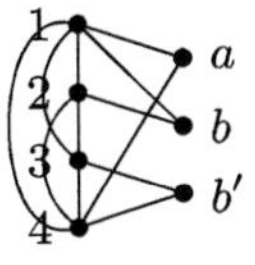

$$w = \underbrace{b1a2b'34}_{u_1}\,\underbrace{b'b}_{v_1}\,\underbrace{12b3a4b'}_{u_2}\,\underbrace{1}_{u_3}\,\underbrace{a}_{v_2}\,\underbrace{234}_{u_4}$$

Fig. 1. Word-representable split graph S and its word-representant w.

Algorithm 1: Constructing a 3-uniform word-representant for a split graph.

Input: A word-representable split graph $G = (I \cup C, E)$ with $1, \ldots, k$ as the labels of the vertices of C.

Output: A 3-uniform word-representant.

1   Initialize p_1, p_2 and p_3 with the permutation $12 \cdots k$ and update them with the vertices of I, as per the following.

2   Initialize $d = 1$.

3   **for** $a \in I$ **do**

4   **if** $a \in A$ **then**

5   **if** $m_a > d$ **then**

6   Assign $d = m_a$.

7   **end**

8   Replace m_a in p_1 with $m_a a$ and n_a in p_2 with $a n_a$.

9   **end**

10   **else**

11   Replace l_a in p_1 with $a l_a$ and r_a in p_2 with $r_a a$.

12   **end**

13 **end**

14 Replace d in p_3 with $d(p_1|_A)^R$.

15 **return** the 3-uniform word $p_1(p_1|_B)^R p_2 p_3$

(i) Let $a \in A$. By line 8 of Algorithm 1, observe that $m_a a (m_a + 1) \ll u_1$ and $(n_a - 1) a n_a \ll u_2$.

(ii) Let $b \in B$. By line 11 of Algorithm 1, observe that $(l_b - 1) b l_b \ll u_1$, if $l_b > 1$ and $b1 \ll u_1$, if $l_b = 1$, and $r_b b (r_b + 1) \ll u_2$, if $r_b < k$ and $kb \ll u_2$, if $r_b = k$.

(iii) Note that $12 \cdots k \ll u_1$, $12 \cdots k \ll u_2$ and $12 \cdots k \ll u_3 u_4$, where $C = \{1, 2, \ldots, k\}$.

Remark 2. (i) For $a \in A$, the first, second and third occurrences of a in the word w appear in the subwords u_1, u_2 and v_2, respectively.

(ii) For $b \in B$, the first, second and third occurrences of b in the word w appear in the subwords u_1, v_1 and u_2, respectively.

(iii) $(12 \cdots k)^3 \ll w$.

3   Word-Representable Split Graphs

In this section, we construct four interval graphs I_1, I_2, I_3 and I_4 on the vertex set $I \cup C$ based on the positions of each letter in the word-representant w (see

Fig. 2) for the split graph G, obtained using Algorithm 1. We show that G is the intersection of these interval graphs, and hence prove the following result.

Theorem 4. *If G is a word-representable split graph, then* box$(G) \leq 4$.

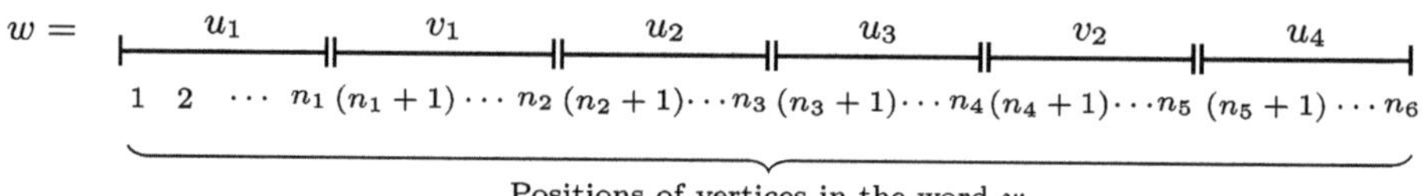

Positions of vertices in the word w

Fig. 2. Decomposition of the word w.

3.1   Interval Graph I_1

The 1-box representation f_1 of the interval graph I_1 on the vertex set $I \cup C$ is as follows. For each $a \in A \cup B \cup C$, assign the following interval

$$f_1(a) = \begin{cases} [(a,w)^3, (a,w)^3], & \text{if } a \in B; \\ [(a,w)^2, (a,w)^3], & \text{otherwise.} \end{cases}$$

Since each vertex holds a distinct position in the word w, the interval assigned to each vertex of I_1 is unique.

Lemma 2. *For $a, b \in V$, if a and b are adjacent in G, then $f_1(a)$ and $f_1(b)$ intersect.*

Proof. Since a and b are adjacent in G, we consider the following cases to show that $f_1(a)$ and $f_1(b)$ intersect.

Case 1: Both $a, b \in C$ or $a \in A$ and $b \in C$. Note that in any case, $f_1(a) = [(a,w)^2, (a,w)^3]$ and $f_1(b) = [(b,w)^2, (b,w)^3]$. Since a and b are adjacent in G, either $ababab \ll w$ or $bababa \ll w$. Without loss of generality, assume that $ababab \ll w$. Clearly, $(a,w)^2 < (b,w)^2 < (a,w)^3 < (b,w)^3$. Thus, $f_1(a)$ and $f_1(b)$ intersect.

Case 2: $a \in C$ and $b \in B$. Then, $a \in N(b) = [l_b, r_b]_z$, $f_1(a) = [(a,w)^2, (a,w)^3]$ and $f_1(b) = [(b,w)^3, (b,w)^3]$. Since $b \in B$, by Remark 2(ii), the third occurrence of b appears in the word u_2 and in view of Remark 1(ii), $r_b b \ll u_2$. Also, since the second occurrence of a appears in u_2 and $a \in [l_b, r_b]_z$, we have $l_b a r_b b \ll u_2$. Furthermore, since the third occurrence of a appears in $u_3 u_4$, we have $(a,w)^2 < (b,w)^3 < (a,w)^3$. Hence, $f_1(b) \subset f_1(a)$ so that $f_1(a)$ and $f_1(b)$ intersect. $\square$

Lemma 3. *For $a, b \in V$, suppose that a and b are non-adjacent in G. Then, $f_1(a)$ and $f_1(b)$ do not intersect if any one of the following holds:*

1. $a \in B$ and $b \in [r_a + 1, k]_z$.

2. $a, b \in B$.

Proof. 1. By Remark 2(ii), the third occurrence of a is in the word u_2 and by Remark 1(ii), $r_a a(r_a + 1) \ll u_2$. Whereas, by Remark 1(iii), the second and third occurrences of b in w appear in the words u_2 and $u_3 u_4$, respectively. Since $b \in [r_a + 1, k]_z$, we have $ab \ll u_2$. Thus, we see that $(a, w)^3 < (b, w)^2 < (b, w)^3$. Hence, $f_1(a)$ and $f_1(b)$ do not intersect.

2. As $a, b \in B$, both $f_1(a)$ and $f_1(b)$ are singletons and the position of any letter in the word w is unique. Clearly, $f_1(a)$ and $f_1(b)$ do not intersect. $\qquad\square$

3.2   Interval Graph I_2

Consider the cyclic shift $u_3 v_2 u_4 u_1 v_1 u_2$ of the word w and call it w_1. Note that by [13, Proposition 5], the word w_1 also represents the graph G.

Remark 3. In the word w_1, the occurrences of each letter are as per the following.

(i) For each $a \in A$, the first, second and third occurrences of a appear in the subwords v_2, u_1 and u_2, respectively.
(ii) For each $a \in B$, the first, second and third occurrences a appear in the subwords u_1, v_1 and u_2, respectively.
(iii) $(12 \cdots k)^3 \ll w_1$.

The 1-box representation f_2 of the interval graph I_2 on the vertex set $I \cup C$ is as follows. For each $a \in A \cup B \cup C$, assign the interval

$$f_2(a) = \begin{cases} [(a, w_1)^1, (a, w_1)^1], & \text{if } a \in B, \\ [(a, w_1)^1, (a, w_1)^2], & \text{otherwise.} \end{cases}$$

Lemma 4. *For $a, b \in V$, if a and b are adjacent in G, then $f_2(a)$ and $f_2(b)$ intersect.*

Proof. Since a and b are adjacent in G, we consider the following cases to show that $f_2(a)$ and $f_2(b)$ intersect.

    Case 1: Both $a, b \in C$ or $a \in A$ and $b \in C$. In any case, we have $f_2(a) = [(a, w_1)^1, (a, w_1)^2]$ and $f_2(b) = [(b, w_1)^1, (b, w_1)^2]$. Since a and b are adjacent in G, either $ababab \ll w_1$ or $bababa \ll w_1$. Without loss of generality, assume that $ababab \ll w_1$. Clearly, $(a, w)^1 < (b, w_1)^1 < (a, w_1)^2 < (b, w)^2$. Thus, $f_2(a)$ and $f_2(b)$ intersect.

    Case 2: $a \in C$ and $b \in B$. We have, $a \in [l_b, r_b]_z$, $f_2(a) = [(a, w_1)^1, (a, w_1)^2]$ and $f_2(b) = [(b, w_1)^1, (b, w_1)^1]$. Since $b \in B$, by Remark 3(ii), the first occurrence of b appears in the word u_1 and by Remark 1(ii), $bl_b \ll u_1$. The first and second occurrences of a in w_1 appear in the words $u_3 u_4$ and u_1, respectively. Since $a \in [l_b, r_b]_z$, we have $(a, w_1)^1 < (b, w_1)^1 < (a, w_1)^2$. Hence, $f_2(a)$ and $f_2(b)$ intersect. $\qquad\square$

Lemma 5. *Suppose a and b are non-adjacent in G such that $a \in B$ and $b \in [1, l_a - 1]_{\mathbb{Z}}$. Then, $f_2(a)$ and $f_2(b)$ do not intersect.*

Proof. By Remark 3(ii), the first occurrence of a in w_1 appears in the words u_1 and $(l_a - 1)al_a \ll u_1$ (by Remark 1(ii)). The first and second occurrences of b in the word w_1 appear in the words $u_3 u_4$ and u_1, respectively. Since $b \in [1, l_a - 1]_{\mathbb{Z}}$, we have $(b, w_1)^1 < (b, w_1)^2 < (a, w_1)^1$. Thus, $f_2(a)$ and $f_2(b)$ do not intersect. $\square$

3.3   Interval Graph I_3

Consider the cyclic shift $u_4 u_1 v_1 u_2 u_3 v_2$ of the word w and call it w_2. By [13, Proposition 5], the word w_2 also represents the graph G.

Remark 4. In the word w_2, the occurrences of each letter are as per the following.

(i) For each $a \in A$, the first, second and third occurrences of a appear in the subwords u_1, u_2 and v_2, respectively.
(ii) For each $b \in B$, the first, second and third occurrences of b appear in the subwords u_1, v_1 and u_2, respectively.
(iii) Recall that $d = \max\{m_c \mid c \in A\}$. We have, $((d+1) \cdots k1 \cdots d)^3 \ll w_2$, where $(d+1) \cdots k1 \cdots d \ll u_4 u_1$, $(d+1) \cdots k1 \cdots d \ll u_1 u_2$ and $(d+1) \cdots k1 \cdots d \ll u_2 u_3$.

The 1-box representation f_3 of the interval graph I_3 on the vertex set $I \cup C$ is as follows. For each $a \in A \cup B \cup C$, assign the following interval

$$f_3(a) = \begin{cases} [(a, w_2)^1, (a, w_2)^1], & \text{if } a \in A; \\ [(a, w_2)^1, (a, w_2)^2], & \text{otherwise.} \end{cases}$$

Lemma 6. *If a and b are adjacent in G, then $f_3(a)$ and $f_3(b)$ intersect.*

Proof. Since a and b are adjacent in G, we consider the following two cases to show that $f_3(a)$ and $f_3(b)$ intersect.

Case 1: Both $a, b \in C$ or $a \in C$ and $b \in B$. In any case, we have $f_3(a) = [(a, w_2)^1, (a, w_2)^2]$ and $f_3(b) = [(b, w_2)^1, (b, w_2)^2]$. Since a and b are adjacent, either $ababab \ll w_2$ or $bababa \ll w_2$. Without loss of generality, let $ababab \ll w_2$. Clearly, $(a, w_2)^1 < (b, w_2)^1 < (a, w_2)^2 < (b, w_2)^2$. Thus, $f_3(a)$ and $f_3(b)$ intersect.

Case 2: $a \in C$ and $b \in A$. We have, $a \in N(b) = [1, m_b]_{\mathbb{Z}} \cup [n_b, k]_{\mathbb{Z}}$. If $a \in [1, m_b]_{\mathbb{Z}}$, then $a \le m_b \le d$, where $d = \max\{m_c \mid c \in A\}$. Thus, the first occurrence of a in the word w_2 appears in u_1. By Remark 4(i), the first occurrence of b appears in u_1 and by Remark 1(i), $m_b b(m_b + 1) \ll u_1$. Thus, we have $ababab \ll w_2$ so that $(a, w_2)^1 < (b, w_2)^1 < (a, w_2)^2$. Hence, $f_3(a)$ and $f_3(b)$ intersect. If $a \in [n_b, k]_{\mathbb{Z}}$, then $d < n_b$. By Remark 4(iii), the first occurrence of a appears in the word u_4 and the second occurrence of a appears in the word u_1. Further, by Remark 1(i), we have $m_b b n_b \ll u_1$. Thus, $(a, w_2)^1 < (b, w_2)^1 < (a, w_2)^2$ so that $f_3(a)$ and $f_3(b)$ intersect. $\square$

Lemma 7. *Suppose a and b are non-adjacent in G. Then, $f_3(a)$ and $f_3(b)$ do not intersect if any one of the following holds:*

1. $a \in A$ and $b \in [m_a + 1, d]_{\mathbb{Z}}$, where $d = \max\{m_c \mid c \in A\}$.
2. $a, b \in A$.
3. $a \in A$ and $b \in B$ such that $m_a < l_b$.

Proof. By Remark 4(i), the first occurrence of a in w_2 appears in u_1 and, by Remark 1(i), we have $m_a a(m_a + 1) \ll u_1$.

1. By Remark 4(iii), the first occurrence of b in the word w_2 appears in u_1 and the second occurrence appears in u_2. Thus, $(m_a, w_2)^1 < (a, w_2)^1 < (b, w_2)^1 < (b, w_2)^2$. Clearly, $f_3(a)$ and $f_3(b)$ do not intersect.
2. As $a, b \in A$, both $f_3(a)$ and $f_3(b)$ are singletons and position of any letter in the word w_2 is unique. Clearly, $f_3(a)$ and $f_3(b)$ do not intersect.
3. By Remark 4(ii), the first occurrence of b in w_2 appears in the word u_1 and $bl_b \ll u_1$ (by Remark 1(ii)). Since $m_a < l_b$, $m_a abl_b \ll u_1$. Thus, $(a, w_2)^1 < (b, w_2)^1 < (b, w_2)^2$. Thus, $f_3(a)$ and $f_3(b)$ do not intersect. $\qquad\square$

3.4   Interval Graph I_4

For the construction of the interval graph I_4, we consider the word w_2 as defined in Sect. 3.3. The 1-box representation f_4 of the interval graph I_4 on the vertex set $I \cup C$ is as follows. For each $a \in A \cup B \cup C$, assign the following interval

$$f_4(a) = \begin{cases} [(a, w_2)^2, (a, w_2)^2], & \text{if } a \in A; \\ [(a, w_2)^2, (a, w_2)^3], & \text{otherwise.} \end{cases}$$

Lemma 8. *If a and b are adjacent in G, then $f_4(a)$ and $f_4(b)$ intersect.*

Proof. Since a and b are adjacent in G, we consider the following two cases to show that $f_4(a)$ and $f_4(b)$ intersect.

Case 1: Both $a, b \in C$ or $a \in C$ and $b \in B$. In any case, we have $f_4(a) = [(a, w_2)^2, (a, w_2)^3]$ and $f_4(b) = [(b, w_2)^2, (b, w_2)^3]$. Without loss of generality, let $ababab \ll w_2$. Clearly, $(a, w_2)^2 < (b, w_2)^2 < (a, w_2)^3 < (b, w_2)^3$. Thus, $f_4(a)$ and $f_4(b)$ intersect.

Case 2: $a \in C$ and $b \in A$. We have, $a \in N(b) = [1, m_b]_{\mathbb{Z}} \cup [n_b, k]_{\mathbb{Z}}$. If $a \in [1, m_b]_{\mathbb{Z}}$, then the second and third occurrences of a in w_2 appear in the subwords u_2 and u_3, respectively (by Remark 4(iii)). Further, the second occurrence of b in w_2 appears in u_2 (by Remark 4(i)). In the word u_2, by Remark 1(i), $m_b bn_b \ll u_2$. Thus, $(a, w_2)^2 < (b, w_2)^2 < (a, w_2)^3$, and hence $f_4(a)$ and $f_4(b)$ intersect. If $a \in [n_b, k]_{\mathbb{Z}}$, then the second and third occurrences of a in w_2 appear in the subwords u_1 and u_2, respectively (by Remark 4(iii)). In the word u_2, by Remark 1(i), $m_b bn_b \ll u_2$. Thus, $(a, w_2)^2 < (b, w_2)^2 < (a, w_2)^3$, and hence $f_4(a)$ and $f_4(b)$ intersect. $\qquad\square$

Lemma 9. *Suppose a and b be non-adjacent in G. Then $f_4(a)$ and $f_4(b)$ do not intersect if any one of the following holds:*

1. *$a \in A$ and $b \in [d + 1, n_a - 1]_z$, where $d = \max\{m_c \mid c \in A\}$.*
2. *$a \in A$ and $b \in B$ such that $m_a \geq l_b$.*

Proof. 1. By Remark 4(iii), the first, second and third occurrences of b in w_2 appear in the subwords u_4, u_1 and u_2, respectively. Whereas, the second occurrence of a in w_2 appears in u_2. In the word u_2, $(n_a - 1)an_a \ll u_2$ (by Remark 1(i)). Thus, $(b, w_2)^2 < (b, w_2)^3 < (a, w_2)^2$, and hence $f_4(a)$ and $f_4(b)$ do not intersect.

2. By Remark 4(i), the second occurrence of a appears in the word u_2 of w_2. By Remark 4(ii), the second and third occurrences of b appear in the words v_1 and u_2, respectively. Since $m_a \geq l_b$, by property (ii) of Theorem 2, $r_b < n_a$. In the word u_2, we have $r_b b a n_a \ll u_2$ (by remark 1(i) and 1(ii)). Thus, $(b, w_2)^2 < (b, w_2)^3 < (a, w_2)^2$, and hence $f_4(a)$ and $f_4(b)$ do not intersect. $\square$

3.5 Proof of Theorem 4 and its Consequences

By Lemmas 2, 4, 6, 8, we see that if a and b are adjacent in G, then a and b are adjacent in each I_j, for all $1 \leq j \leq 4$. Further, if a and b are non-adjacent in G, then, by Lemmas 3, 5, 7 and 9, a and b are non-adjacent in I_j, for some $j \in \{1, 2, 3, 4\}$. Hence, the word-representable split graph G is the intersection of the graphs I_1, I_2, I_3 and I_4 so that $\mathrm{box}(G) \leq 4$.

In view of Theorem 1, we obtain the following corollary of Theorem 4 on the cubicity of word-representable split graphs.

Corollary 1. *If G is a word-representable split graph such that $\alpha(G)$ is the size of the maximum independent set in G, then $\mathrm{cub}(G) \leq 4\lceil \log_2 \alpha(G) \rceil$.*

Corollary 2. *Let $G = (I \cup C)$ be a word-representable split graph. If there is a labeling of the vertices of C such that the set I is either A or B, then $\mathrm{box}(G) \leq 2$.*

Example 1. Consider the word-representable split graph S given in Fig. 1 and its word-representant $w = b1a2b'34b'b12b3a4b'1a234$. Using the positions of second or third occurrences of the vertices $1, 2, 3, 4, a, b, b'$, in w, we construct the interval graph I_1 and its 1-box representation f_1 as depicted in Fig. 3.

Next, consider the cyclic shift $w_1 = 1a234b1a2b'34b'b12b3a4b'$ of w. Based on the first or second occurrences of each vertex of S in w_1, we obtain the interval graph I_2 and its corresponding 1-box representation f_2 (see Fig. 4). Finally, consider the cyclic shift $w_2 = 234b1a2b'34b'b12b3a4b'1a$ of w. Using the positions of first or second occurrences of the vertices in w_2, we construct the interval graph I_3 with its 1-box representation f_3. Similarly, based on the second or third occurrences of the vertices in w_2, we obtain the interval graph I_4 and its 1-box representation f_4 (see Fig. 5). Hence, the graph S is an intersection of the interval graphs I_1, I_2, I_3 and I_4.

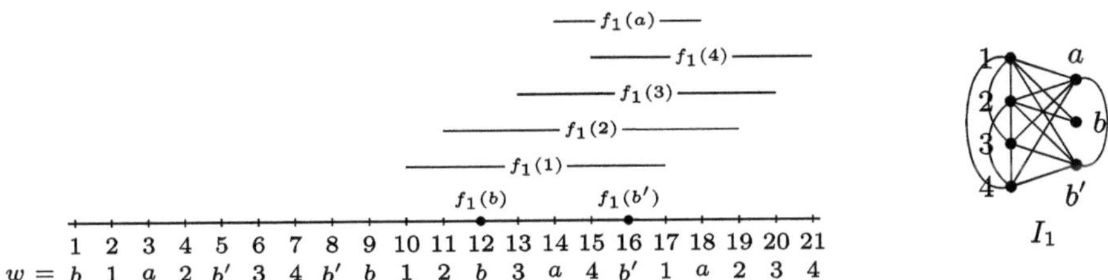

Fig. 3. Interval graph I_1 and its 1-box representation.

4   Split Comparability Graphs

In this section, we show that a split comparability graph can be represented as an intersection of three interval graphs, and hence we obtain the following result.

Theorem 5. *If G is a split comparability graph, then* $\mathrm{box}(G) \le 3$.

In order to prove the above-mentioned theorem, we recall the following characterization of split comparability graphs.

Theorem 6 *bf ([9,16]). Let $G = (I \cup C, E)$ be a split graph. Then, G is a comparability graph if and only if the vertices of C can be labeled from 1 to $k = |C|$ such that the following properties* (i) − (v) *hold: For $a, b \in I$,*

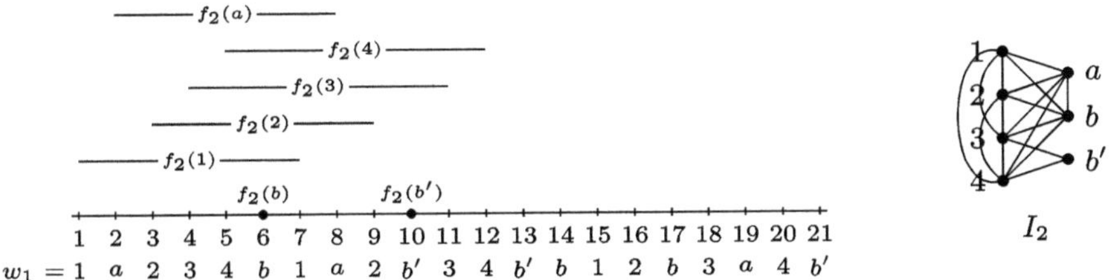

Fig. 4. Interval graph I_2 and its 1-box representation.

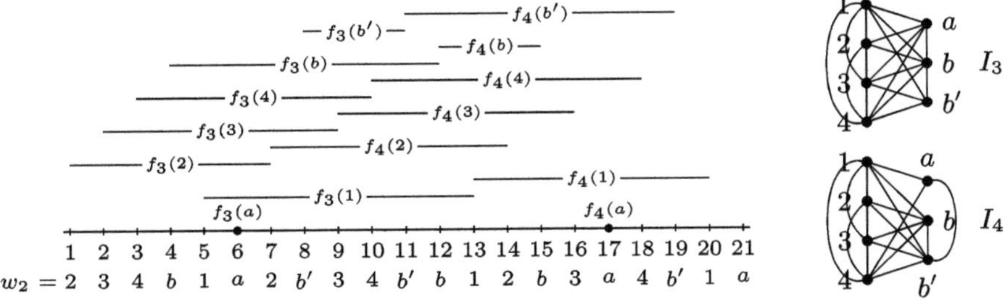

Fig. 5. Interval graphs I_3 and I_4 and their corresponding 1-box representations.

(i) *The neighborhood $N(a)$ has one of the following forms: $[1, m_a]_{\mathbf{z}} \cup [n_a, k]_{\mathbf{z}}$ for $m_a < n_a$, $[1, r_a]_{\mathbf{z}}$ for $r_a < k$, or $[l_a, k]_{\mathbf{z}}$ for $l_a > 1$.*
(ii) *If $N(a) = [1, r_a]_{\mathbf{z}}$ and $N(b) = [l_b, k]_{\mathbf{z}}$, for $r_a < k$ and $l_b > 1$, then $r_a < l_b$.*
(iii) *If $N(a) = [1, m_a]_{\mathbf{z}} \cup [n_a, k]_{\mathbf{z}}$ and $N(b) = [1, r_b]_{\mathbf{z}}$, for $r_b < k$ and $m_a < n_a$, then $r_b < n_a$.*
(iv) *If $N(a) = [1, m_a] \cup [n_a, k]$ and $N(b) = [l_b, k]_{\mathbf{z}}$, for $l_b > 1$ and $m_a < n_a$, then $m_a < l_b$.*
(v) *If $N(a) = [1, m_a]_{\mathbf{z}} \cup [n_a, k]_{\mathbf{z}}$ and $N(b) = [1, m_b]_{\mathbf{z}} \cup [n_b, k]_{\mathbf{z}}$, for $m_a < n_a$ and $m_b < n_b$, then $m_a < n_b$ and $m_b < n_a$.*

In what follows, let $G = (I \cup C, E)$ be a split comparability graph, where the vertices of C are labeled from 1 to $k = |C|$ such that it satisfies the properties of Theorem 6. Note that the labeling of the vertices of C with respect to Theorem 6 satisfies the properties given in Theorem 2. Thus, Algorithm 1 also runs for the split comparability graph G with the labeling of vertices of C satisfying the properties of Theorem 6, and produces a 3-uniform word-representant w for the graph G. Note that Remarks 1 and 2 hold for the word w. Apply cyclic shift on the word w to obtain the word w_2 and the interval graphs I_3 and I_4 as per Sect. 3.3 and Sect. 3.4. In the following, we construct an interval graph I_5, and show that G is the intersection of I_3, I_4 and I_5.

4.1   Interval Graph I_5

Consider a cyclic shift of the word w to obtain the word $u_2 u_3 v_2 u_4 u_1 v_1$ and call it w_3. By [13, Proposition 5], the word w_3 also represents the graph G. Let $d_1 = \min\{l_c \mid c \in B \text{ and } N(c) = [l_c, k]\}$.

Remark 5. In the word w_3, the occurrences of each letter are as per the following.

(i) For each $a \in A$, the first, second and third occurrences of a appear in the subwords u_2, v_2 and u_1, respectively.
(ii) For each $b \in B$, the first, second and third occurrences of b appear in the word u_2, u_1 and v_1, respectively.
(iii) $(12 \cdots k)^3 \ll w_3$.

The 1-box representation f_5 of the interval graph I_5 on the vertex set $I \cup C$ is as follows. For each $a \in A \cup B \cup C$, assign the interval

$$f_5(a) = \begin{cases} [(a, w_3)^1, (a, w_3)^2], & \text{if } a \in C \text{ and } a < d_1; \\ [(a, w_3)^1, (a, w_3)^1], & \text{if } a \in B \text{ and } N(a) = [1, r_a]; \\ [(a, w_3)^2, (a, w_3)^2], & \text{if } a \in B \text{ and } N(a) = [l_a, k]; \\ [(a, w_3)^1, (a, w_3)^3], & \text{otherwise.} \end{cases}$$

Lemma 10. *If a and b are adjacent in G, then $f_5(a)$ and $f_5(b)$ intersect.*

Proof. Since a and b are adjacent in G, either $ababab \ll w_3$ or $bababa \ll w_3$. We prove that $f_5(a)$ and $f_5(b)$ intersect in the following three cases.

Case 1: $a, b \in C$. Without loss of generality, assume that $ababab \ll w_3$. If $a < d_1$ and $b < d_1$, then $f_5(a) = [(a, w_3)^1, (a, w_3)^2]$ and $f_5(b) = [(b, w_3)^1, (b, w_3)^2]$. Thus, $(a, w_3)^1 < (b, w_3)^1 < (a, w_3)^2 < (b, w_3)^2$ so that $f_5(a)$ and $f_5(b)$ intersect. If $a \geq d_1$ and $b \geq d_1$, then $f_5(a) = [(a, w_3)^1, (a, w_3)^3]$ and $f_5(b) = [(b, w_3)^1, (b, w_3)^3]$. Clearly, $(a, w_3)^1 < (b, w_3)^1 < (a, w_3)^3 < (b, w_3)^3$. Thus, $f_5(a)$ and $f_5(b)$ intersect. Finally, if $a < d_1$ and $b \geq d_1$, then $f_5(a) = [(a, w_3)^1, (a, w_3)^2]$ and $f_5(b) = [(b, w_3)^1, (b, w_3)^3]$. We have, $(a, w_3)^1 < (b, w_3)^1 < (a, w_3)^2 < (b, w_3)^3$. Thus, $f_5(a)$ and $f_5(b)$ intersect.

Case 2: $a \in C$ and $b \in B$. If $N(b) = [1, r_b]$, then $f_5(b) = [(b, w_3)^1, (b, w_3)^1]$. Since $b \in B$ and $N(b) = [1, r_b]$, by Remark 5(ii), the first occurrence of b appears in the subword u_2 of w_3 and, by Remark 1(ii), $r_b b \ll u_2$ so that $ab \ll u_2$. Clearly, $ababab \ll w_3$. By property (ii) of Theorem 6, $r_b < d_1$ so that $a \leq r_b < d_1$. Thus, $(a, w_3)^1 < (b, w_3)^1 < (a, w_3)^2$, and hence $f_5(a)$ and $f_5(b)$ intersect. If $N(b) = [l_b, k]$, then $l_b \geq d_1$. By Remark (ii), the second occurrence of b appears in the word u_1 and $b l_b \ll u_1$ (by Remark 1(ii)). We observe that the first and third occurrences of a in the word w_3 appear in the subwords u_2 and u_1, respectively. Since $a \in [l_b, k]$, $ba \ll u_1$. Thus, $(a, w_3)^1 < (b, w_3)^2 < (a, w_3)^3$ so that $f_5(a)$ and $f_5(b)$ intersect.

Case 3: $a \in C$ and $b \in A$. Suppose $a < d_1$. We have $f_5(a) = [(a, w_3)^1, (a, w_3)^2]$ and $f_5(b) = [(b, w_3)^1, (b, w_3)^3]$. Without loss of generality, assume that $ababab \ll w_3$ and then $(a, w)^1 < (b, w_3)^1 < (a, w_3)^2 < (b, w_3)^3$. Thus, $f_5(a)$ and $f_5(b)$ intersect. Suppose $a \geq d_1$. We have $f_5(a) = [(a, w_3)^1, (a, w_3)^3]$ and $f_5(b) = [(b, w_3)^1, (b, w_3)^3]$. Without loss of generality, assume that $ababab \ll w_3$. Then, we have $(a, w)^1 < (b, w_3)^1 < (a, w_3)^3 < (b, w_3)^3$ so that $f_5(a)$ and $f_5(b)$ intersect. $\square$

Lemma 11. *Let a and b be non-adjacent in G. If $a \in C$ and $b \in B$, then $f_5(a)$ and $f_5(b)$ do not intersect.*

Proof. Through the following two cases, we show that $f_5(a)$ and $f_5(b)$ do not intersect.

Case 1: $N(b) = [1, r_b]$. We have $f_5(b) = [(b, w_3)^1, (b, w_3)^1]$. By Remark 5(ii), the first occurrence of b appears in the word u_2 of w_3 and, by Remark 1(ii), we have $r_b b(r_b + 1) \ll u_2$. Since a is non-adjacent to b, a must be greater than r_b. Moreover, the first occurrence of a appears in the word u_2. Thus, $(r_b, w_3)^1 < (b, w_3)^1 < (a, w_3)^1 < (a, w_3)^2$, and therefore $f_5(a)$ and $f_5(b)$ do not intersect.

Case 2: $N(b) = [l_b, k]$. We have $f_5(b) = [(b, w_3)^2, (b, w_3)^2]$. By Remark 5(ii), the second occurrence of b appears in the word u_1 and, by Remark 1(ii), we have $(l_b - 1) b l_b \ll u_1$. Since a is non-adjacent to b, a must be less than l_b. Moreover, the third occurrence of a appears in the word u_1 so that $a b l_b \ll u_1$. Thus, $(a, w_3)^2 < (a, w_3)^3 < (b, w_3)^2$, and therefore $f_5(a)$ and $f_5(b)$ do not intersect. $\square$

4.2 Proof of Theorem 5 and its Consequence

By Lemmas 6, 8 and 10, we see that if a and b are adjacent in G, then a and b are adjacent in each of the interval graph I_j, for $j \in \{3, 4, 5\}$. Further, if a and b

are non-adjacent in G, then, by Lemmas 7, 9 and 11, a and b are non-adjacent in I_j, for some $j \in \{3, 4, 5\}$. Hence, the graph G is the intersection of the graphs I_3, I_4 and I_5 so that $\mathrm{box}(G) \leq 3$.

In view of Theorem 1, we obtain the following corollary of Theorem 5 on the cubicity of split comparability graphs.

Corollary 3. *If G is a split comparability graph, then* $\mathrm{cub}(G) \leq 3\lceil \log_2 \alpha(G) \rceil$.

5   Conclusion

Although the boxicity of a split graph generally depends on the sizes of its clique and independent set, in our work, we established that every word-representable split graph has boxicity at most four. Since word-representable graphs generalize both circle graphs and comparability graphs, this bound immediately applies to split graphs belonging to either of these classes. Moreover, for split comparability graphs, we further proved that their boxicity is at most three. These results lead to the following questions about the tightness: does there exist a word-representable split graph with boxicity four, or a split comparability graph with boxicity three?

A co-bipartite graph is a graph whose vertex set can be partitioned into two disjoint cliques. In [8], a characterization for word-representable co-bipartite graphs was provided in terms of vertex-ordering along with a polynomial-time algorithm to construct a 3-uniform word-representant. A natural direction for future work is to determine the boxicity of word-representable co-bipartite graphs.

References

1. Adiga, A., Bhowmick, D., Chandran, L.S.: The hardness of approximating the boxicity, cubicity and threshold dimension of a graph. Discrete Appl. Math. **158**(16), 1719–1726 (2010)
2. Adiga, A., Bhowmick, D., Chandran, L.S.: Boxicity and poset dimension. SIAM J. Discrete Math. **25**(4), 1687–1698 (2011)
3. Adiga, A., Chandran, L.S.: Cubicity of interval graphs and the claw number. J. Graph Theory **65**(4), 323–333 (2010)
4. Chandran, L.S., Francis, M.C., Sivadasan, N.: Boxicity and maximum degree. J. Combin. Theory Ser. B **98**(2), 443–445 (2008)
5. Chandran, L.S., Sivadasan, N.: Boxicity and treewidth. J. Combin. Theory Ser. B **97**(5), 733–744 (2007)
6. Cozzens, M.B.: Higher and multi-dimensional analogues of interval graphs. ProQuest LLC, Ann Arbor, MI, 1981. Thesis (Ph.D.)–Rutgers The State University of New Jersey - New Brunswick
7. Cozzens, M.B., Roberts, F.S.: Computing the boxicity of a graph by covering its complement by cointerval graphs. Discrete Appl. Math. **6**(3), 217–228 (1983)
8. Das, B., Hariharasubramanian, R.: Representation number of word-representable co-bipartite graph. arXiv:2509.03064 (2025)

9. Dwary, T., Mozhui, K., Krishna, K.V.: Characterization of split comparability graphs. arXiv:2504.19167 (2025)
10. Dwary, T., Mozhui, K., Krishna, K.V.: Representation number of word-representable split graphs. Siberian Advances in Mathematics, To appear (2026). arXiv:2502.00872
11. Kitaev, S., Long, Y., Ma, J., Wu, H.: Word-representability of split graphs. J. Comb. **12**(4), 725–746 (2021)
12. Kitaev, S., Lozin, V.: Words and graphs. Monographs in Theoretical Computer Science. An EATCS Series. Springer, Cham, 2015. With a foreword by Martin Charles Golumbic
13. Kitaev, S., Pyatkin, A.: On representable graphs. J. Autom. Lang. Comb. **13**(1), 45–54 (2008)
14. Kitaev, S., Pyatkin, A.: On semi-transitive orientability of split graphs. Inform. Process. Lett., **184**, Paper No. 106435, 4, (2024)
15. Kratochvíl, J.: A special planar satisfiability problem and a consequence of its NP-completeness. Discrete Appl. Math. **52**(3), 233–252 (1994)
16. Ortiz, C., Villanueva, M.: On split-comparability graphs, pp. 91–105. In II ALIO-EURO Workshop on Pratical Combinatorial Optimization, Valparaiso, Chile (1996)
17. Roberts, F.S.: On the boxicity and cubicity of a graph. In: Recent Progress in Combinatorics (Proc. Third Waterloo Conf. on Combinatorics, 1968), pages 301–310. Academic Press, New York-London (1969)
18. West, D.B.: Introduction to Graph Theory. Prentice Hall (1996)
19. Yannakakis, M.: The complexity of the partial order dimension problem. SIAM J. Algebraic Discrete Methods **3**(3), 351–358 (1982)

Subset Feedback Vertex Set and Subset Vertex Cover on AT-Free Graphs

Joydeep Mukherjee[1] and Tamojit Saha[1,2]([✉])

[1] Ramakrishna Mission Vivekananda Educational and Research Institute, Howrah, India
joydeep.m@gm.rkmvu.ac.in
[2] Institute for Advancing Intelligence, TCG CREST, Kolkata, India
tamojitsaha1@gmail.com
http://rkmvu.ac.in/ , https://www.tcgcrest.org/

Abstract. Given a graph $G = (V, E)$ and a subset $S \subseteq V(G)$, the *Subset Feedback Vertex Set (SFVS)* problem asks to find a minimum vertex set that intersects every cycle containing a vertex from set S. SFVS is known to be NP-hard on general graphs. The computational complexity of SFVS on AT-free graphs was posed as an open problem by Papadopoulos et al. in [25]. We resolve this problem by presenting a polynomial-time algorithm for SFVS on AT-free graphs. Along with this, we also present a polynomial time algorithm for *Subset Vertex Cover (SVC)* problem on AT-free graphs. In SVC, the task is to find a minimum vertex set that contains an endpoint of every edge which contains some vertex from S as its end point. We note here, SVC is used as a subroutine to compute SFVS.

Keywords: AT-free graphs · Graph Algorithm · Optimization Algorithm

1 Introduction

Subset Feedback Vertex Set problem (SFVS) is a fundamental problem in theoretical computer science. Given a graph $G = (V, E)$ and a subset $S \subseteq V(G)$, the objective is to compute a minimum cardinality set $F \subseteq V(G)$ such that the graph $G \setminus F$ contains no cycle passing through any vertex of S. SFVS is a generalization of the classical *Feedback Vertex Set problem (FVS)*, in the sense that when $S = V(G)$, SFVS is equivalent to computing FVS. Although SFVS is known to be NP-hard in general graphs, the computational complexity of SFVS on AT-free graphs was posed as an open problem by Papadopoulos et al. in [25]. We resolve this problem by presenting a polynomial-time algorithm for SFVS on AT-free graphs.

Subset Vertex Cover problem (SVC) generalizes the classical *Vertex Cover problem*. Given a graph $G = (V, E)$ and a subset $S \subseteq V(G)$, define $E_S \subseteq E(G)$ as the set of edges with at least one endpoint in S. The task is to compute a

N. Misra and A. Pandey (Eds.): CALDAM 2026, LNCS 16445, pp. 336–350, 2026.
https://doi.org/10.1007/978-3-032-17156-6_25

minimum vertex set $T \subseteq V(G)$ that intersects every edge in E_S. In this paper we provide a polynomial time algorithm for SVC for AT-free graphs.

Both these problems have significant theoretical and practical motivation. In particular, SFVS not only generalizes the Feedback Vertex Set problem but also generalizes the well-studied *Multi-way Cut problem* [17]. In the Multi-way Cut problem, given a graph $G = (V, E)$ and a subset $S \subseteq V(G)$, the goal is to find a minimum set $K \subseteq V(G)$ such that, in $G \setminus K$, all vertices of S are disconnected. SFVS arises naturally in several domains, including Bayesian inference, constraint satisfaction, and optimization theory [4,5,14,15]. SVC also finds application in many areas like genome research [29].

An *Asteroidal Triple (AT)* in a graph $G = (V, E)$ is a set of three mutually nonadjacent vertices $\{u, v, w\} \subseteq V(G)$ such that, for every pair of vertices in the set, there exists a path connecting them that avoids the neighborhood of the third vertex. A graph containing no asteroidal triple is called *AT-free*. AT-free graphs form a broad graph class that generalizes several well-known families, including cocomparability graphs, interval graphs, permutation graphs, and trapezoid graphs [12]. Each of these subclasses has been widely studied and is known to have diverse practical applications [21].

1.1   Our Contribution

We provide a polynomial time algorithm for SFVS in AT-free graphs.

Theorem 1. *There is a $O(n^{16})$ algorithm that computes SFVS in a given AT-free graph.*

We also provide a polynomial time algorithm for SVC.

Theorem 2. *There is a $O(n^4)$ time algorithm that computes SVC in a given AT-free graph.*

Let $G = (V, E)$ be an AT-free graph and let $S \subseteq V(G)$. Let C^s be a component of $G \setminus N[s]$ where $s \in S$. In this paper, we develop a novel ordering of the vertices in $C^s \cap S$ to facilitate the design of polynomial time algorithms for SFVS and SVC. We call this ordering *inclusion ordering*. Formally, let $\{s_1, \ldots, s_t\}$ be an inclusion ordering of vertices in $C^s \cap S$, then for every $j \in \{1, \ldots, t\}$ we have $I(s, s_j) \cap \{s_{j+1}, \ldots, s_t\} = \emptyset$. Using this ordering, we establish the following lemma, which forms a central component of our algorithms and is also of independent interest. Please see Sect. 2 for undefined notations.

Lemma 1. *For every $\Theta \subseteq V(G)$, containing some vertices from $\{s_1, \ldots, s_t\}$, there exists $i \in \{1, 2, \ldots, t\}$ such that $\Theta \cap I(s, s_i) \cap \{s_1, \ldots, s_t\} = \emptyset$.*

1.2   Organization of the Paper

In the next section we briefly survey related works. Then we define necessary terms and notations in Sect. 2. In Sect. 3, we describe inclusion ordering formally. Section 4 and Sect. 5 contain the algorithms for SVC and SFVS. We note here, SVC is used as a subroutine to compute SFVS.

1.3 Related Work

SFVS has been extensively studied due to its theoretical significance and practical applications. Since the feedback vertex set problem is NP-complete [18], NP-completeness of SFVS follows directly.

The SFVS problem was first introduced by Even et al., who designed a constant-factor approximation algorithm for the weighted version of this problem in general graphs [14]. From the perspective of parameterized complexity, the unweighted version was shown to be fixed-parameter tractable [13] on general graphs, while the fastest known exact algorithm runs in $O(1.87^n)$ time for the weighted version [17], and in $O(1.76^n)$ time for the unweighted version [16].

SFVS has also been studied from the point of view of special graph classes. This problem is NP-complete on split graphs [17], which is a sub-class of chordal graphs. For chordal graphs progressively faster algorithms have been developed, with running times improved to $O(1.6708^n)$ [19] and later to $O(1.6181^n)$ [11]. In fact, the $O(1.62^n)$ algorithm of Chitnis et al. [11] applies to any graph class where the weighted feedback vertex set problem can be solved in polynomial time and the graph class is closed under vertex deletions and edge contractions. From this result, an $O(1.62^n)$ algorithm is known for the SFVS on AT-free graphs. Papadopoulos et al. provided polynomial time algorithms for permutation graph and interval graph [25]. Brettell et al. studied SFVS on $H-$free graphs [7].

More recently, Papadopoulos et al. studied SFVS on chordal graphs via leafage [27], while Bai et al. established fixed-parameter tractability of SFVS on tournaments [1]. Other notable results include SFVS on graphs of bounded independent set size [26], SFVS and Multi-way Cut on graphs of bounded mim-width [6], and improved exact algorithms for restricted versions: $O(1.1550^n)$ on split graphs and $O(1.1605^n)$ on chordal graphs [2]. Furthermore, kernelization results have been obtained for SFVS on chordal graphs [28].

SVC was first introduced by Brettell et al. in [7]. They showed that SVC is polynomial time solvable in $(sP_1 + P_4)-$free graphs for every $s \geq 1$. Brettell et al. in [8] showed that SVC is NP-complete on sub-cubic planar (claw, diamond)-free planar graphs, 2-unipolar graphs. NP-completeness of SVC on 2-unipolar graphs implies that this problem in NP-complete on P_7-free graphs.

AT-free graphs are also very well studied graph class. The Table 1 describes a summary of results on AT-free graphs.

2 Preliminaries

Let $G = (V, E)$ be a simple unweighted graph. We denote the set of vertices by $V(G)$ and the set of edges by $E(G)$. A graph $H = (V', E')$ is a subgraph of G if $V' \subseteq V$ and $E' \subseteq E$. We denote $|V|$ by n and $|E|$ by m.

A subgraph $H = (V', E')$ of G is called an induced subgraph if for all $u, v \in V'$, $(u, v) \in E'$ if and only if $(u, v) \in E$. The induced subgraph on a vertex subset $S \subseteq V$ is denoted by $G[S]$.

The neighbourhood of a vertex v, denoted by $N(v)$, is the set of all vertices adjacent to v. The closed neighbourhood of v is denoted by $N[v] =$

Table 1. Algorithmic studies in AT-free graphs

Problem	Complexity class	Authors
Independent set	Poly time solvable	Broersma et al. [9]
Dominating set	Poly time solvable	Kratsch [23]
Total dominating set	Poly time solvable	Kratsch [23]
Feedback vertex set	Poly time solvable	Kratsch et al. [24]
Connected dominating set	Poly time solvable	Balakrishnan et al. [3]
Induced disjoint path	Poly time solvable	Golovach et al. [20]
Induced matching	Poly time solvable	Jou-Ming Chang [10]
Maximum clique problem	NP-complete	Broersma et al. [9]
Clique cover problem	NP-complete	Broersma et al. [9]
Determination of treewidth	NP-complete	Kloks et al. [22]
Chordal completion	NP-complete	Kloks et al. [22]
Interval completion	NP-complete	Kloks et al. [22]

$\{v\} \cup N(v)$. The neighbourhood of a set of vertices $\{v_1, v_2, \ldots, v_k\}$ is denoted by $N(\{v_1, v_2, \ldots, v_k\}) = \bigcup_{i=1}^{k} N(v_i)$, and the closed neighbourhood is denoted by $N[\{v_1, v_2, \ldots, v_k\}] = \bigcup_{i=1}^{k} N[v_i]$.

Let C be a connected component of G. For a vertex $v \in V(G)$, we denote by $N_C(v)$ the set of neighbours of v that lie in the component C.

A path is a graph, $Y = (V, E)$, such that $V = \{y_1, y_2, \ldots, y_k\}$ and $E = \{y_1 y_2, \ldots, y_{k-1} y_k\}$. We denote a path by the sequence of its vertices, that is $Y = y_1 y_2 \ldots y_k$. Here y_1 and y_k are called endpoints of path Y. The number of vertices present in Y is denoted by $|Y|$. We denote $y_i Y y_j = y_i y_{i+1} \ldots y_j$ where $1 \le i \le j \le k$. A path on k vertices is denoted by Y_k and the length of the path is the number of edges present on the path that is $k - 1$. The distance between two vertices in a graph is the length of the shortest path between them. A cycle is a graph, $C = (V, E)$, such that $V(C) = \{c_1, c_2, \ldots, c_l\}$ and $E(C) = \{c_1 c_2, \ldots, c_{l-1} c_l, c_l c_1\}$. The distance, that is, the length of the shortest path between u and v is denoted by $dist_C(u, v)$ where $u, v \in V(C)$. The number of vertices present in the cycle C is denoted by $|C|$.

Let us consider a graph $G = (V, E)$, where G is an AT-free graph. For a vertex $v \in V(G)$, we denote by $C_1^v, \ldots, C_k^v$ the connected components of the graph obtained by removing the closed neighborhood of v, that is $G \setminus N[v]$.

Definition 1. *The **private neighborhood** of a vertex $x \in X$ with respect to a subset $X \subseteq V(G)$ is defined as the set $N(x) \setminus N(X \setminus x)$.*

Definition 2. *Let $U \subseteq N(v)$. We say that a subset $W \subseteq U$ represents U in G (or is a representative of U in G) if it satisfies the condition: $N(W) \setminus N(v) = N(U) \setminus N(v)$ and W has the minimum possible cardinality among all such subsets.*

Definition 3. *Let C be a connected component of $G \setminus N[v]$. A subset $W' \subseteq U$ is said to be a representative of U in C if: $N(W') \cap C = N(U) \cap C$ and W' is of minimum cardinality among all such subsets.*

We now state some fundamental properties of AT-free graphs, as established in [24].

Lemma 2. *Let $u_1, u_2, u_3 \in N(v)$ be mutually non-adjacent, and let u'_1, u'_2, u'_3 be private neighbors of $u_1, u_2,$ and u_3, respectively, with respect to the set $\{u_1, u_2, u_3\}$.*

 a. *If $u'_i, u'_j \in \{u'_1, u'_2, u'_3\}$ belong to the same connected component of $G \setminus N[v]$, then they are adjacent.*
 b. *The vertices $\{u'_1, u'_2, u'_3\}$ belongs to at most two distinct connected components of $G \setminus N[v]$.*

The algorithms in this paper are motivated by the decomposition technique known as *interval decomposition* introduced by Broersma et al. in [9].

2.1   Interval Decomposition

Let $G = (V, E)$ be an AT-free graph. Denote by $C_1^v, \ldots, C_k^v$ the connected components of the graph $G \setminus N[v]$, for a fixed vertex $v \in V(G)$. Let w be a vertex in one such component, say $w \in V(C_i^v)$.

The interval between two non-adjacent vertices w and v is denoted by $I(w, v) \subseteq V(G)$, where $I(w, v)$ consists of all vertices $s \in V(G)$ such that there exists an $s–v$ path that avoids all neighbors of w, and an $s–w$ path that avoids all neighbors of v. We state the following lemma from [9] and [24]. Please refer to Fig. 1 for illustrations.

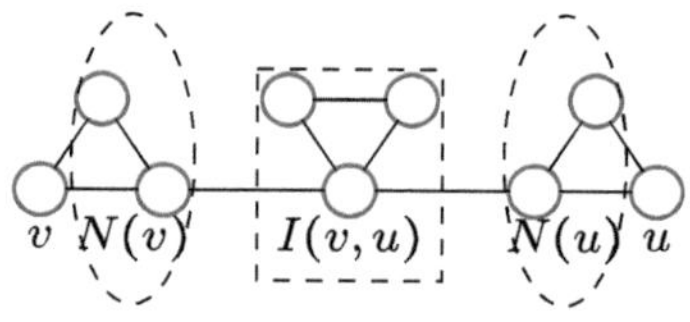

Fig. 1. An example of interval in an AT-free graph.

Lemma 3. *(Broersma et al. [9]) Let D be a component of the graph $C_i^v \setminus N[u]$ where $u \in V(C_i^v)$. Then $N[D] \cap (N[v] \setminus N[u]) = \emptyset$ if and only if D is a component of $G \setminus N[u]$.*

As a consequence of the lemma, we immediately see C_1^u is a component in $G \setminus N[u]$ where, C_1^u plays the role of D in Lemma 3. We note here if D is not the interval $I(u, v)$, in $C_i^v \setminus N[u]$, then D is always a component of $G \setminus N[u]$.

Lemma 4. *Let $G = (V, E)$ be an AT-free graph and $\mathcal{C} = \bigcup_{v \in V(G)} \{C : C$ is a component of $G \setminus N[v]\}$, then $\mathcal{C}$ contains all the non-interval components of G.*

From the above lemma we can see that the total number of non-interval components of an AT-free graph G is $O(n^2)$.

Kratsch et al. ([24]) showed that in AT-free graphs, for any $U \subseteq V(G)$ and for a vertex $v \in U$, there exists a vertex $w \in U \cap C$ where C is a component of $G \setminus N[v]$ and $C \cap U \neq \emptyset$, such that $U \cap I(v, w) = \emptyset$. The lemma is stated below.

Lemma 5. *(Kratsch et al. [24]) Let $G = (V, E)$ be an AT-free graph and $U \subseteq V$. Consider a vertex $v \in U$ and a component C of $G \setminus N[v]$ such that $U \cap C \neq \emptyset$. Then there is a vertex $w \in U \cap C$ such that $U \cap I(v, w) = \emptyset$.*

Due to page restrictions we move some details and proofs in the Appendix. Please refer to the detailed version of the paper for more details.

3   Inclusion Ordering of Vertices in Component

Let $S \subseteq V(G)$ and let C_i^s be a connected component of $G \setminus N[s]$ for some $s \in S$. In this section, we define an ordering of the vertices in $C_i^s \cap S$ that satisfies the following property. Let $\mathcal{A}_{C_i^s}[1, \ldots, t] = \{s_1, \ldots, s_t\}$ be an ordered sequence such that $\{s_1, \ldots, s_t\} = C_i^s \cap S$, and for every $j \in \{1, \ldots, t\}$ we have $I(s, \mathcal{A}_{C_i^s}[j]) \cap \mathcal{A}_{C_i^s}[j+1, \ldots, t] = \emptyset$. We refer to this as the *inclusion ordering* of the vertices. To construct this ordering, we rely on Lemma 5.

Using this lemma we device the following algorithm.

Algorithm 1: Ordering $S \cap C_i^s$

Input: Vertex s and component C_i^s
Output: Inclusion ordering of vertices in $C_i^s \cap S$: $\mathcal{A}_{C_i^s}$

1 Let $S' \leftarrow C_i^s \cap S$
2 Let $\mathcal{A}_{C_i^s}$ be an array of size $|S'|$, which is initially empty.
3 $j \leftarrow 1$
4 **while** $S' \setminus \mathcal{A}_{C_i^s}[1, \ldots, (j-1)] \neq \emptyset$ **do**
5   Find $x \in S' \setminus \mathcal{A}_{C_i^s}[1, \ldots, (j-1)]$ such that
        $I(s, x) \cap (S' \setminus \mathcal{A}_{C_i^s}[1, \ldots, (j-1)]) = \emptyset$
6      $\mathcal{A}_{C_i^s}[j] \leftarrow \{x\}$
7      $j \leftarrow j + 1$
8 **end while**
9 **Return** $\mathcal{A}_{C_i^s}$.

Remark 1. In line 5 of the algorithm, note that, there can be many choices of $x \in S' \setminus \mathcal{A}_{C_i^s}[1, \ldots, (j-1)]$ that satisfies $I(s, x) \cap (S' \setminus \mathcal{A}_{C_i^s}[1, \ldots, (j-1)]) = \emptyset$. Hence the inclusion ordering is not unique. We initially assign a numbering to the vertices of S'. While choosing x we choose the x satisfying the condition and with minimum label according to the initial numbering. In order to avoid notational complication we did not include this detail in the algorithm.

Lemma 6. *Algorithm 1 produces ordering $\mathcal{A}_{C_i^s}[1, \ldots, t] = \{s_1, \ldots, s_t\}$ such that for every $j \in \{1, \ldots, t\}$ we have $I(s, \mathcal{A}_{C_i^s}[j]) \cap \mathcal{A}_{C_i^s}[j+1, \ldots, t] = \emptyset$.*

Proof. We prove this lemma by proving an invariant of Algorithm 1. Let $S' = S \cap C_i^s$.

Invariant: Let j be an arbitrary step of the loop. In the jth step have $\mathcal{A}_{C_i^s}[1, \ldots, j-1]$, and let $U = \{s\} \cup (S' \setminus \mathcal{A}_{C_i^s}[1, \ldots, j-1])$. From Lemma 5, there exists $x \in U \cap C_i^s$ such that $I(s, x) \cap U = \emptyset$, that is, there exists an element $x \in S' \setminus \mathcal{A}_{C_i^s}[1, \ldots, j-1]$ such that $I(s, x) \cap (S' \setminus \mathcal{A}_{C_i^s}[1, \ldots, j-1]) = \emptyset$. By the arbitrariness of the choice of j we conclude that we get an ordering of the vertices in S', which is stored in the array $\mathcal{A}_{C_i^s}$.

Here $\mathcal{A}_{C_i^s}[j]$ denote the vertex inserted in the jth step. Hence all the vertices in set $\{s_{j+1}, \ldots, s_t\}$ belongs to $S' \setminus \mathcal{A}_{C_i^s}[1, \ldots, j]$, that is $\{s_{j+1}, \ldots, s_t\} = S' \setminus \mathcal{A}_{C_i^s}[1, \ldots, j]$. Hence, from the invariant established above, we get, $I(s, \mathcal{A}_{C_i^s}[j]) \cap \{s_{j+1}, \ldots, s_t\} = \emptyset$. $\qquad\square$

In each iteration of the while loop in line 4 of the algorithm one vertex is added to the list $\mathcal{A}_{C_i^s}$ from the set S'. Since there can be at most $O(n)$ vertices in S' the loop runs for $O(n)$ times. For each vertex in $S' \setminus \mathcal{A}_{C_i^s}[1, \ldots, (j-1)]$ it takes $O(n)$ time to check if $I(s, x) \cap (S' \setminus \mathcal{A}_{C_i^s}[1, \ldots, (j-1)]) = \emptyset$. Since there are $O(n)$ such vertices to check from, it takes $O(n^2)$ time to find such a vertex. Hence the running time of this algorithm becomes $O(n^3)$.

Let $s \in S$ and let C^s be a component of $G \setminus N[s]$. Let $\{s_1, \ldots, s_t\}$ be the vertices of $C^s \cap S$ in inclusion ordering as produced in Algorithm 1. We have the following lemma.

Lemma 7. *For every $\Theta \subseteq V(G)$, containing some vertices from $\{s_1, \ldots, s_t\}$, there exist $i \in \{1, 2, \ldots, t\}$ such that $\Theta \cap I(s, s_i) \cap \{s_1, \ldots, s_t\} = \emptyset$.*

Proof. Let $s_l \in \Theta \cap I(s, s_i) \cap \{s_1, \ldots, s_t\}$ be such that $\Theta \cap \{s_1, \ldots, s_{l-1}\} = \emptyset$, that is l is the minimum label in $\{1, \ldots, t\}$ such that $s_l \in \Theta$. From the property of inclusion ordering in Lemma 6, we have $I(s, s_l) \cap \{s_{l+1}, \ldots, s_t\} = \emptyset$. Combining this, with $\Theta \cap \{s_1, \ldots, s_{l-1}\} = \emptyset$, we get $\Theta \cap I(s, s_l) \cap \{s_1, \ldots, s_t\} = \emptyset$. $\qquad\square$

Remark 2. In the proof of Lemma 7, although we have used the minimum labeled vertex to assert the claim, but we note that, there can be other vertices with larger labels for which this lemma holds. Indeed, this follows from the fact that inclusion ordering is not unique, which is stated in Remark 1.

4   Subset Vertex Cover

Consider $G = (V, E)$ to be an AT-free graph. For a given $S \subseteq V(G)$, let $E_S \subseteq E(G)$ be a subset of edges such that each edge contains at least one end-point from S. The subset vertex cover problem asks to find a minimum cardinality vertex set $T \subseteq V(G)$ containing at least one end point of every edge in E_S. This problem can also be seen as finding a maximum induced subgraph H of G

such that every vertex of S present in H is an isolated vertex. We call $V(H)$ an $S-$independent set. We devise a dynamic programming algorithm to find a maximum $S-$independent set in G. Let us denote an $S-$independent set of G by Υ.

We divide the problem in two cases. First case is Υ contains at least one vertex of S and the other case is Υ does not contain any vertex of S. In the second case, the solution is simply $G \setminus S$. To compute the first case we fix a vertex $s \in S$ and take the vertex in the solution and remove all the vertices of $N(s)$ from the solution. Then we compute the maximum $S-$independent set in every component $C_1^s, \ldots, C_k^s$ of $G \setminus N[s]$. For this purpose we do the following observation.

Lemma 8. *If $s, s' \in S$ are two non adjacent vertices, then $\{s, s'\} \cup (I(s, s') \setminus S)$ is an $S-$independent set.*

Let $\mathcal{A}_{C^s} = \{s_1, \ldots, s_t\}$ is an array containing vertices of $S \cap C^s$ in inclusion ordering produced by Algorithm 1.

Lemma 9. *Suppose Υ_{C^s} is an $S-$independent set in C^s, such that, $\Upsilon_{C^s} \cap \mathcal{A}_{C^s} \neq \emptyset$. Then, there exist $w \in \mathcal{A}_{C^s}$ such that $\Upsilon_{C^s} \cap I(s, w) \cap \mathcal{A}_{C^s} = \emptyset$.*

If s_i is the first vertex from $\mathcal{A}_{C^s}$ which is present in the solution in C^s, then the above lemma ensures that $I(s, s_i)$ does not contain any vertex from $\{s_{i+1}, \ldots, s_t\}$. Since, we assumed that, s_i is the first vertex from the ordering to be present in the solution, to compute the optimum contribution from $I(s, s_i)$, we can simply remove all the vertices from the set $I(s, s_i) \cap \{s_1, \ldots, s_{i-1}\}$. We define some required notation.

- $\alpha^S(C^s, \mathcal{A}_{C^s})$: be a maximum $S-$independent set in C^s where C^s is a connected component of $G \setminus N[s]$.
- $MSIS(G, S)$: is a maximum $S-$independent set of G.

As discussed two cases are possible. First at least one vertex of S is present in the solution of C^s and second case is when no vertex of S is present in the solution. We define the following terms.

- Let $s \in S$, and C^s be a component of $G \setminus N[s]$.
- Let s_l denote $\mathcal{A}_{C^s}[l]$.
- Let $C_1^{s_l}, \ldots, C_r^{s_l}$ are the connected components of $C^s \setminus N[s_l]$.

Based on this we have the following recurrence.

$$\alpha^S(C^s, \mathcal{A}_{C^s}) = \max\{A, V(C^s \setminus S)\} \tag{1}$$

A denotes a maximum $S-$independent set in C^s. Equation 1 gives a maximum $S-$independent set in C^s comparing the cases: ($i.$) an $S-$independent set containing at least one vertex from S, ($ii.$) an $S-$independent set containing no vertex from S.

$$A = \max_{l=1}^{t}\{\{s_l\} \cup (I(s, s_l) \setminus S) \cup \bigcup_{i=1}^{r} \alpha^S(C_i^{s_l}, \mathcal{A}_{C_i^{s_l}})\} \tag{2}$$

As suggested by Lemma 9, we iterate over the choices of the vertices $s_l \in \mathcal{A}_{C^s}$. Then following the observation of Lemma 8 we combine s_l with $I(s, s_l) \setminus S$. Disjoint union of S−independent set in smaller components along with $\{s_l\} \cup (I(s, s_l) \setminus S)$ gives our desired solution in C^s.

If $\mathcal{A}_{C^s}$ is empty then,

$$\alpha^S(C^s, \mathcal{A}_{C^s}) = V(C^s) \tag{3}$$

Let $G = (V, E)$ be an AT-free graph and let $S \subseteq V(G)$. Then $MSIS(G, S)$ is computed by the following equation.

$$MSIS(G, S) = \max\{\max_{s \in S}\left(\{s\} \cup \bigcup_{i=1}^{k} \alpha^S(C_i^s, \mathcal{A}_{C^s})\right), V(G \setminus S)\} \tag{4}$$

Correctness of Eq. 4 follows from correctness of recurrence 1-3 stated in Lemma 10. Based on the above recurrences we obtain the following algorithm.

Hence subset vertex cover for given G and $S \subseteq V(G)$ is $G \setminus MSIS(G, S)$.

4.1   Correctness and Complexity

Lemma 10. *The recurrences 1 - 3 compute $\alpha^S(C^s, \mathcal{A}_{C^s})$ correctly.*

Theorem 2. *There is a $O(n^4)$ algorithm to compute minimum subset vertex cover in AT-free graph.*

5   Subset Feedback Vertex Set

5.1   Potential Local Solution and There Compatibility

An induced subgraph H of G is called an S−forest of G, if no cycle in H contains vertices from S. Let F_S be an S−forest of G and let $s \in F_S \cap S$. Let $X_s \subseteq N(s) \cap F_S$ such that vertices in X_s are not leaves of F_S and $Y_s \subseteq N(s) \cap F_S$ is the representative(Definition 2) of L_s where L_s is the set of leaves in F_S that are adjacent to s. We denote a triplet (s, X_s, Y_s) by *potential local solution (PLS)*. In [24], Kratsch et al. proved that, for a forest, $|X_v \cup Y_v| \leq 4$. We mimic their idea in the context of S−forest for the same bound. In fact the steps involved in the proof are also similar, but to make the paper self contained we include their proof. Apart from this observation, we require some novel structural analysis of AT-free graphs to arrive at the desired result.

Lemma 11. *Let C be a component of $G \setminus N[s]$, then, at most one vertex from X_s can have neighbor in $C \cap F_S$.*

Lemma 12. *Let (s, X_s, Y_s) be a PLS, then $|X_v \cup Y_v| \leq 4$.*

Our algorithm uses dynamic programming paradigm to combine solutions in smaller components of a component and the interval therein, in an interval decomposition, to obtain a solution of the bigger component. To maintain the invariant that the union of the solutions of the smaller components is indeed a solution of bigger component we introduce the notion of *compatibility* among solutions in the smaller components.

Consider a PLS (s, X_s, Y_s). Solutions in different components of $G \setminus N[s]$ are vertex disjoint and they are not adjacent to each other. But from Lemma 11, solution in a component may be adjacent to one vertex of X_s. Hence we define compatibility of solutions with PLS.

- Let, Λ denote all PLS (s, X_s, Y_s) such that $s \in S$.

We say two PLS (s, X_s, Y_s) and $(s', X_{s'}, Y_{s'})$ are compatible if $G[\{s, s'\} \cup X_s \cup Y_s \cup X_{s'} \cup Y_{s'}]$ is an $S-$forest and $Y_s \cup Y_{s'}$ are leaves of the $S-$forest. Since size of each PLS are constant, we can find out all the PLSs which are compatible with each other.

- We denote by $\Lambda^{(s, X_s, Y_s)}$ all the PLS that are compatible with the PLS (s, X_s, Y_s).

In the following we define compatibility between a particular solution and a PLS. To be precise, for a given PLS (s, X_s, Y_s), we define compatibility between the PLS and solutions in components of $G \setminus N[s]$, solution in $N(s)$, and solution in interval.

Definition 4. *(Compatibility with a solution in a component of $G \setminus N[s]$)*
Let F be an $S-$forest in C where C is a component of $G \setminus N[s]$. We say F and PLS (s, X_s, Y_s) are compatible if $G[V(F) \cup \{s\} \cup X_s \cup Y_s]$ is an $S-$forest and vertices in Y_s are leaves of that $S-$forest.

Note that, from Lemma 11, at most one vertex from X_s can have neighbor in F.

Definition 5. *(Compatibility with a solution in $N(s)$)* *Let $F_{N(s)}$ be an $S-$forest in $N(s)$ then we say $F_{N(s)}$ is compatible with PLS (s, X_s, Y_s) if:*

- *$X_s \cup Y_s \subseteq V(F_{N(s)})$.*
- *Neighborhood of vertices in $V(F_{N(s)}) \setminus X_s$ is represented by Y_s in G.*
- *$G[\{s\} \cup V(F_{N(s)})]$ is an $S-$forest.*

Definition 6. *(compatibility with a solution in an interval)* *Let F' be an $S-$forest in $I(s, s')$. F' is compatible with PLS (s, X_s, Y_s) and $(s', X_{s'}, Y_{s'})$ if:*

- $G[V(F') \cup \{s, s'\} \cup X_s \cup Y_s \cup X_{s'} \cup Y_{s'}]$ *is an $S-$forest.*
- *Vertices in $Y_s \cup Y_{s'}$ are leaves of the $S-$forest.*

The following lemma is an essential part for correctness proof of the dynamic programming recurrence relations.

Lemma 13. *Let F_i be an $S-$forest in C_i^s compatible with PLS (s, X_s, Y_s), and let $F_{N(s)}$ be an $S-$forest in $N(s)$ compatible with PLS (s, X_s, Y_s). Then $G[\{s\} \cup V(F_{N(s)}) \cup \bigcup_{i=1}^k V(F_i)]$ is an $S-$forest.*

5.2   Dynamic Programming

For a fixed PLS (s, X_s, Y_s) we find solutions in $N[s]$ compatible with the PLS (s, X_s, Y_s). Then we find solutions in every component $C_1^s, \ldots, C_k^s$ of $G \setminus N[s]$ such that the solutions are compatible with (s, X_s, Y_s). We first describe the computation of $S-$forest in $N[s]$. Let F_S be a maximum $S-$forest of G containing the PLS (s, X_s, Y_s).

Now we discuss the recurrence relations to find maximum $S-$forest of G for a given set of vertices S. Let $F_S^*(G, S)$ be a maximum $S-$forest of G with respect to S.

- $\zeta(s, X_s, Y_s, C^s, \mathcal{A}_{C^s})$: is maximum $S-$forest in C^s, that is compatible with PLS (s, X_s, Y_s), where C^s is a connected component of $G \setminus N[s]$, and $\mathcal{A}_{C^s}$ are the vertices of $S \cap C^s$ in inclusion ordering.
- $\iota(s, X_s, Y_s)$: is a maximum $S-$forest in $N(s)$ compatible with PLS (s, X_s, Y_s).
- MaxIForest$(s, X_s, Y_s, s', X_{s'}, Y_{s'})$: the maximum induced S-forest in $I(s, s') \setminus S$ that is compatible with both (s, X_s, Y_s) and $(s', X_{s'}, Y_{s'})$.
- MaxCForest(s, X_s, Y_s, C^s) : is a maximum induced S-forest in $C^s \setminus S$ that is compatible with (s, X_s, Y_s).

A brief overview to compute MaxIForest and MaxCForest are given in Sect. 5.4. A detailed computation please see the detailed version of the paper.

The $F_S \cap N(s)$ is necessarily an independent set otherwise an edge in $F_S \cap N(s)$ creates cycle with s hence contradicting the fact that F_S is an $S-$forest. Let Y' denote the vertices of $N(s)$ that are represented by Y_s in G (see Definition 2), and let $Y = Y' \setminus N[X_s \cup Y_s]$. As pointed out $F_S \cap N(s)$ in independent set, we have the following observation.

Lemma 14. *Consider PLS (s, X_s, Y_s). Then, $\iota(s, X_s, Y_s) = \{s\} \cup X_s \cup Y_s \cup \alpha(G[Y])$.*

Let $\mathcal{A}_{C^s} = \{s_1, \ldots, s_t\}$ is an array containing vertices of $S \cap C^s$ in inclusion ordering produced by Algorithm 1.

Lemma 15. *Suppose F_{C^s} is an $S-$forest in C^s, such that, $F_{C^s} \cap \mathcal{A}_{C^s} \neq \emptyset$. Then, there exist $w \in \mathcal{A}_{C^s}$ such that $F_{C^s} \cap I(s, w) \cap \mathcal{A}_{C^s} = \emptyset$.*

Now we state and prove the recurrence to compute ζ. First, at least one vertex of S is present in the solution of C^s and second case is when no vertex of S is present in the solution. We define the following terms.

- Let (s, X_s, Y_s) be a PLS, and C^s be a component of $G \setminus N[s]$.
- Let s_l denote $\mathcal{A}_{C^s}[l]$.
- Let $C_1^{s_l}, \ldots, C_r^{s_l}$ are the connected components of $C^s \setminus N[s_l]$.

Based on this we have the following recurrence.

$$\zeta(s, X_s, Y_s, C^s, \mathcal{A}_{C^s}) = \max\{A, \mathsf{MaxCForest}(s, X_s, Y_s, C^s)\} \tag{5}$$

A denotes a maximum $S-$forest in C^s. Equation 5 gives a maximum $S-$forest in C^s comparing the cases: $(i.)$ an $S-$forest containing at least one vertex from S, $(ii.)$ an $S-$forest containing no vertex from S.

$$A = \max_{l=1}^{t} \max_{(s_l, X_{s_l}, Y_{s_l}) \in \Lambda^{(s, X_s, Y_s)}} \iota(s_l, X_{s_l}, Y_{s_l}) \cup \mathsf{MaxIForest}(s, X_s, Y_s, s_l, X_{s_l}, Y_{s_l}) \cup$$
$$\bigcup_{j=1}^{r} \zeta(s_l, X_{s_l}, Y_{s_l}, C_j^{s_l}, \mathcal{A}_{C_j^{s_l}}) \tag{6}$$

As suggested by Lemma 15, we iterate over the choices of the vertices $s_l \in \mathcal{A}_{C^s}$. Then following the observation of Lemma 13 we combine s_l with $I(s, s_l) \setminus S$. Disjoint union of $S-$forest in smaller components along with $\{s_l\} \cup (I(s, s_l) \setminus S)$ gives our desired solution in C^s.

If $\mathcal{A}_{C^s}$ is empty then,

$$\zeta(s, X_s, Y_s, C^s, \mathcal{A}_{C^s}) = \mathsf{MaxCForest}(s, X_s, Y_s, C^s) \tag{7}$$

Let $G = (V, E)$ be an AT-free graph and let $S \subseteq V(G)$. Then $F_S^*(G, S)$ is is computed by the following equation. First is if at least one element of S is present in the solution.

$$F_S^*(G, S) = \max\{ \max_{s \in S} \max_{(s, X_s, Y_s) \in \Lambda} \{\iota(s, X_s, Y_s) \cup \bigcup_{j=1}^{r} \zeta(s, X_s, Y_s, C^s, \mathcal{A}_{C^s})\}\} \tag{8}$$

This is the case when no vertices of S is in the solution.

$$|F_S^*(G, S)| = G \setminus S \tag{9}$$

Hence the maximum among these two solutions is the maximum for $F_S^*(G, S)$. Subset feedback vertex set for given G and $S \subseteq V(G)$ is $V(G) \setminus F_S^*(G, S)$.

5.3   Correctness and Complexity

Lemma 16. *The recurrence 5-7 computes $\zeta(s, X_s, Y_s, C, \mathcal{A}_{C^s})$ correctly.*

Theorem 1. *There is a $O(n^{16})$ algorithm that computes SFVS in a given AT-free graph.*

Although we settle the open question posed by Papadopoulos et al. in [25], by providing a polynomial time algorithm for SFVS, the running time of the algorithm is very high. The main bottleneck for high time complexity is due to the fact that, we need to compute MaxIForest for each pair of PLS. Indeed, there can be $O(n^{10})$ possible pairs of PLS and to compute MaxIForest for each pair of PLS requires $O(n^6)$ time. Given this scenario, it will be an interesting question to pursue, if optimum solution can be obtained by computing MaxIForest over a substantially lesser pairs of PLS.

5.4   Computation of MaxIForest and MaxCForest

Consider a fixed $s \in S$ and a fixed PLS $(s, X_s, Y_s) \in \Lambda$ and let C^s be a connected component of $G \setminus N[s]$. To compute a maximum $S-$forest in C^s which is compatible with (s, X_s, Y_s), and which contains some vertices from $\mathcal{A}_{C^s}$, we fix a $w \in \mathcal{A}_{C^s}$ and $(w, X_w, Y_w) \in \Lambda^{(s, X_s, Y_s)}$ and compute a maximum solution F_{C^s} that satisfies the property $F_{C^s} \cap I(s, w) \cap S = \emptyset$. This works due to the fact that Lemma 15 holds. Since we compute a solution where no vertices of $I(s, w) \cap S$ is present in $F_{C^s} \cap I(s, w)$, it is sufficient to consider the set $I(s, w) \setminus S$. Note that, solution in $I(s, w) \setminus S$ must be compatible with both (s, X_s, Y_s) and (w, X_w, Y_w). Based on the adjacencies between vertices of X_s and X_w, we compute the aforementioned solution either by computing a maximum independent set on vertices of $I(s, w) \setminus S$ or computing an $MSIS$ on specific vertices of $I(s, w) \setminus S$. For a detailed explanation please see detailed version of the paper.

References

1. Bai, T., Xiao, M.: A parameterized algorithm for subset feedback vertex set in tournaments. Theoret. Comput. Sci. **975**, 114139 (2023)
2. Bai, T., Xiao, M.: Exact algorithms for restricted subset feedback vertex set in chordal and split graphs. Theoret. Comput. Sci. **984**, 114326 (2024)
3. Balakrishnan, H., Rajaraman, A., Rangan, C.P.: Connected domination and steiner set on asteroidal triple-free graphs. In: Dehne, F., Sack, J.-R., Santoro, N., Whitesides, S. (eds.) WADS 1993. LNCS, vol. 709, pp. 131–141. Springer, Heidelberg (1993). https://doi.org/10.1007/3-540-57155-8_242
4. Bar-Yehuda, R., Geiger, D., Naor, J., Roth, R.M.: Approximation algorithms for the feedback vertex set problem with applications to constraint satisfaction and bayesian inference. SIAM J. Comput. **27**(4), 942–959 (1998)

5. Becker, A., Geiger, D.: Optimization of pearl's method of conditioning and greedy-like approximation algorithms for the vertex feedback set problem. Artif. Intell. **83**(1), 167–188 (1996)
6. Bergougnoux, B., Papadopoulos, C., Telle, J.A.: Node multiway cut and subset feedback vertex set on graphs of bounded mim-width. Algorithmica **84**(5), 1385–1417 (2022)
7. Brettell, N., Johnson, M., Paesani, G., Paulusma, D.: Computing subset transversals in h-free graphs. Theoretical Comput. Sci. **902**, 76–92 (2022)
8. Brettell, N., Oostveen, J.J., Pandey, S., Paulusma, D., Rauch, J., van Leeuwen, E.J.: Computing subset vertex covers in h-free graphs. Theoretical Comput. Sci. **1032**, 115088 (2025)
9. Broersma, H., Kloks, T., Kratsch, D., Müller, H.: Independent sets in asteroidal triple-free graphs. SIAM J. Discrete Math. **12**(2), 276–287 (1999)
10. Chang, J.M.: Induced matchings in asteroidal triple-free graphs. Discrete Appl. Math. **132**(1–3), 67–78 (2003)
11. Chitnis, R., Fomin, F.V., Lokshtanov, D., Misra, P., Ramanujan, M.S., Saurabh, S.: Faster exact algorithms for some terminal set problems. J. Comput. Syst. Sci. **88**, 195–207 (2017)
12. Corneil, D.G., Olariu, S., Stewart, L.: Asteroidal triple-free graphs. SIAM J. Discrete Math. **10**(3), 399–430 (1997)
13. Cygan, M., Pilipczuk, M., Pilipczuk, M., Wojtaszczyk, J.O.: Subset feedback vertex set is fixed-parameter tractable. SIAM J. Discrete Math. **27**(1), 290–309 (2013)
14. Even, G., Naor, J., Zosin, L.: An 8-approximation algorithm for the subset feedback vertex set problem. SIAM J. Comput. **30**(4), 1231–1252 (2000)
15. Festa, P., Pardalos, P.M., Resende, M.G.: Feedback set problems. In: Encyclopedia of Optimization, pp. 1005–1016. Springer (2008)
16. Fomin, F.V., Gaspers, S., Lokshtanov, D., Saurabh, S.: Exact algorithms via monotone local search. J. ACM (JACM) **66**(2), 1–23 (2019)
17. Fomin, F.V., Heggernes, P., Kratsch, D., Papadopoulos, C., Villanger, Y.: Enumerating minimal subset feedback vertex sets. Algorithmica **69**(1), 216–231 (2014)
18. Garey, M.R., Johnson, D.S.: Computers and intractability, vol. 29. wh freeman New York (2002)
19. Golovach, P.A., Heggernes, P., Kratsch, D., Saei, R.: Subset feedback vertex sets in chordal graphs. J. Discrete Algorithms **26**, 7–15 (2014)
20. Golovach, P.A., Paulusma, D., van Leeuwen, E.J.: Induced disjoint paths in AT-free graphs. In: Fomin, F.V., Kaski, P. (eds.) SWAT 2012. LNCS, vol. 7357, pp. 153–164. Springer, Heidelberg (2012). https://doi.org/10.1007/978-3-642-31155-0_14
21. Golumbic, M.C.: Algorithmic graph theory and perfect graphs, vol. 57. Elsevier (2004)
22. Kloks, T., Kratsch, D., Spinrad, J.: On treewidth and minimum fill-in of asteroidal triple-free graphs. Theoret. Comput. Sci. **175**(2), 309–335 (1997)
23. Kratsch, D.: Domination and total domination on asteroidal triple-free graphs. Discret. Appl. Math. **99**(1–3), 111–123 (2000)
24. Kratsch, D., Müller, H., Todinca, I.: Feedback vertex set on AT-free graphs. Discrete Appl. Math. **156**(10), 1936–1947 (2008)
25. Papadopoulos, C., Tzimas, S.: Polynomial-time algorithms for the subset feedback vertex set problem on interval graphs and permutation graphs. Discrete Appl. Math. **258**, 204–221 (2019)
26. Papadopoulos, C., Tzimas, S.: Subset feedback vertex set on graphs of bounded independent set size. Theoret. Comput. Sci. **814**, 177–188 (2020)

27. Papadopoulos, C., Tzimas, S.: Computing a minimum subset feedback vertex set on chordal graphs parameterized by leafage. Algorithmica **86**(3), 874–906 (2024)
28. Philip, G., Rajan, V., Saurabh, S., Tale, P.: Subset feedback vertex set in chordal and split graphs. Algorithmica **81**(9), 3586–3629 (2019)
29. Zhang, P., et al.: An algorithm based on graph theory for the assembly of contigs in physical mapping of dna. Bioinformatics **10**(3), 309–317 (1994)

Outer-Planar Vertex Deletion on AT-Free Graphs

Joydeep Mukherjee[1](✉) and Tamojit Saha[2]

[1] Ramakrishna Mission Vivekananda Educational and Research Institute, Howrah,
India
joydeep.m@gm.rkmvu.ac.in
[2] Institute for Advancing Intelligence, TCG CREST, Kolkata, India
http://rkmvu.ac.in/, https://www.tcgcrest.org/

Abstract. The $\mathcal{F}$-minor-free-deletion problem asks to find minimum number of vertices S in a given graph G such that $G \setminus S$ does not contain any graph from the specified family $\mathcal{F}$ as a minor. In this paper, we consider $\{K_{2,3}, K_4\}$-minor-free-deletion problem in AT-free graphs, which is equivalent to finding a maximum induced outer-planar graph of a given AT-free graph. We show that this problem is solvable in polynomial-time. This problem can also be seen as a natural extension of minimum vertex cover and minimum feedback vertex set problem since they are equivalent to $\{K_2\}$ and $\{K_3\}$ minor free deletion problems respectively and are known to be polynomial time solvable in AT-free graphs [2,24]. Additionally, we provide a polynomial-time algorithm for finding a maximum induced linear forest in AT-free graphs, which serves as a key subroutine in our algorithm to compute a maximum induced outer-planar graph.

Keywords: AT-free graphs · Graph Algorithm · Optimization Algorithm

1 Introduction

Asteroidal Triple (AT) in a graph $G = (V, E)$ is a set of three vertices $\{u, v, w\}$ of $V(G)$ such that these three vertices are mutually nonadjacent, and for any two vertices of this set there exists a path between these two vertices that avoids the neighborhood of the third vertex. A graph that does not contain any asteroidal triple is called *AT-free*. In this work, we study outer-planar vertex deletion problem on AT-free graphs. Recall that an outer-planar graph can be equivalently characterized as a $\{K_{2,3}, K_4\}$–minor free graph [7]. Computing a *maximum induced outer-planar subgraph* (MIOP) is equivalent to finding a vertex set $S \subseteq V(G)$ of minimum cardinality, such that the graph $G \setminus S$ excludes both $K_{2,3}$ and K_4 as minors.

We are motivated to study this problem as a part of answering a more general question on AT-free graphs which we explain next. Consider $\mathcal{F}$ that denotes a fixed family of graphs. Given a graph G, *$\mathcal{F}$-minor-free deletion problem* asks to

© The Author(s), under exclusive license to Springer Nature Switzerland AG 2026
N. Misra and A. Pandey (Eds.): CALDAM 2026, LNCS 16445, pp. 351–364, 2026.
https://doi.org/10.1007/978-3-032-17156-6_26

find a set of vertices S in G of minimum cardinality, such that $G \setminus S$ does not contain any graph from the specified family $\mathcal{F}$ as a minor. In planar $\mathcal{F}$-minor-free deletion problem, $\mathcal{F}$ includes at least one planar graph. Note that, $K_{2,3}$, K_4 being planar graphs, MIOP also forms an instance of planar $\mathcal{F}$-minor-free deletion problem. Other examples include classical *vertex cover* (VC) and *feedback vertex set* (FVS) problems. Indeed, VC corresponds to deleting a minimum number of vertices to obtain a $\{K_2\}$-minor-free graph, while FVS corresponds to $\{K_3\}$-minor-free deletion, where K_2, K_3 are planar graphs. Each of VC, FVS and MIOP admits polynomial sized kernel(for definition of kernel refer [6]) in general graphs [8]. More generally, it is known that planar $\mathcal{F}$-minor-free deletion problems admits polynomial sized kernel in general graphs [11]. On the other hand, VC and FVS are known to be solvable in polynomial time on AT-free graphs [2,24]. Drawing an analogy with the fact: planar $\mathcal{F}$-minor-free deletion problem admits polynomial-sized kernel in general graphs, and VC and FVS being solvable in AT-free graphs, we pose a more general question, **Does planar $\mathcal{F}$-minor-free deletion problem admit polynomial-time algorithm in AT-free graphs?**. As a first step towards answering this question, we study MIOP on AT-free graphs.

Another motivation behind choosing AT-free graphs comes from the fact that it is a fairly large graph class, among the family of special graph classes, as it generalizes several well known graph families which includes cocomparability graphs, interval graphs, permutation graphs, and trapezoid graphs [5], each of which has diverse practical applications [14].

1.1   Our Contribution

In this work, we present a polynomial-time algorithm for the MIOP problem on AT-free graphs, thereby enriching the pool of efficiently solvable problems in this important graph class. We have designed our algorithms by suitably adapting the technique of *interval decomposition* introduced in Broersma et al. [2] and the concept of *representative set* introduced by Kratsch et al. in [24]. To make these techniques work, we make several structural observations, which deepens our understanding about the structural properties of AT-free graphs. Extending the concept of representative set we define *potential local solution* in (Sect. 1.6). We show that there are only polynomially many potential local solutions. Next we use dynamic programming to extend potential local solutions to an MIOP. Since our dynamic programming iterates over all potential local solutions, we could show that our algorithm indeed runs in polynomial time. In fact, we establish the following theorem.

Theorem 1. *There is an $O(n^{19})$ time algorithm to compute MIOP in AT-free graph.*

In a graph $G = (V, E)$ a set $S \subseteq V(G)$ is called a *linear forest* if every vertex in $G[S]$ has degree at most two and $G[S]$ is a forest. As an intermediate step for solving MIOP we provide a polynomial-time algorithm which computes induced linear forest of maximum size in an AT-free graph. We state the problem formally below. In the rest of the paper we denote this problem by *Maximum induced forest of maximum degree two (IFMDT)*. For IFMDT we obtain the following result.

Theorem 2. *There is an $O(n^6)$ time algorithm to compute IFMDT in AT-free graph.*

1.2   Organization of this Paper

In Sect. 1.3, we discuss the known results relevant to our work. We introduce the necessary terminology and notation used throughout the paper, in Sect. 1.4. Section 1.5 describes the interval decomposition technique as introduced in Broersma et al. [2]. An overview of the dynamic programming algorithm is presented in Sect. 1.6. In Sect. 2, we describe the algorithm for IFMDT. Section 3 provides the algorithm for MIOP, and missing proofs are included in the detailed version of the paper.

1.3   Known Results

MIOP can be viewed as a natural extension of VC and FVS. While this problem is NP-complete in general graphs, which follows from the work of Yannakakis et al. [25], it has also been studied from FPT and approximation algorithm point of view: Donkers et al. [8] provides a quartic kernel for MIOP in general graphs. Gupta et al. [15] propose a constant-factor approximation algorithm for same problem in general graphs. Notably, several graph optimization problems that are computationally hard in general graphs become tractable in AT-free graphs, as summarized in Table 1.

MIOP also belong to a broader class of graph modification problems known as vertex deletion problems. A vertex deletion problem for a graph property Π asks to find the minimum number of vertices whose removal makes the remaining graph satisfy property Π. A hereditary graph property is one that holds for any induced sub-graphs. Yannakakis et al. [25] showed that vertex deletion problems for non-trivial hereditary properties are NP-hard on general graphs. Fomin et al. provides a randomized constant factor approximation algorithm and a polynomial kernel for planar $\mathcal{F}$-minor-free deletion [10,11]. Several results have also been established for special graph classes. Fomin et al. provides QPTAS for connected planar $\mathcal{F}$-minor-free deletion on minor free graphs [12]. Yannakakis et al. examined this problem in bipartite graphs [30]. This problem has been investigated in chordal graphs by Cao et al. in [3] and Otachi et al. studied the complexity of vertex deletion problems on graphs with small mim width which include some well known graph classes like interval graphs permutation graphs etc. [27].

Planar vertex deletion or equivalently $\{K_5, K_{3,3}\}$-minor-free deletion, have also been extensively studied. Several results are known for this problem in FPT paradigm, which are due to Jansen et al. [16], Kawarabayashi [18], Marx et al.

Table 1. Algorithmic studies in AT-free graphs

Problem	Complexity class	Authors
Independent set	Poly time solvable	Broersma et al. [2]
Dominating set	Poly time solvable	Kratsch [23]
Total dominating set	Poly time solvable	Kratsch [23]
Feedback vertex set	Poly time solvable	Kratsch et al. [24]
Connected dominating set	Poly time solvable	Balakrishnan et al. [1]
Induced disjoint path	Poly time solvable	Golovach et al. [13]
Induced matching	Poly time solvable	Jou-Ming Chang [4]
Maximum clique problem	NP-complete	Broersma et al. [2]
Clique cover problem	NP-complete	Broersma et al. [2]
Determination of treewidth	NP-complete	Kloks et al. [22]
Chordal completion	NP-complete	Kloks et al. [22]
Interval completion	NP-complete	Kloks et al. [22]

[26]. This problem has also been studied from approximation algorithm point of view by Jansen et al. [17], Kawarabayashi et al. [19,20].

Few structural results are known for IFMDT. Pelsmajer [28] shows that every outer-planar graph contains a linear forest of size $\lceil \frac{4n+2}{7} \rceil$, while Poh [29] shows every planar graph contains a linear forest of at least one-third vertices. Dross et al. [9] provides an upper bound and a lower bound on the number of vertices in maximum induced linear forest of triangle free planar graphs. Klein et al. [21] provides a parallel algorithm for finding linear forest in weighted graphs.

1.4   Preliminaries

Let $G = (V, E)$ be a simple unweighted graph. We denote the set of vertices by $V(G)$ and the set of edges by $E(G)$. A graph $H = (V', E')$ is a subgraph of $G = (V, E)$ if $V' \subseteq V$ and $E' \subseteq E$. We denote $|V|$ by n and $|E|$ by m. A subgraph $H = (V', E')$ of G is an induced subgraph if $V' \subseteq V$ and for $u, v \in V'$, $(u, v) \in E'$ if and only if $(u, v) \in E$. The induced subgraph on any subset $S \subseteq V$ is denoted by $G[S]$. The neighbourhood of a vertex v, denoted by $N(v)$, is the set of all vertices that are adjacent to v. Closed neighbourhood of v is denoted by $N[v] = \{v\} \cup N(v)$. The neighbourhood of a set of vertices $\{v_1, v_2, \ldots, v_k\}$ is denoted by $N(v_1, v_2, \ldots, v_k) = \bigcup_{i=1}^{k} N(v_i)$ and the closed neighbourhood is denoted by $N[v_1, v_2, \ldots, v_k] = \bigcup_{i=1}^{k} N[v_i]$. Assume C is a connected component of G. The set $N_C(v)$ where $v \in V(C)$, denotes the set of neighbour of v that are in the component C. A path is a graph, $Y = (V, E)$, such that $V = \{y_1, y_2, \ldots, y_k\}$ and $E = \{y_1 y_2, y_2 y_3, \ldots, y_{k-1} y_k\}$. We denote a path by the sequence of its vertices, that is $Y = y_1 y_2 \ldots y_k$. Here y_1 and y_k are called endpoints of path Y. The number of vertices present in Y is denoted by $|Y|$. We denote $y_i Y y_j = y_i y_{i+1} \ldots y_j$ where $1 \leq i \leq j \leq k$. A path on k vertices is denoted by Y_k and the length of the path is denoted by the number of edges present on the path that is $k - 1$. The distance between two vertices in a graph is the length of the shortest path between them. A cycle is a graph, $C = (V, E)$, such that $V(C) = \{c_1, c_2, \ldots, c_l\}$ and $E(C) = \{c_1c_2, \ldots, c_{l-1}c_l, c_lc_1\}$. The shortest distance between u and v is denoted by $dist_C(u, v)$ where $u, v \in V(C)$. The number of vertices present in the cycle C is denoted by $|C|$. A graph is planar if the graph can be represented in a two-dimensional plane without any edge crossing. An outer-planar graph is a planar graph such that every vertex of that graph is on the outer face of the graph. It is well known in the literature that a graph is outerplanar if and only if it does not contain any K_4 or $K_{2,3}$ as a minor and a graph is planar if and only if it does not contain any K_5 or $K_{3,3}$ as a minor.

Definition 1. *Let $U \subseteq N(v)$. We call $W \subseteq U$ represents (or representative of) U in G if $N(W) \setminus N(v) = N(U) \setminus N(v)$ and W has minimum possible cardinality. We call $W' \subseteq U$ a representative of U in C, where C is a connected component of $G \setminus N[v]$, if $N(W') \cap C = N(U) \cap C$ and W' is of minimum cardinality. Every vertex in W' must have a private neighbor in component C otherwise we can get a smaller representative set than W'.*

The algorithms in this paper are motivated by the decomposition technique known as *interval decomposition* introduced by Broersma et al. in [2].

1.5   Interval Decomposition

Let $G = (V, E)$ be an AT-free graph. Denote by $C_1^v, \ldots, C_k^v$ the connected components of the graph $G \setminus N[v]$, for a fixed vertex $v \in V(G)$. Let w be a vertex in one such component, say $w \in V(C_i^v)$. The interval between two non-adjacent vertices w and v is denoted by $I(w, v) \subseteq V(G)$, where $I(w, v)$ consists of all vertices $s \in V(G)$ such that there exists an s–v path that avoids all neighbors of w, and an s–w path that avoids all neighbors of v. We state the following lemma from [2,24]. Please refer to Fig. 1 for illustrations.

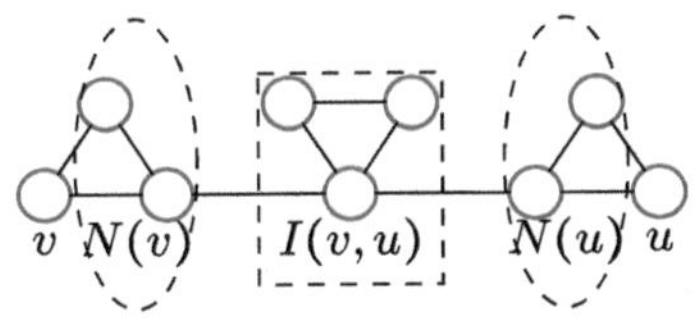

Fig. 1. An example of interval in an AT-free graph.

Lemma 1. *(Broersma et al. [2]) Let D be a component of the graph $C_i^v \setminus N[u]$ where $u \in V(C_i^v)$. Then $N[D] \cap (N[v] \setminus N[u]) = \varnothing$ if and only if D is a component of $G \setminus N[u]$.*

As a consequence of the lemma, we immediately see C_1^u is a component in $G \setminus N[u]$ where, C_1^u plays the role of D in Lemma 1. We note here if D is not the interval $I(u, v)$, in $C_i^v \setminus N[u]$, then D is always a component of $G \setminus N[u]$. Hence we have the following lemma. From this lemma we can see that the total number of non-interval components of an AT-free graph G is $O(n^2)$.

Lemma 2. *Let $G = (V, E)$ be an AT-free graph and $\mathcal{C} = \bigcup_{v \in V(G)} \{C : C \text{ is a}$ component of $G \setminus N[v]\}$, then $\mathcal{C}$ contains all the non-interval components of G.*

From the above lemma we can see that the total number of non-interval components of an AT-free graph G is $O(n^2)$. Kratsch et al. ([24]) showed that in AT-free graphs, for any $U \subseteq V(G)$ and for a vertex $v \in U$, there exists a vertex $w \in U \cap C$ where C is a component of $G \setminus N[v]$ and $C \cap U \neq \varnothing$, such that $U \cap I(v, w) = \varnothing$. The lemma is stated below.

Lemma 3. *(Kratsch et al. [24]) Let $G = (V, E)$ be an AT-free graph and $U \subseteq V$. Consider a vertex $v \in U$ and a component C of $G \setminus N[v]$ such that $U \cap C \neq \phi$. Then there is a vertex $w \in U \cap C$ such that $U \cap I(v, w) = \varnothing$.*

1.6   Dynamic Programming Framework

We employ dynamic programming on interval decomposition of an AT-free graph to formulate our algorithm for each of the problems stated above. We explain this framework by providing the main steps that are required to obtain a solution for MIOP.

Let T be an induced outer-planar graph in G. Assume $v \in V(G) \cap V(T)$. We show that the size of the representative set of $N(v) \cap V(T)$ is at most eight, using AT-freeness of G [Lemma: 6, 7]. Since the number of representative vertices is constant we find all representative sets of $N(v)$ that satisfy the conditions to be present in an MIOP. Such a representative set along with the vertex v is called a *potential local solution (PLS)*. Each PLS being an outer-planar graph and of constant size, they form the base instances of our dynamic programming. We begin our algorithm by fixing a PLS. We find a maximum solution in $N[v]$ containing that PLS. Let $C_1^v, \ldots, C_k^v$ be the components of $G \setminus N[v]$. We show that an MIOP containing a fixed PLS can be obtained as a disjoint union of optimal solutions, under relevant constraints, computed in the components $C_1^v, \ldots, C_k^v$ along with the optimal solution contained in $N[v]$ containing the PLS. The optimal solution is obtained by finding the maximum over all choices of PLS.

A set of induced outer-planar sub-graphs $G[H_1], \ldots, G[H_r]$ of G are said to be compatible if $G[H_1 \cup \ldots \cup H_r]$ is also an outer-planar graph. We fix a PLS X in $N[v]$. We denote an MIOP in $N[v]$ containing X by MIOP_X. Now we discuss how to compute solutions in each of the components $C_1^v, \ldots, C_k^v$, which are compatible with MIOP_X. Consider C_i^v for some $i \in \{1, \ldots, k\}$ and fix a vertex $w \in C_i^v$. Let $C_1^w, \ldots, C_l^w$ be the components of $C_i^v \setminus N[w]$ and $I(v, w)$ be the interval between v and w. We fix a PLS in $N[w]$ say X'. We show that a solution in $N[w]$

containing X' is compatible with a solution in $N[v]$ containing X if and only if X is compatible with X' [Lemma 13]. For an MIOP containing vertex v, it follows from Lemma 3, there exists a vertex $w \in V(C_i^v)$ such that the MIOP does not contain any vertex from $I(v, w)$. As a consequence we restrict our search of an MIOP only within the non-interval components of $C_i^v \setminus N[w]$. Thus for a fixed w in C_i^v, we compute an MIOP in C_i^v containing X', which is compatible with X, and contains no vertex from $I(v, w)$. The maximum such solution over all choices of $X' \in V(C_i^v)$ gives an MIOP in C_i^v which contains w. Iterating over all choices of w we obtain an MIOP in C_i^v which is compatible with MIOP_X. From the above discussion, it is evident that our algorithm proceeds by checking compatibility between a pair of PLS, thereby computing a MIOP in non-interval component. As a result, the running time of this algorithm remains polynomial since there are polynomially many PLS [Lemma: 6, 7], and polynomially many non-interval components [Lemma 2] in G. Given the above outline, we expand upon the following points for each of the problems.

- Description of PLS for each problem.
- Conditions for compatibility between a solution in $N(v)$ containing a PLS and the solutions in each $C_1^v, \ldots, C_k^v$.
- Computation of solution in $N(v)$ containing a particular PLS.
- Recurrence formulation for the problem.

2 Induced Forest of Maximum Degree at Most Two in AT-Free Graphs

2.1 Potential Local Solutions for IFMDT and Their Compatibility

Consider a vertex $v \in V(G)$ and let $C_1^v, \ldots, C_k^v$ are the components of $G \setminus N[v]$. We fix v to be present in IFMDT. The vertex v can have at most two neighbors in IFMDT since every vertex in IFMDT has degree at most two.

Hence a PLS for IFMDT is defined by (v, R_v) where $v \in V(G)$ and $R_v \subseteq N(v)$ such that $G[R_v]$ is independent and $|R_v| \leq 2$. Let Λ be a set containing all possible PLSs of G.

Two PLS (v, R_v) and (w, R_w) are compatible if there is at most one neighbor of R_v in R_w, that is $|N(R_v) \cap R_w| \leq 1$ or the sets R_v and R_w have at most one common element, that is $|R_v \cap R_w| \leq 1$. If R_v contains two vertices then, both vertices in R_v can not have neighbor in the same component. For a solution $S_{C_i^v}$ in C_i^v is compatible with (v, R_v) if $G[\{v\} \cup R_v \cup S_{C_i^v}]$ is a linear forest.

2.2 Dynamic Programming Formulation for IFMDT

Let S be an IFMDT of G containing PLS $(v, R_v) \in \Lambda$. We define following sets for every (v, R_v). Let $\Lambda_{(v,r)}$: All pair $(u, R_u) \in \Lambda$ where $u \in V(C_i^v)$, for some $i \in \{1, \ldots, k\}$ and $r \in R_v$ such that $R_v \cap R_u = r$ or $N(R_u) \cap R_v = r$. That is, r has a neighbor in R_u or r is the common neighbor between R_v and R_u. Let

$\Lambda'_{(v,R_v)}$: All pair $(u, R_u) \in \Lambda$ where $u \in V(C_i^v)$ for some i such that $R_v \cap R_u = \varnothing$ and $N(R_u) \cap R_v = \varnothing$.

Recall, we denote by S an IFMDT of G. Let $\Psi(v, R_v, G)$ denote an IFMDT of $G \setminus N[v]$, which is compatible with the PLS $(v, R_v) \in \Lambda$.

$$|S| = \max_{(v,R_v)\in\Lambda} (|\{v\} \cup R_v \cup \Psi(v, R_v, G)|) \tag{1}$$

In the following we formulate recurrence relations to compute $\Psi(v, R_v, G)$. Let $\Phi(v, r, C_i^v)$ denote an IFMDT in $G[C_i^v]$ such that $r \in R_v$ has a neighbor in $\Phi(v, r, C_i^v)$. Similarly, we define $\Pi(v, R_v, C_i^v)$ to be an IFMDT in $G[C_i^v]$ such that no vertex in R_v has any neighbor in it. We divide the recurrences based on the different possibilities of neighborhood of R_v.

Let A_1 denote the IFMDT compatible with PLS (v, R_v) where $|R_v| = 2$, and both vertices in R_v have a neighbor other than v in the resulting solution. Since vertices of R_v has neighbor in different components, we get,

$$|A_1| = \max_{\{p,q\}\subseteq\{1,\ldots,k\}} (|\Phi(v, r_1, C_p^v) \cup \Phi(v, r_2, C_q^v) \cup (\bigcup_{i\in\{1,\ldots,k\}\setminus\{p,q\}}\Pi(v, R_v, C_i^v))|) \tag{2}$$

Let A_2 denote the IFMDT containing the PLS (v, R_v) with the property that $|R_v| \leq 2$ but exactly one vertex in R_v has a neighbor other than v in the resulting solution.

$$|A_2| = \max_{j\in\{1,2\}} (\max_{p\in\{1,\ldots,k\}} (|\Phi(v, r_j, C_p^v) \cup (\bigcup_{i\in\{1,\ldots,l\}\setminus\{p\}}\Pi(v, R_v, C_i^v)|)) \tag{3}$$

Let A_3 denote the IFMDT containing the PLS (v, R_v) with the property that $|R_v| \leq 2$ but none of the vertex in R_v has a neighbor other than v in the resulting solution.

$$|A_3| = |\bigcup_{i\in\{1,\ldots,k\}} \Pi(v, R_v, C_i^v)| \tag{4}$$

Thus, we get the following recurrence for IFMDT compatible with PLS (v, R_v).

$$|\Psi(v, R_v, G)| = \max(|A_1|, |A_2|, |A_3|) \tag{5}$$

Note that, $|\Phi(v, r, C_i^v)| = 0$ if there is no PLS in C_i^v such that r has exactly one neighbor in it.

$$|\Phi(v, r, C_i^v)| = \max_{\substack{u\in C_i^v, \\ (u,R_u)\in\Lambda_{(v,r)}}} (|\{u\} \cup R_u \cup \Psi(u, R_u \setminus N[r], C_i^v)|) \tag{6}$$

Since r or a neighbor of r is present in R_u, this vertex can not have any other neighbor. Therefore, we remove it from R_u.

$$|\Pi(v, R_v, C_i^v)| = \max_{\substack{u \in C_i^v, \\ (u, R_u) \in \Lambda'_{(v, R_v)}}} (|\{u\} \cup R_u \cup \Psi(u, R_u, C_i^v)|) \qquad (7)$$

Theorem 2. *There is an $O(n^6)$ time algorithm to compute IFMDT in AT-free graph.*

Note: This algorithm can be modified to work for weighted graphs by simply replacing the cardinality of a set by the weight of that set.

3   Outerplanar Vertex Deletion

3.1   Potential Local Solutions for MIOP

In this section we consider T to be an induced outer-planar graph of G and let $v \in V(T)$ and $x \in N_T(v)$. We call x an *extendable vertex* with respect to T and v if x has neighbors in $V(T) \setminus N_T[v]$. We prove that the number of extendable vertices in $N_T(v)$ is upper bounded by four. We begin with Lemma 4 which might be well known in relevant literature, but since we could not find a proper reference, we state as it is required in our algorithm.

Lemma 4. *Suppose $P = V(T) \cap N(v)$. Then $G[P]$ is a forest of maximum degree 2.*

Lemma 5. *Let C be a component of $G \setminus N[v]$. There can be at most two vertices in $N_T(v)$ that can have neighbor(s) in $V(C) \cap V(T)$.*

In Lemma 5 we saw that at most two neighbors of v can have neighbors in a component of $G \setminus N[v]$ when we are considering an induced outerplanar graph.

Lemma 6. *There can be at most 4 extendable vertices in $N_T(v)$.*

Let $L \subseteq N_T(v)$ be those vertices which do not have any neighbor in $V(T) \setminus N_T(v)$. We call these vertices as *non-extendable* vertices. The following lemma shows that at most four vertices in L can represent the neighborhood of L in G.

Lemma 7. *Let $L_r \subseteq L$ be a representative of L in G, that is $N_G(L_r) \setminus N_G(v) = N_G(L) \setminus N_G(v)$. Then $|L_r| \le 4$.*

Using Lemma 6 and Lemma 7 stated above, we define PLS for MIOP. Let (v, X_v, Y_v) be a PLS contained in T'–an induced outer-planar graph, where: $X_v \subseteq N_G(v)$, $|X_v| \le 4$ and vertices of X_v have some neighbor in $V(T') \setminus N_{T'}[v]$, $Y_v \subseteq N_G(v)$, $|Y_v| \le 4$ and vertices of Y_v represents the vertices of $N_{T'}(v)$ that do not have any neighbor in $V(T') \setminus N_{T'}[v]$, and $G[X_v \cup Y_v]$ is cycle free and has maximum degree at most two.

Let Γ be the set of all PLS for MIOP. Recall that T is an induced outer-planar graph of G. Consider $(v, X_v, Y_v) \in \Gamma$ such that (v, X_v, Y_v) contained in T, and let $C_1^v, \ldots, C_k^v$ be the components of $G \setminus N[v]$. From Lemma 5, we know at most two vertices in X_v can have neighbors in a component $C_i^v \cap T$. Suppose $x_1, x_2 \in X_v$ are those two vertices, which have neighbor in $C_i^v \cap T$ for some i. Then Lemma 8 says that, x_1, x_2 both can not have neighbors in $C_j^v \cap T$ for any $j \neq i$.

Lemma 8. *Suppose x_1, x_2 are extendable vertices with respect to T and v such that both x_1 and x_2 have neighbors in $C_i^v \cap T$ for some $i \in \{1, \ldots, k\}$. Then there does not exist any $j \in \{1, \ldots, k\} \setminus \{i\}$ such that both x_1 and x_2 have neighbors in $C_j^v \cap T$.*

Now we study few properties of neighborhood of subset of X_v which will be relevant subsequently.

Definition 2. *A pair of vertices $\{x_1, x_2\} \subseteq X_v$ is called a twin set of X_v, if $N_T(x_1) \cap V(C) \neq \phi$ and $N_T(x_2) \cap V(C) \neq \phi$, where C is a component of $G \setminus N[v]$.*

Lemma 9. *A vertex $x \in X_v$ can be present in at most two twin sets.*

Lemma 10. *Suppose $\{x_1, x_2\} \subseteq X_v$ be a twin set. Then either x_1 and x_2 are adjacent or there is no path between x_1 and x_2 in $G[N_T(v)]$.*

Lemma 11. *Consider a PLS (v, X_v, Y_v), then X_v can contain at most two twin sets.*

Lemma 12. *Assume there exist a path P of length at least two and $V(P) \subseteq X_v$, then the end points of P can not be present in the same twin set. Every other possible pair of vertices in X_v is a valid twin set.*

3.2   Compatibility in MIOP

In this dynamic programming framework, for a particular PLS (v, X_v, Y_v) we find solutions $T_i^v, \ldots, T_k^v$ in every $C_1^v, \ldots, C_k^v$ such that $G[\{v\} \cup X_v \cup Y_v \cup (\cup_{i=1}^k T_i^v)]$ is an outer-planar graph and no vertices in Y_v has neighbor in $\cup_{i=1}^k T_i^v$. We call such solutions compatible with (v, X_v, Y_v).

Definition 3. Compatibility in PLS: *(v, X_v, Y_v) and (w, X_w, Y_w) are compatible if and only if $G[\{v, w\} \cup X_v \cup Y_v \cup X_w \cup Y_w)]$ is an outerplanar graph and no vertices in Y_v has neighbor in $\{w\} \cup X_w \cup Y_w$, and no vertex in Y_w has a neighbor in $\{v\} \cup X_v \cup Y_v$.*

In this section we make observations related to a induced outer-planar graph of an AT-free graph in order to establish the conditions of compatibility between two PLSs. We assume two PLSs (v, X_v, Y_v) and (w, X_w, Y_w) for the following observations. Note that, the vertex w is present in C_i^v where C_i^v is a component of $G \setminus N[v]$. Recall, in Sect. 1.6, we explained that we do not need to consider $I(v, w)$ while computing solution in C_i^v containing the PLS (w, X_w, Y_w). Let T_i^v be such a solution. Then we have the following lemma.

Lemma 13. *T_i^v is compatible with (v, X_v, Y_v) if and only if (w, X_w, Y_w) is compatible with (v, X_v, Y_v).*

From Lemma 5, at most two vertices form X_v can have neighbors in X_w. Based on this fact, we divide our observations into three cases where each case considers the number of vertices in X_v that has neighbor in X_w.

First case is when two vertices of X_v have neighbor in X_w. It is possible that both vertices of X_v are adjacent to a single vertex or two vertices of X_w. Second case is when exactly one vertex from X_v has neighbor in X_w. It is possible that the vertex in X_v is adjacent to exactly one or two vertices in X_w. Third case is when no vertex in X_v is adjacent to any vertex in X_w. In each of the cases we check for all PLS (w, X_w, Y_w) that satisfies the conditions of that case. Along with that we also check, for the given PLS (w, X_w, Y_w), $G[\{v, w\} \cup X_v \cup Y_v \cup X_w \cup Y_w]$ is an outer-planar graph and no vertices in Y_v are adjacent to any vertices in $\{w\} \cup X_w \cup Y_w$. Similarly, we also check no vertices in Y_w are adjacent to any vertices in $\{v\} \cup X_v \cup Y_v$.

- $\Gamma^1_{(X_v^i, Y_v)}$: All triple $(w, X_w, Y_w) \in \Gamma$ which satisfies conditions of first case and compatible with (v, X_v, Y_v) such that X_v^i is a twin set of X_v.
- $\Gamma^2_{(X_v, Y_v)}$: All triple $(w, X_w, Y_w) \in \Gamma$ which satisfies conditions of second and third case and compatible with (v, X_v, Y_v).

3.3 Dynamic Programming for Maximum Induced Outerplanar Graph

MIOP in $N[v]$ Containing Fixed (v, X_v, Y_v): Recall that $N(v) \cap T$ is a linear forest from Lemma 4. Let $H_v \subseteq N(v)$ be set of all vertices that are represented by Y_v in G (Definition 1), that is $H_v = \{h \in N(v) \smallsetminus X_v | N(h) \smallsetminus N(v) \subseteq N(Y_v) \smallsetminus N(v)\}$. From our assumption of T containing fixed PLS (v, X_v, Y_v), no vertices in $N(v) \smallsetminus (X_v \cup Y_v \cup H_v)$ can be present in $T \cap N(v)$. Let $X_v^1 = \{x_1^1, x_2^1\}$ and $X_v^2 = \{x_1^2, x_2^2\}$ be twin sets of X_v in T. Note that, $|X_v^1 \cap X_v^2| \leq 1$ from Lemma 9. We add edges between vertices of X_v^1 and X_v^2. From Lemma 4, $T \cap N(v)$ is the maximum linear forest in $X_v \cup Y_v \cup H_v$ such that X_v and Y_v are contained in it.

Note that the above discussion assumes that both X_v^1 and X_v^2 are twin sets. In general X_v^1 and X_v^2 may not be twin sets. Let $\Theta_{X_v} \in 2^{X_v}$ such that Θ_{X_v} contains only twin sets of X_v in a particular realization of MIOP containing (v, X_v, Y_v) as a PLS. We call Θ_{X_v} a *twin-flag set* of X_v. Hence from Lemma 11, $|\Theta_{X_v}| \leq 2$. Thus we define $\beta(v, X_v, Y_v, \Theta_{X_v})$ to be maximum linear forest in $N(v)$ containing PLS (v, X_v, Y_v) and Θ_{X_v} is the set of twin sets of X_v.

Combining Solutions Using Dynamic Programming: Using the outline in Sect. 1.6 we devise the dynamic programming. Let $\mathcal{C} = \bigcup_{v \in V(G)} \{C : C \text{ is a component of } G \smallsetminus N[v]\}$. Let $\Psi(v, X_v, Y_v, \Theta_{X_v}, C)$ denote an $MIOP$ in $C \smallsetminus N[v]$ where $C \in \mathcal{C}$ is a connected component of G, such that $\Psi(v, X_v, Y_v, \Theta_{X_v}, C)$ is compatible with the PLS (v, X_v, Y_v).

$$|T(G)| = \max_{\substack{(v, X_v, Y_v, \Theta_{X_v}) \\ \text{s.t } (v, X_v, Y_v) \in \Gamma}} (|\beta(v, X_v, Y_v, \Theta_{X_v}) \cup \Psi(v, X_v, Y_v, \Theta_{X_v}, G)|) \qquad (8)$$

We denote by $\Phi(v, X_v^j, Y_v, C_i^v)$ an MIOP in C_i^v such that the twin set X_v^j of X_v w.r.t $C_i^v \cap T$. Similarly, we denote by $\Pi(v, X_v, Y_v, C_i^v)$ an MIOP in C_i^v such that at most one vertex of X_v has neighbors in $C_i^v \cap T$.

Suppose X_v^1 and X_v^2 are twin sets of X_v and $X_v^1 \cup X_v^2 = X_v$. We define,

$$|A_1| = \max_{\{p,q\} \subseteq \{1,\ldots,k\}} |(\Phi(v, X_v^1, Y_v, C_p^v) \cup \Phi(v, X_v^2, Y_v, C_q^v) \cup$$

$$(\bigcup_{i \in \{1,\ldots,k\} \setminus \{p,q\}} \Pi(v, X_v, Y_v, C_i^v)))| \qquad (9)$$

Suppose only one of X_v^1 and X_v^2 are twin sets. We define,

$$|A_2| = \max_{j \in \{1,2\}} (\max_{p \in \{1,\ldots,k\}} |(\Phi(v, X_v^j, Y_v, C_p^v) \cup (\bigcup_{i \in \{1,\ldots,k\} \setminus \{p\}} \Pi(v, X_v, Y_v, C_i^v)))|) \quad (10)$$

Suppose none of X_v^1 and X_v^2 are twin sets. We define,

$$|A_3| = |\bigcup_{i \in \{1,\ldots,k\}} \Pi(v, X_v, Y_v, C_i^v)| \qquad (11)$$

The above four recurrence enumerates all possible solutions. The recurrence for Ψ is as follows.

$$|\Psi(v, X_v, Y_v, \Theta_{X_v}, G)| = \max(|A_1|, |A_2|, |A_3|) \qquad (12)$$

Now we compute of $\Phi(v, X_v^1, Y_v, C_i^v)$ and $\Pi(v, X_v, Y_v, C_i^v)$. Let $Z = N[X_v^1] \cap X_w$. We define two twin-flag sets of X_w, Θ_{X_w} and Θ'_{X_w}. If $|Z| = 2$ and Z satisfies Lemma 12 then we ensure $|\Theta_{X_w}| = 1$ and Θ_{X_w} does not contain Z and $\Theta'_{X_w} = \Theta_{X_w} \cup \{Z\}$. If $|Z| \leq 1$ then Θ_{X_w} is a twin-flag set of X_w such that $|\Theta_{X_w}| \leq 2$ and $\Theta'_{X_w} = \Theta_{X_w}$.

$$\Phi(v, X_v^1, Y_v, C_i^v) = \max_{\substack{w \in C_i^v \\ (w, X_w, Y_w) \in \Gamma^1_{(X_v^1, Y_v)}}} (|\beta(w, X_w, Y_w, \Theta'_{X_w}) \cup \Psi(w, X_w, Y_w, \Theta_{X_w}, C_i^v)|)$$

$$(13)$$

$$\Pi(v, X_v, Y_v, C_i^v) = \max_{\substack{w \in C_i^v \\ (w, X_w, Y_w) \in \Gamma^2_{(X_v, Y_v)}}} (|\beta(w, X_w, Y_w, \Theta'_{X_w}) \cup \Psi(w, X_w, Y_w, \Theta_{X_w}, C_i^v)|)$$

$$(14)$$

Theorem 1. *There is an $O(n^{19})$ time algorithm to compute MIOP in AT-free graph.*

4   Conclusion

In this paper, we present polynomial-time algorithms for MIOP and IFMDT in AT-free graphs. At this point we recall that MIOP admits polynomial kernel for general graphs [8]. It is known that a planar $\mathcal{F}$-minor-free deletion problem admits a polynomial kernel [11]. Drawing an analogy, it would be interesting to explore the algorithmic complexity of planar $\mathcal{F}$-minor-free deletion problem on AT-free graphs.

References

1. Balakrishnan, H., Rajaraman, A., Rangan, C.P.: Connected domination and steiner set on asteroidal triple-free graphs. In: Dehne, F., Sack, J.-R., Santoro, N., Whitesides, S. (eds.) WADS 1993. LNCS, vol. 709, pp. 131–141. Springer, Heidelberg (1993). https://doi.org/10.1007/3-540-57155-8_242
2. Broersma, H., Kloks, T., Kratsch, D., Müller, H.: Independent sets in asteroidal triple-free graphs. SIAM J. Discret. Math. **12**(2), 276–287 (1999)
3. Cao, Y., Ke, Y., Otachi, Y., You, J.: Vertex deletion problems on chordal graphs. Theoret. Comput. Sci. **745**, 75–86 (2018)
4. Chang, J.-M.: Induced matchings in asteroidal triple-free graphs. Discret. Appl. Math. **132**(1–3), 67–78 (2003)
5. Corneil, D.G., Olariu, S., Stewart, L.: Asteroidal triple-free graphs. SIAM J. Discrete Math. **10**(3), 399–430 (1997)
6. Cygan, M., et al.: Parameterized Algorithms, vol. 5. Springer (2015)
7. Diestel, R.: Graph theory. Graduate Texts in Mathematics, Springer-Verlag, New York, Incorporateds vol. 173, p. 107 (2000)
8. Donkers, H., Jansen, B.M.P., Włodarczyk, M.: Preprocessing for outerplanar vertex deletion: an elementary kernel of quartic size. Algorithmica **84**(11), 3407–3458 (2022)
9. Dross, F., Montassier, M., Pinlou, A.: A lower bound on the order of the largest induced linear forest in triangle-free planar graphs. Discret. Math. **342**(4), 943–950 (2019)
10. Fomin, F.V., Lokshtanov, D., Misra, N., Philip, G., Saurabh, S.: Hitting forbidden minors: approximation and kernelization. SIAM J. Discrete Math. **30**(1), 383–410 (2016)
11. Fomin, F.V., Lokshtanov, D., Misra, N., Saurabh, S.: Planar F-deletion: approximation, kernelization and optimal FPT algorithms. In: 2012 IEEE 53rd Annual Symposium on Foundations of Computer Science, pp. 470–479. IEEE (2012)
12. Fomin, F.V., Lokshtanov, D., Saurabh, S., Zehavi, M.: Approximation schemes via width/weight trade-offs on minor-free graphs. In: Chawla, S. (ed.) Proceedings of the 2020 ACM-SIAM Symposium on Discrete Algorithms, SODA 2020, Salt Lake City, UT, USA, 5–8 January 2020, pp. 2299–2318. SIAM (2020). https://doi.org/10.1137/1.9781611975994.141, https://doi.org/10.1137/1.9781611975994.141
13. Golovach, P.A., Paulusma, D., van Leeuwen, E.J.: Induced disjoint paths in AT-free graphs. In: Fomin, F.V., Kaski, P. (eds.) SWAT 2012. LNCS, vol. 7357, pp. 153–164. Springer, Heidelberg (2012). https://doi.org/10.1007/978-3-642-31155-0_14
14. Golumbic, M.C.: Algorithmic Graph Theory and Perfect Graphs, vol. 57. Elsevier (2004)

15. Gupta, A., Lee, E., Li, J., Manurangsi, P., Włodarczyk, M.: Losing treewidth by separating subsets. In: Proceedings of the Thirtieth Annual ACM-SIAM Symposium on Discrete Algorithms, pp 1731–1749. SIAM (2019)
16. Jansen, B.M.P., Lokshtanov, D., Saurabh, S.: A near-optimal planarization algorithm. In: Proceedings of the Twenty-Fifth Annual ACM-SIAM Symposium on Discrete Algorithms, pp. 1802–1811. SIAM (2014)
17. Jansen, B.M.P., Włodarczyk, M.: Lossy planarization: a constant-factor approximate kernelization for planar vertex deletion. In: Proceedings of the 54th Annual ACM SIGACT Symposium on Theory of Computing, pp. 900–913 (2022)
18. Kawarabayashi, K.I.: Planarity allowing few error vertices in linear time. In: 2009 50th Annual IEEE Symposium on Foundations of Computer Science, pp. 639–648. IEEE (2009)
19. Kawarabayashi, K.I., Sidiropoulos, A.: Polylogarithmic approximation for minimum planarization (almost). In: 2017 IEEE 58th Annual Symposium on Foundations of Computer Science (FOCS), pp. 779–788. IEEE (2017)
20. Kawarabayashi, K.I., Sidiropoulos, A.: Polylogarithmic approximation for euler genus on bounded degree graphs. In: Proceedings of the 51st Annual ACM SIGACT Symposium on Theory of Computing, pp. 164–175 (2019)
21. Klein, C., Strzodka, R.: Highly parallel linear forest extraction from a weighted graph on GPUs. In: Proceedings of the 51st International Conference on Parallel Processing, pp. 1–11 (2022)
22. Kloks, T., Kratsch, D., Spinrad, J.: On treewidth and minimum fill-in of asteroidal triple-free graphs. Theoret. Comput. Sci. **175**(2), 309–335 (1997)
23. Kratsch, D.: Domination and total domination on asteroidal triple-free graphs. Discret. Appl. Math. **99**(1–3), 111–123 (2000)
24. Kratsch, D., Müller, H., Todinca, I.: Feedback vertex set on AT-free graphs. Discret. Appl. Math. **156**(10), 1936–1947 (2008)
25. Lewis, J.M., Yannakakis, M.: The node-deletion problem for hereditary properties is NP-complete. J. Comput. Syst. Sci. **20**(2), 219–230 (1980)
26. Marx, D., Schlotter, I.: Obtaining a planar graph by vertex deletion. Algorithmica **62**(3), 807–822 (2012)
27. Otachi, Y., Suzuki, A., Tamura, Y.: Finding induced subgraphs from graphs with small mim-width. In: 19th Scandinavian Symposium and Workshops on Algorithm Theory (SWAT 2024), pp. 38–1. Schloss Dagstuhl–Leibniz-Zentrum für Informatik (2024)
28. Pelsmajer, M.J.: Maximum induced linear forests in outerplanar graphs. Graphs Comb. **20**(1), (2004)
29. Poh, K.S.: On the linear vertex-arboricity of a planar graph. J. Graph Theor. **14**(1), 73–75 (1990)
30. Yannakakis, M.: Node-deletion problems on bipartite graphs. SIAM J. Comput. **10**(2), 310–327 (1981)

The Hardness of Monotone Eccentricity on Polytopes

Krishna Narayanan and Tamon Stephen$^{(\boxtimes)}$

Department of Mathematics, Simon Fraser University,
Burnaby, BC V5A 1S6, Canada
`{krishna_narayanan,tamon}@sfu.ca`

Abstract. We consider the problem of finding the *monotone eccentricity* of a vertex v of a polytope. This is the largest number of pivots through the graph of the polytope required to reach an optimal vertex starting from v. In particular, we study a polytope introduced by Frieze and Teng [3] derived from the exact partition problem. This polytope is simple and nearly 0/1, with at most two fractional components per vertex. We show that Frieze and Teng's result on the complexity of computing lower bounds on diameters of exact partition polytopes can be extended to show that computing monotone eccentricity is also NP-Hard, and in fact D^P-hard, even on simple polytopes.

Keywords: Polytopes · Polyhedral Diameter · Computational Complexity · Linear Optimization · Exact Partition Problem

1   Introduction

The simplex method solves linear programs by moving between adjacent extreme points (vertices) on the feasible region's 1-skeleton. Understanding the lengths of such improving paths on polytopes is therefore crucial.

The *graph diameter* of the 1-skeleton or simply *diameter* of the polytope, defined as the maximum length of a shortest path between any two vertices, is a lower bound for the length of such an improving path.

The computational complexity of diameters of polyhedra was first considered by Frieze and Teng in [3], where they proved that the problem and that of computing the eccentricity of a vertex was *weakly* NP-Hard, and also D^P-Hard [6]. After a long gap, Sanità [7] proved that computing diameters of polytopes is strongly NP-Hard. Wulf, in a recent preprint [8], went further an showed that deciding diameters of polyhedra is Π_2^P-complete.

Since we are motivated by the *improving* paths used in the simplex method, it is natural to ask about the length of such paths. These are characterized by being *monotone* with respect to a linear objective function $\mathbf{c}$, i.e. if $\mathbf{x}_k$ and $\mathbf{x}_{k+1}$ are consecutive vertices on the path, then $\mathbf{c}^T\mathbf{x}_k \leq \mathbf{c}^T\mathbf{x}_{k+1}$. Nöbel and Steiner [5] looked at hardness issues for *monotone* versions of diameter, among other things, getting hardness results even when the constraint matrix of the

© The Author(s), under exclusive license to Springer Nature Switzerland AG 2026
N. Misra and A. Pandey (Eds.): CALDAM 2026, LNCS 16445, pp. 365–377, 2026.
https://doi.org/10.1007/978-3-032-17156-6_27

polytope is totally unimodular. Wulf [8] remarked that his completeness results do not apply to the monotone case.

Our Results. We consider the exact partition polytope as defined in [3], which is a simple polytope and nearly 0/1 in that it allows at most two fractional components per vertex. We provide structural results en route to showing that Frieze and Teng's result that eccentricity is both NP-Hard and D^P-Hard extends to the monotone setting.

Preliminaries. We begin with some definitions, following [9]. An $\mathcal{H}$-*polytope* (henceforth, simply a *polytope*) is the intersection of finitely many closed halfspaces in $\mathbb{R}^d$ that is *bounded*, i.e., it contains no ray $\{\mathbf{x} + t\mathbf{y} : t \geq 0\}$ for $\mathbf{x} \in \mathbb{R}^d$, $\mathbf{y} \neq \mathbf{0}$. The *dimension* of a polytope is the dimension of its affine hull.

A *face* of a polytope is its intersection with a hyperplane such that the polytope lies entirely within one of the two associated halfspaces. The empty set and the polytope itself are trivial faces; all others are *proper* faces. In particular, 0-dimensional faces are the *vertices*, 1-dimensional faces are the *edges*, and the maximal proper faces, of dimension one less than the polytope, are the *facets*. The vertices and edges of a polytope form an undirected graph, called its *1-skeleton*.

A d-dimensional polytope is *simple* if every vertex is adjacent to exactly d edges in the 1-skeleton, or equivalently, contained in exactly d facets. When a vertex is at the intersection of more than d facets, we call it a *degenerate* vertex. If a polytope contains any denegerate vertices, then we call it a *degenerate polytope*.

Throughout this paper, we use the notation $[m]$ to denote the set $\{1, ..., m\}$.

Definition 1. *Let $P \subseteq \mathbb{R}^d$ be a polytope. Let $\mathbf{c} \in \mathbb{R}^d$ be an objective direction. A sequence of adjacent vertices $\mathbf{x}_0, \mathbf{x}_1, \ldots, \mathbf{x}_k$ of P is called $\mathbf{c}$-monotone if $\mathbf{c}^T\mathbf{x}_{i+1} \geq \mathbf{c}^T\mathbf{x}_i$ for every $i \in \{0, \ldots, k-1\}$.*

Definition 2. *Consider a polytope P with an objective function $\mathbf{c}$ that induces a unique maximum $\mathbf{y_c}$. For a vertex $\mathbf{x}$ of polytope P, let $d^\mathbf{c}(\mathbf{x}, \mathbf{y_c})$ denote the length of the shortest $\mathbf{c}$-monotone path from $\mathbf{x}$ to $\mathbf{y_c}$.*[1]

Definition 3 (Monotone Eccentricity). *Consider a polytope $P = \{\mathbf{x} \in \mathbb{R}^d \mid A\mathbf{x} \leq \mathbf{b}\}$. The monotone eccentricity $\eta(x)$ of a vertex $\mathbf{x}$ of P as the longest, over all choices of $\mathbf{c}$, of the shortest $\mathbf{c}$-monotone path from $\mathbf{x}$ to any optimal-cost vertex $\mathbf{y_c}$ for $\mathbf{c}$. That is, $\eta(\mathbf{x}) = \max_\mathbf{c} d^\mathbf{c}(\mathbf{x}, \mathbf{y_c})$*

Though we do not intend to provide any results on the monotone diameter itself, we provide a formal definition for completeness sake.

Definition 4 (Monotone diameter). *Consider a polytope $P = \{\mathbf{x} \in \mathbb{R}^d \mid A\mathbf{x} \leq \mathbf{b}\}$. The monotone diameter of $\delta_\mathcal{E}(P)$ is defined as the maximum*

---

[1] If there exist multiple optimal vertices, then this denotes the shortest path from $\mathbf{x}$ to *any* optimal vertex, where $\mathbf{y_c}$ now denotes a vertex in the optimal set at which this shortest path length is achieved. However, in this case optimal vertex may not be unique, and can depend on the choice of $\mathbf{x}$ as well as $\mathbf{c}$.

of the monotone eccentricities over all vertices $\mathbf{x} \in P$ *and objective functions* $\mathbf{c} \in \mathbb{R}^d$. *Symbolically,*

$$\delta_{\mathcal{E}}(P) = \max_{\mathbf{x} \in P} \eta(\mathbf{x}) = \max_{\mathbf{x} \in P} \max_{\mathbf{c}} d^{\mathbf{c}}(\mathbf{x}, \mathbf{y}_{\mathbf{c}})$$

2   Exact Partition and the Resulting Polytope

We state a variation of the Partition problem from Karp's 21 NP-Complete problems and featured in Garey and Johnson's book [1].

Problem: Exact Partition
Input: A finite set $A = \{s_1, s_2, \ldots, s_{2m}\}$ of integers.
Question: Does there exist a subset $A' \subset A$ with $|A'| = m$ such that

$$\sum_{s \in A'} s = \sum_{t \in A \setminus A'} t?$$

Korte and Schrader in [4] studied this problem via a 0/1 polytope with two knapsack constraints, one limiting the number of elements in the set to m and the other limiting the sum of the selected elements to $\frac{1}{2}S$ where $S := \sum_{i=1}^{2m} s_i$.

Frieze and Teng [3] noted that this polytope, which they call *ILP1*, is highly degenerate. They provide an alternate formulation *ILP2* whose LP-relaxation is simple (non-degenerate).

Let $s_{\max}$ be the largest element in a given problem instance A. Take $M := S + 1$, $d_i := s_{\max} - s_i$ and $\epsilon := \frac{1}{2M}$. Then we have:

$$
\begin{array}{lll}
& \text{Maximize} & \displaystyle\sum_{i=1}^{2m} x_i \text{ with } x_i \in \{0, 1\} \\[3ex]
\textbf{(ILP2)} & \text{subject to} & \displaystyle\sum_{i=1}^{2m} (M + s_i)x_i \leq \tfrac{1}{2}S + mM + \epsilon \qquad \text{(K1)} \\[3ex]
& & \displaystyle\sum_{i=1}^{2m} (M + d_i)x_i \leq \tfrac{1}{2}\sum_{i=1}^{2m} d_i + mM + \epsilon \qquad \text{(K2)}
\end{array}
$$

We see that (K1) limits the sum of elements in the chosen subset, while (K2) limits the number of selected elements. Note this formulation assumes the elements of A are positive. If this does not hold we can translate the problem to this scenario by adding a constant to each element of A. The following key properties are established in [3].

Lemma 5 ([3]). *The optimal value of (ILP2) is either m or $m - 1$. It is m if A has an exact partition, otherwise it is $m - 1$.*

We denote the linear programming relaxation of (ILP2) by P_R. Then P_R is described by $2m$-cube constraints $0 \leq x_i \leq 1$ and two knapsack-type constraints (K1) and (K2) with positive co-efficients.

Lemma 6 ([3]). *When $\epsilon = \frac{1}{2M}$, the polytope defined by the linear programming relaxation of ILP2 is simple (non-degenerate).*

The above lemma ensures that only $2m$ of the constraints are active at any given vertex of P_R.

Following Lemma 5 of [3], we partition the vertices of P_R into three classes based on the number of non-integer components in each: V_0 is the set of integer $(0/1)$ vertices, they do not lie on either of K1 or K2. V_1 is the set of vertices that lie on either K1 or K2, but not both. Lemma 6 implies that there is exactly one fractional component because out of the $4m$-box constraints, exactly $2m - 1$ can be active while either K1 or K2 are active. V_2 is the set of vertices that lie on both K1 and K2. Lemma 6 implies that there are exactly two fractional components by a similar argument as in V_1.

Observation 7 *For any subset of at most $m - 1$ elements of $[2m]$, the indicator vector that is 1 on the subset and 0 otherwise is a vertex of P_R. In particular, the origin is a vertex of V_R.*

Let $I_1(\mathbf{v}) = \{i \in [2m] : v_i = 1\}$, $I_0(\mathbf{v}) = \{i \in [2m] : v_i = 0\}$ and $I_f(\mathbf{v}) = \{i \in [2m] : 0 < v_i < 1\}$. For a vertex $\mathbf{v}$, let us also denote $I^*(\mathbf{v}) = I_1(\mathbf{v}) \cup I_f(\mathbf{v})$; this will be of use in Sect. 2.1.

The following fact is used implicitly by Frieze and Teng [3]; since it is critical to our work we include a proof of it in Appendix A for completeness.

Lemma 8. *For any vertex $\mathbf{v}$ of P_R in V_1 and V_2, $|I_1(\mathbf{v})| \geq m - 1$.*

2.1   Objective Functions for the Exact Partition Polytope

Linear programming asks to find a point $\mathbf{x_0}$ in a polytope P that maximizes a linear function $\mathbf{c^T x}$, that is, such that $\mathbf{c^T x_0} = \max\{\mathbf{c^T x} : \mathbf{x} \in P\}$.

We remark that for any vertex v of a polytope P, a strict convex combination of the normals of the active constraints gives an objective function that is optimized at v. We now do this explicitly for vertices of the exact partition polytope P_R. For a vertex in $\mathbf{v} \in V_0$ the active constraints are all box constraints of the form $0 \leq x_i$ or $x_i \leq 1$. Therefore, the objective function is a positive combination of the outward normals of active box-constraints. In this case, where $v_k = 1$, we have the standard basis vector $e_k = (0, ..., 1, ..., 0)$ with 1 at the kth index and 0 elsewhere, and where $v_k = 0$, we have the standard basis vector $e_k = (0, ..., -1, ..., 0)$ with -1 at the kth index, allowing us to conclude:

Lemma 9. *For a vertex $\mathbf{v} \in V_0$, the objective function:*

$$\mathbf{c} = \sum_{i \in I_1(\mathbf{v})} \lambda_i e_i + \sum_{j \in I_0(\mathbf{v})} \mu_i(-e_j), \quad \lambda_i, \mu_j > 0$$

induces a unique optimum at $\mathbf{v}$.

For a vertex $\mathbf{v} \in V_1$ with one active knapsack constraint $K \in \{K1, K2\}$, we get unique optima at $\mathbf{v}$ from objectives of the form:

$$\mathbf{c} = \delta a^K + \sum_{i \in I_1(\mathbf{v})} \lambda_i e_i + \sum_{j \in I_0(\mathbf{v})} \mu_j(-e_j), \quad \delta, \lambda_i, \mu_j > 0$$

where $\mathbf{a}^K$ is K's normal vector, with all components > 0. Note that $c_i > 0$ for $i \in I_1(\mathbf{v}) \cup I_f(\mathbf{v})$, while the sign of c_j for $j \in I_0(\mathbf{v})$ depends on the relative sizes of δ and the μ_j's. With a relatively large δ they could all be positive, and with a relatively small δ they could all be negative. We summarize this in a Lemma:

Lemma 10. *For a vertex* $\mathbf{v} \in V_1$*, taking* $\mathbf{c} \in \mathbb{R}^{2m}$ *as above induces a unique sink at* $\mathbf{v}$*.*

For a vertex $\mathbf{v} \in V_2$, we note that we add, apart from box constraint normals, the normals from K1 and K2.

The positive linear combination of active constraints at $\mathbf{v}$ is:

$$\mathbf{c} = \delta_1 a^{K1} + \delta_2 a^{K2} + \sum_{i \in I_1} \lambda_i e_i + \sum_{j \in I_0} \mu_j(-e_j), \quad \delta_1, \delta_2, \lambda_i, \mu_j > 0$$

Here we again see that $c_i > 0$ for $i \in I_1(\mathbf{v}) \cup I_f(\mathbf{v})$, while the sign of c_j for $j \in I_0(\mathbf{v})$ depends on the relative sizes of δ_1, δ_2 and the μ_j's. We conclude:

Lemma 11. *The constructed* $\mathbf{c} \in \mathbb{R}^{2m}$ *for a vertex* $\mathbf{v} \in V_2$ *induces a unique sink.*

Remark: While objective functions induce unique sinks at vertices in P_R, they are not necessarily *generic*, in that they don't resolve ties between intermediate non-sink vertices. Achieving full genericity has the benefit of inducing an orientation in the 1-skeleton of P_R. This can be done by perturbing each objective $\mathbf{c} \in \mathbb{R}^{2m}$ to $\mathbf{c}^\Delta = \mathbf{c} + (\Delta, \Delta^2, ..., \Delta^{2m})$ with a small $\Delta > 0$, following Lemma 3.4 in [9].

3 Computational Complexity of Monotone Eccentricity

Here we extend Frieze and Teng's proof [3] that computing eccentricity is NP-hard for $\mathbf{0} \in P_R$ to *monotone* eccentricity. Precisely, we show NP-hardness of the following problem:

EXACT MONOTONE ECCENTRICITY
Instance: Matrix $A \in \mathbb{Q}^{m \times d}$, vector $\mathbf{b} \in \mathbb{Q}^m$ defining polytope $P = \{\mathbf{x} \in \mathbb{R}^d \mid A\mathbf{x} \leq b\}$, integer k, vertex $\mathbf{v} \in P$.
Question: Is k the monotone eccentricity of $\mathbf{v}$ in P?

Any linear objective function achieves its optimum at a vertex in the feasible set polyhedron. The monotone path length from $\mathbf{0}$ to any optimal vertex is at least the length of the shortest non-monotone path to those vertices. It is implicit from the proof of Lemma 5 in [3] that the length of the shortest non-monotone path from $\mathbf{0}$ to any vertex in P_R is exactly the number of non-zero components in a vertex's characteristic vector. We show that the length of the shortest monotone path from $\mathbf{0}$ is tight at its lower bound.

Lemma 12. *The monotone eccentricity of* $\mathbf{0} \in P_R$ *satisfies the following:*

$$\eta(\mathbf{0}) = \begin{cases} m+2 & \textit{if } A \textit{ has an exact partition,} \\ m+1 & \textit{otherwise.} \end{cases}$$

Proof. **Upper Bound**: We first show that $\eta(\mathbf{0}) \leq m+2$ when there exists an exact partition and $\eta(\mathbf{0}) \leq m+1$ otherwise.

Consider any objective function $\mathbf{c}^*$ with a corresponding optimal solution $\mathbf{v}^*$. Note that $\mathbf{c}^*$ must be non-negative on elements of $I_1(\mathbf{v}^*)$. If $\mathbf{v}^*$ is in V_0, then $|I_1(\mathbf{v}^*)| \leq m$, and we can proceed from $\mathbf{0}$ to $\mathbf{v}^*$ via a monotone path by increasing each co-ordinate in $I_1(\mathbf{v}^*)$ from 0 to 1 in turn (in an arbitrary order). It follows from Observation 7 that the proposed steps do indeed traverse vertices and edges.

When $\mathbf{v}^* \in V_1$, we can again follow a path to a point which we will call $\mathbf{v}^I$, defined to be 1 for coordinates in $I_1(\mathbf{v}^*)$ on the remaining coordinates and 0 for coordinates in $I_0(\mathbf{v}^*) \cup I_f(\mathbf{v}^*)$. Here $I_f(\mathbf{v}^*)$ consists of a single coordinate k. The objective coefficient $\mathbf{c}_k^*$ cannot be negative, since in this case we would have $\mathbf{c}^* \mathbf{v}^* < \mathbf{c}^* \mathbf{v}^I$.

Hence we can continue one additional non-decreasing step along this monotone path to arrive at $\mathbf{v}^*$, an adjacent feasible basis.

When $\mathbf{v}^* \in V_2$, we again follow a path to the intermediate vertex $\mathbf{v}^I \in V_0$. From $\mathbf{v}^I$, we can pivot on the kth coordinate, where k corresponds to either of the fractional coordinates of $\mathbf{v}^*$. This bring us to a point $\mathbf{v}'$ which matches $\mathbf{v}^*$ at all coordinates except the remaining fractional coordinate l. Then a final pivot on x_l brings us to $\mathbf{v}^*$ in the required number of steps. Since $\mathbf{c}^*$ is optimal for $\mathbf{v}^*$, $\mathbf{c}_f^* > 0$ for each $f \in I_f(\mathbf{v}^*)$ by Lemma 11, thus both $\mathbf{c}_k > 0$ and $\mathbf{c}_l > 0$. This gives the necessary path of length at most $m+1$ when there is no equal partition and at most $m+2$ when an equal partition is available.

Lower Bound: We refer to the proof of the lower bound in Lemma 5 in [3]. Consider the LP relaxation of ILP2 with slack variables z_1, z_2 and $\mathbf{y}$:

$$\sum_{i=1}^{2m} (M + s_i)\mathbf{x}_i + z_1 = \frac{1}{2}S + mM + \epsilon$$

$$\sum_{i=1}^{2m} (M + d_i)\mathbf{x}_i + z_2 = \frac{1}{2}\sum_{i=1}^{2m} d_i + mM + \epsilon$$

$$x_i + y_i = 1 \quad \forall i \in \{1, \ldots, 2m\}$$

$$\mathbf{x}, \mathbf{y}, z_1, z_2 \geq 0$$

Construction of critical vertex $\mathbf{v}^* \in V_2$: Suppose A has an exact partition. Without loss of generality, assume that indices $(1, \ldots, m)$ correspond to the partition subset of A. Then, there exists a vertex $\mathbf{v}^*$ with basic variables at $\mathbf{v}^*$ are $BV_2 = \{v_1^*, \ldots, v_m^*, v_{m+1}^*, v_{m+2}^*, y_{m+1}, \ldots, y_{2m}\}$. Note that $0 < v_{m+i}^* < 1$, $y_{m+i} = 1 - v_{m+i}^*$ and $z_i = 0$ for $i \in \{1, 2\}$.

The basic variables at $\mathbf{0}$ are $BV_1 = \{y_1, \ldots, y_{2m}, z_1, z_2\}$.

Let distance$(\mathbf{0}, \mathbf{v}^*)$ denote the length of the shortest path between $\mathbf{0}$ and $\mathbf{v}^*$. $\eta(\mathbf{0}) \geq$ distance$(\mathbf{0}, \mathbf{v}^*)$ holds because the length of the shortest monotone path from $\mathbf{0}$ to any other vertex in the polytope is at least that of the shortest undirected path to that vertex. The length of the shortest path from $\mathbf{0}$ to $\mathbf{v}^*$ is determined by the number of basic variable changes:

$$\eta(\mathbf{0}) \geq \text{distance}(\mathbf{0}, \mathbf{v}^*) \geq |BV_1 \setminus BV_2| = |\{y_1, \ldots, y_m, z_1, z_2\}| = m + 2$$

For a problem instance A without an exact partition, an analogous construction of $\mathbf{v}^*$ yields at least $m - 1$ integral v_i^* components and $\eta(\mathbf{0}) \geq m + 1$. This completes our proof. $\square$

As a consequence, we get the following via a reduction from EXACT PARTITION.

Theorem 13. EXACT MONOTONE ECCENTRICITY *is* NP-*Hard on even on simple polytopes.*

3.1 D^P-Hardness and Monotone Eccentricity

In fact we can go further and extend Frieze and Teng's result that ECCENTRICITY is D^P-Hard to the monotone case. The complexity class D^P [6] is:

$$\mathsf{D}^\mathsf{P} = \{L_1 \cap L_2 \mid L_1 \in \mathsf{NP} \ \textbf{and} \ L_2 \in \mathsf{co\text{-}NP}\}$$

Note that D^P is not $\mathsf{NP} \cap \mathsf{coNP}$: the intersection in the definition of $\mathsf{NP} \cap \mathsf{coNP}$ is in the domain of the classes of languages, while a language in D^P is the intersection of two languages. Furthermore, D^P is shown to be a perfectly syntactic class and therefore has complete problems [6].

Many natural and interesting problems are in D^P [6]. One example is the problem of testing whether a given inequality is facet-defining for a the TSP polytope. Many others come from positing *exact* answers to combinatorial optimization problems. For instance, the EXACT TSP problem is: given an $n \times n$ distance matrix c_{ij} for a graph defined on n vertices and an integer k, is the optimal tour for TSP has cost exactly k? This is the intersection of the NP language of problems whose optimal tour has length *at most* k, which can be certified by a short tour, and the coNP language of problems whose optimal tour has length *at least* k, which can be refuted by a short tour.

The problems EXACT DIAMETER, which asks if the value of the diameter of a polytope is exactly some value k, and analogously EXACT ECCENTRICITY are worth considering in this framework. They are shown to be D^P-Hard by Frieze and Teng in [3]. However, certificates are harder to establish as the vertices involved in the longest paths are not easy to identify. Frieze and Teng do manage to show that EXACT DIAMETER is in the complexity class Δ_3 in the polynomial hierarchy which contains D^P and ask if it is complete for the class; this remains unanswered, despite recent interest notably by Wulf [8].

To show the D^P-hardness of MONOTONE ECCENTRICITY, we follow [3] and reduce from a close relative of EXACT PARTITION, namely:

PARTITION-UNPARTITION
Instance: Two sets $A_1 = \{s_1, \ldots, s_{2n_1}\}$ and $A_2 = \{s'_1, \ldots, s'_{2n_2}\}$ of integers
Question: Does A_1 have an exact partition while A_2 does not?

Lemma 14 ([3]). *PARTITION-UNPARTITION is D^P-complete.*

We briefly define the necessary notions and cite the relevant lemmata from [3]. Let $P_1 \subset \mathbb{R}^{m_1}$ and $P_2 \subset \mathbb{R}^{m_2}$ be two polytopes.

Definition 15. *We define the* product of polytopes P_1 and P_2, *denoted* $P_1 \times P_2$, *as the polytope in* $\mathbb{R}^{m_1+m_2}$ *given by:*

$$P_1 \times P_2 :=$$
$$\{(x_1, \ldots, x_{m_1}, y_1, \ldots, y_{m_2}) \mid (x_1, \ldots, x_{m_1}) \in P_1 \text{ and } (y_1, \ldots, y_{m_2}) \in P_2\}.$$

Algebraically, if $P_1 = \{\mathbf{x} : A_1\mathbf{x} \leq b_1\}$ *and* $P_2 = \{\mathbf{y} : A_2\mathbf{y} \leq b_2\}$, *then:*

$$P_1 \times P_2 = \{(\mathbf{x}, \mathbf{y}) : A_1\mathbf{x} \leq b_1, \ A_2\mathbf{y} \leq b_2\}.$$

The product operator $\times$ is associative, and the number of facets satisfies $f(P_1 \times P_2) = f(P_1) + f(P_2)$, where $f(P)$ denotes the number of facets of polytope P.

Definition 16. *Consider two graphs* $G_1 = (V_1, E_1)$ *and* $G_2 = (V_2, E_2)$. *The graph product of* G_1 *and* G_2 *is defined to be* $G_1 \times G_2 = (V_1 \times V_2, E)$ *where*

$$E = \{((\mathbf{u_1}, \mathbf{v_1}), (\mathbf{u_2}, \mathbf{v_2})) \mid$$
$$(\mathbf{u_1}, \mathbf{u_2}) \in E_1 \text{ and } \mathbf{v_1} = \mathbf{v_2} \ OR \ \mathbf{u_1} = \mathbf{u_2} \text{ and } (\mathbf{v_1}, \mathbf{v_2}) \in E_2\}$$

Let $\Delta(G)$ denote the diameter of a graph G and G_P denote the 1-skeleton of a polytope P, where the vertices of G_P are the vertices of the polytope and its edges are the 1-dimensional faces of P.

Proposition 17 ([3]). *Consider two polytopes* P_1 *and* P_2. *Then, we have*

$$G_{P_1 \times P_2} = G_{P_1} \times G_{P_2}.$$

Proposition 18 ([3]). *Consider a pair of graphs* G_1 *and* G_2. $\Delta(G_1 \times G_2) = \Delta(G_1) + \Delta(G_2)$.

Corollary 19 ([3]). $\Delta(P_1 \times P_2) = \Delta(P_1) + \Delta(P_2)$.

Effectively the above and subsequent facts follow from the product structures of $P_1 \times P_2$ and $G_{P_1} \times G_{P_2}$. Pivots in the product project to pivots in one of the two components, with the other component fixed.

Lemma 20. *Consider two pairs of vertices, $v_1, v_1' \in P_1$ and $v_2, v_2' \in P_2$. Then*

$$distance\left((v_1, v_2), (v_1', v_2')\right) = distance(v_1, v_1') + distance(v_2, v_2')$$

with the distances taken in the appropriate polytopes $P_1 \times P_2$, P_1 and P_2.

Denote the eccentricity of a vertex $\mathbf{v}$ in a polytope P as $\beta(\mathbf{v})$. Then we have:

Lemma 21. *Given vertices $\mathbf{v_1} \in P_1$ and $\mathbf{v_2} \in P_2$, for vertex $(\mathbf{v_1}, \mathbf{v_2}) \in P_1 \times P_2$, we have*

$$\beta((\mathbf{v_1}, \mathbf{v_2})) = \beta(\mathbf{v_1}) + \beta(\mathbf{v_2})$$

Consider now objective functions $\mathbf{c_1}$ and $\mathbf{c_2}$ on P_1 and P_2 respectively. These have associated optimal vertices $\mathbf{v_1}$ and $\mathbf{v_2}$ respectively. Then the objective function $(\mathbf{c_1}, \mathbf{c_2})$ on $P_1 \times P_2$ is optimized at $(\mathbf{v_1}, \mathbf{v_2})$. Conversely, if the objective function $\mathbf{c}$ is optimized at vertex $(\mathbf{v_1}, \mathbf{v_2}) \in P_1 \times P_2$, then the objective functions $\mathbf{c_1}$ taken from the first m_1 coordinates of $\mathbf{c}$ and $\mathbf{c_2}$ taken from the last m_2 coordinates of $\mathbf{c}$ are optimized at $\mathbf{v_1}$ and $\mathbf{v_2}$ respectively.

As a consequence, we get the analogue of Lemma 20 for monotone paths, where the distances are taken with respect to objective functions $\mathbf{c_1}$, $\mathbf{c_2}$ and $(\mathbf{c_1}, \mathbf{c_2})$ respectively. This leads to facts on monotone diameter and monotone eccentricity. Recall that $\delta_{\mathcal{E}}(P)$ denotes the monotone diameter of a polytope P.

Proposition 22. $\delta_{\mathcal{E}}(P_1 \times P_2) = \delta_{\mathcal{E}}(P_1) + \delta_{\mathcal{E}}(P_2)$

In one direction, paths in P_1 and P_2 lift to a path in $P_1 \times P_2$. Each individual move improves its component of the concatenated objective function $(\mathbf{c_1}, \mathbf{c_2})$ and therefore, they can be interleaved arbitrarily. The maximum distance occurs when both paths achieve their respective monotone diameters. On the other hand, paths in $P_1 \times P_2$ project to a pair of paths, one in P_1 and one in P_2.

Likewise for eccentricity:

Proposition 23. *Given vertices $\mathbf{v_1} \in P_1$ and $\mathbf{v_2} \in P_2$, for vertex $(\mathbf{v_1}, \mathbf{v_2}) \in P_1 \times P_2$, we have*

$$\eta((\mathbf{v_1}, \mathbf{v_2})) = \eta(\mathbf{v_1}) + \eta(\mathbf{v_2})$$

Theorem 24. Exact Monotone Eccentricity *is* $\mathsf{D^P}$*-Hard*

Proof. Given $(A_1, A_2) \in$ Partition-Unpartition, construct polytopes from the LP-relaxation (ILP2), namely P_1 for A_1 and P_2 for A_2, such that:

$$\eta(\mathbf{0})_{P_i} = \begin{cases} k_i & \text{if } A_i \text{ has an exact partition,} \\ k_i - 1 & \text{otherwise.} \end{cases}$$

where $\eta(\mathbf{0})_{P_i}$ refers to the monotone eccentricity of $\mathbf{0}$ in polytope P_i. Define the product polytope:

$$P := P_1 \times P_1 \times P_2 \subset \mathbb{R}^{4n_1 + 2n_2}$$

Observe that $\eta(\mathbf{0})_P = 2 \cdot \eta(\mathbf{0})_{P_1} + \eta(\mathbf{0})_{P_2}$, and that

$$(A_1, A_2) \in \text{Partition-Unpartition} \iff \eta(\mathbf{0})_P = 2k_1 + (k_2 - 1) = 2k_1 + k_2 - 1$$

If A_1 has partition and A_2 does not: $\eta(0)_{P_1} = k_1$, $\eta(0)_{P_2} = k_2 - 1$. All other cases yield different monotone eccentricities of $\mathbf{0}$ in P. The statement follows. $\square$

Remark: Our reduction for Exact Monotone Eccentricity in Lemma 12 can be extended to threshold versions of the problem. We can reduce the complement of Exact Partition, namely Unpartition, to a version which asks if the monotone eccentricity of a vertex is bounded above by some specified k; this implies coNP-hardness. Likewise, we can reduce Exact Partition to a decision version which asks if the monotone eccentricity of a vertex is bounded below by some specified k, implying NP-hardness.

4   Conclusion

We are able to extended Frieze and Teng's strategy [3] to show the NP-hardness, and indeed D^{P}-hardness, of computing the *monotone* eccentricity of a vertex of a polytope. Notably, the polytopes used in the construction are simple polytopes, while the combinatorial polytopes studied in many recent related results (e.g. [2, 5,8]) have been highly degenerate.

In particular, De Loera, Kafer and Sanità [2] ask about the complexity of finding a shortest monotone path in a simple polytope for a given objective function $\mathbf{c}$. These results do not get us all the way to this result, since it is not clear how to generate a suitable $\mathbf{c}$. But this may be a step in this direction, and the polytope P_R is an appealing object of study.

Our NP-hardness result, like Frieze and Teng's, is a *weak* NP-Hardness result because Exact Partition admits a pseudopolynomial-time algorithm (see [1]). Achieving a strong NP-Hardness result and a full complexity characterization of computing monotone eccentricities and diameters in simple polytopes is an appealing challenge.

Acknowledgments. This research was partially funded by NSERC Discovery Grant RGPIN-2024-06800 to TS. The authors thank the anonymous referees for their suggestions.

A   Appendix

Proof of Lemma 8

We first address the case for a vertex $\mathbf{v} \in V_1$. Suppose that the claim is false. This implies that there exists a vertex $\mathbf{v}$ in V_1 such that $|I_1(\mathbf{v})| \leq m - 2$. Assume without loss of generality that $0 < v_1 < 1$. We now have two cases:

1. K_1 is active. We have:

$$\sum (M + s_i)v_i = \frac{1}{2}S + mM + \epsilon$$

$$(M + s_1)v_1 + \sum_{i \in I_1(\mathbf{v})} (M + s_i)v_i = \frac{1}{2}S + mM + \epsilon$$

$$(|I_1(\mathbf{v})| + v_1)M + \sum_{i \in I_1(\mathbf{v})} s_i + s_1v_1 = \frac{1}{2}S + mM + \epsilon$$

Since $|I_1(\mathbf{v})| \leq m - 2$ by assumption and $v_1 < 1$, $|I_1(\mathbf{v})| + v_1 < m - 1$. Also, $\sum_{i \in I_1(\mathbf{v})} s_i + s_1v_1 < M$, by definition of M. Together, we have:

$$\sum_{i \in I_1(\mathbf{v})} s_i + s_1v_1 + (|I| + v_1)M < (m - 1)M + M = mM$$

Since $\frac{1}{2}S + \epsilon > 0$ by definition of S, we have

$$\sum_{i \in I_1(\mathbf{v})} s_i + s_1v_1 + (|I_1(\mathbf{v})| + v_1)M < mM < \frac{1}{2}S + mM + \epsilon$$

leading to a contradiction.

2. K_2 is active. We have:

$$(M + d_1)v_1 + \sum_{i \in I_1(\mathbf{v})} (M + d_i)v_i = \frac{1}{2}\sum d_i + mM + \epsilon$$

$$(|I_1(\mathbf{v})| + v_1)M + \sum_{i \in I} d_i + d_1v_1 = \frac{1}{2}\sum d_i + mM + \epsilon$$

$$(|I_1(\mathbf{v})| + v_1)M + (|I_1(\mathbf{v})| + v_1)s_{\max} - (\sum_{i \in I_1(\mathbf{v})} s_i + s_1v_1) = m \cdot s_{\max} - \frac{S}{2} + mM + \epsilon$$

As before, $|I| + v_1 < m - 1$. Hence, the left hand side is:

$$(|I_1(\mathbf{v})| + v_1)(M + s_{\max}) - (\sum_{i \in I_1(\mathbf{v})} s_i + s_1v_1)$$

$$< (m - 1)(M + s_{\max}) - (\sum_{i \in I_1(\mathbf{v})} s_i + s_1v_1)$$

And the RHS is:

$$m(M + s_{\max}) - \frac{S}{2} + \epsilon$$

Regardless of whether $\sum_{i \in I_1(\mathbf{v})} s_i + s_1v_1$ is greater than $\frac{S}{2}$ or not, the difference is eclipsed by the $(M + s_{\max})$ in the RHS. Therefore, we have:

$$(m - 1)(M + s_{\max}) - (\sum_{i \in I_1(\mathbf{v})} s_i + s_1v_1) < m(M + s_{\max}) - \frac{S}{2} + \epsilon$$

a contradiction.

We now address the claim for a vertex $\mathbf{v} \in V_2$. Once again, assume that the claim is false. This implies that there exists a vertex $\mathbf{v}$ in V_2 such that $|I_1(\mathbf{v})| \leq m - 2$. Assume without loss of generality that $0 < v_1, v_2 < 1$ by Lemma 6. By definition, K1 and K2 are active. We have:

$$\sum_{i \in I_1(\mathbf{v})} (M + s_i) + (M + s_1)v_1 + (M + s_2)v_2 = \frac{1}{2}S + mM + \epsilon$$

$$\sum_{i \in I_1(\mathbf{v})} (M + d_i) + (M + d_1)v_1 + (M + d_2)v_2 = \frac{1}{2}\sum d_i + mM + \epsilon$$

From K1, we have:

$$Mv_1 + Mv_2 = \frac{S}{2} - \sum_{i \in I_1(\mathbf{v})} s_i v_i - s_1 v_1 - s_2 v_2 + (m - |I_1(\mathbf{v})|)M + \epsilon$$

and from K2, we have:

$$Mv_1 + Mv_2 = (m - |I_1(\mathbf{v})|)M + (m - |I_1(\mathbf{v})|)s_{\max} - \frac{S}{2} + \sum_{i \in I_1(\mathbf{v})} s_i v_i + s_1 v_1 + s_2 v_2 + \epsilon$$

By the equating the two right hand sides, ϵ and $(m - |I_1(\mathbf{v})|)M$ can be canceled out. By rearranging the terms, we get:

$$S - (m - |I_1(\mathbf{v})|)s_{\max} = 2(\sum_{i \in I_1(\mathbf{v})} s_i v_i + s_1 v_1 + s_2 v_2) \tag{1}$$

From the tight equation in K1, we can rearrange the terms to obtain:

$$\sum_{i \in I_1(\mathbf{v})} s_i v_i + s_1 v_1 + s_2 v_2 = \frac{S}{2} - Mv_1 - Mv_2 + (m - |I_1(\mathbf{v})|)(M) + \epsilon$$

By substituting this in (1), we get

$$S - (m - |I_1(\mathbf{v})|)s_{\max} = 2(\frac{S}{2} - Mv_1 - Mv_2 + (m - |I_1(\mathbf{v})|)M + \epsilon)$$

$$2M(v_1 + v_2) = 2(m - |I_1(\mathbf{v})|)M + (m - |I_1(\mathbf{v})|)s_{\max} + 2\epsilon$$

$$v_1 + v_2 = (m - |I_1(\mathbf{v})|) + \frac{(m - |I_1(\mathbf{v})|)s_{\max}}{2M} + \frac{2\epsilon}{2M}$$

Since $|I_1(\mathbf{v})| \leq m - 2$, we have $(m - |I_1(\mathbf{v})|) \geq 2$. As $\epsilon > 0$, $M > 0$ and $s_{\max} > 0$, we have:

$$v_1 + v_2 = (m - |I_1(\mathbf{v})|) + \frac{(m - |I_1(\mathbf{v})|)s_{\max}}{2M} + \frac{2\epsilon}{2M} \geq 2 + \frac{s_{\max} + \epsilon}{M} > 2$$

But $0 < v_1, v_2 < 1$ implies $v_1 + v_2 < 2$, a contradiction.

References

1. Garey, M.R., Johnson, D.S.: Computers and Intractability: A Guide to the Theory of NP-Completeness. W.H. Freeman and Co, New York (1979)
2. De Loera, J.A., Kafer, S., Sanità, L.: Pivot rules for circuit-augmentation algorithms in linear optimization. SIAM J. Optim. **32**, 2156–2179 (2022)
3. Frieze, A.M., Teng, S.-H.: On the complexity of computing the diameter of a polytope. Comput. Complex. **4**, 207–219 (1994)
4. Korte, B., Schrader, R.: On the existence of fast approximation schemes. Nonlinear Program. **4**, 415–437 (1981)
5. Nöbel, C., Steiner, R.: Complexity of polytope diameters via perfect matchings. In: Proceedings of the 2025 Annual ACM-SIAM Symposium on Discrete Algorithms (SODA), pp. 2234–2251 (2025)
6. Papadimitriou, C.H., Yannakakis, M.: The complexity of facets (and some facets of complexity). J. Comput. Syst. Sci. **28**, 244–259 (1984)
7. Sanità, L.: The diameter of the fractional matching polytope and its hardness implications. In: 2018 IEEE 59th Annual Symposium on Foundations of Computer Science (FOCS), pp. 910–921 (2018)
8. Wulf, L.: Computing the polytope diameter is even harder than NP-hard (already for perfect matchings), Preprint: https://arxiv.org/abs/2502.16398. A version of this will appear in the Proceedings of FOCS 2025
9. Ziegler, G.M.: Lectures on Polytopes. Springer, New York, N.Y (1998)

On Odd Coloring of Graphs

Joydeep Patar and I. Vinod Reddy$^{(\boxtimes)}$

Department of Computer Science and Engineering, IIT Bhilai, Bhilai, India
{joydeep,vinod}@iitbhilai.ac.in

Abstract. An odd coloring of a graph G is a proper coloring such that for every non-isolated vertex, at least one color must appear an odd number of times in the open neighborhood of v. The minimum number of colors needed in an odd coloring of a graph G is called odd chromatic number, denoted by $\chi_o(G)$. Given a graph G and a positive integer k, the Odd coloring problem is to check whether $\chi_o(G) \leq k$. It is known that Odd coloring is NP-complete even on bipartite graphs.

- Since the problem is NP-complete on bipartite graphs, we study it on chain graphs and show that the problem can be solved in polynomial time on chain graphs. We also show that the odd chromatic number of convex bipartite graphs can be arbitrarily large.
- Caro et al. (Discrete Appl. Math. 321:392-401, 2022) showed that adding a vertex to a graph increases its odd chromatic number by at most 2. They posed an open question on whether removing a vertex could cause a significant increase in the odd chromatic number. We answer this question by showing that removing a vertex v can increase the odd chromatic number by as much as the degree of v.
- We show that deleting an edge from a graph G increases its odd chromatic number by at most 1, except when the distance between the endpoints of the deleted edge in the resulting graph is 3 or 4, in which case the odd chromatic number can increase by at most 2. We also show that adding an edge between two non-adjacent vertices increases the odd chromatic number by at most 2.
- Finally, we classify all n-vertex graphs whose odd chromatic number equals n or $n - 1$.

1   Introduction

Given a graph G and a positive integer k, a proper k-coloring (or simply a proper coloring) of G is a function $f : V(G) \rightarrow \{1, 2, \ldots, k\}$ such that $f(u) \neq f(v)$ for every edge $uv \in E(G)$. The smallest integer k for which there exists a proper k-coloring of G is called the chromatic number of G, denoted by $\chi(G)$. Proper k-colorings and its variants have been extensively studied from both graph-theoretic [3,10] and complexity-theoretic [6,9] perspectives.

One well-studied generalization of proper coloring is the proper conflict-free coloring (PCFC) [5,8], where the additional constraint requires that for every vertex v, at least one color appears exactly once in the open neighborhood of v. Recently Petruševski and Škrekovski [11] introduced the odd coloring problem,

© The Author(s), under exclusive license to Springer Nature Switzerland AG 2026
N. Misra and A. Pandey (Eds.): CALDAM 2026, LNCS 16445, pp. 378–391, 2026.
https://doi.org/10.1007/978-3-032-17156-6_28

which generalizes PCFC. A proper k-coloring of a connected graph G is called an odd k-coloring, if for every vertex v, there exists at least one color in $N(v)$ that appears an odd number of times, where $N(v)$ denotes the open neighborhood of v. The *odd chromatic number* of G, denoted by $\chi_o(G)$ is the smallest integer k such that G admits an odd k-coloring. Given a graph G and a positive integer k, the ODD k-COLORING asks whether G admits an odd k-coloring. Determining the odd chromatic number of a graph G is NP-hard even when G is bipartite [1].

Petruševski and Škrekovski [11] proved that every planar graph admits an odd 9-coloring and they conjectured that every planar graph admits an odd 5-coloring. The conjecture was later proved for outerplanar graphs [11] and for planar graphs of girth at least seven [7]. Bhyravarapu et al. [2] studied the parameterized complexity of odd coloring. They proved that ODD k-COLORING is fixed-parameter tractable when parameterized by distance to cluster, distance to co-cluster or neighborhood diversity. They also showed that the problem can be solved in polynomial time on cographs and split graphs.

Our Contributions. Since the problem is NP-complete on bipartite graphs, we study its complexity on chain graphs and convex bipartite graphs, which are subclasses of bipartite graphs. In Sect. 3, we show that the odd chromatic number of chain graphs can be computed in polynomial time. In Sect. 4, we show that for a given positive integer k, there exists a convex bipartite graph G_k with odd chromatic number at least k. This shows that the odd chromatic number of convex bipartite graphs can be arbitrarily large.

By deleting an edge from a given graph G, the chromatic number either stays the same or decreases by one. That is $\chi(G) - 1 \leq \chi(G - e) \leq \chi(G)$. However, the odd chromatic number is not monotone under taking subgraphs. If H is a subgraph obtained from G by deleting an edge, then $\chi_o(H)$ may be either larger or smaller than $\chi_o(G)$. In Sect. 5, we show that deleting an edge between two vertices of a graph G increases its odd chromatic number by at most one, except when the distance between the endpoints of the deleted edge in the resulting graph is three or four, in which case the odd chromatic number can increase by at most two. We also show that adding an edge between two non-adjacent vertices increases the odd chromatic number by at most two.

In [11], the authors showed that adding a new vertex v and connecting it to all or some vertices can increase the odd chromatic number by at most two. The case of removing a vertex was left open. In Sect. 6, we show that removing a vertex from a graph can significantly increase (up to its degree) the odd chromatic number.

In the case of proper coloring, it is easy to see that $\chi(G) = n$ if and only if $G = K_n$. In Sect. 7, we investigate the class of graphs for which the odd chromatic number equals n or $n - 1$. In particular we show that $\chi_o(G) = n$ if and only if $G = P_3, C_4, C_5, K_n$, where $n = |V(G)|$. We also present a characterization of graphs having odd chromatic number $n - 1$, when $n \geq 6$. Let $\mathcal{F} = \{I_3, 2K_2, P_4, H_1, H_2, H_3, H_4\}$ be the set of graphs as shown in the Fig. 5. We show that if G is a connected and non-complete graph on $n \geq 6$ vertices. Then $\chi_o(G) = n - 1$ if and only if G is $\mathcal{F}$-free.

2   Preliminaries

For a positive integer k, we use $[k]$ to denote the set $\{1, 2, \ldots, k\}$. All graphs considered in this paper are undirected, connected and simple. For a graph $G = (V, E)$, we denote its vertex set and edge set by $V(G)$ and $E(G)$ respectively. An edge between vertices u and v is denoted as uv for simplicity. For a subset $S \subseteq V(G)$, the graph $G[S]$ denotes the subgraph of G induced by vertices in S. For a vertex $v \in V(G)$, the open neighborhood of v is $N(v) := \{u \in V(G) \mid vu \in E(G)\}$ and the closed neighborhood is $N[v] = N(v) \cup \{v\}$. For a subset $S \subseteq V(G)$, the neighborhood of S is defined as $N(S) = \left(\bigcup_{v \in S} N(v) \right) \setminus S$.

If $f : V(G) \to [k]$ is a coloring of G, then we use $f^{-1}(i) = \{u \in V(G) | f(u) = i\}$ to denote the subset of vertices colored i under f. For a subset $X \subseteq V(G)$, we use $f(X)$ to denote set $\{f(u) \mid u \in X\}$. We say that a vertex $v \in V(G)$ has a color $c \in [k]$ as its *odd color* with respect to f, if the number of neighbors of v assigned color c is odd. Formally, c is an odd color of v if $|f^{-1}(c) \cap N(v)|$ is odd.

For a vertex set $X \subseteq V(G)$, we denote $G \setminus X$, the graph obtained from G by deleting all vertices of X and their incident edges. Given two graphs G and H, we say that G *contains* H if some induced subgraph of G is isomorphic to H. A graph G is said to be H-free if it does not contain H. For a family of graphs $\mathcal{H}$, the graph G is $\mathcal{H}$-free if it is H-free for every $H \in \mathcal{H}$. For more details on standard graph-theoretic notation and terminology, we refer the reader to [12].

3   Chain Graphs

In this section, we study the odd coloring problem on chain graphs. First, we show that the odd chromatic number of chain graphs is at most four. Then, we characterize chain graphs with odd chromatic numbers two and three.

Definition 1 (Bipartite graph). *A graph G is* bipartite *if its vertex set $V(G)$ can be partitioned into two disjoint sets X and Y such that no two vertices within the same set are adjacent. We denote such a graph by $G = (X, Y, E)$.*

Definition 2 (Chain graph). *A bipartite graph $G = (X, Y, E)$ is called a* chain graph *if there exist orderings $(x_1, x_2, \ldots, x_r)$ of the vertices of X and $(y_1, y_2, \ldots, y_s)$ of the vertices of Y such that $N(x_r) \subseteq N(x_{r-1}) \subseteq \ldots \subseteq N(x_1)$ and $N(y_s) \subseteq N(y_{s-1}) \subseteq \ldots \subseteq N(y_1)$.*

If $G = (X, Y, E)$ is a chain graph, then from the definition it follows that $\deg(x_1) \geq \deg(x_2) \geq \cdots \geq \deg(x_r)$ and $\deg(y_1) \geq \deg(y_2) \geq \cdots \geq \deg(y_s)$.

Lemma 1. *If $G = (X, Y, E)$ is a chain graph, then $\chi_o(G) \leq 4$.*

Proof. Let $G = (X, Y, E)$ be a chain graph. We can order the vertices of X and Y as

$$X = \{x_1, x_2, \ldots, x_r\}, \qquad Y = \{y_1, y_2, \ldots, y_s\},$$

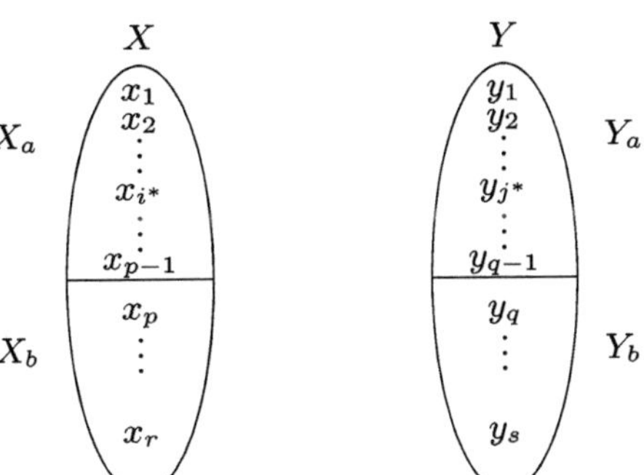

Fig. 1. $G = (X \cup Y, E)$ is a chain graph. The vertex x_1 (resp. y_1) is adjacent to every vertex of Y (resp. X). The edges between the vertices of X and Y are not shown in the figure due to space constraints.

such that $N(x_{i+1}) \subseteq N(x_i)$ for all $i \in [r-1]$, $N(y_{j+1}) \subseteq N(y_j)$ for all $j \in [s-1]$. Here $r + s = n$.

Define a coloring $f : V(G) \to \{1, 2, 3, 4\}$ as follows:

$$f(x_1) = 1, \quad f(y_1) = 2, \quad f(x_i) = 3 \text{ for } i \geq 2, \quad f(y_j) = 4 \text{ for } j \geq 2.$$

It is easy to see that f is an odd 4-coloring of G, hence $\chi_o(G) \leq 4$. $\qquad\square$

Next, we characterize chain graphs with odd chromatic numbers two, three and four. The following result from [4] provides the characterization of chain graphs with odd chromatic number two.

Lemma 2. *[4] Let G be a connected chain graph with at least two vertices. Then $\chi_o(G) = 2$ if and only if every vertex of G has odd degree.*

From now on, we assume that the chain graph $G = (X, Y, E)$ contains at least one even-degree vertex. If all vertices in either X or Y have odd degree vertices, then $\chi_o(G) = 3$. Therefore, we further assume that both X and Y contain at least one even-degree vertex.

Lemma 3. *Let $G = (X, Y, E)$ be a chain graph such that both X and Y contain at least one even-degree vertex. Let $i^* = \max\{i : \deg(x_i) \text{ is even}\}$ and $j^* = \max\{j : \deg(y_j) \text{ is even}\}$. Define $p := \deg(y_{j^*})$, $q := \deg(x_{i^*})$.*

Then $\chi_o(G) = 3$ if and only if x_p and y_q are nonadjacent (equivalently, $\deg(x_p) < q$ and $\deg(y_q) < p$).

Proof. Forward direction. Assume $\chi_o(G) = 3$. Let $X = \{x_1, \ldots, x_r\}$ and $Y = \{y_1, \ldots, y_s\}$ be the bipartition of the chain graph, ordered by neighborhoods. Let $i^* = \max\{i : \deg(x_i) \text{ is even}\}$ and $j^* = \max\{j : \deg(y_j) \text{ is even}\}$. Since both X and Y contain at least one even-degree vertex, we have $i^*, j^* \geq 1$ and $p, q \geq 1$ (Fig. 1). We show that x_p and y_q are nonadjacent.

Let $f : V(G) \to \{1, 2, 3\}$ be an odd 3-coloring of G. Without loss of generality, assume $f(x_1) = 1$ and $f(y_1) = 2$. Since $N(x_1) = Y$ and $N(y_1) = X$, this implies $1 \notin f(Y)$ and $2 \notin f(X)$. Define

$$S := \{x \in X : f(x) = 3\}, \qquad T := \{y \in Y : f(y) = 3\}.$$

Since both X and Y contain at least one even-degree vertex, both S and T are non-empty.

Let ℓ (resp. t) be the smallest index such that $x_\ell \in S$ (resp. $y_t \in T$). Thus all vertices $x_1, \ldots, x_{\ell-1}$ are colored 1 and all $y_1, \ldots, y_{t-1}$ are colored 2 under f.

If $\ell \leq p$ and $t \leq q$, then $N(x_p) \subseteq N(x_\ell)$ and $N(y_q) \subseteq N(y_t)$. Since both x_ℓ and y_t are colored with 3, we have $x_\ell y_t \notin E(G)$ which implies $x_p y_t \notin E(G)$. Since $t \leq q$, we have $x_p y_q \notin E(G)$.

If $\ell > p$, then $deg(y_{j*}) = p$, which is even and $N(y_{j*}) = \{x_1, \ldots, x_p\}$. Since every vertex in $N(y_{j*})$ is assigned color 1 under f, the vertex y_{j*} does not satisfy the odd coloring condition. This contradicts the assumption that f is an odd coloring of G. Hence, the case $\ell > p$ cannot occur. By a symmetric argument, the case $t > q$ also cannot occur. Therefore x_p and y_q are nonadjacent in G.

Backward Direction. Assume x_p and y_q are nonadjacent. We show that $\chi_o(G) = 3$. Since both X and Y contain at least one even-degree vertex, we have $\chi_o(G) \geq 3$. Next, we give an odd 3-coloring of G.

Let $X_a = \{x_1, \ldots, x_{p-1}\}$, $X_b = \{x_p, \ldots, x_r\}$ and $Y_a = \{y_1, \ldots, y_{q-1}\}$ and $Y_b = \{y_q, \ldots, y_s\}$. Since $x_p y_q \notin E(G)$, we have $deg(x_p) < q = deg(x_{i*})$, hence $p > i^*$. Similarly, $\deg(y_q) < p = \deg(y_{j*})$, hence $q > j^*$.

Define a coloring $f : V(G) \to \{1, 2, 3\}$ as

$$f(v) = \begin{cases} 1, & v \in X_a, \\ 2, & v \in Y_a, \\ 3, & v \in X_b \cup Y_b. \end{cases}$$

First, note that the coloring is proper: there are no edges inside X or inside Y, and there are no edges between X_b and Y_b because $\deg(x_i) \leq \deg(x_p) < q \leq j$ for every $x_i \in X_b$ and every $y_j \in Y_b$.

It remains to verify the odd coloring condition at every vertex.

- Consider a vertex $x_k \in X_a$. If $\deg(x_k)$ is odd, then the odd coloring condition is trivially satisfied at x_k. If $\deg(x_k)$ is even, then $deg(x_k) \geq deg(x_{i*}) = q$. Since $q - 1$ is odd, color 2 appears an odd number of times in $N(x_k)$ and the odd coloring condition is satisfied at x_k.
- Consider a vertex $x_k \in X_b$. Since every vertex of X_b has odd degree, the odd coloring condition is trivially satisfied at x_k.
- Similar to the above case, we can show that every vertex in $Y_a \cup Y_b$ also satisfies the odd coloring condition. $\qquad\square$

Theorem 1. ODD COLORING *can be solved in polynomial time on chain graphs.*

Proof. Follows from Lemmas 1, 2 and 3 .

4   Convex Bipartite Graphs

In this section, we show that the odd chromatic number of convex bipartite graphs can be arbitrarily large.

Definition 3 (Convex bipartite graph). *A bipartite graph $G = (X, Y, E)$ is called* convex *with respect to X, if there exists an ordering of the vertices of X such that for every $y \in Y$ the vertices adjacent to y are consecutive with respect to the ordering on X. A bipartite graph is called* convex bipartite *if it is convex with respect to either X or Y.*

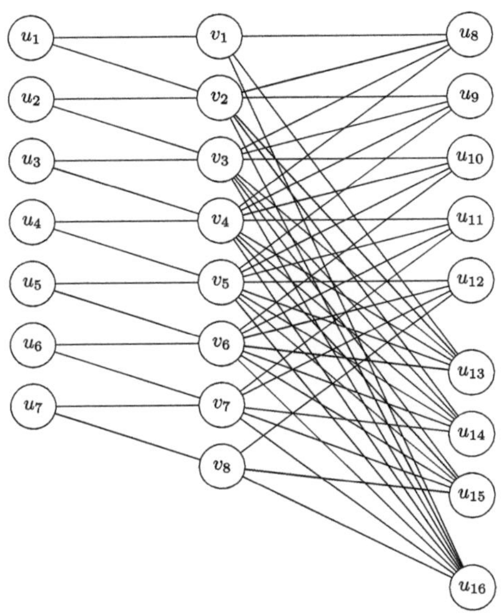

Fig. 2. The convex graph $G_4 = (X, Y, E)$ with bipartition $X = \{v_1, \ldots, v_8\}$ and $Y = \{u_1, \ldots, u_{16}\}$.

Fix $k \in \mathbb{N}$, where $k \geq 2$. Let $G_k = (X, Y, E)$ be the bipartite graph defined as follows.

- $X = \{v_1, \ldots, v_p\}$, $Y = \{u_1, \ldots, u_q\}$, where $p = 2^{k-1}$ and $q = 2^{2k-4}$.
- Define $s_1 = 1$, and for $j = 2, \ldots, k$:

$$s_j = s_{j-1} + 2^{k-1} - (2(j-1) - 1), \quad \text{and } s_{k+1} = q + 1.$$

- For $u_r \in Y$, if $s_j \leq r < s_{j+1}$, set

$$N(u_r) = \{v_{r-s_j+1}, \ldots, v_{r-s_j+2j}\}.$$

For $k = 4$, the graph G_4 is shown in Fig. 2. Since each vertex $u_r \in Y$ is adjacent to a consecutive block of $2j$ vertices in X, where j depends on the index r. Thus, G_k is convex with respect to X.

Theorem 2. $\chi_o(G_k) \geq k$.

Proof. Let $P_r = \{v_1, \ldots, v_{2r}\}$, $r \in [k]$. By construction of the graph G_k, for each $r \in [k]$ there exists a $u \in Y$ such that $N(u) = P_r$. Hence, in any odd coloring f of G_k, each P_r must contain some color appearing an odd number of times.

Suppose, for contradiction, that there exists an odd coloring f of G_k using only $t \leq k-1$ colors. Let the colors used by f are $1, 2 \ldots, t$. For each $j \in [k]$ and color $c \in [t]$, define

$$p_{j,c} = |\{v \in P_j \mid f(v) = c\}| \quad (\mathrm{mod}\ 2)$$

Arrange these bits into a $k \times t$ matrix A, viewed as elements of $\mathbb{F}_2$. The j-th row of A is a t-bit vector $(p_{j,1}, p_{j,2}, \ldots, p_{j,t})$. Observe that

1. No row of A is a zero row: If a row j of A contains all zeros, then no color appears odd number of times in P_j, which is a contradiction to the odd coloring f.
2. As $k > t$, the k row vectors must be linearly dependent over $\mathbb{F}_2$. That is, there exists a non-empty subset $J \subseteq [k]$ such that

$$\sum_{j \in J} p_{j,c} \equiv 0 \quad (\mathrm{mod}\ 2) \quad \text{for all colors } c.$$

Let $M := \triangle_{j \in J} P_j$, the symmetric difference of the P_j's. That is, a vertex belongs to M if and only if it appears in an odd number of P_j's with $j \in J$.

Observe that the P_j's are nested: $P_1 \subset P_2 \subset \cdots \subset P_k$, with $P_\ell \setminus P_{\ell-1} = \{v_{2\ell-1}, v_{2\ell}\}$. Thus, the symmetric difference M is a union of disjoint intervals of the form $S_{i_r, j_r} = \{v_{i_r}, v_{i_r+1} \ldots, v_{j_r}\}$, each of even length. For example, if $J = \{2, 5, 9, 15\}$, then $M = S_{5,10} \cup S_{19,30}$. Also, note that for every interval S_{i_r, j_r} there exists a vertex $u_r \in Y$ such that $N(u_r) = S_{i_r j_r}$.

From the linear dependence of rows of A, whose indices present in J, for each color c, $\sum_{j \in J} p_{j,c} \equiv 0$ (mod 2), which implies that each color appears an even number of times in M. Therefore, in each interval S_{i_r, j_r} of M, all colors appear an even number of times. This contradicts the odd-coloring property of f, since each interval (as a neighborhood) must contain some color appearing an odd number of times. Hence, $\chi_o(G_k) \geq k$. $\square$

5   Addition and Deletion of One Edge

In this section, we study how the odd chromatic number of a graph G is affected by the removal of an existing edge or the addition of a new edge between two non-adjacent vertices.

Theorem 3. *Let G be a graph and H be the graph obtained from G by deleting an edge $e = uv$. Let $d := d_H(u, v)$ be the distance between the vertices u and v in the graph H. Then*

$$\chi_o(H) \leq \begin{cases} \chi_o(G) + 2 & \textit{if } d \in \{3, 4\}; \\ \chi_o(G) + 1 & \textit{otherwise;} \end{cases}$$

Proof. Given an odd coloring f of G, we can obtain an odd coloring of H using at most two new colors: recolor one vertex of $N_H(u)$ and one vertex of $N_H(v)$ with new colors. Hence we have $\chi_o(H) \leq \chi_o(G) + 2$.

Next we prove that the above bound can be improved, if $d \notin \{3, 4\}$.

Case 1: $d = 2$

Since the distance between u and v in graph H is two, we have $N_H(u) \cap N_H(v) \neq \emptyset$. Given an odd coloring f of G, we can obtain an odd coloring of H by recoloring an arbitrary vertex in $N_H(u) \cap N_H(v)$ with a new color. Therefore, $\chi_o(H) \leq \chi_o(G) + 1$.

Case 2: $d \geq 5$

Let $S_u = \{w \mid d_H(u, w) \leq 2\}$ and $S_v = \{w \mid d_H(v, w) \leq 2\}$. Since $d \geq 5$, we have $S_u \cap S_v = \emptyset$. Let $w_1 \in N_H(u)$ and $w_2 \in N_H(v)$ be two arbitrary vertices. Given an odd coloring f of G, we define a coloring g of H as follows.

$$g(z) = \begin{cases} \chi_o(G) + 1 & \text{if } z \in \{w_1, w_2\}; \\ f(z) & \text{otherwise;} \end{cases}$$

We now argue that g is an odd coloring of H.

- For every vertex $w \in V(H) \setminus (N_H(w_1) \cup N_H(w_2))$ the odd color of w with respect to f is the same as the odd color of w with respect to g.
- Since $N_H(w_1) \cap N_H(w_2) = \emptyset$, the color $\chi_o(G) + 1$ acts as an odd color for each vertex in the set $N_H(w_1) \cup N_H(w_2)$.

Therefore, we have $\chi_o(H) \leq \chi_o(G) + 1$. $\square$

For $d \in \{3, 4\}$, the bound given in the above theorem is tight, as shown by the examples in Fig. 3.

Theorem 4. *Let H be a graph and G be the graph obtained from H by adding an edge $e = uv$ between two non-adjacent vertices u and v. Then $\chi_o(G) \leq \chi_o(H) + 2$.*

Proof. Let g be an odd coloring of H with $\chi_o(H)$ colors. We construct an odd coloring f of G using at most $\chi_o(H) + 2$ colors as follows: Assign two new colors c_1 and c_2 (not used in g) to the vertices u and v, respectively. Define

$$f(w) = \begin{cases} c_1 & \text{if } w = u, \\ c_2 & \text{if } w = v, \\ g(w) & \text{otherwise.} \end{cases}$$

We now show that f is an odd coloring of G. Since f differs from g only at u and v, and these two vertices are now assigned distinct new colors, their neighbors still satisfy the odd coloring condition. Since $uv \in E(G)$, $f(u) \neq f(v)$ and they are not in $\{1, 2, \ldots, \chi_o(H)\}$, both u and v satisfy odd coloring condition. Hence, f is a valid odd coloring of G using $\chi_o(H) + 2$ colors.

To show that the bound is tight, consider $H = P_5$ (the path on five vertices), and let $G = C_5$, the cycle on five vertices, obtained by adding an edge between the end vertices of P_5. It is known that $\chi_o(P_5) = 3$ and $\chi_o(C_5) = 5$, so equality is achieved in the bound. $\square$

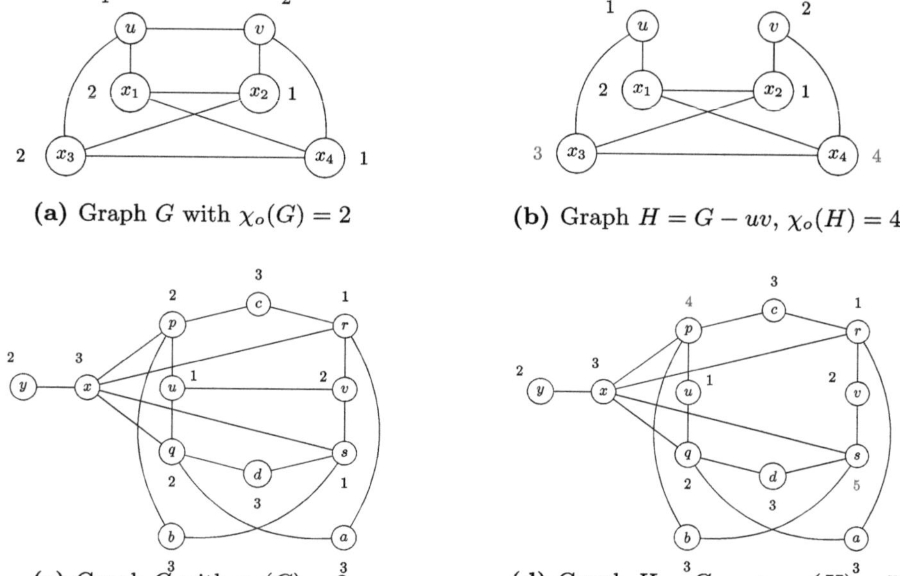

(a) Graph G with $\chi_o(G) = 2$

(b) Graph $H = G - uv$, $\chi_o(H) = 4$

(c) Graph G with $\chi_o(G) = 3$

(d) Graph $H = G - uv$, $\chi_o(H) = 5$

Fig. 3. Examples showing that the upper bound in Theorem 3 is tight when $d \in \{3, 4\}$. The top row shows the case where the distance between u and v in H is 3, and the odd chromatic number increases from 2 to 4. The bottom row shows the case where the distance between u and v in H is 4, and the odd chromatic number increases from 3 to 5.

6   Addition and Deletion of One Vertex

In the paper [4], the authors showed that adding a new vertex v and connecting it to all or some vertices can increase the odd chromatic number by at most two. The case of removing a vertex was left open. We show that removing a vertex from a graph can significantly increase the odd chromatic number.

Theorem 5. *Let G be a graph and H be the graph obtained from G by deleting a vertex v. Then, $\chi_o(H) \leq \chi_o(G) + \deg(v)$, where $\deg(v)$ denotes the degree of the vertex v in G.*

Proof. Let f be an odd coloring of G with $\chi_o(G)$ colors. Let $N(v) = \{u_1, \ldots, u_{\deg(v)}\}$ denote the neighbors of v in G. Let $H = G - v$ be the graph obtained by deleting vertex v.

We construct an odd coloring of H by restricting f to H. However, for each neighbor u_i of v, the removal of v may cause the odd coloring condition at u_i to fail. To restore the odd coloring condition at each u_i we can assign a new color to one of its neighbors (other than v) in H. In the worst case, we may need to introduce one new color per affected neighbor of v, resulting in at most $\deg(v)$ additional colors. Therefore, we obtain $\chi_o(H) \leq \chi_o(G) + \deg(v)$. The bound is tight as shown in the Fig. 4. $\qquad\square$

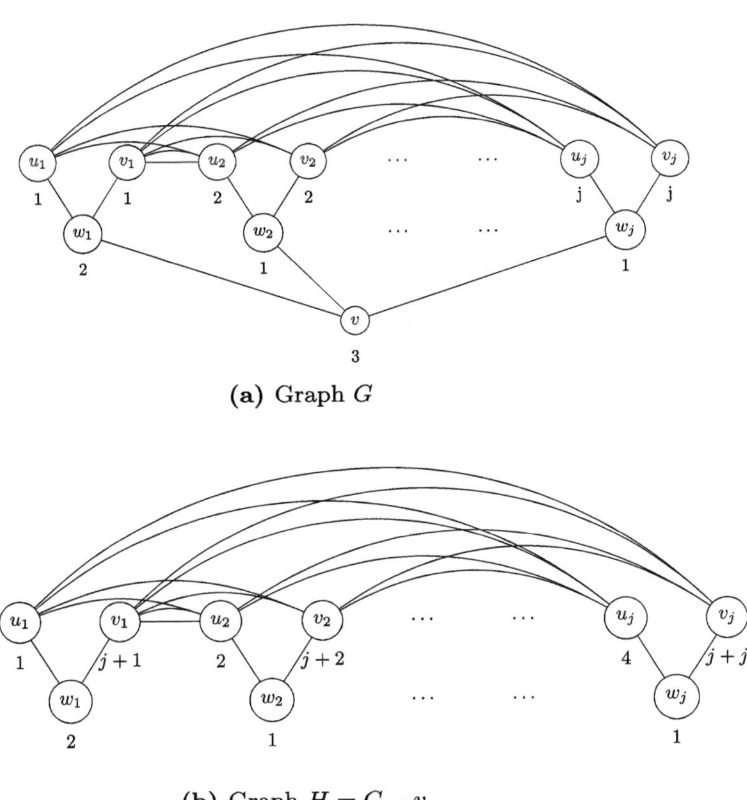

(a) Graph G

(b) Graph $H = G - v$

Fig. 4. In this example $\chi_o(G) = j$ and $\chi_o(H) = \chi_o(G - v) = 2j$, where $\deg(v) = j$.

7   Graphs with $\chi_o(G) = n$ or $n - 1$

In the case of proper coloring, it is easy to see that $\chi(G) = n$ if and only if $G = K_n$. In this section, we investigate graphs for which the odd chromatic number equals n or $n - 1$.

Theorem 6. *Let G be a graph with n vertices. Then $\chi_o(G) = n$ if and only if $G \in \{P_3, C_4, C_5, K_n\}$*

Proof. Forward direction. Let G be a graph with n vertices such that $\chi_o(G) = n$.
Case 1: $n \geq 6$

Suppose $G \not\cong K_n$. Then, there exists two non-adjacent vertices v_i and v_j in G. We will show that $\chi_o(G) \leq n-1$ by giving an odd coloring f of G using $n-1$ colors.

Define $S = \{u \in V(G) \mid \{v_i, v_j\} \subsetneq N(u)\}$ and $S^2 = \{u \in V(G) \mid N(u) = \{v_i, v_j\}\}$
Subcase 1(a): $S^2 = \emptyset$.

Define a coloring $f : V(G) \to [n-1]$ such that $f(v_i) = f(v_j) = 1$ and assign distinct colors from the remaining $n-2$ colors to other $n-2$ vertices. It is easy to verify that f is an odd coloring of G. Therefore, $\chi_o(G) \leq n-1$, which is a contradiction to the fact that $\chi_o(G) = n$.

Subcase 1(b): $S^2 \neq \emptyset$.

Let $z \in S^2$. Define a coloring $f : V(G) \to [n-1]$ such that $f(v_i) = 1$, $f(v_j) = 2$, $f(z) = 3$, and other remaining $n-3$ vertices are assigned distinct colors from the set $\{3, 4, \ldots, n-1\}$ ensuring that for both v_i and v_j, at least one of their neighbors (other than z, if such a neighbor exists) is colored with a color different from 3. It may happen that both v_i and v_j have exactly one another neighbor (other than z), and these neighbors are adjacent. In that case, we assign them distinct colors. Such a coloring is possible because $n \geq 6$. Since color 3 is used twice by f, the coloring f uses at most $n-1$ colors. We now argue that f is an odd coloring of G.

- For every $w \in S^2$, since $N(w) = \{v_i, v_j\}$ and $f(v_i) \neq f(v_j)$, w has an odd color with respect to f.
- Since the coloring f ensures that one neighbor of v_i (resp. v_j), other than z, is assigned a color $c(\neq 3)$, this color c acts as an odd color for v_i (resp. v_j). If no such neighbor exists, then the color 3 will act as an odd color for both v_i and v_j.
- For every $w \in V(G) \setminus (S^2 \cup \{v_i, v_j\})$, each color is used exactly once in $N(w)$, so w has an odd color with respect to f.

Therefore, f is an odd coloring of G, implying $\chi_o(G) \leq n-1$, which is a contradiction. Hence $G \cong K_n$.

Case 2: $n \leq 5$

In this case, we verified through a tedious case-by-case analysis that the only graphs for which $\chi_o(G) = n$ are P_3, C_4, C_5 and K_n for $n \leq 5$.

Backward Direction. It is easy to verify that $\chi_o(G) = n$ for each of the graphs P_3, C_4, C_5, and K_n for $n \leq 5$. $\qquad\qquad\square$

Next, we characterize the graphs having odd chromatic number $n-1$, for $n \geq 6$. Let $\mathcal{F} = \{I_3, 2K_2, P_4, H_1, H_2, H_3, H_4\}$ be the set of graphs shown in the Fig. 5.

Theorem 7. *Let G be a connected and non-complete graph on $n \geq 6$ vertices. Then $\chi_o(G) = n-1$ if and only if G is $\mathcal{F}$-free.*

We prove the contrapositive of the above theorem, stated below.

Theorem 8. *Let G be a connected and non-complete graph on $n \geq 6$ vertices. Then $\chi_o(G) \leq n-2$ if and only if G contains an induced subgraph from $\mathcal{F}$.*

Proof. Due to space constraints, we present the forward direction of the proof in the lemma below. The reverse direction is given in the full version of the paper.

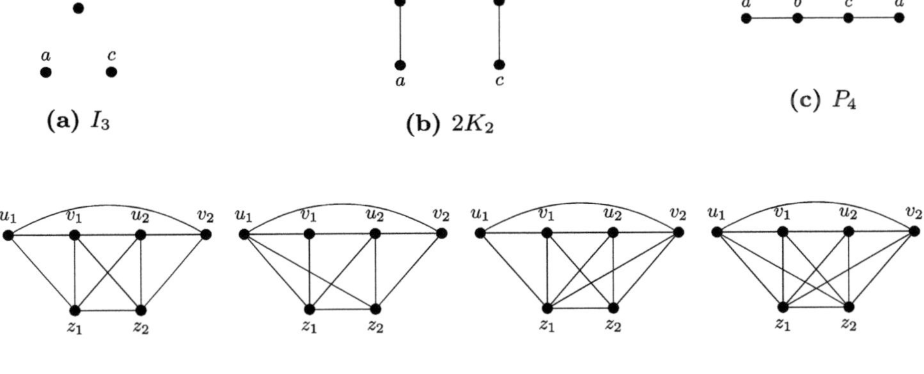

(a) I_3 **(b)** $2K_2$ **(c)** P_4

(d) H_1 **(e)** H_2 **(f)** H_3 **(g)** H_4

Fig. 5. Forbidden induced subgraphs for the class of graphs with $\chi_o(G) = n - 1$.

Lemma 4. *Let G be a connected graph on $n \geq 6$ vertices. If $\chi_o(G) \leq n - 2$, then G contains a graph from $\mathcal{F}$ as an induced subgraph.*

Proof. Let G be a graph such that $\chi_o(G) \leq n - 2$. Let $f : V(G) \to [k]$ be an odd k-coloring of G, where $k \leq n - 2$. We have two cases based on the number of times a color is used in the coloring f.

Case 1: There exists an $i \in [k]$ such that $|f^{-1}(i)| \geq 3$.

In this case the set $f^{-1}(i)$ is an independent set of size at least three. Hence, G contains I_3, which is a graph from $\mathcal{F}$.

Case 2: There exists i, j such that $|f^{-1}(i)| = 2$, $|f^{-1}(j)| = 2$ and $|f^{-1}(\ell)| = 1$ for every $\ell \in [n - 2] \setminus \{i, j\}$.

Let $f^{-1}(i) = \{u_1, u_2\}$ and $f^{-1}(j) = \{v_1, v_2\}$ and let $S = \{u_1, u_2, v_1, v_2\}$. If $G[S]$ contains less than four edges then G contains one of I_3, $2K_2$ or P_4. Therefore, we assume that $G[S]$ contains at least four edges. It is easy to see that $G[S]$ cannot have more than four edges and $G[S] \cong C_4$. Let the edges of C_4 be $u_1v_1, v_1u_2, u_2v_2, v_2u_1$.

Let $T = V(G) \setminus S$ denote the other $n - 4$ vertices of G which are colored with $n - 4$ distinct colors by f. Since $n \geq 6$, we have $|T| \geq 2$.

Case A. $|T| = 2$.

Let $T = \{z_1, z_2\}$.

Case A.1. If $z_1 z_2 \notin E(G)$.

- If either z_1 or z_2 has degree one, say z_i, then either $\{u_1, u_2, z_i\}$ or $\{v_1, v_2, z_i\}$ forms an I_3.
- If a $z_i \in \{z_1, z_2\}$ has degree two:
  - If its two neighbors are adjacent, then the graph contains a P_4.
  - If its two neighbors are non-adjacent, then z_i cannot have an odd color (since both neighbors have the same color), a contradiction to the fact that f is an odd coloring of G.

– If both z_1 and z_2 have degree three:
  • If $N(z_1) = N(z_2)$, then there is a vertex y in S of degree two and both its neighbors colored with the same color, violating the assumption that f is an odd coloring. Hence, this is not allowed.
  • If $N(z_1) \neq N(z_2)$, then we can find vertices $a \in N(z_1) \setminus N(z_2)$ and $b \in N(z_2) \setminus N(z_1)$. If $ab \in E(G)$ are adjacent then $z_1 - a - b - z_2$ forms a P_4. Otherwise, z_1a and z_2b forms a $2K_2$.
– Neither z_1 nor z_2 can have degree four, since f is an odd coloring.

Case A.2. If $z_1z_2 \in E(G)$

– If either z_1 or z_2 has degree at most two, say z_1, then either $\{u_1, u_2, z_1\}$ or $\{v_1, v_2, z_1\}$ forms an I_3.
– If $z_i \in \{z_1, z_2\}$ has degree three:
  • If its two neighbors in S are adjacent, then the graph contains a P_4.
  • If its two neighbors in S are non-adjacent, then the graph contains an I_3.
– If both z_1 and z_2 have degree four:
  • If $N[z_1] = N[z_2]$, then there exists a vertex $y \in S$ of degree two with both neighbors colored the same, contradicting the assumption that f is an odd coloring.
  • If $N[z_1] \neq N[z_2]$, then G is isomorphic to H_1 or H_2.
– If $\deg(z_1) = 5$ and $\deg(z_2) = 4$, then G is isomorphic to H_3.
– If $\deg(z_1) = \deg(z_2) = 5$, then G is isomorphic to H_4.

Therefore, if $|T| = 2$, then in all possible cases, G contains a graph from the family $\mathcal{F}$.

Case B. $|T| \geq 3$.

We know that G has at least seven vertices, with $V(G) = T \cup S$, where $T = \{z_1, \ldots, z_{n-4}\}$ and $|T| \geq 3$. We know that f is an $(n-2)$-odd coloring of G satisfying: $f(u_1) = f(u_2) = i$, $f(v_1) = f(v_2) = j$, all other vertices in T are colored with distinct colors by f.

– If no two vertices in T are adjacent, then G contains an I_3.
– If some $z_i \in T$ has exactly one neighbor in S, then G contains an I_3 formed by z_i and two non-adjacent vertices from S.
– If $z_i \in T$ has exactly two neighbors in S:
  • If those neighbors are adjacent, then G contains a P_4.
  • If they are not adjacent, then G contains an I_3.
– Therefore, every vertex in T must have at least three neighbors in S.
– Let $z_1, z_2 \in T$ be two adjacent vertices T.
  • If both z_1 and z_2 have exactly three neighbors in S, and $N[z_1] \neq N[z_2]$, then (as shown in previous cases) G contains one of H_1 or H_2 as an induced subgraph.

- If both z_1 and z_2 have exactly three neighbors in S and $N[z_1] = N[z_2] = \{u_1, v_1, u_2\}$. As f is an odd coloring of G, the vertex v_2 must have a neighbor in T. Let $z_p \in T$ be a neighbor of v_2. The vertex z_p must be adjacent to both z_1 and z_2 otherwise G contains either a $2K_2$ or P_4 as an induced subgraph. Since z_p must have at least three neighbors in S and $N[z_1] \neq N[z_p]$, by the above case, G again contains one of H_1 or H_2 as an induced subgraph.
- If one of z_1 and z_2 have exactly four neighbors in S and the other has three neighbors then G contains H_3.
- If both z_1 and z_2 have exactly four neighbors in S and the other has three neighbors then G contains H_4.

Therefore, in all possible cases when $|T| \geq 3$, the graph G contains one of the graphs from $\mathcal{F}$ as an induced subgraph. $\qquad\qquad\square$

References

1. Ahn, J., Im, S., Oum, S.i.: The proper conflict-free k-coloring problem and the odd k-coloring problem are np-complete on bipartite graphs. Discrete Appl. Math. **377**, 10–17 (2025)
2. Bhyravarapu, S., Kumari, S., Vinod Reddy, I.: On the parameterized complexity of odd coloring. In: Conference on Algorithms and Discrete Applied Mathematics, pp. 60–72. Springer (2025)
3. Bohman, T., Holzman, R.: On a list coloring conjecture of reed. J. Graph Theor. **41**(2), 106–109 (2002)
4. Caro, Y., Petruševski, M., Škrekovski, R.: Remarks on odd colorings of graphs. Discret. Appl. Math. **321**, 392–401 (2022)
5. Caro, Y., Petruševski, M., Škrekovski, R.: Remarks on proper conflict-free colorings of graphs. Discret. Math. **346**(2), 113221 (2023)
6. de CM Gomes, G., Lima, C.V., dos Santos, V.F.: Parameterized complexity of equitable coloring. Discrete Math. Theor. Comput. Sci. **21** (2019)
7. Cranston, D.W., Lafferty, M., Song, Z.X.: A note on odd colorings of 1-planar graphs. Discret. Appl. Math. **330**, 112–117 (2023)
8. Fabrici, I., Lužar, B., Rindošová, S., Soták, R.: Proper conflict-free and unique-maximum colorings of planar graphs with respect to neighborhoods. Discret. Appl. Math. **324**, 80–92 (2023)
9. Kratochvil, J., Tuza, Z.: Algorithmic complexity of list colorings. Discret. Appl. Math. **50**(3), 297–302 (1994)
10. Meyer, W.: Equitable coloring. Am. Math. Monthly **80**(8), 920–922 (1973)
11. Petruševski, M., Škrekovski, R.: Colorings with neighborhood parity condition. Discret. Appl. Math. **321**, 385–391 (2022)
12. West, D.B., et al.: Introduction to Graph Theory, vol. 2. Prentice hall Upper Saddle River (2001)

Matching Extendability in Cartesian Product of Hypercubes and Paths

A. A. Pereira$^{(\boxtimes)}$ and C. N. Campos

Universidade Estadual de Campinas, Av. Albert Einstein, 1251,
Campinas-SP 13083-852, Brazil
`alessandra.pereira@ic.unicamp.br`, `cnc@unicamp.br`

Abstract. A matching M in a graph G is said to be extendable to a perfect matching if there exists a perfect matching M^* of G such that $M \subseteq M^*$. In this work, we study the extendability of matchings under a neighbourhood condition: no unsaturated vertex has all of its neighbours M-saturated. Vandenbussche and West showed that, in the hypercube Q_n, any matching of size at most $2n - 4$ is extendable to a perfect matching if and only if it satisfies this condition. We extend their result to the Cartesian product $Q_n \square P_m$ by proving that every matching of size at most $2n - 2$ is extendable to a perfect matching if and only if it does not saturate the neighbourhood of any unsaturated vertex. Furthermore, we demonstrate that this bound is tight by constructing a matching of size $2n + 2\delta(H) - 3$ in $Q_n \square H$ that satisfies the neighbourhood condition but is not extendable to a perfect matching.

Keywords: Matching extendability · k-extendability · Perfect matchings · Hypercubes · Cartesian product

1 Introduction

Let G be a finite, undirected and simple graph with vertex set $V(G)$ and edge set $E(G)$. Let $N_G(v) = \{u \in V(G) : uv \in E(G)\}$ be the *(open) neighbourhood* of a vertex $v \in V(G)$. Each $u \in N_G(v)$ is called a *neighbour* of v. Similarly, for a subset $S \subseteq V(G)$, the *neighbourhood* of S is given by $N_G(S) = \cup_{v \in S} N_G(v)$. When the context is clear, we omit the subscript and simply write $N(v)$ and $N(S)$. The *degree* of a vertex $v \in V(G)$ is $d(v) = |N(v)|$. The *minimum degree* of G is denoted by $\delta(G) = \min_{v \in V(G)}\{d(v)\}$ and the *maximum degree* by $\Delta(G) = \max_{v \in V(G)}\{d(v)\}$. For $u, v \in V(G)$, the *distance* $\mathrm{dist}(u, v)$ between u and v is the number of edges in the shortest path connecting them; $\mathrm{dist}(u, v) = \infty$ if such a path does not exist. A *matching* M in G is a set of pairwise non-adjacent edges. A vertex $v \in V(G)$ is said to be *M-saturated* if it is incident with some edge in M; otherwise, v is said to be *M-unsaturated*; we drop M when it is

This research was supported by CAPES, grants 88887.520851/2020-00 and 88887.600764/2021-00.

clear in the context. The matching M is called *perfect* when every vertex of G is M-saturated and we say that M *extends* to a perfect matching if M is a subset of some perfect matching of G.

Matching theory has played a fundamental role in Graph Theory since its origins in the late 19th century. Over the years, significant progress has been made in characterizing their existence and enumerating perfect matchings [27,29], and also analysing their interaction with other structural properties of graphs [20,28]. In particular, numerous studies have introduced decomposition techniques for graphs in terms of their matchings [7,8,12].

In one of such study, Lovász [18] introduced the concept of a bicritical graph. A graph is *bicritical* if $G - u - v$ has a perfect matching for every $u, v \in V(G)$, $u \neq v$. Motivated by this concept, Plummer [22] introduced in 1980 the notion of k-extendable graphs. Let G be a graph on $2p$ vertices and let k be an integer such that $1 \leq k \leq p - 1$. Then, G is said to be k-*extendable* if every matching of size k in G extends to a perfect matching of G. Note that, for $k = 1$, this definition is equivalent to stating that G is a *matching-covered graph* (a graph such that every edge belongs to some perfect matching), which has been extensively studied in the literature [5,19]. In the same article, Plummer [22] showed that every 2-extendable graph is either bipartite or a 3-connected bicritical graph. Furthermore, the author proved that every k-extendable graph is also $(k-1)$-extendable and $(k+1)$-connected.

Since then, the theory of k-extendability has been further developed through numerous contributions, addressing necessary and sufficient conditions for k-extendable graphs [15], their behaviour on surfaces [2] and its criticality [3]. Additionally, several studies have addressed the relationship between extendability and graph parameters such as independence number, binding number, toughness, among others [6,17,21,32]. A comprehensive overview of the foundational concepts and early developments in the theory of k-extendable graphs can be found in three extensive surveys authored by Plummer [24–26].

In 1998, Lakhal and Litzler [13] showed that deciding whether a bipartite graph is k-extendable can be done in polynomial time. For an arbitrary graph, Hackfeld and Koster [10] proved that the problem belongs to co-NP-complete. This naturally leads to the study of k-extendability in classes of graphs. Some examples include regular graphs [1], triangle-free graphs [16] and planar graphs [23]. In addition, the problem has been investigated for graphs obtained from certain operations, such as Cartesian product. For instance, Györi and Plummer [9] proved that the Cartesian product of a k-extendable and an l-extendable graph is a $(k + l + 1)$-extendable graph. Other operations, such as the lexicographic product [4] and the complementary prism of graphs [11], have also been explored.

In the study of k-extendability of graphs, several properties have been shown to be useful. Among them, the neighbourhood of unsaturated vertices is often analysed, as it provides immediate insight into the matching extendability of graphs.

Property 1. Let G be a graph and M a non-perfect matching of G. If there exists an unsaturated vertex $u \in V(G)$ such that every neighbour $v \in N(u)$ is M-saturated, then M is not extendable to a perfect matching of G.

Some studies have investigated the extendability of matchings in the absence of this type of neighbourhood obstruction. Limaye and Sarvate [14] proved that in a hypercube Q_n, $n \geq 2$, a matching M of size at most n can be extended to a perfect matching of Q_n if and only if M does not saturate the neighbourhood of any unsaturated vertex. Since any matching M of Q_n with $1 \leq |M| \leq n-1$ does not saturate the neighbourhood of any unsaturated vertex, this result implies that the hypercube is k-extendable for all $1 \leq k \leq n-1$. The authors also showed that, for the hypercube Q_4, there exists a matching of size $n+1 = 5$ that does not saturate the neighbourhood of any unsaturated vertex and does not extend to a perfect matching. For $n \geq 5$, Vandenbussche and West [30] strengthened these results for matchings of size up to $2n - 4$, as stated in Theorem 1. More recently, the problem was revisited by Vandenbussche and Westlund [31], who extended the analysis to the n-fold Cartesian product of cycles $G(2m)_d \cong C_{2m} \,\square\, \ldots \,\square\, C_{2m}$.

Theorem 1. (Vandenbussche and West [30]). *Let M be a matching of the hypercube Q_n, $n \geq 5$, such that $|M| \leq 2n - 4$. Then, M is extendable to a perfect matching of Q_n if and only if M does not saturate the neighbourhood of any unsaturated vertex.* $\qquad\square$

In our work, we focus on the Cartesian product $Q_n \,\square\, P_m$, $n \geq 4$ and $m \geq 2$, and prove that a matching of size at most $2n - 2$ is extendable to a perfect matching if and only if it does not saturate the neighbourhood of any unsaturated vertex. Furthermore, we show that this bound is tight. We also provide some insights about the extendability of matchings that satisfy the neighbourhood condition for the Cartesian product of hypercubes with arbitrary graphs.

2   Results

An n-dimensional hypercube, with $n \geq 1$, denoted by Q_n, is the graph whose vertex set consists of the 2^n binary strings of length n. That is, if $v \in V(Q_n)$, then $v = b_1 b_2 \ldots b_n$ such that every *coordinate* $b_i \in \{0, 1\}$, $1 \leq i \leq n$. Moreover, two vertices are adjacent if and only if their strings differ in exactly one coordinate. For a positive integer i, $1 \leq i \leq n$, we define the *i-th decomposition* of Q_n such that $V(Q_n) = \{V_0^i, V_1^i\}$, with V_0^i containing all vertices with 0 in coordinate i, and V_1^i containing all vertices with 1 in coordinate i. Note that $Q_n[V_0^i]$ and $Q_n[V_1^i]$ are both isomorphic to the hypercube Q_{n-1}.

Given the Cartesian product $G \cong Q_n \,\square\, H$, we define the *canonical notation* of G as follows. Let $V(Q_n) = \{v_0, v_1, \ldots, v_{2^n-1}\}$, such that the index of each vertex v_j, $0 \leq j \leq 2^n - 1$, corresponds to the decimal value of the binary representation of its binary string. Let $V(H) = \{u_1, u_2, \ldots, u_m\}$. We construct the graph G from m copies of the hypercube Q_n, denoted $G^1, G^2, \ldots, G^m$. Moreover, we define the vertex set of each copy G^i as $V(G^i) = \{v_1^i, v_2^i, \ldots, v_{2^n}^i\}$, for

$1 \leq i \leq m$. These copies are connected in such a way that the induced subgraph $G[v_j^1, v_j^2, \ldots, v_j^m]$ is isomorphic to the graph H, for every $1 \leq j \leq 2^n$. We say that v_j^i and v_j^k, $i \neq k$, are *corresponding vertices*.

Given the copies $G^1, G^2, \ldots, G^m$ of Q_n in the canonical notation, we can apply the i-th decomposition of each G^k, such that $V(G^k) = \{V_0^{k,i}, V_1^{k,i}\}$. Thus, $G[\cup_{k=1}^m V_1^{k,i}] \cong G[\cup_{k=1}^m V_2^{k,i}] \cong Q_{n-1} \,\square\, H$. Observe that, by the definition of the hypercube, $G \cong (Q_{n-1} \,\square\, H) \,\square\, K_2$. In other words, G can be built from two subgraphs G' and G'', both isomorphic to $Q_{n-1} \,\square\, H$, such that G' and G'' are connected by a perfect matching. Finally, since there are n possible i-th decompositions of each G^k, we conclude that for any two distinct vertices u and v in G, if they are not corresponding vertices, then it is possible to built G in such a way that u belongs to G' and v belongs to G''.

Vandenbussche and West [30] showed that the bound in Theorem 1 is tight; that is, for every $n \geq 4$, there exists a matching in Q_n of size $2n-3$ that does not saturate the neighbourhood of any unsaturated vertex and cannot be extended to a perfect matching. The idea behind the construction involves the i-th decomposition of Q_n. Let $Q^k \cong Q_{n-1}$, $k \in \{1,2\}$, be the two $(n-1)$-dimensional hypercubes obtained in this decomposition. Let u and v be two vertices in Q^1 such that $d(u,v) = 2$. Note that $|N(u) \cap N(v)| = 2$. Let $N(u) \cap N(v) = \{x,y\}$. Let $R = N(u) \setminus \{x,y\}$ and $S = N(v) \setminus \{x,y\}$. The authors argue that it is possible to construct a matching M with $2n-3$ edges that saturates all vertices in $R \cup S \cup \{x\}$. Figure 1 illustrates the construction for $n = 4$. In this example, vertex u can only be matched with y. However, v can also only be matched with y and, thus, there is no perfect matching in Q_n that contains M.

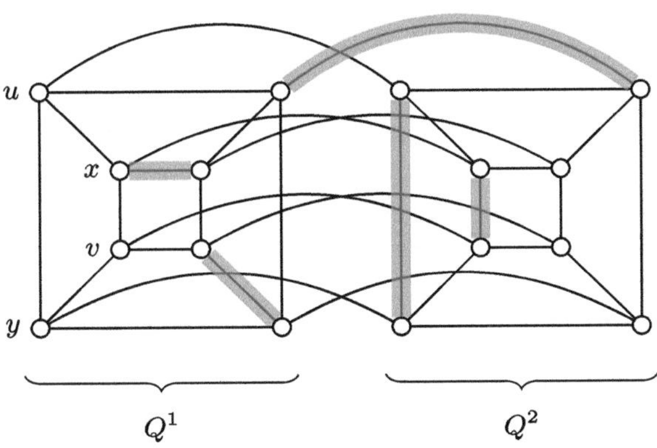

Fig. 1. A matching of Q_4 that does not saturate the neighbourhood of any unsaturated vertex and cannot be extended to a perfect matching.

In our first result, presented in Corollary 1, we adapted the construction proposed by Vandenbussche and West [30] to build non-extendable matchings in other graph classes related to hypercubes. Consider graph $G \cong Q_n \,\square\, H$, with $n \geq 4$ and $|V(H)| = m$, using the canonical notation. By applying the construction of Vandenbussche and West, it is possible to obtain a matching of size $2n - 3$ in a copy G^i that does not saturate the neighbourhood of any unsaturated vertex. Let $w \in V(H)$ such that $d(w) = \delta(H)$ and let G^i be the copy of Q_n corresponding to w. Let u, v and y be the vertices as in the aforementioned construction. Note that $\{u, v, y\} \subseteq V(G^i)$. Since u and v have each $\delta(H)$ neighbours in $V(G) \setminus V(G^i)$, we can obtain a matching in $G[V(G) \setminus V(G^i)]$ with size $2\delta(H)$ that does not saturate the neighbourhood of any unsaturated vertex. Again, u and v can only be matched with y and thus there is no perfect matching in G that contains M. Figure 2 shows an example.

Corollary 1. *Let $G \cong Q_n \,\square\, H$ with $n \geq 4$. Then, there exists a matching M in G with $|M| = 2n + 2\delta(H) - 3$ such that M does not saturate the neighbourhood of any unsaturated vertex and M is not extendable to a perfect matching of G.*□

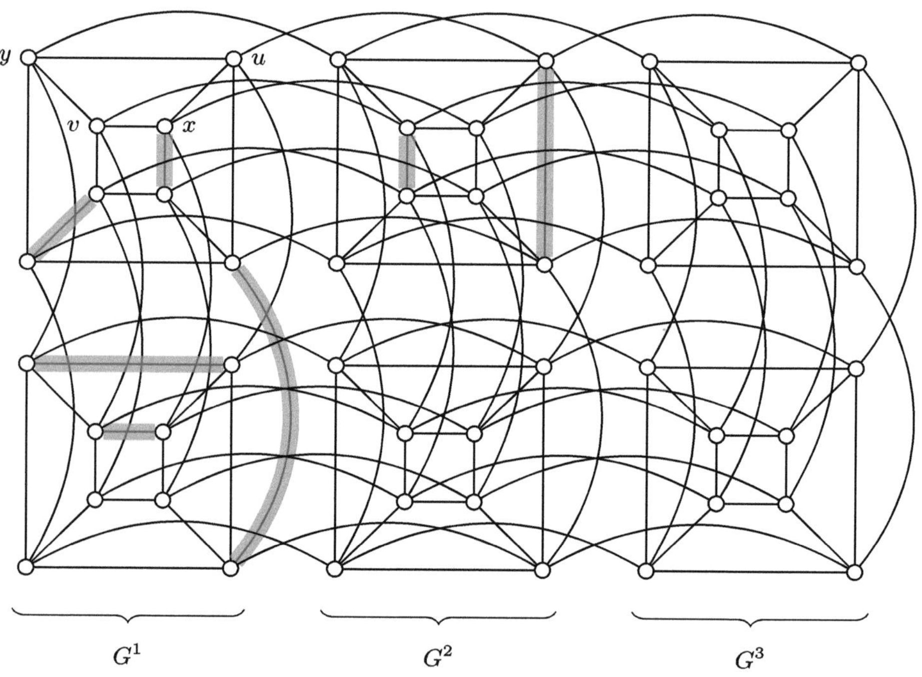

Fig. 2. A matching of $Q_4 \,\square\, P_3$ that does not saturate the neighbourhood of any unsaturated vertex and cannot be extended to a perfect matching.

In the special case in which H is a path P_m, the bound provided by Corollary 1 implies that G admits a matching of size $2n - 1$ that does not saturate the neighbourhood of any unsaturated vertex and is not extendable to a perfect matching. We shall prove that this bound is tight for the graph $Q_n \,\square\, P_m$, $n \geq 4$

and $m \geq 2$, by showing that every matching of size at most $2n - 2$ that does not saturate the neighbourhood of any unsaturated vertex is extendable to a perfect matching.

Our main result, presented in Theorem 2, states that a matching M in a graph $G \cong Q_n \,\square\, P_m$, $n \geq 4$ and $m \geq 2$, with $|M| \leq 2n - 2$, is extendable to a perfect matching if and only if M does not saturate the neighbourhood of any unsaturated vertex. In order to prove it, we have to establish auxiliary lemmas that use the well-known Hall's Theorem: in a bipartite graph with bipartition $\{X, Y\}$, there exists a perfect matching saturating X if and only if $|N(S)| \geq |S|$ for every subset $S \subseteq X$. We show that Hall's condition remains satisfied even after the removal of the vertices saturated by the initial matching. In particular, we show that the surplus $|N(S)| - |S|$ is sufficiently large to guarantee the existence of a perfect matching in the remaining graph.

We proceed by proving the first auxiliary lemma. This lemma is then applied in Lemma 2, where we establish a bound on the neighbourhood sizes when the set S is small and contained in one part of the bipartition.

Lemma 1. *Let $G \cong Q_n \,\square\, P_m$, $n \geq 1$ and $m \geq 2$. Let $S \subseteq V(G)$. If $S \neq \emptyset$, then $|N(S) \setminus S| > (n + 1)|S| - \binom{|S|+1}{2}$.*

Proof. Let $G \cong Q_n \,\square\, P_m$, $n \geq 1$ and $m \geq 2$, with its canonical notation. First, note that if $|S| = 1$, then $N(S) \cap S = \emptyset$ and $N(S) \setminus S = N(S)$. This implies $|N(S) \setminus S| = |N(S)| \geq n + 1 > n = (n + 1) - 1$. Now, suppose $|S| \geq 2$. The proof proceeds by induction on n. Consider $n = 1$. In this case, we want to show that $|N(S) \setminus S| > 2|S| - \binom{|S|+1}{2}$. Suppose $2|S| - \binom{|S|+1}{2} < 0$. Since $|N(S) \setminus S| \geq 0$, we have $|N(S) \setminus S| \geq 0 > 2|S| - \binom{|S|+1}{2}$ and the result follows. Suppose $2|S| - \binom{|S|+1}{2} \geq 0$. If $2|S| - \binom{|S|+1}{2} = 0$, then $|S| = 3$; otherwise, $|S| = 2$.

Consider $|S| = 3$. First, note that if $|N(S) \setminus S| = 0$, then $N(S) = S$. Therefore, since $d(u) \geq n + 1 = 2$ for every $u \in S$, we conclude that $G[S] \cong K_3$. However, this contradicts the fact that G is bipartite. Suppose $|N(S) \setminus S| \geq 1$. In this case, $|N(S) \setminus S| \geq 1 > 2|S| - \binom{|S|+1}{2} = 0$ as required.

Consider $|S| = 2$. Let $S = \{x, y\}$. If $N(S) \cap S = \emptyset$, then $|N(S) \setminus S| = |N(S)| \geq n + 1 = 2 > 2|S| - \binom{|S|+1}{2} = 1$, satisfying the inequality. Consider $N(S) \cap S \neq \emptyset$. This implies $N(S) \cap S = \{x, y\}$. As $d(x) \geq n + 1$ and $x \notin N(x)$, there exists $u_x \in N(x)$ such that $u_x \notin \{x, y\}$. Similarly, there exists $u_y \in N(y)$ such that $u_y \notin \{x, y\}$. Since G is bipartite, we must have $u_x \neq u_y$. Therefore, $|N(S) \setminus S| \geq 2$. Then again, we have $|N(S) \setminus S| \geq 2 > 2|S| - \binom{|S|+1}{2} = 1$ and the result follows.

Now, consider $n \geq 2$. Suppose that G cannot be partitioned into $H_1 \cong Q_{n-1} \,\square\, P_m$ and $H_2 \cong Q_{n-1} \,\square\, P_m$ such that $S \cap V(H_1) \neq \emptyset$ and $S \cap V(H_2) \neq \emptyset$. Then each G^i contains at most one vertex of S and, for every $\{u, v\} \subseteq S$, u and v are corresponding vertices. Since each $v_k^i \in S$ has n neighbours in G^i, we conclude that $|N(S) \setminus S| \geq n|S|$. It remains to show that $n|S| > (n + 1)|S| - \binom{|S|+1}{2}$. Note that $n|S| > n|S| + \frac{|S|-|S|^2}{2}$ since $|S| \geq 2$. This implies

$$n|S| > n|S| + |S| - |S| + \frac{|S| - |S|^2}{2}$$
$$= (n+1)|S| - \binom{|S|+1}{2}$$

and the result follows.

Suppose that there exists a partition of G into H_1 and H_2 such that $S \cap V(H_1) \neq \emptyset$ and $S \cap V(H_2) \neq \emptyset$. Let $S_i = S \cap V(H_i)$. By the induction hypothesis, $|N_{H_i}(S_i) \setminus S_i| \geq n|S_i| - \binom{|S_i|+1}{2} + 1$. Since $S_1 \cap S_2 = \emptyset$ and $N_{H_1}(S_1) \cap N_{H_2}(S_2) = \emptyset$, $|N(S) \setminus S| \geq n(|S_1|+|S_2|) - \left[\binom{|S_1|+1}{2} + \binom{|S_2|+1}{2}\right] + 2$. Moreover, as $|S_1|+|S_2| = |S|$ and $|S_i| \geq 1$, the expression $\binom{|S_1|+1}{2} + \binom{|S_2|+1}{2}$ is maximized when $|S_1| = |S| - 1$ and $|S_2| = 1$. Therefore, $|N(S) \setminus S| \geq n|S| - \left[\binom{(|S|-1)+1}{2} + \binom{1+1}{2}\right] + 2$. Thus,

$$n|S| - \left[\binom{(|S|-1)+1}{2} + \binom{1+1}{2}\right] + 2 = n|S| - \binom{|S|}{2} + 1$$
$$= n|S| + |S| - |S| - \binom{|S|}{2} + 1$$
$$= (n+1)|S| - |S| - \frac{|S|(|S|-1)}{2} + 1$$
$$= (n+1)|S| - \binom{|S|+1}{2} + 1,$$

which concludes the proof. $\qquad\qquad\square$

Lemma 2. *Let $G \cong Q_n \,\square\, P_m$, $n \geq 3$ and $m \geq 2$, and let $\{X, Y\}$ be a bipartition of G. Let $S \subseteq X$. If $2 \leq |S| \leq n$, then $|N(S)| - |S| \geq 2n - 2$.*

Proof. Let G, $\{X, Y\}$ and S be as stated in the hypothesis. Suppose $2 \leq |S| \leq n$. Initially, note that, since $S \subseteq X$, $N(S) \subseteq Y$ and, therefore, $|N(S) \setminus S| = |N(S)|$. By Lemma 1, $|N(S) \setminus S| \geq (n+1)|S| - \binom{|S|+1}{2} + 1$. Thus, $|N(S)| - |S| \geq (n+1)|S| - \binom{|S|+1}{2} + 1 - |S| = n|S| - \binom{|S|+1}{2} + 1$. So, to prove the result, it suffices to show that $n|S| - \binom{|S|+1}{2} + 1 \geq 2n - 2$.

We proceed by induction on n. Consider $n = 3$. Then, $|S| \in \{2, 3\}$. Note that, in both cases, $3|S| - \binom{|S|+1}{2} + 1 = 4 \geq 2n - 2 = 4$. For $n \geq 4$, assume by the induction hypothesis that $(n-1)|S| - \binom{|S|+1}{2} + 1 \geq 2(n-1) - 2$. Observe that $n|S| - \binom{|S|+1}{2} + 1 = \left((n-1)|S| - \binom{|S|+1}{2} + 1\right) + |S|$. Then,

$$n|S| - \binom{|S|+1}{2} + 1 = (n-1)|S| - \binom{|S|+1}{2} + 1 + |S|$$
$$\geq 2(n-1) - 2 + |S|$$
$$= 2n - 4 + |S|.$$

Since $|S| \geq 2$, we have $2n - 4 + |S| \geq 2n - 2$, concluding the proof. $\qquad\square$

Lemma 3 establishes bounds for the neighbourhood of larger vertex sets, using the following proposition.

Proposition 1. *Let G be a bipartite graph and let $\{X, Y\}$ be a bipartition of G such that $|X| = |Y|$. Then, for every $S \subseteq X$, $|S| - |N(S)| \leq |T| - |N(T)|$, with $T = Y \setminus N(S)$.*

Proof. Since $T = Y \setminus N(S)$, it follows that $|Y| = |N(S)| + |T|$. Moreover, by the definition, $N(T) \cap S = \emptyset$. Therefore, $|N(T)| + |S| \leq |X|$. As $|X| = |Y|$, we have $|N(T)| + |S| \leq |N(S)| + |T|$ and, hence, $|S| - |N(S)| \leq |T| - |N(T)|$. $\qquad\square$

Lemma 3. *Let $G \cong Q_n \,\square\, P_m$, $m, n \geq 3$, and let $\{X, Y\}$ be a bipartition of G. For $S \subseteq X$, if $2 \leq |S| \leq m2^{n-1} - 2n + 1$, then $|N(S)| - |S| \geq 2n - 2$.*

Proof. Let G, $\{X, Y\}$ and S be as stated in the hypothesis. Suppose $2 \leq |S| \leq m2^{n-1} - 2n + 1$. By Lemma 2, if $2 \leq |S| \leq n$, the claim holds. Suppose $|S| > n$. First, suppose $|S| \leq m2^{n-2}$. Note that, by the definition of S, it holds that $N(S) \cap S = \emptyset$. Also, since G is a bipartite graph that admits a perfect matching, by Hall's Theorem, we have that $|N(S)| \geq |S|$ for every $S \subseteq X$. Therefore, $|N(S)| - |S| \geq 0$.

We proceed by induction on n. Consider $n = 3$. We show that $|N(S)| - |S| \geq 4$. Assume G with its canonical notation and let $S_{G^i} = S \cap G^i$. Since $S \subseteq X$, it follows that $S_{G^i} \subseteq X$ and $|S_{G^i}| \leq 4$, for every $1 \leq i \leq m$. Furthermore, if $|S_{G^i}| = 1$, then $|N_{G^i}(S_{G^i})| - |S_{G^i}| = 2$; if $|S_{G^i}| = 2$, then $|N_{G^i}(S_{G^i})| - |S_{G^i}| = 2$; if $|S_{G^i}| = 3$, then $|N_{G^i}(S_{G^i})| - |S_{G^i}| = 1$; and if $|S_{G^i}| = 4$, then $|N_{G^i}(S_{G^i})| - |S_{G^i}| = 0$. Observe that $|N(S)| - |S| \geq \sum_{i=1}^{m} |N_{G^i}(S_{G^i})| - |S_{G^i}|$, since the sum of the neighbourhoods within the isolated copies does not account for inter-copy neighbours.

Suppose $|S_{G^i}| \neq \emptyset$ for every $1 \leq i \leq m$. Since $m \geq 3$ and $|S| \leq 2m$, we conclude that there exist at least two distinct sets S_{G^i} and S_{G^j} such that $1 \leq |S_{G^i}| \leq 2$ and $1 \leq |S_{G^j}| \leq 2$. Hence, $|N_{G^i}(S_{G^i})| - |S_{G^i}| + |N_{G^j}(S_{G^j})| - |S_{G^j}| \geq 4$. As $|N(S)| - |S| \geq \sum_{i=1}^{m} |N_{G^i}(S_{G^i})| - |S_{G^i}| \geq 4$, the result follows. This conclusion is used implicitly in the remainder of the proof.

Now, suppose that there exists i such that $S_{G^i} = \emptyset$. We can assume, adjusting notation if necessary, that there exists an index i such that $S_{G^i} = \emptyset$ and $S_{G^{i+1}} \neq \emptyset$. There are $|S_{G^{i+1}}|$ neighbours of $S_{G^{i+1}}$ in G^i, that is, $|N_{G^i}(S_{G^{i+1}})| = |S_{G^{i+1}}|$. So, if $|S_{G^{i+1}}| \geq 2$, we conclude that $|N(S)| - |S| \geq |N_{G^{i+1}}(S_{G^{i+1}})| - |S_{G^{i+1}}| + |N_{G^i}(S_{G^{i+1}})| \geq 4$.

Consider $|S_{G^{i+1}}| = 1$. Then, $|N_{G^{i+1}}(S_{G^{i+1}})| - |S_{G^{i+1}}| + |N_{G^i}(S_{G^{i+1}}) \geq 3$. This implies that, if there exists j, for $1 \leq j \leq m$ and $j \notin \{i, i+1\}$, such that $1 \leq |S_{G^j}| \leq 3$, then $|N(S)| - |S| \geq (|N_{G^{i+1}}(S_{G^{i+1}})| - |S_{G^{i+1}}| + |N_{G^i}(S_{G^{i+1}})|) + (|N_{G^j}(S_{G^j})| - |S_{G^j}|) \geq 4$ and the result follows. Consider then $|S_{G^j}| \in \{0, 4\}$ for every $j \notin \{i, i+1\}$. In particular, this implies $|S| \geq 5$. Suppose $i + 1 = m$. Since $m \geq 3$ and $|S| \geq 5$, there exists j, for $1 \leq j \leq i - 1$, such that $|S_{G^j}| = 4$. Choose G^j such that $|S_{G^j}| = 4$ and j is maximum. Thus, $S_{G^{j+1}} = \emptyset$. Therefore, $|N_{G^j}(S_{G^j})| - |S_{G^j}| + |N_{G^{j+1}}(S_{G^j})| \geq 4$ and the result follows. Suppose $i + 1 < m$. If $S_{G^{i+2}} = \emptyset$, then $|N_{G^i}(S_{G^{i+1}})| + |N_{G^{i+1}}(S_{G^{i+1}})| - |S_{G^{i+1}}| + |N_{G^{i+2}}(S_{G^{i+1}})| \geq 4$

and we are done. Suppose $|S_{G^{i+2}}| = 4$. Since each vertex of $S_{G^{i+2}}$ has a distinct neighbour in G^{i+1} and $|S_{G^{i+1}}| = 1$, there must be a vertex of $N_{G^{i+1}}(S_{G^{i+2}})$ that is not adjacent to the vertex of $S_{G^{i+1}}$. Therefore, $|N_{G^{i+1}}(S_{G^{i+1}})| - |S_{G^{i+1}}| + |N_{G^{i+2}}(S_{G^{i+2}})| - |S_{G^{i+2}}| + |N_{G^{i+1}}(S_{G^{i+2}})| + |N_{G^i}(S_{G^{i+1}})| \geq 4$. This concludes the base case.

Consider $n \geq 4$. Suppose that G cannot be partitioned into $H_1 \cong Q_{n-1} \square P_m$ and $H_2 \cong Q_{n-1} \square P_m$ such that $S \cap V(H_1) \neq \emptyset$ and $S \cap V(H_2) \neq \emptyset$. Hence, each G^i has at most one vertex in S and, for every $\{u, v\} \subseteq S$, u and v are corresponding vertices. Since every $v_k^i \in S$ has n neighbours in G^i, we obtain $|N(S)| - |S| \geq n|S| - |S| = (n-1)|S|$. Because $n \geq 4$, the inequality $(n-1)|S| \geq 2(n-1)$ holds whenever $|S| \geq 2$. Since in this case $|S| \geq 5$, we conclude that $|N(S)| - |S| \geq 2n - 2$.

Suppose that there exists a partition of G into H_1 and H_2 such that $S \cap V(H_1) \neq \emptyset$ and $S \cap V(H_2) \neq \emptyset$. Let $S_i = S \cap V(H_i)$, $i \in \{1, 2\}$. We can assume, without loss of generality, $|S_1| \geq |S_2|$. As $|S| \geq 5$, we have $|S_1| \geq 3$. Suppose $|S_1| \leq m2^{n-2} - 2(n-1) + 1$. By the induction hypothesis, $|N_{H_1}(S_1)| - |S_1| \geq 2(n-1) - 2$. To conclude the result, we analyse the cases $|S_2| = 1$ and $|S_2| \geq 2$.

First, suppose $|S_2| = 1$. Then, $|N_{H_2}(S_2)| \geq (n-1) + 1$. Since $|N(S)| - |S| \geq (|N_{H_1}(S_1)| - |S_1|) + (|N_{H_2}(S_2)| - |S_2|)$, we conclude that $|N(S)| - |S| \geq 3n - 4$. As $n \geq 4$, it holds that $3n - 4 \geq 2n - 2$ and the result follows. Suppose $|S_2| \geq 2$. By the induction hypothesis, $|N_{H_2}(S_2)| - |S_2| \geq 2(n-1) - 2$. Similarly, we have $|N(S)| - |S| \geq 4n - 8 \geq 2n - 2$.

Now, suppose $|S_1| \geq m2^{n-2} - 2(n-1) + 2 = m2^{n-2} - 2n + 4$. Let $S_1' \subseteq S_1$ such that $|S_1'| = m2^{n-2} - 2(n-1) + 1 = m2^{n-2} - 2n + 3$. By the induction hypothesis, $|N_{H_1}(S_1')| - |S_1'| \geq 2(n-1) - 2 = 2n - 4$. Then,

$$|N_{H_1}(S_1')| - |S_1'| \geq 2n - 4$$
$$|N_{H_1}(S_1')| \geq |S_1'| + 2n - 4$$
$$\geq m2^{n-2} - 2n + 3 + 2n - 4$$
$$= m2^{n-2} - 1.$$

Since $|N_{H_1}(S_1)| \geq |N_{H_1}(S_1')|$, it follows that $|N_{H_1}(S_1)| \geq m2^{n-2} - 1$. Moreover, note that each vertex of S_1 has a neighbour in H_2. Therefore, $|N(S)| \geq |N_{H_1}(S_1)| + |S_1| \geq m2^{n-2} - 1 + |S_1|$. Since $|S_1| = |S| - |S_2|$, we have $|N(S)| \geq m2^{n-2} - 1 + |S| - |S_2|$, which implies $|N(S)| - |S| \geq m2^{n-2} - 1 - |S_2|$.

If $|S_2| \leq m2^{n-2} - 2n + 1$, then $|N(S)| - |S| \geq 2n - 2$ as required. Suppose $|S_2| \geq m2^{n-2} - 2n + 2$. Since $|S| \leq m2^{n-2}$, $|S_1| \geq m2^{n-2} - 2n + 4$ and $|S_2| = |S| - |S_1|$, it follows that $|S_2| \leq 2n - 4$. Thus, we obtain $2n - 4 \geq |S_2| \geq m2^{n-2} - 2n + 2$, which implies $4n - 6 \geq m2^{n-2}$. First, observe that $4n - 6 < m2^{n-2}$ for $n \geq 5$ and $m \geq 3$. Since by hypothesis $n \geq 4$, we must have $n = 4$, which implies that the inequality $4n - 6 \geq m2^{n-2}$ is satisfied only when $m \leq 2$. However, this contradicts the fact that $m \geq 3$.

Now, consider $|S| > m2^{n-2}$. Let $T = Y \setminus N(S)$. Then, $|T| = |Y| - |Y \cap N(S)|$ and, since $N(S) \subseteq Y$, it follows that $|T| = |Y| - |N(S)|$. As $|Y| = m2^{n-1}$ and $|N(S)| \geq |S| > m2^{n-2}$, we conclude that $|T| < m2^{n-2}$. By Proposition 1,

$|S| - |N(S)| \leq |T| - |N(T)|$ and thus $|N(S)| - |S| \geq |N(T)| - |T|$. To conclude the proof, we consider the cases where $|T| \leq 1$ and $|T| \geq 2$.

Suppose $|T| \leq 1$. As $|T| = |Y| - |N(S)|$, we conclude that $|N(S)| \geq m2^{n-1} - 1$. Since, by hypothesis, $|S| \leq m2^{n-1} - 2n + 1$, it follows that $|N(S)| - |S| \geq 2n - 2$. Suppose $|T| \geq 2$. Since $|T| < m2^{n-2}$ and $m, n \geq 3$, we have $2 \leq |T| \leq m2^{n-1} - 2n + 1$. Therefore, as we proved before, $|N(T)| - |T| \geq 2n - 2$. Hence $|N(S)| - |S| \geq |N(T)| - |T| \geq 2n - 2$. This concludes the proof. $\square$

Using the previous lemmas, we prove now our main result.

Theorem 2. *Let $G \cong Q_n \,\square\, P_m$, $n \geq 4$ and $m \geq 2$. Let M be a matching of G such that $|M| \leq 2n - 2$. Then, M is extendable to a perfect matching of G if and only if M does not saturate the neighbourhood of any unsaturated vertex.*

Proof. Suppose that M is extendable to a perfect matching of G. Then, by Property 1, M does not saturate the neighbourhood of any unsaturated vertex. Conversely, suppose M, with $|M| \leq 2n - 2$, does not saturate the neighbourhood of any unsaturated vertex. For $m = 2$, since $G \cong Q_n \,\square\, P_m \cong Q_{n+1}$, we conclude, by Theorem 1, that M is extendable to a perfect matching of G. Consider $m \geq 3$.

Let V_M be the set of M-saturated vertices of G and let $\{X, Y\}$ be a bipartition of G. To prove the result, we show that $G[V(G) \setminus V_M]$ admits a perfect matching. It suffices, by Hall's Theorem, to show that $|N(S) \setminus V_M| \geq |S \setminus V_M|$ for every $S \subseteq X$.

Since M does not saturate the neighbourhood of any unsaturated vertex, if follows that $|N(S) \setminus V_M| \geq 1$. Therefore, if $|S| = 1$, the result follows. Suppose $2 \leq |S| \leq m2^{n-1} - 2n + 1$. Then, by Lemma 3, $|N(S)| - |S| \geq 2n - 2$ and thus $|N(S)| \geq |S| + 2n - 2$. On the other hand, $N(S) \subseteq Y$ and $|N(S)| = |N(S) \cap V_M| + |N(S) \setminus V_M|$. Note that $|V_M \cap Y| = |M| \leq 2n - 2$ and $|N(S) \cap V_M| \leq |V_M \cap Y|$. Thus, we have $|N(S) \setminus V_M| \geq |S|$. As $|S| \geq |S \setminus V_M|$, the result follows.

Suppose $|S| \geq m2^{n-1} - 2n + 2$. Let $S' \subseteq S$ such that $|S'| = m2^{n-1} - 2n + 1$. Then, by Lemma 3, it follows that $|N(S')| \geq |S'| + 2n - 2 = m2^{n-1} - 1$. Since $S' \subseteq S$, $|N(S)| \geq m2^{n-1} - 1$. Let $T = Y \setminus N(S) \setminus (V_M \cap Y)$. As $T \subseteq Y \setminus N(S)$ and $|Y| = m2^{n-1}$, we conclude that $|T| \leq m2^{n-1} - (m2^{n-1} - 1) = 1$. Moreover, since M does not saturate the neighbourhood of any unsaturated vertex, $|N(T) \setminus V_M| \geq |T|$. Therefore, $|N(T) \setminus V_M| - |T| \geq 0$.

Let $G' = G \setminus V_M$. Note that $N_{G'}(T) = N(T) \setminus V_M$. Then, $|N_{G'}(T)| - |T| = |N(T) \setminus V_M| - |T| \geq 0$. Since G' is bipartite , by Proposition 1, we have $|N_{G'}(S)| - |S \setminus V_M| \geq |N_{G'}(T)| - |T| \geq 0$. Since $N_{G'}(S) = N(S) \setminus V_M$, it follows that $|N(S) \setminus V_M| - |S \setminus V_M| \geq 0$ and, therefore, $|N(S) \setminus V_M| \geq |S \setminus V_M|$. $\square$

3   Concluding Remarks

The results presented in this work shed light on the extendability of matchings that do not saturate the neighbourhood of unsaturated vertices in Cartesian products involving hypercubes. In particular, for $Q_n \,\square\, H$ when H is an arbitrary graph, we proved in Corollary 1 that there exists a matching of size $2n + 2\delta(H) - 3$

that satisfies the neighbourhood condition but is not extendable to a perfect matching. This motivates the following conjecture.

Conjecture 1. Let $G \cong Q_n \,\square\, H$, $n \geq 4$, and let M be a matching of G such that $|M| \leq 2n + 2\delta(H) - 4$. Then, M is extendable to a perfect matching of G if and only if M does not saturate the neighbourhood of any unsaturated vertex.

We have verified this conjecture when H is isomorphic to a path. The techniques used in this case are based on Hall's Theorem and structural properties of the hypercube, suggesting that the results may be generalized to cases where H is a bipartite graph. However, further efforts are needed to establish the conjecture for an arbitrary graph H.

References

1. Aldred, R., Holton, D.A., Lou, D.: N-extendability of symmetric graphs. J. Graph Theor. **17**(2), 253–261 (1993)
2. Aldred, R., Kawarabayashi, K., Plummer, M.: On the matching extendability of graphs in surfaces. J. Comb. Theor. Ser. B **98**(1), 105–115 (2008)
3. Aldred, R., Plummer, M.: Extendability and criticality in matching theory. Graphs Comb. **36**(3), 573–589 (2020)
4. Bai, B., Wu, Z., Yang, X., Yu, Q.: Lexicographic product of extendable graphs. Bull. Malaysian Math. Sci. Soc. Second Ser. **33**(2), 197–204 (2010)
5. de Carvalho, M.H., Lucchesi, C.L., Murty, U.S.R.: On a conjecture of Lovász concerning bricks: I. the characteristic of a matching covered graph. J. Comb. Theor. Ser. B **85**(1), 94–136 (2002)
6. Chen, C.: Binding number and toughness for matching extension. Discret. Math. **146**(1–3), 303–306 (1995)
7. Edmonds, J.: Paths, trees, and flowers. Can. J. Math. **17**, 449–467 (1965)
8. Gallai, T.: Kritische graphen II. Magyar Tudományos Akadémia Matematikai Kutató Intézetének Közleményei **8**, 373–395 (1963)
9. Györi, E., Plummer, M.: The cartesian product of a k-extendable and an l-extendable graph is (k + l + 1)-extendable. Julius Petersen Graph Theor. Centennial **6**, 417 (1992)
10. Hackfeld, J., Koster, A.: The matching extension problem in general graphs is co-NP-complete. J. Comb. Optim. **35**, 853–859 (2018)
11. Janseana, P., Ananchuen, N.: Extendability of the complementary prism of extendable graphs. Thai J. Math. **13**(3), 703–721 (2015)
12. Kotzig, A.: On the theory of finite graphs with a linear factor I. Fyzikálny Časopis Slovenskej Akademie Vied **9**(10) (1959)
13. Lakhal, J., Litzler, L.: A polynomial algorithm for the extendability problem in bipartite graphs. Inf. Process. Lett. **65**(1), 11–16 (1998)
14. Limaye, N.B., Sarvate, D.G.: On r-extendability of the hypercube Q_n. Math. Bohem. **122**(3), 249–255 (1997)
15. Lou, D.: Some conditions for n-extendable graphs. Australasian J. Comb. **9**, 123–136 (1994)
16. Lou, D.: A local independence number condition for n-extendable graphs. Discret. Math. **195**(1–3), 263–268 (1999)
17. Lou, D.: On matchability of graphs. Australas. J. Comb. **21**, 201–210 (2000)

18. Lovász, L.: On the structure of factorizable graphs. Acta Math. Hungar. **23**(1–2), 179–195 (1972)
19. Lovász, L.: Ear-decompositions of matching-covered graphs. Combinatorica **3**, 105–117 (1983)
20. Lovász, L., Plummer, M.: Matching theory, vol. 367. American Mathematical Soc. (2009)
21. Maschlanka, P., Volkmann, L.: Independence number in n-extendable graphs. Discret. Math. **154**(1–3), 167–178 (1996)
22. Plummer, M.: On n-extendable graphs. Discret. Math. **31**(2), 201–210 (1980)
23. Plummer, M.: A theorem on matchings in the plane. Ann. Discrete Math. **41**, 347–354 (1988)
24. Plummer, M.: Extending matchings in graphs: a survey. Discret. Math. **127**(1–3), 277–292 (1994)
25. Plummer, M.: Extending matchings in graphs: an update. Congr. Numer. **116**, 3 (1996)
26. Plummer, M.: Recent progress in matching extension. Build. Bridges: Between Math. Comput. Sci. 427–454 (2008)
27. Propp, J.: Enumeration of matchings: problems and progress. New Perspect. Geom. Combinatorics **38**, 255–291 (1999)
28. Raman, V., Ramanujan, M.S., Saurabh, S.: Paths, Flowers and vertex cover. In: Demetrescu, C., Halldórsson, M.M. (eds.) ESA 2011. LNCS, vol. 6942, pp. 382–393. Springer, Heidelberg (2011). https://doi.org/10.1007/978-3-642-23719-5_33
29. Tutte, W.: The factorization of linear graphs. J. Lond. Math. Soc. **1**(2), 107–111 (1947)
30. Vandenbussche, J., West, D.B.: Matching extendability in hypercubes. SIAM J. Discret. Math. **23**(3), 1539–1547 (2009)
31. Vandenbussche, J., Westlund, E.: Matching extendability in Cartesian products of cycles. Australas. J. Comb. **82**(3), 317 (2022)
32. Wu, J., Wang, J., Kang, L.: Spectral conditions for matching extension. Appl. Math. Comput. **483**, 128982 (2024)

Extending Unlocked Matchings in Graphs

A. A. Pereira[✉][iD] and C. N. Campos[iD]

Universidade Estadual de Campinas, Av. Albert Einstein, 1251, Campinas, SP
13083-852, Brazil
`alessandra.pereira@ic.unicamp.br`, `cnc@unicamp.br`

Abstract. Let M be a matching of a graph G and let $S(M)$ and $U(M)$ be the sets of M-saturated and M-unsaturated vertices of G, respectively. A vertex $v \in U(M)$ is unlocked if there exists a vertex $u \in N(v) \cap U(M)$. The matching M is unlocked if every vertex in $U(M)$ is unlocked. We say that a graph G is k-fully-extendable if every unlocked matching M in G of size k can be extended to a perfect matching of G. In this work, we study k-fully-extendable graphs of order $2n$, characterizing the cases $k = n-1$ and $k = n-2$. For the case $k = n-2$, we provide additional necessary conditions, based on degree bounds, as well as sufficient conditions, based on classes of graphs.

Keywords: Matching extendability · Unlocked matching · Perfect matching · k-fully-extendable graphs

1 Introduction

Let G be a finite, undirected and simple graph with vertex set $V(G)$ and edge set $E(G)$. Let $N(v) = \{u \in V(G) : uv \in E(G)\}$ be the *(open) neighbourhood* of a vertex $v \in V(G)$. Each $u \in N(v)$ is called a *neighbour* of v. The *degree* of a vertex $v \in V(G)$ is $d_G(v) = |N(v)|$. The *minimum degree* of G is denoted $\delta(G)$ and the *maximum degree* is denoted $\Delta(G)$. Given $S \subseteq V(G)$, we denote $\bar{S} = V(G) \setminus S$. For a subgraph H of G, we analogously define $\bar{H} \cong G[V(G) \setminus V(H)]$. A *matching* M in G is a set of pairwise non-adjacent edges. A vertex $v \in V(G)$ is said to be *M-saturated* if it is incident with some edge in M; otherwise, v is said to be *M-unsaturated*. Denote $S(M)$ and $U(M)$ as the sets of M-saturated and M-unsaturated vertices of G, respectively. Set M is a *perfect matching* of a graph G when $S(M) = V(G)$. A graph is *matchable* if it admits a perfect matching. We say that M *extends* to a perfect matching if M is a subset of some perfect matching of G.

Matching theory has long been a focal point of Graph Theory, due both to its significant structural results [8,10,14,19,32] and its wide-ranging applications [9, 30,35,36]. As the theory developed, researchers investigated conditions under

This research was supported by CAPES, grants 88887.520851/2020-00 and 88887.600764/2021-00.

which matchings extend to perfect matchings. In this direction, Plummer [23] introduced the concept of k-extendable graphs in 1980.

We say that a graph G is *k-extendable* if every matching M of G with $|M| = k$ can be extended to a perfect matching of G. In the special case $k = 1$, k-extendability has been widely studied in the literature [15, 20] under the concept of a *matching-covered graph* – a graph in which every edge lies in some perfect matching. The problem has also been studied for other fixed values of k.

For $k = n - 1$, Anunchuen and Caccetta [3] proved that a matchable graph of order $2n$ is $(n-1)$-extendable if and only if it is isomorphic to K_{2n} or $K_{n,n}$. In a subsequent work [4], the same authors provided a complete characterization for $k = n - 2$. They showed that a matchable graph G of order $2n$, with $n \geq 5$, is $(n-2)$-extendable if and only if: either $G \cong K_{n,n}$ or $G \cong K_{2n}$; or G is bipartite with minimum degree $n - 1$; or G has minimum degree $2n - 3$ and contains a maximum independent set of order at most two; or, finally, G has minimum degree $2n - 2$.

The theory of k-extendable graphs has been explored in many directions, such as the establishment of necessary and sufficient conditions based on forbidden subgraphs [25], the analysis of criticality [2], the study of relationships with other graph parameters [6, 18, 21] and the investigation of their behaviour under graph operations [5, 11]. In addition, since Hackfeld and Koster [12] proved that the original problem is co-NP-complete, it has also been studied in classes of graphs, including bipartite, regular and planar graphs [1, 16, 24]. These and other related results are presented in three surveys by Plummer [26–28].

A key obstacle to extending a matching arises from the neighbourhood of an unsaturated vertex. In particular, for a non-perfect matching M of G, if there exists $u \in U(M)$ such that $N(u) \subseteq S(M)$, then M is not extendable to a perfect matching of G. Motivated by this property, we analyse extendable matchings without this type of neighbourhood obstruction.

Let M be a matching of a graph G. We say that a vertex $v \in U(M)$ is *unlocked* if there exists a vertex $u \in N(v) \cap U(M)$. We extend this notion to matchings by saying that M is *unlocked* if every vertex in $U(M)$ is unlocked. Observe that $U(M)$ may be empty, in which case M is unlocked. Moreover, a graph G is *k-fully-extendable* if every unlocked matching M of G with $|M| = k$ can be extended to a perfect matching of G. Observe that every matchable graph of order $2n$ is trivially 0-fully-extendable and n-fully-extendable. For a given graph G, we define $\mathcal{M}_k^f = \{M : M$ is an unlocked matching of G and $|M| = k\}$.

Using a different terminology, in 1997, Limaye and Sarvate [17] proved that every hypercube Q_r, $r \geq 2$, is k-fully-extendable, $1 \leq k \leq r$. Furthermore, the authors showed that, for Q_4, there exists an unlocked matching of size 5 (i.e. $r + 1$) that cannot be extended to a perfect matching, thereby confirming the tightness of the bound. For $r \geq 5$, Vandenbussche and West [33] strengthened these results by showing that the hypercube Q_r is k-fully-extendable for $1 \leq k \leq 2r - 4$. Recently [22], we proved that the cartesian product $Q_r \mathbin{\Box} P_n$, $r \geq 4$ and $n \geq 2$, is k-fully-extendable, $1 \leq k \leq 2r - 2$.

One can note that, if a graph G is k-extendable, then every matching M of G with $|M| = k$ is unlocked. Therefore, a k-extendable graph is also k-fully-extendable. However, the converse is not necessarily true. For instance, every hypercube Q_r, $r \geq 3$, is r-fully-extendable but not r-extendable. In fact, for $r \geq 3$, there exists a matching M of Q_r with $|M| = r$ such that, for some $u \in U(M)$, $N(u) \subseteq M$. Consequently, M is not extendable.

In this work, we study the extendability of unlocked matchings by investigating the concept of k-fully-extendable graphs of order $2n$ for $k \in \{n - 1, n - 2\}$. First, we prove that every matchable graph of order $2n$ is $(n-1)$-fully-extendable. Then, we address the case of $(n-2)$-fully-extendable graphs, for which we provide a characterization in terms of forbidden subgraphs and derive additional necessary and sufficient conditions involving classes of graphs and degree bounds.

2   Preliminaries

In this section, we present some results on matchable and Hamiltonian graphs used to establish our main results. Observe that every Hamiltonian graph of even order is matchable.

We begin with the classical theorem from Tutte [32]. Let G be a graph and $S \subseteq V(G)$. Denote $o(G \setminus S)$ the number of connected components of $G \setminus S$ having odd order.

Theorem 1 (Tutte [32]). *A graph G is matchable if and only if $o(G \setminus S) \leq |S|$ for every $S \subseteq V(G)$.* $\qquad\square$

A *claw* is a graph isomorphic to $K_{1,3}$. Let H be a claw. We call the vertex $c \in V(H)$ with $d_H(c) = 3$ the *central vertex* of H. An *induced claw* of a graph G is an induced subgraph of G isomorphic to a claw. A graph G is *claw-free* if it does not contain induced claws. The following theorem states that every connected claw-free graph of even order is matchable.

Theorem 2 (Sumner [31], Las Vergnas [34]). *Let G be a connected graph of even order. If G is claw-free, then G is matchable.* $\qquad\square$

We conclude this section by presenting a theorem that unifies some sufficient conditions for Hamiltonian graphs. For a graph G, let the *connectivity* $\kappa(G)$ of G be the maximum value of k for which G is k-connected.

Theorem 3 (Dirac [7], Pósa [29], Häggkvist and Nicoghossian [13]). *Let G be a graph of order $n \geq 3$. Then, G is Hamiltonian if any of the following holds:*

1. *$\delta(G) \geq \frac{n}{2}$;*
2. *for every integer $1 \leq k < \frac{n}{2}$, the number of vertices of degree at most k is strictly less than k;*
3. *G is 2-connected and $\delta(G) \geq \frac{n + \kappa(G)}{3}$.* $\qquad\square$

3   Results

Now, we present our results on fully-extendable graphs. We begin by proving, in Proposition 1, that a matchable graph contains unlocked matchings of every possible size.

Proposition 1. *Let G be a matchable graph of order $2n$, $n \geq 1$. Then, for every $0 \leq k \leq n$, there exists an unlocked matching of G with $|M| = k$.*

Proof. Let M be a perfect matching of G and let $0 \leq k \leq n$. Let $\{M', N\}$ be a bipartition of M with $|M'| = k$. We claim that M' is unlocked. In order to see this, consider $v \in U(M')$. By construction, $U(M') = S(N)$. Therefore, $v \in S(N)$ and there exists $uv \in N$, which implies $u \in S(N)$. Therefore, $u \in U(M')$ and v is unlocked. $\qquad\square$

In Theorem 4, our first result, we show that G is also $(n-1)$-fully-extendable. This result is an immediate consequence of Proposition 1.

Theorem 4. *Every matchable graph G of order $2n$, $n \geq 1$, is $(n - 1)$-fully-extendable.*

Proof. Let G be a matchable graph with $|V(G)| = 2n$. By Proposition 1, it follows that $\mathcal{M}^f_{n-1} \neq \emptyset$. Let $M \in \mathcal{M}^f_{n-1}$ and $U(M) = \{u, v\}$. Since M is unlocked, we have $N(u) \cap U(M) \neq \emptyset$ and $N(v) \cap U(M) \neq \emptyset$, which implies $N(u) \setminus S(M) = \{v\}$ and $N(v) \setminus S(M) = \{u\}$. Therefore, $uv \in E(G)$ and $M \cup \{uv\}$ is a perfect matching of G. $\qquad\square$

Our primary focus in this work is the study of $(n - 2)$-fully-extendability in graphs. This case is more subtle and requires additional care, in contrast to the relative straightforwardness of the $n - 1$ case discussed in Theorem 4.

In Theorem 5, we establish a characterization of $(n - 2)$-fully-extendable graphs in terms of induced claws, which provides the foundation for the subsequent results. We then strengthen this characterization in Theorems 6 and 7, by deriving further necessary and sufficient conditions based on classes of graphs and minimum degree bounds.

Theorem 5. *Let G be a matchable graph of order $2n$, $n \geq 2$. Then, G is $(n - 2)$-fully-extendable if and only if for every induced claw H of G, $\overline{H}$ is non-matchable.*

Proof. Let G be as stated in the hypothesis. Suppose first that G is $(n-2)$-fully-extendable. Let H be an induced claw of G. Suppose $\overline{H}$ is matchable. Let M be a perfect matching of $\overline{H}$. Then, M is unlocked and $|M| = n - 2$. By definition, H is non-matchable. Therefore, M cannot be extended to a perfect matching of G, a contradiction.

Conversely, suppose that, for every induced claw H of G, $\overline{H}$ is non-matchable. Since G is matchable, by Proposition 1, $\mathcal{M}^f_{n-2} \neq \emptyset$. Let $M \in \mathcal{M}^f_{n-2}$. Then, $|U(M)| = 4$. Moreover, for every $u \in U(M)$, we have $N(u) \cap U(M) \neq \emptyset$. Since

$G[V(G) \setminus U(M)] \cong G[S(M)]$ is matchable, by hypothesis, $G[U(M)] \not\cong K_{1,3}$. If $G[U(M)]$ is not connected, then it has exactly two connected components, both isomorphic to K_2, and the result follows. Suppose $G[U(M)]$ is connected. Since $G[U(M)] \not\cong K_{1,3}$, by Theorem 2, $G[U(M)]$ is matchable and the result follows. $\square$

Theorem 6. *Let G be a matchable graph of order $2n$, $n \geq 2$. If G is claw-free or bipartite, then G is $(n-2)$-fully-extendable.*

Proof. Let G be as stated in the hypothesis. Suppose that G is claw-free. By Proposition 1, $\mathcal{M}^f_{n-2} \neq \emptyset$. Let $M \in \mathcal{M}^f_{n-2}$ and $U(M) = \{u, v, x, y\}$. Since M is unlocked, $N(u) \cap U(M) \neq \emptyset$. Hence, we may assume, without loss of generality, $uv \in E(G)$. If $xy \in E(G)$, then $M \cup \{uv, xy\}$ is a perfect matching of G and the result follows. Otherwise, since $N(x) \cap U(M) \neq \emptyset$, we may assume $ux \in E(G)$. Considering that G is claw-free and $N(y) \cap U(M) \neq \emptyset$, we may assume $vy \in E(G)$. Therefore, $M \cup \{ux, vy\}$ is a perfect matching of G and the result follows.

Now, consider that G is bipartite and non-claw-free. Let H be an induced claw of G and let $V(H) = \{c, u, v, w\}$ with c being its central vertex. Note that, since G is bipartite, $\overline{H}$ is also bipartite. Let $\{X, Y\}$ be a bipartition of G. Adjust notation so that $c \in X$ and $\{u, v, w\} \subseteq Y$. Then, the bipartition of $\overline{H}$ is given by $X' = X \setminus \{c\}$ and $Y' = Y \setminus \{u, v, w\}$, with $|X'| = |X| - 1$ and $|Y'| = |Y| - 3$. This implies that $\overline{H}$ is non-matchable and, by Theorem 5, G is $(n-2)$-fully-extendable. $\square$

Theorem 7. *Let G be a connected graph of order $2n$, $n \geq 4$. If G is $(n-2)$-fully-extendable and G is non-claw-free, then $\delta(G) \leq n$.*

Proof. Let G be a connected graph of order $2n$, $n \geq 4$. Suppose that G is $(n-2)$-fully-extendable. Suppose also that G is non-claw-free. Let $\mathcal{H} = \{H : H$ is an induced claw of $G\}$. Note that, for every $H \in \mathcal{H}$, we have $\delta(\overline{H}) \geq \delta(G) - 4$ and $|V(\overline{H})| = 2n - 4$.

Suppose $\delta(G) \geq n + 2$. Let $H \in \mathcal{H}$. Then, $\delta(\overline{H}) \geq n - 2 \geq \frac{|V(\overline{H})|}{2}$. Applying Theorem 3, $\overline{H}$ is Hamiltonian and, hence, matchable. By Theorem 5, we conclude that G is not $(n-2)$-fully-extendable.

Suppose $\delta(G) = n + 1$. If there exists $H \in \mathcal{H}$ such that $\delta(\overline{H}) \geq n - 2$, then, by the previous reasoning, G is not $(n-2)$-fully-extendable. Hence, we may assume that, for every $H \in \mathcal{H}$, $\delta(\overline{H}) = n - 3$.

Let $H \in \mathcal{H}$ and $V_\delta = \{v \in V(\overline{H}) : d_{\overline{H}}(v) = n - 3\}$. Since $\delta(\overline{H}) = n - 3$, the only integer $1 \leq l < \frac{|V(\overline{H})|}{2} = n - 2$ for which $\overline{H}$ has vertices of degree at most l is $l = n - 3$. Thus, if $|V_\delta| \leq n - 4$, it follows from Theorem 3 that $\overline{H}$ is Hamiltonian, which implies that G is not $(n-2)$-fully-extendable. We conclude that $|V_\delta| \geq n - 3$. Let $u \in V(H)$ be the central vertex of H and $V(H) \setminus \{u\} = \{v_1, v_2, v_3\}$. Note that every $w \in V_\delta$ is adjacent to all vertices of H and $G[\{w, v_1, v_2, v_3\}] \cong K_{1,3}$. Thus, we can choose $H \in \mathcal{H}$ so that $d_G(u) = n + 1$. This implies $|V_\delta| \leq n - 2$.

We now consider the following cases based on the connectivity of $\overline{H}$.

Case 1. $\kappa(\overline{H}) = 0$.

Let G_1 and G_2 be two distinct connected components of $\overline{H}$. Considering that $|V(\overline{H})| = 2n - 4$, we may assume $|V(G_1)| \le n - 2$. Moreover, considering that $\delta(\overline{H}) = n - 3$, it follows that $|V(G_1)| \ge n - 2$ and $|V(G_2)| \ge n - 2$. Hence, $|V(G_1)| = |V(G_2)| = n - 2$. Thus, each G_i, $i \in \{1, 2\}$, contains exactly $n - 2$ vertices of degree $n - 3$ in $\overline{H}$. Furthermore, since $\delta(G) \ge n + 1$, each vertex of G_i is adjacent in G to every vertex of H. This implies $|V_\delta| = 2n - 4$, contradicting $|V_\delta| \le n - 2$.

Case 2. $\kappa(\overline{H}) = 1$.

In this case, $\overline{H}$ has a cut vertex x. Let G_1' and G_2' denote two distinct connected components of $G' \cong \overline{H}[V(\overline{H}) \setminus \{x\}]$. Without loss of generality, assume that $|V(G_1')| \le |V(G_2')|$. Note that $|V(G')| = 2n - 5$ and $\delta(G') \in \{n - 3, n - 4\}$. If $\delta(G') = n - 3$, then, since $n \ge 4$, we would have $|V(G_1')| \ge n - 2$ and $|V(G_2')| \ge n - 2$, implying $|V(G')| \ge 2n - 4$, a contradiction. Hence, $\delta(G') = n - 4$.

First, consider $n = 4$. In this case, $\delta(G') = 0$ and $\overline{H}$ must be isomorphic to one of the two graphs shown in Fig. 1. Since $n - 3 \le |V_\delta| \le n - 2$, it follows that $\overline{H}$ can only be isomorphic to the graph in Fig. 1(a). Thus $\overline{H}$ is matchable, and by Theorem 5, G is not $(n - 2)$-fully-extendable.

Fig. 1. Graph $\overline{H}$ for $n = 4$ when $\delta(G') = n - 4$.

Now suppose $n \ge 5$. Recall that $|V(G')| = 2n - 5$. Therefore, we may assume $|V(G_1')| \le n - 3$. Since $\delta(G') = n - 4$, every vertex of G_1' has at least $n - 4$ neighbours in G_1' and, thus, $|V(G_1')| \ge (n - 4) + 1 = n - 3$. Hence, $|V(G_1')| = n - 3$. This implies $|V(G_2')| = n - 2$. Note that $G_1' \cong K_{n-3}$. Suppose $\delta(G_2') = n - 3$. Then, similarly to the previous case, we conclude that $G_2' \cong K_{n-2}$. Thus, $\overline{H}$ is a connected claw-free graph. By Theorem 2, $\overline{H}$ is matchable and, hence, G is not $(n - 2)$-fully-extendable. Therefore, we can assume $\delta(G_2') = n - 4$.

Suppose, first, $n = 5$. Recalling that $|V(G_1')| = 2$ and $|V(G_2')| = 3$, we conclude that $G_1' \cong K_2$ and $G_2' \cong P_3$. If y is a vertex of degree one in G_2', then y is adjacent to x; otherwise $d_G(y) \le 5 < n + 1 = 6$. Then, $\overline{H}$ contains a spanning path P of six vertices consisting of the vertices of G_2', followed by x, and then the vertices of G_1'. Since P is matchable, by Theorem 5, G is not $(n - 2)$-fully-extendable.

Suppose $n \geq 6$. Then, $|V(G_2')| \geq 3$ and $n-4 \geq \frac{n-2}{2}$. Therefore, by Theorem 3, G_2' is Hamiltonian. Considering that $|V(G')| = 2n-5$, $|V(G_1')|$ and $|V(G_2')|$ have opposite parity. Let $x_1 \in V(G_1')$ and $x_2 \in V(G_2')$ be neighbours of x. Suppose $|V(G_1')|$ even. Then, since $G_1' \cong K_{n-3}$, G_1' has a perfect matching M_1. Moreover, since $|V(G_2')|$ is odd and G_2' is Hamiltonian, $G_2'[V(G_2') \setminus \{x_2\}]$ has a perfect matching M_2. We conclude that $M_1 \cup M_2 \cup \{xx_2\}$ is a perfect matching of $\overline{H}$ and G is not $(n-2)$-fully-extendable. The case $|V(G_1')|$ odd and $|V(G_2')|$ even is analogous.

Case 3. $\kappa(\overline{H}) \geq 2$.

If $\kappa(\overline{H}) \leq n-5$, then $\frac{2n-4+\kappa(\overline{H})}{3} \leq n-3$. Hence, by Theorem 3, $\overline{H}$ is Hamiltonian since $\delta(\overline{H}) = n-3$, and we are done again. Suppose, now, $\kappa(\overline{H}) \geq n-4$. If $\overline{H}$ is non-matchable, then, by Theorem 1, there exists $X \subseteq V(\overline{H})$ such that for $G' \cong \overline{H}[V(\overline{H}) \setminus X]$, $o(G') > |X|$. It follows that $|X| \leq n-3$ since $|V(\overline{H})| = 2n-4$. On the other hand, $|X| \geq n-4$ since $\kappa(\overline{H}) \geq n-4$. We conclude that $n-4 \leq |X| \leq n-3$.

Suppose $|X| = n-3$. Since $o(G') > |X|$, we conclude that $o(G') \geq n-2$. On the other hand, $|V(G')| = |V(\overline{H})| - |X| = n-1$. Hence, $o(G') \leq n-1$. If $o(G') = n-2$, then $|V(G')| - o(G') = 1$, a contradiction since a connected component of even order requires at least two vertices. Therefore, $o(G') = n-1$ and G' consists of $n-1$ isolated vertices. The equalities $|X| = n-3$ and $\delta(\overline{H}) = n-3$ imply that every vertex of G' has degree $n-3$ in $\overline{H}$, contradicting the fact that $|V_\delta| \leq n-2$.

Suppose $|X| = n-4$. Then, since $\overline{H}$ is 2-connected, $|X| \geq 2$ and, thus, $n \geq 6$. Moreover, $n-3 \leq o(G') \leq n$. If G' has an isolated vertex z, then $d_{\overline{H}}(z) \leq n-4$, a contradiction. Therefore, for every connected component G_i' of G' with odd order, we have $|V(G_i')| \geq 3$. Hence, $o(G') \leq \frac{n}{3}$. However, $\frac{n}{3} \geq o(G') > |X| = n-4$ only when $n < 6$, another contradiction.

We conclude that if G is $(n-2)$-fully-extendable and G is non-claw-free, then $\delta(G) \leq n$. $\qquad\square$

It is well known that connected bipartite matchable graphs of order $2n$ have $\delta(G) \leq n$. Moreover, equality is attained by the complete bipartite graphs. By Theorem 6, these graphs are $(n-2)$-fully-extendable and, from the perspective of Theorem 7, they demonstrate that the bound stated therein is tight. Next, we show that the bound remains tight even when considering non-bipartite graphs.

Theorem 8. *For every integer $n \geq 5$, there exists a matchable, non-bipartite, and non-claw-free graph G of order $2n$ with $\delta(G) = n$ that is $(n-2)$-fully-extendable.*

Proof. Let n be an integer such that $n \geq 5$. The proof is constructive and divided into cases depending on the parity of n.

Case 1. n is even.

Let G_1, G_2, and G_3 be three pairwise disjoint graphs such that $G_1 \cong K_{n-2}$, $G_2 \cong K_{n-1}$, $V(G_3) = \{x, y, z\}$ and $E(G_3) = \emptyset$. We construct G as follows:

- $V(G) = V(G_1) \cup V(G_2) \cup V(G_3)$;
- $E(G) = E(G_1) \cup E(G_3) \cup E_{13} \cup E_{23}$ such that:
  - $E_{13} = \{ux, uy, uz : u \in V(G_1)\}$;
  - given a bipartition $\{V_{xy}, V_{yz}\}$ of $V(G_2)$ such that $|V_{xy}| = \lfloor \frac{n-1}{2} \rfloor$ and $|V_{yz}| = \lceil \frac{n-1}{2} \rceil$, we define $E_{23} = \{vx, vy : v \in V_{xy}\} \cup \{vy, vz : v \in V_{yz}\}$.

This construction is illustrated for $n = 6$ in Fig. 2. Next, we show that $\delta(G) = n$ and that G is a non-bipartite, non-claw-free, matchable, and $(n-2)$-fully-extendable graph.

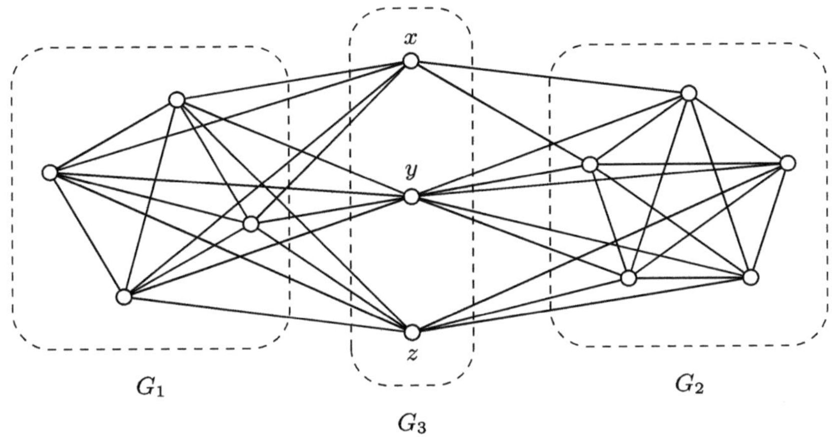

Fig. 2. Construction of a matchable, non-bipartite and non-claw-free graph G of order $2n = 12$ with $\delta(G) = n = 6$ that is $(n-2)$-fully-extendable.

First, note that $|V(G)| = (n-2) + 3 + (n-1) = 2n$. For $u \in V(G_1)$, we have $d_G(u) = (n-3) + 3 = n$ and for $u \in V(G_2)$, $d_G(u) = (n-2) + 2 = n$. For the vertices of G_3, $d_G(y) = (n-2) + (n-1) = 2n - 3 \geq n$, $d_G(x) = (n-2) + \lfloor \frac{n-1}{2} \rfloor$ and $d_G(z) = (n-2) + \lceil \frac{n-1}{2} \rceil$. Since n is even and $n \geq 6$, we have $d_G(x) \geq n$ and $d_G(z) \geq n$. We conclude that $\delta(G) = n$.

Since $G_1 \cong K_{n-2}$, G is non-bipartite. Moreover, for every $u \in V(G_1)$, $G[\{u, x, y, z\}] \cong K_{1,3}$. We now prove that G is matchable. As $G_1 \cong K_{n-2}$ and $n - 2$ is even, G_1 has a perfect matching M_1. Let $\{u_x, u_y, u_z\} \subseteq V(G_2)$ be a set of distinct vertices such that $\{u_x x, u_y y, u_z z\} \subseteq E(G)$. Then, $G[V(G_2) \setminus \{u_x, u_y, u_z\}] \cong K_{n-4}$ has a perfect matching M_3. Thus, $M = M_1 \cup M_3 \cup \{u_x x, u_y y, u_z z\}$ is a perfect matching of G.

It remains to show that G is $(n-2)$-fully-extendable. Let H be an induced claw of G such that $V(H) = \{c, v_1, v_2, v_3\}$ with c being its central vertex. Since $G_1 \cong K_{n-2}$ and $G_2 \cong K_{n-1}$, we have $|\{v_1, v_2, v_3\} \cap V(G_1)| \leq 1$ and $|\{v_1, v_2, v_3\} \cap V(G_2)| \leq 1$. Furthermore, since y is adjacent to all vertices of G_2 and every vertex of G_2 is adjacent to exactly two vertices of G_3, it follows that $c \notin V(G_2)$. Also, note that $E(G_3) = \emptyset$. Therefore, $c \notin V(G_3)$. Thus, we conclude that $c \in V(G_1)$. Consequently, $\overline{H}$ consists of two connected components isomorphic to K_{n-3} and K_{n-1}. Since n is even, K_{n-3} and K_{n-1} are non-matchable. Therefore, by Theorem 5, G is $(n-2)$-fully-extendable.

Case 2. n is odd.

Let G_1, G_2, and G_3 be three pairwise disjoint graphs such that $G_1 \cong K_{n-2}$, $G_2 \cong K_{n-2}$, $V(G_3) = \{w, x, y, z\}$ and $E(G_3) = \{wx, wy, wz\}$. We construct G as follows:

- $V(G) = V(G_1) \cup V(G_2) \cup V(G_3)$;
- $E(G) = E(G_1) \cup E(G_2) \cup E(G_3) \cup E_w \cup E_x \cup E_y \cup E_z$, such that, for fixed vertices $v_1 \in V(G_1)$ and $v_2 \in V(G_2)$:
  - $E_w = \{vw : v \in V(G_1) \cup V(G_2)\}$;
  - $E_x = \{vx : v \in V(G_1) \cup \{v_2\}\}$;
  - $E_y = \{vy : v \in V(G_2) \cup \{v_1\}\}$;
  - $E_z = \{vz : v \in V(G_1) \cup V(G_2) \setminus \{v_1, v_2\}\}$.

This construction is illustrated for $n = 7$ in Fig. 3. Next, we show that $\delta(G) = n$ and that G is a non-bipartite, non-claw-free, matchable, and $(n-2)$-fully-extendable graph.

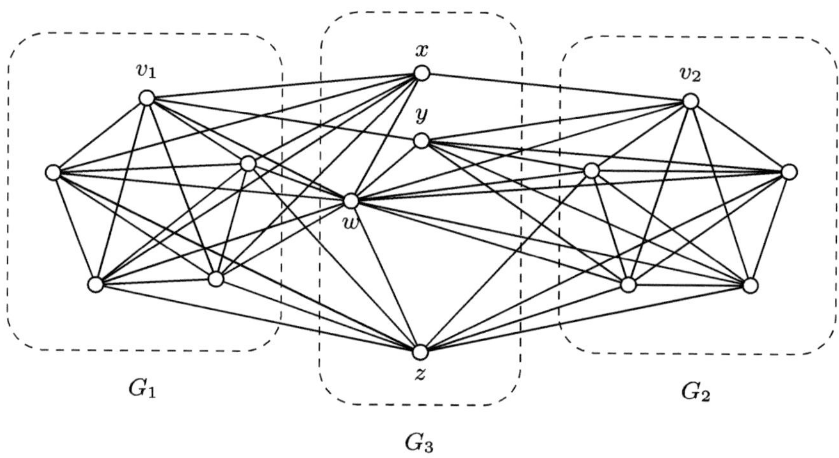

Fig. 3. Construction of a matchable, non-bipartite and non-claw-free graph G of order $2n = 14$ with $\delta(G) = n = 7$ that is $(n-2)$-fully-extendable.

First, observe that $|V(G)| = (n-2) + 4 + (n-2) = 2n$. For $u \in V(G_1) \setminus \{v_1\}$, $d_G(u) = (n-3) + 1 + 1 + 1 = n$ and $d_G(v_1) = (n-3) + 1 + 1 + 1 = n$. Similarly, for $u \in V(G_2) \setminus \{v_2\}$, $d_G(u) = (n-3) + 1 + 1 + 1 = n$ and $d_G(v_2) = (n-3) + 1 + 1 + 1 = n$. For the vertices of G_3, $d_G(w) = (n-2) + (n-2) + 3 = 2n - 1 \geq n$, $d_G(x) = d_G(y) = (n-2) + 1 + 1 = n$ and $d_G(z) = (n-3) + (n-3) + 1 = 2n - 5$. Since $n \geq 5$, we have $d_G(z) \geq n$. Hence, $\delta(G) = n$.

Once again, since $G_1 \cong K_{n-2}$, G is non-bipartite. Moreover, G contains an induced claw since $G_3 \cong K_{1,3}$. In order to prove that G is matchable, note that $G[V(G_1) \setminus \{v_1\}] \cong K_{n-3}$ and $G[V(G_2) \setminus \{v_2\}] \cong K_{n-3}$ have perfect matchings M_1 and M_2, respectively. Thus, $M = M_1 \cup M_2 \cup \{v_1 y, v_2 x, wz\}$ is a perfect matching of G.

It remains to show that G is $(n-2)$-fully-extendable. Let H be an induced claw of G such that $V(H) = \{c, t_1, t_2, t_3\}$ with c being its central vertex. Since $G_1 \cong K_{n-2}$ and $G_2 \cong K_{n-2}$, we have $|\{t_1, t_2, t_3\} \cap V(G_1)| \leq 1$ and $|\{t_1, t_2, t_3\} \cap V(G_2)| \leq 1$. Suppose $c \in V(G_1)$. Then, $\{t_1, t_2, t_3\} \subseteq V(G_1) \cup V(G_3)$. By construction, every vertex of $V(G_1) \setminus \{v_1\}$ is adjacent to w, x and z, and v_1 is adjacent to w, x and y. Therefore, since w is also adjacent to x, y and z, it follows that $|\{t_1, t_2, t_3\} \cap V(G_3)| \leq 2$. This implies $|\{t_1, t_2, t_3\} \cap V(G_1)| = 1$, $|\{t_1, t_2, t_3\} \cap V(G_3)| = 2$ and $w \notin \{t_1, t_2, t_3\}$. If $c = v_1$, then we can assume $\{t_1, t_2\} = \{x, y\}$ and $t_3 \in V(G_1) \setminus \{v_1\}$. If $c \in V(G_1) \setminus \{v_1\}$, then we can assume $\{t_1, t_2\} = \{x, z\}$ and $t_3 \in V(G_1) \setminus \{c\}$. However, this is a contradiction since in both cases $x \in \{t_1, t_2\}$ and x is adjacent to every vertex of G_1, including t_3. Therefore, $c \notin V(G_1)$. Analogously, $c \notin V(G_2)$. We conclude that $c \in V(G_3)$. Considering that w is adjacent to all the other vertices of G and $|\{t_1, t_2, t_3\} \cap V(G_1)| \leq 1$ and $|\{t_1, t_2, t_3\} \cap V(G_2)| \leq 1$, it follows that $c = w$. Consequently, either $V(H) = \{w, x, y, z\}$ or $V(H) = \{w, v_1, v_2, z\}$. In both cases, $\overline{H}$ consists of two connected components isomorphic to K_{n-2}. Therefore, again by Theorem 5, G is $(n-2)$-fully-extendable. $\qquad\square$

4 Concluding Remarks

The problem of k-fully-extendability has previously appeared in the literature as the study of extendability of matchings of size k that do not saturate the neighbourhood of any unsaturated vertex. Motivated by this, we introduce the new terminology of k-fully-extendable graphs and investigate their structural properties.

In particular, we have shown, in Theorem 4, that every matchable graph of order $2n$ is $(n-1)$-fully-extendable. In Theorem 5, we present a general characterization of $(n-2)$-fully-extendable graphs in terms of induced claws. Building on this characterization, Theorem 6 shows that claw-free and bipartite graphs are $(n-2)$-fully-extendable. In Theorem 7, we prove that a graph that is $(n-2)$-fully-extendable and non-claw-free must satisfy $\delta(G) \leq n$. Moreover, in Theorem 8, we prove that this bound is tight even when considering non-bipartite graphs.

In order to further characterize $(n-2)$-fully-extendable graphs using Theorems 6 and 7, we need to identify the conditions under which matchable, non-bipartite, non-claw-free graphs of order $2n$ with $\delta(G) \leq n$ are $(n-2)$-fully-extendable. Another challenging open problem is to study k-fully-extendable graphs for $1 \leq k \leq n-3$. Moreover, it is also interesting to explore extremal aspects by characterizing maximal and minimal k-fully-extendable graphs, as well as to investigate the relationship with other parameters and their behaviour under graph operations.

References

1. Aldred, R., Holton, D.A., Lou, D.: N-extendability of symmetric graphs. J. Graph Theory **17**(2), 253–261 (1993)
2. Aldred, R., Plummer, M.: Extendability and criticality in matching theory. Graphs Combin. **36**(3), 573–589 (2020)
3. Anunchuen, N., Caccetta, L.: On minimally k-extendable graphs. Aust. J. Combin. **9**, 153–168 (1994)
4. Anunchuen, N., Caccetta, L.: On $(n-2)$-extendable graphs. J. Combin. Math. Combin. Comput. **16**, 115–128 (1994)
5. Bai, B., Wu, Z., Yang, X., Yu, Q.: Lexicographic product of extendable graphs. Bull. Malay. Math. Sci. Soc. Second Ser. **33**(2), 197–204 (2010)
6. Chen, C.: Binding number and toughness for matching extension. Discret. Math. **146**(1–3), 303–306 (1995)
7. Dirac, G.A.: Some theorems on abstract graphs. Proc. Lond. Math. Soc. **3**(1), 69–81 (1952)
8. Edmonds, J.: Paths, trees, and flowers. Can. J. Math. **17**, 449–467 (1965)
9. Gale, D., Shapley, L.: College admissions and the stability of marriage. Am. Math. Mon. **69**(1), 9–15 (1962)
10. Gallai, T.: Kritische graphen II. Magyar Tud. Akad. Mat. Kutato Int. Kozl. **8**, 373–395 (1963)
11. Györi, E., Plummer, M.: The Cartesian product of a k-extendable and an l-extendable graph is $(k+l+1)$-extendable. Julius Petersen Graph Theory Centennial **6**, 417 (1992)
12. Hackfeld, J., Koster, A.: The matching extension problem in general graphs is co-NP-complete. J. Comb. Optim. **35**, 853–859 (2018)
13. Häggkvist, R., Nicoghossian, G.: A remark on hamiltonian cycles. J. Combin. Theory Ser. B **30**(1), 118–120 (1981)
14. Hall, P.: On representatives of subsets. In: Classic Papers in Combinatorics, pp. 58–62 (1987)
15. He, J., Wei, E., Ye, D., Zhai, S.: On perfect matchings in matching covered graphs. J. Graph Theory **90**(4), 535–546 (2019)
16. Lakhal, J., Litzler, L.: A polynomial algorithm for the extendability problem in bipartite graphs. Inf. Process. Lett. **65**(1), 11–16 (1998)
17. Limaye, N.B., Sarvate, D.G.: On r-extendability of the hypercube Q_n. Math. Bohem. **122**(3), 249–255 (1997)
18. Lou, D.: On matchability of graphs. Aust. J. Combin. **21**, 201–210 (2000)
19. Lovász, L.: On the structure of factorizable graphs. Acta Math. Hungar. **23**(1–2), 179–195 (1972)
20. Lu, F., Kothari, N., Feng, X., Zhang, L.: Equivalence classes in matching covered graphs. Discret. Math. **343**(8), 111945 (2020)
21. Maschlanka, P., Volkmann, L.: Independence number in n-extendable graphs. Discret. Math. **154**(1–3), 167–178 (1996)
22. Pereira, A.A., Campos, C.N.: Matching extendability in cartesian product of hypercubes and paths. Manuscript (2025)
23. Plummer, M.: On n-extendable graphs. Discret. Math. **31**(2), 201–210 (1980)
24. Plummer, M.: A theorem on matchings in the plane. Ann. Discret. Math. **41**, 347–354 (1988)
25. Plummer, M.: Extending matchings in claw-free graphs. Discret. Math. **125**(1–3), 301–307 (1994)

26. Plummer, M.: Extending matchings in graphs: a survey. Discret. Math. **127**(1–3), 277–292 (1994)
27. Plummer, M.: Extending matchings in graphs: an update. Congr. Numer. **116**, 3 (1996)
28. Plummer, M.: Recent progress in matching extension. In: Building Bridges: Between Mathematics and Computer Science, pp. 427–454 (2008)
29. Pósa, L.: A theorem concerning Hamilton lines. Magyar Tud. Akad. Mat. Kutató Int. Közl **7**, 225–226 (1962)
30. Roth, A.: Online and Matching-Based Market Design. Cambridge University Press (2023)
31. Sumner, D.P.: Graphs with 1-factors. Proc. Am. Math. Soc. **42**(1), 8–12 (1974)
32. Tutte, W.: The factorization of linear graphs. J. Lond. Math. Soc. **1**(2), 107–111 (1947)
33. Vandenbussche, J., West, D.B.: Matching extendability in hypercubes. SIAM J. Discret. Math. **23**(3), 1539–1547 (2009)
34. Vergnas, L.: A note on matchings in graphs. Cahiers Cent. Études Recherche Opér. **17**, 257 (1975)
35. Vukičević, D.: Applications of perfect matchings in chemistry. In: Dehmer, M. (ed.) Structural Analysis of Complex Networks, pp. 463–482. Springer, Heidelberg (2010). https://doi.org/10.1007/978-0-8176-4789-6_19
36. Witwer, C., Hofacker, I., Stadler, P.: Prediction of consensus RNA secondary structures including pseudoknots. IEEE/ACM Trans. Comput. Biol. Bioinf. **1**(2), 66–77 (2004)

A Note on Reconfiguration Graphs of Cliques

Nhat-Quan Lam[1], Huu-An Phan[2(✉)], and Duc A. Hoang[3]

[1] Université Gustave Eiffel, Champs-sur-Marne, France
`nhat-quan.lam@edu.univ-eiffel.fr`
[2] Nanyang Technological University, Singapore, Singapore
`HUUAN002@e.ntu.edu.sg`
[3] VNU University of Science, Vietnam National University, Hanoi, Vietnam
`hoanganhduc@hus.edu.vn`

Abstract. In a reconfiguration setting, each clique of a graph G is viewed as a set of tokens placed on vertices of G such that no vertex has more than one token and any two tokens are adjacent. Three well-known reconfiguration rules have been studied in the literature: Token Jumping (TJ), Token Sliding (TS), and Token Addition/Removal (TAR). Given a graph G and a reconfiguration rule $\mathsf{R} \in \{\mathsf{TS}, \mathsf{TJ}, \mathsf{TAR}\}$, a reconfiguration graph of k-cliques of G, denoted by $\mathsf{R}_k(G)$, is the graph whose vertices are cliques of G of size k and two vertices are adjacent if one can be obtained from the other by applying R exactly once. In this paper, we initiate the study of structural properties of reconfiguration graphs of cliques, proving several interesting results primarily under TS and TJ rules. In particular, we establish a formula relating the clique number of G and that of $\mathsf{TS}_k(G)$, and bound the chromatic number of $\mathsf{TS}_k(G)$ via that of an appropriate Johnson graph. Additionally, we present an algorithm to construct $\mathsf{TS}_{\omega(G)-1}(G)$ from $\mathsf{TJ}_{\omega(G)}(G)$ and derive structural properties of $\mathsf{TJ}_{\omega(G)}(G)$ graphs, where $\omega(G)$ denotes the clique number of G. Finally, we show that $\mathsf{TS}_k(G)$ is planar whenever G is planar and establish bounds on the number of 3- and 4-cliques based on results concerning $\mathsf{TS}_k(G)$ graphs. In particular, we prove that any planar graph G with n vertices can contain at most $3n - 8$ triangles, which aligns with the classical bound on maximal planar graphs.

Keywords: combinatorial reconfiguration · reconfiguration graph · clique · structural properties
2020 MSC: 05C99

1 Introduction

1.1 Combinatorial Reconfiguration

Recently, *combinatorial reconfiguration* has emerged in different areas of computer science, including recreational mathematics (e.g., games and puzzles), com-

---

N.-Q. Lam—Work completed while at University of Science, Ho Chi Minh City, Vietnam.

putational geometry (e.g., flip graphs of triangulations), constraint satisfaction (e.g., solution space of Boolean formulas), and even quantum complexity theory (e.g., ground state connectivity). Given a *source problem* $\mathcal{P}$ (e.g., SATISFIABILITY, VERTEX-COLORING, INDEPENDENT SET, etc.) and a prescribed *reconfiguration rule* R that usually describes a "small" change in a feasible solution of $\mathcal{P}$ (e.g., satisfying truth assignments, proper vertex-colorings, independent sets, etc.) without affecting its feasibility (e.g., flipping one bit of a satisfying truth assignment, recoloring one vertex of a proper vertex-coloring, adding/removing one member of an independent set, etc.), one can define the corresponding *reconfiguration graph of $\mathcal{P}$ under the rule* R as follows. In such a graph, each *feasible solution* of $\mathcal{P}$ is a node, and two nodes are *adjacent* if one can be obtained from the other by applying the rule R exactly once. In particular, the reconfiguration graph of satisfying truth assignments of a Boolean formula on n variables under the "one-bit-flipping rule", also known as the *solution graph* of a Boolean formula, is an induced subgraph of the well-known *hypercube* graph Q_n—the graph whose nodes are length-n binary strings and two nodes are adjacent if they differ in exactly one bit. (See Fig. 1).

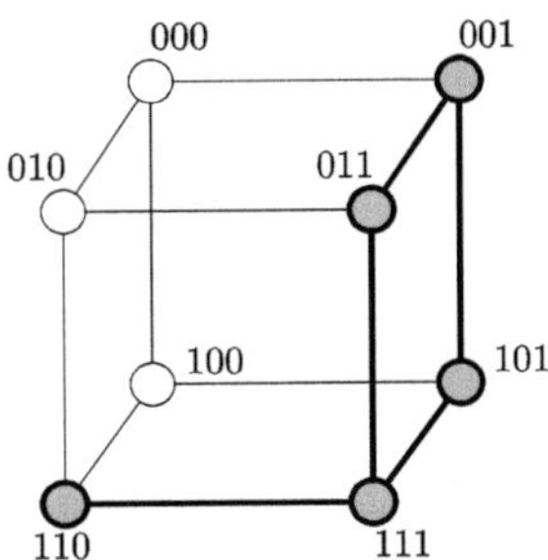

Fig. 1. The reconfiguration graph of satisfying truth assignments of the Boolean formula $(x \wedge y) \vee z$ under the "one-bit-flipping rule" is an induced subgraph of Q_3. Each binary string corresponds to a truth assignment of the variables x, y, and z, respectively.

To the best of our knowledge, reconfiguration graphs have been studied from both *algorithmic* and *graph-theoretic* viewpoints. From the *algorithmic perspective*, the main goal is to understand whether certain algorithmic questions, such as finding a (shortest) path between two given nodes of the reconfiguration graph, can be answered efficiently, and if so, design an algorithm to do it. A major challenge is that the size of a reconfiguration graph is often quite huge, and thus the whole graph can never be a part of the input. From the *graph-theoretic perspective*, the main goal is to study the structural properties (e.g., connectedness, bipartitedness, Hamiltonicity, etc.) of reconfiguration graphs as well as classify them based on known graph classes (which graphs are reconfiguration graphs). With respect to several well-known source problems, reconfiguration graphs have been extensively studied from the algorithmic viewpoint [2,13,16].

On the other hand, from the *graph-theoretic perspective*, reconfiguration graphs have been well-characterized only for a limited number of source problems— namely those whose "feasible solutions" are satisfying truth assignments of a Boolean formula [14], or general vertex subsets [11], (maximum) matchings [4], dominating sets, or proper vertex-colorings [12] of a graph. We refer readers to the surveys [2,12,13,16] for more details on recent advances in this research area.

1.2   Reconfiguration of Cliques

In this paper, we consider CLIQUE as the source problem. A *clique* in a graph G is a subset of vertices that are all adjacent. Each clique can be viewed as a set of tokens on the vertices of G, with no vertex holding more than one token. The following reconfiguration rules are well-known:

- **Token Jumping (TJ):** Two cliques are *adjacent under* TJ if one can be transformed into the other by moving a token to an unoccupied vertex.
- **Token Sliding (TS):** Two cliques are *adjacent under* TS if one can be transformed into the other by moving a token to an adjacent unoccupied vertex.
- **Token Addition/Removal (TAR):** Two cliques are *adjacent under* TAR_k *(resp. TAR^k)* if one can be transformed into the other by adding a token to an unoccupied vertex or removing one from an occupied vertex, ensuring the resulting set has at least (resp. at most) k tokens. (Here, $\emptyset$ is considered a clique of size 0).

Given a graph G and possibly an integer $k \geq 0$, one can define different reconfiguration graphs of cliques of G as follows.

- $\mathsf{TJ}(G)$ (resp. $\mathsf{TJ}_k(G)$) is the graph whose nodes are cliques (resp. cliques of size k) of G and edges are defined under TJ. Similar definitions hold for $\mathsf{TS}(G)$ and $\mathsf{TS}_k(G)$.
- $\mathsf{TAR}_k(G)$ (resp. $\mathsf{TAR}^k(G)$) is the graph whose nodes are cliques of size *at least* (resp. *at most*) k of G and edges are defined under TAR_k (resp. TAR^k). For convenience, we denote by $\mathsf{TAR}(G)$ the graph $\mathsf{TAR}_0(G) \cong \mathsf{TAR}^{\omega(G)}(G)$, where $\omega(G)$ is the maximum size of a clique of G. By definition, $\mathsf{TAR}(G)$ is indeed the union of $\mathsf{TAR}_k(G)$ and $\mathsf{TAR}^k(G)$.

Naturally, for $\mathsf{R} \in \{\mathsf{TS}, \mathsf{TJ}, \mathsf{TAR}\}$ and a given graph G, we call G a R-*reconfiguration graph* (or simply R-*graph*) if there exists a graph H such that $G \cong \mathsf{R}(H)$. We define a R_k-*reconfiguration graph* (or simply R_k-*graph*) and a R^k-*reconfiguration graph* (or simply R^k-*graph*) similarly.

We review related results on reconfiguration graphs of cliques from both algorithmic and graph-theoretic perspectives. Algorithmically, Ito et al. [8,9] initiated the study of CLIQUE RECONFIGURATION (CR) under $\mathsf{R} \in \{\mathsf{TS}, \mathsf{TJ}, \mathsf{TAR}_k\}$, asking whether a path exists between two given cliques in $\mathsf{R}(G)$. They showed that TS, TJ, and TAR_k are equivalent regarding polynomial-time solvability, with CR being PSPACE-complete for perfect graphs but polynomial-time solvable for even-hole-free graphs and cographs. They also designed polynomial-time

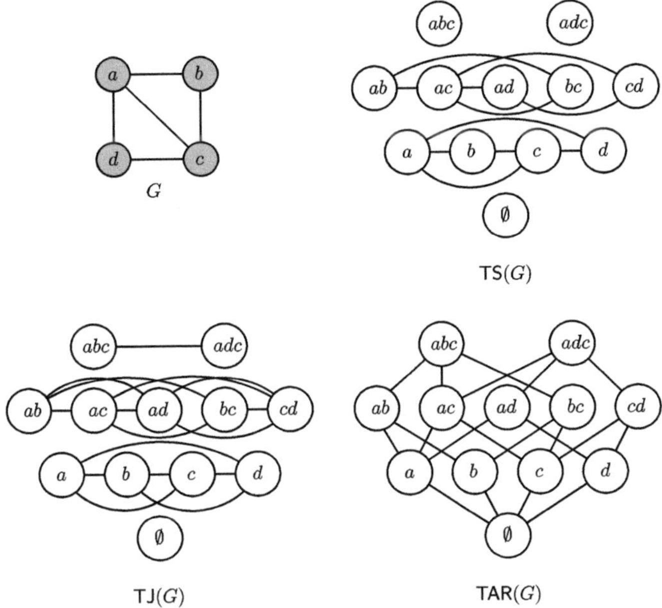

Fig. 2. An example of a graph G and some reconfiguration graphs of cliques of G. Each label inside a white-colored vertex (e.g., abc) indicates a clique of G (e.g., $\{a, b, c\}$).

algorithms for the shortest path variant when G is chordal, bipartite, planar, or has bounded treewidth (Fig. 2).

A different way of looking at a clique is to consider it as a collection of edges instead of vertices. With this viewpoint, the *edge-variants* of CR can be defined similarly under all above rules. Hanaka et al. [6] initiated the study of these edge-variants (as restricted cases of a more generalized problem called SUBGRAPH RECONFIGURATION) from the algorithmic viewpoint by showing that the problems can be solved in linear time under any of TS and TJ.

From the graph-theoretic viewpoint, the TAR-graph (of a graph) was first introduced in 1989 by Bandelt and van de Vel [1] under the name *simplex graph*[1]. The TS_k-graph of a complete graph K_n is closely related to the so-called *token graphs* [11] and *Johnson graphs* [7]. A *token graph* of a graph G, denoted by $F_k(G)$, is a graph whose vertices are size-k vertex-subsets of G and two vertices are adjacent if one can be obtained from the other by applying a single TS-move. A *Johnson graph* $J(n, k)$ is a graph whose vertices are size-k subsets of an n-element set and two vertices are adjacent if their intersection is of size exactly $k - 1$. One can readily verify that $\mathsf{TS}_k(K_n) \cong F_k(K_n) \cong J(n, k)$. To the best of

---

[1] Apparently, in some contexts, a *simplex* is a vertex subset where any two members form an edge (which is the same as our definition of a clique) and a *clique* is a maximal simplex which is not contained in any larger simplex.

our knowledge, reconfiguration graphs of cliques under TS or TJ have not yet been systematically studied in the literature.

1.3 Our Problem and Results

In this paper, we consider reconfiguration graphs of cliques and study their structural properties. We prove a number of interesting results, most of which are under TS and TJ rules.

In Sect. 3, we study the maximum clique size and chromatic number of $\mathsf{TS}_k(G)$. We establish a formula relating $\omega(G)$ and $\omega(\mathsf{TS}_k(G))$, and bound $\chi(\mathsf{TS}_k(G))$ via the chromatic number of an appropriate Johnson graph.

In Sect. 4, we focus on TJ_k-graphs with $k = \omega(G)$. We give an algorithm to construct $\mathsf{TS}_{\omega(G)-1}(G)$ from $\mathsf{TJ}_{\omega(G)}(G)$ and derive structural properties of maximum TJ graphs, i.e., graphs H for which $H \cong \mathsf{TJ}_{\omega(G)}(G)$ for some G.

Finally, in Sect. 5, we prove that $\mathsf{TS}_k(G)$ is planar whenever G is planar. We also derive bounds on the number of 3- and 4-cliques in terms of $|E(G)|$. Specifically, for 3-cliques, we prove that any planar graph G with $n = |V(G)|$ vertices can have at most $3n - 8$ copies of K_3, which matches the classical bound on maximal planar graphs [5]. Due to space constraints, several proofs are deferred to the full version [10].

2   Preliminaries

For the concepts and notations not defined here, we refer readers to [3]. Unless otherwise mentioned, all graphs in this paper are simple, connected, and undirected. We denote by $V(G)$ and $E(G)$ the vertex-set and edge-set of a graph G, respectively. The *chromatic number* of G, denoted by $\chi(G)$, is the smallest number of colors that can be used to color vertices of G such that no two adjacent vertices receive the same color. For two sets X, Y, we sometimes write $X + Y$ and $X - Y$ to respectively indicate $X \cup Y$ and $X \setminus Y$. When $Y = \{y\}$, we simply write $X + y$ and $X - y$ instead of $X + \{y\}$ and $X - \{y\}$, respectively. We denote by $X \triangle Y$ the *symmetric difference* of X and Y, i.e., $X \triangle Y = (X - Y) + (Y - X)$. For a vertex-subset X, we denote by $G[X]$ the subgraph of G induced by vertices in X. For two distinct graphs H and G, we denote by $H + G$ the *disjoint union* of H and G and by cH the disjoint union of c copies of H, where c is a positive integer.

We now formally define the TS, TJ, and TAR rules. Two cliques C and C' of G are *adjacent under* TJ if there exist $u, v \in V(G)$ such that $C - C' = \{u\}$ and $C' - C = \{v\}$. We say that they are *adjacent under* TS if one additional constraint $uv \in E(G)$ is satisfied. Two cliques C and C' are *adjacent under* TAR_k *(resp. TAR^k)* if $|C \triangle C'| = 1$ and $\min\{|C|, |C'|\} \geq k$ (resp. $\max\{|C|, |C'|\} \leq k$).

3   Token Sliding

We prove a useful observation regarding cliques in TS_k-graphs. We remark that similar observations for Johnson graphs have been proved in [15].

Lemma 1. *Let G be a graph. Suppose that $\mathsf{TS}_k(G)$ contains a complete subgraph K_n ($n \geq 3$) whose vertices $A_1, \ldots, A_n$ are k-cliques of G. There exist either a clique $\mathsf{Uni} \subseteq V(G)$ of size $k+1$ and pairwise distinct vertices $a_1, \ldots, a_n \in \mathsf{Uni}$ such that $A_i = \mathsf{Uni} - a_i$ or a clique $\mathsf{Int} \subseteq V(G)$ of size $k-1$ and pairwise distinct vertices $a_1, \ldots, a_n \in V(G) \setminus \mathsf{Int}$ such that $A_i = \mathsf{Int} + a_i$, where $1 \leq i \leq n$. In particular, when $n > k+1$, a clique Int satisfying the above conditions exists.*

We now prove a relationship between the clique number of G and that of $\mathsf{TS}_k(G)$. A similar relationship has been established for Johnson graphs in [15].

Theorem 1. *Let G be a graph.*

(a) If $k > \omega(G)$ then $\omega(\mathsf{TS}_k(G)) = 0$.
(b) If $k = \omega(G)$ then $\omega(\mathsf{TS}_k(G)) = 1$.
(c) If $k < \omega(G)$ then $\omega(\mathsf{TS}_k(G)) = \max\{k+1, \omega(G) - k + 1\}$.

In the next propositions, we will link the chromatic number of $\mathsf{TS}_k(G)$ with that of a Johnson graph. For a set X, we denote by $P(X)$ the *power set* of X, i.e., the set of all subsets of X.

Proposition 1. *Given an arbitrary graph G and number k, we have $\chi(\mathsf{TS}_k(G)) \leq \chi(J(\chi(G), k))$.*

Proposition 2. *Given an arbitrary graph G and number k, we have $\chi(\mathsf{TS}_k(G)) \geq \chi(J(\omega(G), k))$.*

Proof. Let H be a clique of size $\omega(G)$ in G. We know that $\mathsf{TS}_k(H)$ is a subgraph of $TS_k(G)$ and $J(\omega(G), k) \cong TS_k(H)$. Hence $\chi(\mathsf{TS}_k(G)) \geq \chi(J(\omega(G), k))$.

4   Token Jumping

In this section, we consider TJ_k-graphs. As $\mathsf{TS}_k(G)$ is a subgraph of $\mathsf{TJ}_k(G)$, we immediately have the following consequence of Theorem 1.

Corollary 1. *If $k < \omega(G)$ then $\omega(\mathsf{TJ}_k(G)) \geq \max\{k+1, \omega(G) - k + 1\}$.*

For a graph G, we now consider $\mathsf{TJ}_{\omega(G)}(G)$ and prove some of its properties.

Lemma 2. *Suppose that the $\omega(G)$-cliques A, B, C of a graph G form a triangle in $\mathsf{TJ}_{\omega(G)}(G)$. Then, $A \cap B = B \cap C = A \cap C$.*

Using Lemma 2, we can prove the following proposition. A *diamond* $K_4 - e$ is a graph obtained from K_4 by removing exactly one edge.

Proposition 3. *For any graph G, the graph $\mathsf{TJ}_{\omega(G)}(G)$ does not contain any diamond as an induced subgraph.*

Proposition 4. *Let G be any graph and let $k = \omega(G)$. Let U be a vertex of $\mathsf{TJ}_k(G)$. One can partition the set $N_{\mathsf{TJ}_k(G)}(U)$ of neighbors of U in $\mathsf{TJ}_k(G)$ into k disjoint (possibly empty) subsets $S_1(U), S_2(U), \ldots, S_k(U)$ such that each $S_i(U)$ $(1 \leq i \leq k)$, if it is not empty, is a clique of $\mathsf{TJ}_k(G)$. Moreover, there is no edge in $\mathsf{TJ}_k(G)$ joining a vertex of $S_i(U)$ and that of $S_j(U)$, for distinct pairs $i, j \in \{1, \ldots, k\}$.*

Next, we prove an interesting relationship between the two graphs $\mathsf{TJ}_{\omega(G)}(G)$ and $\mathsf{TS}_{\omega(G)-1}(G)$, for any graph G. The following lemma is useful.

Lemma 3. *Let G be a graph and k be an integer satisfying $2 \leq k \leq \omega(G)$. Let $A, B \in V(\mathsf{TS}_{k-1}(G))$. Then, $AB \in E(\mathsf{TS}_{k-1}(G))$ if and only if $A \cup B \in V(\mathsf{TJ}_k(G))$.*

We remark that for two graphs G and T satisfying $T \cong \mathsf{TJ}_k(G)$, one can indeed label vertices of T by k-cliques of G. We call such a labelling a *set label* of T. Given a graph $T \cong \mathsf{TJ}_k(G)$ and one of its set labels, using Lemma 3, one can indeed construct a graph H such that $\mathsf{TS}_{k-1}(G) \cong H + cK_1$ for some integers $k \geq 2$ and $c \geq 0$: vertices of H are exactly the size-$(k-1)$ subsets of each k-clique of G (i.e., vertex in $\mathsf{TJ}_k(G)$), and two vertices of H are adjacent if their union is a k-clique of G (Lemma 3).

We now consider a more generalized case when T is given without any set label. Now for an integer k and graph G, we call G a *k-good* graph if it satisfies the following *k-good conditions*: for every vertex u of G, the set $N_G(u)$ can be partitioned into k disjoint subsets (possibly empty) $S_1(u), S_2(u), \ldots, S_k(u)$ such that each $S_i(u)$ $(1 \leq i \leq k)$, if it is not empty, is a clique in G and there is no edge joining a vertex of $S_i(u)$ and that of $S_j(u)$, for a distinct pair $i, j \in \{1, \ldots, k\}$.

Proposition 5. *For a k-good graph $T = (V, E)$, one can find a multiset $\mathbf{Msets}$ of subsets of V such that the following conditions are satisfied.*

(a) For each pair of different sets $X, Y \in \mathbf{Msets}$, $|X \cap Y| < 2$.
(b) For each $X \in \mathbf{Msets}$, X is a clique in T.
(c) For each $u \in V(T)$, there are exactly k sets $X \in \mathbf{Msets}$ such that $u \in X$.
(d) For each $uv \in E(T)$, there exists exactly one set $X \in \mathbf{Msets}$ such that $u, v \in X$.

From Proposition 4, $\mathsf{TJ}_{\omega(G)}$ is $\omega(G)$-good. Therefore, the following corollary directly follows from Proposition 5.

Corollary 2. *Let G be any graph and let $k = \omega(G)$. There exists a multiset $\mathbf{Msets}(\mathsf{TJ}_k(G))$ containing subsets of $V(\mathsf{TJ}_k(G))$ such that the following conditions are satisfied.*

(a) For each different sets $X, Y \in \mathbf{Msets}(\mathsf{TJ}_k(G))$, $|X \cap Y| < 2$
(b) X form a clique in $T J_k(G)$, $(\forall X \in \mathbf{Msets}(\mathsf{TJ}_k(G)))$
(c) For each $u \in V(\mathsf{TJ}_k(G))$, there are exactly k sets $X \in \mathbf{Msets}(\mathsf{TJ}_k(G))$ such that $u \in X$

(d) For each $uv \in E(\mathsf{TJ}_k(G))$, there exists one set $X \in \mathbf{Msets}(\mathsf{TJ}_k(G))$ such that $u, v \in X$

Next, we will examine the relation between graphs $\mathsf{TS}_{k-1}(G)$ and $\mathsf{TJ}_k(G)$, where $k = \omega(G)$. For each set v in $V(\mathsf{TS}_{k-1}(G))$ let $\mathbf{Expand}(v)$ be the set of u in $V_{\mathsf{TJ}_k(G)}$ such that $v \subset u$, then $\mathbf{Expand}(v)$ is a clique in $\mathsf{TJ}_k(G)$ (since each set pair in $\mathbf{Expand}(v)$ will have intersection v).

Proposition 6. *For each w, r in $V(\mathsf{TS}_{k-1}(G))$: $\mathbf{Expand}(w), \mathbf{Expand}(r)$ has a common element if and only if wr is in $E(\mathsf{TS}_{k-1}(G))$.*

Corollary 3. *The vertices $w \in V(TS_{k-1}(G))$ such that $\mathbf{Expand}(w) = \emptyset$ are isolated vertices.*

Proposition 7. *Given $k = \omega(G)$. Let $\mathbf{Msets}$ be the multiset of non-empty sets $\mathbf{Expand}(w)$, $(\forall w \in V(TS_{k-1}(G))$ then $Msets = \mathbf{Msets}(TJ_k(G))$.*

For the next proposition, we use $H + cK_1$ to denote the graph obtained from H by adding c isolated vertices.

Proposition 8. *Given $k = \omega(G)$ and $\mathsf{TJ}_k(G)$ as input without knowing the set label of $\mathsf{TJ}_k(G)$, one can construct a graph H in $k^2 \cdot poly(size(\mathsf{TJ}_k(G)))$ time such that $\mathsf{TS}_{k-1}(G) \cong H + cK_1$ for some integer $c \geq 0$.*

Proof. We call graph $T = (V, E)$ for short for our input graph $\mathsf{TJ}_k(G)$. Our algorithm will be split into 2 phases. First phase: We want to check if the graph is k-good, that is, for each $u \in V$, we can partition $N_T(u)$ into k sets $S_1(u), S_2(u), \ldots, S_k(u)$.

Second phase: We initialize multiset $\mathbf{Msets}$ and insert each of $M_i(u)$ if it has only one element or is not contained. Next, we will create a new graph H with vertex set being the multiset $\mathbf{Msets}$ and the edges will be between those two set that have common elements.

Evaluation: The running time of first algorithm is at most $|N_T(u)|^2$ for each vertex $u \in V$ and $i \in [1, k]$. This means, in total, the running time is at most $k \cdot \sum_{u \in V} |N_T(u)|^2$ which is $k \cdot poly(size(T))$. For the second algorithm, the checking if $M_i(u) \in Msets(u \in V, i \in [1, k])$ can be done in $O(size(T))$ since we can check if there is any vertex in $M_i(u)$ that has been iterated and also then building the E_H take at most $k^2 \cdot poly(size(T))$ since $Msets$ has at most $k \cdot poly(size(T))$ elements. Hence, in total, we need a running time of $O(k^2 \cdot poly(size(T)))$.

Justification: Firstly, we prove that Algorithm 1 can be used to check for k-good graph and also split $N_T(u)$ into k disjoint sets that form a clique.

- If Algorithm 1 runs successfully, we can split $N_T(u)(\forall u \in V)$ into k disjoint sets $(S_1(u), S_2(u), \cdots, S_k(u))$ that form a clique and there is no edge between different sets. Hence, T is a k-good graph and the partition of it is valid.
- If Algorithm 1 fails somewhere, we consider the following cases.

Algorithm 1: Algorithm to check if a graph is k-good and partition the neighbour set of each vertex.

Input: A graph $T = \mathsf{TJ}_k(G)$

Output: For each $u \in V(T)$, a collection of k sets $S_1(u), S_2(u), \ldots, S_k(u)$ such that each $S_i(u)$, if non-empty, is a clique of T, and no edge of T joining a vertex of $S_i(u)$ and that of $S_j(u)$, where $i \neq j$ and $1 \leq i, j \leq k$

```
 1  for u ∈ V do
 2  │   Remain ← N_T(u)
 3  │   for i = 1 to k do
 4  │   │   if Remain is empty then
 5  │   │   │   S_i(u) ← ∅
 6  │   │   else
 7  │   │   │   v be an element of Remain
 8  │   │   │   S_i(u) be the sets of elements in Remain that are incident to v and v
 9  │   │   │   Remain ← Remain − S_i(u)
10  │   │   │   if S_i(u) does not form a clique then
11  │   │   │   │   report T is not k-good
12  │   │   │   else if exist an edge from S_i(u) to Remain then
13  │   │   │   │   report T is not k-good
14  │   if Remain is not empty then
15  │   │   report T is not k-good
```

Algorithm 2: Algorithm to obtain a graph H satisfying Proposition 8.

Input: A k-good graph $T = \mathsf{TJ}_k(G)$ for some graph G without any set label, and for each $u \in V(T)$, a collection of k sets $S_1(u), S_2(u), \ldots, S_k(u)$ obtained from Algorithm 1

Output: A graph H satisfying Proposition 8

```
 1  Msets ← ∅
 2  for u ∈ V do
 3  │   for i = 1 to k do
 4  │   │   M_i(u) ← S_i(u) ∪ {u}
 5  │   │   if M_i(u) = {u} then
 6  │   │   │   Msets ← Msets + {u}
 7  │   │   else
 8  │   │   │   if M_i(u) ∉ Msets then
 9  │   │   │   │   Msets ← Msets + M_i(u)
10  Edge set E_H ← ∅
11  for U, V ∈ Msets do
12  │   if |U ∩ V| ≠ ∅ then
13  │   │   E_H ← E_H ∪ {U, V}
14  return graph H = (Msets, E_H)
```

- If it fails because $S_i(u)$ does not form a clique. This means there are u, v, a, b in V such that $uv, ua, ub, va, vb \in E$ but not ab. This means that T is not k-good since if it is then a, b cannot be in the same clique in the partition of $N_T(u)$ but a, v and v, b must be in the same clique which creates a contradiction.
- If it fails because $S_i(u), Remain$ have an edge between them. This means there are u, v, a, b in V such that $uv, ua, ub, va, ab \in E$ but not vb. This means that T is not k-good because of the same explanation as above.
- If it fails because after the loop $Remain$ is not empty. Then the induced subgraph of T with vertex set $N_T(u)$ have at least $k + 1$ connected components which does not apply for k-good graph. Hence, T is not k-good.

Next, the second algorithm follows the idea of Proposition 7 and Proposition 6 to build $\mathsf{TS}_{k-1}(G)$. However, there are several isolated points because of Corollary 3.

In the next part, for some integer k, recall that a graph T satisfying that $T \cong \mathsf{TJ}_k(G)$ is called a TJ_k-*graph*. Additionally, we call T a *maximum* TJ_k-*graph* if it is a TJ_k-graph for some graph G satisfying $k = \omega(G)$.

Proposition 9. *For an integer k, if T is a maximum TJ_k-graph then it is a (maximum) TJ_n-graph for any $n \geq k$.*

Proposition 10. *If for infinitely many integers n, a non-empty graph T is a TJ_n-graph, then there exists k so that T is a maximum TJ_k-graph.*

Thus, because of Proposition 9 and Proposition 10, we have the following corollary.

Corollary 4. *A non-empty graph T is a TJ_n-graph for infinitely many integers n if and only if there exists k so that graph T is a maximum TJ_k-graph.*

Proposition 11. *For $k \leq 3$, if a non-empty graph T is a TJ_k-graph and T does not contain any diamond $(K_4 - e)$ as an induced subgraph then T is a maximum TJ_k-graph.*

Corollary 5. *For an integer $k \leq 3$, if a graph is T is a TJ_n-graph for all $n \geq k$, then T is a maximum TJ_k-graph.*

Proof. Because of Proposition 10, then there exist a number N such that T is a maximum TJ_N-graph. Because of Proposition 3, T does not contain a diamond as its induced subgraph. By Proposition 11, T is a maximum TJ_k-graph.

5   Planar TS_k-Graphs

This section is devoted to proving the following theorem.

Theorem 2. *If a graph G is planar, then so is $\mathsf{TS}_k(G)$, where $1 \leq k \leq \omega(G) \leq 4$.*

We remark that Theorem 2 is trivial when either $k = 1$ or $k = 4$. When $k = 1$, $\mathsf{TS}_k(G) \cong G$ and therefore it is planar. When $k = 4$, no two vertices of $\mathsf{TS}_k(G)$ are adjacent (otherwise, G must have a K_5, which contradicts the assumption that it is planar), and thus it is again planar.

In the next lemma, we consider the case $k = 2$.

Lemma 4. *If a graph G is planar, then so is $\mathsf{TS}_2(G)$.*

To conclude our proof of Theorem 2, we consider the case $k = 3$.

Lemma 5. *If a graph G is planar, so is $\mathsf{TS}_3(G)$.*

Corollary 6. *For a planar graph $G = (V, E)$, let F_3, F_4 be the number of K_3, K_4 in graph G then we have $F_3 \leq |E| - 2$ and $2 \cdot F_4 \leq F_3 - 2$. Moreover, $F_3 \leq 3 \cdot |V| - 8$.*

Proof. For a planar graph $T = (V_T, E_T)$, the bound for the number of edge is : $|E_T| \leq 3|V_T| - 6$. Apply this bound to $TS_2(G)$ and $TS_3(G)$. Finally, apply the bound to G to get $F_3 \leq 3 \cdot |V| - 8$.

Acknowledgment. This work began during the 2023–2024 edition of the Vietnam Polymath REU (VPR) program. We thank the organizers and participants of VPR for creating a fruitful research environment. Duc A. Hoang's research was partially supported by the Vietnam Institute for Advanced Study in Mathematics (VIASM), and the Vietnam National University, Hanoi under the project QG.25.07 "A study on reconfiguration problems from algorithmic and graph-theoretic perspectives". Finally, we would like to thank the anonymous reviewers of the International Conference on Algorithms and Discrete Applied Mathematics for their valuable comments and suggestions that helped improve this paper.

References

1. Bandelt, H.J., van de Vel, M.: Embedding topological median algebras in products of dendrons. Proc. Lond. Math. Soc. **3**(3), 439–453 (1989). https://doi.org/10.1112/plms/s3-58.3.439
2. Bousquet, N., Mouawad, A.E., Nishimura, N., Siebertz, S.: A survey on the parameterized complexity of the independent set and (connected) dominating set reconfiguration problems. arXiv preprint (2022). https://arxiv.org/abs/2204.10526
3. Diestel, R.: Graph Theory, Graduate Texts in Mathematics, vol. 173, 5th edn. Springer, Cham (2017). https://doi.org/10.1007/978-3-662-53622-3
4. Eroh, L., Schultz, M.: Matching graphs. J. Graph Theory **29**(2), 73–86 (1998). https://doi.org/10.1002/(SICI)1097-0118(199810)29:2⟨73::AID-JGT3⟩3.0.CO;2-9
5. Hakimi, S.L., Schmeichel, E.F.: On the number of cycles of length k in a maximal planar graph. J. Graph Theory **3**(1), 69–86 (1979). https://doi.org/10.1002/jgt.3190030108
6. Hanaka, T., et al.: Reconfiguring spanning and induced subgraphs. Theoret. Comput. Sci. **806**, 553–566 (2020). https://doi.org/10.1016/j.tcs.2019.09.018
7. Holton, D.A., Sheehan, J.: 8. The Johnson graphs and even graphs. In: The Petersen Graph. Australian Mathematical Society Lecture Series, vol. 7, p. 300. Cambridge University Press (1993). https://doi.org/10.1017/CBO9780511662058.010

8. Ito, T., Ono, H., Otachi, Y.: Reconfiguration of cliques in a graph. In: Jain, R., Jain, S., Stephan, F. (eds.) TAMC 2015. LNCS, vol. 9076, pp. 212–223. Springer, Cham (2015). https://doi.org/10.1007/978-3-319-17142-5_19

9. Ito, T., Ono, H., Otachi, Y.: Reconfiguration of cliques in a graph. Discret. Appl. Math. **333**, 43–58 (2023). https://doi.org/10.1016/j.dam.2023.01.026

10. Lam, Q.N., Phan, H.A., Hoang, D.A.: A note on reconfiguration graphs of cliques. arXiv preprint (2025). https://arxiv.org/abs/2506.07821

11. Monroy, R.F., Flores-Peñaloza, D., Huemer, C., Hurtado, F., Urrutia, J., Wood, D.R.: Token graphs. Graphs Combin. **28**(3), 365–380 (2012). https://doi.org/10.1007/s00373-011-1055-9

12. Mynhardt, C., Nasserasr, S.: Reconfiguration of colourings and dominating sets in graphs. In: Chung, F., Graham, R., Hoffman, F., Mullin, R.C., Hogben, L., West, D.B. (eds.) 50 years of Combinatorics, Graph Theory, and Computing, 1st edn., pp. 171–191. CRC Press (2019). https://doi.org/10.1201/9780429280092-10

13. Nishimura, N.: Introduction to reconfiguration. Algorithms **11**(4), 52 (2018). https://doi.org/10.3390/a11040052

14. Scharpfenecker, P.: On the structure of solution-graphs for Boolean formulas. In: Kosowski, A., Walukiewicz, I. (eds.) FCT 2015. LNCS, vol. 9210, pp. 118–130. Springer, Cham (2015). https://doi.org/10.1007/978-3-319-22177-9_10

15. Shuldiner, P., Oldford, R.W.: The clique structure of Johnson graphs. arXiv preprint (2022). https://arxiv.org/abs/2208.12710

16. van den Heuvel, J.: The complexity of change. In: Surveys in Combinatorics. London Mathematical Society Lecture Note Series, vol. 409, pp. 127–160. Cambridge University Press (2013). https://doi.org/10.1017/cbo9781139506748.005

Improved Upper Bounds on Color Reversal by Local Inversions

Kumud Singh Porte[(✉)], R B Sandeep[iD], and Kamal Santra[iD]

Department of Computer Science and Engineering, Indian Institute of Technology Dharwad, Dharwad 580011, Karnataka, India
{cs24dp012,sandeep,ra.kamal.santra}@iitdh.ac.in

Abstract. We study the problem of color reversal in bicolored graphs under local inversions. A *bicoloration* of a graph $G = (V, E)$ is a mapping $\beta : V \to \{-1, 1\}$. A *local inversion* at a vertex $v \in V$ consists of reversing the colors of all neighbors of v and replacing the subgraph induced by these neighbors with its complement, while leaving v and the rest of G unchanged. Sabidussi (Discrete Mathematics, 1987) showed that any bicolored graph on n vertices without isolated vertices can be color-reversed (that is, all vertex colors flipped while preserving the underlying graph) in at most $6n + 3$ local inversions, and that any bicolored graph can be transformed into another bicolored graph on the same underlying graph in at most $9n$ local inversions. We improve both bounds: we prove that the first task can be accomplished in at most $4n - 3$ local inversions, and the second in at most $\lfloor \frac{11n-3}{2} \rfloor$ local inversions. Furthermore, we show that for stars and complete graphs, color reversal can be performed with at most $3n$ local inversions.

Keywords: local inversion · local complementation · bicolored graph · color reversal

1   Introduction

Local transformations of graphs form a common language across structural graph theory, algebraic graph theory, and quantum information. Among these, *local complementation*—toggling adjacency within the open neighborhood of a chosen vertex–plays a central role. It appears in Bouchet's theory of isotropic systems and the structure of circle graphs [2,3], underpins the vertexminor relation and rank-width [10], and captures equivalences between quantum graph states under local Clifford operations [8,9]. Closely related, Seidel switching is well known for its applications to combinatorial and algebraic aspects of graphs [13]. Local complementation falls under the broad umbrella of graph modification problems, which have been extensively studied in algorithmic and parameterized complexity [1,5,7].

Supported by SERB/ANRF MATRICS grant MTR/2022/000692: *Algorithmic study on hereditary graph properties*

We work with simple, undirected, labelled graphs $G = (V, E)$ whose vertices carry a binary color (encoded by $\beta : V \to \{-1, +1\}$). A single move, called *local inversion*, at a vertex $v \in V$ applies local complementation at v to the underlying graph and simultaneously flips the colors of all neighbors of v. Iterating moves along a string $w \in V^*$ transforms the bicolored graph $B = (G, \beta)$ into B_w, where V^* denotes the set of all finite sequences of vertices from V. For $S \subseteq V$, we write B^S for the bicolored graph that keeps the underlying graph G and flips the colors of vertices in S only; in particular, B^V performs a *global color reversal*. Our central quantity is the *color reversal number $cr(G)$*: the minimum ℓ such that, for every initial coloring β, there exists a string w with $|w| \leq \ell$ and $B_w = B^V$. Note that if there exists a string w such that $B_w = B^V$, then for every bicolored graph B' on G, we also have $B'_w = (B')^V$. Therefore, it is independent of the initial coloring.

In 1987, Sabidussi [11] introduced the color-reversal problem under local inversions. He proved that a bicolored graph without isolated vertices can be color-reversed using at most $6n + 3$ local inversions. He also studied a more general problem: given two bicolored graphs B and B' with the same underlying graph G, can one transform B into B' using local inversions? He showed that, for any such pair, this can be done in at most $9n$ local inversions. Brijder and Hoogeboom [4] applied these results in their study of the group structure of pivot and loop complementation on graphs and set systems.

Our Contributions. We improve both bounds of Sabidussi [11].

- For every connected n-vertex graph G with $n \geq 2$,

$$cr(G) \leq \begin{cases} 4n - 4, & \text{if } n \text{ is even,} \\ 4n - 3, & \text{if } n \text{ is odd,} \end{cases}$$

  obtained via parity-aware decompositions and repeated applications of constant-length strings (Theorem 1).
- For any two bicolored graphs $B = (G, \beta)$ and $B' = (G, \beta')$ on the same connected n-vertex graph G with $n \geq 2$, there exists a string w with $|w| \leq \lfloor \frac{11n-3}{2} \rfloor$ such that $B_w = B'$ (Theorem 2).
- For star graphs and complete graphs G with at least two vertices, we have $cr(G) \leq 3n$ (Theorem 3, Theorem 4).

Theorem 1 and Theorem 2 crucially use the perfect forest theorem of Scott [12], which states that every graph of even order has a spanning forest in which each tree is an induced subgraph of the graph and every vertex has odd degree within its tree. See Caro et al. [6] for two shorter proofs of the theorem. All constructions that we provide are explicit and run in polynomial time.

The rest of the paper is organized as follows. Section 2 introduces the preliminaries, providing definitions and notations for graphs, bicolored graphs, local complementation, and local inversion. It also states several known results and recalls the Perfect Forest Theorem. Section 3 presents the proof of Theorem 1, which establishes an improved bound on $cr(G)$, and of Theorem 2, which bounds the number of moves required to transform one bicolored graph into another. Section 4 focuses on special graph families, namely star graphs and complete graphs, and shows that stronger, family-specific bounds can be obtained through explicit constructions, leading to Theorem 3 and Theorem 4. Finally, Section 5 discusses a few open problems.

2   Preliminaries

We consider only simple, undirected, labelled graphs. The vertex and edge sets of a graph G are denoted by $V(G)$ and $E(G)$, respectively. Two graphs G and G' on the same vertex set are *equivalent* if and only if $E(G) = E(G')$; equivalently, for all distinct $u, v \in V(G)$, $\{u, v\} \in E(G)$ if and only if $\{u, v\} \in E(G')$. For a subset $S \subseteq V(G)$, we write $G[S]$ for the subgraph of G induced by S. The *open neighborhood* of a vertex v in G is denoted by $N_G(v)$. A path, a complete graph, and a star on t vertices are denoted by P_t, K_t, and S_t, respectively.

A *bicoloration* of a graph $G = (V, E)$ is a mapping $\beta : V \to \{-1, 1\}$. A *bicolored graph* is a pair $B = (G, \beta)$, where G is a graph and β is a bicoloration of G. Let $G = (V, E)$ be a graph and let $a \in V$. The *local complement* of G at a is the graph $G_a = (V, E_a)$ obtained by toggling adjacency among the neighbors of a; formally, for distinct $x, y \in V$,

$$\{x, y\} \in E_a \iff \begin{cases} \{x, y\} \in E \text{ and } (x \notin N_G(a) \text{ or } y \notin N_G(a)), \\ \text{or} \\ \{x, y\} \notin E \text{ and } x, y \in N_G(a). \end{cases}$$

Figure 1 illustrates the construction of G_a from G.

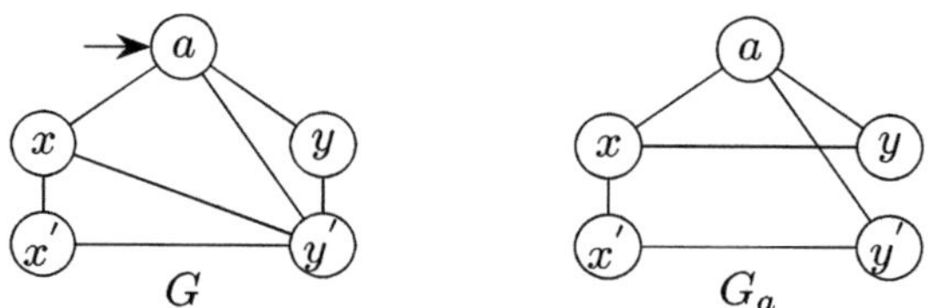

Fig. 1. Local complement of a graph G with respect to vertex a.

For a graph G, a *string* on the alphabet $V(G)$ is a finite sequence of vertices of $V(G)$. The set $V(G)^*$ denotes the set of all finite sequences over $V(G)$. We use ε to denote the *empty string*. Let $w \in V(G)^*$ be any string. We define the local complement of G with respect to w, denoted G_w, inductively as follows:

- Base case: $G_\varepsilon = G$.
- Inductive case: if $w = w'a$ with $a \in V(G)$, then $G_w = (G_{w'})_a$.

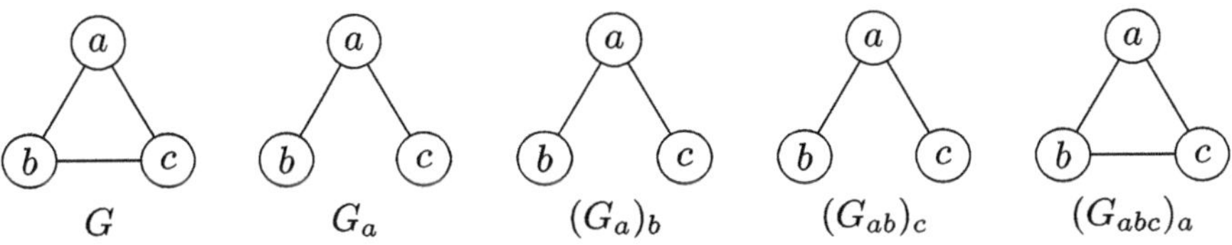

Fig. 2. Local complement of a graph with respect to the string $abca$.

Figure 2 shows an example of local complementation with respect to the string $w = abca$.

Given a bicoloration β, the bicoloration with respect to a vertex a, denoted β_a, is defined by inverting the color of each neighbor of a:

$$\beta_a(x) = \begin{cases} -\beta(x), & \text{if } x \in N_G(a), \\ \beta(x), & \text{otherwise.} \end{cases}$$

Let $B = (G, \beta)$ be a bicolored graph and let $a \in V(G)$. Writing G_a and β_a as above, the *local inversion* of B at a, denoted B_a, is (G_a, β_a). Let $B = (G, \beta)$ be a bicolored graph and let $w \in V(G)^*$ be a string. The local inversion of B with respect to w is defined inductively as follows:

- Base case: $B_\varepsilon = B$.
- Inductive case: if $w = w'a$ with $a \in V(G)$, then $B_w = (B_{w'})_a$.

For example, see Fig. 3.

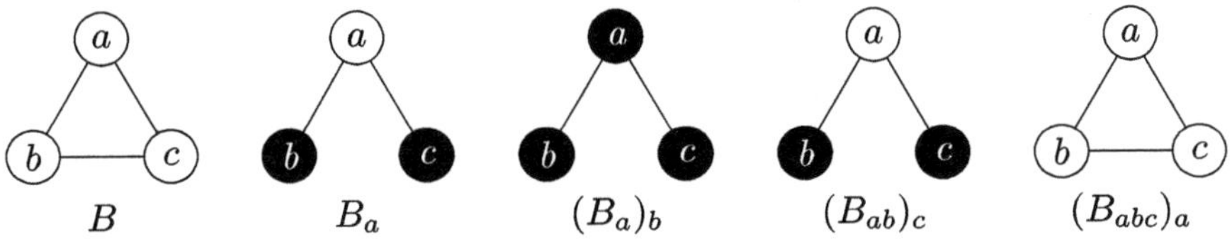

Fig. 3. Local inversion of a graph with respect to the string $abca$.

We now state some known results that will be used in the following sections. Proposition 1 means that it does not make sense to repeatedly apply local inversion on a vertex.

Proposition 1 ([11]). *For any bicolored graph $B = (G, \beta)$, $B_{aa} = B$, where a is any vertex in G.*

Let $B = (G, \beta)$ be a bicolored graph with $\beta : V(G) \to \{-1, 1\}$, and let $A \subseteq V(G)$. Define the operator h_A by

$$(h_A(\beta))(v) = \begin{cases} -\beta(v), & \text{if } v \in A, \\ \beta(v), & \text{if } v \notin A. \end{cases}$$

We write $B^A = (G, h_A(\beta))$ for the bicolored graph obtained from B by flipping the colors of all vertices in A and leaving all others unchanged. We write B^a for $B^{\{a\}}$.

Proposition 2 states that in a bicolored graph, the endpoints of an edge can be color-reversed using 6 local inversions, while the underlying graph remains unchanged.

Proposition 2 ([11]). *Let $B = (G, \beta)$ be a bicolored graph. If $ab \in E(G)$ and $w = ababab$ (of length 6), then $B_w = B^{\{a,b\}}$.*

Proposition 3 states that one vertex of a triangle in a bicolored graph can be color-reversed in 7 moves.

Proposition 3 ([11]). *Let $B = (G, \beta)$ be a bicolored graph. If $a, b, c \in V(G)$ form a triangle and $w = abacbac$ (of length 7), then $B_w = B^a$.*

Proposition 4 handles the base cases when G has 2 or 3 vertices.

Proposition 4 ([11]). *A bicolored graph on K_2 with vertices a, b can be color-reversed using 2 local inversions, for example, with the string ab. A bicolored graph on a triangle abc can be color-reversed using 9 local inversions, for example, with the string abababcac. Similarly, a bicolored graph on a path P_3 with vertices a, b, c, where b is the degree-2 vertex, can be color-reversed using 9 local inversions, for example, with the string ababacacb.*

An *odd tree* is a tree in which every vertex has odd degree. A *perfect forest* of a graph G is a spanning forest of G in which each tree is an induced subgraph of G and is an odd tree.

Proposition 5 (Perfect Forest Theorem [6,12]**).** *Let G be a connected graph on n vertices, where n is even. Then G has a perfect forest, and such a forest can be computed in polynomial time.*

3   General bounds

In this section, we prove Theorem 1 and Theorem 2. We start with some simple lemmas derivable from the results by Sabidussi [11]. Lemma 1 states that in a bicolored graph, the endvertices of an induced P_3 can be color-reversed in 8 local inversions, without changing the underlying graph.

Lemma 1. *Let $B = (G, \beta)$ be a bicolored graph, and let $a, b, c \in V(G)$ with $a, b \in N_G(c)$ and $ab \notin E(G)$. Let $w = cabababc$ (of length 8). Then $B_w = B^{\{a,b\}}$.*

Proof. Let $w' = ababab$ (of length 6). Then $w = cw'c$. In G_c, since $a, b \in N_G(c)$ and $ab \notin E(G)$, the edge ab appears, so a, b, c form a triangle in G_c. In B_c, relative to B, the colors of the neighbors of c are flipped, while the colors of all other vertices remain unchanged. By Proposition 2, $(G_c)_{w'} = G_c$, and in $(B_c)_{w'}$ the colors of a and b are flipped (relative to B_c). Thus $B_{cw'}$ has underlying graph

G_c, with the colors of all vertices in the neighborhood of c flipped except those of a and b, while the colors of the remaining vertices stay unchanged (relative to B). Finally, $(B_{cw'})_c$ has underlying graph $G_{cc} = G$ (by Proposition 1), and the colors of all neighbors of c except a and b revert to their original values. Therefore, only the colors of a and b are flipped relative to B.   □

Lemma 2 states that an endvertex of an induced P_3 in a bicolored graph can be color-reversed using 7 local inversions.

Lemma 2. *Let* $B = (G, \beta)$ *be a bicolored graph, and let* $a, b, c \in V(G)$ *with* $a, b \in N_G(c)$ *and* $ab \notin E(G)$. *Let* $w = cabacba$ *(of length 7). Then* $B_w = B^a$.

Proof. Let $w' = abacbac$ (of length 7). Then

$$cw'c = (c)(abacbac)(c) = (cabacba)(cc) \sim cabacba = w,$$

where the simplification $cc \sim \epsilon$ follows from Proposition 1.

In G_c, since $a, b \in N_G(c)$ and $ab \notin E(G)$, the edge ab appears, so a, b, c form a triangle in G_c. In B_c, relative to B, the colors of the neighbors of c are flipped, while the colors of all other vertices remain unchanged. By Proposition 3, $(G_c)_{w'} = G_c$, and in $(B_c)_{w'}$ the color of a is flipped (relative to B_c).

Thus, in $B_{cw'}$, the underlying graph is G_c, with the colors of all vertices in $N_G(c)$ flipped, except for a, while the colors of all other vertices remain unchanged (relative to B).

Finally, $(B_{cw'})_c$ has underlying graph $G_{cc} = G$ (by Proposition 1), and the colors of all neighbors of c are flipped again. Therefore, only the color of a is flipped relative to B, yielding $B_w = B^a$.   □

Proposition 3 and Lemma 2 together imply Lemma 3, which states that only 7 moves are sufficient to color-reverse any vertex in a nontrivial connected graph.

Lemma 3. *Let* G *be a connected graph on* $n \geq 3$ *vertices. Let* a *be any vertex in* G. *Let* B *be any bicolored graph of* G. *Then there is a string* w *of length 7 such that* $B_w = B^a$.

Lemma 4 states that the edges of an odd tree can be partitioned into copies of P_3 together with one K_2, with certain useful properties.

Lemma 4. *Let* T *be a rooted odd tree on* n *vertices. Then there exists a partition* $\mathcal{P}$ *of the edges of* T *into* $(n-2)/2$ *copies of* P_3 *together with one* K_2, *such that the following conditions are satisfied:*

   i. In each P_3 *of* $\mathcal{P}$, *the two endvertices are children (with respect to the given root) of its center.*
   ii. One of the endvertices of the K_2 *in* $\mathcal{P}$ *is the root.*
   iii. Each vertex of T *is an endvertex of exactly one path in* $\mathcal{P}$.
   iv. The partition can be computed in polynomial time.

Proof. Clearly, n is even. We proceed by induction on n. The base case $n = 2$ is trivial.

Assume $n \geq 4$ and that the claim holds for all smaller even values of n. Let r be the root of T, and let v be a vertex farthest from r that is adjacent to a leaf. Then v has no non-leaf descendant. Since v is not a leaf and every degree in T is odd, we have $\deg(v) \geq 3$. Hence, v has at least two leaf neighbors; let u and w be two of them.

Delete the path u–v–w (i.e., remove the edges uv and vw) and then delete the isolated vertices u and w to obtain a tree T' on $n - 2$ vertices. All degrees in T' remain odd (the degree of v decreases by 2, and the degrees of all other vertices are unchanged). Moreover, T' has an even number of vertices. By the induction hypothesis, $E(T')$ can be partitioned into $(n - 4)/2$ copies of P_3 together with one K_2. Adding the P_3 u–v–w to this partition yields the desired partition $\mathcal{P}$.

Since v is the center of the P_3 $u - v - w$, statement (i) follows inductively. For every P_3 in $\mathcal{P}$ that contains r, by (i), r must be its center. Since r has odd degree, it must also be an endvertex of the K_2 in $\mathcal{P}$, proving (ii). At each step, the process deletes the endvertices of the newly created P_3. Therefore, each vertex is the endvertex of at most one path in $\mathcal{P}$. As no vertex remains at the end, each vertex is the endvertex of exactly one path in $\mathcal{P}$, establishing statement (iii). Finally, the construction of $\mathcal{P}$ can clearly be carried out in polynomial time, proving statement (iv). See Fig. 4 for a demonstration. $\square$

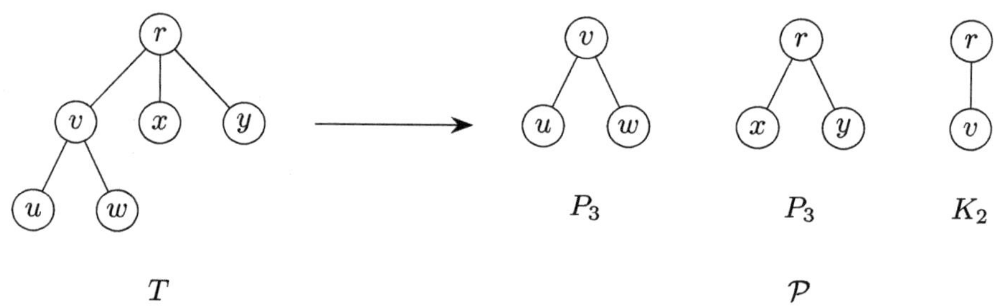

Fig. 4. Partition of T into P_3s and K_2

The next two lemmas state that the vertices of an induced odd tree T in a graph can be color-reversed using $4|V(T)| - 4$ local inversions, by means of a string w that ends (Lemma 5) or starts (Lemma 6) with a designated vertex r of T.

Lemma 5. *Let G be a graph on n vertices, and let T be an induced odd tree of G on at least 4 vertices. Let r be any vertex of T, and let $B = (G, \beta)$ be any bicolored graph of G. Then there exists a string $w \in V(T)^*$ such that the following conditions hold:*

i. $|w| = 4|V(T)| - 4,$

ii. $B_w = B^{V(T)}$,

iii. w ends with r.

Proof. Root T at r. By Lemma 4, there exists a partition $\mathcal{P}$ of $E(T)$ into $(n-2)/2$ copies of P_3 together with one K_2, such that in each P_3 the endvertices are children of its center, and every vertex of T is an endvertex of exactly one path in $\mathcal{P}$. Moreover, r is one of the endvertices of the K_2 in $\mathcal{P}$. Let rv denote this K_2.

Since T has at least 4 vertices, either r or v has degree greater than 1 in T.

First, assume that r has degree greater than 1. Then r also belongs to some P_3 in $\mathcal{P}$. Since r can be an endvertex of at most one path in $\mathcal{P}$, it must serve as the center of every P_3 containing it. Let xry be one such P_3.

Apply Lemma 1 on B successively with each P_3 in $\mathcal{P}$, except xry. Let the resulting bicolored graph be B', and let w' be the corresponding string used. In B', every vertex of T is color-flipped, except x, y, r, and v.

Define

$$w_1 = vrvrvr, \quad w_2 = rxyxyxyr, \quad w'' = (vrvrv)(xyxyxyr).$$

By Proposition 1,
$$w_1 w_2 = (vrvrvr)(rxyxyxyr) \sim (vrvrv)(rr)(xyxyxyr) \sim (vrvrv)(xyxyxyr)$$
$$= w''.$$

By Proposition 2, in B'_{w_1} the colors of r and v (with respect to B') are flipped without altering the underlying graph G. Then, by Lemma 1, in $(B'_{w_1})_{w_2}$ the colors of x and y are flipped, again without changing G. Hence

$$B'_{w_1 w_2} = B'_{w''}$$

is a bicolored graph in which the colors of all vertices of T are flipped with respect to B. Note that $w = w'w''$ ends with w'', which in turn ends with r.

Now, assume that r has degree 1. Then v has degree greater than 1 in T. Let xvy be a P_3 containing v in $\mathcal{P}$. In this case, the same argument works with

$$w_1 = vxyxyxyv, \quad w_2 = vrvrvr, \quad w'' = (vxyxyxy)(rvrvr).$$

Finally, we count the length of w. We applied w'' with $|w''| = 12$, and we used Lemma 1 on $(|V(T)|-4)/2$ paths, contributing $8 \cdot (|V(T)|-4)/2$. Therefore,

$$|w| = 8 \cdot \frac{|V(T)| - 4}{2} + 12 = 4|V(T)| - 4.$$

This completes the proof.   □

Lemma 6, where the only difference from Lemma 5 is that w starts with r, can be proved in a similar fashion.

Lemma 6. *Let G be a graph on n vertices, and let T be an induced odd tree of G on at least 4 vertices. Let r be any vertex of T, and let $B = (G, \beta)$ be any bicolored graph of G. Then there exists a string $w \in V(T)^*$ such that the following conditions hold:*

   i. $|w| = 4|V(T)| - 4,$
  ii. $B_w = B^{V(T)},$
iii. w *starts with* r.

The next two lemmas state that if G' is a nontrivial connected induced subgraph of a graph G with $n' = |V(G')|$ even, then the vertices of G' can be color-reversed (in G) using $4n' - 4$ local inversions, by means of a string that ends (Lemma 7) or starts (Lemma 8) with a designated vertex $v \in V(G')$.

Lemma 7. *Let G be a graph and let G' be any connected induced subgraph of G with $n' \geq 4$ vertices, where n' is even. Let v be any vertex of G', and let $B = (G, \beta)$ be any bicolored graph of G. Then there exists a string $w \in V(G')^*$ such that the following conditions hold:*

   i. $|w| \leq 4n' - 4,$
  ii. $B_w = B^{V(G')},$
iii. w *ends with* v.

Proof. By Proposition 5, G' admits a perfect forest F. Let $T_1, T_2, \ldots, T_t$ (with $t \geq 1$) be the trees in F. Without loss of generality, assume that $v \in T_t$.

For each tree T_i, we construct a string w_i as follows. If T_i is a K_2, say xy, then set $w_i = xyxyxy$. In particular, if T_t is a K_2, say uv, then $w_t = uvuvuv$. If T_i has more than two vertices, then w_i is the string obtained from Lemma 5. In particular, if T_t is a tree with more than two vertices, then w_t is the string from Lemma 5, which ends in v.

Let $w = w_1 w_2 \cdots w_t$. Clearly, w ends with v. By Proposition 2 and Lemma 5, the underlying graph of B_w is still G, and the colors of all vertices in G' are flipped relative to B, while the colors of all other vertices remain unchanged.

It remains to prove the bound on $|w|$. Suppose t' trees in F are K_2s, for some $0 \leq t' \leq t$. Then, by Proposition 2 and Lemma 5,

$$
\begin{aligned}
|w| &\leq 6t' + 4(n' - 2t') - 4(t - t') & & t'K_2 \text{ covers } 2t' \text{ vertices} \\
&= 4n' - 4t + 2t' \\
&\leq 4n' - 4t + 2t & & \text{since } t' \leq t \\
&= 4n' - 2t.
\end{aligned}
$$

If $t \geq 2$, we obtain $|w| \leq 4n' - 4$ as required. If $t = 1$, then G' itself is a tree. Since G' has at least four vertices, T_1 cannot be a K_2, and the bound follows from Lemma 5. $\qquad\square$

Lemma 8, where the only difference from Lemma 7 is that w starts with v, can be proved in a similar fashion.

Lemma 8. *Let G be a graph and let G' be any connected induced subgraph of G with $n' \geq 4$ vertices, where n' is even. Let v be any vertex of G', and let $B = (G, \beta)$ be any bicolored graph of G. Then there exists a string $w \in V(G')^*$ such that the following conditions hold:*

   i. $|w| \leq 4n' - 4$,
  ii. $B_w = B^{V(G')}$,
iii. w *starts with* v.

Lemma 9 states that if G' is a nontrivial connected induced subgraph of G with an odd number n' of vertices, then the vertices of G' can be color-reversed in G using at most $4n' - 3$ local inversions.

Lemma 9. *Let G be a graph and let G' be any connected induced subgraph of G with $n' \geq 5$ vertices, where n' is odd. Let $B = (G, \beta)$ be any bicolored graph of G. Then there exists a string $w \in V(G')^*$ such that the following conditions hold:*

   i. $|w| \leq 4n' - 3$,
  ii. $B_w = B^{V(G')}$.

Proof. Let a be any vertex of G' such that $G'' = G' - a$ is connected.

First, assume that a lies on a triangle, say abc in G'. Let $w_1 = abacbac$. Applying w_1 to B yields a bicolored graph B_1. By Proposition 3, B_1 has the same underlying graph G, and compared to B, only the color of a is flipped.

Since G'' is a connected induced subgraph of G with $n' - 1 \geq 4$ vertices (which is even), we can apply Lemma 8 to obtain a string w_2 starting with c that flips the colors of all vertices of G'' in B. Therefore, $B_{w_1 w_2}$ differs from B exactly in that all vertices of G' have their colors flipped, while the underlying graph remains G.

Let w be the string obtained from $w_1 w_2$ by removing the final c of w_1 and the initial c of w_2. By Proposition 1, $B_{w_1 w_2} = B_w$. Furthermore,

$$\begin{aligned}
|w| &= |w_1| + |w_2| - 2 \\
&\leq 7 + (4(n' - 1) - 4) - 2 \qquad \text{by Lemma 8} \\
&= 4n' - 3.
\end{aligned}$$

Now assume that a does not lie on any triangle in G'. Since G'' is connected and has at least five vertices, there exist vertices $b, c \in V(G'')$ such that acb is an induced P_3. In this case, let $w_1 = cabacba$. By arguments analogous to the previous case, together with Lemma 2 and Lemma 7, we obtain the desired string w satisfying the two conditions. $\qquad\square$

We are now ready to prove Theorem 1.

Theorem 1. *Let G be a connected graph with $n \geq 2$ vertices. If n is even, then $cr(G) \leq 4n - 4$, and if n is odd, then $cr(G) \leq 4n - 3$.*

Proof. If $n = 2$, then by Proposition 4, we have $cr(G) \leq 2 \leq 4 \cdot 2 - 4$. If $n = 3$, then by Proposition 4, we have $cr(G) \leq 9 \leq 4 \cdot 3 - 3$. If $n \geq 4$ and n is even, then the statement follows from Lemma 7. If $n \geq 5$ and n is odd, then the statement follows from Lemma 9. $\qquad\square$

Corollary 1 follows immediately from Theorem 1, noting that each component of a graph with even order may itself have odd order.

Corollary 1. *Let G be a graph with n vertices and $t \geq 2$ components, and suppose G has no isolated vertices. Then $cr(G) \leq 4n - 3t$.*

We now state and prove our second main result.

Theorem 2. *Let G be a connected graph on $n \geq 2$ vertices. Let $B = (G, \beta)$ and $B' = (G, \beta')$ be two bicolored graphs on G. Then there exists a string w of length at most $\lfloor \frac{11n-3}{2} \rfloor$ such that $B_w = B'$.*

Proof. Let $V_0 = \{v \in V(G) : \beta(v) = \beta'(v)\}$ with $n_0 = |V_0|$, and let $V_1 = \{v \in V(G) : \beta(v) = -\beta'(v)\}$ with $n_1 = |V_1| = n - n_0$. If $n_1 = 0$, then $B = B'$ and we are done. If $n_0 = 0$, then $\beta = -\beta'$, and by Theorem 1 there is a string of length at most $4n - 3 \leq \frac{11n-3}{2}$ (since $n \geq 2$).

Assume henceforth $n_0, n_1 \geq 1$. Let $G_0 = G[V_0]$ and $G_1 = G[V_1]$. Let I_0 (resp. I_1) be the set of isolated vertices in G_0 (resp. G_1), and write $i_0 = |I_0|$, $i_1 = |I_1|$.

We consider two strategies and take the cheaper.

(1) Fix V_1 directly. Use Lemma 3 on each vertex of I_1 (cost 7 per isolated vertex), then apply Theorem 1 on $G_1 - I_1$ (cost at most $4(n_1 - i_1) - 3$ if $n_1 - i_1 > 0$, else 0). Thus, the total p cost satisfies

$$p \leq 7i_1 + \max\{\, 4(n_1 - i_1) - 3,\, 0 \,\} \leq 7n_1.$$

(2) Flip V_0, then flip all of G. First use Lemma 3 on I_0 (cost 7 per isolated vertex), then apply Theorem 1 on $G_0 - I_0$ (cost at most $4(n_0 - i_0) - 3$ if $n_0 - i_0 > 0$, else 0), and finally apply Theorem 1 on G (cost at most $4n - 3$). Hence, we can bound the total cost q

$$q \leq 7i_0 + \max\{\, 4(n_0 - i_0) - 3,\, 0 \,\} + (4n - 3) \leq 7n_0 + 4n - 3.$$

For the given partition (n_0, n_1) we can realize B' from B with length at most $\min\{p, q\} \leq \min\{\, 7(n - n_0),\, 7n_0 + 4n - 3 \,\}$. Maximizing this upper bound over $0 \leq n_0 \leq n$ occurs when the two arguments are equal:

$$7(n - n_0) = 7n_0 + 4n - 3 \implies 14n_0 = 3n + 3,$$

which yields

$$\min\{p, q\} \leq 7(n - n_0) = \frac{14(n - n_0)}{2} = \frac{14n - 3n - 3}{2} = \frac{11n - 3}{2}.$$

Therefore, there exists a string w of length at most $\lfloor \frac{11n-3}{2} \rfloor$ such that $B_w = B'$.
$\square$

Corollary 2 follows immediately.

Corollary 2. *Let G be a graph without isolated vertices, and let $B = (G, \beta)$ and $B' = (G, \beta')$ be two bicolored graphs on G. Then there exists a string w of length at most $\lfloor \frac{11n-3t}{2} \rfloor$ such that $B_w = B'$, where t is the number of components of G.*

4 Stars and complete graphs

In this section, we obtain tighter bounds for stars and complete graphs.

Theorem 3. *Let S_n be the star graph on $n \geq 2$ vertices. Then $cr(S_n) \leq 3n$.*

Proof. Let c_0 be the center and $c_1, c_2, \ldots, c_{n-1}$ be the leaves of S_n.
Consider the bicolored graph $B = (S_n, \beta)$. Define

$$w_n = (c_1 c_0 c_1 c_0 c_1)(c_2 c_0 c_2)(c_3 c_0 c_3) \cdots (c_{n-1} c_0 c_{n-1})(c_0).$$

We claim that $B_{w_n} = (S_n, -\beta)$.

For $2 \leq i \leq n$, let $A_i = \{c_0, c_1, \ldots, c_{i-1}\}$. We prove that $B_{w_i} = B^{A_i}$. For $i = 1$, $w_1 = c_1 c_0 c_1 c_0 c_1 c_0$, and this case follows from Proposition 2. Assume that the statement holds for some $i = n - 1$. We now prove it for $i = n$.

Observe that

$$w = (w_{n-1})(c_0 c_{n-1} c_0 c_{n-1} c_0 c_{n-1})(c_{n-1}) \qquad \text{assume}$$
$$\sim (c_1 c_0 c_1 c_0 c_1)(c_2 c_0 c_2)(c_3 c_0 c_3) \cdots (c_{n-2} c_0 c_{n-2})(c_0)(c_0 c_{n-1} c_0 c_{n-1} c_0 c_{n-1})$$
$$(c_{n-1})$$
$$= (c_1 c_0 c_1 c_0 c_1)(c_2 c_0 c_2)(c_3 c_0 c_3) \cdots (c_{n-2} c_0 c_{n-2})(c_0 c_0)(c_{n-1} c_0 c_{n-1})(c_0)(c_{n-1}$$
$$c_{n-1})$$
$$\sim (c_1 c_0 c_1 c_0 c_1)(c_2 c_0 c_2)(c_3 c_0 c_3) \cdots (c_{n-2} c_0 c_{n-2})(c_{n-1} c_0 c_{n-1})(c_0)$$
$$\text{by Proposition 1}$$
$$= w_n.$$

Therefore, it is enough to show that $B_w = (S_n, -\beta)$.

By the induction hypothesis, $B_{w_{n-1}} = B^{A_{n-1}}$; that is, $B_{w_{n-1}}$ has underlying graph S_n, where the colors of all vertices except c_{n-1} are flipped. Let $w' = c_0 c_{n-1} c_0 c_{n-1} c_0 c_{n-1}$. By Proposition 2, $(B_{w_{n-1}})_{w'}$ has the same underlying graph S_n, and the colors of c_0 and c_{n-1} are flipped relative to $B_{w_{n-1}}$. Thus, $B_{w_{n-1} w'}$ has all vertex-colors flipped except c_0, with respect to B. Finally, applying c_{n-1} to $(B_{w_{n-1} w'})$ preserves the underlying graph (since c_{n-1} has degree 1) and flips the color of c_0. Hence, all vertex colors are flipped, completing the proof. $\square$

Theorem 4. *For $n \geq 2$, we have $cr(K_n) \leq 3n$.*

Proof. Let $V(K_n) = \{c_0, c_1, \ldots, c_{n-1}\}$, and let $B = (K_n, \beta)$ be a bicolored graph. Define
$$w = (c_0)(w')(c_0),$$
where
$$w' = (c_1 c_0 c_1 c_0 c_1)(c_2 c_0 c_2)(c_3 c_0 c_3) \cdots (c_{n-1} c_0 c_{n-1})(c_0),$$
which is the string used for the color-reversal of S_n in Theorem 3.

In B_{c_0}, the underlying graph is a star with center c_0, and the colors of all vertices except c_0 are flipped. By Theorem 3, $(B_{c_0})_{w'}$ has the same underlying graph (the star centered at c_0), and the colors of all vertices are flipped relative to B_{c_0}. Therefore, in $B_{c_0 w'}$, only the color of c_0 is flipped relative to B. Finally, applying c_0 once more, $(B_{c_0 w'})_{c_0}$ has underlying graph K_n, with all vertex colors flipped relative to B.

The claim follows since

$$w = (c_0)(w')(c_0)$$
$$= (c_0)(c_1 c_0 c_1 c_0 c_1)(c_2 c_0 c_2)(c_3 c_0 c_3) \cdots (c_{n-1} c_0 c_{n-1})(c_0)(c_0)$$
$$\sim (c_0)(c_1 c_0 c_1 c_0 c_1)(c_2 c_0 c_2)(c_3 c_0 c_3) \cdots (c_{n-1} c_0 c_{n-1}) \qquad \text{by Proposition 1.}$$

This completes the proof. $\qquad\qquad\square$

5 Concluding remarks

We conclude with a few open problems. We have shown that any bicolored graph can be color-reversed in at most $4n - 4$ local inversions when n is even, and in at most $4n - 3$ inversions when n is odd. Although this bound is tight for some small graphs, such as P_3, K_3, and S_4, we do not believe it is tight for larger graphs. With the aid of a computer program, we determined $cr(G)$ for all graphs with at most 5 vertices and found that the value never exceeds $3n$.

Problem 1: Is it true that $cr(G) \leq 3n$?

We proved that Problem 1 has an affirmative answer for stars and complete graphs. But is it tight for them?

Problem 2: Is it true that $cr(G) = 3n$ for all stars and complete graphs of at least 3 vertices?

As mentioned earlier, all our proofs are constructive, and the corresponding strings can be obtained in polynomial time. Nevertheless, the complexity of computing $cr(G)$ remains unknown.

Problem 3: Is the following problem solvable in polynomial time? Given a graph G and an integer k, decide whether $cr(G) \leq k$.

Note that no lower bounds are currently known for $cr(G)$, which is another direction worth exploring. Some good lower bounds may help us to get a good approximation algorithm. In particular:

Problem 4: Does our algorithm (implied by Theorem 1) serve as an approximation algorithm with a good approximation factor for the optimization version of the problem?

The bound we obtained for the number of local inversions required to transform one bicolored graph into another (Theorem 2) is weaker than the color-reversal bound. However, it is unclear whether selective color-reversal inherently requires more local inversions than global color reversal.

Problem 5: Is it true that the minimum number of local inversions required to transform B into B' is at most $cr(G)$, where G is the underlying graph of B and B'?

A trivial brute-force algorithm for the decision version runs in $n^{O(k)}$ time, implying that the problem lies in XP.

Problem 6: Does the problem admit an FPT algorithm parameterized by k?

References

1. Bodlaender, H.L., Heggernes, P., Lokshtanov, D.: Graph modification problems (dagstuhl seminar 14071). Dagstuhl Rep. **4**(2), 38–59 (2014). https://doi.org/10.4230/DAGREP.4.2.38
2. Bouchet, A.: Graphic presentations of isotropic systems. J. Comb. Theory B **45**(1), 58–76 (1988). https://doi.org/10.1016/0095-8956(88)90055-X
3. Bouchet, A.: Circle graph obstructions. J. Comb. Theory B **60**(1), 107–144 (1994). https://doi.org/10.1006/JCTB.1994.1008
4. Brijder, R., Hoogeboom, H.J.: The group structure of pivot and loop complementation on graphs and set systems. Eur. J. Comb. **32**(8), 1353–1367 (2011). https://doi.org/10.1016/j.ejc.2011.03.002
5. Cai, L.: Fixed-parameter tractability of graph modification problems for hereditary properties. Inf. Process. Lett. **58**(4), 171–176 (1996). https://doi.org/10.1016/0020-0190(96)00050-6
6. Caro, Y., Lauri, J., Zarb, C.: Two short proofs of the perfect forest theorem. Theor. Appl. Graphs **4**(1), Article 4 (2017). https://doi.org/10.20429/tag.2017.040104
7. Crespelle, C., Drange, P.G., Fomin, F.V., Golovach, P.A.: A survey of parameterized algorithms and the complexity of edge modification. Comput. Sci. Rev. **48**, 100556 (2023). https://doi.org/10.1016/j.cosrev.2023.100556
8. Dehaene, J., Nest, M., Moor, B.D.: Graphical description of the action of local Clifford transformations on graph states. Phys. Rev. A **69**(2), 022316 (2004). https://doi.org/10.1103/PhysRevA.69.022316
9. Hein, M., Eisert, J., Briegel, H.J.: Multiparty entanglement in graph states. Phys. Rev. A **69**(6), 062311 (2004). https://doi.org/10.1103/PhysRevA.69.062311
10. Oum, S.: Rank-width and vertex-minors. J. Comb. Theory B **95**(1), 79–100 (2005). https://doi.org/10.1016/J.JCTB.2005.03.003
11. Sabidussi, G.: Color-reversal by local complementation. Discret. Math. **64**(1), 81–86 (1987). https://doi.org/10.1016/0012-365X(87)90240-8
12. Scott, A.D.: On induced subgraphs with all degrees odd. Graphs Combinatorics **17**(3), 539–553 (2001). https://doi.org/10.1007/s003730170028
13. Seidel, J.: A survey of two-graphs. In: Corneil, D., Mathon, R. (eds.) Geometry and Combinatorics: Selected Works of J.J. Seidel, pp. 146–176. Academic Press (1991). https://doi.org/10.1016/B978-0-12-189420-7.50018-9

Hardness and Algorithmic Results for Roman {3}-Domination

Sangam Balchandar Reddy[(✉)] [iD]

School of Computer and Information Sciences, University of Hyderabad,
Hyderabad, India
21mcpc14@uohyd.ac.in

Abstract. A Roman {3}-dominating function on a graph $G = (V, E)$ is a function $f : V \rightarrow \{0, 1, 2, 3\}$ such that for each vertex $u \in V$, if $f(u) = 0$ then $\sum_{v \in N(u)} f(v) \geq 3$ and if $f(u) = 1$ then $\sum_{v \in N(u)} f(v) \geq 2$. The weight of a Roman {3}-dominating function f is $\sum_{u \in V} f(u)$. The objective of ROMAN {3}-DOMINATION is to compute a Roman {3}-dominating function of minimum weight. The problem is known to be NP-complete on chordal graphs, star-convex bipartite graphs and comb-convex bipartite graphs. In this paper, we study the complexity of ROMAN {3}-DOMINATION and show that the problem is NP-complete on split graphs. In addition, we prove that the problem is W[2]-hard parameterized by weight. On the positive front, we present a polynomial-time algorithm for block graphs, thereby resolving an open question posed by Chaudhary and Pradhan [Discrete Applied Mathematics, 2024].

Keywords: domination · Roman {3}-domination · split graphs · block graphs · parameterized complexity

1 Introduction

Given a graph $G = (V, E)$, a Roman dominating function $f : V \rightarrow \{0, 1, 2\}$ is a labeling of vertices such that each vertex $u \in V$ with $f(u) = 0$ has a vertex $v \in N(u)$ with $f(v) = 2$. The weight of a Roman dominating function f is the sum of $f(u)$ over all $u \in V(G)$, that is, $\sum_{u \in V(G)} f(u)$. The objective of ROMAN DOMINATION (RD) is to compute a Roman dominating function of minimum weight.

RD is a fascinating concept in graph theory that draws inspiration from the strategic positioning of legions to protect the Roman Empire. The concept of RD was introduced by Cockayne et al. [10] in 2004. Liu et al. [18] later proved that RD is NP-complete on split graphs and bipartite graphs. They also provided a linear-time algorithm for strongly chordal graphs. RD also admits linear-time algorithms for cographs, interval graphs, graphs of bounded clique-width and a polynomial-time algorithm for AT-free graphs [17].

Fernau et al. [14] studied RD in the realm of parameterized complexity and proved that the problem is W[2]-complete parameterized by weight. They also provided an FPT algorithm that runs in time $\mathcal{O}^*(5^t)$, where t is the treewidth of the input graph. In addition, on planar graphs, they showed that RD can be solved in $\mathcal{O}^*(3.3723^k)$ time. Ashok et al. [2] provided an $\mathcal{O}^*(4^d)$ algorithm for RD where d is the distance to cluster. Recently, Mohanapriya et al. [19] proved that RD parameterized by solution size is W[1]-hard, even on split graphs.

Many variants of RD, such as ROMAN {2}-DOMINATION, DOUBLE ROMAN DOMINATION, TOTAL ROMAN DOMINATION, INDEPENDENT ROMAN DOMINATION and WEAK ROMAN DOMINATION were studied from an algorithmic standpoint. Now, we discuss the problems ROMAN {2}-DOMINATION and DOUBLE ROMAN DOMINATION, as well as their computational complexities.

Chellali et al. [8] introduced the concept of ROMAN {2}-DOMINATION in graphs. For a graph $G = (V, E)$, a Roman {2}-dominating function $f : V \rightarrow \{0, 1, 2\}$ is a labeling of vertices such that for each vertex $u \in V$ with $f(u) = 0$, either there exist two vertices $v_1, v_2 \in N(u)$ with $f(v_1) = 1$ and $f(v_2) = 1$ or there exists a vertex $v \in N(u)$ with $f(v) = 2$. The results related to the complexity of the problem can be found in [9, 13, 21].

The study on DOUBLE ROMAN DOMINATION was initiated by Beeler et al. [4]. A double Roman dominating function $f : V \rightarrow \{0, 1, 2, 3\}$ is a labeling of vertices such that for each vertex $u \in V$ with $f(u) = 0$, either there exist two vertices $v_1, v_2 \in N(u)$ with $f(v_1) = 2$ and $f(v_2) = 2$ or there exists a vertex $v \in N(u)$ with $f(v) = 3$. For each vertex $u \in V$ with $f(u) = 1$ there exists a vertex $v \in N(u)$ with $f(v) \geq 2$. More details on the problem complexity can be found in [1, 3, 23, 24].

A Roman {3}-dominating function on a graph $G = (V, E)$ is a function $f : V \rightarrow \{0, 1, 2, 3\}$ such that for each vertex $u \in V$, if $f(u) = 0$ then $\sum_{v \in N(u)} f(v) \geq 3$, else if $f(u) = 1$ then $\sum_{v \in N(u)} f(v) \geq 2$. The weight of a Roman {3}-dominating function f is $\sum_{u \in V(G)} f(u)$. The minimum weight over all Roman {3}-dominating functions for G is denoted by $\gamma_{R3}(G)$. The objective of ROMAN {3}-DOMINATION (R3D) is to compute $\gamma_{R3}(G)$.

Mojdeh et al. [20] initiated the study on R3D by showing that the problem is NP-complete on bipartite graphs. Later, Chakradar et al. [6] proved that the problem is NP-complete on chordal graphs, planar graphs, star-convex bipartite graphs and comb-convex bipartite graphs. They provided linear-time algorithms for bounded treewidth graphs, chain graphs and threshold graphs. They also showed that the problem is APX-complete on graphs with maximum degree 4. Goyal et al. [15] proposed an $\mathcal{O}(\ln \Delta(G))$ approximation algorithm. Recently, Chaudhary et al. [7] proved that the problem is NP-complete on chordal bipartite graphs, planar graphs and star-convex bipartite graphs. They also presented linear-time algorithms for chain graphs and cographs. In addition, they provided several approximation-related results.

In this paper, we extend the study on the complexity aspects of R3D and obtain the following results.

1. As the problem is known to be NP-complete on chordal graphs, we study the problem complexity on split graphs, a subclass of chordal graphs and prove that the problem is NP-complete.

2. We initiate the study on parameterized complexity of R3D and prove that the problem is W[2]-hard parameterized by weight.
3. In addition, we present a polynomial-time algorithm for block graphs.

2   Preliminaries

Let $G = (V, E)$ be a graph with V as the vertex set and E as the edge set, such that $n = |V|$ and $m = |E|$. The set of vertices that belong to $N(u)$ and $N[u]$, respectively, are referred to as the neighbours and closed neighbours of u. The open neighbourhood of a set $T \subseteq V$ is denoted by $N(T)$ and the closed neighbourhood by $N[T]$. $N(T) = \bigcup_{u \in T} N(u)$ and $N[T] = \bigcup_{u \in T} N[u]$. For any vertex $u \in V$, we use the term **label** to denote the value assigned to u by the function f, i.e., $f(u)$. Similarly, for any vertex $u \in V$, we define the term **labelSum** to be the sum of the labels of all the vertices in the closed neighbourhood of u, i.e., $\sum_{v \in N[u]} f(v)$. For a vertex subset $T \subseteq V$, the **weight** of T, denoted by $f(T)$, is the sum of the labels of all the vertices in T, i.e., $\sum_{u \in T} f(u)$. We say that a vertex $u \in V$ with $f(u) \in \{0, 1\}$ is *dominated* if u satisfies the Roman $\{3\}$-dominating function constraint. A vertex $u \in V$ with $f(u) \in \{0, 1\}$ is *dominated* if and only if u has a labelSum of at least three.

A problem is considered to be *fixed-parameter tractable* with respect to a parameter k, if there exists an algorithm that runs in time $\mathcal{O}(f(k) \cdot n^{\mathcal{O}(1)})$, where f is some computable function. The class FPT consists of all *fixed-parameter tractable* problems. In parameterized complexity, there exists a hierarchy of complexity classes: FPT $\subseteq$ W[1] $\subseteq$ W[2] $\subseteq$ ... $\subseteq$ XP. Under the assumption that FPT $\neq$ W[i], a problem is said to be W[i]-hard with respect to a parameter if it is believed not to be *fixed-parameter tractable*. This is because W[i]-hard problems are at least as hard as all problems in W[i], making it unlikely they admit *fixed-parameter tractable* algorithms. For more information on *graph thery* and *parameterized complexity*, we refer the reader to [22] and [11], respectively.

Now, we define the graph classes that are used in this paper. The graph class *split graphs* is defined as follows.

Definition 1. (Split Graphs). *A graph $G = (V, E)$ is a split graph if the vertex set V can be partitioned into two sets V_1 and V_2, such that the subgraph induced by V_1, denoted by $G[V_1]$, is a complete graph and the subgraph induced by V_2, denoted by $G[V_2]$, is an independent set.*

The graph class *block graphs* is defined as follows.

Definition 2. (Block Graphs). *A vertex u is a cut-vertex of G if the number of components in $G - \{u\}$ is higher than that of G. A block of G is a maximal connected subgraph of G that does not have a cut-vertex. A block graph is a simple graph in which every block is a complete graph.*

Let $\{c_1, c_2, \ldots, c_\ell\}$ be the set of cut-vertices and $\{\mathcal{B}_1, \mathcal{B}_2, \ldots, \mathcal{B}_r\}$ be the set of blocks of a block graph G. The *cut-tree* of G, denoted by T_G, is defined as the graph with vertex set $V(T_G) = \{\mathcal{B}_1, \mathcal{B}_2, \ldots, \mathcal{B}_r, c_1, c_2, \ldots, c_\ell\}$ and an edge set $E(T_G) = \{\mathcal{B}_i c_j : c_j \in \mathcal{B}_i, \ i \in [r], \ j \in [\ell]\}$. A block graph G and its corresponding *cut-tree* T_G are given in Fig. 1. An *end block* of a block graph is a block that contains exactly one cut-vertex.

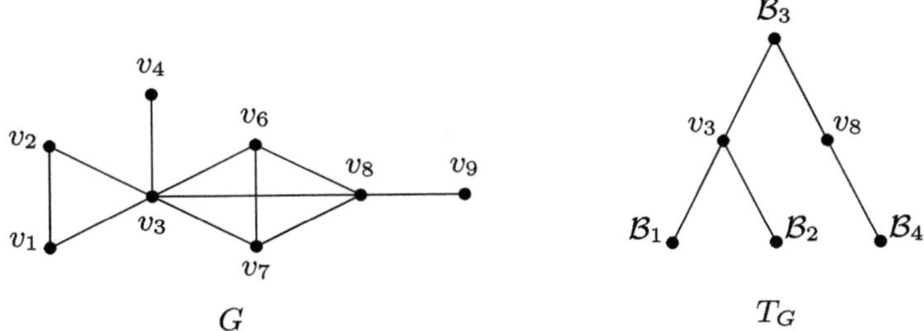

Fig. 1. Block graph G and the corresponding *cut-tree* T_G.

3   NP-Complete on Split Graphs

R3D was proved to be NP-complete on chordal graphs [6,7]. In this section, we strengthen this result by showing that R3D remains NP-complete on split graphs, a subclass of chordal graphs. We provide a polynomial-time reduction from the EXACT 3-COVER (X3C) problem to prove that R3D is NP-complete on split graphs. The X3C problem is defined as follows.

EXACT 3-COVER: *Input:* A set $\overline{X}$ with $|\overline{X}| = 3q$ and a collection $\overline{C}$ of 3-element subsets of $\overline{X}$.
Output: YES, if there exists a collection of sets $\overline{C'}$ of $\overline{C}$ such that every element in $\overline{X}$ occurs in exactly one member of $\overline{C'}$; NO otherwise.

The complexity result related to X3C is given as follows.

Theorem 1. ([16]).   *X3C is NP-complete.*

Consider an instance $\langle \overline{X}, \overline{C} \rangle$ of X3C, where $\overline{X} = \{\overline{x}_1, \overline{x}_2, \ldots, \overline{x}_{3q}\}$ and $\overline{C} = \{\overline{C}_1, \overline{C}_2, \ldots, \overline{C}_t\}$. For the sake of generality, we assume that q is even. This is valid since any odd value of q can be made even by appending three dummy elements $\overline{x}_{3q+1}, \overline{x}_{3q+2}$ and $\overline{x}_{3q+3}$ to the set $\overline{X}$ and by augmenting $\overline{C}$ with an additional 3-element subset $\overline{C}_{t+1} = \{\overline{x}_{3q+1}, \overline{x}_{3q+2}, \overline{x}_{3q+3}\}$. From an instance $\langle \overline{X}, \overline{C} \rangle$ of X3C, we construct an instance $I = (G, k)$ of R3D as follows.

- For each $\overline{x}_i \in \overline{X}$, we introduce a vertex x_i in G. Similarly, for each $\overline{C}_j \in \overline{C}$, we introduce a vertex c_j in G. Let $X = \{x_1, x_2, \ldots, x_{3q}\}$ and $C = \{c_1, c_2, \ldots, c_t\}$.
- We create two additional copies of the vertex set X, denoted by, A and B. A copy of vertex $x_i \in X$ is denoted by a_i in A and b_i in B.

446   S. B. Reddy

- We create two sets Y and Z of size $10q$ each. Let $Y = \{y_1, y_2, ..., y_{10q}\}$ and $Z = \{z_1, z_2, ..., z_{10q}\}$.
- For each $x_i \in X$ and $c_j \in C$, we add an edge between x_i and c_j if and only if $\overline{x}_i \in \overline{C}_j$.
- We make vertex x_i adjacent to both a_i and b_i, for each $i \in [3q]$.
- We construct the adjacency between the vertices of the sets A and Z as follows. We arbitrarily select a subset $A_i \subseteq A$ of six vertices, and a subset $Z_j \subseteq Z$ of twenty vertices. We make each vertex in Z_j adjacent to a unique triplet of vertices in A_i. This process is repeated for each such pair of subsets A_i and Z_j chosen in successive iterations. Importantly, in each iteration, the sets A_i and Z_j should not include any vertices from A and Z, respectively, that were selected in previous iterations. In other words, vertices in the sets $A \cup Z$ must participate exactly once in this adjacency construction. In a similar manner, we establish the adjacency between the sets B and Y.
- Additionally, we make the set $A \cup B \cup C$ a clique.

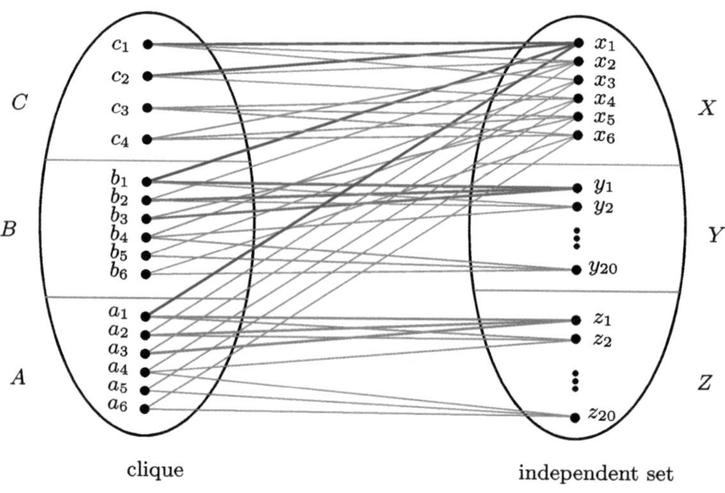

Fig. 2. Reduced instance G for an instance of X3C where $\overline{X} = \{1, 2, 3, 4, 5, 6\}$ and $\overline{C} = \{\overline{C_1} = \{1, 2, 3\}, \overline{C_2} = \{1, 2, 4\}, \overline{C_3} = \{4, 5, 6\}$ and $\overline{C_4} = \{2, 5, 6\}\}$. Edges between the vertices of the clique $A \cup B \cup C$ are not shown in the figure. The edges incident on the vertices x_1, y_1 and z_1 are colored in blue, red and green respectively.

This concludes the construction of G. See Fig. 2 for an illustration.

Observe that G is a split graph, where the subgraph induced by $A \cup B \cup C$ forms a clique and the subgraph induced by $X \cup Y \cup Z$ forms an independent set.

Lemma 1. *Let f be a minimum weighted Roman $\{3\}$-dominating function for G. Then $f(A) \geq 3q$ and $f(B) \geq 3q$.*

Proof. Let $A_i = \{a_1, a_2, a_3, a_4, a_5, a_6\}$ be a six-vertex subset of A. By construction, set A_i is associated with $Z_j = \{z_1, z_2, ..., z_{20}\}$ in Z. Each vertex in Z_j is adjacent to a distinct triple of vertices from A_i. Importantly, the vertices of Z are adjacent only to the vertices of A. Therefore, optimally, no vertex from Z gets a label from $\{1, 2, 3\}$. To ensure the required labelSum for the vertices in Z_j, every triple (a_i, a_j, a_k) from A_i must satisfy $f(a_i) + f(a_j) + f(a_k) \geq 3$. The argument holds for each triplet in every six-vertex subset chosen from A and B. Therefore, we conclude that $f(A) \geq 3q$ and similarly, we have $f(B) \geq 3q$. $\square$

Lemma 2. $\langle \overline{X}, \overline{C} \rangle$ *is a yes-instance of X3C if and only if I has a Roman {3}-dominating function of weight $7q$.*

Proof. [$\Rightarrow$] Let $\overline{C'} \subseteq \overline{C}$ be a solution to a yes-instance $\langle \overline{X}, \overline{C} \rangle$ of X3C. We construct a Roman {3}-dominating function f as follows.

$$
f(v) = \begin{cases}
0, & \text{if } v \in X \cup Y \cup Z \cup \bigcup_{\overline{C}_j \notin \overline{C'}} \{c_j\}, \\
1, & \text{if } v \in A \cup B \cup \bigcup_{\overline{C}_j \in \overline{C'}} \{c_j\}, \\
2, & \text{if } v \in \emptyset, \\
3, & \text{if } v \in \emptyset
\end{cases}
$$

Each vertex in $X \cup Y \cup Z$ is dominated because it has three neighbours in $A \cup B \cup C$ with a label of 1. Since $A \cup B \cup C$ forms a clique, every vertex in it is dominated. The weight of $A \cup B$ is $6q$ and the weight of C is q. Hence, we conclude that f is a Roman {3}-dominating function of weight $7q$.

[$\Leftarrow$] From Lemma 1, we have that $f(A \cup B) \geq 6q$. At this point, each vertex of the set $A \cup B \cup C \cup Y \cup Z$ is dominated. The vertices of X are yet to be dominated. Given that the overall weight of the Roman {3}-dominating function is $7q$, this leaves at most a weight of q for the remaining vertices in $X \cup C$. To ensure that all $3q$ vertices in X are dominated, we must increase the labelSum of each vertex by at least one. As each vertex $x_i \in X$ is adjacent to exactly one vertex from both A and B, we cannot afford to increase the label of any vertex from $A \cup B$ to dominate the vertices of X. This is true because we have $3q$ vertices in X to dominate and only a weight of q. Hence, each vertex in $A \cup B$ gets a label of 1. Thus, the remaining weight available for the set $X \cup C$ is exactly q.

Note that each vertex $c_j \in C$ is adjacent to exactly three vertices in X and labeling c_j with 1, will increase the labelSum of all three of its neighbors by one. Since we have a weight of q left to consume over $X \cup C$, and we need to increase the labelSum of all $3q$ vertices in X by one. The only way is to label exactly q vertices in C with 1. This ensures that all vertices in X get the required labelSum of three. As a result, each vertex in X gets a label of 0, and exactly q vertices in C receive a label of 1. The vertices that get a label of 1 from C will correspond to $\overline{C'} \subseteq \overline{C}$ that forms a solution to the X3C instance. $\square$

Hence, by Theorem 1 and Lemma 2, we arrive at the following theorem.

Theorem 2. *R3D on split graphs is NP-complete.*

4   W[2]-Hard Parameterized by Weight

It was asked in [7] regarding the parameterized complexity of R3D. In this section, we investigate the parameterized complexity of R3D and prove that

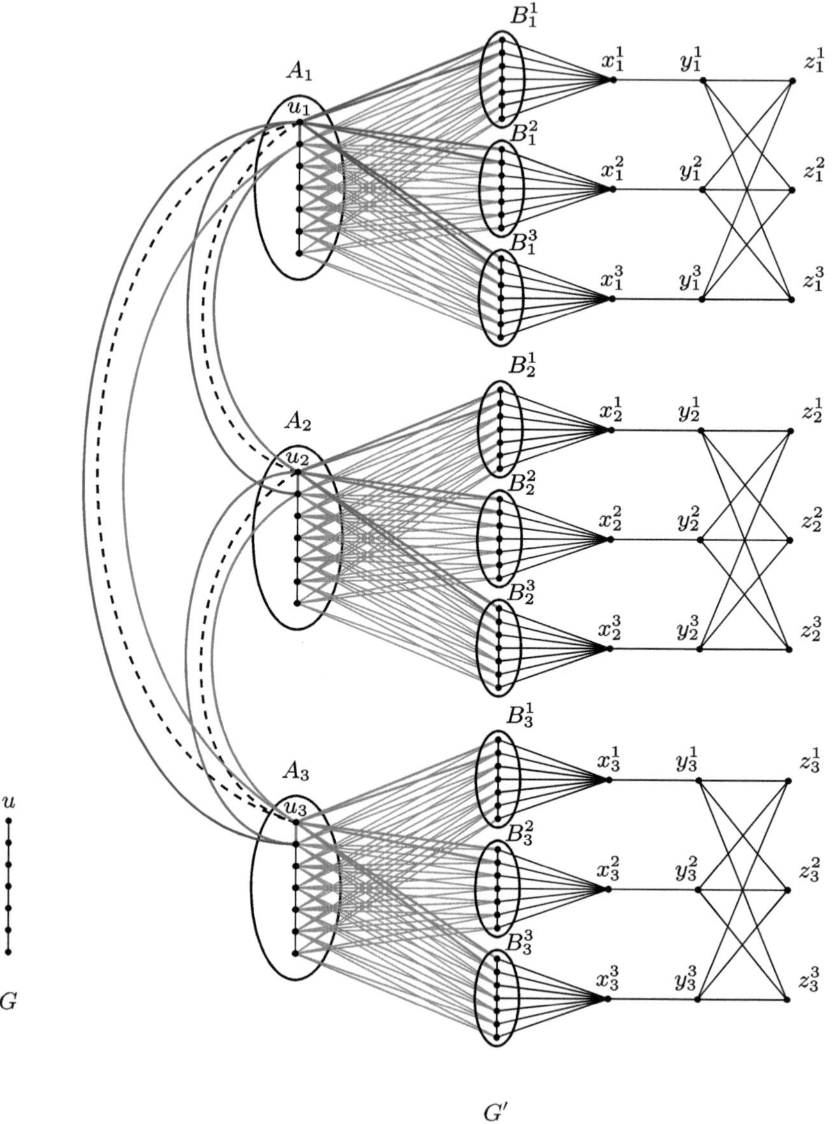

Fig. 3. Graph G' constructed from G and $k = 3$. In the induced subgraph $G[A_1 \cup A_2 \cup A_3]$, only the edges incident on u_1, u_2 and u_3 are shown in the figure. Dashed edges indicate the edges between the vertices of the set $\{u_1, u_2, u_3\}$. The edges incident on u_1, u_2 and u_3 (except dashed) are colored in blue, red and green, respectively.

the problem is W[2]-hard parameterized by weight. We achieve this result by providing a parameterized reduction from DOMINATING SET which is known to be W[2]-hard parameterized by solution size [12, Theorem 13.21]. Given a graph $G = (V, E)$, a set $S \subseteq V$ is a dominating set if every vertex $u \in V \setminus S$ has a neighbour in S. The decision version of DOMINATING SET (DS) asks whether there exists a dominating set of size at most k.

Given an instance $I = (G, k)$ of DS, we construct an instance $I' = (G', 12k)$ of R3D as follows. Without loss of generality, we assume that k in DS instance I is a multiple of 3 (as this can be achieved by adding the isolated vertices to G).

- We begin by creating three disjoint copies of G, denoted by A_1, A_2 and A_3, and include them in G'. Vertex $u \in V(G)$ is denoted by u_1 in A_1, u_2 in A_2 and u_3 in A_3.
- Additionally, for every $i \in [3]$ and $j \in [k]$, we create a copy of the vertex set $V(G)$, denoted B_i^j. Thus, there are $3k$ such sets in total. Vertex $v \in V(G)$ is denoted by v_i^j in B_i^j, for each $i \in [3]$ and $j \in [k]$.
- For each set B_i^j, we introduce three new vertices x_i^j, y_i^j and z_i^j in G'.
- For every $i \in [3]$ and $j \in [k]$, we make x_i^j adjacent to y_i^j and also we make x_i^j adjacent to each vertex of B_i^j.
- We make y_i^{3j+1}, y_i^{3j+2} and y_i^{3j+3} adjacent to each vertex among z_i^{3j+1}, z_i^{3j+2} and z_i^{3j+3}, for every $i \in [3]$ and $j \in \{0\} \cup [\frac{k}{3} - 1]$.
- For every $i \in [3]$ and $j \in [k]$, and for every pair of vertices $u_i \in A_i$ and $v_i^j \in B_i^j$, we add an edge between u_i and v_i^j if and only if $u \in N_G(v)$.
- Similarly, for every $i \in [3]$ and $j \in [k]$, and for every pair of vertices $u_i \in A_i$ and $u_i^j \in B_i^j$, we add an edge between u_i and u_i^j.
- For every $i, j \in [3]$, and for each pair of vertices $u_i \in A_i$ and $v_j \in A_j$, we add an edge between u_i and v_j if and only if $u \in N_G(v)$.
- Finally, for every $i, j \in [3]$ with $i \neq j$, and for each pair of vertices $u_i \in A_i$ and $u_j \in A_j$, we add an edge between u_i and u_j.

This completes the construction of G'. See Fig. 3 for more details. The size of G' is $\mathcal{O}(n^2)$.

Lemma 3. *There exists a dominating set of size at most k if and only if there exists a Roman {3}-dominating function of weight at most $12k$.*

Proof. [$\Rightarrow$] Let S be a dominating set of size at most k. We define a Roman {3}-dominating function $f : V \to \{0, 1, 2, 3\}$ as follows.

$$f(v) = \begin{cases} 0, & \text{if } v \in \bigcup_{u \notin S}\{u_1, u_2, u_3\} \cup \bigcup_{i \in [3]}\bigcup_{j \in [k]}(B_i^j \cup \{z_i^j\}), \\ 1, & \text{if } v \in \bigcup_{u \in S}\{u_1, u_2, u_3\} \cup \bigcup_{i \in [3]}\bigcup_{j \in [k]}\{y_i^j\}, \\ 2, & \text{if } v \in \bigcup_{i \in [3]}\bigcup_{j \in [k]}\{x_i^j\}, \\ 3, & \text{if } v \in \emptyset \end{cases}$$

- Every vertex in $\bigcup_{i \in [3]}\bigcup_{j \in [k]}\{y_i^j, z_i^j\}$ has a labelSum of three from its closed neighbours in $\bigcup_{i \in [3]}\bigcup_{j \in [k]}\{x_i^j, y_i^j\}$.

- Every vertex in $\bigcup_{i\in[3]}\bigcup_{j\in[k]} B_i^j$ has a labelSum of three from its neighbour x_i^j and its neighbours in A_i.
- Every vertex in $A_1 \cup A_2 \cup A_3$ has a labelSum of three from its three closed neighbours with a label of 1 in $A_1 \cup A_2 \cup A_3$ itself.

The weight of $\bigcup_{i\in[3]}\bigcup_{j\in[k]}\{x_i^j, y_i^j\}$ is $9k$ and the weight of $A_1 \cup A_2 \cup A_3$ is at most $3k$. Hence, we conclude that f is a Roman $\{3\}$-dominating function for G' with weight at most $12k$.

[$\Leftarrow$] Let f be a Roman $\{3\}$-dominating function of G' with weight at most $12k$.

- Since each z_i^j is adjacent to exactly three vertices in $\bigcup_{i\in[3]}\bigcup_{j\in[k]} y_i^j$ and to no other vertices, the optimal labeling assigns label 0 to each z_i^j, as their neighbours in $\bigcup_{i\in[3]}\bigcup_{j\in[k]} y_i^j$ will get positive labels.
- For each $i \in [3]$ and $j \in [k]$, y_i^j must be labeled at least 1, in order to dominate its neighbours in $\bigcup_{i\in[3]}\bigcup_{j\in[k]} z_i$.
- Since the vertex x_i^j is adjacent y_i^j and to each vertex in the set B_i^j, we label x_i^j with 2 and fix the label of 1 to y_i^j. After these assignments, the labelSum of y_i^j becomes three and each vertex of the set B_i^j has a labelsum of two.
- At this point, each vertex of the set $\bigcup_{i\in[3]}\bigcup_{j\in[k]}\{y_i^j, z_i^j\}$ has a labelSum of three and each vertex of the set $\bigcup_{i\in[3]}\bigcup_{j\in[k]}\{x_i^j\}$ is labeled 2.
- Every vertex in $\bigcup_{i\in[3]}\bigcup_{j\in[k]} B_i^j$ has a labelSum of two from its neighbour x_i^j. So far, the vertices of A have a labelSum of zero.
- The weight of $\bigcup_{i\in[3]}\bigcup_{j\in[k]}\{x_i^j, y_i^j, z_i^j\}$ is $9k$ and we are left with weight at most $3k$.
- We label k vertices from each of A_1, A_2 and A_3 with a label of 1 such that each vertex in $\bigcup_{i\in[3]}\bigcup_{j\in[k]} B_i^j$ is adjacent to at least one of them.
- Similarly, every vertex in $A_1 \cup A_2 \cup A_3$ must have a labelSum of three from its closed neighborhood within $A_1 \cup A_2 \cup A_3$. This condition is satisfied only if the k vertices that are labeled 1 in each of A_1, A_2, and A_3 are adjacent to all the other vertices within their respective sets.
- The k vertices that are labeled 1 in A_1 (or A_2, or A_3) form a dominating set S of size k in G.

$\square$

Hence, by Lemma 3 and the W[2]-hardness of DS parameterized by solution size, we obtain the following result.

Theorem 3. *R3D is W[2]-hard parameterized by weight.*

5   Polynomial-Time Algorithm for Block Graphs

The complexity of R3D on block graphs was posed as an open question in [7]. In this section, we study the complexity of R3D on block graphs and present an algorithm that runs in $\mathcal{O}(n^3)$ time.

Given a graph $G = (V, E)$ and let $R(G)$ be a set of functions (not necessarily Roman {3}-dominating) on G. For a vertex $u \in V(G)$ and $i \in \{0\} \cup [8]$, we define the sets $D^i(G, u)$ as follows.

1. $D^i(G, u) = \{f \in R(G) : f$ is a Roman {3}-dominating function on G such that $f(u) = i\}$, for each $i \in \{0\} \cup [3]$.
2. $D^4(G, u) = \{f \in R(G) : f_{G-u}$ is a Roman {3}-dominating function on $G - u$, either there exists exactly one vertex $v \in N_G(u)$ with $f(v) = 2$ and for each $w \in N_G[u] \setminus \{v\}$, $f(w) = 0$ or there exist exactly two vertices $v_1, v_2 \in N_G(u)$ with $f(v_1) = f(v_2) = 1$ and for each $w \in N_G[u] \setminus \{v_1, v_2\}$, $f(w) = 0\}$.
3. $D^5(G, u) = \{f \in R(G) : f_{G-u}$ is a Roman {3}-dominating function on $G - u$, there exists exactly one vertex $v \in N_G(u)$ with $f(v) = 1$ and $f(w) = 0$ for each $w \in N_G[u] \setminus \{v\}\}$.
4. $D^6(G, u) = \{f \in R(G) : f_{G-u}$ is a Roman {3}-dominating function on $G - u$, $f(w) = 0$ for each $w \in N_G[u]\}$.
5. $D^7(G, u) = \{f \in R(G) : f_{G-u}$ is a Roman {3}-dominating function on $G - u$, $f(u) = 1$ and there exists exactly one vertex $v \in N(u)$ with $f(v) = 1$ and $f(w) = 0$ for each $w \in N_G(u) \setminus \{v\}\}$.
6. $D^8(G, u) = \{f \in R(G) : f_{G-u}$ is a Roman {3}-dominating function on $G - u$, $f(u) = 1$ and $f(w) = 0$ for each $w \in N_G(u)\}$.

Let $\gamma_{R3}^i(G, u) = \min(w(f) : f \in D^i(G, u))$ for each $i \in \{0\} \cup [8]$.

Lemma 4 ($\star$). [1]   *Consider a graph G and let v be an arbitrary vertex of G. Then, $\gamma_{R3}(G) = \min(\gamma_{R3}^0(G, v), \gamma_{R3}^1(G, v), \gamma_{R3}^2(G, v), \gamma_{R3}^3(G, v))$.*

Lemma 5 ($\star$).   *Let f be a Roman {3}-dominating function on G and let G' be an induced subgraph of G such that there exists a vertex $v \in V(G')$ such that $N_G(v) \setminus N_{G'}(v) \neq \emptyset$ and $N_G(u) = N_{G'}(u)$ for each vertex $u \in V(G') \setminus \{v\}$. If $f(v) = 0$, then the following hold:*

(i) If there exists exactly one vertex $v_a \in N_G(v) \setminus N_{G'}(v)$ such that $f(v_a) = 1$ and $f(u) = 0$ for each vertex $u \in N_G(v) \setminus (N_{G'}(v) \cup \{v_a\})$, then $f_{G'} \in D^0(G', v) \cup D^4(G', v)$.

(ii) If there exists exactly one vertex $v_a \in N_G(v) \setminus N_{G'}(v)$ such that $f(v_a) = 2$ and $f(u) = 0$ for each vertex $u \in N_G(v) \setminus (N_{G'}(v) \cup \{v_a\})$, then $f_{G'} \in D^0(G', v) \cup D^4(G', v) \cup D^5(G', v)$.

(iii) If there exist exactly two vertices $v_a, v_b \in N_G(v) \setminus N_{G'}(v)$ such that $f(v_a) = f(v_b) = 1$ and $f(u) = 0$ for each vertex $u \in N_G(v) \setminus (N_{G'}(v) \cup \{v_a, v_b\})$, then $f_{G'} \in D^0(G', v) \cup D^4(G', v) \cup D^5(G', v)$.

(iv) If there exist three vertices $v_a, v_b, v_c \in N_G(v) \setminus N_{G'}(v)$ such that $f(v_a) \geq 1, f(v_b) \geq 1, f(v_c) \geq 1$ or there exist two vertices $v_a, v_b \in N_G(v) \setminus N_{G'}(v)$ such that $f(v_a) \geq 2$ and $f(v_b) \geq 1$ or there exists a vertex $v_a \in N_G(v) \setminus N_{G'}(v)$ such that $f(v_a) = 3$, then $f_{G'} \in D^0(G', v) \cup D^4(G', v) \cup D^5(G', v) \cup D^6(G', v)$.

---

[1] Due to the space limit, the proofs of the statements marked with the $\star$ are presented in the full version of this paper.

Lemma 6 ($\star$).   *Let f be a Roman $\{3\}$-dominating function on G and let G' be an induced subgraph of G such that there exists a vertex $v \in V(G')$ such that $N_G(v) \setminus N_{G'}(v) \neq \emptyset$ and $N_G(u) = N_{G'}(u)$ for each vertex $u \in V(G') \setminus \{v\}$. If $f(v) = 1$, then the following hold:*

(i) If there exists exactly one vertex $v_a \in N_G(v) \setminus N_{G'}(v)$ such that $f(v_a) = 1$ and $f(u) = 0$ for each vertex $u \in N_G(v) \setminus (N_{G'}(v) \cup \{v_a\})$, then $f_{G'} \in D^1(G', v) \cup D^7(G', v)$.

(ii) If there exists a vertex $v_a \in N_G(v) \setminus N_{G'}(v)$ such that $f(v_a) = 2$ or there exist two vertices $v_a, v_b \in N_G(v) \setminus N_{G'}(v)$ such that $f(v_a) = f(v_b) = 1$, then $f_{G'} \in D^1(G', v) \cup D^7(G', v) \cup D^8(G', v)$.

Observation 1 *Let f be a Roman $\{3\}$-dominating function on G and let G' be an induced subgraph of G such that there exists a vertex $v \in V(G')$ such that $N_G(v) \setminus N_{G'}(v) \neq \emptyset$ and $N_G(u) = N_{G'}(u)$ for each vertex $u \in V(G') \setminus \{v\}$. If $f(v) = 2$ (resp. $f(v) = 3$), then $f_{G'} \in D^2(G', v)$ (resp. $f_{G'} \in D^3(G', v)$).*

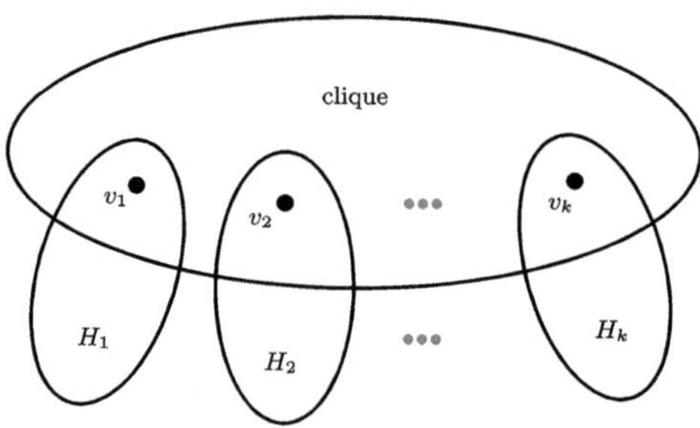

Fig. 4. Graph H.

To design our algorithm, we assume that each graph is rooted at a designated vertex. The Lemma 7 forms the foundation of our polynomial-time algorithm for block graphs. Our approach relies on a specific graph composition technique. Let $H_1, H_2, \ldots, H_k$ (with $k \geq 2$) be the graphs rooted at vertices $v_1, v_2, \ldots, v_k$, respectively. We define a graph H to be the composition of $H_1, H_2, \ldots, H_k$ if it is obtained by taking their disjoint union and then adding edges to make the set $\{v_1, v_2, \ldots, v_k\}$ a clique in H. Note that by starting with trivial graphs and repeatedly applying this composition operation, we can construct any connected block graph (refer to [5]). See Fig. 4 for more details.

Given a graph H rooted at v_1, we compute the values of $\gamma_{R3}^i(H, v_1)$, for $i \in \{0\} \cup [8]$ as follows.

Lemma 7 ($\star$).   *Let $H_1, H_2, \ldots, H_k (k \geq 2)$ be the graphs rooted at $v_1, v_2, \ldots, v_k$, respectively and H be the graph rooted at v_1. We obtain the values of $\gamma_{R3}^i(H, v_1)$, for $i \in \{0\} \cup [8]$ as follows.*

$$(a)\gamma_{R3}^0(H,v_1) = \min \begin{cases} \sum_{1\leq r\leq k} \gamma_{R3}^0(H_r,v_r); \\ \min\{\gamma_{R3}^i(H_1,v_1) : i = 0,4\} + C_1; \\ \min\{\gamma_{R3}^i(H_1,v_1) : i = 0,4,5\} + \min\{B_1,B_2\}; \\ \min\{\gamma_{R3}^i(H_1,v_1) : i = 0,4,5,6\} + \min\{A_1,A_2,A_3\}. \end{cases}$$

$$(b)\gamma_{R3}^1(H,v_1) = \min \begin{cases} \gamma_{R3}^1(H_1,v_1) + \sum_{r\in\{2,\ldots,k\}} \min\{\gamma_{R3}^i(H_r,v_r) : i = 0,4\}; \\ \min\{\gamma_{R3}^i(H_1,v_1) : i = 1,7\} + C_2; \\ \min\{\gamma_{R3}^i(H_1,v_1) : i = 1,7,8\} + \min\{B_3,B_4\}. \end{cases}$$

$$(c)\gamma_{R3}^2(H,v_1) =$$
$$\min \begin{cases} \gamma_{R3}^2(H_1,v_1) + \sum_{r\in\{2,\ldots,k\}} \min\{\gamma_{R3}^i(H_r,v_r) : i = 0,4,5\}; \\ \gamma_{R3}^2(H_1,v_1) + C_3; \end{cases}$$

$$(d)\gamma_{R3}^3(H,v_1) = \gamma_{R3}^3(H_1,v_1) + \sum_{r\in\{2,\ldots,k\}} \min\{\gamma_{R3}^i(H_r,v_r) : i = 0,1,2,3,4,5,6,7,8\}.$$

$$(e)\gamma_{R3}^4(H,v_1) = \min \begin{cases} \gamma_{R3}^4(H_1,v_1) + \sum_{r\in\{2,\ldots,k\}} \gamma_{R3}^0(H_r,v_r); \\ \gamma_{R3}^5(H_1,v_1) + C_1; \\ \gamma_{R3}^6(H_1,v_1) + \min\{B_1,B_2\}. \end{cases}$$

$$(f)\gamma_{R3}^5(H,v_1) = \min \begin{cases} \gamma_{R3}^5(H_1,v_1) + \sum_{r\in\{2,\ldots,k\}} \gamma_{R3}^0(H_r,v_r); \\ \gamma_{R3}^6(H_1,v_1) + C_1. \end{cases}$$

$$(g)\gamma_{R3}^6(H,v_1) = \gamma_{R3}^6(H_1,v_1) + \sum_{r\in\{2,\ldots,k\}} \gamma_{R3}^0(H_r,v_r).$$

$$(h)\gamma_{R3}^7(H,v_1) = \min \begin{cases} \gamma_{R3}^7(H_1,v_1) + \sum_{r\in\{2,\ldots,k\}} \gamma_{R3}^0(H_r,v_r); \\ \gamma_{R3}^8(H_1,v_1) + C_2. \end{cases}$$

$$(i)\gamma_{R3}^8(H,v_1) = \gamma_{R3}^8(H_1,v_1) + \sum_{r\in\{2,\ldots,k\}} \gamma_{R3}^0(H_r,v_r).$$

where $A_1 = \min_{a\in\{2,\ldots,k\}}\{\gamma_{R3}^3(H_a,v_a) + \sum_{r\in\{2,\ldots,k\}\setminus\{a\}} \min\{\gamma_{R3}^i(H_r,v_r) : i = 0,1,2,3,4,5,6,7,8\}\}$,

$A_2 = \min_{a,b\in\{2,\ldots,k\}}\{\gamma_{R3}^2(H_a,v_a) + \min\{\gamma_{R3}^i(H_b,v_b) : i = 1,7,8\} + \sum_{r\in\{2,\ldots,k\}\setminus\{a,b\}} \min\{\gamma_{R3}^i(H_r,v_r) : i = 0,1,2,3,4,5,6,7,8\}\}$,

$A_3 = \min_{a,b,c\in\{2,\ldots,k\}}\{\min\{\gamma_{R3}^i(H_a,v_a) : i = 1,7,8\} + \min\{\gamma_{R3}^i(H_b,v_b) : i = 1,7,8\} + \min\{\gamma_{R3}^i(H_c,v_c) : i = 1,7,8\} + \sum_{r\in\{2,\ldots,k\}\setminus\{a,b,c\}} \min\{\gamma_{R3}^i(H_r,v_r) : i = 0,1,2,3,4,5,6,7,8\}\}$,

$B_1 = \min_{a\in\{2,\ldots,k\}}\{\gamma_{R3}^2(H_a,v_a) + \sum_{r\in\{2,\ldots,k\}\setminus\{a\}} \min\{\gamma_{R3}^i(H_r,v_r) : i = 0,1,4,5,7,8\}\}$,

$B_2 = \min_{a,b\in\{2,\ldots,k\}}\{\min\{\gamma_{R3}^i(H_a,v_a) : i = 1,7\} + \min\{\gamma_{R3}^i(H_b,v_b) : i = 1,7\} + \sum_{r\in\{2,\ldots,k\}\setminus\{a,b\}} \min\{\gamma_{R3}^i(H_r,v_r) : i = 0,1,4,5,7,8\}\}$,

$B_3 = \min_{a\in\{2,\ldots,k\}}\{\gamma_{R3}^2(H_a,v_a) + \sum_{r\in\{2,\ldots,k\}\setminus\{a\}} \min\{\gamma_{R3}^i(H_r,v_r) : i = 0,1,4,5,6,7,8\}\}$,

$B_4 = \min_{a,b\in\{2,\ldots,k\}}\{\min\{\gamma_{R3}^i(H_a,v_a) : i = 1,7,8\} + \{\min\{\gamma_{R3}^i(H_b,v_b) : i = 1,7,8\} + \sum_{r\in\{2,\ldots,k\}\setminus\{a,b\}} \min\{\gamma_{R3}^i(H_r,v_r) : i = 0,1,4,5,6,7,8\}\}$,

$C_1 = \min_{a\in\{2,\ldots,k\}}\{\gamma_{R3}^1(H_a,v_a) + \sum_{r\in\{2,\ldots,k\}\setminus\{a\}} \min\{\gamma_{R3}^i(H_r,v_r) : i = 0,4\}\}$,

$C_2 = \min_{a\in\{2,\ldots,k\}}\{\min\{\gamma_{R3}^i(H_a,v_a) : i = 1,7\} + \sum_{r\in\{2,\ldots,k\}\setminus\{a\}} \min\{\gamma_{R3}^i(H_r,v_r) : i = 0,1,4,5,7,8\}\}$,

---

Algorithm 1: R3DN-BLOCK(G)

---

1 Input: block graph G;

2 Output: $\gamma_{R3}(G)$;

3 $G' = G$;

4 $S = \emptyset$;

5 **foreach** $v \in V(G)$ **do**

6 $\quad$ $\gamma_{R3}^i(G[\{v\}], v) = \infty$ for $i \in \{0,1\}$;

7 $\quad$ $\gamma_{R3}^2(G[\{v\}], v) = 2$;

8 $\quad$ $\gamma_{R3}^3(G[\{v\}], v) = 3$;

9 $\quad$ $\gamma_{R3}^i(G[\{v\}], v) = \infty$ for $i \in \{4,5\}$;

10 $\quad$ $\gamma_{R3}^6(G[\{v\}], v) = 0$;

11 $\quad$ $\gamma_{R3}^7(G[\{v\}], v) = \infty$;

12 $\quad$ $\gamma_{R3}^8(G[\{v\}], v) = 1$;

13 **while** G' *has more than one vertex* **do**

14 $\quad$ Choose an end block $\mathcal{B}$ in G', where $V(\mathcal{B}) = \{v_1, v_2, ..., v_k\}$ and v_1 is the only cut-vertex when $\mathcal{B}$ has a cut-vertex;

15 $\quad$ $S = S \cup V(\mathcal{B})$;

16 $\quad$ Let H_i be the component in $G[S] - E(\mathcal{B})$ such that $V(\mathcal{B}) \cap V(H_i) = \{v_i\}$ for each $i \in [k]$;

17 $\quad$ Let H be the graph constructed from $H_1, H_2, ..., H_k$ by adding edges to make $\{v_1, v_2, ..., v_k\}$ a clique in H;

18 $\quad$ Determine $\gamma_{R3}^i(H, v_1)$ for all $i \in \{0\} \cup [8]$ using Lemma 7;

19 $\quad$ $G' = G' - \{v_2, v_3, ..., v_k\}$

20 return $\min\{\gamma_{R3}^0(G, v_r), \gamma_{R3}^1(G, v_r), \gamma_{R3}^2(G, v_r), \gamma_{R3}^3(G, v_r)\}$, where v_r is the only vertex in G'.

$$C_3 = \min_{a \in \{2,...,k\}} \{\min\{\gamma_{R3}^i(H_a, v_a) : i = 1, 7, 8\} + \sum_{r \in \{2,...,k\} \setminus \{a\}} \min\{\gamma_{R3}^i(H_r, v_r) : i = 0, 1, 2, 3, 4, 5, 6, 7, 8\}\}.$$

We present the algorithm R3DN-BLOCK(G) to compute $\gamma_{R3}(G)$ for a given connected block graph G, using a dynamic programming approach. The algorithm maintains a working graph G', which is initially set to G. The values of $\gamma_0, \gamma_1, \gamma_2, \gamma_3, \gamma_4, \gamma_5, \gamma_6, \gamma_7$ and γ_8 are initialized at each vertex $v \in V(G)$, corresponding to the subgraph $G[\{v\}]$. At each iteration, the algorithm selects an end block of G'. Suppose at a particular iteration, the block $\mathcal{B}$ is selected, where $V(\mathcal{B}) = \{v_1, v_2, \ldots, v_k\}$ and v_1 is the cut-vertex connecting $\mathcal{B}$ to the rest of G'. The algorithm computes the values of $\gamma_0, \gamma_1, \gamma_2, \gamma_3, \gamma_4, \gamma_5, \gamma_6, \gamma_7$ and γ_8 for this block using Lemma 7, and stores them at the cut-vertex v_1. The block $\mathcal{B}$ is then removed from G'. This process is repeated until only one block remains in G', say $\mathcal{B}_q$. At this point, the algorithm selects an arbitrary vertex $v_r \in V(\mathcal{B}_q)$ to store the final computed values. Finally, using Lemma 4, we determine $\gamma_{R3}(G)$ from the values stored at v_r.

Lemma 8 ($\star$). *Algorithm* R3DN-BLOCK(G) *computes* $\gamma_{R3}(G)$ *of a block graph in* $\mathcal{O}(n^3)$ *time.*

Hence, from Lemma 8, we obtain the following theorem.

Theorem 4. *R3D on block graphs is polynomial-time solvable.*

References

1. Ahangar, H.A., Chellali, M., Sheikholeslami, S.M.: On the double roman domination in graphs. Discret. Appl. Math. **232**, 1–7 (2017)
2. Ashok, P., Das, G.K., Pandey, A., Paul, K., Paul, S.: (independent) roman domination parameterized by distance to cluster. In: International Conference on Combinatorial Optimization and Applications, pp. 157–169. Springer (2024)
3. Banerjee, S., Henning, M.A., Pradhan, D.: Algorithmic results on double roman domination in graphs. J. Comb. Optim. **39**(1), 90–114 (2020)
4. Beeler, R.A., Haynes, T.W., Hedetniemi, S.T.: Double roman domination. Discret. Appl. Math. **211**, 23–29 (2016)
5. Chain-Chin, Y., Lee, R.C.: The weighted perfect domination problem and its variants. Discret. Appl. Math. **66**(2), 147–160 (1996)
6. Chakradhar, P., Reddy, P.V.S.: Algorithmic aspects of roman {3}-domination in graphs. RAIRO-Operations Res. **56**(4), 2277–2291 (2022)
7. Chaudhary, J., Pradhan, D.: Roman {3}-domination in graphs: complexity and algorithms. Discret. Appl. Math. **354**, 301–325 (2024)
8. Chellali, M., Haynes, T.W., Hedetniemi, S.T., McRae, A.A.: Roman 2-domination. Discret. Appl. Math. **204**, 22–28 (2016)
9. Chen, H., Lu, C.: Roman {2}-domination problem in graphs. Discussiones Math. Graph Theory **42**(2), 641–660 (2022)
10. Cockayne, E.J., Dreyer, P.A., Hedetniemi, S.M., Hedetniemi, S.T.: Roman domination in graphs. Discret. Math. **278**(1), 11–22 (2004)
11. Parameterized Algorithms. Springer, Cham (2015). https://doi.org/10.1007/978-3-319-21275-3_15
12. Downey, R.G., Fellows, M.R.: Parameterized Complexity. Springer (1999)
13. Fernández, L., Leoni, V.: New complexity results on roman {2}-domination. RAIRO-Operations Res. **57**(4), 1905–1912 (2023)
14. Fernau, H.: ROMAN DOMINATION: a parameterized perspective. In: Wiedermann, J., Tel, G., Pokorný, J., Bieliková, M., Štuller, J. (eds.) SOFSEM 2006. LNCS, vol. 3831, pp. 262–271. Springer, Heidelberg (2006). https://doi.org/10.1007/11611257_24
15. Goyal, P., Panda, B.: Hardness and approximation results of roman {3}-domination in graphs. In: Computing and Combinatorics: 27th International Conference, COCOON 2021, Tainan, Taiwan, October 24–26, 2021, Proceedings 27, pp. 101–111. Springer (2021)
16. Johnson, D.S., Garey, M.R.: Computers and intractability: a guide to the theory of NP-completeness. WH Freeman (1979)
17. Liedloff, M., Kloks, T., Liu, J., Peng, S.L.: Efficient algorithms for roman domination on some classes of graphs. Discret. Appl. Math. **156**(18), 3400–3415 (2008)
18. Liu, C.H., Chang, G.J.: Roman domination on strongly chordal graphs. J. Comb. Optim. **26**(3), 608–619 (2013)
19. Mohanapriya, A., Renjith, P., Sadagopan, N.: Roman k-domination: hardness, approximation and parameterized results. In: International Conference and Workshops on Algorithms and Computation, pp. 343–355. Springer (2023)
20. Mojdeh, D.A., Volkmann, L.: Roman {3}-domination (double italian domination). Discret. Appl. Math. **283**, 555–564 (2020)

21. Poureidi, A., Rad, N.J.: On the algorithmic complexity of roman {2}-domination (italian domination). Iranian J. Sci. Technol. Trans. A Sci. **44**(3), 791–799 (2020)
22. West, D.B.: Introduction to graph theory, 2nd edn. Pearson, Chennai (2015)
23. Yue, J., Wei, M., Li, M., Liu, G.: On the double roman domination of graphs. Appl. Math. Comput. **338**, 669–675 (2018)
24. Zhang, X., Li, Z., Jiang, H., Shao, Z.: Double roman domination in trees. Inf. Process. Lett. **134**, 31–34 (2018)

Hardness and Approximation Results on the Total Dominating Set Problem

Sasmita Rout[1,2]([✉]) and Gautam Kumar Das[1]

[1] Indian Institute of Technology Guwahati, Guwahati 781039, Assam, India
sasmitarout20.84@gmail.com, gkd@iitg.ac.in
[2] SRM University-AP, Amaravati 522240, Andhra Pradesh, India

Abstract. Let $G = (V, E)$ be a simple undirected graph with no isolated vertex. A set $D_t \subseteq V$ is a total dominating set of G if (i) D_t is a dominating set, and (ii) the set D_t induces a subgraph with no isolated vertex. The total dominating set of the minimum cardinality is called the minimum total dominating set, and the size of the minimum total dominating set is called the total domination number $(\gamma_t(G))$. Given a graph G, the total dominating set (TDS) problem is to find a total dominating set of minimum cardinality. In this paper, we enhance the hardness of the TDS problem by proving that it is NP-complete on grid-aligned UDGs, a subclass of unit disk graphs (UDGs). Furthermore, we present a 6.29 factor approximation algorithm for the TDS problem in UDGs.

Keywords: Total Dominating Set · Unit Disk Graphs · Grid Graphs · NP-complete · Approximation Algorithm

1 Introduction

In a simple undirected graph G, $V(G)$ and $E(G)$ denote the sets of vertices and edges, respectively. Given a vertex v of G, $N_G(v)$ and $N_G[v]$ represent the open and closed neighborhood of v, and are defined as: $N_G(v) = \{u \in V : uv \in E(G)\}$ and $G[v] = N_G(v) \cup \{v\}$, respectively. For a subset $S \subseteq V(G)$, $G[S]$ denotes the induced subgraph[1] in G. Given a graph G, a subset $S \subseteq V(G)$ is said to be a dominating set (DS) of G if each vertex $v \in V(G)$ is either present in S or adjacent to a vertex in S. The dominating set with minimum size is the minimum dominating set, and the size of the minimum dominating set is the domination number $(\gamma(G))$.

The dominating set problem has numerous applications and, consequently, many variants. Several of these can be found in studies such as [2, 6, 12, 13, 20, 21], and others. This paper focuses on a well-studied variant of the dominating set problem known as the total dominating set (TDS) which has several important applications in networking, particularly in areas such as network monitoring,

---

[1] For each $u, v \in S$, $uv \in E(G[S])$ if and only if $uv \in E(G)$.

N. Misra and A. Pandey (Eds.): CALDAM 2026, LNCS 16445, pp. 457–470, 2026.
https://doi.org/10.1007/978-3-032-17156-6_34

fault tolerance, energy efficiency, and resource placement.

A subset $D_t \subseteq V(G)$ is a TDS if (i) D_t is a dominating set (in this paper, we refer to it as *domination* property), and (ii) $G[D_t]$ induces a graph with no isolated vertex (in this paper, we refer to it as *total* property). The total dominating set with minimum cardinality is the minimum total dominating set, and the cardinality of the set is the total domination number $(\gamma_t(G))$. Given a graph G, the objective of the minimum total dominating set problem is to find the total domination number of G. The concept of *total dominating set* was introduced by Cockayne et al. [5]. They showed that for any connected graph G with at least 3 vertices, $\gamma_t \leq \frac{2}{3}n$, where n is the order of the graph. In [1], Atapour and Soltankhah showed that for any isolate free graph G, $\gamma_t \leq n - \Delta + 1$, where n is the number of vertices and Δ is the maximum degree. They also characterized the bipartite graphs and trees that achieve this upper bound. Thomassé and Yeo [22] proved that every graph with minimum degree at least 3 (respectively, 4) has total domination number at most $\frac{n}{2}$ (respectively, $\frac{3n}{7}$). In [9], DeLaViña et al. Showed that in a connected graph with $n \geq 1$, $\gamma_t \geq r$, where r is the radius of the graph[2]. They also proved that the total domination number of any connected graph equals the total domination number of a spanning tree of the same graph. Another interesting aspect of trees with respect to total domination is that it is possible to characterize some vertices that are in every total dominating set or not in any total dominating set [4]. Furthermore, in [11], Haynes and Henning established three equivalent conditions to have a unique minimum total dominating set in a tree. The authors also gave a constructive characterization of such trees in the same paper. In [3], Chellali and Haynes characterized the trees for which $\gamma_t = \frac{n+2-l_f}{2}$ and proved that for a non-trivial tree, $\gamma_t \geq \frac{n+2-l_f}{2}$, where n and l_f are the order and number of leaves of the tree, respectively. In [10], the authors showed that planar graphs with diameter 3 and radius 2 have total domination number at most 5. As far as complexity is concerned, the decision version of total dominating set problem is NP-complete, when restricted to bipartite graphs [19]. However, there is a linear-time algorithm for computing the total dominating set of a tree [16]. In [16], Lasker et al. Also showed that the problem remains NP-complete in undirected path graphs. In 1993 [15], Keil proved that the total dominating set problem is NP-complete in circle graphs. In the same article, they observed that the total domination number can be obtained in polynomial time for cycles and paths. They also established a set of relations between (i) γ_t and the maximum degree, and (ii) γ_t and the cut vertices of the graph. Peter Damaschke et al. showed that the TDS problem is polynomially solvable in chordal bipartite graphs [8].

In studies on the applications of domination and its variants, unit disk graphs (UDGs) are widely used to model wireless and mobile networks. In this paper, we focus on the total domination problem in UDGs and grid-aligned UDGs, which are UDGs embedded on grids. In 2021, Jena and Das [14] proved that

---

[2] Given a graph G, radius is the minimum eccentricity taken over all the vertices of G.

the total dominating set (TDS) problem is NP-complete in UDGs and presented an 8-factor approximation algorithm. We strengthen this result by proving that the problem remains NP-complete even for grid-aligned UDGs, and we further propose a 6.29-factor approximation algorithm for UDGs.

1.1 Organisation

The remaining part of this paper is organized as follows. In Sect. 2, we introduce the required preliminaries and notations. In Sect. 3, we prove that the TDS problem is NP-complete in grid-aligned UDGs. Next, in Sect. 4, we propose a 6.29-factor approximation algorithm for the total dominating set problem in unit disk graphs. At the end, we conclude the paper in Sect. 5.

2 Preliminaries

In this section, we define some notation and definitions relevant to the paper. For the completeness of the paper, we revisit some of the already known facts and properties of the unit disk graphs (UDGs). A graph $G = (V, E)$ is said to be a geometric unit disk graph (UDG) corresponding to the set of points P if there exists a one-to-one correspondence between each $v_i \in V(G)$ and $p_i \in P$, and an edge $v_i v_j \in E(G)$ if and only if the Euclidean distance $\delta(p_i, p_j) \leq 1$. Here, $\delta(.,.)$ signifies the Euclidean distance between two points in $\mathbb{R}^2$. Let $U(p)$ denote a disk of radius 1 centered at a point $p \in V$ and $U(P)$ denote the unit disks centered at the points in a set $P \subseteq V$, that is, $U(P) = \{U(p) : p \in P\}$. The set of disks $U(P)$ is said to be independent if for any pair $p, q \in P$, $p \notin U(q)$, that is, $\delta(p, q) > 1$. We further investigate the effect of an additional constraint on grid graphs, where the graph is viewed as a UDG in which all disks have radius 1 and their centers are positioned exclusively at integer coordinates. We refer to this class of graphs as grid-aligned UDGs. In this section, we also revisit some of the previously established lemmas that are relevant to this work.

Lemma 1. *[18] Let $\mathcal{P}$ be a unit disk centered at point p and let S be a set of independent unit disks such that each disk in S contains the point p. Then, $|S| \leq 5$.*

Lemma 2. *[14] Consider two points $p, q \in \mathbb{R}^2$ such that $\delta(p, q) \leq 1$. Let S be the set of independent unit disks such that each disk in S contains the points p and/or q. Then, $|S| \leq 8$.*

Lemma 3. *[23] Let $G = (V, E)$ be a planar graph of degree at most 3. The graph G can be embedded in a grid of area $O(|V|^2)$ such that each $v \in V$ is positioned at a grid point with coordinates $(10i, 10j)$, where i and j are integers, and each edge $e \in E$ is a finite sequence of consecutive segments, each of length 10 units, aligned along the grid lines.*

Observation 1. *Let G be a graph with no isolated vertex. Then, $\gamma(G) \leq \gamma_t(G)$.*

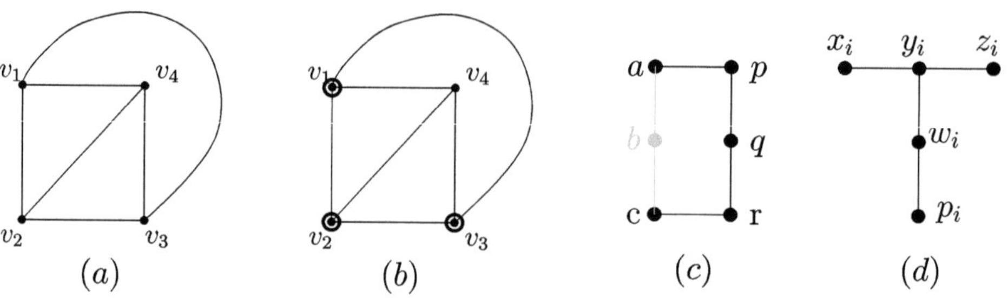

Fig. 1. (a) $G = (V, E)$, (b) $S_{vc} = \{v_1, v_2, v_3\}$, VC of G, (c) path, and (d) gadget.

Theorem 1. *[7] The minimum set cover (MSC) problem[3] can be approximated with an approximation factor $H(max\{|\mathbb{S}_i| : \mathbb{S}_i \in \mathbb{S}\})$ using GreedySetCover$(\mathbb{U}, \mathbb{S})$ in time $O(n^2 m)$, where $H(m)$ is the m-th harmonic number.*

3 Hardness of TDS Problem

In this section, we aim to establish the TDS problem's hardness result by demonstrating that its decision version is NP-complete in grid-aligned UDGs. To achieve this, we employ a reduction from the decision version of the *vertex cover* (VC) problem in planar graphs with degree at most 3 to the decision version of the TDS problem in grid-aligned unit disk graphs. Since the reduction is based on geometry, we often refer to a node or vertex as a point. Formally, the decision versions of the above stated problems are as follows:

Decision version of the VC problem in planar graphs of degree at most 3 (D-VC-$PGD3$): Given a planar graph G of degree at most 3 and a positive integer K, does G have a VC of size at most K?

Decision version of the TDS problem in grid-aligned unit disk graphs (D-TDS-$GAUDGs$): Given a grid aligned UDG G and a positive integer K, does G have a TDS of size at most K?

Lichtenstein and David [17] established the NP-completeness of **D-VC-PGD3** by reducing the planar 3SAT problem to the planar vertex cover problem. We prove the hardness result of the TDS problem in grid-aligned UDGs by making a polynomial time reduction from an arbitrary instance of **D-VC-PGD3** to an instance of **D-TDS-GAUDGs**. To demonstrate this assertion, we use Lemma 3, given by Valiant.

---

[3] The minimum set cover (MSC) problem is an optimization problem that finds the smallest number of subsets from a given collection that together cover all elements of a universe.

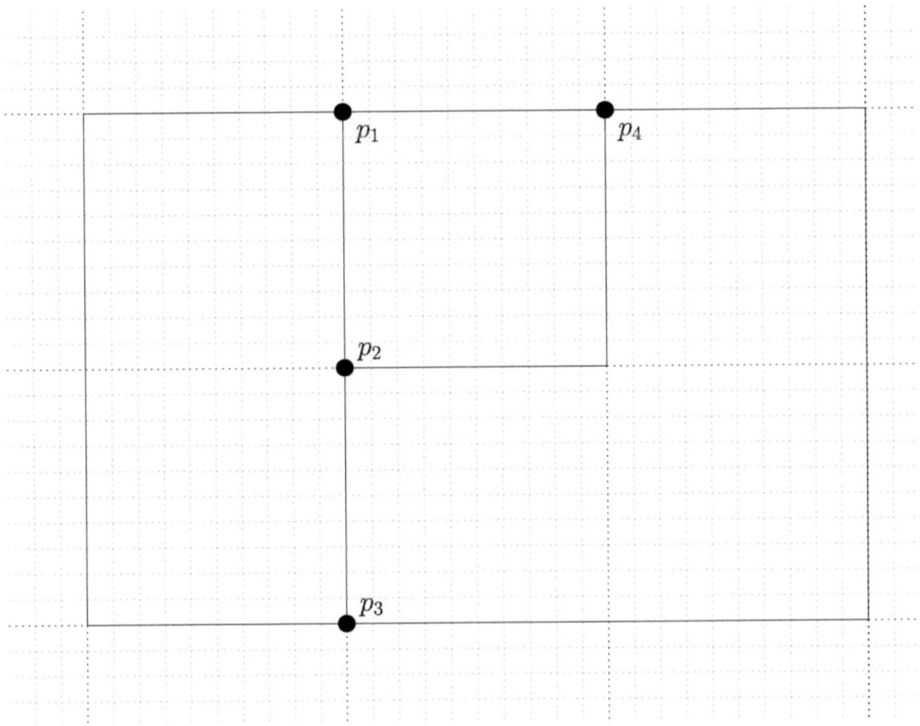

Fig. 2. Embedding of G on a grid of cell size 10×10.

Lemma 4. *If $G = (V, E)$ is an instance of **D-VC-PGD3** without any isolated vertex, then an instance $G' = (V', E')$ of **D-TDS-GAUDGs** can be constructed from G in polynomial time.*

Proof. Let $V = \{v_1, v_2, \ldots, v_n\}$ and $E = \{e_1, e_2, \ldots, e_m\}$ be the vertex set and edge set of the instance G. We construct a graph $G' = (V', E')$ from G by the four steps as given below:

Step 1 (Embedding): The graph G (see Fig. 1(a)) is initially embedded onto a grid with cells of size 10×10 (see Fig. 2), utilizing Valiant's algorithm (Lemma 3) as described in [23]. Each edge $e \in E$ is represented as a consecutive sequence of line segments on the grid, where each segment has a length of 10 units. Let ℓ_{ij} denote the total number of line segments in $p_i p_j$. For every vertex $v_i \in V$, a point is positioned at a grid coordinate of the form $(10i, 10j)$, where i and j are integers. Let p_i represent the point located in the grid corresponding to the vertex $v_i \in V$, where $1 \leq i \leq n$. We refer to these points as embedding nodes. Let N be the set of these nodes and $|N| = |V| = n$.

Step 2 (Inclusion of auxiliary nodes): In this step, we augment each line segment $p_i p_j$ with additional auxiliary nodes. Given that each $p_i p_j$ consists of ℓ_{ij}

segments, there will be $10\ell_{ij} - 1$ intermediary grid points along $p_i p_j$ (when the cell size is 1×1), excluding p_i and p_j. Consequently, a node is added at each grid point, as illustrated in Fig. 3. A modification is made if the count of nodes added along each $p_i p_j$ does not conform to the pattern $4k_{ij} + 1$, where k_{ij} is an integer. Specifically, three consecutive grid points, denoted as a, b, and c (excluding those adjacent to either p_i or p_j), are selected. Subsequently, point b and its incident edges are removed, and three new nodes, p, q, and r, are introduced to $p_i p_j$ such that the path $apqrc$ is formed (refer to Fig. 1(c), Fig. 3 and Fig. 4). This adjustment ensures that the count of grid points on $p_i p_j$ becomes $4k_{ij} + 1$, where k_{ij} is an integer. Let A be the set representing the auxiliary nodes added in this step. We have that $|A| = \sum_{v_i v_j \in E}(4k_{ij}+1) = 4\sum_{v_i v_j \in E} k_{ij} + |E| = 4\sum_{v_i v_j \in E} k_{ij} + m$.

Step 3 **(Inclusion of gadgets):** In this stage, considering that each node within the planar graph has a degree at most 3 and is embedded within a grid, there exists at least one feasible position at every embedded node to accommodate an additional edge. Consequently, we introduce a special construction called gadget (refer to Fig. 1(d)), at each embedded node p_i. This gadget at p_i comprises four distinct nodes: x_i, y_i, z_i, and w_i. Let S represent the set of nodes added in this process. Given that each gadget comprises four nodes (excluding node point), it follows that $|S| = 4|N| = 4n$.

Step 4 **(Construction of the grid-aligned UDGs):** Let $G' = (V', E')$ be the grid-aligned UDGs constructed after applying the above 3 steps on graph G, where $V' = N \cup A \cup S$ and $E' = \{uv : u, v \in V'$ and $\delta(u,v) = 1\}$, where $\delta(u,v)$ is the Euclidean distance between u and v.

From Lemma 3, we conclude that the number of segments $\ell = \sum_{v_i v_j \in E} \ell_{ij} = O(n^2)$. Therefore, the upper bound on the number of vertices and the number of edges in G' is $O(n^2)$. Hence, G' can be constructed from G in polynomial time (Fig. 5).

Theorem 2. ***D-TDS-GAUDGs*** *belongs to the class NP-complete.*

Proof. Let $G = (V, E)$ be a grid-aligned UDGs. Given a subset $S \subseteq V$ and a positive integer K, we can verify whether S is a TDS of G of size at most K or not in polynomial time. Therefore, $D\text{-}TDS\text{-}GAUDGs \in NP$.

To establish the NP-hardness of $D\text{-}TDS\text{-}GAUDGs$, we employ a polynomial time reduction from $D\text{-}VC\text{-}PGD3$ to $D\text{-}TDS\text{-}GAUDGs$. Utilizing Lemma 4, we construct an instance $G' = (V', E')$ of $D\text{-}TDS\text{-}GAUDGs$ from an arbitrary instance $G = (V, E)$ of $D\text{-}VC\text{-}PGD3$ in polynomial time. From the following claim, we have $D\text{-}TDS\text{-}GAUDGs \in NP\text{-}hard$. Therefore, $D\text{-}TDS\text{-}GAUDGs \in NP\text{-}complete$. $\square$

Claim. G has a vertex cover S_{vc} of size at most K if and only if G' has a total dominating set D_t of size at most $K + 2n + 2\sum_{v_i v_j \in E} k_{ij}$.

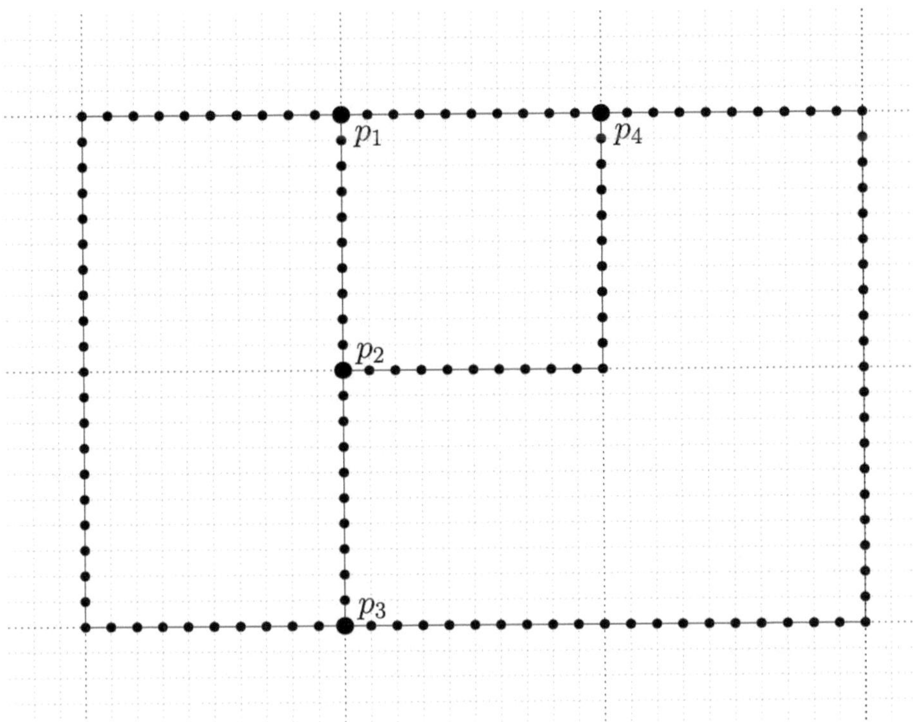

Fig. 3. Inclusion of auxiliary points.

Proof. ($\Longrightarrow$) Let S_{vc} be a *vertex cover* of G such that $|S_{vc}| \leq K$, and let T_{vc} be the set of vertices in G' corresponding to the vertices in S_{vc}, i.e., $T_{vc} = \{p_i \in V' : v_i \in S_{vc}\}$. Now, we construct two sets $T_a \subseteq A$ and $T_g \subseteq S$ such that $D_t = T_{vc} \cup T_a \cup T_g$ forms a *total dominating set* with cardinality less than or equal to $K + 2n + 2\sum_{v_i v_j \in E} k_{ij}$.

The construction of T_a and T_g proceeds as follows: since S_{vc} is a *vertex cover* of G (see Fig. 1(b)), at least one endpoint of every edge in G is inside S_{vc}. Since every edge in G corresponds to a sequence of segments in G', we start from the endpoint inside T_{vc}. For each $p_i p_j$ in G' corresponding to each $v_i v_j \in E$, at least one out of p_i and p_j is in T_{vc}. Without loss of generality, let $p_i \in T_{vc}$ (if $p_i, p_j \in T_{vc}$, then break the tie by picking either). We traverse from p_i towards p_j. While traversing, we leave two vertices next to p_i, add the next two vertices to T_a, then leave the next two vertices and select the next two vertices for T_a again. This process is repeated until we reach p_j. For each $v_i v_j \in E$, we apply this process to the corresponding $p_i p_j$ in G'. Since $p_i p_j$ contains $4k_{ij} + 1$ intermediate grid points, it requires at least $2k_{ij}$ points from $p_i p_j$ in T_a for total domination. Therefore, $|T_a| = 2\sum_{v_i v_j \in E} k_{ij}$, where k_{ij} is the integer associated to each $v_i v_j \in E$ derived in Step 2 in the proof of Lemma 4. In Fig. 6, the red cross points represent T_a.

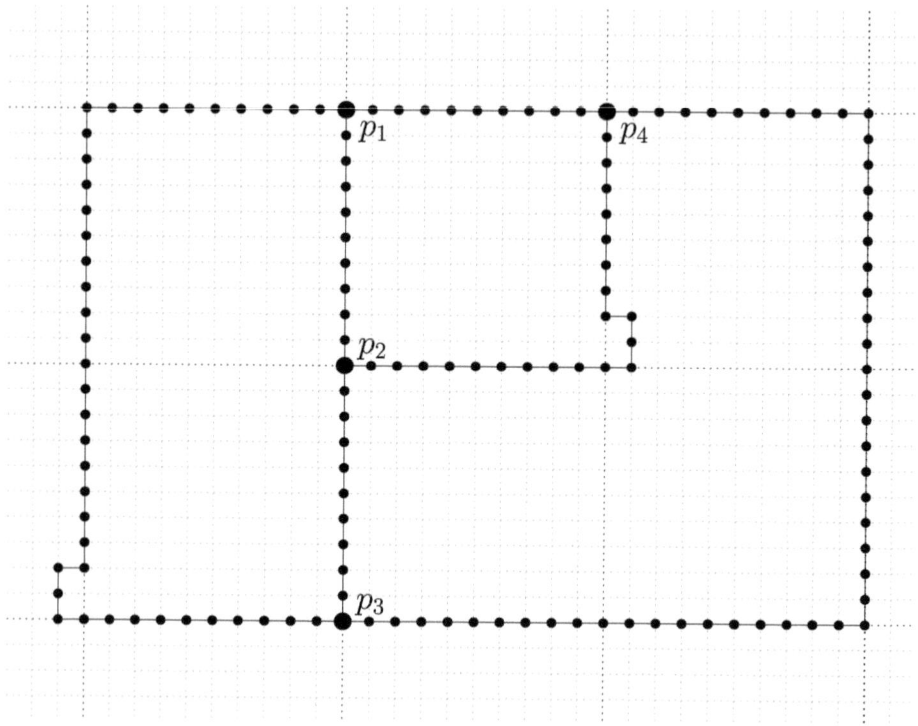

Fig. 4. Inclusion of an additional path if needed.

From each $p_i \in V'$, we choose y_i and w_i for T_g. Hence, $|T_g| = 2n$. In Fig. 6, the blue circles depict T_g.

Now, we need to demonstrate that the set D_t is a *total dominating set* (TDS) of G', satisfying two conditions: (i) every vertex in G' is dominated, and (ii) there is no isolated vertex in $G[D_t]$. Firstly, since w_i and y_i from $T_g \subseteq D_t$ dominate w_i, x_i, y_i, z_i, and p_i, and these are adjacent vertices, $G[T_g]$ does not have any isolated vertex. Secondly, $G[T_g \cup T_{vc}]$ does not have an isolated vertex since $y_i w_i p_i$ is a path. Now, each $p_i p_j$ corresponding to $v_i v_j$ has $4k_{ij}$ vertices. These vertices are dominated by $2k_{ij}$ vertices in T_a since out of $4k_{ij} + 1$ vertices one vertex is already dominated by a vertex in T_{vc} (as $T_{vc} = \{p_i \in V' : v_i \in S_{vc}\}$). Since the construction alternates between adding two vertices to T_a and skipping two vertices, no vertex in $G[T_a]$ is isolated. Hence, the set $D_t = T_{vc} \cup T_g \cup T_a$ is indeed a TDS of G' and if $S_{vc} \le K$, then $|D_t| \le K + 2n + 2\sum_{v_i v_j \in E} k_{ij}$.

($\Longleftarrow$) Let D_t be a TDS of G' such that $|D_t| \le K + 2n + 2\sum_{v_i v_j \in E} k_{ij}$. Then, we will show that G has a vertex cover S_{vc} of size at most K. To prove this, we prove the following observations.

(i) Out of four points in the gadget associated with each p_i in G' (i.e., x_i, y_i, z_i, w_i), at least two points belong to D_t.

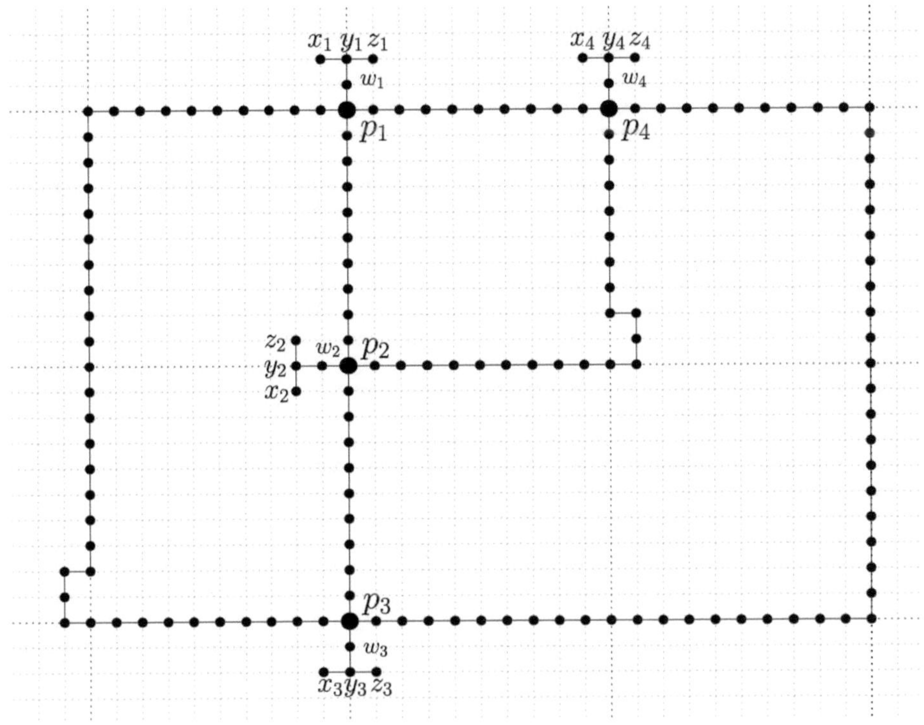

Fig. 5. Inclusion of gadget at each node point.

(ii) If p_i and p_j in G' correspond to the end vertices of an edge $v_i v_j \in E$ and none of p_i and p_j are in D_t, then at least $2k_{ij} + 1$ vertices of $p_i p_j$ are in D_t.

Observation (i): Corresponding to each $p_i \in E'$, there are 4 points (i.e., x_i, y_i, z_i and w_i) in the corresponding gadget. Since x_i and z_i are pendant vertices and $G[D_t]$ cannot have an isolated vertex, the points y_i and w_i must be in the total dominating set D_t. Hence, in G', $|S \cap D_t| \geq 2n$.

Observation (ii): Since $p_i, p_j \notin D_t$, there are $4k_{ij} + 1$ points on $p_i p_j$ for domination. Since two consecutive points can dominate four vertices, $4k_{ij} + 1$ number of vertices can be dominated by $\lceil \frac{4k_{ij}+1}{4} \times 2 \rceil = 2k_{ij} + 1$.

Given a TDS D_t of G' with a size at most $K + 2n + 2\sum_{v_i v_j \in E} k_{ij}$, we need to demonstrate that by deleting and/or replacing some vertices from D_t, we can obtain a *vertex cover* S_{vc} of G with a size at most K. The vertices of the set S account for at least $2n$ vertices in D_t due to Observation (i). Let us define a set $S'_{vc} = D_t \setminus S$ and $S_{vc} = \{v_i \in V : p_i \in S'_{vc}\}$. By Observation (i), we have $|S'_{vc}| \leq K + 2\sum_{v_i v_j \in E} k_{ij}$. Then, for each edge $v_i v_j \in E$, if $p_i, p_j \notin S_{vc}$, then the corresponding $p_i p_j$ in G' has $2k_{ij} + 1$ points in D_t instead of $2k_{ij}$ (refer to Observation (ii)). For each such edge, we add either v_i or v_j to S_{vc}. Since

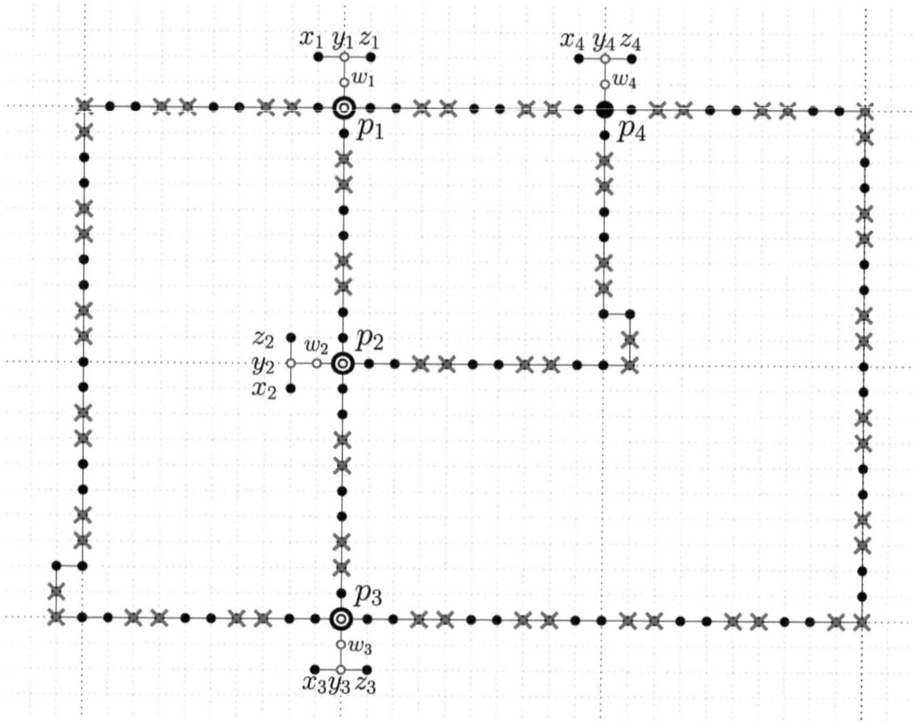

Fig. 6. $G' = (V', E')$, where the red crosses are in T_a and the blue circles are in T_g. (Color figure online)

every $p_i p_j$ corresponding to each $v_i v_j$ contributes at least $2k_{ij}$ to D_t, $|D_t \cap T_a| \geq 2\sum_{v_i v_j \in E} k_{ij}$. Therefore, $|S_{vc}| \leq K$. This proves that $D\text{-}TDS\text{-}GAUDGs \in NP\text{-}hard$. $\qquad\square$

4   Approximation Algorithm for the TDS Problem

In this section, we propose a 6.29-factor approximation algorithm for the TDS problem in UDGs.

4.1   Algorithm

The algorithm aims to find a TDS in a unit disk graph $G = (V, E)$, where $V = \{p_1, p_2, \ldots, p_n\}$ represents the centers of unit disks in $\mathbb{R}^2$. To begin, the algorithm identifies a maximal independent set $D \subseteq V$, which satisfies the *domination property* (see Lines 2–6 of Algorithm 1). Next, to fulfill the *total property*, the algorithm selects a subset $T \subseteq V$ such that for each $v \in D$, there exists a vertex $u \in V \setminus D$ where u is adjacent to v. To find the set T, the algorithm constructs a set cover instance $< D, S >$, where D serves as the universal set, and S consists of subsets $S_i = N_G(u_i) \cap D$ for each $u_i \in V \setminus D$ (see Lines 8–12 of Algorithm 1). Here, $N_G(u_i) \cap D$ represents the neighbors of u_i within D. Subsequently, a greedy set cover algorithm is applied to $< D, S >$ to find a minimal subset $S' \subseteq S$ such that $\bigcup_{S_i \in S'} S_i = D$ (see Line 13 of Algorithm 1). For each $S_i \in S'$, T contains the corresponding u_i of $V \setminus D$ (see Line 13 of Algorithm 1). Then, it reports $D_t = D \cup T$ as a TDS of G.

Algorithm 1. TDS-UDG-SC

Input: A unit disk graph, $G = (V, E)$, with known disk centers
Output: A TDS D_t for G

1: $V' = V, D = \emptyset, S = \emptyset$
2: **while** $V' \neq \emptyset$ **do**   ▷ *domination* property of TDS
3:     choose a vertex $v \in V'$
4:       $D = D \cup \{v\}$
5:       $V' = V' \setminus N_G[v]$
6: **end while**
7: $i = 1$
8: **for** each $u \in V \setminus D$ **do**
9:       $S_i = N_G(u) \cap D$
10:      $S = S \cup S_i$
11:        $i = i + 1$
12: **end for**
13: $S' = GreedySetCover(D, S)$
14: $T = \{u_i | S_i \in S'\}$ ▷ *total* property of TDS
15: $D_t = D \cup T$
16: **return** $|D_t|$

Lemma 5. *Given a maximal independent set D of a geometric unit disk graph $G = (V, E)$ such that $S = \{S_i\}$, where $S_i = D \cap N_G(u_i)$, for each $u_i \in V \setminus D$. If C^* is an optimal* set cover *of the set cover instance $< D, S >$ and D^* is an optimal dominating set of G, then $|C^*| \leq |D^*|$.*

Proof. Given that D is a *maximal independent set* in the unit disk graph G, for any vertex $v \in V \setminus D$, the Lemma 1 ensures that $|N_G(v) \cap D| \leq 5$. Let D^* denote an *optimal dominating set* of G, then either (i) $D^* \subseteq D$ or (ii) $D^* \subseteq V \setminus D$ or (iii) $D^* = D_D^* \cup D_{V \setminus D}^*$, where $D_D^* = D^* \cap D$ and $D_{V \setminus D}^* = D^* \cap (V \setminus D)$. We need to demonstrate that for each of the given scenarios $|C^*| \leq |D^*|$.

(i) $D^* \subseteq D$: If $D^* \subseteq D$, then $D^* = D$; otherwise, there exists at least one vertex $v \in D$, such that v is not dominated by D^* (since D is an *independent set*). This leads to a contradiction that D^* is a dominating set. Since D is an independent set, in the worst case, D requires at most $|D|$ subsets (i.e., vertices) from S (i.e., $V \setminus D$) to cover D. If C^* is an optimal cover of the set cover instance $< D, S >$, then $|C^*| \leq |D| = |D^*|$.

(ii) $D^* \subseteq V \setminus D$: We prove this case by contradiction, so let us assume $C^* > D^*$. Since D^* is a dominating set, each vertex in $D \cup V \setminus D$ (i.e., V) is either in D^* or is adjacent to at least one vertex in D^*. If so, a set $D' \subseteq D^*$ exists such that D' dominates D (since D is an independent set). This implies $|D'|$ subsets (i.e., vertices) from S (i.e., $V \setminus D$) are sufficient to cover D, which is less than or equal to $|C^*|$. This leads to a contradiction to the fact that C^* is an optimal cover of D.

(iii) $D^* = D_D^* \cup D_{V \setminus D}^*$, where $D_D^* = D^* \cap D$ and $D_{V \setminus D}^* = D^* \cap (V \setminus D)$: Since $D_D^* \subseteq D$, D_D^* is an independent set. In the worst case, at most $|D_D^*|$ number of subsets (vertices) from S are required to cover vertices in D_D^*. Since $D_{V \setminus D}^*$ dominates the set $D \setminus D_D^*$, $|D_{V \setminus D}^*|$ number of subsets (vertices) are sufficient for the coverage of remaining vertices in $D \setminus D_D^*$. Therefore, $|C^*| \leq |D_D^*| + |D_{V \setminus D}^*| = |D^*|$. $\qquad\square$

Lemma 6. *The set D_t in Algorithm 1 is a TDS of G.*

Proof. In the first phase, Algorithm 1 finds a *maximal independent set D* of G that satisfies the domination property (see Lines 2–6 in Algorithm 1). Next, the algorithm runs $GreedySetCover(D, S)$ to find a subset T such that $T = \{u_i | S_i \in S'\}$ (see Line 13 and Line 14 in Algorithm 1). The set T ensures that for each vertex $v \in D$, there exists a vertex $u \in T$ such that $uv \in E$. Hence, the set T, when combined with D, confirms that none of the vertices in D_t is isolated. Therefore, the nominated points in D and T satisfy both the properties of TDS. Consequently, the set D_t forms a TDS of G. $\qquad\square$

4.2   Analysis

The set D_t in *TDS-UDG-SC* is a TDS of G, where $D_t = D \cup T$ (see Lemma 6). Let D^* and D_t^* be an optimal DS and an optimal TDS of G, respectively. Since 2 vertices of an edge in $G[D_t^*]$ dominate at most 8 vertices of D (Lemma 2), the following equation holds.

$$|D| \leq 4|D_t^*| \tag{1}$$

The set T in *TDS-UDG-SC* satisfies the total property when added to the independent set D, where $T = \{u_i | S_i \in S'\}$ and S' is the set cover of the instance $< D, S >$. Let C^* be an *optimal set cover* of $< D, S >$. Then from Lemma 5, the following equation holds.

$$|C^*| \leq |D^*| \tag{2}$$

Therefore, from Lemma 6, we conclude the following:

$$\begin{aligned}
|D_t| &= |D| + |T| \\
&\leq 4|D_t^*| + H(5) \times |C^*| \text{ (Equation 1 and Theorem 1)} \\
&\leq 4|D_t^*| + \frac{137}{60} \times |D^*| \text{ (Lemma 5)} \\
&\leq 4|D_t^*| + \frac{137}{60} \times |D_t^*| \text{ (Observation 1)} \\
&= \frac{377}{60} \times |D_t^*| \leq 6.29 \times |D_t^*|
\end{aligned}$$

$$(3)$$

5 Conclusion

In conclusion, we have shown that the Total Dominating Set (TDS) problem is NP-complete in grid-aligned UDGs and have proposed a 6.29-approximation algorithm for addressing the TDS problem in unit disk graphs (UDGs).

References

1. Atapour, M., Soltankhah, N.: On total dominating sets in graphs. Int. J. Contemp. Math. Sciences **4**(6), 253–257 (2009)
2. Chain-Chin, Y., Lee, R.C.: The weighted perfect domination problem and its variants. Discret. Appl. Math. **66**(2), 147–160 (1996)
3. Chellali, M., Haynes, T.W.: A note on the total domination number of a tree. J. Comb. Math. Comb. Comput. **58**, 189–193 (2006)
4. Cockayne, E.J., Henning, M.A., Mynhardt, C.M.: Vertices contained in all or in no minimum total dominating set of a tree. Discret. Math. **260**(1–3), 37–44 (2003)
5. Cockayne, E.J., Dawes, R., Hedetniemi, S.T.: Total domination in graphs. Networks **10**(3), 211–219 (1980)
6. Cockayne, E.J., Dreyer, P.A., Jr., Hedetniemi, S.M., Hedetniemi, S.T.: Roman domination in graphs. Discret. Math. **278**(1–3), 11–22 (2004)
7. Cormen, T.H., Leiserson, C.E., Rivest, R.L., Stein, C.: Introduction to Algorithms. MIT Press, 4 edn. (2022)
8. Damaschke, P., Müller, H., Kratsch, D.: Domination in convex and chordal bipartite graphs. Inf. Process. Lett. **36**(5), 231–236 (1990)
9. DeLaVina, E., Liu, Q., Pepper, R., Waller, B., West, D.B.: Some conjectures of graffiti.pc on total domination. Congressus Numerantium **185**, 81 (2007)
10. Dorfling, M., Goddard, W., Henning, M.A.: Domination in planar graphs with small diameter ii. Ars Combin. **78**, 237–255 (2006)
11. Haynes, T.W., Henning, M.A.: Trees with unique minimum total dominating sets. Discussiones Mathematicae Graph Theory **22**(2), 233–246 (2002)
12. Haynes, T.W., Henning, M.A., Howard, J.: Locating and total dominating sets in trees. Discret. Appl. Math. **154**(8), 1293–1300 (2006)
13. Jallu, R.K., Jena, S.K., Das, G.K.: Liar's dominating set in unit disk graphs. In: International Computing and Combinatorics Conference, pp. 516–528. Springer (2018)

14. Jena, S.K., Das, G.K.: Total domination in geometric unit disk graphs. In: Proceedings of the 33rd Canadian Conference on Computational Geometry, 2021, August 10-12, 2021, pp. 219–227 (2021)
15. Keil, J.M.: The complexity of domination problems in circle graphs. Discret. Appl. Math. **42**(1), 51–63 (1993)
16. Laskar, R., Pfaff, J., Hedetniemi, S.M., Hedetniemi, S.T.: On the algorithmic complexity of total domination. SIAM J. Algebraic Discrete Methods **5**(3), 420–425 (1984)
17. Lichtenstein, D.: Planar formulae and their uses. SIAM J. Comput. **11**(2), 329–343 (1982)
18. Marathe, M.V., Breu, H., Hunt, H.B., III., Ravi, S.S., Rosenkrantz, D.J.: Simple heuristics for unit disk graphs. Networks **25**(2), 59–68 (1995)
19. Pfaff, J., Laskar, R., Hedetniemi, S.: NP-completeness of total and connected domination and irredundance for bipartite graphs. technicial report 428, dept. math. Sciences, Clemson University (1983)
20. Rout, S., Das, G.K.: Semi-total domination in unit disk graphs. In: Conference on Algorithms and Discrete Applied Mathematics, pp. 117–129. Springer (2024)
21. Rout, S., Mishra, P.K., Das, G.K.: Total Roman domination and total domination in unit disk graphs. Commun. Combinatorics Optim. (2024)
22. Thomassé, S., Yeo, A.: Total domination of graphs and small transversals of hypergraphs. Combinatorica **27**(4), 473–487 (2007)
23. Valiant, L.G.: Universality considerations in VLSI circuits. IEEE Trans. Comput. **100**(2), 135–140 (1981)

Area-Universality in Outerplanar Graphs

Ravi Suthar[✉], Raveena Chahar, and Krishnendra Shekhawat

Department of Mathematics, Birla Institute of Technology and Science, Pilani, Vidya Vihar, Pilani Campus, Pilani, Rajasthan 333031, India
`p20240055@pilani.bits-pilani.ac.in`

Abstract. A rectangular floorplan is a partition of a rectangle into smaller rectangles such that no four rectangles meet at a single point. Rectangular floorplans arise naturally in a variety of applications, including VLSI design, architectural layout, and cartography, where efficient and flexible spatial subdivisions are required. A central concept in this domain is that of area-universality: a floorplan (or more generally, a rectangular layout) is area-universal if, for any assignment of target areas to its constituent rectangles, there exists a combinatorially equivalent layout that realizes these areas.

In this paper, we investigate the structural conditions under which an outerplanar graph admits an area-universal rectangular layout. We establish a necessary and sufficient condition for area-universality in this setting, thereby providing a complete characterization of admissible outerplanar graphs. Furthermore, we present an algorithmic construction that guarantees that the resulting layout is always area-universal.

Keywords: Area-Universal · Graph Theory · Outerplanar Graphs · Algorithms

1 Introduction

The rectangular layout is a fundamental concept that involves strategic arrangements of non-overlapping rectangles (modules) within a rectangular space. Additionally, it ensures that there are no intersecting joints formed by any four non-overlapping modules within the layout. The rectangular layout has broad application in cartography, architecture, VLSI circuit design, and graph drawings. This paper explores the idea of area-universal rectangular layouts using a graph-theoretic approach. The notion of area-universal layouts [2] emerges as an essential instrument for generating diverse rectangular arrangements that accommodate specific area assignments while preserving the consistent adjacency configurations defined by the adjacency graphs. From a mathematical perspective, the area-universal layouts can be represented through graphs, wherein each room or module within the layout is represented as a vertex, and the adjacency (i.e., the physical connection or proximity between rooms) is denoted by an edge connecting the corresponding vertices. The central challenge lies in determining

N. Misra and A. Pandey (Eds.): CALDAM 2026, LNCS 16445, pp. 471–484, 2026.
https://doi.org/10.1007/978-3-032-17156-6_35

whether the graph that captures these adjacency constraints can accommodate every possible assignment of areas to its rectangles while preserving the adjacencies and connectivity among the modules. Area-universal rectangular layouts of graphs serve a vital role in various fields, particularly in VLSI design, where the rectangles represent circuit components and their common boundaries model adjacency requirements. During the early stages of VLSI design, when chip component areas are not yet specified, only the relative positions of the components matter. However, later design stages require specific component areas, which an area-universal layout accommodates, simplifying the design process. Past works [11] have presented heuristic algorithms for computing rectangular layouts based on given area assignments.

1.1   Preliminary

A graph $\mathcal{G} = (V, E)$ is a fundamental combinatorial structure consisting of a finite set of vertices V and a set of edges E, where each edge connects a pair of distinct vertices in V. A graph $\mathcal{G}$ is *biconnected* if, for every pair of distinct vertices $u, v \in V(\mathcal{G})$, there exist at least two vertex-disjoint paths between u and v. A graph is called planar if it can be drawn on the plane in such a way that its edges intersect only at their shared endpoints, thereby allowing a crossing-free representation. Such a drawing of a planar graph in the plane without edge crossings is referred to as a plane graph or a plane embedding. In this paper, the *order* of a graph $\mathcal{G}$ denotes $|V|$, the number of vertices of $\mathcal{G}$.

Definition 1. *In a plane graph, a separating triangle [10] is a cycle of three edges (u, v, w) such that at least one vertex of the graph lies in the interior region bounded by this cycle (see Fig. 1a).*

Definition 2. *A properly triangulated plane graph (PTPG) [6] is a connected plane graph which satisfies the following properties: (see Fig. 1b)*

1. *Exterior face has length ≥ 4,*
2. *Every face (except the exterior) is a triangle (bounded by three edges),*
3. *It does not contain a separating triangle (ST).*

Definition 3. *A graph $\mathcal{G}$ is said to be* outerplanar *if it admits a plane embedding in which all vertices lie on the boundary of the exterior face.*

The *four–completion* [4] of a PTPG $\mathcal{G}$ is obtained by introducing four new vertices N, E, S, W (*cardinal vertices*) on the outer boundary, connecting them to form the outer four-cycle $(N - E - S - W)$, and partitioning the outer boundary of $\mathcal{G}$ into four contiguous paths, each vertex on respective paths is then made adjacent to one of the cardinal vertices. The resulting graph is planar, triangulated, free of separating triangles, and has (N, E, S, W) as its outer face (see Fig. 1c). An outerplanar graph $\mathcal{G}$ is referred to as an *extended graph* after it undergoes the *four-completion process* (see Fig. 2a).

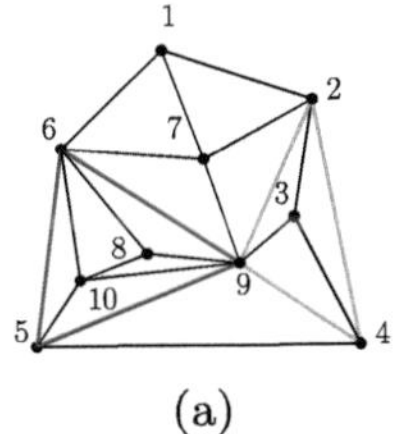 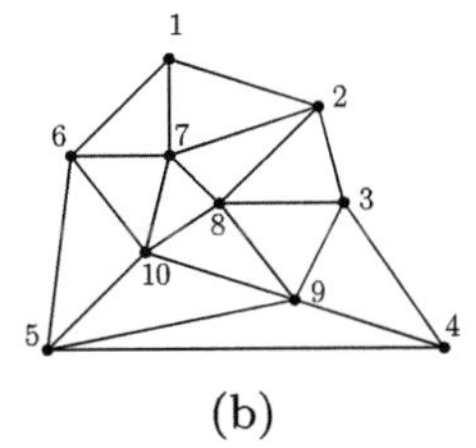 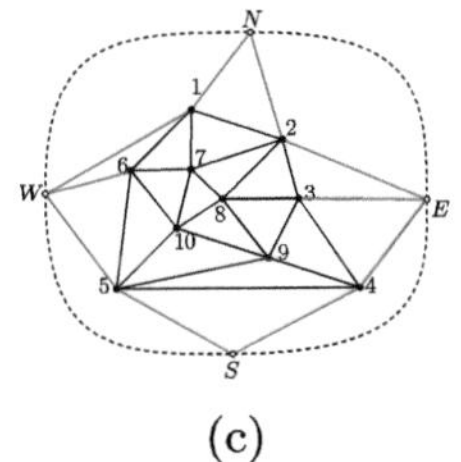

(a)   (b)   (c)

Fig. 1. (a) Presence of separating triangles $(2, 4, 9)$, $(6, 9, 10)$, and $(6, 9, 5)$. (b) A properly triangulated plane graph (PTPG). (c) A four-completion of a PTPG.

Definition 4. *Regular edge labeling [REL] [5]: A regular edge labeling for a biconnected PTPG $\mathcal{G}$ with four outer vertices N, W, S, E (ordered counterclockwise), constitutes a partition and orientation of its interior edges into two disjoint subsets, T_1 (represented using blue color) and T_2 (represented using red color), satisfying the following conditions:*

(I) For every interior vertex v, the incident edges are arranged counterclockwise around v in the sequence:
- *Directed toward v: incoming T_1 edges.*
- *Directed away from v: outgoing T_2 edges*
- *Directed away from v: outgoing T_1 edges*
- *Directed toward v: incoming T_2 edges*

(ii) Edges incident to vertex N are included in T_1 and are directed toward N. Conversely, edges incident to W are part of T_2 and are directed away from W. Edges incident to S are contained in T_1 and are directed away from S, while edges incident to E lie in the set T_2 and are directed toward E.

Definition 5. *A flippable edge [3] in a REL of a graph $\mathcal{G}$ is an edge e that is not incident to any degree-four vertex, and must be a diagonal of a four-cycle whose edges are alternately labeled in the REL. Furthermore, if an edge is flippable, its label can be changed without affecting the rest of the labeling. In Fig. 2b, edges $(5, 7)$ and $(4, 8)$ are flippable, and the REL obtained after flipping these edges is shown in Fig. 2c.*

Definition 6. *Floor plan $(\mathcal{F})$ [9]: A floor plan (layout) decomposes a polygon into smaller component polygons via straight-line segments. The outer polygon is called boundary of the floor plan, while the smaller component polygons are termed modules. Two modules are adjacent if they share a wall segment; mere point contact (four joint where four line segments meet) does not constitute adjacency. A special class of floor plans is the rectangular floor plan (RFP), in which the boundary and all modules are rectangular.*

Definition 7. *A rectangular layout $\mathcal{F}$ is said to be area-universal [2] if for every assignment of positive areas to the rectangles of $\mathcal{F}$, there exists a drawing of $\mathcal{F}$ in the plane in which each rectangle realizes its assigned area. In such a drawing, only the sizes of the rectangles may change; the combinatorial structure of $\mathcal{F}$ (i.e., the adjacencies between rectangles) must remain exactly the same as in the original layout.*

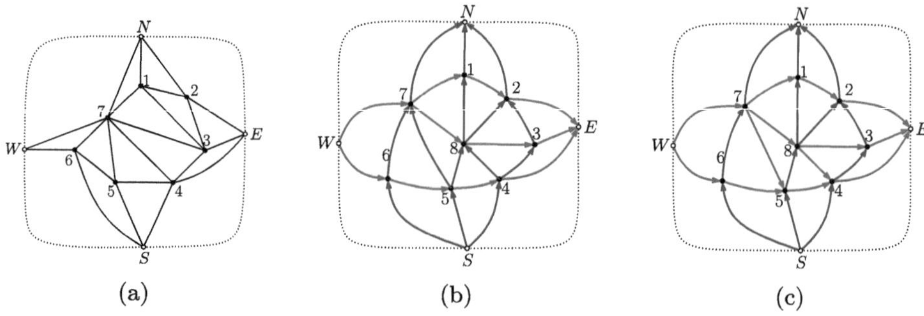

Fig. 2. (a) An extended outerplanar graph. (b) Presence of flippable edges $(5,7)$ and $(4,8)$. (c) Altered labels of the flippable edges $(5,7)$ and $(4,8)$.

1.2   Motivation and Contributions

A planar graph admits multiple rectangular layouts, not all of which are area-universal. Rinsma [8] demonstrated explicit examples where specific outerplanar graphs with given area assignments could not be realized by any rectangular layout. Eppstein et al. [3] established a geometrical characterization, which is both necessary and sufficient, for a rectangular layout to be area-universal (see Theorem 1).

In a rectangular layout $\mathcal{F}$, a line segment is defined as a sequence of consecutive inner edges of $\mathcal{F}$. A line segment is said to be maximal if it is not contained within any other line segment. Moreover, a maximal line segment is one-sided if it is a side of at least one rectangle in a rectangular layout.

Theorem 1. *A rectangular layout $\mathcal{F}$ is said to be area-universal if and only if every internal maximal line segment of $\mathcal{F}$ is one-sided. [3]*

The rectangular layout depicted in Fig. 3a is not area-universal, since the maximal internal line segment o (highlighted in bold) does not align with the side of any rectangle. In contrast, the rectangular layout shown in Fig. 3b is area-universal, as every maximal line segment aligns with the side of some rectangle.

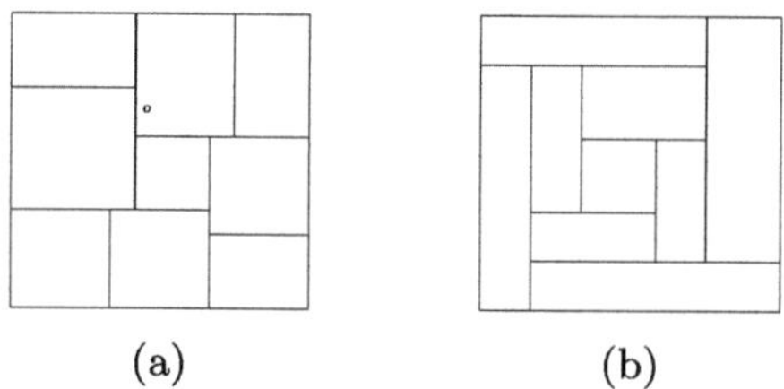

Fig. 3. (a) A rectangular layout that is not area-universal. (b) An area-universal rectangular layout.

Eppstein et al. [3] proposed an algorithm to construct area-universal layouts for certain graph families. The method, however, suffers from very high worst-case complexity, specifically $O\!\left(2^{O(K^2)} n^{O(1)}\right)$, where K represents the maximum number of degree-4 vertices in any minimal separation component. As a result, the problem remains computationally difficult in the general case. Later studies have therefore emphasized restricted graph classes. For instance, Chang and Yen [1] in 2017 showed that biconnected outerplanar graphs admit area-universal convex polygonal drawings, thus offering constructive realizations of area-universality within that class. More recently, a polynomial-time algorithm is proposed in [12], leveraging insights into the properties of area-universal layouts and their relationship to regular edge labelling construction. Subsequently, Kumar et al. [7] identified a family of rectangularly dualizable graphs (RDGs) for which every RDG can be realized as an area-universal rectangular layout in polynomial time.

Our study delves into area-universal layouts using graph theory concepts and aims to identify the class of outerplanar graphs that can accommodate them. Outerplanar graphs, which have all their vertices on the boundary of a single face when embedded in the plane, are fascinating due to their structural properties.

2 Necessary and Sufficient Condition

We endeavour to establish a comprehensive understanding of outerplanar graphs that can accommodate area-universal layouts while also identifying those that cannot. The illustration presented in Fig. 4 showcases the smallest graph known to lack an area-universal layout. It has been established in [2] that a layout qualifies as area-universal if and only if it adheres to the criterion of being one-sided. An area-universal layout is characterized by the property that every maximal line segment (a line segment within any layout is termed maximal if it cannot be entirely contained within another line segment) serves as a side of at least one rectangle in the layout. Theorem 2 outlines the essential requirements needed to determine whether an outerplanar graph can have an area-universal layout, which is also known as a one-sided layout. Before formally stating Theorem 2, we first prove an essential lemma that will be used throughout the paper.

Lemma 1. *Let $\mathcal{G}$ be a plane graph and $r(\mathcal{G})$ be any REL of $\mathcal{G}$. If $r(\mathcal{G})$ contains a flippable edge, then the rectangular layout of G corresponding to $r(\mathcal{G})$ is not area-universal. [3] (for proof, see the full version of this paper).*

Theorem 2. *Let $\mathcal{G}$ be a biconnected outerplanar triangulated graph of order $n(n > 3)$. An area-universal layout corresponding to $\mathcal{G}$ exists if and only if it contains exactly two vertices of degree two.*

Proof. Considering $\mathcal{G}$ as a biconnected outerplanar triangulated graph implies that the minimum degree of $\mathcal{G}$ is at least two. Since every bounded face of $\mathcal{G}$ is a triangle, we define its weak dual T by introducing one vertex for each interior

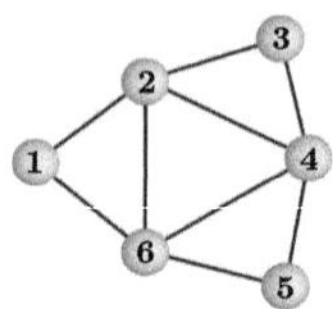

Fig. 4. The smallest outerplanar graph G not possessing any area-universal layout.

triangular face and joining two such vertices whenever the corresponding faces share an internal edge (see Fig. 5). It follows that T is a tree. Each leaf of T corresponds to a triangular face of $\mathcal{G}$ that shares exactly one internal edge, where the opposite boundary vertex in $\mathcal{G}$ has degree two. Thus, the degree-2 vertices of $\mathcal{G}$ are in one-to-one correspondence with the leaves of T. Moreover, since every tree with $|V(T)| > 1$ has at least two leaves, $\mathcal{G}$ must contain at least two degree-2 vertices whenever $n > 3$.

First, we assume that $\mathcal{G}$ has exactly two degrees-2 vertices. We aim to demonstrate the existence of a corresponding area-universal layout. We present an algorithm (refer to Algorithm 1) for constructing such a layout. By proving that this algorithm consistently generates an area-universal layout for any biconnected outerplanar graph having two degree-2 vertices, we establish the validity of this statement.

Conversely, assuming that an area-universal layout exists corresponding to $\mathcal{G}$, then we need to show that it has exactly two degree-2 vertices. Here, we will prove the contrapositive statement. For that, we assume that $\mathcal{G}$ has at least three (say $k(k > 2)$) vertices of degree two, then its weak dual T has at least three leaves and therefore, T must contain at least one vertex of degree three. In T, a vertex of degree three corresponds to a interior triangular face (a triangular face $F = \{v_1, e_1, v_2, e_2, v_3, e_3, v_1\}$, each of whose three edges is shared with another triangular face) in $\mathcal{G}$ (see Fig. 5). Now consider a biconnected outerplanar graph $\mathcal{G}$ that contains an interior triangle (v_1, v_2, v_3) that corresponds to a degree-3 vertex in the weak dual. When constructing a rectangular layout corresponding to graph $\mathcal{G}$, the four-completion step requires partitioning the outer boundary into four contiguous vertex paths (P_1, P_2, P_3, P_4), while ensuring that no separating triangle is created. For any partition of the outer boundary of $\mathcal{G}$ into four contiguous paths in the four-completion process, if a degree-2 vertex lies on exactly one boundary path, then in the extended graph its degree increases to 3 and it becomes the interior vertex of a separating triangle. Thus, no single path can contain a degree-2 vertex, each degree-2 vertex is necessarily shared by at least two distinct boundary paths. It follows that the vertices v_1, v_2, v_3 of the triangular face must lie on different paths in the four-completion, and consequently, the associated modules are positioned along different cardinal directions (N, W, S, E). Without loss of generality, let v_1, v_2, and v_3 lie on the boundary paths P_1, P_2, and P_3, respectively. Consequently, the modules associated with v_1, v_2, and v_3 are positioned along the N, E, and S cardinal directions, respectively. Since v_1, v_2, and v_3 are pairwise adjacent and lie on three distinct boundary paths, two of the corresponding modules appear collinear in the rectangular layout, while the third module is orthogonal to these two. In the REL, there are precisely three possible labeling patterns among v_1, v_2, and v_3 that realize this configuration (see Fig. 5).

(i) $(v_1, v_2), (v_3, v_2) \in T_2$ and $(v_3, v_1) \in T_1$.
(ii) $(v_3, v_1), (v_2, v_1) \in T_1$ and $(v_3, v_2) \in T_2$.
(iii) $(v_3, v_1), (v_3, v_2) \in T_1$ and $(v_1, v_2) \in T_2$.

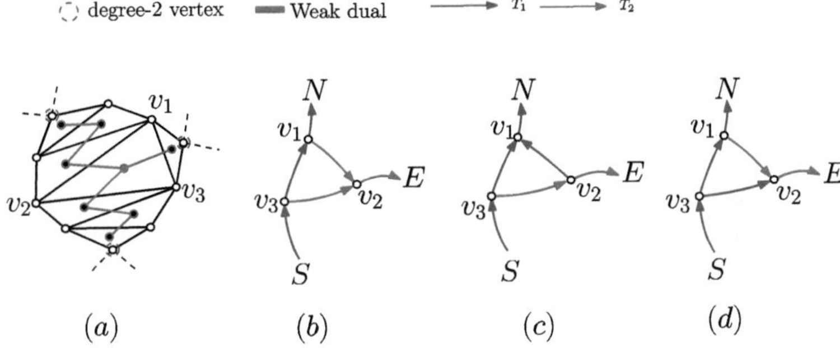

Fig. 5. (a) Paths are separated by degree-2 vertices. (b-d) Three possible labeling patterns among v_1, v_2, and v_3.

Let $\alpha_1 (\neq v_3)$ be a common neighbour of the vertices v_1 and v_2. Depending on the adjacency in the graph $\mathcal{G}$, either α_1 lies on the path P_1 or on the path P_2. In both cases, the edge (v_1, v_2) is the diagonal of the alternately labeled four-cycle $(v_2 - \alpha_1 - v_1 - v_3)$ in the REL, and hence (v_1, v_2) is a flippable edge (see Fig. 6). Moreover, in this configuration the edge (v_3, v_2) is also flippable. Let $\alpha_2 (\neq v_1)$ be a common neighbour of the vertices v_3 and v_2. Depending on the adjacency in the graph, either α_2 lies on the path P_2 or on the path P_3. In both scenarios, the edge (v_3, v_2) is the diagonal of the alternately labeled four-cycle $(v_3 - \alpha_2 - v_2 - v_1)$, and is therefore flippable (see Fig. 6).

Case (ii) can be derived directly from Case (i). In Case (i), the edge (v_1, v_2) belongs to T_2; after flipping its label so that $(v_2, v_1) \in T_1$, we obtain the configuration of Case (ii), in which (v_2, v_1) remains a flippable edge. Similarly, Case (iii) can be derived from Case (i) by starting with $(v_3, v_2) \in T_2$ and flipping its label so that $(v_3, v_2) \in T_1$, yielding the configuration of Case (iii), where (v_3, v_2) is still a flippable edge.

Therefore, there always exists a flippable edge among (v_1, v_2), (v_1, v_3), and (v_2, v_3), where v_i, $1 \le i \le 3$, are the vertices of the triangular face corresponding to a degree-3 vertex of the weak dual, depending on the partition of the outer boundary and the adjacencies in the outerplanar graph $\mathcal{G}$. $\qquad \square$

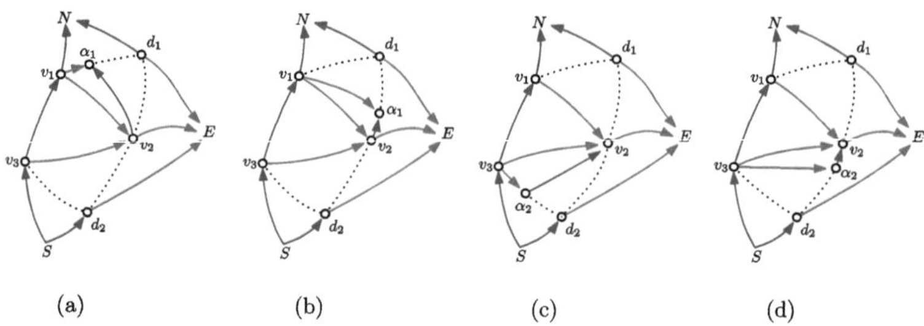

Fig. 6. (a-b) edge (v_1, v_2) is flippable, (c-d) edge (v_3, v_2) is flippable.

Remark: An outerplanar graph with exactly two vertices of degree two corresponds to an area-universal layout. However, not every four-completion of such graph leads to an area-universal layout. As shown in Fig. 7, an outerplanar graph is associated with two different ways with respect to four-completion: one yields an area-universal layout (see Fig. 7b). At the same time, the other fails to do so (see Fig. 7a). Hence, it is essential to consider properties of the *extended* outerplanar graph. Theorem 3 specifies the particular four-completions that must be avoided in order to obtain an area-universal rectangular layout.

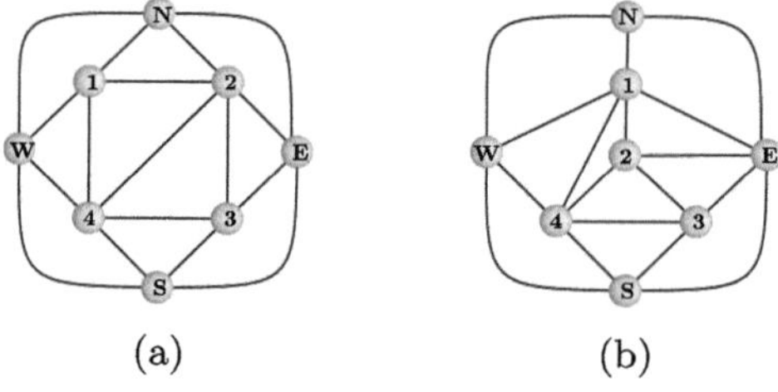

Fig. 7. (a) An extended outerplanar graph that does not admit an area-universal layout. (b) An extended outerplanar graph that admits an area-universal layout.

Theorem 3. *Let $\mathcal{G}$ be an outerplanar graph having exactly two vertices of degree two, and $\mathcal{G}'$ be its extended graph obtained by a four-completion. If four-completion leads to area-universal layout, then a vertex must be adjacent to exactly three cardinal vertices in $\mathcal{G}'$.*

Proof. The full proof is given in the full version of this paper.

3   Constructing Area-Universal Layouts for Outerplanar Graphs: Algorithm and Correctness

3.1   Overview

To address the areauniversal layout problem for outerplanar graphs, we first need a mechanism that identifies whether the given input graph satisfies the necessary structural condition for such a layout to exist. Specifically, an outerplanar graph admits an areauniversal layout only if it contains exactly two vertices of degree two (see Theorem 2). This condition can be checked in linear time by simply counting degree-2 vertices.

The purpose of OUTER4COMPLETION is to transform a given outerplanar graph into an *extended* outerplanar graph such that the REL obtained from this four-completion has no flippable edges, and hence yields an area-universal layout.

Notation and helper functions:

- $B = (b_1, \ldots, b_n)$: ordered outer-face cycle.
- d_1, d_2: the two degree-two vertices.
- $\mathrm{CARD} = [N, E, S, W]$: array of the four new cardinal vertices.
- $\mathrm{indexOf}(v, B)$: returns the position of v in B.
- $\mathrm{addEdge}(u, w)$: inserts an edge (u, w) while preserving planarity.
- For $1 \leq i \leq 4$, P_i denote the set of vertices selected from the outer cycle B, listed in the same ordering as they appear in B, and determined by the vertices v, d_1, and d_2.
- E_c: set of edges introduced during the four-completion process.

This augmentation procedure is formally described in Algorithm 1. Given a biconnected outerplanar graph $\mathcal{G}$ with outer–face cycle $B = (b_1, \ldots, b_n)$, and exactly two degree–2 vertices d_1 and d_2, the algorithm augments $\mathcal{G}$ by introducing four new vertices N, E, S, W arranged in clockwise order along the boundary. These vertices, collectively denoted as $\mathrm{CARD} = [N, E, S, W]$, serve as the cardinal vertices. A pivot vertex $v \in V$ with degree greater than two is then selected. The outer boundary cycle B is traversed clockwise starting from the successor of v, and is partitioned into four sets P_1, P_2, P_3, and P_4 determined by the locations of d_1 and d_2 on B.

P_1 : consisting solely of the vertex v,

P_2 : the vertices encountered from v to the first degree–2 vertex d_1 in clockwise order along B,

P_3 : the vertices from d_1 to the second degree–2 vertex d_2 in clockwise order along B,

P_4 : the remaining vertices from d_2 back to v.

The vertices in P_i, $1 \leq i \leq 4$ are then made adjacent to their designated cardinal vertices, with vertex v itself adjacent to three consecutive cardinal vertices (W, N, E). Finally, the four cardinal vertices are connected by edges into the four-cycle (N, E, S, W) so that the new exterior face of the augmented graph is a quadrilateral. The output is the augmented graph $\mathcal{G}_v$ that serves as a valid four-completion of $\mathcal{G}$, guaranteeing that the resulting extended outerplanar graph can admit an area–universal rectangular layout.

Algorithm 1 Outer4Completion($\mathcal{G}, B, d_1, d_2$)

1: **Input:** biconnected outerplanar graph $\mathcal{G} = (V, E)$, outer-face cycle $B = (b_1, \ldots, b_n)$ (clockwise), degree-2 vertices d_1, d_2

2: **Output:** augmented graph $\mathcal{G}_v = (V \cup \{N, E, S, W\}, E \cup E_C)$

3: $V_{>2} \leftarrow \{u \in V : \deg_{\mathcal{G}_p}(u) > 2\}$

4: add vertices N, E, S, W on outer boundary; CARD$= [N, E, S, W]$ $\quad\quad\triangleright$ clockwise indices 0..3

5: choose a vertex $v \in V_{>2}$

6: $\mathcal{G}_v \leftarrow \mathcal{G}$

7: addEdge(v, W), addEdge(v, N), addEdge(v, E)

8: $i \leftarrow (\text{indexOf}(v, B) + 1) \bmod n$,

9: $P_2 = \phi$, $P_3 = \phi$, $P_4 = \phi$

10: **while** true **do**

11: $\quad$ append $B[i]$ to P_2;

12: $\quad$ **if** $B[i] = d_1$ or $B[i] = d_2$ **then**

13: $\quad\quad$ append $B[i]$ to P_3

14: $\quad\quad$ $i \leftarrow (i + 1) \bmod n$

15: $\quad\quad$ **break**

16: $\quad$ **end if**

17: $\quad$ $i \leftarrow (i + 1) \bmod n$

18: **end while**

19: **for all** $u \in P_2$ **do** addEdge(u, E)

20: **end for**

21: **while** true **do**

22: $\quad$ append $B[i]$ to P_3;

23: $\quad$ **if** $B[i] = d_1$ or $B[i] = d_2$ **then**

24: $\quad\quad$ append $B[i]$ to P_4

25: $\quad\quad$ $i \leftarrow (i + 1) \bmod n$

26: $\quad\quad$ **break**

27: $\quad$ **end if**

28: $\quad$ $i \leftarrow (i + 1) \bmod n$

29: **end while**

30: **for all** $u \in P_3$ **do** addEdge(u, S)

31: **end for**

32: **while** true **do**

33:

34: $\quad$ **if** $B[i] = v$ **then** **then** **Break**

35: $\quad$ **end if**

36: $\quad$ append $B[i]$ to P_4

37: $\quad$ $i \leftarrow (i + 1) \bmod n$

38: **end while**

39: **for all** $u \in P_4$ **do** addEdge(u, W)

40: **end for**

41: addEdge($\mathcal{G}_v, (N, E)$); $\quad\quad$ addEdge($\mathcal{G}_v, (E, S)$); $\quad\quad$ addEdge($\mathcal{G}_v, (S, W)$); addEdge($\mathcal{G}_v, (W, N)$)

42: **return** $\mathcal{G}_v$ $\quad\quad\quad\quad\quad\quad\quad\quad\triangleright$ success: augmented graph

---

3.2   Correctness and Time Complexity

To establish that the OUTER4COMPLETION procedure always produces an extended outerplanar graph that admits an area–universal layout, independent of the particular chosen REL, we rely on the following theorem. Specifically, we show that for any biconnected outerplanar graph G, if there exists a vertex v with $\deg(v) > 2$ that can be made adjacent to three consecutive cardinal vertices, then the resulting extended outerplanar graph serves as a valid four–completion of G. Moreover, such a four–completion guarantees the existence of an area–universal layout corresponding to every possible REL. This claim is formally stated in Theorem 4.

Theorem 4. *Let $\mathcal{G}$ be a biconnected outerplanar graph of order $n > 3$, and let v be a vertex of degree $k > 2$. It is sufficient for $\mathcal{G}$ to admit a one-sided rectangular layout (area-universal layout) if v is adjacent to exactly three cardinal vertices during the four-completion process.*

Proof. Let $\mathcal{G}$ be a biconnected outerplanar graph embedded in the plane with its vertices listed in clockwise order on the outer face. Choose a vertex $v \in V(\mathcal{G})$ and make it adjacent to exactly three cardinal vertices; denote the four cardinals in clockwise order by N, E, S, W. Then the outer face of $\mathcal{G}$ is partitioned uniquely into four contiguous boundary paths P_1, P_2, P_3, P_4. The uniqueness follows from outerplanarity together with the fact that v ($deg(v) > 2$) is incident to three distinct cardinals (see Figs. 8a and 8b). We adopt the convention $P_1 = \{v\}$ and take P_2, P_3, P_4 in clockwise order from v, so that P_2 begins at v and ends at the first degree-2 vertex d_1, P_3 runs from d_1 to the other degree-2 vertex d_2, and the remaining boundary vertices (including d_2) lie on P_4. Any other partition of the outer face into four paths would produce a separating triangle and hence would not correspond to a valid RFP.

Since v belongs to both P_2 and P_4, realizing the third (non-cardinal) adjacency of v by connecting it to a vertex of either P_2 or P_4 would create a forbidden local configuration (a complex/separating triangle). Therefore, the third neighbor that witnesses $\deg(v) > 2$ must lie on P_3; denote this vertex by $w \in P_3$. With this choice, no adjacency between the vertices of P_2 and the vertices of P_4 is possible (see Fig. 8c).

Now, vertices in path p_3 can only be adjacent to vertices in p_2, and similarly, vertices in path p_1 can only be adjacent to vertices in p_2. Under this partition, any edge (v_i, v_j) with $v_i, v_j \in p_k$ receives the same label. For a four-cycle to be alternating, there are only two possible configurations: either two vertices from p_i and two from p_j, or one vertex from p_i and three from p_j. In both cases, an alternating four-cycle cannot occur. It follows that no alternating four-cycle can be formed in any REL obtained from these four-completions. Consequently, no flippable edge arises in any REL produced by this construction. Therefore, any REL derived from this four-completion will ultimately yield an area-universal layout. $\square$

The time complexity of Algorithm 1 can be analyzed as follows. Constructing the set $V_{>2}$ of vertices of degree greater than two requires a single pass over all

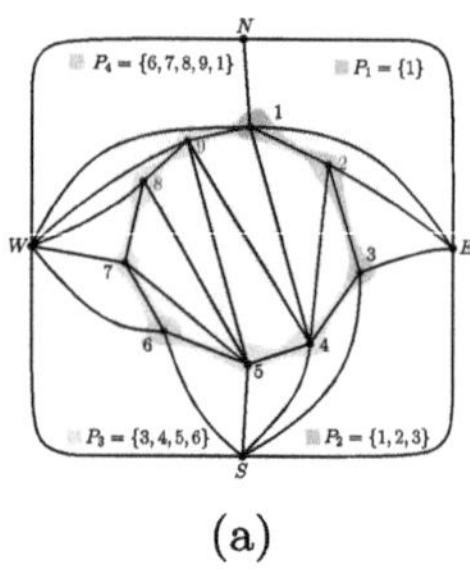 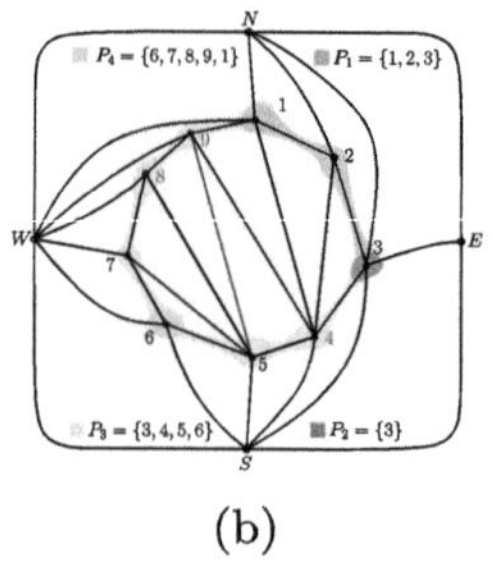 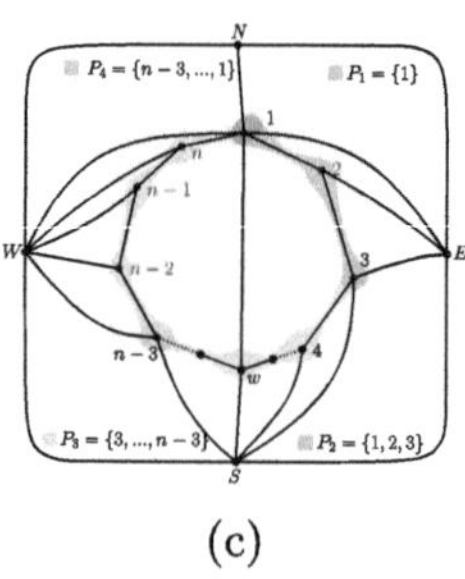

(a)   (b)   (c)

Fig. 8. (a) Unique partition of paths when vertex 1 is adjacent to all three cardinal vertices. (b) Multiple partitions are possible in an outerplanar graph, even when vertex 3 is adjacent to three cardinal vertices. (c) The adjacency (u, v) does not exist, where $u \in P_2$ and $v \in P_4$.

vertices of $\mathcal{G}$, which takes $O(n)$ time, where $n = |V|$. The algorithm then selects one pivot vertex v and makes a single call to $\mathtt{indexOf}(v, B)$, which involves scanning the boundary cycle B once and hence costs $O(n)$ time. The subsequent partitioning of the boundary into three contiguous sets P_2, P_3, P_4 is carried out by advancing a pointer around the outer cycle, and therefore each vertex on B is processed at most once, yielding another $O(n)$ contribution. Attaching vertices of each partition to the designated cardinal nodes incurs a constant number of edge insertions per vertex, also bounded by $O(n)$. Finally, the algorithm adds four edges to close the cycle (N, E, S, W), which is constant time. Thus the total running time for constructing a four-completion with respect to a single pivot is bounded by $O(n)$, dominated by linear scans of the boundary and degree computations.

4   Conclusion and Future Work

In this work we have addressed the problem of constructing area–universal layouts for outerplanar graphs. Our main contribution is the identification of both necessary and sufficient conditions under which an outerplanar graph admits area-universal layout. In particular, we established that while the presence of exactly two degree–two vertices is a necessary requirement. However, not every four-completion of such a graph leads to an area-universal layout. To overcome this limitation, we introduced the concept of an extended outerplanar (a special four-completion) graph and characterized the sufficient condition under which the four–completion indeed produces an area–universal layout corresponding to every regular edge labeling (see Fig. 9).

The results presented here provide a structural foundation for constructing area–universal layouts in outerplanar graphs and contribute to a deeper understanding of the interplay between four–completions and rectangular layouts. There are several potential avenues for future research based on this work. A significant direction would be to generalize the concept of area-universal layouts to a broader class of graphs. Specifically, identifying and characterizing the entire class of graphs for which an area-universal layout exists remains an open problem. Additionally, developing efficient algorithms to construct such layouts for these graphs is another challenging yet promising area for future exploration. Extending the approach to handle more complex graph structures and refining the algorithm to optimize layout properties such as compactness and symmetry can further enhance its applicability to real-world problems.

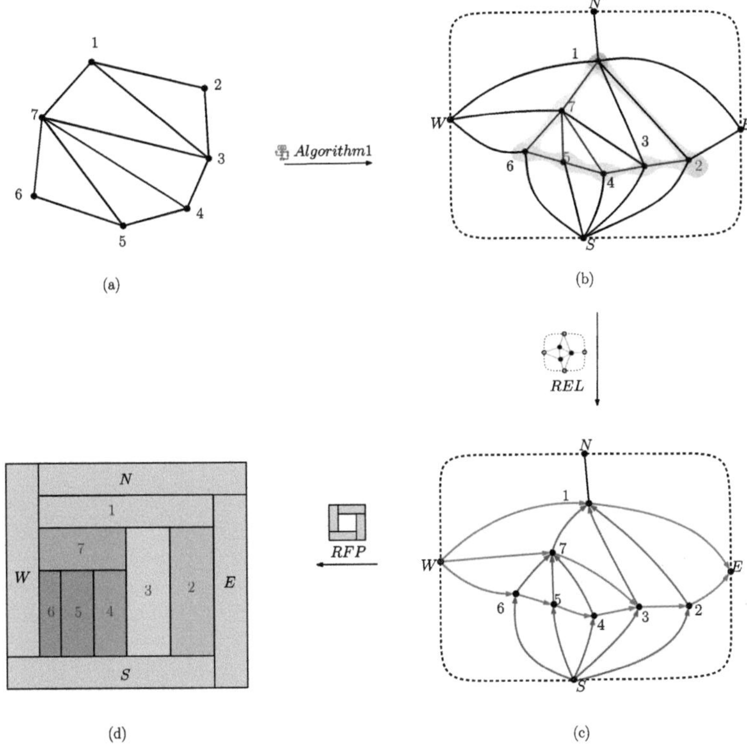

Fig. 9. (a) An outerplanar graph with exactly two vertices of degree-2, (b) An extended outerplanar graph, with four contiguous paths separated with different colors, obtained from Algorithm 1, (c) A REL of the extended outerplanar graph, (d) An area-universal layout corresponding to input outerplanar graph.

References

1. Chang, Y.J., Yen, H.C.: Area-universal drawings of biconnected outerplane graphs. Inf. Process. Lett. **118**, 1–5 (2017)
2. Eppstein, D., Mumford, E., Speckmann, B., Verbeek, K.: Area-universal rectangular layouts. Proc. Twenty-Fifth Ann. Symp. Comput. Geometry **41**(3), 267–276 (2009)

3. Eppstein, D., Mumford, E., Speckmann, B., Verbeek, K.: Area-universal and constrained rectangular layouts. SIAM J. Comput. **41**(3), 537–564 (2012)
4. He, X.: On finding the rectangular duals of planar triangular graphs. SIAM J. Comput. **22**(6), 1218–1226 (1993). https://doi.org/10.1137/0222072
5. Kant, G., He, X.: Regular edge labeling of 4-connected plane graphs and its applications in graph drawing problems. Theoret. Comput. Sci. **172**(1–2), 175–193 (1997)
6. Koźmiński, K., Kinnen, E.: Rectangular duals of planar graphs. Networks **15**(2), 145–157 (1985)
7. Kumar, V., Shekhawat, K.: Transformations of rectangular dualizable graphs. Theoretical Comput. Sci. (2021)
8. Rinsma, I.: Nonexistence of a certain rectangular floorplan with specified areas and adjacency. Environ. Plann. B. Plann. Des. **14**(2), 163–166 (1987)
9. Rinsma, I.: Existence theorems for floor-plans. Bull. Aust. Math. Soc. **37**(3), 473–475 (1988)
10. Sun, Y., Yeap, K.H.: Edge covering of complex triangles in rectangular dual floor-planning. J. Circuits Syst. Comput. **3**(03), 721–731 (1993)
11. Van Kreveld, M., Speckmann, B.: On rectangular cartograms. Comput. Geom. **37**(3), 175–187 (2007)
12. Wang, J.J.: A polynomial time algorithm for finding area-universal rectangular layouts (2016). https://arxiv.org/abs/1302.3672

Abstracts for Posters

Eccentric Connectivity Index of Strongly Connected Digraphs

Vysakh Chakooth[1]([✉]), Prasanth G. Narasimha-Shenoi[2,4],
and Prakash G. Narasimha-Shenoi[3,4]

[1] Department of Mathematics, NSS College Ottappalam, Palakkad 679103, Kerala,
India
v2sakh@gmail.com
[2] Department of Mathematics, Government College Chittur, Palakkad 678104,
Kerala, India
prasanthgns@gmail.com
[3] Department of Mathematics, Maharajas College Ernakulam, Ernakulam 682011,
Kerala, India
prakashgn@gmail.com
[4] Department of Collegiate Education, Government of Kerala, Thiruvananthapuram
695033, Kerala, India

1 Introduction

Topological indices are numerical invariants that represent structural features of graphs and are widely applied in chemical graph theory for predicting physico-chemical and biological properties. The *eccentric connectivity index*, of simple graphs introduced by Sharma et al., in [2], is a distance-cum-degree based descriptor defined as

$$\xi^C(G) = \sum_{u \in V} d_u \, ecc(u)$$

where d_u is the degree of the vertex u of and $ecc(u)$ is the eccentricity of the vertex u of the graph $G = (V, E)$.

2 Eccentric Connectivity Index of Digraphs

Definition 1. *Let $D = (V, A)$ be a digraph, the* eccentric connectivity index *of D is defined as*

$$\xi^C(D) = \frac{1}{2} \sum_{u \in V} (d_u^+ + d_u^-) \, mecc(u)$$

Chakooth V.—Is also a part-time research scholar in the Department of Mathematics, Government College Chittur, Palakkad, Kerala, India - 678104, affiliated to University of Calicut, Thenhipalam, Kerala, India - 673635.

N. Misra and A. Pandey (Eds.): CALDAM 2026, LNCS 16445, pp. 487–488, 2026.
https://doi.org/10.1007/978-3-032-17156-6

with respect to the metric, *maximum distance* defined by $md(u,v) = \max\{\vec{d}(u,v), \vec{d}(v,u)\}$, where d_u^+ and d_u^- are the out-degree and in-degree, $mecc(u)$ is the m-eccentricity of the vertex $u \in V$.

Theorem 1. *If G be any connected graph, then the eccentric connectivity index of G and its symmetric digraph $\widehat{G}$ are same. That is $\xi^C(G) = \xi^C(\widehat{G})$.*

Theorem 2. *If D is a strongly connected digraph with n vertices and a arcs, then $a.mrad(D) \leq \xi^C(D) \leq a.mdiam(D)$.*

2.1   Eccentric Connectivity Index of Orientations of K_n

Different orientations of K_n may have different eccentric connectivity index.

Theorem 3. *If $D = (V, A)$ be a strongly connected digraph of order $n, n \geq 4$ then $\xi^C(D) \geq 3(n-1)$, and equality holds if and only if D is the biorientation of star graph $\overleftrightarrow{S}_n$.*

3   Maximum Eccentric Connectivity Index of Digraphs

Many studies where done to find the upperbounds of eccentric connectivity index of graphs. Theorem 9 of [1] found maximum eccentric connectivity index of graphs on n vertices.

If $P_n = u_1 u_2 \ldots u_n$ is any (simple) path, then P_n^* is the digraph obtained from the biorientation of P_n by adding exactly one of the arcs (u_1, u_n) or (u_n, u_1).

Theorem 4. *If P_n is any simple path, then $\xi^C(P_n) = \xi^C(\overleftrightarrow{P_n}) \leq \xi^C(P_n^*)$.*

The Theorem 4 imply, one can increase the eccentric connectivity index of a digraph by adding arcs without changing the the eccentricities of the vertices. So the problem here is to find the maximum number of arcs that can be added to $\overleftrightarrow{P}_n$ so that meccentricity remains the same but as number of arcs increases the eccentric connectivity index become maximum.

References

1. Hauweele, P., Hertz, A., Mélot, H., Ries, B., Devillez, G.: Maximum eccentric connectivity index for graphs with given diameter. Discrete Appl. Math. **268**, 102–111 (2019)
2. Sharma, V., Goswami, R., Madan, A.K.: Eccentric connectivity index: a novel highly discriminating topological descriptor for structure- property and structure-activity studies. J. Chem. Inf. Comput. Sci. **37**(2), 273–282 (1997)

On the Hardness of Hop Domination and Total Hop Domination

Sobyasachi Chatterjee[✉]

The Institute of Mathematical Sciences (HBNI), Chennai, India
`sobyasachic@imsc.res.in`

Abstract. Problems related to domination in graphs play a central role in graph algorithms. For every fixed integer $r \in \mathbb{N}$, the r-HOP DOMINATION problem is defined as follows: given a graph G and a positive integer k, the task is to decide whether there exists a set $S \subseteq V(G)$ of size at most k such that for every vertex $v \in V(G) \setminus S$, there is a vertex $u \in S$ with distance between u and v being exactly r. The case $r = 2$ was studied by Karthika et al. in their recent work at CALDAM 2025, where they proved that 2-HOP DOMINATION parameterized by the solution size is W[1] hard and therefore unlikely to admit a fixed parameter tractable algorithm. We strengthen this result by proving that r-HOP DOMINATION parameterized by the solution size is W[2]-hard even on 6 degenerate graphs. Furthermore, we study a closely related problem, namely the TOTAL r-HOP DOMINATION problem. In this problem, we are given a graph G and a positive integer k, and the task is to decide whether there exists a set $S \subseteq V(G)$ of size at most k such that for every vertex $v \in V(G)$, there exists a vertex $u \in S$ with the distance between u and v being exactly r. We prove that TOTAL r-HOP DOMINATION is W[2]-hard even on 6 degenerate graphs. Moreover, we show that when parameterized by the vertex cover number of the input graph, TOTAL 2-HOP DOMINATION does not admit a polynomial compression unless NP $\subseteq$ coNP/poly.

1 Introduction

Domination at exact distances has received recent attention [3]. In HOP DOMINATION, each vertex outside the solution must be at distance two from a chosen vertex. TOTAL HOP DOMINATION strengthens this by requiring that every vertex, (including solution vertices) be hop dominated, i.e., vertices cannot dominate themselves. The r-HOP DOMINATION and TOTAL r-HOP DOMINATION variants generalize this to distance r. This paper studies hardness results for these problems on sparse graphs.

One of the earliest studies of exact distance based domination was carried out by Chidambaram and Ayyaswamy [1], who formally introduced the HOP DOMINATION problem. The HOP DOMINATION problem was further explored by Henning, Pal, and Pradhan [3], who proved that it is NP-hard even on bipartite and chordal graphs.

N. Misra and A. Pandey (Eds.): CALDAM 2026, LNCS 16445, pp. 489–490, 2026.
https://doi.org/10.1007/978-3-032-17156-6

2 Hardness Results

In this section, we present the hardness results; the full proofs are omitted due to space constraints. A sketch of the proofs will be provided at the end of the section.

Theorem 1. (Hop DominationHardness). *For every fixed integer $r \geq 2$, the r-HOP DOMINATION problem parameterized by the solution size k is W[2]-hard even on 6 degenerate graphs.*

Theorem 2. (Total Hop DominationHardness). *For every fixed integer $r \geq 2$, the TOTAL r-HOP DOMINATION problem parameterized by the solution size k is W[2]-hard even on 6 degenerate graphs.*

Theorem 3. (Kernelization Lower Bound). TOTAL HOP DOMINATION *parameterized by the vertex cover number of the input graph does not admit a polynomial compression unless NP $\subseteq$ coNP/poly.*

Proof Overview. The hardness results are obtained through reductions from the well known $(k, r - 1)$-CENTER problem for any $r > 2$, which is W[2]-hard even on 2 degenerate graphs [2]. The starting point of each reduction is an instance of $(k, r - 1)$-CENTER, and the challenge is to convert it into an instance of hop domination while preserving the parameter. This is achieved by constructing layered gadgets that enforce exact distances and ensure that domination of specific layers is possible only through designated centerpiece vertices.

In the reduction for r-HOP DOMINATION, the gadget contains a sequence of layers that correspond to copies of the input graph. Edges between consecutive layers are created so that a vertex in a higher layer can dominate a vertex in a lower layer only when their corresponding original vertices are equal or adjacent in the input instance. This forces any small hop dominating set to select vertices that encode a valid solution of the original $(k, r - 1)$-CENTER problem. The construction also includes special vertices that guarantee that all remaining vertices are dominated without altering the correspondence with the center solution. A careful ordering of the vertices shows that every vertex has at most six later neighbors, which proves that the constructed graph is six degenerate. For the total variant, the argument is similar to that of r-HOP DOMINATION.

The kernelization lower bound is obtained by a parameter preserving reduction from RED BLUE DOMINATING SET.

References

1. Chidambaram, N., Ayyaswamy, S.K.: Hop Domination in Graphs, pp. 1–15. Springer, Cham (2015)
2. Golovach, P.A., Villanger, Y.: Parameterized Complexity for Domination Problems on Degenerate Graphs. In: Broersma, H., Erlebach, T., Friedetzky, T., Paulusma, D. (eds.) WG 2008. LNCS, vol. 5344, pp. 195–205. Springer, Heidelberg (2008). https://doi.org/10.1007/978-3-540-92248-3_18
3. Henning, M.A., Pal, S., Pradhan, D.: Algorithm and hardness results on hop domination in graphs. Inf. Process. Lett. **160**, 1–6 (2020)

Towards the Burning Number Conjecture

Sandip Das[1], Sk Samim Islam[1], Ritam Manna Mitra[1(✉)], and Sanchita Paul[2]

[1] Indian Statistical Institute, Kolkata, India
sandipdas@isical.ac.in, samimislam08@gmail.com, rmmitra98@gmail.com
[2] Department of Mathematics, Techno Main Salt Lake, Kolkata, India
sanchitajumath@gmail.com

1 Introduction

Given a simple connected graph $G(V, E)$ with $|V| = n$, all vertices are initially unburnt. In each step, one new unburnt vertex is selected to burn, and in the following steps the fire spreads from this vertex to all its unburnt neighbors . Once a vertex is burned, it stays burned. This process stops when all vertices are burned. The burning number of a graph G, denoted by $b(G)$, is the minimum number of steps required to burn the entire graph. N. Alon [1] first worked on this problem in the class of graphs called n-dimensional hypercubes. Later this problem was formally introduced by A. Bonato, J. Janssen, and E. Roshanbin [3] under the name GRAPH BURNING.

Burning Number Conjecture (BNC) [3]: Let G be a connected graph with n vertices. Then $b(G) \leq \lceil \sqrt{n} \rceil$.

In this paper we provide meaningful progress towards proving the conjecture and we sharpen some previous known bounds (Table 1).

Table 1. List of our contributions.

Previous works	Their results	Our improved results	Comments
Murakami [5]	BNC is true when $n_2 = 0$	BNC is true when, $n_2 \leq 4\lceil \sqrt{n} \rceil - 9$	n_2 is the number of degree-2 vertices of T
Bonato et al. [4]	$b(T) \leq \sqrt{\frac{4}{3}n} + O(1)$	$b(T) \leq \lceil \sqrt{\frac{4}{3}n + 8} \rceil - 1$	When T has number of degree-2 vertices less than $n/3$.
Bessy et al. [2]	$b(T) \leq \lceil \sqrt{n + n_2 + 1/4} + 1/2 \rceil$	$b(T) \leq \lceil \sqrt{n + n_2 + 8} \rceil - 1$	Our results are better when $n \geq 50$
	$b(T) \leq \lceil \sqrt{2n + \frac{1}{4}} - \frac{1}{2} \rceil$	$b(T) \leq \lceil \sqrt{2n + 6} \rceil - 1$	

N. Misra and A. Pandey (Eds.): CALDAM 2026, LNCS 16445, pp. 491–493, 2026.
https://doi.org/10.1007/978-3-032-17156-6

2 Our Results

We define a special type of k-ary tree, where the root of the tree has exactly 2 children and no parents. All other nodes except the root and the leaves can have at least 2 children and at most k children. We call this tree a $(2, k)$-*ary tree*. For any vertex v of T, the subtree rooted at v is denoted by T_v.

Lemma 1. *For a given tree T of size n and n_2 number of degree-2 vertices, if $\Delta(T) = k + 1$ then a $(2, k)$-ary tree T' of size $n + n_2 - 1$ can be created where T is a subgraph of T'.*

2.1 Algorithm 1

We have the following results:

Lemma 2. *Let T be a $(2, k)$-ary tree with n vertices. There exists an algorithm which burns T in $\lceil \sqrt{n} \rceil$ steps in $O(n^{1.5})$ time.*

Consequently, using Lemma 1, and Lemma 2, we have the following result:

Theorem 1. *A general tree T of size n and with n_2 number of degree-2 vertices can be burnt in $\lceil \sqrt{n + n_2 - 1} \rceil$ steps.*

2.2 Algorithm 2

We have the following results:

Lemma 3. *Let T be a $(2, k)$-ary tree with n vertices. There exists an algorithm which burns T in $\lceil \sqrt{n + 9} \rceil - 1$ steps in $O(n^{1.5})$ time.*

Consequently, using Lemma 1, and Lemma 3, we have the following result:

Theorem 2. *A general tree T of size n and with n_2 number of degree-2 vertices can be burnt in $\lceil \sqrt{n + n_2 + 8} \rceil - 1$ steps.*

2.3 Main Result

Theorem 3. *Let T be a tree of size n and n_2 be the number of degree-2 vertices. When $n_2 \leq 4\lceil \sqrt{n} \rceil - 9$,*

$$b(T) \leq \lceil \sqrt{n} \rceil.$$

References

1. Alon, N.: Transmitting in the n-dimensional cube. Discrete Appl. Math. **37**, 9–11 (1992)
2. Bessy, S., Bonato, A., Janssen, J., Rautenbach, D., Roshanbin, E.: Bounds on the burning number. Discrete Appl. Math. **235**, 16–22 (2018)

3. Bonato, A., Janssen, J., Roshanbin, E.: How to burn a graph. Internet Math. **12**(1–2), 85–100 (2016)
4. Bonato, A., Kamali, S.: An improved bound on the burning number of graphs. arXiv preprint arXiv:2110.01087 (2021)
5. Murakami, Y.: The burning number conjecture is true for trees without degree-2 vertices. Graphs Comb. **40**(4), 82 (2024)

On the Complexity of Maximum Component Minus Vertex Cut (Extended Abstract)

K. Mohamed Harith[1]([✉]), Sounaka Mishra[1], and B. V. Raghavendra Rao[2]

[1] Department of Mathematics, IIT Madras, Chennai, India
harry1496@gmail.com, sounak@iitm.ac.in
[2] Department of Computer Science and Engineering, IIT Madras, Chennai, India
bvrr@iitm.ac.in

A classical example of a graph separation problem is the minimum k-cut problem, which seeks to delete a minimum set of edges to partition the graph into at least k connected components. The `k-way-vertex-cut` problem, the vertex analogue of minimum k-cut problem, asks for a minimum vertex set whose removal creates at least k connected components. Formally, given a graph $G = (V, E)$ and positive integers s and k, the problem seeks a vertex set S with $|S| \leq s$ such that $c_G(S) \geq k$, where $c_G(S)$ denotes the number of connected components in the induced subgraph $G[V \setminus S]$. Marx [4] established that this problem is W[1]-hard when parameterized by k. Subsequently, Berger et al. [2] proved NP-hardness of approximation within a factor of $n^{1-\epsilon}$ for general graphs.

Cornaz et al. [3] studied the `Vertex-k-Cut` problem, demonstrating that finding a minimum cardinality vertex-k-cut is NP-hard for all $k \geq 3$. In contrast to fixing the number of components (the parameter k) and minimizing the cut size, we consider the `Max-Vertex-Cut` problem, which maximizes the number of connected components without constraint on the cut set size:

From [2], `Max-Vertex-Cut` can be deduced to be NP-hard to approximate. Building upon this foundation, we introduce a novel optimization problem that balances the competing objectives of maximizing components while penalizing the size of the deletion set:

Our first result establishes a strong inapproximability for `Max-CV-Cut`, using the hardness of `Max-Vertex-Cut`.

Theorem 1. *For any $\varepsilon > 0$, `Max-Vertex-Cut` cannot be approximated within a factor of $n^{1-\varepsilon}$, unless P = NP.*

We prove the hardness of `Max-CV-Cut` by establishing an approximation preserving reduction from `Max-Vertex-Cut` and using Theorem 1.

Theorem 2. *For any $\varepsilon > 0$, `Max-CV-Cut` cannot be approximated within a factor of $\frac{1}{2\sqrt{3}}n^{\frac{1}{2}-\varepsilon}$, unless P = NP.*

We restrict attention to graphs with maximum degree 4 and establish APX-hardness. The reduction exploits the APX-completeness of vertex cover on cubic graphs [1].

Theorem 3. *Max-CV-Cut-4 (the restriction of Max-CV-Cut to graphs with maximum degree 4) is APX-hard.*

Despite the hardness results, we show that Max-CV-Cut becomes tractable for graphs of bounded treewidth through dynamic programming on a nice tree decomposition.

Theorem 4. *Let $G = (V, E)$ be a graph of treewidth at most k. Then Max-CV-Cut can be solved in time $\mathcal{O}(n \cdot 2^{k+1} \cdot \mathcal{B}(k+1)^2)$ for G, where n is the number of vertices and $\mathcal{B}(k+1)$ is the $(k+1)$-th Bell number.*

We consider the parameterized version CV-Cut with two parameters: k (upper bound on $|S|$) and l (lower bound on $f_G(S)$).

We obtain the following result by establishing a reduction from Maximum Independent Set problem, which is known to be W[1]-hard.

Theorem 5. *CV-Cut is W[1]-hard with parameters k and l.*

Acknowledgements. The first author thanks the Council of Scientific and Industrial Research (CSIR), India, for financial support. The second author would like to thank SERB for the grant under MATRICS (MTR/2023/000364).

References

1. Alimonti, P., Kann, V.: Hardness of approximating problems on cubic graphs. In: Bongiovanni, G., Bovet, D.P., Di Battista, G. (eds.) CIAC 1997. LNCS, vol. 1203, pp. 288–298. Springer, Heidelberg (1997). https://doi.org/10.1007/3-540-62592-5_80
2. Berger, A., Grigoriev, A., van der Zwaan, R.: Complexity and approximability of the k-way vertex cut. Networks **63**(2), 170–178 (2014)
3. Cornaz, D., Furini, F., Lacroix, M., Malaguti, E., Mahjoub, A.R., Martin, S.: The vertex k-cut problem. Discrete Optim. **31**, 8–28 (2019)
4. Marx, D.: Parameterized graph separation problems. Theoret. Comput. Sci. **351**(3), 394–406 (2006)

On the Partitions Associated with Christoffel Words

Abhishek Krishnamoorthy$^{(\boxtimes)}$ [ORCID], Robinson Thamburaj [ORCID],
and Durairaj Gnanaraj Thomas [ORCID]

Madras Christian College, Chennai, Tamil Nadu, India
abhishek@mcc.edu.in, robinsonthamburaj@gmail.com, gnanarajthomas@gmail.com

Abstract. In this paper, we have studied some of the properties of the partitions associated with a subclass of binary words called Christoffel words and have shown how these properties can help deduce a Christoffel word of any slope solely using its associated partition and its continued fraction representation. Additionally, an application of one of the properties of the partition of a Christoffel word has been used to derive some results regarding Wiener type indices.

1 Core Words and Christoffel Words

Definition 1. *[2] The core of a word w over the alphabet $A = \{a, b\}$, denoted by $core(w)$, is the unique subword w_0 of w with smallest possible length satisfying $w = b^* w_0 a^*$.*

Definition 2. *[2] A word $w \in A^*$ is said to be a core word if and only if $core(w) = w$.*

Teh and Kwa showed in [2] that any core word can be associated with a partition of a natural number and that the converse also holds true. Since if w is a core word, it is of the form

$$w = a^{n_1} b\, a^{n_2} b \cdots a^{n_{|w|_b}} b$$

Thus, w can be identified with the partition

$$|w|_{ab} = n_1 + (n_1 + n_2) + \cdots + (n_1 + n_2 + \cdots + n_k)$$

The converse is also true.

Example 1. The partition associated with the core word $abbab$ is $1 + 1 + 2$.

Definition 3. *[1] Let x, y be coprime positive integers and $n = x + y$. Define $r_i \equiv ix \pmod{n}$. The Christoffel word of slope $\frac{x}{y}$ is the word w over the alphabet A of length n with $w[i] = a$ if $r_{i-1} < r_i$ and $w[i] = b$ otherwise.*

Example 2. For the Christoffel word $C(\frac{4}{7})$, the values $r_i \equiv 4i \pmod{11}$ are $0, 4, 8, 1, 5, 9, 2, 6, 10, 3, 7, 0$. Hence $C(\frac{4}{7}) = aabaabaabab$.

N. Misra and A. Pandey (Eds.): CALDAM 2026, LNCS 16445, pp. 496–497, 2026.
https://doi.org/10.1007/978-3-032-17156-6

2   Main Results

Theorem 1. *Let w be the Christoffel word of slope $\frac{x}{y}$. Then,*
$m_1 + m_2 + \cdots + m_{x-1} + y$ *is a partition of the natural number* $\left(\frac{x+1}{2}\right) y + \frac{x-1}{2}$.

Proposition 1. *Let w be the Christoffel word of slope $\frac{x}{y}$ and (w_1, w_2) be its standard factorization, where w_1 and w_2 are the Christoffel words of slope $\frac{u}{v}$ and $\frac{k}{l}$ respectively. Then the m_{x-u} th term in the partition of w is given by $y + 1 - v$.*

Proposition 2. *If w_1 and w_2 are two core words with partitions $m_1 + m_2 + \cdots + m_k$ and $m_1' + m_2' + \cdots + m_l'$ respectively, then the core word $w_1 w_2$ has the partition $m_1 + m_2 + \cdots + m_k + m_1' + m_k + m_2' + m_k + \cdots + m_l' + m_k$.*

Remark 1. Using the above results one can recursively compute the partition associated with a Christoffel word of any slope.

We extend the study of a few Wiener-type indices defined on binary core words carried out by Nobin Thomas et al. [3], to Christoffel words.

Proposition 3. *The Wiener index of the Parikh word representable graph of the Christoffel word w of slope $\frac{x}{y}$ is*
$(x + y)^2 - (x + y) + xy - (x + 1)y - x + 1.$

Proposition 4. *Let $w = a^{n_1} b a^{n_2} b \cdots a^{n_x} b$ be the Christoffel word of slope $\frac{x}{y}$, then*
$\sum_{i=1}^{x} \left((x + 2i - 3)n_i \right) = 1 + 2xy - (x + 2y).$

Proposition 5. *The Wiener polynomial of the Parikh word representable graph of the Christoffel word w of slope $\frac{x}{y}$ is* $\left(xy - \frac{(x+1)}{2}y + \frac{(x-1)}{2} \right) z^3 + \frac{1}{2}\big(y(y - 1) + x(x - 1) \big) z^2 + \left(\frac{(x+1)}{2}y - \frac{(x-1)}{2} \right) z.$

Proposition 6. *The multiplicative Wiener index of the Parikh word representable graph of the Christoffel word w of slope $\frac{x}{y}$ is* $2^{\frac{y(y-1)+x(x-1)}{2}} 3^{\frac{xy-(x+y)+1}{2}}.$

References

1. Berstel, J., Lauve, A., Reutenauer, C., Saliola, F.V.: Combinatorics on Words: Christoffel Words and Repetitions in Words. CRM Monograph Series, vol. 27. American Mathematical Society, Providence, R.I. (2008)
2. Teh, W.C., Kwa, K.H.: Core words and Parikh matrices. Theor. Comput. Sci. **582**, 60–69 (2015). https://doi.org/10.1016/j.tcs.2015.03.037
3. Thomas, N., Mathew, L., Sriram, S., Subramanian, K.G.: Wiener-type indices of Parikh word representable graphs. Ars Math. Contemp. **20**(2), 243–260 (2021). https://doi.org/10.26493/1855-3974.2359.a7b

Efficient Approximate Priority Queues: Soft Heaps in the PEM Model

G. Sajith[iD] and Ranajit Senko[(✉)][iD]

Department of Computer Science and Engineering, Indian Institute of Technology, Guwahati, India
`sajith@iitg.ac.in`, `ranajit@iitg.ac.in`

Abstract. We present PEMSH, the first deterministic soft heap for the Parallel External Memory (PEM) model. Soft heaps permit controlled key corruption: after N insertions, at most ϵN elements may have their keys increased. While soft heaps exist in the RAM and EM models, no parallel, I/O-efficient variant previously existed. PEMSH adapts the EM soft heap to the PEM framework by using rank-partitioned processors and $(\sqrt{m}/\log\sqrt{m})$-way trees with per-rank buckets. We prove correctness and provide amortized bounds identical to those of the EM soft heap, with natural parallel speedup across ranks.

1 Introduction

Soft heaps, introduced by Chazelle [1], allow limited key corruption, enabling several optimal or near-optimal algorithms. In hierarchical-memory settings, the EM soft heap of Bhushan and Sajith [2] provides I/O-efficient approximate priority queues.

Modern algorithms increasingly rely on multicore systems that support parallel disk I/O, modeled by the Parallel External Memory (PEM) model [3]. However, no approximate priority queue existed for the PEM setting.

Contribution. We design PEMSH, the first PEM soft heap. It supports INSERT, FINDMIN, DELETEMIN, MELD, and preserves the invariant that at most ϵN elements become corrupt.

Amortized I/O bounds:

$$\text{Insert: } O\left(\frac{1}{B}\log_{\frac{\sqrt{m}}{\log\sqrt{m}}}\frac{1}{\epsilon}\right), \qquad \text{FINDMIN, DELETEMIN, MELD : non-positive.}$$

2 Structure of PEMSH

PEMSH organizes items into $(\sqrt{m}/\log\sqrt{m})$-way trees stored in external memory. Ranks 0 to r hold *PNODES* containing sorted buffers, while ranks $> r$ hold *CNODES* containing list-nodes with potentially corrupt keys. For each rank i,

N. Misra and A. Pandey (Eds.): CALDAM 2026, LNCS 16445, pp. 498–499, 2026.
https://doi.org/10.1007/978-3-032-17156-6

the following are maintained: (i) Information bucket IB_i: tree count and root pointers, (ii) Data bucket DB_i: smallest block of elements or list-node.

Processor p_i maintains all trees of rank i, enabling independent parallel restoration during melds or deletions.

3   Operations

Insert: Elements are added to an input buffer; when full, the buffer becomes a new rank-0 node and melds into the heap.
Findmin: Zero I/O; minimal element stored in memory.
Deletemin: Extracts from the DB_i having the global minimum. Empty DB_is trigger a local FILLUP by processor p_i.
Meld: Rank-wise merge using parallel loops over processors $p_0, \ldots, p_r$. Root sets are merged and invariants restored.
Auxiliary: SIFT restores PNI/CNI (PNODE or CNODE invariants). EXTRACT gathers smallest elements from children. FILLUP replenishes DB_is.

4   Correctness and I/O Analysis

Correctness relies on the top-down SIFT procedure: PNODES extract smallest child elements to restore PNI; CNODES update their ckeys via list-node merges to restore CNI; Recursive repairs maintain heap order and corruption bounds.

Amortized analysis uses a charge-accounting scheme over buckets, nodes, and elements. Insert performs most of the work; all other operations release enough charge to offset their I/O cost.

5   Conclusion

PEMSH is the first approximate priority queue for the PEM model. It preserves soft-heap properties while enabling parallel I/O efficiency. The structure provides a foundation for scalable graph algorithms and data-intensive computation in multicore external-memory environments.

References

1. Chazelle, B.: The soft heap: an approximate priority queue with optimal error rate. J. ACM **47**(6), 1012–1027 (2000)
2. Bhushan, A., Gopalan, S.: External Memory Soft Heap, and Hard Heap, a Meldable Priority Queue. In: Gudmundsson, J., Mestre, J., Viglas, T. (eds.) COCOON 2012. LNCS, vol. 7434, pp. 360–371. Springer, Heidelberg (2012). https://doi.org/10.1007/978-3-642-32241-9_31
3. Arge, L., Goodrich, M.T., Nelson, M., Sitchinava, N.: Fundamental parallel algorithms for private-cache chip multiprocessors. In: Meyer auf der Heide, F., Shavit N. (eds.) Proceedings of the 20th Annual ACM Symposium on Parallelism in Algorithms and Architectures (SPAA 2008), pp. 197–206, ACM, Munich, Germany (2008)

Dominating Set Knapsack: Profit Optimization on Dominating Sets

Sipra Singh[✉]

Indian Institute of Technology Kharagpur, Kharagpur, India
`sipra.singh@iitkgp.ac.in`

Abstract. In a large-scale network, we want to choose some influential nodes to make a profit by paying some cost within a limited budget so that we do not have to spend more budget on some nodes adjacent to the chosen nodes; our problem is the graph-theoretic representation of it. We define our problem DOMINATING SET KNAPSACK by attaching KNAPSACK PROBLEM with DOMINATING SET on graphs. Each vertex $v(\in \mathcal{V})$ is associated with a cost factor $w(v)$ and a profit amount $\alpha(v)$. We aim to choose some vertices within a fixed budget (s) that give maximum profit so that we do not need to choose their 1-hop neighbors. We show that the DOMINATING SET KNAPSACK problem is strongly NP-complete even when restricted to bipartite graphs, but weakly NP-complete for star graphs. We present a pseudo-polynomial time algorithm for trees in time $\mathcal{O}(n \cdot min\{s^2, (\alpha(\mathcal{V}))^2\})$. We show that DOMINATING SET KNAPSACK is unlikely to be *fixed parameter tractable* by proving that it is W[2]-hard parameterized by the solution size. We developed pseudo-FPT algorithms with running time $\mathcal{O}(4^{tw} \cdot n^{\mathcal{O}(1)} \min\{s^2, \alpha(\mathcal{V})^2\})$ and $\mathcal{O}(2^{vck-1} \cdot n^{\mathcal{O}(1)} \min\{s^2, \alpha(\mathcal{V})^2\})$, where tw represents the TREEWIDTH of the given graph $\mathcal{G}(\mathcal{V}, \mathcal{E})$, vck is the solution size of the VERTEX COVER KNAPSACK, s is the capacity or size of the knapsack and $\alpha(\mathcal{V}) = \sum_{v \in \mathcal{V}} \alpha(v)$.

Keywords: DOMINATING SET PROBLEM · KNAPSACK PROBLEM · Parameterized algorithm

1 Introduction

In our problem, we include a graph constraint on KNAPSACK PROBLEM, where we represent each item as a vertex of a graph, and each vertex has weight and profit values. Based on a graph constraint, we choose some vertices that give the maximum profit within the fixed capacity of the knapsack. There are some existing variants of the KNAPSACK PROBLEM with graph constraints. Pferschy and Schauer showed NP-hardness and approximation results for some special graph classes for the KNAPSACK PROBLEM with the independent set as graph property [4]. Dey et al. studied KNAPSACK PROBLEM with graph connectivity, path, shortest path [2], and VERTEX COVER [1]. This paper considers another graph property as a constraint, the DOMINATING SET [3], with the

N. Misra and A. Pandey (Eds.): CALDAM 2026, LNCS 16445, pp. 500–502, 2026.
https://doi.org/10.1007/978-3-032-17156-6

Knapsack Problem. A Dominating Set of an undirected graph is a set of vertices such that all other vertices are adjacent to at least one element of that set. We use the name Dominating Set Knapsack for our problem. The formal definition is as follows:

Definition 1. (Dominating Set Knapsack). *Given an undirected graph* $\mathcal{G} = (\mathcal{V}, \mathcal{E})$, *weight and profit of vertices are defined as the functions* $w : \mathcal{V} \to \mathbb{Z}_{\geq 0}$ *and* $\alpha : \mathcal{V} \to \mathbb{R}_{\geq 0}$ *respectively, size* $s(\in \mathbb{Z}^+)$ *of the knapsack and target value* $d(\in \mathbb{R}^+)$, *compute if there exists a subset* $\mathcal{U} \subseteq \mathcal{V}$ *of vertices such that: (I)* $\mathcal{U}$ *is a* Dominating Set, *(II)* $\sum_{u \in \mathcal{U}} w(u) \leq s$, *(III)* $\sum_{u \in \mathcal{U}} \alpha(u) \geq d$. *We denote an arbitrary instance of* Dominating Set Knapsack *by* $(\mathcal{G}, (w(u))_{u \in \mathcal{V}}, (\alpha(u))_{u \in \mathcal{V}}, s, d)$.

We refer to our full paper [5] for all the details. The technical preliminaries and detailed literature survey are also available in our full paper.

2 Our Contributions

The main contributions of this paper are on Classical NP-Hardness and Parameterized Complexity. The list of results are in the Table 1. For the proofs of the theorems in Table 1, please refer to full version of our paper [5].

Table 1. List of results. n: The number of vertices; tw: Treewidth of the input graph; s: Target size; d: Target profit; k: The number of vertices in the solution; vck: The solution size of the Vertex Cover Knapsack [1]; $\alpha(\mathcal{V}) = \sum_{v \in \mathcal{V}} \alpha(v)$ where $\alpha(v)$ is the value of the vertex v in $\mathcal{V}$.

Theorem 1.	Dominating Set Knapsack is strongly NP-complete general graphs.
Theorem 2.	Dominating Set Knapsack is strongly NP-complete even when the instances are restricted to a bipartite graphs.
Theorem 3.	Dominating Set Knapsack is weakly NP-complete for stars.
Theorem 4.	There is a pseudo-polynomial time algorithm of Dominating Set Knapsack for trees with complexity $\mathcal{O}(n \cdot min\{s^2, (\alpha(V))^2\})$
Theorem 5.	Dominating Set Knapsack is W[3]-complete parameterized by the solution size k.
Theorem 6.	There is an algorithm of Dominating Set Knapsack with running time $\mathcal{O}(2^{vck-1} \cdot n^{\mathcal{O}(1)} \min\{s^2, \alpha(\mathcal{V})^2\})$ parameterized by solution size of the Vertex Cover Knapsack vck
Theorem 7.	There is an algorithm of Dominating Set Knapsack with running time $\mathcal{O}(4^{tw} \cdot n^{\mathcal{O}(1)} \min\{s^2, \alpha(\mathcal{V})^2\})$ parameterized by treewidth tw

References

1. Dey, P., Hota, A., Kolay, S., Singh, S.: Knapsack with vertex cover, set cover, and hitting set. arXiv preprint arXiv:2406.01057 (2024)
2. Dey, P., Kolay, S., Singh, S.: Knapsack: connectedness, path, and shortest-path. In: Soto, J.A., Wiese, A. (eds.) LATIN 2024. LNCS, vol. 14579, pp. 162–176. Springer, Cham (2024). https://doi.org/10.1007/978-3-031-55601-2_11
3. Garey, M.R., Johnson, D.S., Stockmeyer, L.: Some simplified np-complete problems. In: Proceedings of the Sixth Annual ACM Symposium on Theory of Computing, pp. 47–63 (1974)
4. Pferschy, U., Schauer, J.: The knapsack problem with conflict graphs. J. Graph Algorithms Appl. **13**(2), 233–249 (2009)
5. Singh, S.: Dominating set knapsack: profit optimization on dominating sets. arXiv preprint arXiv:2506.24032 (2025)

Author Index